compressible-fluid dynamics

advanced engineering series

Irving H. Shames, *Consulting Editor*

Hodge: *Continuum Mechanics*
Leigh: *Nonlinear Continuum Mechanics*
Meirovitch: *Methods of Analytical Dynamics*
Oden: *Finite Elements of Nonlinear Continua*
Thompson: *Compressible-Fluid Dynamics*

compressible-fluid dynamics

philip a. thompson

Division of Fluid, Chemical, and Thermal Processes
Rensselaer Polytechnic Institute

McGRAW-HILL BOOK COMPANY

New York St. Louis San Francisco Düsseldorf Johannesburg
Kuala Lumpur London Mexico Montreal New Delhi
Panama Rio de Janeiro Singapore Sydney Toronto

compressible-fluid dynamics

Library of Congress Catalog Card Number 79–148996
07–064405–5

1234567890MAMM7987654321

This book was set in Times New Roman, and was printed and bound by The Maple Press Company. The designer was Paula Tuerk. The editors were B. J. Clark and J. W. Maisel. Les Kaplan supervised production.

for Jasper Banks

contents

preface

This book introduces the fundamentals of compressible-fluid motion, or gasdynamics. It is intended as a text for senior and graduate students in engineering, physics, and applied mathematics. It may also be useful as a source for applied researchers in various fields.

The first three chapters are introductory in nature and are included to provide a self-contained treatment of fluid mechanics. Chapters 1 and 2 concisely present the relevant continuum mechanics and thermodynamics; Chapter 3 is a description of dimensional reasoning as it applies to compressible-fluid flow. A student who encounters this material for the first time in these chapters will likely find it heavy going: on this account some previous preparation in the form of a first course in fluid mechanics and in thermodynamics is desirable. Those students who already enjoy adequate preparation in these areas may wish to skip this material altogether.

This book contains more material than can reasonably be covered in a normal one-semester or one-quarter course. The selection of material suitable to a particular course can safely be entrusted to the instructor.

There are already several excellent books in the field of gasdynamics, in particular Shapiro's *Compressible Fluid Flow* and Liepmann and Roshko's *Elements of Gasdynamics*. I have enjoyed the considerable advantage of reading these works as a student and as a teacher, and acknowledge my debt to them. Merely to retread the ground already covered by these existing works would, however, be wasteful and self-defeating. What is offered here is not a redigest of some venerable work but a book whose merits and deficiencies are peculiarly its own. They can best be assessed by reading the book itself.

In selecting the material to be covered, I have tried to emphasize fundamental topics such as acoustics, shock waves, and the nature of compressible flow itself. In short, the selection has favored fundamentals over techniques. In choosing specific problems for consideration, I have preferred those which seem to be accessible to ordinary experience: along these lines, no special effort has been made to treat somewhat esoteric subjects such as high-temperature gases or fluids far from thermodynamic equilibrium.

Many individuals have given valuable help along the road to publication. The manuscript has been read and criticized in its entirety by Louis Solomon. It has been read and criticized in its various parts by several of my colleagues at Rensselaer, in particular Steven Ball, Henrik Hagerup, Gerald Kliman, Howard Littman, Charles Muckenfuss, Euan Somerscales, and Hendrik Van Ness. The thorough review of the manuscript by Richard Corlett and the many corrections to it by Howard Cyphers have been especially helpful. Assistance in calculation has been given by E. T. Laskaris, Michael Liu, Dean Nairn, Antonio Artiles, and Pedro Porrello. Editorial help at a very practical level has come from Bruce, Claudia, Stephen, and Jean Thompson. The typing has been most effectively handled by Joanne Margosian. To all these individuals, and several others unnamed, my thanks.

This book could be written only in a situation where individual enterprises of this kind are encouraged: the support of the School of Engineering at Rensselaer Polytechnic Institute, and of Hendrik Van Ness in particular, has been invaluable. Finally, I would like to acknowledge the influence of a zealous advocate of vectors, Kenneth Bisshopp, and of a fine teacher, Ascher Shapiro, who first introduced me to this subject.

Philip A. Thompson

list of symbols[1]

a	Helmholtz function, $a \equiv e - Ts$
A	Cross-sectional area; amplitude coefficient; surface area
$\mathbf{b}$	Boundary or shock-front velocity
B	Constant in Tait equation
c	Sound speed, $c^2 \equiv \left(\frac{\partial P}{\partial \rho}\right)_s$, or phase velocity
c_p	Specific heat at constant pressure, $c_p \equiv \left(\frac{\partial h}{\partial T}\right)_p$
c_v	Specific heat at constant volume, $c_v \equiv \left(\frac{\partial e}{\partial T}\right)_v$
C^+, C^-	Labels for characteristics in xt plane
D_{ik}	Component of rate-of-deformation tensor
e	Base of natural logarithm; specific internal energy
$\mathbf{e}$	Unit vector
E	Acoustic energy per unit volume, $E \equiv \frac{p^2}{2\rho_0 c_0{}^2} + \frac{\rho_0 u^2}{2}$; voltage (electrical potential)
f	Wave function; general function; number of degrees of freedom; number of fundamental dimensions
F	Wave function; force; thermodynamic function, $F \equiv \int \frac{dP}{\rho c}$
g	Wave function; acceleration of gravity; Gibbs function, $g \equiv h - Ts$

[1] A few minor symbols, which have been used only briefly, are omitted from this list.

$\mathbf{G}$	Specific body-force vector
h	Specific enthalpy, $h \equiv e + Pv$; Planck's constant; liquid depth in shallow-water theory
H	Scale height of the atmosphere, $H \equiv \dfrac{RT_0}{g_0}$; Bernoulli constant, $H \equiv h + \dfrac{u^2}{2} + \Psi$
i	Imaginary unit, $i \equiv \sqrt{-1}$
$\mathbf{i}$	Cartesian unit vector
I	Integral
I_s	Specific impulse
$\mathcal{I}m$	Imaginary part of following expression
$\mathbf{j}$	Cartesian unit vector
J	Mass flux, $J \equiv \rho u$; molecular flux
J^+, J^-	Riemann invariants
k	Wave number, $k \equiv \dfrac{2\pi\nu}{c} = \dfrac{\omega}{c}$; Boltzmann constant
$\mathbf{k}$	Wave-number vector, $\mathbf{k} \equiv k\mathbf{e}$
K	Constant; bulk modulus; spring constant
l	Distance along acoustic ray; spacing distance
L	Length; inductance
m	Particle mass
m^+, m^-	Labels for Mach waves (characteristics); distances along such waves
$\dot{m}$	Rate of mass flow
M	Mach number, $M \equiv \dfrac{u}{c}$; total mass
$\tilde{M}$	Molecular weight
n	Distance normal to a streamline; index of refraction, $n \equiv \dfrac{c_0}{c}$; number density of molecules; number of dimensional variables
$\mathbf{n}$	Unit normal vector
N	Frequency of oscillation in the atmosphere ($N^2 > 0$ for stability)
$\tilde{N}$	Avogadro's number
$\mathcal{N}$	Number of moles

O	Abbreviation for order of magnitude
p	Acoustic (perturbation) pressure, $p \equiv P - P_0$; number of dimensionless variables
P	Pressure (absolute)
Pr	Prandtl number, $\mathrm{Pr} \equiv \mu c_p/\kappa$
q	Traffic flow rate, $q \equiv \rho u$
$\mathbf{q}$	Heat-flux vector
r	Spherical or cylindrical coordinate; number of variables with independent dimensions
R	Radius; specific gas constant, $R \equiv \dfrac{\tilde{R}}{\tilde{M}}$; electrical resistance
$\tilde{R}$	Universal gas constant, $\tilde{R} = \tilde{N}k$
$\mathscr{R}$	Acoustic impedance, $\mathscr{R} \equiv \rho c$
Re	Reynolds number, $\mathrm{Re} \equiv \dfrac{Lu}{\nu}$
$\mathscr{R}e$	Real part of following expression
s	Specific entropy; distance along streamline
S	Condensation, $S \equiv \dfrac{\rho - \rho_0}{\rho_0}$; surface
t	Time
T	Absolute temperature
$\mathbf{T}$	Surface-traction vector
u	Velocity magnitude; velocity component in x direction
$\mathbf{u}$	Vector velocity
U	Velocity; volume velocity $U \equiv uA$
v	Relative velocity component parallel to shock front; velocity component in y direction
$\mathbf{v}$	Molecular velocity
$\mathscr{v}$	Specific volume, $\mathscr{v} \equiv \dfrac{1}{\rho}$
V	Volume; peculiar molecular velocity
w	Relative velocity component normal to shock front
$\mathbf{w}$	Wind velocity
$\mathbf{x}$	Vector space coordinate
X	General variable

x, y, z	Cartesian space coordinates
z	Cylindrical coordinate; altitude above sea level
α	Attenuation coefficient; constant; amplitude function; thermal diffusivity, $\alpha \equiv \kappa/\rho c_p$
β	Bulk coefficient of thermal expansion, $\beta \equiv \dfrac{1}{v}\left(\dfrac{\partial v}{\partial T}\right)_p$; shock-front angle; coefficient in linearized equation, $\beta = \lvert\sqrt{M_\infty^2 - 1}\rvert$
γ	Ratio of specific heats, $\gamma \equiv \dfrac{c_p}{c_v}$; exponent in Tait equation
Γ	Dimensionless thermodynamic variable, $\Gamma \equiv \dfrac{\rho^3 c^4}{2}\left(\dfrac{\partial^2 v}{\partial P^2}\right)_s$
Γ_c	Circulation, $\Gamma_c \equiv \oint \mathbf{u} \cdot d\mathbf{l}$
δ	Small change; diffusivity, $\delta \equiv \dfrac{4\mu/3 + \mu_v}{\rho} + \dfrac{\gamma - 1}{\mathrm{Pr}}\dfrac{\mu}{\rho}$; ratio of specific heats, $\delta \equiv \dfrac{c_{pp}}{c_{pg}}$; boundary-layer thickness
δ_{ik}	Kronecker delta
Δ	Change; decibel level; shock-front thickness
ε	Small quantity; deviation of shock-front angle from Mach angle, $\varepsilon \equiv \beta - \mu$; energy per molecule
ε_{ijk}	Levi-Civita triple-index tensor
η	Similarity variable; mass fraction
θ	Flow angle; temperature
κ	Thermal conductivity
λ	Wavelength
Λ	Molecular mean free path
μ	Viscosity (ordinary shear viscosity); Mach angle, $\mu \equiv \sin^{-1}\dfrac{1}{M}$
μ_v	Bulk viscosity
ν	Kinematic viscosity, $\nu \equiv \dfrac{\mu}{\rho}$; frequency
ξ	Fluid-particle displacement; mass displacement
Ξ	Entropy flux
π	Pi, $\pi = 3.14159\ldots$

Π	Dimensionless pressure jump, $\Pi \equiv \dfrac{[P]}{\rho_1 c_1^{\,2}}$; general dimensionless quantity
Π_{ik}	Molecular momentum flux
ρ	Density
σ	Molecular diameter
σ_{ik}	Component of stress tensor
Σ	Sum; function of spatial coordinates
Σ_{ik}	Component of viscous stress tensor
τ	Relaxation time; dimensionless time
Υ	Dissipation function, $\Upsilon \equiv \Sigma_{ik} D_{ik}$
ϕ	Velocity potential
φ	Intermolecular potential
Φ_e	Acoustic energy flux (intensity)
Φ_m	Acoustic momentum flux
χ	Mole fraction
ψ	Polar coordinate; stream function
Ψ	Force potential
ω	Angular frequency, $\omega = 2\pi\nu$; Prandtl-Meyer function
$\mathbf{\Omega}$	Vorticity vector, $\mathbf{\Omega} \equiv \nabla \times \mathbf{u}$
Ω_{ik}	Component of spin tensor

compressible-fluid dynamics

one
descriptions of fluid motion

1.1 Introduction

Fluid mechanics is the study of the motions of gases and liquids. *Gas-dynamics*, or *compressible-fluid flow*, is the study of those motions for which changes in fluid density play an essential role.

Although there is some density change in every physical flow, it is often possible to neglect such changes and to treat the flow according to the idealization that the fluid is incompressible. This approximation may be applicable to gases, e.g., in low-speed flow around an airplane or flow through a vacuum cleaner, as well as to liquids.

On the other hand, the very small density changes associated with acoustic motions in liquids and gases cannot be neglected. Discussion of the precise conditions under which density changes must be considered will require the development of a little analytical apparatus and will be postponed to Chap. 3. Anticipating the results of that discussion, we simply set down the main practical categories of motion for which fluid compressibility plays a crucial role:

1 Wave propagation within the fluid

2 Steady flow in which the fluid speed is of the same order of magnitude as the speed of sound

3 Convection driven by body forces, e.g., gravity, acting on fluid subject to thermal expansion

4 Large-scale convection of gases in the presence of body forces

This list is not necessarily exhaustive, and some motions may fit into more than one category. This book is devoted mostly to the first two categories of motion, which have in common a relation to the fluid sound speed.

The principal equations of motion are developed in fairly general form in this chapter. This level of generality is not required for many of the applications, e.g., viscous forces can be neglected in many problems, but will permit us to arrive rationally at the various simplifying approximations. Some of the detailed steps will be omitted from the derivations given in this chapter. For a more complete treatment, see *Batchelor* [1967, chaps. 1–3] or *Aris* [1962, chaps. 1–6].

The equations of motion are developed from the concept that the fluid is a *continuum*; i.e., the fluid is considered to be matter which exhibits no structure, however finely it may be divided. This model makes it possible to treat fluid properties (such as density, temperature, and velocity) at a *point* in space and mathematically as continuous functions of space and time. The application of such a model to fluid motion is due principally to Leonhard Euler (1755). Treatments of fluid mechanics and solid mechanics from the continuum viewpoint have much in common (many of the equations in this chapter are applicable to fluids and solids indifferently, in fact), and the subjects taken together are called *continuum mechanics*.

There is an alternative way of proceeding, which begins with the particulate view of matter and by averaging over large numbers of molecules arrives finally at the continuum equations. This method, while perhaps more general in principle, is limited by practical difficulties to the Boltzmann equation applied to a dilute gas and will not be pursued here.

The continuum model may be expected to fail when the size of the fluid region of interest is of the same order as a characteristic dimension of the molecular structure. A suitable characteristic dimension for gases is the mean free path Λ (of the order of 10^{-7} m for air at standard conditions). For liquids, a corresponding characteristic molecular dimension is not clearly defined but may be taken to be a distance equal to several intermolecular spacings (for water, the intermolecular spacing L is of the order of 10^{-10} m). These dimensions are so small that the continuum model is violated only in extreme cases; two examples are the motion of dust or smoke particles of very small diameter d in the atmosphere (with $d \sim \Lambda$) and the propagation of high-frequency sound of wavelength λ in gases (with $\lambda \sim \Lambda$) or even in liquids (with $\lambda \sim L$). On the other hand, the mean free path Λ varies inversely with the density of a gas, so that under conditions of very low pressure, e.g., at high altitudes or in a vacuum chamber, even relatively large fluid regions cannot be described by the continuum model. We will be concerned only slightly with such extreme cases and adopt the continuum view henceforth. Occasional use will be made, however, of results from the elementary kinetic theory of gases, i.e., the *particulate* view, where such results enhance our physical insight for the problem of fluid motions.

1.2 Dynamical laws of motion

A natural concept within the continuum model is that of the *material volume*. This is an arbitrary collection of matter of fixed identity enclosed by a *material surface* (or boundary) every point of which moves with the local fluid velocity. This surface is purely hypothetical and in general does not correspond to any physical boundary in the flow: it may be helpful to imagine it as a perfectly flexible and extensible membrane of zero mass. As the material volume moves through space, it is deformed in shape and changed in volume, as sketched in Fig. 1.1. We will refer to the material volume as $V(t)$ and the material surface as $S(t)$. If the volume $V(t)$ is shrunk to a point, the resulting material point is called a *fluid particle*.

By definition, the surface $S(t)$ is impenetrable to matter: interdiffusion of chemical species thus cannot be accounted for by this particular model.

The dynamical laws of motion, from which most of the equations in this book are deduced, are stated for a material volume as follows:

1 *Conservation of mass* (*continuity*): The mass of a material volume is constant.

2 *Balance of linear momentum* (*Newton's second law*): The rate of change of the material-volume momentum is equal to the sum of the surface forces (due to pressure and viscous stresses) and body forces (such as gravity) acting upon it.

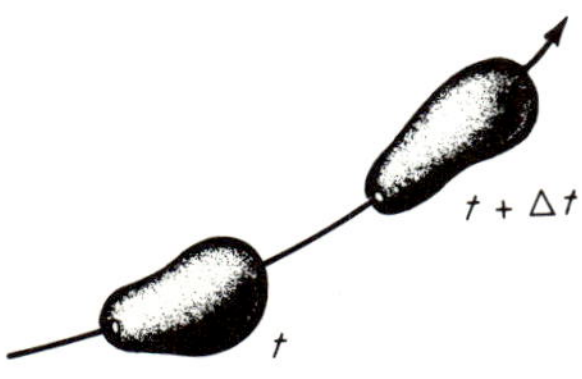

Figure 1.1
Material volume at time t and at time $t + \Delta t$.

3 *Balance of energy (first law of thermodynamics)*: The rate of change of the material-volume energy (internal plus kinetic) is equal to the rate at which forces do work upon it plus the rate at which heat is transferred to it.

4 *Creation of entropy (second law of thermodynamics)*: The (positive) rate of entropy creation within the material volume is equal to the rate of increase of the material entropy plus the rate at which entropy flows out through the material surface.

The first two laws are those of classical mechanics. We will have occasion also to make use of the balance of moment of momentum to obtain one special result, that the viscous stresses display a particular kind of symmetry.

These laws are axiomatic or primitive in that they are not derivable from a smaller number of other (alternative) primitive axioms and are ultimately justified only by experiment.

At very high temperatures, energy, momentum, and entropy may be stored in a thermal radiation field which is superimposed on the flow. These effects, which may for example be important in a plasma at very high temperature, are not considered here.

The only consistent view of continuum mechanics is that which completely ignores the existence of molecular structure. The molecular aggregation of matter is extraneous to the continuum model. We are nevertheless conceptually burdened by our knowledge of the existence of molecules, and it may be useful to comment briefly on the connection between continuum and molecular viewpoints. It has already been noted that interdiffusion of chemical species cannot be accounted for by the present material-volume description; but even if the fluid under consideration is chemically pure (nitrogen gas, say), there is an exchange of matter at the molecular level across a hypothetical material surface. For example, if all the nitrogen molecules in a certain region $V(t)$ could somehow be tagged (conceptually, by coloring them red, say) at some instant, the material surface thus defined would with time become fuzzy by virtue of molecular diffusion.[1] After a sufficient time, the material volume would lose its identity completely. This "paradox" results from a dual model of matter; it is "resolved" by introducing continuum phenomena which account for the molecular exchange of matter. The exchange of *mass* at the molecular level is automatically accounted for by the con-

[1] Though often very slowly. If a stream of dye is introduced into a *laminar* flow of liquid or gas at ordinary density, it retains a smooth filamentous configuration far downstream from the point of introduction.

tinuum velocity as the mass-average velocity. The exchange of *momentum* at the molecular level is accounted for by continuum surface forces, e.g., the pressure. The exchange of *energy* at the molecular level is accounted for by the continuum heat flux. These correspondences can be made precise within the framework of kinetic theory and need not be considered further here.

1.3 Kinematics

The spatial coordinates will be represented by the three Cartesian components x_1, x_2, x_3 (equivalent to x, y, z). These are the components of the position vector[1] $\mathbf{x}$. Let the velocity of a fluid particle be represented by $\mathbf{u}$. Then a *velocity field* is specified by the function

$$\mathbf{u}(\mathbf{x},t) = \mathbf{u}(x_1,x_2,x_3,t) \tag{1.1}$$

That the velocity is defined at every point in space is a property of the continuum model. The specification of such a function would constitute a complete kinematic[2] description of a flow. The discovery of such a function may be the formal objective in a particular problem; it is not in general known in advance. It can of course be manipulated in analysis whether it is known or not; in the following discussion, it may be helpful to imagine that it is a known function.

In addition to the velocity field (1.1) we will also be concerned with the distribution of other flow properties, e.g., the density field $\rho(\mathbf{x},t)$. The point of view taken, with the spatial coordinates and time regarded as independent variables, is usually called *Eulerian*. The eulerian description, e.g., Eq. (1.1), may be thought of as giving a picture of the distribution of properties at each instant of time. An alternative description, called *Lagrangian*, consists in labeling each fluid particle by its position $\mathbf{x}_0$ at some instant t_0, giving functions such as $\mathbf{u} = \mathbf{u}(\mathbf{x}_0,t)$. Although the Lagrangian, or "material," description is well suited to certain calculations, we will not be concerned with it in this book (for a brief treatment, see *Serrin* [1959, pp. 128ff]).

[1] Vector quantities are represented by boldface symbols, a convention introduced by J. W. Gibbs of thermodynamics fame.

[2] From κίνημα = movement.

Indicial Notation

In this section we consider the kinematic (purely geometric) aspects of the fluid motion, as characterized by the velocity field (1.1). It is convenient to make use of *indicial* notation which may be introduced as follows: let the unit vectors along the coordinate axes x_1, x_2, and x_3 respectively be $\mathbf{e}_1$, $\mathbf{e}_2$, and $\mathbf{e}_3$, as shown in Fig. 1.2. Then by vector addition the position vector $\mathbf{x}$ is

$$\mathbf{x} = \mathbf{e}_1 x_1 + \mathbf{e}_2 x_2 + \mathbf{e}_3 x_3 \tag{1.2}$$

For conciseness we introduce the *Einstein summation convention*, whereby the above is written

$$\mathbf{x} = \mathbf{e}_i x_i \qquad i = 1, 2, 3 \tag{1.3}$$

The convention is simply that an index (subscript i in this case) which is *repeated* stands for a summation, with the index being replaced successively by 1, 2, and 3. Thus (1.2) and (1.3) are identical statements. Note that the particular indicial symbol used is of no significance; thus, for example, $a_i b_i$ and $a_k b_k$ have exactly the same meaning.

Scalar dot-product multiplication of (1.3) with the unit vector $\mathbf{e}_k$ gives

$$\mathbf{e}_k \cdot \mathbf{x} = \mathbf{e}_k \cdot \mathbf{e}_i x_i \tag{1.4}$$

Because the unit vectors $\mathbf{e}_i$ are mutually orthogonal, $\mathbf{e}_k \cdot \mathbf{e}_i$ is zero ($i \neq k$) or unity ($i = k$). A concise notation is the *Kronecker delta*[1] δ_{ik}, defined by

$$\delta_{ik} = \begin{cases} 1 & i = k \\ 0 & i \neq k \end{cases} \tag{1.5}$$

[1] After Leopold Kronecker (1823–1891), German mathematician.

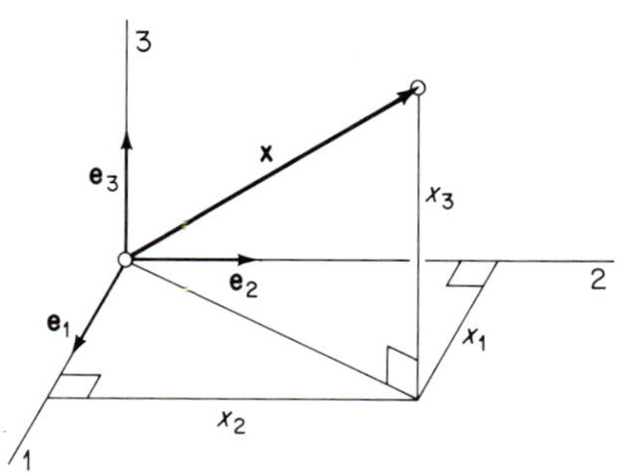

Figure 1.2

Then (1.4) can be written

$$\mathbf{e}_k \cdot \mathbf{x} = \delta_{ik} x_i \tag{1.6}$$

The only surviving term on the right-hand side is the one for which $i = k$, and so this becomes

$$\mathbf{e}_k \cdot \mathbf{x} = x_k \tag{1.7}$$

which is a concise way of writing the *three* equations

$$x_1 = \mathbf{e}_1 \cdot \mathbf{x} \qquad x_2 = \mathbf{e}_2 \cdot \mathbf{x} \qquad x_3 = \mathbf{e}_3 \cdot \mathbf{x}$$

That is, the index k can take on any of the values 1, 2, or 3 at will.

Exactly analogous operations apply to the velocity $\mathbf{u}$ and its Cartesian components u_1, u_2, and u_3. In particular,

$$\mathbf{u} = \mathbf{e}_i u_i \tag{1.8}$$

and taking the scalar product with $\mathbf{e}_k$ gives

$$\mathbf{e}_k \cdot \mathbf{u} = u_k \tag{1.9}$$

Occasionally it is desired to suppress the summation implied by the Einstein convention, in which case the repeated index is enclosed in parentheses. For example,

$$a_n = \lambda_n b_{(n)} \tag{1.10}$$

represents the three distinct relations,

$$a_1 = \lambda_1 b_1 \qquad a_2 = \lambda_2 b_2 \qquad a_3 = \lambda_3 b_3$$

Material Derivative

The material (or substantial) derivative represents the time rate of change as seen by an observer moving with a fluid particle. Let F be any differentiable function of space and time (such as the fluid density ρ or velocity $\mathbf{u}$), $F = F(\mathbf{x},t)$. For arbitrary changes in the four independent variables, making use of the summation convention, we write

$$dF = \frac{\partial F}{\partial t}\,dt + \frac{\partial F}{\partial x_i}\,dx_i$$

where each of the partial differentiations is performed with the three "inactive" independent variables held constant. If the spatial variables are now *restricted* to be some particular functions of time, $x_i = x_i(t)$, which may be viewed as specifying the motion of an observer in space, we can write

$$\frac{dF}{dt} = \frac{\partial F}{\partial t} + \frac{\partial F}{\partial x_i}\frac{dx_i}{dt}$$

In the important special case for which $x_i(t)$ is restricted to be the position of any infinitesimal fluid particle, dx_i/dt is just the velocity component u_i. For this case we use by convention the special notation D/Dt (first used by G. G. Stokes, ca. 1850) in place of d/dt; then

$$\frac{DF}{Dt} \equiv \frac{\partial F}{\partial t} + u_i \frac{\partial F}{\partial x_i} \tag{1.11}$$

This is the *material derivative* or *substantial derivative* of F. The first term is called the *unsteady part*, and the last three terms are called the *convective part*. In vector notation this is written

$$\frac{DF}{Dt} = \frac{\partial F}{\partial t} + (\mathbf{u} \cdot \nabla)F \tag{1.12}$$

The material derivative has the following "physical" interpretation: an observer traveling with the fluid particle, located at the point $\mathbf{x}$ at some instant t, observes a time rate of change of the quantity F given by DF/Dt. Part of this change is due to local unsteadiness, and part is due to motion of the particle to a location where F has a different value.

In particular, if F is the fluid velocity $\mathbf{u}$, we have the *material acceleration*

$$\frac{D\mathbf{u}}{Dt} = \frac{\partial \mathbf{u}}{\partial t} + u_i \frac{\partial \mathbf{u}}{\partial x_i} \tag{1.13}$$

This has the same meaning as acceleration in ordinary particle mechanics. An example may be helpful: suppose a small stick is dropped into a river, where it flows through a narrow constriction, and that the stick faithfully follows the motion of surrounding water. The measured acceleration of the stick is just the material (fluid) acceleration at that instant, at the position of the stick. Furthermore if the stick is imagined to be equipped with a device for continuously measuring the local water temperature, then the measured rate of change of the temperature at some point is just the local value of the material derivative of temperature.

The derivative operator D/Dt has the usual properties; e.g., the reader can directly verify that

$$\frac{D}{Dt}ab = a\frac{Db}{Dt} + b\frac{Da}{Dt}$$

However, it does not necessarily *commute* with other operators. For example, the operations D/Dt and the curl are not in general commutative; the relation

$$\frac{D}{Dt}(\nabla \times \mathbf{u}) = \nabla \times \frac{D\mathbf{u}}{Dt}$$

is not valid in general.

Local Deformation and Rotation of the Fluid

A small material volume deforms with time according to the relative motion of neighboring fluid particles. The velocity of the particle located at $\mathbf{x} + d\mathbf{x}$ relative to that at $\mathbf{x}$ is

$$d\mathbf{u} = \frac{\partial \mathbf{u}}{\partial x_k} dx_k \tag{1.14}$$

as illustrated in Fig. 1.3. The relative velocity $d\mathbf{u}$ has three components du_i,

$$du_i = \frac{\partial u_i}{\partial x_k} dx_k \tag{1.15}$$

There are nine possible values (called *components*) for the coefficient $\partial u_i/\partial x_k$, corresponding to three different values for i and three for k.

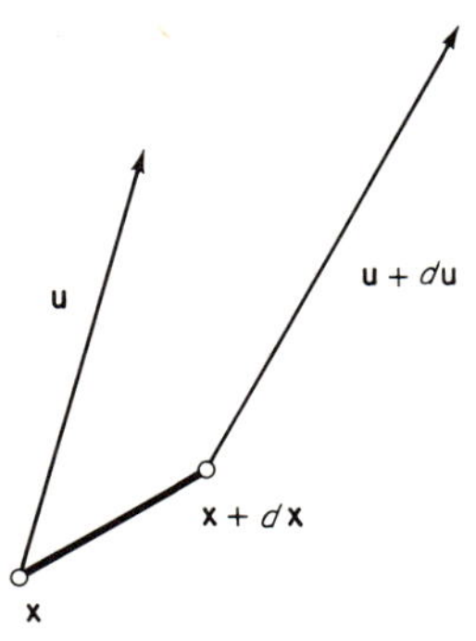

Figure 1.3

Collectively these components form the *velocity-gradient tensor* at the point $\mathbf{x}$. Clearly, if each of the nine values is known, it is possible to calculate the motion of *all* neighboring points a distance dx away. Such a calculation would thus show how an infinitesimal fluid particle was deforming with time and (with the velocity $\mathbf{u}$) amounts to a complete description of the state of motion in the neighborhood of the point $\mathbf{x}$.

It is useful to split the velocity gradient tensor into symmetric and antisymmetric parts; formally,

$$\frac{\partial u_i}{\partial x_k} = \frac{1}{2}\left(\frac{\partial u_i}{\partial x_k} + \frac{\partial u_k}{\partial x_i}\right) + \frac{1}{2}\left(\frac{\partial u_i}{\partial x_k} - \frac{\partial u_k}{\partial x_i}\right) \tag{1.16}$$

The first tensor on the right is the *rate-of-deformation tensor* (hereafter called simply the deformation tensor and called the rate-of-strain tensor by some authors). This tensor is denoted

$$D_{ik} \equiv \frac{1}{2}\left(\frac{\partial u_i}{\partial x_k} + \frac{\partial u_k}{\partial x_i}\right) \tag{1.17}$$

and is symmetric with respect to interchange of the two indices, for example, $D_{13} = D_{31}$. The second tensor is the *spin tensor*,

$$\Omega_{ik} \equiv \frac{1}{2}\left(\frac{\partial u_i}{\partial x_k} - \frac{\partial u_k}{\partial x_i}\right) \tag{1.18}$$

which is antisymmetric with respect to interchange of the two indices, for example, $\Omega_{13} = -\Omega_{31}$.

The above tensors are called second order because the components are labeled by two indices (by this nomenclature a vector is a first-order tensor and a scalar is a zero-order tensor). It is usual to represent a second-order tensor by a matrix,[1] with the value of the first index corresponding to the row number and the value of the second index correspond-

[1] A Cartesian tensor is distinguished from an ordinary matrix in that certain combinations of the components are *invariants* under a rotation of axes, and the components themselves transform according to a certain law for a rotation of axes, viz. (for a second-order tensor) $\bar{A}_{ik} = l_{ji}l_{mk}A_{jm}$, where $\bar{A}_{ik}$ is the component in the rotated coordinate system and l's are direction cosines. Convenient rules to remember are that any product of vectors, for example, $u_i x_k$ or $u_i u_k$, is a tensor and that any spatial derivative of a vector field, for example, $\partial u_i/\partial x_k$, is a tensor.

ing to the column number. For example, the deformation tensor D_{ik} is written

$$D_{ik} = \begin{bmatrix} D_{11} & D_{12} & D_{13} \\ D_{21} & D_{22} & D_{23} \\ D_{31} & D_{32} & D_{33} \end{bmatrix} \tag{1.19}$$

The components for which $i = k$ form the *main diagonal* (shown as a dashed line) of the tensor.

For conciseness, the *comma notation* for derivatives is introduced, viz.,

$$u_{i,k} \equiv \frac{\partial u_i}{\partial x_k}$$

In this notation the deformation tensor is written out explicitly

$$D_{ik} = \begin{bmatrix} u_{1,1} & \tfrac{1}{2}(u_{1,2} + u_{2,1}) & \tfrac{1}{2}(u_{1,3} + u_{3,1}) \\ \tfrac{1}{2}(u_{2,1} + u_{1,2}) & u_{2,2} & \tfrac{1}{2}(u_{2,3} + u_{3,2}) \\ \tfrac{1}{2}(u_{3,1} + u_{1,3}) & \tfrac{1}{2}(u_{3,2} + u_{2,3}) & u_{3,3} \end{bmatrix} \tag{1.20}$$

which shows by inspection the symmetric property. The spin tensor may be written out similarly, and will show zeros along the main diagonal.

We have decomposed the tensor $u_{i,k}$ of Eq. (1.15) into the sum of a symmetric tensor D_{ik} and an antisymmetric tensor Ω_{ik}. Such a step is formally correct but at first sight may appear to be a needless elaboration of a simple equation. It remains to discuss briefly the geometric significance of the deformation tensor D_{ik} and the *vector* of the spin tensor, which justifies their introduction.

Because of the antisymmetry property, the spin tensor has only three distinct nonzero components, which are in fact the components of a vector. We will take advantage of this and replace the spin tensor by the appropriate vector. The *vorticity* vector $\mathbf{\Omega}$ (sometimes called the rotation) is defined to be the curl of the velocity,

$$\mathbf{\Omega} \equiv \nabla \times \mathbf{u} \tag{1.21}$$

The components of $\mathbf{\Omega}$ are just twice the appropriate components of the spin tensor; writing out the Cartesian components of (1.21) by the usual

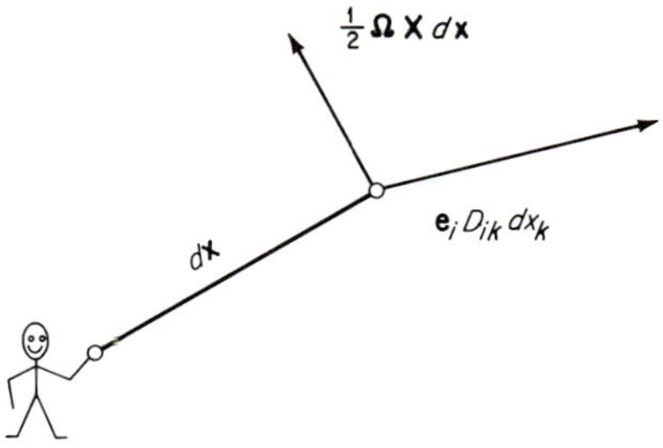

Figure 1.4

determinant expansion for the curl and comparing with (1.18) gives for the vector components[1] Ω_i

$$\begin{aligned}\Omega_1 &= 2\Omega_{32} = -2\Omega_{23}\\ \Omega_2 &= 2\Omega_{13} = -2\Omega_{31}\\ \Omega_3 &= 2\Omega_{21} = -2\Omega_{12}\end{aligned} \tag{1.22}$$

The fundamental deformation equation (1.15) is rewritten

$$du_i = D_{ik}\, dx_k + \Omega_{ik}\, dx_k \tag{1.23}$$

Multiplying through by $\mathbf{e}_i$ gives, in mixed notation,

$$d\mathbf{u} = \mathbf{e}_i\, D_{ik}\, dx_k + \tfrac{1}{2}\boldsymbol{\Omega} \times d\mathbf{x} \tag{1.24}$$

The second term can be verified by writing out all the components explicitly. The relative velocity $d\mathbf{u}$ is thus resolved into two components, as indicated in Fig. 1.4. The component $\frac{1}{2}\boldsymbol{\Omega} \times d\mathbf{x}$ is necessarily perpendicular to $d\mathbf{x}$, but the deformation component $\mathbf{e}_i\, D_{ik}\, dx_k$ can in general have any direction. There are, however, three special and mutually perpendicular choices for $d\mathbf{x}$ for which $\mathbf{e}_i\, D_{ik}\, dx_k$ is just parallel to $d\mathbf{x}$: these directions define the *principal axes* at the point in question.[2]

[1] Formally, these relations are given by

$$\Omega_{ik} = \tfrac{1}{2}\varepsilon_{ijk}\Omega_j$$

where ε_{ijk} is the *Levi-Civita permutation symbol* defined by

$$\varepsilon_{ijk} = \begin{cases} 0 & \text{if any two indices are alike}\\ +1 & \text{if indices are in cyclic order}\\ -1 & \text{if indices are not in cyclic order}\end{cases}$$

and cyclic order is 123, 312, or 231. Thus, for example, $\varepsilon_{133} = 0$, $\varepsilon_{123} = +1$, $\varepsilon_{132} = -1$. The expansion for the curl is given in Appendix B.

[2] For a discussion of principal axes, see *Aris* [1962, p. 92].

EXAMPLE 1.1 RIGID-BODY MOTION

The simplest nontrivial case of motion is that of rigid-body motion, with a velocity field given by

$$\mathbf{u} = \mathbf{u}_0 + \boldsymbol{\omega} \times \mathbf{x} \tag{1.25}$$

where $\mathbf{u}_0$ is the velocity of the center of rotation and $\boldsymbol{\omega}$ is the angular velocity. The center of rotation has been put instantaneously at the origin without loss of generality. An example of a *fluid* motion of this kind would be a bucket of water (or a vessel of gas) rotated about an axis, say the axis of symmetry; if the angular velocity $\boldsymbol{\omega}$ is constant and $\mathbf{u}_0$ is constant, all fluid motion *relative* to the bucket will eventually cease, and the fluid is kinematically a rigid body. The velocity components are, from (1.25),

$$\begin{aligned} u_1 &= u_{01} + \omega_2 x_3 - \omega_3 x_2 \\ u_2 &= u_{02} + \omega_3 x_1 - \omega_1 x_3 \\ u_3 &= u_{03} + \omega_1 x_2 - \omega_2 x_1 \end{aligned}$$

The components of the velocity gradient tensor are then, by direct differentiation, $u_{1,1} = 0$, $u_{1,2} = -\omega_3$, $u_{2,1} = +\omega_3$, and so on. We find thus that *all* components of the deformation tensor are zero,

$$D_{ik} \equiv 0$$

which will always be true for rigid-body motion. Therefore the relative velocity of neighboring points is, from (1.24),

$$d\mathbf{u} = \tfrac{1}{2}\boldsymbol{\Omega} \times d\mathbf{x}$$

But from (1.25)

$$d\mathbf{u} = \boldsymbol{\omega} \times d\mathbf{x}$$

Thus, the vorticity vector $\boldsymbol{\Omega} = 2\boldsymbol{\omega}$, or is everywhere twice the angular velocity. Every infinitesimal fluid line $d\mathbf{x}$ rotates with the angular velocity $\boldsymbol{\Omega}/2$. These results provide some justification for the decomposition of the velocity gradient tensor and for the naming of the deformation tensor and the vorticity (rotation) vector.

That $D_{ik} = 0$ for rigid-body motion has the important consequence that the local value of D_{ik} for an *arbitrary* fluid motion is *independent of the motion of an observer*, i.e., is independent of the motion of the coordinate frame in which D_{ik} is found. This follows from the argument (which can be made quite formal; see Prob. 1.15) that the effect of the observer's motion is just to add a rigid-body velocity, i.e., Eq. (1.25), to the arbitrary velocity field. Quantities which have a value independent of the observer's motion are said to be *objective*. While the deformation tensor is objective, the vorticity vector cannot be, because the effect of the observer's motion (with angular velocity $\boldsymbol{\omega}$) is to add $-2\boldsymbol{\omega}$ to the vorticity. In the preceding example the vorticity becomes zero only for an observer rotating with the fluid.

The instantaneous local state of motion is completely described via (1.24) by the local values of the deformation tensor D_{ik} and the vorticity vector $\boldsymbol{\Omega}$. The more important properties of the deformation tensor and the vorticity are summarized below:

1 The components D_{ik} of the deformation tensor are individually zero for rigid-body motion. The deformation tensor represents essentially *fluid* deformations.

2 The sum of the diagonal components of the deformation tensor, $D_{ii} = u_{i,i} = \nabla \cdot \mathbf{u}$, represents the relative rate of volume growth of an infinitesimal fluid particle and is invariant with respect to the choice of coordinate directions. (The sum of the diagonal components of a tensor is often called the *trace*.)

3 At every point in the velocity field there are three mutually perpendicular directions (the *principal* directions); if the coordinate axes are aligned with these directions, the deformation tensor reduces to *diagonal form*, in which all the off-diagonal components ($i \neq k$) are zero. The set of three infinitesimal fluid lines (a *fluid triad*) instantaneously aligned with the principal directions rotates like a solid body with an extensive deformation along each of the lines (Fig. 1.5).

4 The deformation tensor is objective while the vorticity vector is not.

5 The vorticity vector $\boldsymbol{\Omega} = \nabla \times \mathbf{u}$ represents a local rate of rotation of a

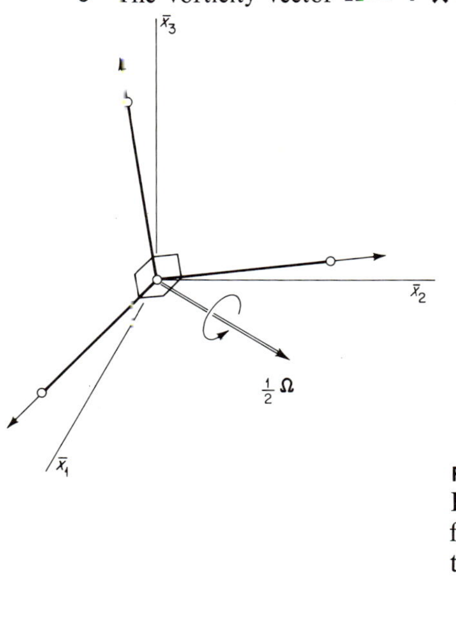

Figure 1.5
Rotation and extension of a principal fluid triad instantaneously aligned with the principal axes $\bar{x}_i$.

fluid particle. In particular, $\mathbf{\Omega}$ is just twice the angular velocity of rigid-body motion and just twice the angular velocity of a principal fluid triad.

A large body of gasdynamics is based on the assumption $\mathbf{\Omega} = \nabla \times \mathbf{u} = 0$ (this approximation is discussed in Chap. 2). Such flows are appropriately called *irrotational.* Classical hydrodynamics is based on this and the further approximation $\nabla \cdot \mathbf{u} = 0$.†

Reynolds' Transport Theorem

One further kinematical result is needed. Let $V^*(t)$ be a moving volume (not necessarily a material volume) with bounding surface $S^*(t)$ and outward unit normal vector $\mathbf{n}$, as shown in Fig. 1.6. The local boundary velocity is $\mathbf{b}$ and may vary over the surface S^*. Such an arbitrary moving volume is often called a *control volume*; the boundary need not in general be identified with any physical boundaries. Let $\chi(\mathbf{x},t)$ be any summable continuous function, such as the density ρ. The "content" of V^* is $\int \chi \, dV$. It is desired to find an expression for the rate of change with time of this integral,

$$\frac{d}{dt} \int_{V^*(t)} \chi \, dV$$

Two contributions to the rate of change are recognized: (1) the value of χ may be changing with time within the volume, giving a rate of change $\partial\chi/\partial t \, dV$ in each volume element dV; (2) the moving surface

† The condition $\nabla \times \mathbf{u} = 0$ is automatically satisfied by the introduction of a velocity potential ϕ, defined by $\mathbf{u} = \nabla\phi$. If the fluid is incompressible, $\nabla \cdot \mathbf{u} = 0$, giving Laplace's equation,

$$\nabla^2\phi = 0$$

which is the basic equation of classical hydrodynamics. This equation is not of interest in gasdynamics since it neglects compressibility, but it will appear in modified form when $\nabla \cdot \mathbf{u} \neq 0$.

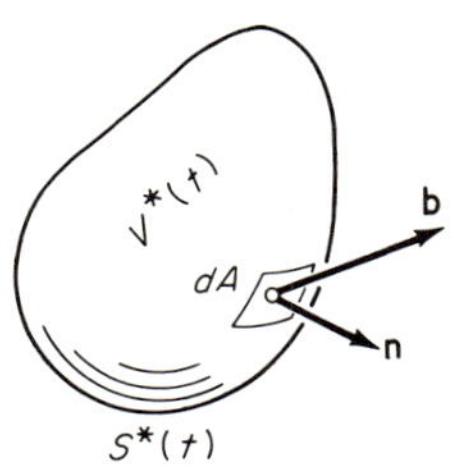

Figure 1.6
Control volume.

envelops new regions of space with time, giving a rate of change $\chi\mathbf{b}\cdot\mathbf{n}\,dA$ at each surface element dA. Adding these contributions gives[1]

$$\frac{d}{dt}\int_{V^*(t)}\chi\,dV = \int_{V^*(t)}\frac{\partial\chi}{\partial t}\,dV + \int_{S^*(t)}\chi\mathbf{b}\cdot\mathbf{n}\,dA \tag{1.26}$$

This statement, Reynolds' transport theorem, is analogous to *Leibniz' formula* for the derivative of an integral with variable limits. In the present case the variable limits correspond of course to the moving boundary.

It is emphasized that (1.26) is a kinematical statement quite independent of the usual laws of physics. The only "physical" interpretation is that (Euclidean) volume is conserved.

Often the transport theorem is applied to a *material* volume, in which case the boundary velocity $\mathbf{b}$ is just the fluid velocity $\mathbf{u}$ and the statement becomes

$$\frac{d}{dt}\int_{V(t)}\chi\,dV = \int_{V(t)}\frac{\partial\chi}{\partial t}\,dV + \int_{S(t)}\chi\mathbf{u}\cdot\mathbf{n}\,dA \tag{1.27}$$

As a very simple example, let $\chi = 1$. Then this becomes

$$\frac{dV}{dt} = \int_{S(t)}\mathbf{u}\cdot\mathbf{n}\,dA$$

Application of the divergence theorem[2] gives

$$\frac{dV}{dt} = \int_{V(t)}\nabla\cdot\mathbf{u}\,dV$$

Taking the limit as $V \to 0$ then gives

$$\lim_{V\to 0}\frac{1}{V}\frac{dV}{dt} = \nabla\cdot\mathbf{u} \tag{1.28}$$

This formula is due to Euler (1755). The local divergence $\nabla\cdot\mathbf{u}$ is thus interpreted as the relative rate of volume growth of an infinitesimal fluid particle, a result which has already been referred to in the summary of deformation properties. By definition, an incompressible fluid has constant density; it follows that $\nabla\cdot\mathbf{u} = 0$ for an incompressible fluid.

[1] The preceding derivation is heuristic and equivalent to the original version of Osborne *Reynolds* [1903, arts. 13 and 14]. A comprehensive derivation is given by *Truesdell and Toupin* [1960, p. 347].

[2] The divergence theorem, together with related formulas, is given in Appendix B.

1.4 Constitutive equations

The relations between observable effects and the internal constitution of matter are described by *constitutive equations.* Specifically, we consider in this section the relations between stress and rate of deformation and between heat flux and temperature gradient.

By hypothesis, only two kinds of physical forces need to be reckoned. The first kind consists of *body force* acting on matter in bulk, with an overall force proportional to the amount of mass (or if electric and magnetic forces are considered, to the net charge; this case will not be specifically considered). The most important example of body force is the gravitational attraction $\mathbf{g}$ per unit mass.

The second kind of force consists of *surface force*, in which the overall force is proportional to the amount of surface acted upon. The surface forces considered in fluid mechanics need not act on a *physical* surface, e.g., a material interface, but may be applied to an imaginary surface, in particular the boundary of a material volume. It should be noted that the forces under consideration are *not* the membrane-type forces called *surface tension*, which can act only at phase boundaries. The most familiar example of a surface force is simple hydrostatic pressure. Let $\mathbf{T}$ be the vector surface force per unit area, sometimes called the surface traction or stress vector. The local orientation of the material surface is defined by the unit normal vector $\mathbf{n}$ (Fig. 1.7). By hypothesis, the value of $\mathbf{T}$ depends on the orientation of the surface, as well as position and time; thus,

$$\mathbf{T} = \mathbf{T}(\mathbf{n},\mathbf{x},t) \tag{1.29}$$

At a given point in space and time we then write $\mathbf{T} = \mathbf{T}(\mathbf{n})$ (it may be helpful to recall that hydrostatic pressure has this character).

Consider the case for which the surface normal $\mathbf{n}$ is aligned with one of the Cartesian axes; then $\mathbf{n} = \mathbf{e}_i$, say. The corresponding surface traction $\mathbf{T}(\mathbf{e}_i)$, which is not necessarily parallel to $\mathbf{e}_i$, may be broken into

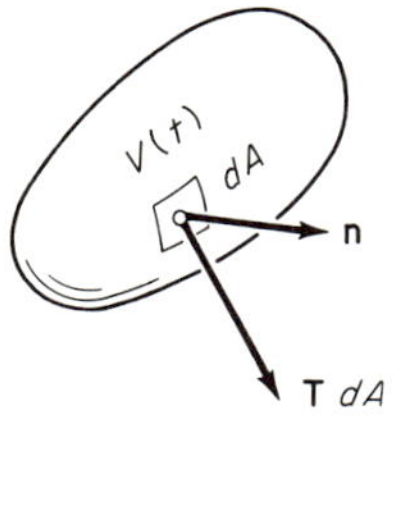

Figure 1.7

three Cartesian components; specifically, let σ_{ik} be the kth component of $\mathbf{T}(\mathbf{e}_i)$, which may then be written

$$\mathbf{T}(\mathbf{e}_i) = \mathbf{e}_k\sigma_{ik} \tag{1.30}$$

Defines plane (the i in σ_{ik}); Defines direction of component (the k in σ_{ik})

The (ensemble) σ_{ik} is called the *stress tensor*. At a point in space the nine components are of physical interest.

A material volume of infinitesimal size must be in equilibrium from the surface forces alone, because the surface forces are proportional to area while the body force and momentum change are proportional to volume. This result allows the traction $\mathbf{T}(\mathbf{n})$ on a surface with arbitrary orientation $\mathbf{n}$ to be expressed in terms of σ_{ik} (see Prob. 1.16) as follows:

$$\boxed{T_k(\mathbf{n}) = n_i\sigma_{ik}} \tag{1.31}$$

A similar argument applied to the moment of momentum for a small material volume shows that for all fluids of interest the stress tensor is symmetric,[1] viz.,

$$\sigma_{ik} = \sigma_{ki} \tag{1.32}$$

For a fluid at rest or in solid-body motion [as discussed in Example 1.1 (page 13)] the stress is expected to reduce to simple hydrostatic pressure. This observation motivates the decomposition of the stress tensor into two parts,

$$\sigma_{ik} = -P\delta_{ik} + \Sigma_{ik} \tag{1.33}$$

The first part, $-P\delta_{ik}$, may be called the *pressure stress*. The second part, Σ_{ik}, is the *viscous stress* (or deviatoric stress) and will be required to vanish when $D_{ik} = 0$, that is, for a fluid at rest or in solid-body motion. It is noted that an arbitrary decomposition of this kind is always possible. Since σ_{ik} and $P\delta_{ik}$ are symmetric, the viscous-stress tensor Σ_{ik} is also necessarily symmetric.

In the hydrostatic case, with $\Sigma_{ik} = 0$, the stress is $\sigma_{ik} = -P\delta_{ik}$, and (1.31) gives for an arbitrary internal surface

$$T_k(\mathbf{n}) = n_i\sigma_{ik} = -n_iP\delta_{ik} = -Pn_k$$

[1] This result fails only for fluids with *inherent* angular momentum due to the rotation of individual molecules, with their angular momentum vectors not randomly distributed so that there is a net contribution. There is apparently little evidence for the existence of this phenomenon, which is the subject of current research.

or

$$\mathbf{T}(\mathbf{n}) = -P\mathbf{n} \tag{1.34}$$

The surface traction is a compressive force acting normal to the surface, consistent with our ideas about pressure.

When there is relative motion, "hydrostatic" stress has no clear significance. In what follows, we will *define* P to be the thermodynamic pressure, i.e., the pressure given by an equation of state, such as $P(\rho,T)$, where ρ is density and T the absolute temperature.[1]

Relation between Stress and Rate of Deformation

The local viscous stress Σ_{ik} should depend on the local fluid motion through the deformation D_{ik} but not through the vorticity (rotation) $\mathbf{\Omega}$. This follows formally from the requirement that stress be objective and therefore dependent only on objective quantities. Physically, it is consistent with our idea of a *fluid* as a substance for which nonhydrostatic stress vanishes in the absence of relative motion.

The most general *linear* dependence of viscous stress on deformation is thus

$$\Sigma_{ik} = \alpha_{ikmn} D_{mn} \tag{1.35}$$

in which each of the Σ's depends on all nine of the D's. There are thus $3^4 = 81$ *coefficients of viscosity* α.

Such a large number of material parameters would be extremely awkward for experiment and theory. Fortunately, arguments of symmetry and isotropy reduce the 81 parameters to only 2. These arguments are: (1) The viscous-stress tensor and the deformation tensor are symmetric. This requires that $\alpha_{ikmn} = \alpha_{kimn} = \alpha_{iknm} = \alpha_{kinm}$. These relations reduce the 81 coefficients to 36 distinct values. (2) The fluid is *isotropic*, i.e., has no preferred directions (as would be the case, for example, if the fluid had the structure of aligned filaments). This requires that a tensile

[1] Like many other ideas in this chapter, that of a *pressure* acting on a surface internal to the fluid is due to Leonhard *Euler*, who introduced it with the symbol p (*pressus*) in 1753. [The famous equation of Daniel *Bernoulli* (approximately 1730), now written $P + \rho u^2/2 = \text{const}$, was never put in such terms by Bernoulli himself.] The generalization to internal forces not necessarily normal to the surface is due to Augustin Louis *Cauchy* (1823), though *Newton* (1687) had considered tangential internal resistance "from want of slipperiness" in connection with vortex motions. Much of the historical commentary in this book is based on the splendid summary of *Truesdell* [1954].

stress can result only in a deformation in the direction of the tension and that the *principal axes of deformation are also the principal axes of stress.* The 36 coefficients are thereby reduced to only 2. Omitting further details, the resulting stress law is

$$\Sigma_{ik} = 2\mu(D_{ik} - \tfrac{1}{3}\delta_{ik}D_{mm}) + \mu_v\delta_{ik}D_{mm} \tag{1.36}$$

where the *shear viscosity* μ and the *bulk viscosity* μ_v are coefficients dependent only on the local thermodynamic state. Then the full stress σ_{ik} is given by

$$\boxed{\sigma_{ik} = -P\delta_{ik} + 2\mu(D_{ik} - \tfrac{1}{3}\delta_{ik}D_{mm}) + \mu_v\delta_{ik}D_{mm}} \tag{1.37}$$

The nine components of the stress tensor are given for each combination of the indices i and k. This is the required constitutive equation for stress in a linearly viscous fluid.[1] Almost all fluids in ordinary experience are very nearly of this type, including air, water, whole blood, and a host of others.[2] The exceptions, called non-Newtonian fluids, include high polymers, printing inks, and certain lubricants.

The average normal stress, i.e., the average of the stress components normal to the coordinate planes, $\bar{\sigma}$ is, from (1.37),

$$\bar{\sigma} = \tfrac{1}{3}\sigma_{ii} = -P + \mu_v D_{mm} \tag{1.38}$$

in which $D_{mm} = \nabla \cdot \mathbf{u}$ is the relative rate of volume increase of an infinitesimal fluid particle. Then for incompressible fluids, with $\nabla \cdot \mathbf{u} = 0$ by virtue of (1.28), the average normal stress is just the (negative of) pressure. For real, compressible fluids the deviation of the average normal stress from the pressure is just the bulk viscosity times the rate of volume dilatation.

The normal stresses $\sigma_{i(i)}$ are by convention positive if they are tensile; e.g., simple hydrostatic stress corresponds to a negative value for each of the diagonal components of σ_{ik}. Real fluids, however, cannot sustain

[1] The nomenclatures for this relation are numerous. In recognition of Newton's conjectures on tangential friction in the *Principia*, fluids satisfying (1.37) are often called *Newtonian*. According to Truesdell, the name best justified historically is the *Navier-Poisson law*. Occasionally, fluids satisfying (1.37) are called *Stokesian*, though the relation proposed by G. G. Stokes is slightly more general.

[2] Blood may display non-Newtonian behavior at very low deformation rates or in very small capillaries but for most purposes is described by (1.37).

tensile stresses under ordinary circumstances, hence the average normal stress $\bar{\sigma} < 0$.

According to *Stokes' hypothesis* (G. G. Stokes, 1845), which was not, however, put forward very strongly by Stokes himself, the bulk viscosity μ_v should be set equal to zero. Stokes reasoned that in an isotropic fluid expansion (see Example 1.5, page 28), the local stress should be independent of the dilatation rate $\nabla \cdot \mathbf{u}$, adding that since $\nabla \cdot \mathbf{u}$ is normally small, it should not make very much difference whether μ_v is zero or not. In many treatments of fluid mechanics Stokes' hypothesis is accepted without further ado, so that (1.37) contains only the "ordinary" shear viscosity μ.

Experimentally, however, μ_v is in general found to be other than zero. In some cases it may in fact be two *orders of magnitude* greater than the ordinary shear viscosity μ (see Table 1.1). Only for dilute monatomic gases (helium, argon, etc.) is μ_v found to be zero, both on the basis of experiments and the Boltzmann kinetic theory. For many fluid motions $\nabla \cdot \mathbf{u}$ is small enough so that, as Stokes suggested, the effect of bulk viscosity is negligible. In gasdynamics (specifically, acoustic motions and the interior of shock waves), however, this effect is important. The bulk viscosity is further discussed in Chap. 2.

Four simple examples of stress-deformation fields follow.

EXAMPLE 1.2 PLANE LAMINAR SHEARING

A possible velocity distribution, with $u_2 = u_3 = 0$ and the velocity gradient in the x_2 direction, is shown in Fig. 1.8. Such a motion occurs in a *laminar boundary layer*, at least approximately, and in *Couette flow*. As shown in the figure, there is a solid plane boundary at $x_2 = 0$ where the velocity vanishes, $u_2 = 0$; this is the *no-slip condition*, that the tangential component of velocity vanish at a solid boundary (this condition, however, plays no part in the calculation to follow).

At some typical point a in the velocity field we presume a given value for

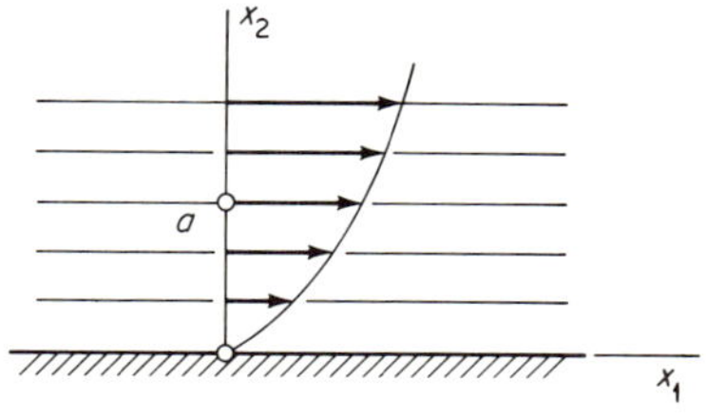

Figure 1.8

$\partial u_1/\partial x_2 = u_{1,2}$ (all other gradients are zero). Then the components of the deformation tensor are

$$D_{12} = \tfrac{1}{2}(u_{1,2} + u_{2,1}) = \tfrac{1}{2}u_{1,2}$$
$$D_{21} = \tfrac{1}{2}(u_{2,1} + u_{1,2}) = \tfrac{1}{2}u_{1,2}$$

and all other components D_{ik} are identically zero. Note that $D_{mm} = \nabla \cdot \mathbf{u} = 0$ and the motion is incompressible, though this does *not* preclude a density gradient in the x_2 direction.

From (1.37) the corresponding stresses are

$$\sigma_{21} = \sigma_{12} = \mu u_{1,2}$$
$$\sigma_{11} = \sigma_{22} = \sigma_{33} = -P$$

and all other stresses are zero. At the solid boundary there is a tangential stress $\sigma_{21} = \mu(u_{1,2})_{x_2=0}$, which tends to pull the boundary in the direction of fluid motion (this stress is called skin friction or viscous drag in aeronautics), as well as a normal stress $-P$.

The simple shearing stress σ_{21} is what is commonly meant when one speaks of viscous stress. It will be of interest to find some numerical values. Water flowing in a capillary tube of diameter 0.1 mm behaves as a Newtonian fluid at shear rates as large as $u_{1,2} = 10^6\ \text{s}^{-1}$. With the shear viscosity μ of water equal to 10^{-3} kg/(m)(s), the corresponding shear stress at this extreme deformation rate is $\sigma_{21} = (10^{-3})(10^6) = 10^3\ \text{N/m}^2$, or about 0.01 atm. In *air* the extreme shear rate of $2 \times 10^7\ \text{s}^{-1}$ is found in certain gas bearings. With $\mu = 1.8 \times 10^{-5}$ kg/(m)(s), the corresponding shear stress is $\sigma_{21} = (1.8 \times 10^{-5}) \times (2 \times 10^7) \approx 4 \times 10^2\ \text{N/m}^2$, or about 0.004 atm.

The values given above are for extreme situations yet are still very small compared (say) to atmospheric pressure. It would, however, be a mistake to assume that viscous stresses were negligible on this account! An appropriate estimate of the importance of viscous stress can be made by comparing the magnitude of the viscous terms with other terms in the equations of motion (Chap. 3). One profound influence of the viscous stresses which should be mentioned is their effect on the *separation* of a flow from a solid surface.

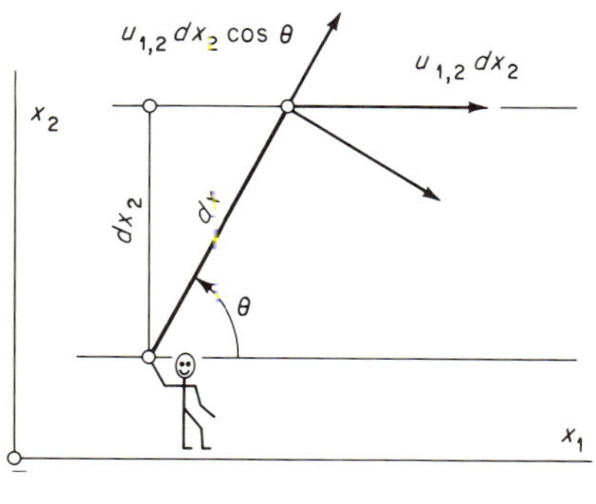

Figure 1.9

The vorticity $\boldsymbol{\Omega}$ is found to be

$$\boldsymbol{\Omega} = \nabla \times \mathbf{u} = -\mathbf{e}_3 u_{1,2} \tag{1.39}$$

The vorticity vector is everywhere perpendicular to the plane of the motion (as will always be the case for plane flows). Even though individual fluid particles move along straight lines, the flow is *rotational*. This simple shearing motion is, in fact, the archetype of rotational flows.

The principal directions for stress and deformation can be simply determined from the condition that the (logarithmic) rate of elongation $\dot{E}$ of a fluid line is *stationary* when the line is aligned with one of the principal directions (Fig. 1.9). This condition will simply be used here, and no formal proof of its validity will be given (for a proof, see *Aris* [1962]).

The relative velocity between the end points is $u_{1,2}\,dx_2$, and the extensive component of this is $u_{1,2}\,dx_2 \cos\theta$. Then, with $dx_2 = dx \sin\theta$, the rate of elongation per unit length is

$$\dot{E} = u_{1,2} \sin\theta \cos\theta \tag{1.40}$$

and this is a stationary value if

$$\frac{d\dot{E}}{d\theta} = u_{1,2}(\cos^2\theta - \sin^2\theta) = 0$$

Then $\sin\theta = \pm\cos\theta$, giving the principal directions

$$\theta = \pm 45^\circ$$

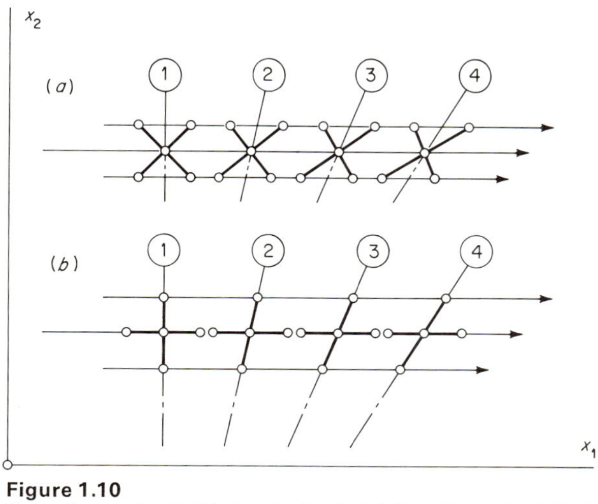

Figure 1.10
Rotation of a fluid dyad: (*a*) initially aligned with the principal axes; (*b*) initially aligned with the original axes.

Equation (1.40) then gives for the principal deformations, with axis $\bar{x}_2$ at 45° (say) and $\bar{x}_1$ at $-45°$,

$$\bar{u}_{1,1} = \bar{D}_{11} = -\tfrac{1}{2}u_{1,2} \qquad \bar{u}_{2,2} = \bar{D}_{22} = +\tfrac{1}{2}u_{1,2}$$

since $\dot{E}_i$ along a principal axis i is just the derivative $\bar{u}_{i,(i)}$. Figure 1.10 shows the motion of a fluid dyad initially aligned with the principal axes and another initially aligned with the original axes (the construction assumes a steady flow). Note that in the former case (Fig. 1.10*a*) the dyad initially rotates like a solid body (the motion is rotational!). The associated principal stresses are given by (1.37).

EXAMPLE 1.3 PURELY EXTENSIVE MOTION

If the fluid motion is in one direction only and the velocity gradient is in the same direction, the flow may be said to be purely one-dimensional (Fig. 1.11). Such flows are typical of *acoustic* motions and longitudinal waves in general. With $u_2 = u_3 = 0$ and all gradients other than $\partial u_1/\partial x_1 = u_{1,1}$ equal to zero, the only nonzero component of the deformation tensor is

$$D_{11} = u_{1,1}$$

The reader should be able to convince himself that the chosen axes are principal and that the motion is irrotational! The volume dilatation is

$$D_{mm} = \nabla \cdot \mathbf{u} = u_{1,1}$$

so that the fluid is necessarily compressible.

From (1.37) the stresses are

$$\begin{aligned} \sigma_{11} &= -P + (\mu_v + \tfrac{4}{3}\mu)u_{1,1} \\ \sigma_{22} = \sigma_{33} &= -P + (\mu_v - \tfrac{2}{3}\mu)u_{1,1} \\ \sigma_{ik} &= 0 \qquad i \neq k \end{aligned} \tag{1.41}$$

so that the bulk viscosity plays an important role in the motion. An extreme value for the viscous normal stress $\Sigma_{11} = (\mu_v + \tfrac{4}{3}\mu)u_{1,1}$ is represented by the flow of air through a normal shock, where the deformation rate $u_{1,1} \sim 10^{10}\ \mathrm{s}^{-1}$, giving $\Sigma_{11} \sim 5$ atm.

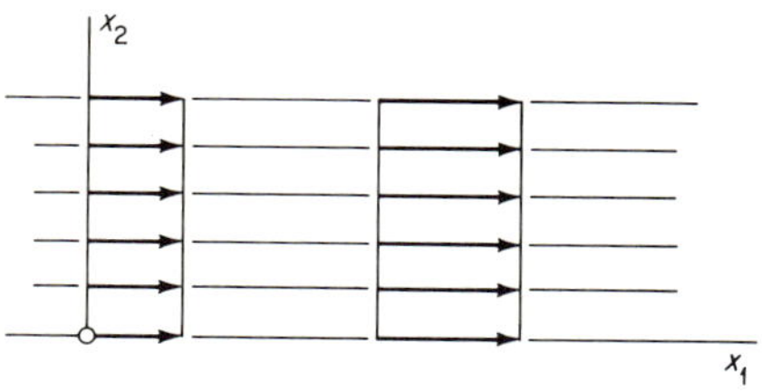

Figure 1.11

EXAMPLE 1.4 POTENTIAL VORTEX

Still another plane motion of some practical importance is the *potential vortex* (Fig. 1.12). Working approximations to such a flow field are found in tornadoes whirlpools, and flow in curved channels.

In cylindrical coordinates (appropriate to the circular motion of the vortex) the velocity field is given by

$$\mathbf{u} = \mathbf{e}_\theta \frac{K}{r} \tag{1.42}$$

where K is a constant. First we establish the irrotational character of the vortex. The *circulation* Γ_C is defined as the line integral around any closed curve C,

$$\Gamma_C \equiv \oint_C \mathbf{u} \cdot \mathbf{t}\, dl \tag{1.43}$$

For any curve not enclosing the origin (Fig. 1.12*b*), with $\mathbf{u} \cdot \mathbf{t}\, dl = ur\, d\theta = K\, d\theta$

$$\Gamma_C = \int K\, d\theta = 0$$

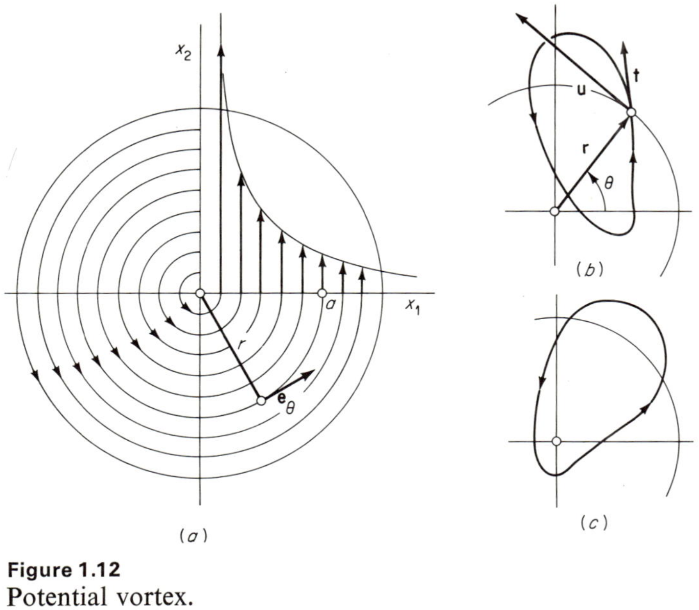

Figure 1.12
Potential vortex.

since the net change of θ for one loop is zero. By Stokes' theorem this is

$$\Gamma_C = \int_S (\nabla \times \mathbf{u}) \cdot \mathbf{n}\, dA = 0$$

where both $\mathbf{n}$ and $\boldsymbol{\Omega} = \nabla \times \mathbf{u}$ are perpendicular to the plane of the motion. Thus

$$\Gamma_C = \int_S \Omega\, dA = 0$$

Since the curve C (and therefore the enclosed region S) is arbitrary, the integrand is necessarily identically zero everywhere within C. If, however, the curve encloses the origin (Fig. 1.12c), the net change of θ for one loop is just 2π and (1.43) gives

$$\Gamma_C = 2\pi K \tag{1.44}$$

This value is often called the *strength* of the vortex and holds for any curve enclosing the origin, in particular for a tiny loop just around the origin. From Stokes' theorem, we conclude that the vorticity Ω is necessarily infinite at the origin itself. Thus

$$\Omega = \begin{cases} 0 & r \neq 0 \\ \infty & r = 0 \end{cases}$$

The vortex is "potential" because the velocity field is then derivable from a velocity potential, $\mathbf{u} = \nabla\phi$, everywhere except at the origin.

The (Cartesian) deformation tensor can be determined for a typical point a. The velocity components are given by

$$u_1 = -\frac{K}{r}\sin\theta = -\frac{Kx_2}{x_1^2 + x_2^2}$$

$$u_2 = \frac{K}{r}\cos\theta = \frac{Kx_1}{x_1^2 + x_2^2}$$

By differentiating and *then* setting x_2 to zero, we obtain for point a

$$u_{1,1} = u_{2,2} = 0$$

$$u_{1,2} = -\frac{K}{x_1^2}$$

$$u_{2,1} = -\frac{K}{x_1^2}$$

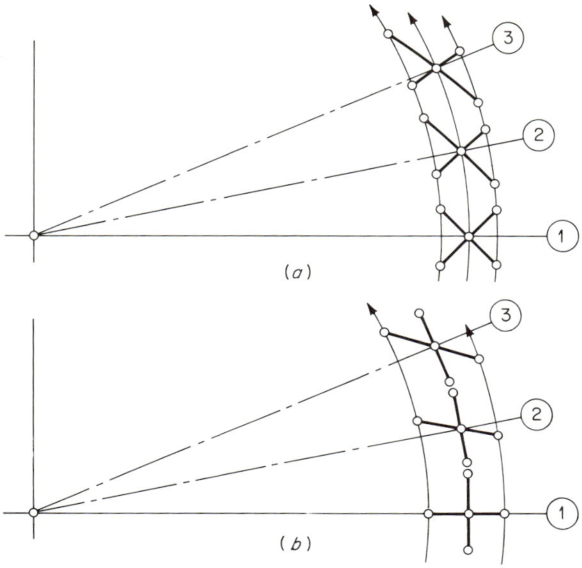

Figure 1.13
Irrotational motion of a fluid dyad: (*a*) initially aligned with the principal axes; (*b*) initially aligned with the x_1 and x_2 axes.

Thus

$$D_{12} = \tfrac{1}{2}(u_{1,2} + u_{2,1}) = -\frac{K}{x_1^2}$$

$$D_{21} = \tfrac{1}{2}(u_{2,1} + u_{1,2}) = -\frac{K}{x_1^2}$$

and all other components D_{ik} are identically zero. The deformation tensor then has exactly the same form as in plane laminar shearing (Example 1.1, page 13) and the principal directions are at $\pm 45°$ to the local radius vector. Note that $D_{mm} = \nabla \cdot \mathbf{u} = 0$, though there may be a radial density gradient (the "compressible" vortex is of practical interest).

To emphasize the irrotational character of the flow, the motion of fluid dyads is shown in Fig. 1.13.

The ideal potential vortex specified by (1.42) has infinite velocity at the origin $r = 0$. Real vortices cannot behave in this way. A very simple and often used approximation to reality is the *solid-core vortex*,

$$u = \begin{cases} \dfrac{K}{r_0}\dfrac{r}{r_0} & r \le r_0 \\[2ex] \dfrac{K}{r} & r \ge r_0 \end{cases}$$

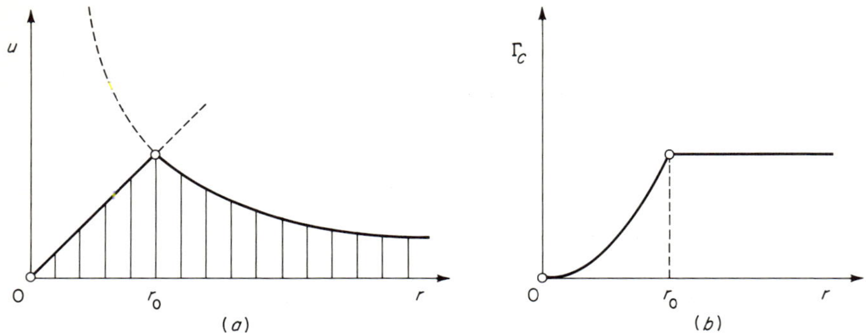

Figure 1.14
Solid-core vortex: (*a*) velocity distribution; (*b*) distribution of circulation.

as shown in Fig. 1.14*a*. The corresponding distribution of circulation is

$$\Gamma_C = \begin{cases} 2\pi K \dfrac{r^2}{{r_0}^2} & r \leq r_0 \\ 2\pi K & r \geq r_0 \end{cases}$$

as shown in Fig. 1.14*b*.

EXAMPLE 1.5 ISOTROPIC EXPANSION

The simplest possible three-dimensional motion is spherically symmetric isotropic expansion (Fig. 1.15). The velocity field is given by

$$\mathbf{u} = K\mathbf{r} \tag{1.45}$$

where K is a constant. Such a motion is a Euclidean description of the expanding universe; it has the remarkable property that the *same* motion is viewed by all material observers in the field. In effect, every material point is an origin for the motion (see Prob. 1.6). One can imagine a fluid motion like (1.45) in response to uniform heating or the radial movement of a spherical enclosure.

The Cartesian velocity components are

$$u_i = Kx_i \tag{1.46}$$

By differentiation we obtain the deformation tensor

$$D_{ik} = K\delta_{ik} \tag{1.47}$$

which written out is

$$D_{ik} = \begin{bmatrix} K & 0 & 0 \\ 0 & K & 0 \\ 0 & 0 & K \end{bmatrix}$$

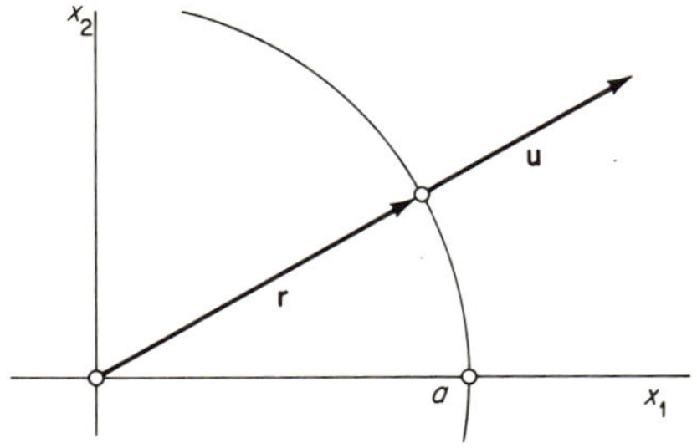

Figure 1.15

A tensor which is unchanged by a rotation of coordinate axes is said to be *isotropic*; this tensor is isotropic, as can be seen from the fact that (1.46) holds for any choice of axes. Other examples of isotropic tensors are the pressure, δ_{ik} itself, and the Levi-Civita index ε_{ijk}.

The resulting stresses are, from (1.37),

$$\sigma_{ik} = (-P + 3\mu_v K)\delta_{ik} \tag{1.48}$$

This represents isotropic stress and depends on the bulk viscosity μ_v but not the shear viscosity μ.

The Fourier Conduction Law

According to accepted thermodynamical ideas, transfer of energy may be classified as either *work* or *heat*, the latter depending on differences of temperature. Let the vector $\mathbf{q}$ represent the heat flux, with the units of energy per unit time per unit area, and let T represent absolute temperature.

A linear constitutive relation between heat flux and temperature gradient is then

$$q_i = \beta_{ik} \frac{\partial T}{\partial x_k} \tag{1.49}$$

For an isotropic fluid, however, the heat-flux vector is necessarily in the same direction as the temperature gradient, giving $\beta_{ik} = -\kappa\delta_{ik}$ and

$$\boxed{\mathbf{q} = -\kappa\,\nabla T} \tag{1.50}$$

This is the Fourier heat conduction law, and the coefficient κ is called the *coefficient of thermal conductivity*, or simply the thermal conductivity. With $\kappa > 0$, (1.50) requires that heat flows from regions of higher temperature toward regions of lower temperature. Like the Navier-Poisson law

Table 1.1 Transport Properties at 300 K, 1 Atm†

Fluid	$\mu \times 10^5$, kg/(m)(s)	μ_v/μ	$\kappa \times 10^2$, J/(m)(s)(K)	$\frac{\mu}{\rho} \times 10^5$, m²/s	Pr
He	1.98	0	15.0	12.2	0.67
Ar	2.27	0	1.77	1.40	0.67
H_2	0.887	32	17.3	10.8	0.71
N_2	1.66	0.8	2.52	1.46	0.71
O_2	2.07	0.4	2.58	1.59	0.72
CO_2	1.50	1,000	1.66	0.837	0.75
Air	1.85	0.6	2.58	1.57	0.71
H_2O (*liquid*)	85.7	3.1	61	0.0857	6.0
Ethyl alcohol	110	4.5	18.3	0.14	15
Glycerine	134,000	0.4	29	109	11,000

† Source references for transport properties are given in Appendix F.

for stress, this linear relation between heat flux and temperature gradient has been justified by many experiments.

The shear viscosity μ, bulk viscosity μ_v, and thermal conductivity κ are collectively called the *transport coefficients* and are material properties depending on the local thermodynamic state.[1] With the local state described by two independent variables, say pressure and temperature, the transport properties are given in principle by $\mu(P,T)$, $\mu_v(P,T)$, and $\kappa(P,T)$. Some specific fluids will be discussed in Chap. 2.

In Table 1.1 are listed some values for the transport properties at 1 atm and 300 K. Also listed is the *Prandtl number* Pr,

$$\text{Pr} \equiv \frac{\mu c_p}{\kappa} \tag{1.51}$$

a dimensionless combination of properties (with c_p the specific heat at constant pressure) which is an indicator of the relative importance of

[1] Thermodynamics traditionally is concerned with *equilibrium* states. Since the transport properties measure departures from equilibrium, it is not clear that they *should* behave like thermodynamic properties, though they are invariably treated in this way. This question has been resolved by *Coleman and Mizel* [1964], who give a systematic treatment for the local state of a linearly viscous, linearly conducting fluid.

viscous stress and heat transfer, as discussed in Chap. 3. In the equations of motion it is often found that the coefficient of the viscous-force terms is μ/ρ, rather than μ, so this quantity, which is called the *kinematic viscosity* ν, is listed. It may be noted that in terms of kinematic viscosity air is about 15 times more viscous than water.

It has already been noted that a large part of gasdynamics is based on the approximation of *irrotational* flow, $\nabla \times \mathbf{u} = 0$. An important and related approximation is that of *inviscid* flow, with $\Sigma_{ik} = \mathbf{q} = 0$; that is, the fluid is treated as nonviscous and nonconducting. The relation between these approximations is discussed in Chap. 2.

Another class of constitutive equations, the thermodynamical equations of state, is also considered in Chap. 2.

1.5 Equations of motion

Moving Reference Frames

The fluid motion is assumed to be described within a *Galilean reference frame*, i.e., within a coordinate system which has only uniform translation (at constant velocity) with respect to a primitive system embedded in the nearby "fixed" stars. In such a frame, sometimes called an inertial frame, the equations of Newtonian mechanics hold.

An example of a frame which is not *strictly* Galilean is one fixed with respect to the surface of the earth. For most fluid motions of interest, however, the accelerations within the frame are far greater than the accelerations associated with the motion of the coordinate system; a notable exception is the large-scale circulation of the atmosphere. An example of a frame which is not likely to be even approximately Galilean is one fixed with respect to a rotating turbine rotor.

It is always possible, however, to write the equations of motion applicable to a non-Galilean frame by making use of transformation formulas. If $\mathbf{u}$ is the (apparent) fluid velocity within a moving frame which is translating with velocity $\mathbf{v}$ and rotating with angular velocity $\boldsymbol{\omega}$ with respect to some "fixed" Galilean frame, then the absolute velocity $\mathbf{U}$ and absolute acceleration $\dot{\mathbf{U}}$ are, respectively, with $\mathbf{x}$ the (apparent) position in the moving frame,

$$\mathbf{U} = \mathbf{u} + \mathbf{v} + \boldsymbol{\omega} \times \mathbf{x} \tag{1.52}$$

$$\dot{\mathbf{U}} = \frac{D\mathbf{u}}{Dt} + \dot{\mathbf{v}} + \dot{\boldsymbol{\omega}} \times \mathbf{x} + 2\boldsymbol{\omega} \times \mathbf{u} + \boldsymbol{\omega} \times (\boldsymbol{\omega} \times \mathbf{x}) \tag{1.53}$$

where D/Dt is the (apparent) material derivative as seen in the moving frame.[1] For all Galilean frames, $\dot{\mathbf{v}} = \boldsymbol{\omega} = \dot{\boldsymbol{\omega}} = 0$, conditions which are henceforth assumed.

As has already been noted, any quantity which is unchanged between moving coordinate frames (put differently, which has a value independent of the velocity and acceleration of the observer) is said to be *objective*. In particular, quantities unchanged between Galilean frames are called *Galilean invariants*. Examples of Galilean invariants are acceleration, stress, vorticity, pressure, temperature, and heat flux. Examples of quantities which are not Galilean invariants are velocity, kinetic energy, and molecular mean free path.

Integral Equations of Motion

The equations of motion are obtained from the dynamical laws applied to a material volume, as stated in Sec. 1.2.

For a material volume the mass is constant, so that conservation of mass takes the form[2]

$$\frac{d}{dt}\int_{V(t)} \rho \, dV = 0 \tag{1.54}$$

The balance of momentum equates the rate of change of the material-volume momentum (with the momentum per unit mass just the velocity $\mathbf{u}$) to the net body force plus the net surface force

$$\frac{d}{dt}\int_{V(t)} \rho\mathbf{u} \, dV = \int_{V(t)} \rho\mathbf{G} \, dV + \int_{S(t)} \mathbf{T} \, dA \tag{1.55}$$

The specific energy (energy per unit mass) of a fluid is $e + u^2/2$, where e is the specific internal energy, a thermodynamic quantity, and $u^2/2$ is the specific kinetic energy. It is just the quantity e which is introduced to satisfy our ideas about the *balance* of energy (net rate of energy input via work and heat equals rate of accumulation). Equating the rate of energy increase for a material volume to the rate at which energy is transferred to the volume via work and heat gives

[1] See, for example, *Fox* [1967, p. 37].

[2] The natural temptation to use the material derivative D/Dt for the rate of change of an integral over a material volume is resisted, and d/dt used instead. The material derivative as defined in (1.13) applies only at a point in space.

$$\frac{d}{dt}\int_{V(t)} \rho\left(e + \frac{u^2}{2}\right) dV = \int_{V(t)} \rho \mathbf{G} \cdot \mathbf{u}\, dV + \int_{S(t)} \mathbf{T} \cdot \mathbf{u}\, dA - \int_{S(t)} \mathbf{q} \cdot \mathbf{n}\, dA \tag{1.56}$$

The last (heat) term has a minus sign because $\mathbf{q} \cdot \mathbf{n} > 0$ for an outward-directed heat flux $\mathbf{q}$; that is, it represents a loss of energy. The final dynamical equation is the condition (from the second law of thermodynamics) that entropy can be created but not destroyed, i.e., that any real process must result in a net production of entropy. For a material volume this statement is

$$\frac{d}{dt}\int_{V(t)} \rho s\, dV + \int_{S(t)} \mathbf{\Xi} \cdot \mathbf{n}\, dA \geq 0 \tag{1.57}$$

where s is the specific entropy and $\mathbf{\Xi}$ is the entropy flux (which will later be identified as $\mathbf{q}/T$) written in the same spirit as the heat flux $\mathbf{q}$. All the entropy created within $V(t)$ must appear as an increase in the material entropy, represented by the first integral, or as a net outflow from the boundary $S(t)$, as represented by the second integral; that the sum must be positive means that entropy is created.

Differential Equations of Motion

For conciseness, we will use the *comma notation* for spatial partial derivatives. With the application of the divergence theorem, and making use of the fact that the material volume is arbitrary, the preceding integral statements for mass, momentum, energy, and entropy, respectively, reduce to the corresponding point statements

$$\frac{\partial \rho}{\partial t} + (\rho u_k)_{,k} = 0 \tag{1.58}$$

$$\rho \frac{Du_i}{Dt} = \rho G_i + \sigma_{ki,k} \tag{1.59}$$

$$\rho \frac{D}{Dt}\left(e + \frac{u^2}{2}\right) = (\sigma_{ik} u_k)_{,i} + \rho G_k u_k - q_{k,k} \tag{1.60}$$

$$\rho \frac{Ds}{Dt} + \Xi_{k,k} \geq 0 \tag{1.61}$$

In this form the equations are very general and applicable, for example, to solids as well as fluids. With the insertion of the linear constitutive relations already given for fluids

$$\sigma_{ik} = -P\delta_{ik} + 2\mu(D_{ik} - \tfrac{1}{3}\delta_{ik}D_{mm}) + \mu_v\delta_{ik}D_{mm}$$

$$q_k = -\kappa T_{,k}$$

these equations [and the momentum equation (1.59) in particular] are the celebrated *Navier-Stokes equations*.[1] Alternative forms and special cases are considered below.

Mass

Equation (1.58) is often referred to as the *continuity equation*. Expanding the second term gives, in vector notation,

$$\frac{1}{\rho}\frac{D\rho}{Dt} + \nabla \cdot \mathbf{u} = 0 \tag{1.62}$$

Introducing the *specific volume* $v \equiv 1/\rho$ (the volume occupied by a unit mass), this becomes

$$\frac{1}{v}\frac{Dv}{Dt} = \nabla \cdot \mathbf{u} \tag{1.63}$$

where $\nabla \cdot \mathbf{u} = u_{k,k} = D_{mm}$ is the trace of the deformation tensor [this result is *almost* the same as Eq. (1.28)]. For an incompressible fluid this gives of course $\nabla \cdot \mathbf{u} = 0$.

With the help of the continuity equation we can obtain a useful special form of Reynolds' theorem: the integral of interest is

$$\frac{d}{dt}\int_{V(t)} \rho\chi \, dV$$

Applying Reynolds' theorem (1.27) to this (with the integrand $\rho\chi$ replacing the former χ) and then applying the divergence theorem gives

$$\frac{d}{dt}\int_{V(t)} \rho\chi \, dV = \int_{V(t)} \left\{\chi\left[\frac{\partial\rho}{\partial t} + \nabla \cdot (\rho\mathbf{u})\right] + \rho\left[\frac{\partial\chi}{\partial t} + \mathbf{u} \cdot \nabla\chi\right]\right\} dV$$

[1] After L. M. H. Navier (1785–1836), French military engineer and professor at the École Polytechnique and the École des Ponts et Chausées (bridges and dikes), and G. G. Stokes (1819–1903), Lucasian professor at Cambridge. Navier gave (1.59) in a form appropriate to incompressible fluids (1822), while Stokes developed the more general form in a famous paper (1845). Where (1.59) appears without a specified stress law it is known as *Cauchy's equation* (A. L. Cauchy, 1827).

The terms in square brackets are respectively zero (by continuity) and $D\chi/Dt$. Thus, finally,

$$\frac{d}{dt}\int_{V(t)} \rho\chi \, dV = \int_{V(t)} \rho \frac{D\chi}{Dt} \, dV \tag{1.64}$$

follows from the constancy of $\int \rho \, dV$.

Remark on Steady Flow

We will define a *steady flow* to be one for which, at any fixed point $\mathbf{x}$ in the flow field, all the flow variables (velocity, pressure, density, etc.) are independent of time. Thus, in a steady flow all partial derivatives with respect to time vanish; for example, $\partial\mathbf{u}/\partial t = 0$, $\partial P/\partial t = 0$, $\partial\rho/\partial t = 0$.

If we specify that the velocity field is $\mathbf{u} = \mathbf{u}(\mathbf{x})$, independent of time, this in itself does not guarantee that other variables are independent of time. As a simple example, consider fluid moving at *constant* velocity through a constant-area pipe (Fig. 1.16). The continuity equation (1.62) can be written

$$\frac{D\rho}{Dt} + \rho\frac{\partial u}{\partial x} = 0$$

But $\partial u/\partial x = 0$ by assumption; thus $D\rho/Dt = 0$: the density of each fluid particle is constant. The density may vary from one particle to another, however, because

$$\frac{D\rho}{Dt} = \frac{\partial\rho}{\partial t} + u\frac{\partial\rho}{\partial x} = 0$$

and

$$\frac{\partial\rho}{\partial t} = -u\frac{\partial\rho}{\partial x}$$

is not necessarily zero. A possible density distribution of this kind is illustrated in Fig. 1.17. The fluid motion may be described as a density

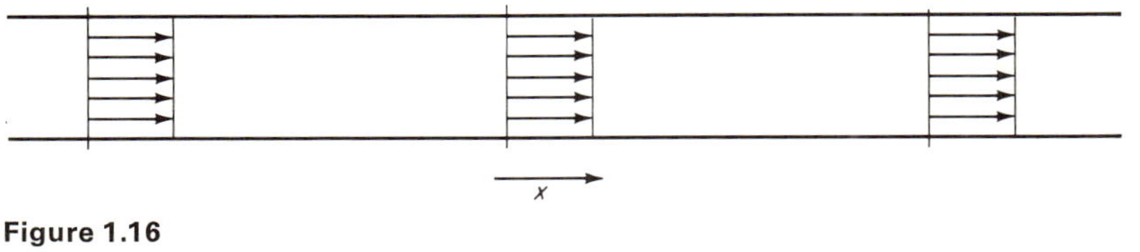

Figure 1.16

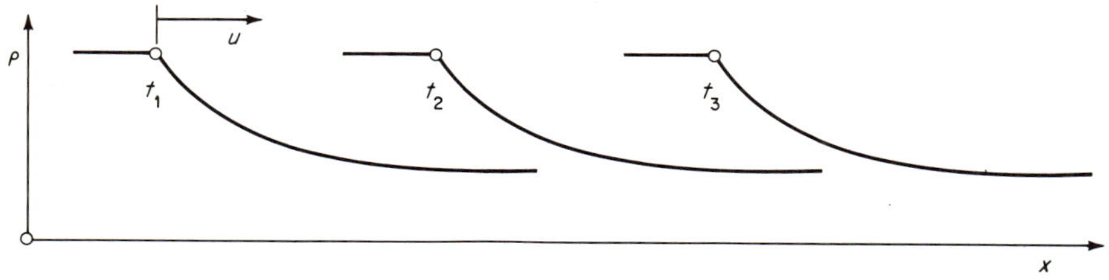

Figure 1.17
Density wave at various times t: $t_3 > t_2 > t_1$.

wave propagating with the fluid velocity u. Such an unsteady motion might occur, for example, in a pipe carrying gas of nonuniform temperature at low speed.

Momentum

Splitting the stress tensor into pressure and viscous terms, $\sigma_{ki} = -P\delta_{ki} + \Sigma_{ki}$, the momentum equation (1.59) becomes

$$\rho \frac{Du_i}{Dt} + P_{,i} = \rho G_i + \Sigma_{ki,k} \tag{1.65}$$

If μ and μ_v can be treated as constants (an approximation which is not however always justified), it is convenient to put (1.65) into vector form; multiplying through by $\mathbf{e}_i$ and replacing $\mathbf{e}_i\, \partial/\partial x_i$ by ∇, and so forth, gives

$$\rho \frac{D\mathbf{u}}{Dt} + \nabla P = \rho \mathbf{G} + \mu \nabla^2 \mathbf{u} + (\mu_v + \tfrac{1}{3}\mu)\nabla(\nabla \cdot \mathbf{u}) \tag{1.66}$$

In many problems in gasdynamics and acoustics the viscous terms in (1.66) or (1.65) can be neglected to a fair approximation (as discussed in Chaps. 2 and 3), and the equation becomes *Euler's equation* (L. Euler, 1755)

$$\rho \frac{D\mathbf{u}}{Dt} = -\nabla P + \rho \mathbf{G} \tag{1.67}$$

For the problem of *hydrostatics*, the viscous stresses will always vanish (because there is no motion), and

$$\nabla P = \rho \mathbf{G} \tag{1.68}$$

in which ρ is not necessarily constant.

Multiplying Cauchy's equation (1.59) by u_i and interchanging the dummy indices for future convenience gives an expression for the creation of kinetic energy

$$\rho \frac{D}{Dt}\frac{u^2}{2} = \sigma_{ik,i}u_k + \rho G_k u_k \tag{1.69}$$

Energy

Subtracting the kinetic-energy expression (1.69) from the energy equation (1.60) gives

$$\rho \frac{De}{Dt} = \sigma_{ik}u_{k,i} - q_{k,k}$$

This can be put in a more convenient form. The double sum can be rewritten

$$\sigma_{ik}u_{k,i} = \tfrac{1}{2}\sigma_{ik}u_{k,i} + \tfrac{1}{2}\sigma_{ki}u_{i,k}$$

and since the stress tensor is symmetric, this reduces to

$$\sigma_{ik}\tfrac{1}{2}(u_{i,k} + u_{k,i}) = \sigma_{ik}D_{ik}$$

and we obtain

$$\rho \frac{De}{Dt} = \sigma_{ik}D_{ik} - q_{k,k} \tag{1.70}$$

This is the first alternative form of the energy equation. This result is exactly equivalent to the first law of thermodynamics (sometimes written $dE = \delta W + \delta Q$). The term $\sigma_{ik}D_{ik}$ represents the rate at which surface forces do work on a fluid element, including pressure forces producing a volume change and viscous forces producing deformations, as will be discussed in Chap. 2.

A second alternative form is obtained by splitting the stress tensor, $\sigma_{ik} = -P\delta_{ik} + \Sigma_{ik}$. Then the work term in (1.60) becomes, with $u_{i,i} = \dfrac{1}{v}\dfrac{Dv}{Dt}$ by continuity,

$$(\sigma_{ik}u_k)_{,i} = -\frac{P}{v}\frac{Dv}{Dt} - u_iP_{,i} + (\Sigma_{ik}u_k)_{,i}$$

Inserting this into (1.60) gives, after some algebra,

$$\rho \frac{D}{Dt}\left(e + Pv + \frac{u^2}{2}\right) = \frac{\partial P}{\partial t} + (\Sigma_{ik}u_k)_{,i} + \rho G_k u_k - q_{k,k} \tag{1.71}$$

Finally, if the body force field has a potential (as for example in the case of gravitational force),

$$G_k = -\Psi_{,k} \tag{1.72}$$

where Ψ is the force potential or potential energy per unit mass. Then if $\Psi(\mathbf{x})$ is independent of time, as is usually the case,

$$\rho G_k u_k = -\rho u_k \Psi_{,k} = -\rho \frac{D\Psi}{Dt}$$

Defining the *specific enthalpy* h (a thermodynamic property)

$$h \equiv e + Pv \tag{1.73}$$

the energy equation (1.71) can now be written

$$\rho \frac{D}{Dt}\left(h + \frac{u^2}{2} + \Psi\right) = \frac{\partial P}{\partial t} + (\Sigma_{ik}u_k)_{,i} - q_{k,k} \tag{1.74}$$

This is the second alternative form of the energy equation and the prototype for all the various equations which are called *Bernoulli's equation*.

The essential function of pumps, compressors, and turbines is to change the *Bernoulli constant* $(h + u^2/2 + \Psi)$ of a flow; one conclusion from (1.74) is that such a device cannot operate without viscous forces and heat transfer in the absence of unsteadiness.

If, as often happens, the viscous stresses and the heat transfer are *simultaneously* negligible, the flow is called *inviscid*. If the flow is both steady and inviscid, (1.74) reduces to

$$\mathbf{u} \cdot \nabla\left(h + \frac{u^2}{2} + \Psi\right) = 0$$

That is, the quantity $h + u^2/2 + \Psi$ has no variation in the direction of $\mathbf{u}$. Thus under these conditions an integral is

$$h + \frac{u^2}{2} + \Psi = \text{const} \tag{1.75}$$

on any given streamline[1] (the constant may in general vary from one

[1] A streamline in steady flow is the path traversed by an infinitesimal fluid particle. Its equation is in general $dx_i = u_i\,dt$, or

$$\frac{dx_1}{u_1} = \frac{dx_2}{u_2} = \frac{dx_3}{u_3}$$

streamline to another). We will make considerable use of this result in steady gasdynamics. For an *incompressible* fluid, it will be shown in Chap. 2 that (1.75) reduces to

$$P + \tfrac{1}{2}\rho u^2 + \rho\Psi = \text{const} \tag{1.76}$$

Entropy

Unlike the other dynamical laws, the entropy equation (1.61) is in the form of an *inequality* and rules out as physically impossible any process which violates this inequality. Further applications, in particular the limiting case $Ds/Dt = 0$, are discussed in Chap. 2.

The equations of motion (1.58) to (1.60) are five in number (there are three momentum equations). We can now verify that there are just five unknowns, so that the set of equations (with appropriate boundary conditions) is determinate. Anticipating the *state principle* of thermodynamics, the thermodynamic state is fixed by two thermodynamic variables, say P and ρ. Then ρ, P, μ, κ, μ_v, and T are all known in terms of P and ρ. There are then five dependent variables in the equations of motion: P, ρ, and the three velocity components u_i.

In principle, then, the equations of motion can be solved for a particular physical problem with particular boundary conditions. Yet in practice this is almost never possible, not even numerically with the help of large computers, without making drastic simplifying approximations beyond those already inherent in the equations themselves. The remaining chapters of this book are devoted to problems permitting such simplification. In most cases it will be possible to extract useful information, e.g., by dimensional analysis, even without formal mathematical solutions.

.6 Equations of balance for control volumes

In many practical problems it is convenient to apply the laws of motion to macroscopic regions which are not necessarily material volumes. The control volume is a region in space $V^*(t)$ bounded by a control surface $S^*(t)$. These must be specified in some way for each problem. The material volume is a special case of the control volume.

Possible examples of control volumes are the cylinder of an engine, with the top of the (moving) piston forming part of the boundary, and a region enclosing a portion of an advancing wave. Problems in which control volumes are useful include flow through jet engines and rockets, flow through chemical and nuclear reactors, pipe flow, and analysis of shock-type discontinuities.

The Nature of Balance Statements

A conceptual model for the physical balance statements is a bag (the control volume) containing a variable number of jelly beans. A proposed balance statement might be: "The rate at which the number of jelly beans contained is increasing is equal to the net rate at which jelly beans are put into the bag." This might be called the principle of conservation of jelly beans. The principle fails, however, if there is a jelly-bean factory or a goblin with a weakness for jelly beans inside the bag. We can allow for such a possibility by adding a production term to the principle, which then reads

Rate of increase of contents = net inflow + production

Such a balance statement can be written for any *countable* (summable) quantity. It is usefully applied to the quantities mass, momentum, energy, and entropy, since for these quantities a precise statement can be made about the production term. The special forms of the inflow and production terms are listed in Table 1.2. The convection terms represent inflow carried by matter crossing the control surface.

An arbitrary control volume $V^*(t)$ is shown in Fig. 1.6. The formal balance statements are obtained by the application of Reynolds' transport theorem to Eqs. (1.54) to (1.57) for a material volume $V(t)$. For balance of mass,

$$\frac{d}{dt}\int_{V(t)} \rho \, dV = 0$$

Table 1.2

Fluid Property	Inflow	Production
Mass	Convection	
Momentum	Convection	Applied force
Energy	Convection, heat, work	
Entropy	Convection, heat ÷ absolute temp.	Dissipation Ψ, temperature gradient (sum positive)

By Reynolds' theorem (1.27) this is expanded to give

$$\int_{V(t)} \frac{\partial \rho}{\partial t} \, dV + \int_{S(t)} \rho \mathbf{u} \cdot \mathbf{n} \, dA = 0 \tag{1.77}$$

Reynolds' theorem can also be applied to a control volume $V^*(t)$ to give

$$\frac{d}{dt} \int_{V^*(t)} \rho \, dV = \int_{V^*(t)} \frac{\partial \rho}{\partial t} \, dV + \int_{S^*(t)} \rho \mathbf{b} \cdot \mathbf{n} \, dA \tag{1.78}$$

The (arbitrary) material volume is now chosen to be instantaneously coincident with the desired control volume, so that integrals over $V^*(t)$ are identical to those over $V(t)$ and those over $S^*(t)$ identical to those over $S(t)$ (this does not apply, however, to the time derivatives of the integrals). Then equating the first right-hand term in (1.78) to the equivalent term in (1.77) gives

$$\boxed{\frac{d}{dt} \int_{V^*(t)} \rho \, dV + \int_{S^*(t)} \rho (\mathbf{u} - \mathbf{b}) \cdot \mathbf{n} \, dA = 0} \tag{1.79}$$

the balance of mass for the control volume. Proceeding similarly with (1.55) to (1.57) gives for momentum, energy, and entropy:

$$\frac{d}{dt} \int_{V^*(t)} \rho \mathbf{u} \, dV + \int_{S^*(t)} \rho \mathbf{u} (\mathbf{u} - \mathbf{b}) \cdot \mathbf{n} \, dA = \int_{V^*(t)} \rho \mathbf{G} \, dV + \int_{S^*(t)} \mathbf{T} \, dA \tag{1.80}$$

$$\frac{d}{dt} \int_{V^*(t)} \rho \left(e + \frac{u^2}{2} \right) dV + \int_{S^*(t)} \rho \left(e + \frac{u^2}{2} \right) (\mathbf{u} - \mathbf{b}) \cdot \mathbf{n} \, dA = \int_{V^*(t)} \rho \mathbf{G} \cdot \mathbf{u} \, dV + \int_{S^*(t)} \mathbf{T} \cdot \mathbf{u} \, dA - \int_{S^*(t)} \mathbf{q} \cdot \mathbf{n} \, dA \tag{1.81}$$

$$\frac{d}{dt} \int_{V^*(t)} \rho s \, dV + \int_{S^*(t)} \rho s (\mathbf{u} - \mathbf{b}) \cdot \mathbf{n} \, dA + \int_{S^*(t)} \frac{1}{T} \mathbf{q} \cdot \mathbf{n} \, dA \geq 0 \tag{1.82}$$

The entropy flux $\boldsymbol{\Xi}$ has been written as $\mathbf{q}/T$, anticipating a result in Chap. 2. The convection terms each contain the relative flux as seen by an observer on the boundary (who sees a fluid velocity $\mathbf{u} - \mathbf{b}$). For the special case in which the control volume is chosen to be a material volume, these terms disappear (because $\mathbf{u} \equiv \mathbf{b}$ on the boundary) and the above statements are identical to (1.54) to (1.57). For a control volume at rest, of course, $\mathbf{b} = 0$.

Later use will be made of the above theorems, particularly for shock waves, and illustrative examples are given below.

EXAMPLE 1.6 STEADY-FLOW ENERGY EQUATIONS

Consider a piece of fluid machinery, such as a pump, turbine, or chemical reactor, fixed in inertial space ($\mathbf{b} = 0$), as shown in Fig. 1.18. The integrated heat *input* rate in (1.81) is

$$\dot{Q} \equiv -\int_{S^*} \mathbf{q} \cdot \mathbf{n}\, dA$$

The surface work term in (1.81) is zero at the stationary boundaries, but there is a contribution where the rotating shaft enters and at the fluid inlet section 1 and fluid exit section 2. We set the former equal to $\dot{W}$ (the *shaft work*) and assume the surface traction $\mathbf{T} = -P\mathbf{n}$ at the inlet and exit sections. Then

$$\int_{S^*} \mathbf{T} \cdot \mathbf{u}\, dA = \dot{W} - \int_{1,2} P\mathbf{n} \cdot \mathbf{u}\, dA$$

The body force $\mathbf{G}$ is assumed to have a potential Ψ. Then the integrand of the body-force work term in (1.81) is

$$\rho\mathbf{G} \cdot \mathbf{u} = \mathbf{G} \cdot (\rho\mathbf{u}) = -\nabla\Psi \cdot (\rho\mathbf{u}) = -\nabla \cdot (\Psi\rho\mathbf{u})$$

because $\nabla \cdot (\rho\mathbf{u}) = 0$ for the assumed *steady* flow (in a practical case, the flow may be steady only in the sense of an average over time, however). Then upon use of the divergence theorem, this work integral becomes

$$\int_{V^*} \rho\mathbf{G} \cdot \mathbf{u}\, dV = -\int_{V^*} \nabla \cdot (\Psi\rho\mathbf{u})\, dV = -\int_{S^*} \Psi\rho\mathbf{u} \cdot \mathbf{n}\, dA$$

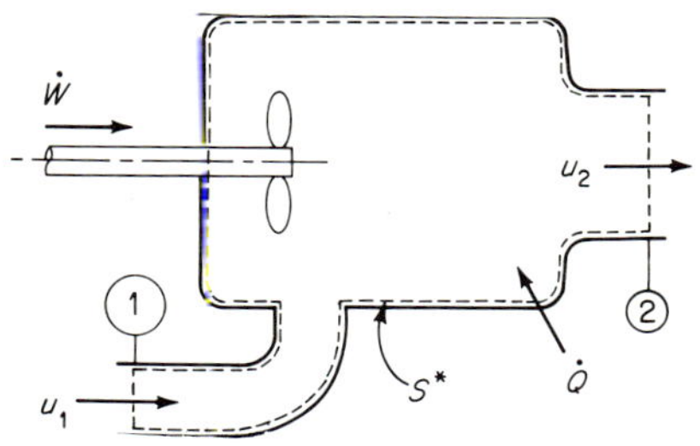

Figure 1.18
Control volume for the energy equations.

which has a contribution only at sections 1 and 2. Substituting into (1.81), with the first (storage) term zero for steady flow, gives the energy balance

$$\int_{S^*}\left(h+\frac{u^2}{2}+\Psi\right)\rho\mathbf{u}\cdot\mathbf{n}\,dA=\dot{W}+\dot{Q} \tag{1.83}$$

From the mass balance (1.79)

$$\int_{S^*}\rho\mathbf{u}\cdot\mathbf{n}\,dA=0 \tag{1.84}$$

because there is no storage of mass inside. Defining the *mass throughput* $\dot{m}$,

$$\dot{m}\equiv-\int_1\rho\mathbf{u}\cdot\mathbf{n}\,dA=+\int_2\rho\mathbf{u}\cdot\mathbf{n}\,dA$$

and assuming uniform conditions at the inlet section 1 and exit section 2, the energy balance (1.81) reduces to

$$\dot{m}\left[h+\frac{u^2}{2}+\Psi\right]_1^2=\dot{W}+\dot{Q} \tag{1.85}$$

which is the *steady-flow energy equation* of elementary thermodynamics [compare Eq. (1.75)]. This has an obvious generalization to the case where fluid crosses the boundary at more than two points.

The "rigid" walls in this example can be considered to be formed by the streamlines of a steady three-dimensional flow; the walls are then referred to as a *stream tube*. Then, with negligible viscous stress and negligible heat conduction at these boundaries, (1.85) reduces to

$$h+\frac{u^2}{2}+\Psi=\text{const}\equiv H \tag{1.86}$$

along a streamline. This result [which we found earlier in the form of Eq. (1.75)] is sometimes called the *adiabatic energy equation.* The constant H is often called the *Bernoulli constant;* in gas dynamics it is usually called the *stagnation enthalpy*. We will discuss this result at length in Chap. 5.

EXAMPLE 1.7 ROCKET MOTION

The balance statements (1.79) to (1.82) apply to an accelerated-rocket control volume provided that the fluid and boundary velocities are taken with respect to an inertial coordinate frame. The rocket velocity $\mathbf{b}(t)$ and the fluid velocity $\mathbf{u}$ are thus "absolute" velocities. It will be assumed that $\mathbf{b}(t)$ is uniform over the control surface $S^*(t)$, which implies that the rocket is in rectilinear motion, i.e., not tumbling. The fluid velocity as seen by an observer on the rocket is then

$$\mathbf{v}\equiv\mathbf{u}-\mathbf{b}$$

Fluid is discharged only at the exit plane with relative velocity $\mathbf{v}_e$. With the total (instantaneous) mass of the rocket $M = \int \rho \, dV$, the balance (1.79) just gives for the rate of change of mass

$$\dot{M} = -\int_{S_e^*} \rho \mathbf{v}_e \cdot \mathbf{n} \, dA \tag{1.87}$$

The overall external surface forces are represented by $\mathbf{F}$ (drag, lift, etc.); for a rocket traversing a vacuum[1] the only surface force is the pressure in the exit plane, and $\mathbf{F}$ will reduce to $-P_e A_e \mathbf{n}_e$. Then with the gravitational body force just the acceleration of gravity $\mathbf{g}$ the momentum balance (1.80) becomes

$$\frac{d}{dt} M\mathbf{b} + \frac{d}{dt}\int_{V^*} \rho \mathbf{v} \, dV + \int_{S_e^*} \rho_e \mathbf{v}_e \mathbf{v}_e \cdot \mathbf{n} \, dA - \mathbf{b}\dot{M} = M\mathbf{g} + \mathbf{F}$$

Assuming that $\mathbf{v}_e$ is uniform over the exit plane, this simplifies to

$$M\dot{\mathbf{b}} + \frac{d}{dt}\int_{V^*} \rho \mathbf{v} \, dV - \dot{M}\mathbf{v}_e = M\mathbf{g} + \mathbf{F} \tag{1.88}$$

In most situations of practical importance the second term on the left is negligible. In order to estimate this term we assume a simplified model for the generation of propellant fluid. Let the source of the propellant be a stationary (with respect to the rocket) region V_s with material of density ρ_s, for example, solid fuel. The generated propellant, normally a gas, has density ρ_0 in the reservoir region V_0 and travels toward the rocket nozzle with (small) velocity $\mathbf{v}_0$. Flow conditions in the nozzle are assumed to be completely steady in the frame of the rocket. The rate at which the total rocket mass is changing is

$$\dot{M} = \frac{d}{dt}(\rho_0 V_0 + \rho_s V_s)$$

[1] It was not always obvious that a rocket can propel itself through vacuum. We quote from an editorial in *The New York Times*, Jan. 13, 1920: "It is when one considers the multiple-charge rocket as a traveler to the moon that one begins to doubt . . . for after the rocket quits our air and really starts on its longer journey, its flight would be neither accelerated nor maintained by the explosion of the charges it then might have left . . . . That Professor Goddard, with his "chair" in Clark College and the countenancing of the Smithsonian Institution, does not know the relation of action to reaction, and of the need to have something better than a vacuum against which to react—to say that would be absurd. Of course he only seems to lack the knowledge ladled out daily in high schools."

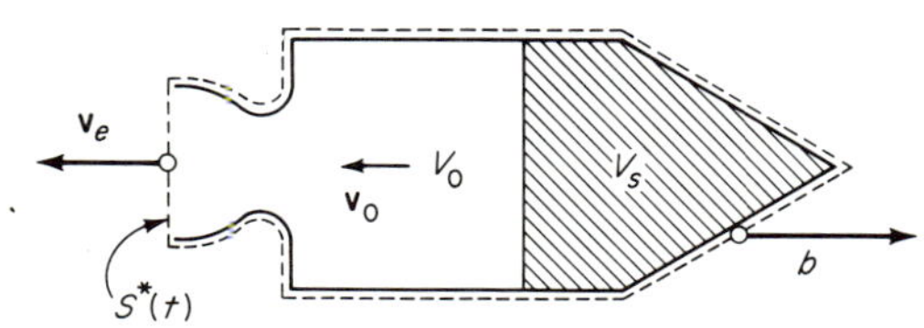

Figure 1.19

With the sum $V_0 + V_s$ constant, this gives

$$\dot{M} = -(\rho_s - \rho_0)\frac{dV_0}{dt}$$

Then the term of interest in (1.88) becomes, with $\mathbf{v}_s = 0$,

$$\frac{d}{dt}\int_{V^*} \rho\mathbf{v}\, dV = \frac{d}{dt}\rho_0\mathbf{v}_0 V_0 = -\dot{M}\frac{\rho_0}{\rho_s - \rho_0}\mathbf{v}_0 \tag{1.89}$$

This term is normally much smaller than the corresponding term $\dot{M}\mathbf{v}_e$ because $v_e \gg v_0$ by several orders of magnitude and also because $\rho_0/(\rho_s - \rho_0)$ is small compared to unity. Then (1.88) becomes

$$M\dot{\mathbf{b}} = \dot{M}\mathbf{v}_e + M\mathbf{g} + \mathbf{F} \tag{1.90}$$

This is the desired equation of rocket motion. It is just of the form of Newton's second law for a body of fixed mass, viz., $M\mathbf{a} = \sum \mathbf{F}$, *provided* that the term $\dot{M}\mathbf{v}_e$ is treated as a force. This term is in fact the *thrust force* of rocket terminology,

$$\mathbf{F}_t \equiv \dot{M}\mathbf{v}_e \tag{1.91}$$

A crucial figure of merit in rocketry is the *specific impulse*, the thrust force per unit rate of mass discharge. From (1.91) the (scalar) specific impulse is just the exit speed v_e,

$$I_s = v_e \tag{1.92}$$

It is desirable that this quantity be as large as possible,[1] thereby reducing the rate of propellant discharge required to maintain a given thrust level and consequently reducing the amount of propellant which must be carried.

By inspection, the dimensions of specific impulse are simply L/T. Strangely enough, however, it has become common practice to report specific impulse figures in units of seconds, for example, $I_s = 400$ s. This arcane convention presumably arose when someone divided lb_f by lb_m/s and arrived at seconds. To convert such a figure to rational units it is necessary to multiply by the standard acceleration of gravity, $g_0 = 32.17\ \text{ft/s}^2 = 9.80\ \text{m/s}^2$.

A rocket is a very inefficient lifting device. Consider a vertical rocket which just has sufficient thrust to support itself against gravity. Then from (1.90)

$$-\dot{M}v_e = Mg$$

[1] From a momentum standpoint; not, however, from an energy standpoint, in terms of the power which must be supplied.

which integrates to

$$\frac{M}{M_0} = e^{-gt/v_e} \tag{1.93}$$

or for small exponent,

$$\frac{M}{M_0} \approx 1 - \frac{gt}{v_e}$$

For a typical specific impulse, $v_e = (300 \text{ s})g$, and the mass is decreased by $1/300$ in 1 s. For a large rocket with (say) $M_0 = 3 \times 10^6 \text{ lb}_m$, 10^4 lb_m must be discharged in the first second just to support the rocket.

EXAMPLE 1.8 ROCKET MOTION WITH INTERNAL MOMENTUM CHANGE

As an example of a rocket for which the internal momentum change, i.e., the second term in (1.88), may not be negligible, consider the toy rocket shown in Fig. 1.20. The propellant is essentially liquid water, under pressure from the gas above. The term in question is

$$\frac{d}{dt}\int_{V^*} \rho \mathbf{v}\, dV \approx \frac{d}{dt}\rho_0 \mathbf{v}_0 V_0 = \mathbf{v}_0 \dot{M}$$

With $\mathbf{v}_e A_e = \mathbf{v}_0 A_0$, this becomes

$$\frac{d}{dt}\int_{V^*} \rho \mathbf{v}\, dV = \frac{A_e}{A_0}\dot{M}\mathbf{v}_e$$

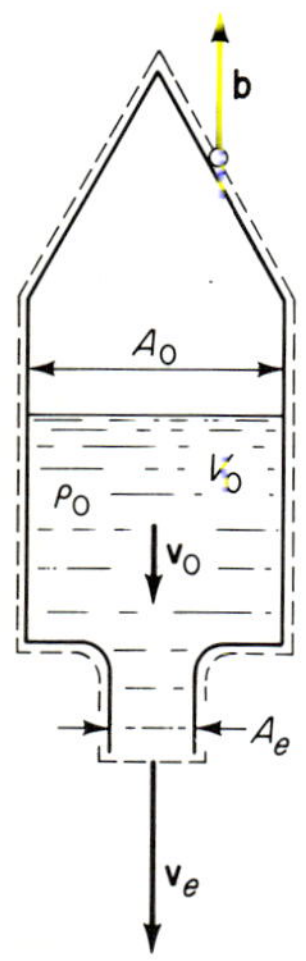

Figure 1.20

Equation (1.88) then gives the rocket equation

$$M\dot{\mathbf{b}} = \dot{M}\left(1 - \frac{A_e}{A_0}\right)\mathbf{v}_e + M\mathbf{g} + \mathbf{F} \tag{1.94}$$

which can be compared to (1.90). Unless the exit area A_e is very small compared to A_0, the equation is significantly modified.

It should be remarked that this is a very inefficient rocket, with very small specific impulse. Assuming that the rocket acceleration is small enough for Bernoulli's equation (1.76) to hold within the rocket frame, we find for the exit velocity

$$v_e = \sqrt{\frac{2(P_0 - P_a)}{\rho_0[1 - (A_e/A_0)^2]}}$$

where P_0 is the gas pressure in the chamber and P_a is atmospheric pressure. This gives the following values for specific impulse, assuming $A_e \ll A_0$,

$P_0 - P_a$, lb_f/in^2	v_e, ft/s	v_e/g_0, s
100	122	3.8
1,000	386	12.0
10,000	1,219	37.9

Problems

1.1 Evaluate the sums δ_{ii}, $\delta_{ik}\delta_{ik}$, $\varepsilon_{ijk}\varepsilon_{ijk}$, and $\varepsilon_{ijk}\varepsilon_{ikj}$.

Answers $+3$; $+3$; $+6$; -6

1.2 Find the value of the divergence $\nabla \cdot \mathbf{u}$ and the vorticity $\nabla \times \mathbf{u}$ for each of the following velocity fields. The factors A and B are constants.

(*a*) $\mathbf{u} = \mathbf{e}_1 A$
(*b*) $\mathbf{u} = \mathbf{e}_1 A x_1$
(*c*) $\mathbf{u} = \mathbf{e}_r f(r,t)$, where r is a spherical coordinate
(*d*) $\mathbf{u} = \mathbf{e}_1 A x_2 + \mathbf{e}_2 B x_1$
(*e*) $\mathbf{u} = A(\mathbf{e}_1 \times \mathbf{x})$
(*f*) $\mathbf{u} = A\left(\frac{\mathbf{e}_1}{x_1} + \frac{\mathbf{e}_2}{x_2} + \frac{\mathbf{e}_3}{x_3}\right)$

Answers (*a*) 0; 0 (*b*) $+A$; 0

(*c*) $\frac{1}{r^2}\frac{\partial}{\partial r}(r^2 f)$; 0 (*d*) 0; $\mathbf{e}_3(B - A)$

(*e*) 0; $2A\mathbf{e}_1$ (*f*) $-A\left(\frac{1}{x_1^2} + \frac{1}{x_2^2} + \frac{1}{x_3^2}\right)$; 0

1.3 Suggest physical means for producing a flow with $\nabla \cdot \mathbf{u} > 0$.

1.4 Show that the rate of elongation of a fluid line $d\mathbf{x}$ is

$d\mathbf{x} \cdot d\mathbf{u} = dx_i\, D_{ik}\, dx_k$

where $d\mathbf{u}$ is the velocity change between $\mathbf{x}$ and $\mathbf{x} + d\mathbf{x}$.

1.5 For an irrotational flow adjacent to a plane wall, how does the tangential component of the velocity vary with the (normal) distance from the wall? Note that the tangential component does not vanish at the wall (in the absence of a no-slip condition) but the normal component does.

1.6 Specify any simple velocity field for which $\mathbf{\Omega}$ and $\mathbf{u}$ are parallel, at least at some point in the field.

1.7 Show formally that the velocity field $\mathbf{u}(\mathbf{x},t)$ derived from a potential $\phi(\mathbf{x},t)$

$u = \nabla\phi$

is irrotational.

1.8 For the velocity field $\mathbf{u} = K\mathbf{r}$ in Fig. 1.15, show that *any* observer moving with the local velocity sees exactly the same velocity field. For example, an observer located instantaneously at $\mathbf{r}_0$ sees the field $\mathbf{u} - \mathbf{u}_0 = K(\mathbf{r} - \mathbf{r}_0)$.

1.9 Consider a spherical surface of radius $R(t)$ enclosing a region $V^*(t)$. Let $\chi(r,t)$ be some spherically symmetric function. Formally evaluate the derivative

$$\frac{d}{dt}\int_{V^*(t)} \chi(r,t)\, dV$$

by Reynolds' transport theorem and also by Leibniz' rule and verify that the results are identical.

1.10 Find the material acceleration $D\mathbf{u}/Dt$ for the velocity field

$$\mathbf{u} = \mathbf{e}_1 \alpha \frac{x_1}{t}$$

where α is a constant, and interpret the result for the special case $\alpha = 1$.

Answer $$\frac{D\mathbf{u}}{Dt} = \mathbf{e}_1 \alpha(\alpha - 1)\frac{x_1}{t^2}$$

1.11 For a potential vortex the velocity field is

$$\mathbf{u} = \mathbf{e}_\theta \frac{K}{r}$$

For an infinitesimal fluid particle located at $r = R$ let the angular displacement in time Δt be $\Delta\theta_0$. Show that the relative angular displacement $\Delta\theta - \Delta\theta_0$ of a particle located at $r = R + y$, where $y \ll R$, is given by

$$\frac{\Delta\theta - \Delta\theta_0}{\Delta\theta_0} = -\frac{2y}{R} + 3\left(\frac{y}{R}\right)^2 \cdots$$

1.12 Show that the Euler momentum equation

$$\rho\frac{Du_i}{Dt} + P_{,i} = 0$$

can be put in the form

$$\frac{\partial}{\partial t}\rho u_i + (\rho u_i u_k + P\delta_{ik})_{,k} = 0$$

The quantity $\rho u_i u_k$ is the *momentum tensor.*

1.13 For an incompressible fluid at rest, with $e = c_v T + \text{const}$, show that the energy equation (1.71) reduces to the *heat equation*

$$\rho c_v \frac{\partial T}{\partial t} = \kappa\nabla^2 T$$

where the thermal conductivity κ has been assumed constant.

1.14 Consider a small circle of radius a lying in a plane with orientation defined by the unit vector $\mathbf{n}$. Show from Stokes' theorem that the average tangential component of velocity on the circle, $\bar{u}_t = \overline{\mathbf{u}\cdot\mathbf{t}}$, is given by $\bar{u}_t = (a/2)\boldsymbol{\Omega}\cdot\mathbf{n}$. (This provides an interpretation of the vorticity $\boldsymbol{\Omega}$.)

1.15 Show that at a rigid boundary with orientation $\mathbf{n}$, bounding an incompressible fluid, the surface traction is given by

$$\mathbf{T} = -P\mathbf{n} - \mu(\mathbf{n}\times\boldsymbol{\Omega})$$

1.16 Given that the small stressed tetrahedron shown is in equilibrium from surface forces alone and that $\mathbf{T}(-\mathbf{e}_i) = -\mathbf{T}(\mathbf{e}_i)$, show that the traction on the surface dA (with arbitrary orientation $\mathbf{n}$) is

$$T_k(\mathbf{n}) = n_i\sigma_{ik}$$

which is Eq. (1.31).

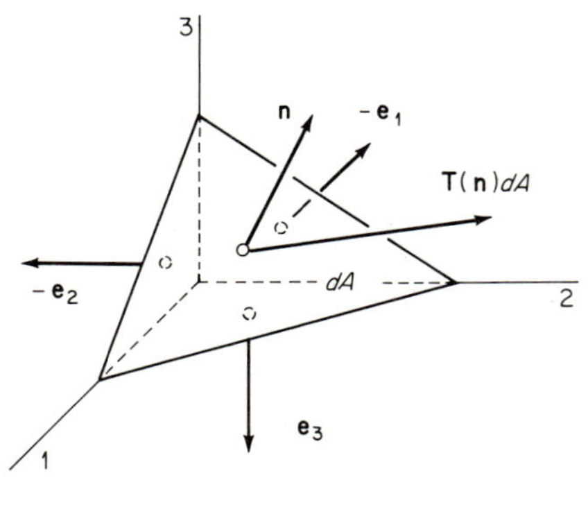

1.17 A coordinate frame translates with velocity $\mathbf{v}_0(t)$ and rotates with angular velocity $\boldsymbol{\omega}_0(t)$. Let $\mathbf{U}(\mathbf{x},t)$ be absolute fluid velocity and $\mathbf{u}(\mathbf{x},t)$ the fluid velocity relative to the moving frame, with

$$\mathbf{U} = \mathbf{u} + \mathbf{v}_0 + \boldsymbol{\omega}_0 \times \mathbf{x}$$

Show that the deformation tensor D_{ik} has the same value in the moving and absolute frames.

1.18 Two fluid jets, 1 and 2, combine to form a common jet as shown. The discharge velocities u_1, u_2 are known along with $\dot{m}_1$, $\dot{m}_2$. The pressure is everywhere equal to P_a. Find the angle θ of the combined jet from the known quantities by the use of the momentum balance (1.80) with suitable assumptions.

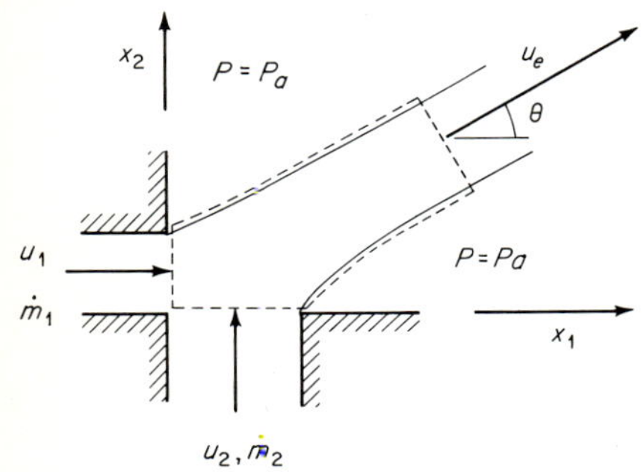

1.19 (*Archimedes' principle*). A body V of density ρ is submerged in a fluid of density ρ_a. With the help of Eq. (1.68) for hydrostatic equilibrium, show that the net force on the body is given by

$$\int_V \rho \mathbf{G}\, dV + \int_S \mathbf{T}\, dA = \int_V (\rho - \rho_a)\mathbf{G}\, dV$$

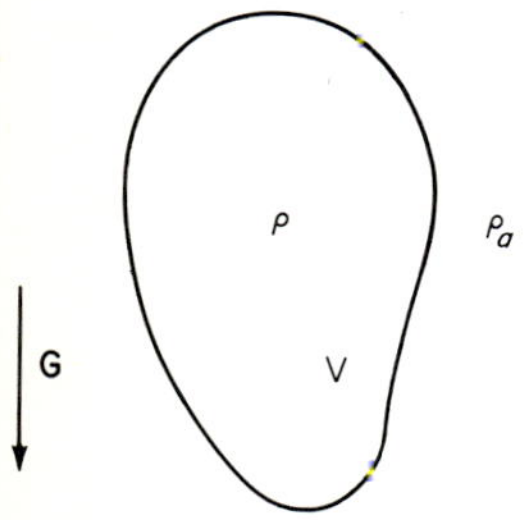

1.20 Show that the scalar product of a symmetric with an antisymmetric tensor, for example, $D_{ik}\Omega_{ik}$, is identically zero.

1.21 Verify that the specific potential energy $\Psi = gx_3$ in a constant gravity field, with x_3 the vertical coordinate, is consistent with the force potential defined by Eq. (1.72).

1.22 The equation of *static* equilibrium for the earth's atmosphere is

$$\nabla P = \rho \mathbf{g}$$

With the gravitational force derivable from a potential, show that this equation requires that *both* pressure and density be constant on an equipotential surface. Also discuss the motion induced by local heating of the atmosphere.

1.23 A very simple model of a chimney has constant fluid density ρ_o in the outside atmosphere and constant density ρ_i inside the chimney, with $\rho_o > \rho_i$. For an

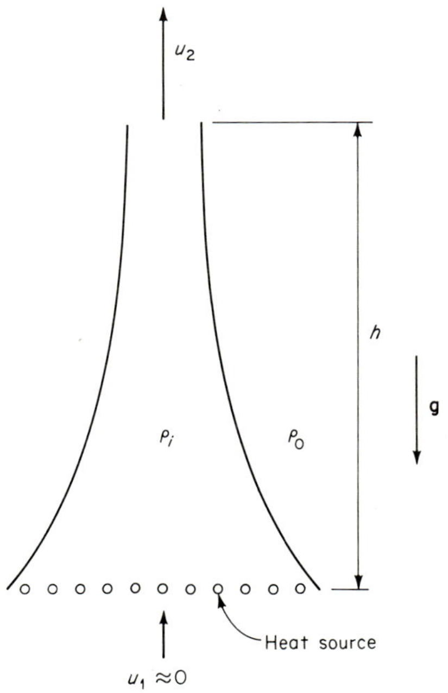

idealized frictionless chimney, find the exit velocity u_2 in terms of the quantities ρ_0, ρ_i, g, and h. The flow is steady, and the pressures are matched at the entrance and exit; that is, $(P_1)_i = (P_1)_o$ and $(P_2)_i = (P_2)_o$.
Hint: Use Bernoulli's equation and hydrostatics.

Answer $\quad u_2 = \sqrt{2 \dfrac{\rho_0 - \rho_i}{\rho_i} gh}$

two
thermodynamics of motion

2.1 Introduction

Thermodynamics as usually presented deals with relatively permanent states, called *equilibrium states*, within uniform matter, i.e., matter in which there are no spatial gradients in quantities such as temperature, pressure, and concentration. Any changes are assumed to take place with extreme slowness. The fluid motions of interest in gasdynamics do not appear to be slow; nevertheless, it is found that the usual results of classical thermodynamics and statistical mechanics can be directly applied to such motions provided that the instantaneous *local* thermodynamic state is considered and that rates of change are not too large. A more precise statement is that equilibrium thermodynamics is locally applicable to fluids which are described by the *linear* stress and heat conduction relations, i.e., the constitutive relations given in Chap. 1.

The thermodynamical results of interest are the first and second laws as represented by Eqs. (1.70) and (1.61), respectively, viz.,

$$\rho \frac{De}{Dt} = \sigma_{ik} D_{ik} - q_{k,k}$$

$$\rho \frac{Ds}{Dt} + \Xi_{k,k} \geq 0$$

along with the state principle, which fixes the number of independent thermodynamic variables at two; the Gibbs equation, which connects the specific entropy to other variables; and equation-of-state data. These results are put into various useful forms by use of the calculus of two independent variables.

2.2 The local thermodynamic state

In the equations of motion, the following thermodynamic variables have been introduced:

P = pressure
T = absolute temperature
ρ = density
$v = 1/\rho$ = specific volume
e = specific internal energy
$h = e + Pv$ = specific enthalpy
s = specific entropy

In addition are the transport coefficients, which behave like thermodynamic variables:

μ = shear viscosity
μ_v = bulk viscosity
κ = thermal conductivity

Each of the above variables is conventionally termed *intensive* (as opposed to extensive) because a value obtains at each point in space, independent of the quantity of surrounding matter. According to the field description of flow, each of these variables may be considered a function of position and time; thus $P = P(\mathbf{x},t)$, $e = e(\mathbf{x},t)$, and so on.

According to the *state principle*, the local thermodynamic state is fixed by any two independent thermodynamic variables[1] provided that the chemical composition of the fluid is not changed by mixing or diffusion, i.e., that the material volume is a valid model. This is an empirical law, like the first and second laws (even though it has not been awarded a number). In most situations any two of the above variables may be selected as the independent pair, with the obvious exception of ρ and v.

[1] In general, the state principle stipulates that the number of independent variables is equal to the number of reversible (not contributing to entropy production) work modes plus 1. We will find that there is one reversible work mode, corresponding to $P\ Dv/Dt$.

If, for example, s and v are selected as the independent pair, the other thermodynamic variables are in principle known in terms of them,

$$
\begin{aligned}
e &= e(s,v) \\
T &= T(s,v) \\
P &= P(s,v) \\
&\cdots\cdots
\end{aligned}
$$

Relations of this type are known as *equations of state* and exist in principle for every fluid. Then the equations of motion can be rewritten such that only two thermodynamic quantities appear.

Of particular interest, because directly measurable quantities are involved, is the *thermal* equation of state,

$$P = P(v,T) \tag{2.1}$$

which can be written in alternative forms such as $f(P,v,T) = 0$ or $v = v(P,T)$. An especially important example is the thermal equation of state for an ideal gas,

$$Pv = RT \tag{2.2}$$

in which R is the specific gas constant, $R = \tilde{R}/\tilde{M}$, where $\tilde{R}$ is the universal gas constant and $\tilde{M}$ is the molecular weight of the gas in question. In general, (2.1) can exist in the form of an algebraic expression or in the form of tabulated data.

By itself, the thermal equation of state (2.1) does not contain sufficient information to allow determination of all properties at a given vT state; for example, $s(v,T)$ cannot be determined from (2.1) alone. An additional relation must be given. One possibility is to specify a so-called *caloric* equation of state,

$$e = e(v,T) \tag{2.3}$$

It is easy to show that a knowledge of (2.1) and (2.3) for the fluid in question allows determination of any thermodynamic property for the fluid in question. As we will see later, however, these equations are not independent of each other.

Alternatively, a *potential* (or canonical) equation of state such as

$$e = e(s,v) \tag{2.4}$$

contains implicitly a complete specification of all thermodynamic variables, as will be illustrated in Example 2.1 (page 62). In any case, it will be assumed that complete equation-of-state data are in principle known for fluids of interest.

In a changing environment, fluid does not instantaneously achieve an equilibrium state. Suppose, for example, that a partially dissociated diatomic gas is locally in equilibrium at some point a (where $T = T_a$, $P = P_a$); then the fractional dissociation α is fixed by the local conditions, $\alpha_a{}^0 = \alpha(P_a,T_a)$. If the gas flows rapidly to point b (where $T = T_b$, $P = P_b$), the fractional dissociation will not in general achieve the new equilibrium value $\alpha_b{}^0 = \alpha(P_b,T_b)$ but will have some intermediate value between $\alpha_a{}^0$ and $\alpha_b{}^0$. The required dissociation or recombination reactions take time and do not occur fast enough to keep up with the predicted equilibrium value for the fractional dissociation. Such a situation could arise physically if a hot gas were rapidly expanded through a small supersonic nozzle or in passing through a shock wave.

In general, lagging internal processes such as dissociation, ionization, evaporation, chemical reaction, and transfer of energy between molecular modes (translation, rotation, vibration) are called *relaxation processes.* When such processes are important and can be characterized by rate equations, the time enters the (nonequilibrium) thermodynamic state explicitly; e.g., we might write for the specific internal energy e

$$e = e(P,T,t) \tag{2.5}$$

Fluids so described are referred to as fluids with *memory.* In the preceding example of the dissociating gas, the gas "remembers" that it came from a region a of high fractional dissociation (say), and this is reflected in its actual state at point b.

In many internal gas processes, it is possible to assign a *relaxation time* τ, which is a measure of the time required for a disturbed quantity to regain its equilibrium value. If the time Δt required for a given fluid change, e.g., the time required to pass through a small nozzle, satisfies the inequality

$$\Delta t \gg \tau \tag{2.6}$$

it will be a good approximation to treat the fluid as though it were in equilibrium. In most of the problems considered in this book (2.6) is implicitly assumed. That is, from the standpoint of departure from thermodynamic equilibrium, the flows involved are in most cases *slow.* Some representative relaxation times are shown in Table 2.1.[1]

[1] For discussion of relaxation in gases and data, see *Zel'dovich and Raizer* [1967, chap. 6], *Bradley* [1962], and *Gaydon and Hurle* [1963].

Table 2.1 Relaxation Times at $P = 1$ Atm

Gas	Mode	T, K	τ, s
N_2	Translation	300	1.6×10^{-10}
N_2	Rotation	300	1.2×10^{-9}
N_2	Vibration	3,000	3.0×10^{-5}
CF_4	Vibration	373	6.6×10^{-7}
H_2O	Vibration	486	3.7×10^{-8}
O_2	Dissociation	3,100	2.3×10^{-5}

The fluids considered in this book are described by a linear viscosity law (1.37), a linear conduction law (1.50), and a state fixed by two thermodynamic variables. Such fluids are called *simple*.

2.3 Entropy

Entropy is a useful parameter in fluid mechanics primarily because it may be taken as constant in many flows. This results in the considerable simplification that the thermodynamic state is fixed by a single thermodynamic variable; for example, $e = e(s,v)$ becomes with the specific entropy held constant $e = e(v)$. By similar reasoning the thermal equation of state $P = P(v,T)$ reduces to

$$P = P(v) \qquad \text{or} \qquad P = P(\rho)$$

under the same conditions. Fluids for which the pressure depends only on the density are said to be *barotropic*.[1] This leads to a simplification in the equations of motion almost as great as if the density itself were taken constant. Accordingly, we consider the production and transfer of entropy, its relation to the other flow quantities, and the conditions under which it may be assumed constant.

[1] The term *barotropic* originated in meteorology, where its application to the atmosphere implies that a surface of constant pressure coincides with a surface of constant density (in this restricted sense, an incompressible fluid may be barotropic). In the same usage, the term *baroclinic* means that surfaces of constant pressure do not coincide with surfaces of constant density but are mutually inclined. Finally it should be mentioned that a fluid will be barotropic if any thermodynamic variable (the temperature T, for example) is held constant; the case of practical interest here, however, is that of constant entropy.

The Production of Entropy

The energy equation (first law) was obtained in the form (1.70), viz.,

$$\rho \frac{De}{Dt} = \sigma_{ik} D_{ik} - q_{k,k}$$

and will be put into a form equivalent to the *second* law. With $\sigma_{ik} = -P\delta_{ik} + \Sigma_{ik}$ and the use of (1.63) this becomes

$$\frac{De}{Dt} + P \frac{Dv}{Dt} = v\Sigma_{ik} D_{ik} - v q_{k,k} \tag{2.7}$$

The scalar $\Sigma_{ik} D_{ik}$ has the property that it is *always positive* for a linearly viscous fluid, as we now show. With (1.36)

$$\Sigma_{ik} D_{ik} = 2\mu D_{ik} D_{ik} + (\mu_v - \tfrac{2}{3}\mu)(D_{mm})^2 \tag{2.8}$$

After some algebra this reduces to

$$\begin{aligned}\Sigma_{ik} D_{ik} = {} & \tfrac{2}{3}\mu[(D_{11} - D_{22})^2 + (D_{22} - D_{33})^2 + (D_{33} - D_{11})^2] \\ & + 4\mu(D_{12}{}^2 + D_{13}{}^2 + D_{23}{}^2) + \mu_v(D_{11} + D_{22} + D_{33})^2\end{aligned} \tag{2.9}$$

which is necessarily positive (or zero) if the viscosity coefficients μ and μ_v are positive. Simple stability arguments show that the viscosities are indeed positive; in a negative-viscosity fluid, for example, a simple shearing motion as indicated in Fig. 1.8 would be accelerated without limit, and a shaft supported by oil-lubricated journal bearings would accelerate continuously if given a slight initial rotation.

The positive quantity $\Sigma_{ik} D_{ik}$ is called the *dissipation function* and will be indicated by the (suggestively drooping) symbol Υ

$$\Upsilon \equiv \Sigma_{ik} D_{ik} \geq 0 \tag{2.10}$$

This represents that part of the viscous work going into deformation (as opposed to acceleration) of a fluid particle.

It is also possible to find a positive definite part in the heat-conduction term of (1.70). With the Fourier conduction law $q_k = -\kappa T_{,k}$, and the expansion

$$\left(\frac{\kappa T_{,k}}{T}\right)_{,k} = -\frac{\kappa}{T^2} T_{,k} T_{,k} + \frac{1}{T}(\kappa T_{,k})_{,k} \tag{2.11}$$

yields a quadratic term. In vector notation this expansion is

$$\nabla \cdot \frac{\kappa \nabla T}{T} = -\frac{\kappa}{T^2}(\nabla T)^2 + \frac{1}{T}\nabla \cdot (\kappa \nabla T) \tag{2.12}$$

Solving for the last term and putting into (2.7) gives

$$\frac{De}{Dt} + P\frac{Dv}{Dt} - vT\nabla \cdot \frac{\kappa \nabla T}{T} = v\Upsilon + \frac{v\kappa(\nabla T)^2}{T} \tag{2.13}$$

The last term is positive if T and κ are positive; κ is positive from a thermal stability argument, since a local perturbation in temperature in an otherwise uniform medium would grow without limit by (1.50) if κ were negative. We assume T positive, and the terms on the right side of (2.13) are necessarily individually positive.

Equation (2.13) is now almost the second law of thermodynamics. We now introduce the *Gibbs equation*,[1]

$$T\,ds = de + P\,dv \tag{2.14}$$

which connects the entropy to the other properties. Then (2.10) can be written

$$\frac{Ds}{Dt} - v\nabla \cdot \frac{\kappa \nabla T}{T} = \frac{v\Upsilon}{T} + v\kappa\left(\frac{\nabla T}{T}\right)^2 \geq 0 \tag{2.15}$$

which is to be compared with the formal statement of the second law already given in Eq. (1.61), viz.,

$$\frac{Ds}{Dt} + v\nabla \cdot \boldsymbol{\Xi} \geq 0$$

These statements are identical[2] if the entropy flux $\boldsymbol{\Xi}$ is equated to $\mathbf{q}/T$, where $\mathbf{q} = -\kappa \nabla T$.

The second-law statement (2.15) can now be interpreted as follows: the quantity on the right-hand side of the equal sign is the (always positive) *production* of entropy per unit mass, the divergence term represents the

[1] After J. Willard Gibbs (1839–1903), greatest of American theoretical physicists and founder of equilibrium thermodynamics. Gibbs was the first American to receive a Ph.D. in engineering, later becoming professor of physics at Yale: "Mr. Josiah Willard Gibbs, of New Haven, was appointed Professor of Mathematical Physics, without salary, in the Department of Philosophy and the Arts." (Minutes of the Yale Corporation Meeting, July 13, 1871.)

[2] A more systematic development, in which the Gibbs equation appears as a *result* rather than an axiom, is given by *Coleman and Mizel* [1964].

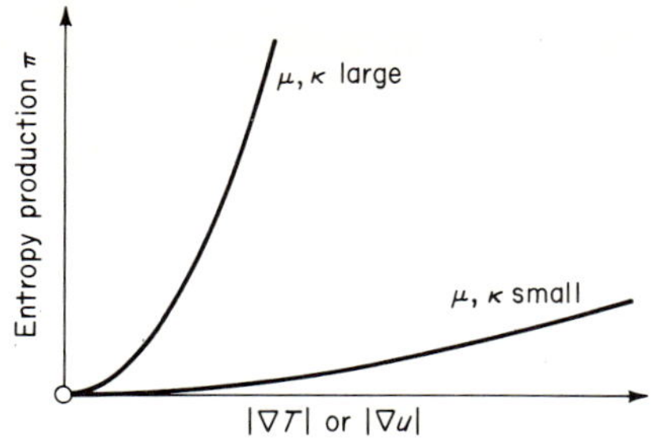

Figure 2.1

net outflow of entropy per unit mass, and Ds/Dt is the *rate of increase* of entropy in a unit mass.

A *reversible* process is one for which there is no production of entropy, i.e., for which the right side of (2.15) vanishes and entropy becomes a conserved quantity. Such processes are approximated for small gradients because the production is proportional to the *square* of temperature and velocity gradients, as sketched in Fig. 2.1. Thus, if the gradients are of order ε, the production $\pi \sim \varepsilon^2$.

The overall entropy balance (2.15) can be rewritten finally as

$$\boxed{\rho T \frac{Ds}{Dt} = \Upsilon - \nabla \cdot \mathbf{q}} \tag{2.16}$$

where $\mathbf{q} = -\nabla T$. In thermodynamics books this relation is often written in the form of the *Clausius inequality*, $T ds \geq \delta q$, where $\delta q = -v \nabla \cdot \mathbf{q}\, dt$ is the heat transfer to a unit mass of fluid.

Isentropic and Homentropic Idealizations

If both heat conduction $\mathbf{q}$ and dissipation Υ can be neglected altogether, (2.16) becomes simply

$$\frac{Ds}{Dt} = 0$$

and the entropy of a given fluid particle is constant. Such a process is called *reversible adiabatic* or *isentropic*. This approximation is tenable in gasdynamics provided that the gradients are not large, this requirement usually being satisfied outside of boundary layers, shock layers, and wakes.

Flows which satisfy the further condition that the entropy is the same for *all* fluid particles will be called *homentropic*:

$$\text{Isentropic flow:} \qquad \frac{Ds}{Dt} = 0$$

$$\text{Homentropic flow:} \qquad \frac{Ds}{Dt} = \nabla s = 0$$

In general a flow can be isentropic or homentropic only if the effects of viscosity and heat conduction are negligibly small. In gasdynamics fluids which may be considered to have neither viscosity nor thermal conductivity are called *inviscid*, as has already been noted, and this term will be often used. In hydrodynamics fluids without viscosity are called *perfect* (they are perhaps perfect from the point of view of the theoretical hydrodynamicist; perfect fluids of this kind have no connection with the *perfect gas* discussed in Sec. 2.5).

Remark on the Distinction between the Adiabatic and Isentropic Idealizations

It is unfortunate that the terms adiabatic and isentropic are occasionally used as though they were perfectly interchangeable. This remark constitutes a reminder that they are *not*, in fact, synonymous.

By definition, an *adiabatic*[1] flow has $\mathbf{q} = 0$ everywhere (certain flows are adiabatic in some mean sense if $\mathbf{q} = 0$ on material boundaries; such a flow is treated in Sec. 6.4). Also by definition, an *isentropic* flow has $Ds/Dt = 0$ everywhere. By virtue of (2.16), an adiabatic flow is isentropic only if the dissipation Υ is also zero, i.e., if the fluid may be treated as inviscid.

Relations between Thermodynamic Variables

Rewriting the Gibbs equation in terms of the internal energy e, enthalpy $h \equiv e + Pv$, Helmholtz function $a \equiv e - Ts$, and Gibbs function $g \equiv h - Ts$ gives the alternative (equivalent) forms:

$$de = T\,ds - P\,dv \tag{2.17a}$$

$$dh = T\,ds + v\,dP \tag{2.17b}$$

$$da = -s\,dT - P\,dv \tag{2.17c}$$

$$dg = -s\,dT + v\,dP \tag{2.17d}$$

[1] From the Greek ἀδιάβατος, not passing across; i.e., the heat does not flow across a surface within the fluid.

(The quantities a and g will not be of further interest but are introduced only to obtain the following relations.) Comparing (2.17a) with the differential of $e(s,v)$,

$$de = \left(\frac{\partial e}{\partial s}\right)_v ds + \left(\frac{\partial e}{\partial v}\right)_s dv$$

gives the identities

$$T = \left(\frac{\partial e}{\partial s}\right)_v \qquad P = -\left(\frac{\partial e}{\partial v}\right)_s$$

If the surface $e(s,v)$ is smooth, the crossed second derivatives are equal; i.e.,

$$\frac{\partial^2 e}{\partial s\, \partial v} = \frac{\partial^2 e}{\partial v\, \partial s}$$

Thus $(\partial T/\partial v)_s = -(\partial P/\partial s)_v$. Applying the same reasoning further, we thus obtain the identities resulting from the identification of all coefficients in (2.17)

$$\begin{aligned} T &= \left(\frac{\partial e}{\partial s}\right)_v & -P &= \left(\frac{\partial e}{\partial v}\right)_s \\ T &= \left(\frac{\partial h}{\partial s}\right)_P & v &= \left(\frac{\partial h}{\partial P}\right)_s \\ -s &= \left(\frac{\partial a}{\partial T}\right)_v & -P &= \left(\frac{\partial a}{\partial v}\right)_T \\ -s &= \left(\frac{\partial g}{\partial T}\right)_P & v &= \left(\frac{\partial g}{\partial P}\right)_T \end{aligned} \tag{2.18}$$

and making use of the equality of crossed derivatives gives the *Maxwell relations,*[1]

$$\begin{aligned} \left(\frac{\partial T}{\partial v}\right)_s &= -\left(\frac{\partial P}{\partial s}\right)_v \\ \left(\frac{\partial T}{\partial P}\right)_s &= \left(\frac{\partial v}{\partial s}\right)_P \\ \left(\frac{\partial s}{\partial v}\right)_T &= \left(\frac{\partial P}{\partial T}\right)_v \\ \left(\frac{\partial s}{\partial P}\right)_T &= -\left(\frac{\partial v}{\partial T}\right)_P \end{aligned} \tag{2.19}$$

[1] After James Clerk Maxwell (1831–1879), British physicist known particularly for work in kinetic theory and electromagnetics. For an account of Maxwell's use of the signature $(\partial P/\partial T)_v$ in private correspondence, see *Klein* [1970].

By manipulation of (2.17) and (2.19) and making use of the rules of calculus it is possible to arrive at a large number of identities. Generally the motive in seeking such identities is to express more or less obscure quantities (as may be needed in analysis) in terms of known or directly measurable quantities. In addition to pressure, volume, and temperature, two such quantities are the *specific heats* (sometimes called the heat capacities) at constant volume and pressure, respectively,

$$c_v \equiv \left(\frac{\partial e}{\partial T}\right)_v \qquad c_p \equiv \left(\frac{\partial h}{\partial T}\right)_p \tag{2.20}$$

The ratio of these will occur frequently and is indicated by the symbol γ

$$\gamma \equiv \frac{c_p}{c_v} \tag{2.21}$$

Several examples involving the preceding thermodynamic relations are given in the following.

EXAMPLE 2.1 POTENTIAL EQUATION OF STATE

We wish to show the determination of all thermodynamic properties from a potential equation of state, such as

$$e = e(s,v) \tag{2.4}$$

An example of an equation of state in this form is the expression for a perfect gas

$$\frac{e}{e_0} = \left(\frac{v}{v_0}\right)^{-R/c_v} e^{(s-s_0)/c_v} \tag{2.22}$$

where R and c_v are constants and e_0, v_0, and s_0 are constants corresponding to a fixed reference state.

Given a function of the form (2.4), the temperature and pressure can be found with (2.18) by direct differentiation, yielding $T(s,v)$ and $P(s,v)$. Then the enthalpy is found from its definition

$$h = e(s,v) + vP(s,v)$$

and similarly for the Gibbs and Helmholtz functions. The specific heat at constant volume is

$$c_v = \left(\frac{\partial e}{\partial T}\right)_v = \left(\frac{\partial e}{\partial s}\right)_v \left(\frac{\partial s}{\partial T}\right)_v = T\left(\frac{\partial s}{\partial T}\right)_v$$

and $(\partial s/\partial T)_v$ can be found from $T(s,v)$. The specific heat at constant pressure is found similarly. Thus all properties are found essentially by differentiation.

The other potential equations contain the variables appearing as differentials in each of Eqs. (2.17); for example, $h(s,P)$ is a potential equation.

If the operations just described are applied to Eq. (2.22) for a perfect gas, we find

$$T = \left(\frac{\partial e}{\partial s}\right)_v = \frac{e}{c_v}$$

$$P = -\left(\frac{\partial e}{\partial v}\right)_s = \frac{RT}{v}$$

$$c_v = \text{const}$$

which are well-known properties of a perfect gas, to be discussed in Sec. 2.5.

EXAMPLE 2.2 RELATION BETWEEN ISENTROPIC AND ISOTHERMAL COMPRESSIBILITIES

As an example of a thermodynamic identity of frequent use, we wish to show that

$$\left(\frac{\partial P}{\partial v}\right)_s = \gamma\left(\frac{\partial P}{\partial v}\right)_T \tag{2.23}$$

By definition

$$\gamma \equiv \frac{c_p}{c_v} = \frac{\left(\frac{\partial h}{\partial T}\right)_p}{\left(\frac{\partial e}{\partial T}\right)_v} = \frac{\left(\frac{\partial h}{\partial s}\right)_p\left(\frac{\partial s}{\partial T}\right)_p}{\left(\frac{\partial e}{\partial s}\right)_v\left(\frac{\partial s}{\partial T}\right)_v} = \frac{\left(\frac{\partial s}{\partial v}\right)_p\left(\frac{\partial v}{\partial T}\right)_p}{\left(\frac{\partial s}{\partial P}\right)_v\left(\frac{\partial P}{\partial T}\right)_v}$$

By calculus,

$$\left(\frac{\partial s}{\partial v}\right)_p\left(\frac{\partial v}{\partial P}\right)_s\left(\frac{\partial P}{\partial s}\right)_v = -1 \qquad \text{and} \qquad \left(\frac{\partial v}{\partial T}\right)_p\left(\frac{\partial T}{\partial P}\right)_v\left(\frac{\partial P}{\partial v}\right)_T = -1$$

Using these relations, the above equation becomes

$$\gamma = \frac{\left(\frac{\partial P}{\partial v}\right)_s}{\left(\frac{\partial P}{\partial v}\right)_T} = \frac{\frac{1}{v}\left(\frac{\partial P}{\partial v}\right)_s}{\frac{1}{v}\left(\frac{\partial P}{\partial v}\right)_T}$$

which is the desired identity [the quantities $\frac{1}{v}\left(\frac{\partial v}{\partial P}\right)_T$ and $\frac{1}{v}\left(\frac{\partial v}{\partial P}\right)_s$ are respectively the *isothermal* compressibility and the *isentropic* compressibility]. Equation (2.23) is useful, for example, in the calculation of the speed of sound when PvT data are known. Further identities are given in Appendix B.

EXAMPLE 2.3 BERNOULLI'S EQUATION FOR ISENTROPIC FLOW

We will reduce Bernoulli's equation (1.74) to its simplest form. For steady flow with negligible viscous stress and heat conduction we found that (1.74) reduces to

$$d\left(h + \frac{u^2}{2} + \Psi\right) = 0$$

along a streamline. Now in the absence of viscous stress and heat conduction, the flow is *isentropic* by (2.15). Then Gibbs' equation (2.17*b*) gives $dh = v\,dP$, and the above becomes

$$v\,dP + d\left(\frac{u^2}{2} + \Psi\right) = 0$$

or

$$\int_{P_0}^{P} v\,dP + \frac{u^2}{2} + \Psi = \text{const} \tag{2.24}$$

where P_0 is any reference state and the integral has a definite value dependent only on P because $v = v(P)$ with the entropy constant. The result (2.24) can also be obtained from Euler's equation (1.67). For an incompressible fluid $v = 1/\rho =$ const along a streamline and (2.24) reduces to the famous equation

$$P + \tfrac{1}{2}\rho u^2 + \rho\Psi = \text{const} \tag{2.25}$$

The reader who is not familiar with applications of this equation may refer to any elementary physics text.

EXAMPLE 2.4 ATMOSPHERIC STABILITY

The determination of the condition for stability of a stationary atmosphere is an application of simple thermodynamic identities. For an atmosphere at rest with vertical coordinate z and (downward) gravitational acceleration g, Eq. (1.68) of hydrostatic equilibrium is just

$$\frac{dP}{dz} = -\rho_a g \tag{2.26}$$

where ρ_a is the local atmospheric density. Consider the small vertical displacement ξ of a fluid particle δV, as sketched in Fig. 2.2. The net upward force on the particle is $g(\rho_a - \rho)\,\delta V$ from Archimedes' principle (see Prob. 1.19). If this force is negative, i.e., if the particle density is greater than that of the local atmosphere, the displaced particle will tend to return to its initial position and

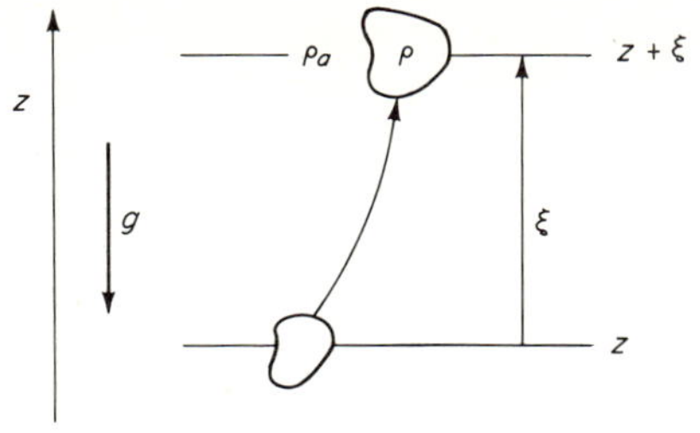

Figure 2.2

the atmosphere may be considered stable. Setting the force equal to mass times acceleration gives

$$g(\rho_a - \rho) = \rho\ddot{\xi} \tag{2.27}$$

To calculate ρ_a and ρ we expand each in a Taylor series from the initial state $\rho(z) = \rho_a(z)$. For the atmosphere,

$$\rho_a = \rho_a(z) + \left(\frac{\partial\rho}{\partial P}\right)_s [P - P(z)] + \left(\frac{\partial\rho}{\partial s}\right)_p [s - s(z)] + \cdots$$

It is assumed that the displaced particle experiences no entropy change, i.e., dissipation and conduction are negligible, and the Taylor expansion for the particle is just

$$\rho = \rho(z) + \left(\frac{\partial\rho}{\partial P}\right)_s [P - P(z)] + \cdots$$

Now $P - P(z)$ is the same in both expressions, since the particle is assumed to be in pressure equilibrium with its environment. Then subtraction gives, with $s - s(z) = \xi\, ds/dz$,

$$\rho_a - \rho = \left(\frac{\partial\rho}{\partial s}\right)_p \frac{ds}{dz}\xi$$

For convenience we rewrite the thermodynamic derivative in terms of $v = 1/\rho$ and

$$\rho_a - \rho = -\frac{1}{v^2}\left(\frac{\partial v}{\partial s}\right)_p \frac{ds}{dz}\xi$$

The equation of motion for the particle (2.27) becomes

$$\ddot{\xi} + N^2\xi = 0 \tag{2.28}$$

where

$$N^2 \equiv \frac{g}{v}\left(\frac{\partial v}{\partial s}\right)_p \frac{ds}{dz} \tag{2.29}$$

If $N^2 > 0$, the particle will oscillate about its initial position, for example, $\xi = \sin Nt$, and the atmosphere is *stable*. If $N^2 < 0$, the particle motion is accelerated, for example, $\xi = e^{|N|t}$, and the atmosphere is unstable. The parameter N is called the *Brunt-Väisälä*, or *buoyancy frequency*, and the preceding derivation is due to *Väisälä* [1925], as discussed in *Eckart* [1960].

The frequency N can be put in terms of conveniently observable quantities as follows:

$$\left(\frac{\partial v}{\partial s}\right)_p = \left(\frac{\partial v}{\partial T}\right)_p \left(\frac{\partial T}{\partial h}\right)_p \left(\frac{\partial h}{\partial s}\right)_p$$

Substituting $c_p \equiv (\partial h/\partial T)_p$ and $T = (\partial h/\partial s)_p$ from (2.18) gives

$$\left(\frac{\partial v}{\partial s}\right)_p = \frac{T}{c_p}\left(\frac{\partial v}{\partial T}\right)_p$$

Expanding $s(T,P)$, we obtain for the entropy gradient in the atmosphere

$$\frac{ds}{dz} = \left(\frac{\partial s}{\partial T}\right)_p \frac{dT}{dz} + \left(\frac{\partial s}{\partial P}\right)_T \frac{dP}{dz}$$

Now

$$\left(\frac{\partial s}{\partial T}\right)_p = \left(\frac{\partial s}{\partial h}\right)_p \left(\frac{\partial h}{\partial T}\right)_p = \frac{c_p}{T}$$

as above and $dP/dz = -\rho_a g$ from (2.26). We also have the Maxwell relation $(\partial s/\partial P)_T = -(\partial v/\partial T)_p$. Then

$$\frac{ds}{dz} = \frac{c_p}{T}\frac{dT}{dz} + \frac{g}{v}\left(\frac{\partial v}{\partial T}\right)_p$$

Equation (2.29) becomes

$$N^2 = \frac{g}{v}\left(\frac{\partial v}{\partial T}\right)_p \left[\frac{dT}{dz} + \frac{gT}{vc_p}\left(\frac{\partial v}{\partial T}\right)_p\right] > 0 \tag{2.30}$$

for stability. This relation contains of course only the thermodynamic properties of the stationary atmosphere. The thermal-expansion coefficient $\frac{1}{v}\left(\frac{\partial v}{\partial T}\right)_p$ is positive for almost all fluids (exceptions being water near 4°C and liquid bismuth near its freezing point), and $N^2 > 0$ if the sum in square brackets is positive. In particular, for an ideal gas (2.30) becomes, with $v = RT/P$ and hence $(\partial v/\partial T)_p = v/T$,

$$N^2 = \frac{g}{T}\left(\frac{dT}{dz} + \frac{g}{c_p}\right) > 0 \tag{2.31}$$

Thus for this case

$$\frac{dT}{dz} > -\frac{g}{c_p} \tag{2.32}$$

is the condition for stability.

The value of g/c_p for dry air is conveniently about 1 K per 100 m. Thus -0.01 K/m represents the metastable condition corresponding to equality in (2.32) or $ds/dz = 0$ from (2.29) and is called the *dry adiabatic lapse rate* in meteorology. At this condition a rising or sinking bubble of air changes temperature such that the temperature (and therefore the density) of the bubble always matches that of the local atmosphere.

Atmospheric *inversions* for which (by definition) $dT/dz > 0$ can be observed on the morning following a clear winter's night. Such atmospheres are "extremely" stable and by observation often tranquil. The stable inversion is of course a major factor in holding smog over cities. The stable *stratosphere* makes high-altitude flight attractive.

Violent convective instability is observed when air near the earth's surface is strongly heated, resulting for example in the updraft *thermals* utilized by gliders and soaring birds and in the formation of powerful "dust devils" in desert regions. Convective instability is also a factor in thunderstorms, with condensation playing a vital role.

In real oscillatory atmospheric motions, notably the spectacular *lee waves* sometimes observed behind mountains, the Brunt-Väisälä frequency N given by (2.30) is not necessarily the observed frequency of the motion. In such cases the assumptions of the preceding derivation, e.g., local pressure equilibrium, may not be fulfilled. The frequency N, however, will enter as a parameter for the real motion.

Elementary *models* of the earth's atmosphere (distribution of pressure, temperature, etc.) are of interest. The simplest possible models are the *isentropic* atmosphere and the *isothermal* atmosphere, considered separately below.

The stationary one-dimensional atmosphere satisfies the condition (2.26) of hydrostatic equilibrium

$$\frac{dP}{dz} = -\rho g \tag{2.33}$$

in which we will take $g = g_0$ to be a constant. To find, say, $\rho(z)$ from this equation it is necessary to know how P varies with ρ. If we take the atmosphere to be *isentropic*, P and ρ are related by

$$P\rho^{-\gamma} = \text{const} \tag{2.34}$$

where $\gamma = c_p/c_v$ is the ratio of specific heats (this equation will be derived in Sec. 2.5). This can be written in the form

$$\frac{P}{P_0} = \left(\frac{\rho}{\rho_0}\right)^\gamma \tag{2.35}$$

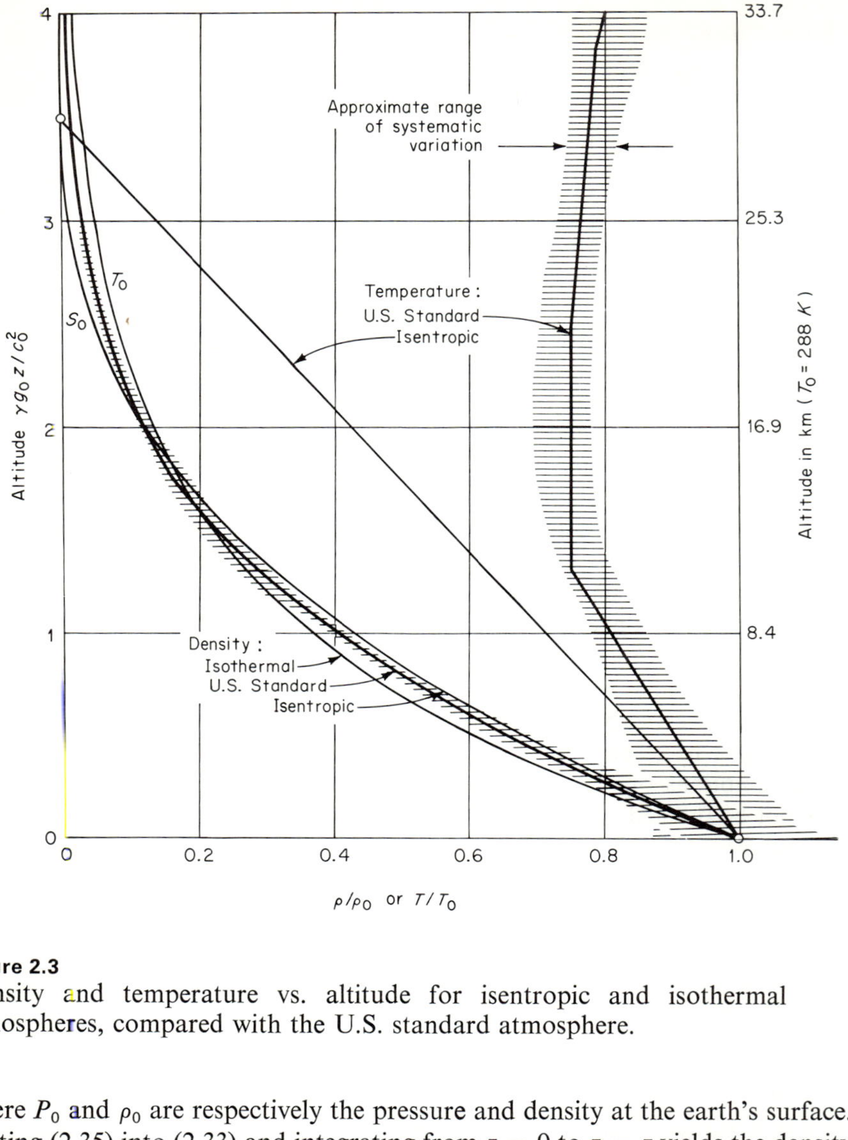

Figure 2.3
Density and temperature vs. altitude for isentropic and isothermal atmospheres, compared with the U.S. standard atmosphere.

where P_0 and ρ_0 are respectively the pressure and density at the earth's surface. Putting (2.35) into (2.33) and integrating from $z = 0$ to $z = z$ yields the density distribution

$$\frac{\rho}{\rho_0} = \left[1 - \frac{(\gamma - 1)g_0 z}{c_0^2}\right]^{1/(\gamma-1)} \tag{2.36}$$

where we have written $\gamma P_0/\rho_0 \equiv c_0^2$. This density distribution is compared to that of the U.S. standard atmosphere[1] (the "real" atmosphere) in Fig. 2.3.

[1] An extensive collection of data is given in *The U.S. Standard Atmosphere 1962*, U.S. Government Printing Office, Washington, 1962.

With the ideal gas equation of state $P = \rho RT$ (where R is constant), Eqs. (2.35) and (2.36) give the temperature distribution

$$\frac{T}{T_0} = 1 - \frac{(\gamma - 1)g_0 z}{c_0^2} \tag{2.37}$$

The isentropic model atmosphere described by (2.36) and (2.37) has the peculiar feature of a distinct top surface at $z = c_0^2/[(\gamma - 1)g_0]$, where the atmosphere simply ends, with zero density and zero temperature. This model is thus unrealistic at higher altitudes.

The alternative simple model is the *isothermal* atmosphere. With $P = \rho RT$ and $T = T_0 =$ const, (2.33) integrates to

$$\frac{\rho}{\rho_0} = e^{-\gamma g_0 z/c_0^2} \tag{2.38}$$

where $c_0^2 \equiv \gamma P_0/\rho_0 = \gamma RT_0$ as before. From Fig. 2.3 we see that this distribution is fairly realistic, except at high altitudes. On the basis of (2.38) we can define the *scale height of the atmosphere H* as

$$H \equiv \frac{c_0^2}{\gamma g_0} = \frac{P_0}{\rho_0 g_0} = \frac{RT_0}{g_0} \tag{2.39}$$

a value which depends on temperature. For $T_0 = 288.15$ K (the U.S. standard-atmosphere value), $H = 8.435$ km.

The properties of the standard atmosphere are given in Table D.4, Appendix D.

EXAMPLE 2.5 DEPENDENCE OF INTERNAL ENERGY ON TEMPERATURE

For certain fluids the internal energy depends only on temperature. To investigate such a possibility we begin by assuming $e = e(P,T)$ or

$$de = \left(\frac{\partial e}{\partial P}\right)_T dP + \left(\frac{\partial e}{\partial T}\right)_p dT$$

and determine the conditions under which $(\partial e/\partial P)_T$ vanishes, so that $e = e(T)$ alone. From the Gibbs equation (2.17*a*) we form this derivative

$$\left(\frac{\partial e}{\partial P}\right)_T = T\left(\frac{\partial s}{\partial P}\right)_T - P\left(\frac{\partial v}{\partial P}\right)_T \tag{2.40}$$

But

$$\left(\frac{\partial s}{\partial P}\right)_T = -\left(\frac{\partial v}{\partial T}\right)_p$$

from a Maxwell relation.

Thus

$$\left(\frac{\partial e}{\partial P}\right)_T = -T\left(\frac{\partial v}{\partial T}\right)_p - P\left(\frac{\partial v}{\partial P}\right)_T \tag{2.41}$$

which expresses the dependence of e on P via the thermal equation of state. The right-hand side vanishes for two cases:

1 *Incompressible fluid:* We *define* an incompressible fluid by the relation $v = \text{const}$ for a fluid particle; that is, $(\partial v/\partial T)_p = (\partial v/\partial P)_T = 0$. Then (2.41) gives immediately $(\partial e/\partial P)_T = 0$, and $e = e(T)$ only. From the Gibbs equation (2.17*a*) it also follows that $s = s(T)$ only. It may be noted that e cannot depend on P by physical reasoning, because a "compression" of the fluid cannot result in an increase in its internal energy as no work is done. It should be remarked that incompressibility is an *idealization*; all real fluids are compressible to some extent.

2 *Fluid for which v depends only on P/T:* With $v = f(P/T)$, differentiation gives

$$\left(\frac{\partial v}{\partial T}\right)_p = -\frac{P}{T^2}f'\left(\frac{P}{T}\right)$$

$$\left(\frac{\partial v}{\partial P}\right)_T = \frac{1}{T}f'\left(\frac{P}{T}\right)$$

where f' indicates differentiation with respect to the argument P/T. Substitution in (2.41) then gives $(\partial e/\partial P)_T = 0$ and $e = e(T)$ only. The important practical case is the *ideal gas*, for which $v = RT/P$.

2.4 Entropy and vorticity

The connection between the vorticity $\mathbf{\Omega} \equiv \nabla \times \mathbf{u}$ and the gradient of the specific entropy is established by reformulation of the momentum equation. If the transport coefficients are constant, the momentum equation is (1.66),

$$\rho\frac{D\mathbf{u}}{Dt} + \nabla P = \rho\mathbf{G} + \mu\,\nabla^2\mathbf{u} + (\mu_v + \tfrac{1}{3}\mu)\nabla(\nabla\cdot\mathbf{u})$$

With the standard vector identities (from those listed in Appendix B),

$$(\mathbf{u}\cdot\nabla)\mathbf{u} = (\nabla \times \mathbf{u}) \times \mathbf{u} + \nabla\frac{u^2}{2}$$

$$\nabla^2\mathbf{u} = \nabla(\nabla\cdot\mathbf{u}) - \nabla \times (\nabla \times \mathbf{u})$$

the above form of the momentum equation can be rewritten

$$\rho\left(\frac{\partial \mathbf{u}}{\partial t} + \boldsymbol{\Omega} \times \mathbf{u} + \nabla \frac{u^2}{2}\right)$$
$$= \rho \mathbf{G} - \nabla P - \mu(\nabla \times \boldsymbol{\Omega}) + (\mu_v + \tfrac{4}{3}\mu)\nabla(\nabla \cdot \mathbf{u}) \qquad (2.42)$$

Note that an irrotational incompressible flow experiences no net viscous forces. The Gibbs equation (2.17*b*) gives $\nabla P = \rho\, \nabla h - \rho T\, \nabla s$, and it is assumed that the force has a potential, $\mathbf{G} = -\nabla \Psi$. Then (2.42) becomes

$$\frac{\partial \mathbf{u}}{\partial t} + \boldsymbol{\Omega} \times \mathbf{u} = T\, \nabla s - \nabla\left(h + \frac{u^2}{2} + \Psi\right)$$
$$- \mu v(\nabla \times \boldsymbol{\Omega}) + (\mu_v + \tfrac{4}{3}\mu) v \nabla(\nabla \cdot \mathbf{u}) \qquad (2.43)$$

This equation in its inviscid form is due to Crocco (1936) and Vazsonyi (1945). For steady inviscid flow this is

$$\boldsymbol{\Omega} \times \mathbf{u} = T\, \nabla s - \nabla H \qquad (2.44)$$

where $H = h + u^2/2 + \Psi$ is the *Bernoulli constant.* In many steady flows of practical interest, H and s are constant over the flow field; in this case the momentum equation assumes the peculiar form

$$\boldsymbol{\Omega} \times \mathbf{u} = 0 \qquad (2.45)$$

This can be satisfied by any of the conditions

(i) $\mathbf{u} \equiv 0$
(ii) $\boldsymbol{\Omega}$ everywhere parallel to $\mathbf{u}$
(iii) $\boldsymbol{\Omega} \equiv 0$

The solution (i) is trivial. Flows for which (ii) holds are called *Beltrami flows*: the simplest example is perhaps swirling motion of flow through a circular pipe. The *irrotational-flow* solution (iii) is of greatest practical interest and occurs often in gasdynamics. One very simple illustration of such a flow was given in Example 1.4 (page 25).

An alternative and useful form of the Crocco-Vazsonyi equation is found by taking the curl of (2.43). With the help of several vector identities and considerable algebra the result reduces to

$$\frac{D\boldsymbol{\Omega}}{Dt} = (\boldsymbol{\Omega} \cdot \nabla)\mathbf{u} - \boldsymbol{\Omega}(\nabla \cdot \mathbf{u}) + \nabla T \times \nabla s$$
$$+ (\mu_v + \tfrac{4}{3}\mu)[\nabla v \times \nabla(\nabla \cdot \mathbf{u})]$$
$$+ \mu[v \nabla^2 \boldsymbol{\Omega} - \nabla v \times (\nabla \times \boldsymbol{\Omega})] \qquad (2.46)$$

The term $\nabla T \times \nabla s$ will play an important role in the following discussion. By suitable expansion, this term can be written as the cross product of the gradients of any two thermodynamic variables; e.g., with $T = T(P,v)$ and $s = s(P,v)$ we find

$$\nabla T \times \nabla s = \left[\left(\frac{\partial T}{\partial P}\right)_v\left(\frac{\partial s}{\partial v}\right)_P - \left(\frac{\partial T}{\partial v}\right)_P\left(\frac{\partial s}{\partial P}\right)_v\right](\nabla P \times \nabla v)$$

Thus this term will vanish if *any* thermodynamic variable, e.g., the density or the temperature, is constant over the flow field. In gasdynamics the only case of practical interest is that in which the entropy is constant: in this (barotropic) case, P is a unique function of v.

For inviscid ($\mu = \mu_v = 0$) and homentropic ($\nabla s = 0$) flow (2.46) then is

$$\frac{D\mathbf{\Omega}}{Dt} = (\mathbf{\Omega} \cdot \nabla)\mathbf{u} - \mathbf{\Omega}(\nabla \cdot \mathbf{u}) \tag{2.47}$$

This equation has the important consequence that *fluid initially without vorticity will remain forever without vorticity* in the absence of entropy gradients or viscous forces. This follows from the fact that the right-hand side of (2.47) is initially zero; then $D\mathbf{\Omega}/Dt = 0$, vorticity cannot develop, and the vorticity of the fluid particle remains zero.

Some special forms of the general equation (2.46) are of interest. For the flow of a uniform *incompressible* fluid $\nabla v = \nabla \cdot \mathbf{u} = 0$, and with $s = s(T)$ only [as found in Example 2.5 (page 69)], $\nabla T \times \nabla s = 0$. This result also follows from the arguments given following Eq. (2.46). Then

$$\frac{D\mathbf{\Omega}}{Dt} = (\mathbf{\Omega} \cdot \nabla)\mathbf{u} + \nu\, \nabla^2\mathbf{\Omega} \tag{2.48}$$

where $\nu = \mu/\rho$ is the kinematic viscosity. For plane flow, $\mathbf{\Omega}$ is perpendicular to the plane of the velocity, and this reduces to

$$\frac{D\Omega}{Dt} = \nu\, \nabla^2\Omega \tag{2.49}$$

It is remarkable that this has exactly the form of the energy equation in incompressible flow provided that the dissipation is negligible. In that case the energy equation (2.7) becomes, with $e = e(T)$ and constant thermal conductivity,

$$\frac{DT}{Dt} = \alpha\, \nabla^2 T \tag{2.50}$$

where $\alpha = \kappa/\rho c_v$ is the thermal diffusivity. The correspondence between (2.49) and (2.50) is the basis for an analogy between the convection of heat and the convection of vorticity (see Example 2.9, page 75). We can go even further: with $s = s(T)$, the specific entropy will be found to satisfy a similar diffusion equation.

For an *inviscid* fluid with compressibility, Eq. (2.46) can with the help of continuity be written

$$\frac{D}{Dt} v\boldsymbol{\Omega} = (v\boldsymbol{\Omega} \cdot \boldsymbol{\nabla})\mathbf{u} + v(\boldsymbol{\nabla}T \times \boldsymbol{\nabla}s) \tag{2.51}$$

which is the vorticity equation of Vazsonyi. This has interesting consequences, but we forego further discussion (see, for example, *Serrin* [1959]).

The following examples should help to make clear the conditions under which the considerable simplification afforded by setting $\boldsymbol{\Omega} = 0$ is possible and when it is not.

EXAMPLE 2.6 SUBSONIC FLOW OVER A BODY

The streaming flow over a body is sketched in Fig. 2.4. Conditions far upstream (to the left of the figure) are supposed to be uniform: $\mathbf{u} =$ const, $P =$ const, etc. Such flows are said to have a *uniform free stream* and occur for example in aeronautical problems, where the free stream corresponds to the nominally uniform atmosphere as seen by an observer, e.g., an airplane pilot, traveling with the moving body.

For sufficiently large free-stream velocities, it is found that viscous forces are effectively restricted to a relatively thin region, called the *boundary layer*, near the surface of the body (in consequence of the *no-slip condition* there is no relative motion between the fluid and the body surface, as indicated in Fig. 1.8, and relatively high shear rates occur within the boundary layer). Outside of the boundary layer viscous forces are negligible, and the flow is inviscid; this external flow is then *irrotational* ($\Omega = 0$) via Eq. (2.47), because the initial free-stream vorticity is zero and there is no entropy gradient.

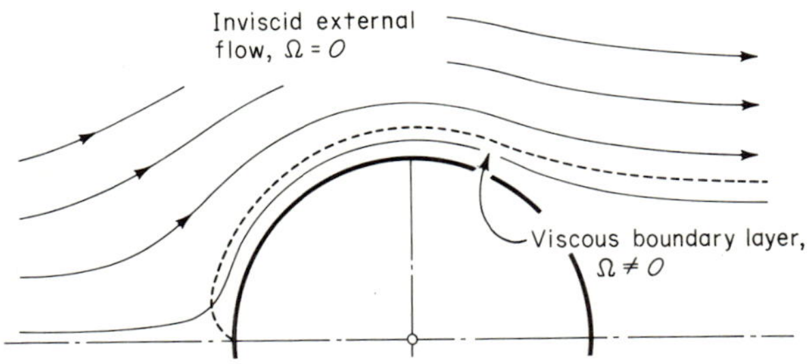

Figure 2.4

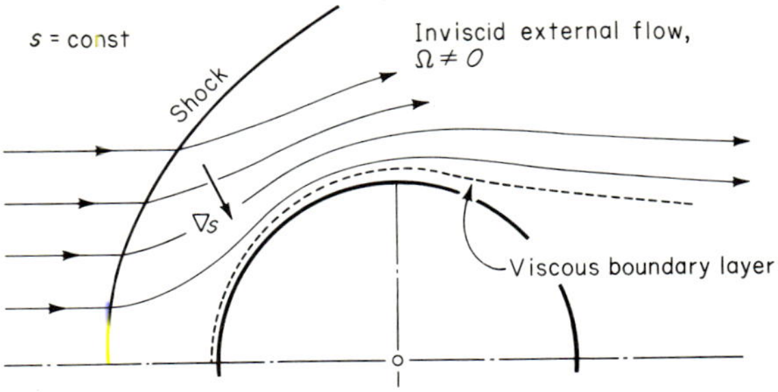

Figure 2.5

EXAMPLE 2.7 SUPERSONIC FLOW OVER A BODY

If the magnitude of the free-stream velocity exceeds the speed of sound in the fluid, the flow is said to be supersonic and the situation is significantly different from the preceding example. In general a *shock*, a surface across which all flow properties are discontinuous, will form ahead of the body. Shocks will be discussed in detail in Chap. 7; the feature of interest at present is that the specific *entropy* is discontinuous across the shock. This entropy jump will vary with the local strength of the shock. In the situation shown in Fig. 2.5, the entropy jump is greatest near the centerline of the flow (where the shock is strongest) and diminishes away from the centerline. Since the free-stream entropy is uniform, an entropy gradient in the direction indicated necessarily occurs behind the shock.

It is convenient to discuss the vorticity by the use of the Crocco-Vazsonyi theorem (2.44). We will later show that the *Bernoulli constant H* has no jump across the shock; in consequence, H has a constant value within the region of inviscid external flow, and $\nabla H = 0$. The entropy gradient ∇s is necessarily perpendicular to streamlines, since $s =$ const along each streamline. Thus $\boldsymbol{\Omega}$ is nonzero and is perpendicular to ∇s. The external flow is inviscid and *rotational*.

The result that the flow downstream of a stationary shock of nonuniform strength is rotational is sometimes referred to as *Hadamard's theorem*. It is sometimes roughly stated "flow downstream of a curved shockfront is rotational."

EXAMPLE 2.8 ACOUSTIC WAVES

Consider fluid at rest adjacent to a heated surface. If the surface were first heated or "switched on" only a short time before the instant of consideration, the heated fluid would be confined to a relatively thin layer near the surface (Fig. 2.6). Within this layer the temperature gradient, and therefore the entropy gradient, may be large. Outside of this layer the entropy is very nearly constant.

Suppose that the fluid is now set into motion by acoustic (sound) waves

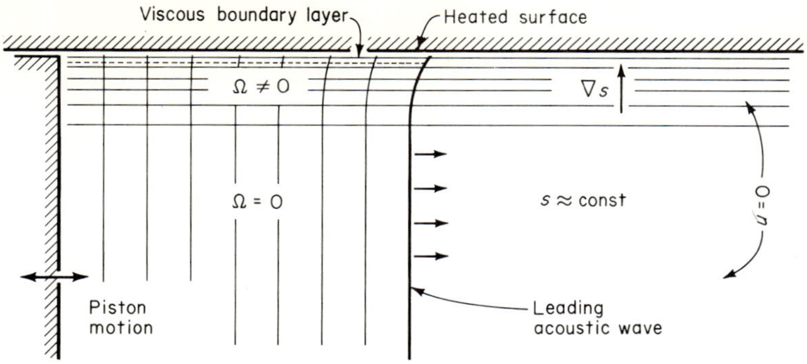

Figure 2.6

incident from the left, as by a piston which initiates a high-frequency oscillation at a certain time. The only feature of the acoustic waves of present interest is that they establish pressure gradients and fluid motion in the horizontal direction.

As we shall see in Chap. 4, the acoustic motion is inviscid to a good approximation, except very near the solid surface. Consider Eq. (2.51): there is a vorticity-producing term $\nabla T \times \nabla s$. It may be helpful to rewrite this term; with $T = T(P,s)$

$$\nabla T = \left(\frac{\partial T}{\partial P}\right)_s \nabla P + \left(\frac{\partial T}{\partial s}\right)_p \nabla s$$

and

$$\nabla T \times \nabla s = \left(\frac{\partial T}{\partial P}\right)_s \nabla P \times \nabla s \tag{2.52}$$

because the cross product of ∇s with itself is zero. Now the pressure gradient ∇P associated with the acoustic waves is (very nearly) perpendicular to ∇s, and the term (2.52) is nonzero. A fluid particle within the heated layer acquires vorticity by (2.51) after the passage of the leading acoustic wave (the initial vorticity of the fluid at rest is of course zero). Outside of the heated layer the motion is irrotational.

XAMPLE 2.9 ANALOGY BETWEEN HEAT AND VORTICITY CONVECTION

It has been pointed out that there is a strict analogy between vorticity convection and heat convection for plane incompressible flow, as represented by Eqs. (2.49) and (2.50), viz.,

$$\frac{D\Omega}{Dt} = \nu \nabla^2 \Omega$$

$$\frac{DT}{Dt} = \alpha \nabla^2 T$$

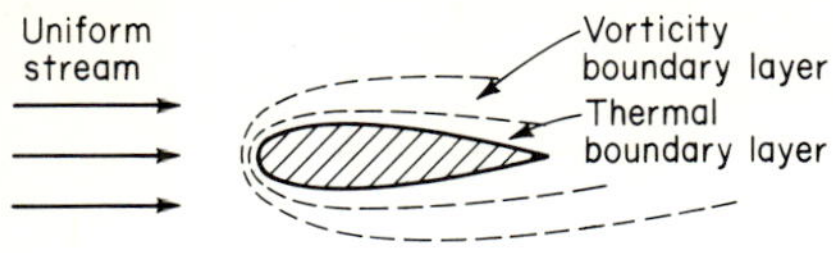

Figure 2.7

On the assumption that the spreading out of temperature is more firmly grounded in intuition than the spreading out of vorticity, this analogy provides a useful basis for visualizing the latter process.

Specifically, for the streaming flow shown in Fig. 2.7 consider that the two-dimensional, i.e., infinitely long in the direction perpendicular to the plane of the paper, body is heated to some uniform temperature. The surrounding region of heated fluid is swept downstream by the moving fluid to form the thermal boundary layer shown. The vorticity which is generated at the surface by virtue of the no-slip condition spreads out similarly to form the vorticity boundary layer (or viscous boundary layer, or momentum boundary layer). The analogy is not quite exact, for while the airfoil surface may have constant temperature, the vorticity is not in general constant on this surface, though it is generally greatest there. The relative thickness of the vorticity and thermal layers is governed by the relative values of the kinematic viscosity ν (momentum diffusivity) and thermal diffusivity α in Eqs. (2.49) and (2.50), i.e., by the value

$$\frac{\nu}{\alpha} = \frac{\mu/\rho}{\kappa/\rho c_v} = \frac{\mu c_v}{\kappa} = \frac{\mu c_p}{\kappa} = \mathrm{Pr}$$

where $c_v = c_p$ for incompressible fluid (see, for example, *Schlichting* [1968, chap. 14]).

As the speed u_∞ of the free stream is increased, the vorticity and thermal layers are swept increasingly downstream and lie closer to the surface of the body. In Chap. 3 it will be indicated that this effect depends on the free-stream speed in the *dimensionless* form

$$\mathrm{Re} = \frac{Lu_\infty}{\nu} \tag{2.53}$$

where L is a characteristic dimension of the body, e.g., the "chord" or left-to-right dimension in the figure, and Re is an abbreviation for *Reynolds number*.

2.5 Ideal gas

Many of the compressible fluids of practical interest can be approximately treated as ideal gases. We *define* an ideal gas to be a substance with the equation of state

$$Pv = RT \tag{2.54}$$

where the specific gas constant R is

$$R = \frac{\tilde{R}}{\tilde{M}} \tag{2.55}$$

in which $\tilde{M}$ is the molecular weight of the gas and $\tilde{R}$ is the universal gas constant with a numerical value

$$\tilde{R} = \begin{cases} 8{,}314 \dfrac{\text{J}}{(\text{kg mol})(\text{K})} \\[2ex] 8{,}314 \dfrac{\text{m}^2/\text{s}^2}{\text{K}} \dfrac{\text{kg}}{\text{kg mol}} \\[2ex] 49{,}720 \dfrac{\text{ft}^2/\text{s}^2}{{}^\circ\text{R}} \dfrac{\text{lb}_m}{\text{lb mol}} \\[2ex] 1.987 \dfrac{\text{cal}}{(\text{g mol})(\text{K})} \\[2ex] 1.987 \dfrac{\text{Btu}}{(\text{lb mol})({}^\circ\text{R})} \\[2ex] 1{,}545 \dfrac{\text{ft} \cdot \text{lb}_f}{(\text{lb mol})({}^\circ\text{R})} \end{cases}$$

(We will also have some occasion to make use of the related Boltzmann constant $k = \tilde{R}/\tilde{N}$, where $\tilde{N}$ is Avogadro's number.) Note that there is a definite value of the gas constant $R = \tilde{R}/\tilde{M}$ for any chemical substance, quite independent of whether the substance is an ideal gas or not; e.g., this value plays a role in the theory of perfect crystals.

Gases of relatively low density (sometimes called dilute gases) satisfy (2.54) approximately. The degree of approximation can be determined from Fig. 2.8, in which Pv/RT (which is of course unity for an ideal gas) is plotted against *reduced pressure* (the actual pressure normalized with respect to the critical-point pressure) for various values of *reduced temperature* (temperature normalized with respect to the critical-point temperature). Some representative values for nitrogen ($P_{\text{crit}} = 33.5$ atm, $T_{\text{crit}} = 126$ K) are

P, atm	T, K	Pv/RT
1	300	0.9998
100	300	1.005
10	150	0.942

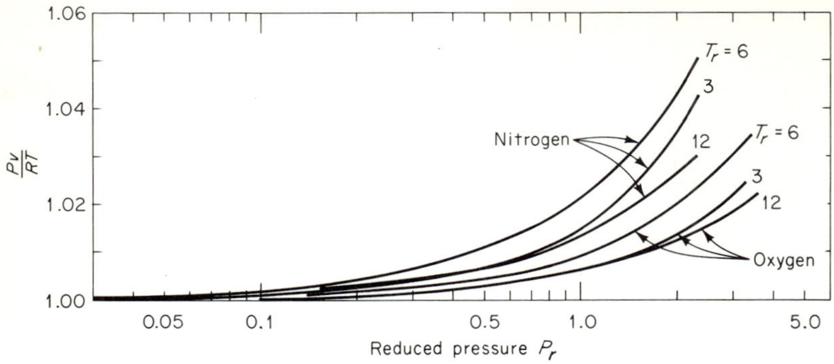

Figure 2.8
Correlation of PvT data for nitrogen and oxygen.

Nature has generously provided us with gases which, over a considerable range of pressures, satisfy to good approximation the very simple equation of state (2.54). There are no other substances of interest described with comparable algebraic simplicity.

Specific Heats and Ratio of Specific Heats

In the absence of chemical reaction, e.g., dissociation of a diatomic gas, the molecular weight $\tilde{M}$ and therefore the specific gas constant R are actual constants. For this case we found in Example 2.5 (page 69) that the specific internal energy e depends only on temperature. Since $h = e - Pv = e + RT$, it follows that the specific enthalpy h also depends only on temperature. Thus, it is a consequence of (2.54) that

$$e = e(T) \qquad h = h(T) \tag{2.56}$$

The corresponding differential forms are

$$de = \frac{de}{dT}dT = c_v(T)\,dT \qquad dh = \frac{dh}{dT}dT = c_p(T)\,dT \tag{2.57}$$

There is a simple relation between c_p and c_v:

$$c_p - c_v = \frac{dh}{dT} - \frac{de}{dT} = \frac{d}{dT}(h - e) = \frac{d}{dT}RT = R$$

$$c_p - c_v = R \tag{2.58}$$

Introducing the ratio of specific heats $\gamma \equiv c_p/c_v$, we obtain from (2.58)

$$c_p = \frac{\gamma}{\gamma - 1} R \qquad c_v = \frac{1}{\gamma - 1} R \tag{2.59}$$

where in general $\gamma = \gamma(T)$.

An ideal gas for which γ is constant, and therefore c_p and c_v are constant, will be called a *perfect gas* (in the nomenclature of aeronautics, such a gas is called *thermally and calorically perfect*). Thus

$$Pv = RT \qquad \leftrightarrow \text{ideal gas}$$

$$Pv = RT, \gamma = \text{const} \leftrightarrow \text{perfect gas}$$

In molecular terms, the validity of (2.54) is dependent on a molecular number density n (number of molecules per unit volume) sufficiently small for the average intermolecular spacing to be large compared to the distance over which intermolecular force is effective. An ideal gas thus has negligible molecular interaction (except during *collisions*, which occupy a small fraction of the lifetime of any given molecule), and there is a negligible contribution to the internal energy of the gas due to the potential energy associated with intermolecular forces. That is, all the energy of the gas resides in the molecules themselves. Under these conditions, the equipartition theorem of classical statistical mechanics predicts an average molecular energy of $\frac{1}{2}kT$ for each degree of freedom with quadratic energy dependence. The number of degrees of freedom f is just the number of coordinates required to fix the energy state of the molecule; e.g., for a monatomic molecule $f = 3$, corresponding to the molecular position coordinates x_1, x_2, x_3, with associated energies $m\dot{x}_1^2/2$, $m\dot{x}_2^2/2$, and $m\dot{x}_3^2/2$. In a general case, with f degrees of freedom, the average molecular energy $\bar{\varepsilon}$ is

$$\bar{\varepsilon} = \frac{f}{2} kT \tag{2.60}$$

and the internal energy per mole is $\tilde{N}\bar{\varepsilon} = f\tilde{R}T/2$, giving a specific internal energy

$$e = \frac{f}{2} RT$$

and specific enthalpy

$$h = \frac{f + 2}{2} RT$$

Then the specific heats are $c_v = fR/2$ and $c_p = (f + 2)R/2$, giving for the ratio of specific heats

$$\gamma = \frac{c_p}{c_v} = \frac{f + 2}{f} \tag{2.61}$$

An ideal gas which satisfies equipartition of energy is thus perfect because $\gamma =$ const. For a monatomic gas (He, Ne, Ar, etc.), the simplest possible molecular structure, $f = 3$, and (2.61) gives $\gamma = 5/3$. This prediction is very well confirmed by experiment. At the other extreme of molecular complexity, very complicated molecules have a large number of degrees of freedom, and γ may approach unity, which represents the minimum possible value, since $c_p \geq c_v$ by virtue of a general thermodynamic argument [see Eq. (B.6) in Appendix B]. Then γ necessarily has a range of values,

$$5/3 \geq \gamma \geq 1 \tag{2.62}$$

For *diatomic* gases the number of degrees of freedom is reckoned as follows:

Translation of center of mass	3
Rotation about two principal axes	2
Vibration about center of mass (quadratic KE and PE)	2
	$f = 7$

This gives a ratio of specific heats $\gamma = 9/7$. Experimentally most diatomic gases, nitrogen and oxygen in particular, have $\gamma = 7/5$ at room temperature, gradually increasing to $\gamma = 9/7$ at a few thousand degrees Kelvin (see Fig. 2.9). Equipartition thus gives an incorrect result, except at quite high temperatures. The reason is that classical statistical mechanics does not properly account for the quantization of molecular vibration (neither does

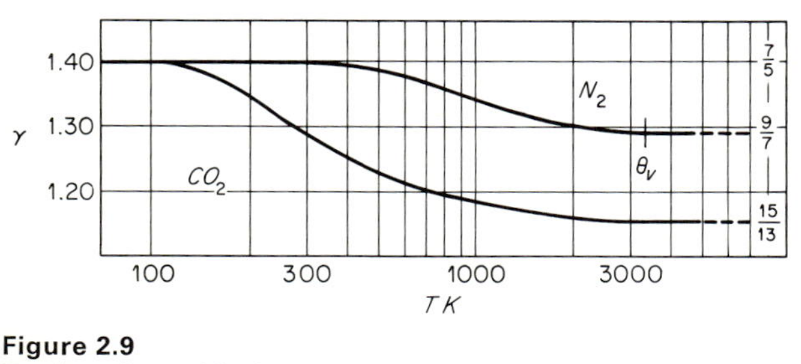

Figure 2.9
Ratio of specific heats vs. temperature.

Table 2.2 Physical Constants for Diatomic Molecules

	M	θ_r, K	θ_v, K	θ_d, K
H_2	2.016	87.5	6,325	52,000
N_2	28.02	2.89	3,393	113,400
O_2	32.00	2.08	2,273	59,000
CO	28.01	2.78	3,122	129,000
HCl	36.47	15.2	4,300	51,950

it account for the quantization of rotation, but this introduces an error only at very low temperatures). We define a characteristic temperature for vibration

$$\theta_v = \frac{h\nu}{k} \tag{2.63}$$

where ν is the natural frequency of vibration of the diatomic oscillator (which is easily measured because ν is also the light frequency of the principal lines in the vibrational spectrum of the molecules, a consequence of the quantization of vibrational energy) and h is Planck's constant. In terms of this characteristic temperature, the specific heat is found to be

$$c_v = \left[\frac{5}{2} + \frac{(\theta_v/T)^2 e^{\theta_v/T}}{(e^{\theta_v/T} - 1)^2}\right] R \tag{2.64}$$

provided that the gas temperature T is large compared to the characteristic temperature θ_r for rotation ($\theta_r \equiv h^2/8\pi^2 Ik$, where I is the molecular moment of inertia). With $c_p = c_v + R$, the ratio of specific heats γ can be found accordingly. Equation (2.64) yields the values

$$T \gg \theta_v \qquad c_v = \tfrac{7}{2}R \qquad \gamma = \tfrac{9}{7}$$
$$T \ll \theta_v \qquad c_v = \tfrac{5}{2}R \qquad \gamma = \tfrac{7}{5}$$

which are in agreement with the curve in Fig. 2.9. Some characteristic temperatures are listed in Table 2.2.

Dissociation and Chemical Reaction

A chemically reactive mixture of ideal gases is still described by (2.54), but R is no longer constant (because $R \equiv \tilde{R}/\tilde{M}$ and the average molecular weight $\tilde{M}$ is now variable), and the conclusions $e = e(T)$ and $h = h(T)$ no longer follow. There is, however, no difficulty in principle

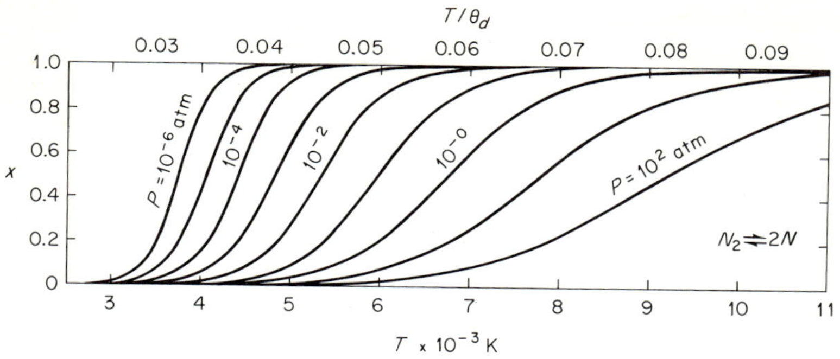

Figure 2.10
Dissociation equilibrium of nitrogen. Mole fraction of atomic nitrogen vs. temperature at various pressures.

in predicting the equilibrium concentrations of reactants, and the corresponding properties such as $h(P,T)$ (see, for example, *Fay* [1965, chap. 14]). Of particular interest is the dissociation of a diatomic gas with increasing temperature. The mole fraction χ of atomic nitrogen for the dissociation reaction $N_2 \rightleftharpoons 2N$ is shown in Fig. 2.10. The dissociation energy ε_d can be expressed as a characteristic temperature $\theta_d \equiv \varepsilon_d/k$. For diatomic gases other than nitrogen, the fraction dissociated χ can be estimated by reading χ as a function of the dimensionless quantity T/θ_d; that is, to a very rough approximation, $\chi = \chi(P, T/\theta_d)$. Ionization, which for an (initially) diatomic gas is important at even higher temperatures, shows a temperature-pressure dependence like that of dissociation.

We will not carry this highly abbreviated discussion of chemical reaction further. The important point is that in the absence of diffusion the reactive mixture behaves like a simple fluid, with the state fixed by two thermodynamic quantities such as the pressure and the temperature.

Nonreactive Mixtures of Ideal Gases

Frequently, gases of interest may not be chemically pure but made up of a mixture of chemical species. The most important example is of course air. The thermodynamic properties of such a mixture can be found from a knowledge of the properties of the constituent species.

A particular chemical species will be represented by subscript i. The mass m_i of species i is

$$m_i = \tilde{M}_i \mathscr{N}_{(i)} \tag{2.65}$$

where $\mathcal{N}_i$ is the number of moles, a quantity which is *defined* by this equation. For example, a quantity of (diatomic) oxygen $m_1 = 20$ kg, with $\tilde{M}_1 = 32.00$ kg/kg mol contains $\mathcal{N}_1 = 20/32 = 0.625$ kg mol. Then making use of the summation convention, the total mass m is

$$m = \tilde{M}_i \mathcal{N}_i \tag{2.66}$$

The total number of moles $\mathcal{N}$ is

$$\mathcal{N} = \sum_i \mathcal{N}_i \tag{2.67}$$

It is convenient to define the mole fraction χ_i as

$$\chi_i = \frac{\mathcal{N}_i}{\mathcal{N}} \tag{2.68}$$

and (2.67) gives

$$\sum_i \chi_i = 1 \tag{2.69}$$

The molecular weight $\tilde{M}$ of the mixture is, by definition,

$$\tilde{M} = \frac{m}{\mathcal{N}} = \frac{\tilde{M}_i \mathcal{N}_i}{\mathcal{N}} = \tilde{M}_i \chi_i \tag{2.70}$$

Then the specific gas constant for the mixture is

$$R = \frac{\tilde{R}}{\tilde{M}} = \frac{\tilde{R}}{\tilde{M}_i \chi_i} \tag{2.71}$$

The preceding equations (2.65) to (2.71) are simply definitions and apply to *any* mixture, whether it is made up of ideal gases or not.

A conceptual model for a mixture of ideal gases is a container filled with molecules of various species, in which molecules do not interact. In the volume V shown in Fig. 2.11 the white molecules (species 1) are "unaware" of the existence of the black molecules (species 2). We write

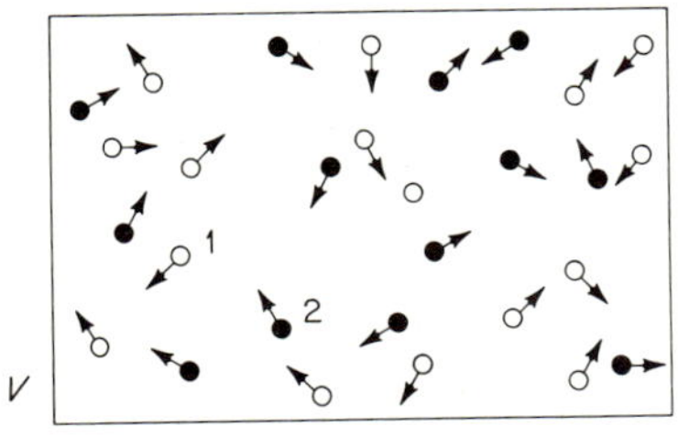

Figure 2.11

the ideal-gas law for each species separately; multiplying (2.54) by the mass m_1 gives, for example,

$$P_1V = m_1R_1T_1$$

where P_1 is the *partial pressure* of species 1, that is, the contribution to the pressure at the wall (say) due to the impact of species 1. The model of an ideal gas as an aggregation of noninteracting molecules is not quite correct because the molecules do interact via infrequent collisions and the overall effect of these collisions is to ensure that the *temperature* of all species is equal provided that the mixture is in equilibrium. Thus in the above example, $T_1 = T_2 = T$, and we can write

$$P_1V = m_1R_1T$$
$$P_2V = m_2R_2T$$

In general, for the ith species in a mixture, with $m_iR_{(i)} = m_i\tilde{R}/\tilde{M}_{(i)} = \mathcal{N}_i\tilde{R}$,

$$P_iV = \mathcal{N}_i\tilde{R}T \tag{2.72}$$

The partial pressure due to species i is proportional to the number of molecules of that species. Summing this over all species i yields

$$PV = \mathcal{N}\tilde{R}T \tag{2.73}$$

where the total pressure $P = \sum P_i$. Dividing this into (2.72) gives simply

$$P_i = \chi_iP \tag{2.74}$$

Dividing the gas law for the mixture (2.73) by the total mass $m = \mathcal{N}\tilde{M}$ gives

$$Pv = RT$$

which is identical to (2.54), and the specific gas constant R is the appropriate average given by (2.71).

Because the internal energy resides in the molecules themselves, a definite energy e_i is assignable to each species. Thus the total energy in a mixture is

$$me = m_ie_i$$

Then the mixture specific internal energy e is

$$e = \frac{m_i}{m}e_i \tag{2.75}$$

where m_i/m is the *mass fraction*, related to the mole fraction χ_i by

$$\frac{m_i}{m} = \frac{\tilde{M}_i\chi_{(i)}}{\tilde{M}} \tag{2.76}$$

Differentiating (2.75) with respect to temperature yields, with constant mass fraction m_i/m (in the absence of chemical reaction or condensation),

$$c_v = \frac{m_i}{m} c_{vi} \tag{2.77}$$

Similarly,

$$c_p = \frac{m_k}{m} c_{pk} \tag{2.78}$$

Then the ratio of specific heats γ for the mixture is the ratio of sums,

$$\gamma = \frac{(m_k/m)c_{pk}}{(m_i/m)c_{vi}} \tag{2.79}$$

Alternatively, this can be put in the form

$$\gamma = \frac{\chi_k \dfrac{\gamma_k}{\gamma_k - 1}}{\chi_i \dfrac{1}{\gamma_i - 1}} \tag{2.80}$$

EXAMPLE 2.10 THE PROPERTIES OF AIR

The mole fractions of dry air are, approximately, nitrogen 78.0 percent, oxygen 21.0 percent, and argon 1.0 percent. Find the properties of the mixture at 273 K.

It is noted that the above figures are often quoted as *volume fractions*, which amount to the same thing, for if a mixture is separated into pure constituents at the pressure and temperature of the original mixture, each constituent will by (2.73) occupy a volume proportional to its mole fraction. The properties of air of interest, from which others may be derived, are the molecular weight, gas constant, and ratio of specific heats. The constituent properties are shown in Table 2.3.

Table 2.3

	$\tilde{M}$	γ	χ
N_2	28.02	1.400	0.780
O_2	32.00	1.400	0.210
Ar	39.95	1.667	0.010

The molecular weight is from (2.70)

$$\tilde{M} = 0.78(28.02) + 0.21(32) + 0.01(39.95) = 28.98$$

and the gas constant is from (2.71)

$$R = \frac{8314}{28.98} = 286.9\ \text{J/(kg)(K)}$$

The ratio of specific heats is from (2.80)

$$\gamma = \frac{0.78(3.5) + 0.21(3.5) + 0.01(2.5)}{0.78(2.5) + 0.21(2.5) + 0.01(1.5)} = 1.402$$

and the specific heats are then given by (2.59).

From the gas law (2.54) the specific volume at 1 atm and 273 K is

$$v = \frac{(286.9)(273)}{1.013 \times 10^5} = 0.7732\ \text{m}^3/\text{kg}$$

and the density is

$$\rho = \frac{1}{v} = 1.293\ \text{kg/m}^3$$

The properties of moist air will differ slightly from the above. In particular the density

$$\rho = \frac{P}{RT} = \frac{\tilde{M}P}{\tilde{R}T}$$

will be slightly *less*, because the addition of water vapor ($\tilde{M} = 18$) lowers the mixture molecular weight.

A complete list of the constituents of air is given in the *U.S. Standard Atmosphere*. The constituents given there lead to a molecular weight $\tilde{M} = 28.964$.

The Entropy of an Ideal Gas

The Gibbs equation (2.17*b*) is

$$T\,ds = dh - v\,dP$$

For an ideal gas, $dh = c_p\,dT$ and $v = RT/P$ so that this becomes

$$ds = c_p \frac{dT}{T} - R\frac{dP}{P}$$

Let P_0 and T_0 define any reference state, with entropy $s_0 = s(P_0, T_0)$. Then integration gives

$$s - s_0 = \int_{T_0}^{T} c_p(T) \frac{dT}{T} - R \ln \frac{P}{P_0} \tag{2.81}$$

Similar expressions can be found for $s(P,v)$ and $s(T,v)$. If the gas is considered *perfect*, c_p is constant and (2.81) yields

$$\frac{P}{P_0} = e^{(s_0 - s)/R} \left(\frac{T}{T_0}\right)^{\gamma/(\gamma - 1)} \tag{2.82}$$

Substituting $P = R\rho T$ on the left-hand side and then on the right-hand side gives, respectively,

$$\frac{\rho}{\rho_0} = e^{(s_0 - s)/R} \left(\frac{T}{T_0}\right)^{1/(\gamma - 1)} \tag{2.83}$$

$$\frac{P}{P_0} = e^{(s - s_0)/c_v} \left(\frac{\rho}{\rho_0}\right)^{\gamma} \tag{2.84}$$

For *isentropic* changes of state the last equation can be simply written

$$Pv^{\gamma} = \text{const} \tag{2.85}$$

Alternative statements are, from (2.82) to (2.84),

$$\frac{P}{P_0} = \left(\frac{T}{T_0}\right)^{\gamma/(\gamma - 1)} \qquad \frac{\rho}{\rho_0} = \left(\frac{T}{T_0}\right)^{1/(\gamma - 1)} \qquad \frac{P}{P_0} = \left(\frac{\rho}{\rho_0}\right)^{\gamma} \tag{2.86}$$

Molecular Motion in an Ideal Gas

About the year 1620 the Flemish alchemist van Helmont boldly postulated a new class of matter:

> This spirit, hitherto unknown,
> I call by a new name, gas.

(The word "gas" was not soon accepted but was reintroduced by Lavoisier some 150 years later.) According to Helmont himself, he took the name from the Greek χάος (chaos). This etymology is remarkably appropriate to the modern view of the molecular motion within a gas.

A statistical description of this chaotic motion is the *Maxwell distribution* of molecular velocities. For a dilute gas, i.e., a gas of low density, at rest, let the velocity of a molecule be $\mathbf{v}$, with components v_1, v_2, and v_3, and the spatial number density of molecules be n. Then the Maxwellian *distribution function* f^0 is

$$f^0 = n\left(\frac{m}{2\pi kT}\right)^{3/2} e^{-(m/2kT)(v_1^2+v_2^2+v_3^2)} \tag{2.87}$$

where m is the mass of one molecule.[1] The number of molecules per unit volume with velocities within the range $v_1 \pm \delta v_1/2$, $v_2 \pm \delta v_2/2$, $v_3 \pm \delta v_3/2$ is $f^0(\mathbf{v})\ \delta v_1\ \delta v_2\ \delta v_3$.

The distribution function can be interpreted for a uniform gas as a density in *velocity space*. Suppose that the velocity of each of the n molecules within a unit volume of physical space is measured at some instant of time and that each measurement is recorded as a representative point in the three-dimensional space of the velocity components. Such a plot is shown in Fig. 2.12 and is a representation of the molecular velocity

[1] A derivation of (2.87), which is a special case of a Maxwell-Boltzmann distribution, can be found in almost any book which treats statistical mechanics, e.g., *Fay* [1965]. We follow here the usual practice in using the molecular constants m and k; note, however, that $k/m = \tilde{R}/\tilde{M} = R$.

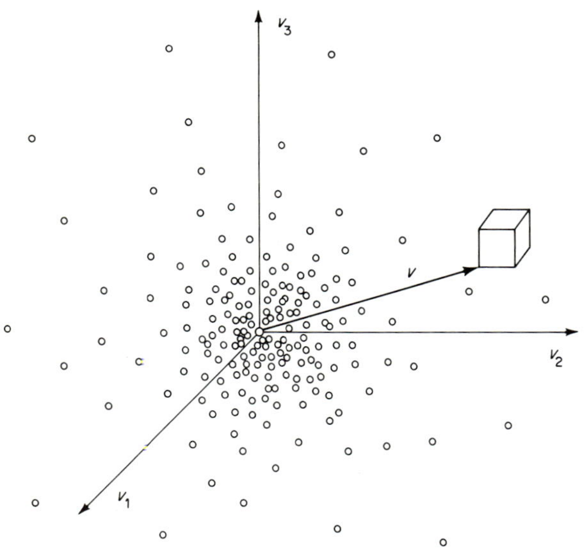

Figure 2.12
Distribution of molecular-representative points in velocity space (schematic).

distribution. In particular, f^0 is the number density of points in the velocity space, and we assume that the number of points is sufficiently large so that $f^0(\mathbf{v})$ may be taken as a continuous function. Because there is a total of n representative points, the sum over all velocity space must be just n,

$$n = \iiint_{-\infty}^{\infty} f^0 \, dv_1 \, dv_2 \, dv_3 \tag{2.88}$$

which the reader may verify by evaluating the definite integral.

The distribution function f^0 depends only on the molecular speed $v = (v_i v_i)^{1/2}$ in Eq. (2.87). The distribution of representative points in velocity space thus depends only on the (scalar) distance from the origin and is spherically symmetric. This corresponds to the absence of any preferred direction for molecular motion. The exponent in (2.87) contains the physical properties of the gas only in the form of a factor m/T. The effect of decreasing the value of this factor is to move the representative points in Fig. 2.12 further away from the origin, i.e., to spread out the distribution to higher molecular speeds, and this effect can be achieved either by increasing the temperature T or decreasing the molecular mass m.

The number of representative points in a small volume $\delta v_1 \, \delta v_2 \, \delta v_3$ is just $f^0(v) \, \delta v_1 \, \delta v_2 \, \delta v_3$. By integrating this in various ways we arrive at alternative distributions. The number of representative points in a thin slab between the planes $v_1 = \text{const}$ and $v_1 + dv_1 = \text{const}$ is

$$dv_1 \iint_{-\infty}^{\infty} f^0 \, dv_2 \, dv_3 \equiv n_{v_1} \, dv_1 \tag{2.89}$$

The quantity n_{v_1} is thus the number of representative points (number of molecules) per unit component velocity. With f^0 given by (2.87) this expression becomes

$$n_{v_1} = \frac{n}{\pi} \left(\frac{m}{2\pi kT}\right)^{1/2} e^{-mv_1^2/2kT} \left(\int_{-\infty}^{\infty} e^{-\eta^2} \, d\eta\right)^2$$

where $\eta^2 = mv_i^2/2kT$. The definite integral has a value of $\sqrt{\pi}$; thus

$$n_{v_1} = n \left(\frac{m}{2\pi kT}\right)^{1/2} e^{-mv_1^2/2kT} \tag{2.90}$$

is the distribution in one component of velocity, as shown in dimensionless form in Fig. 2.13. Precisely equivalent distributions apply of course to the velocity components v_2 and v_3.

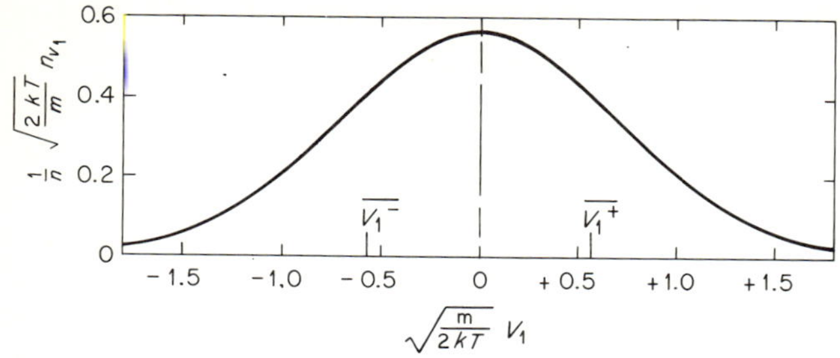

Figure 2.13
Maxwell distribution in one component of velocity.

As a simple application, we can calculate the *number flux of molecules* crossing a stationary surface $x_1 =$ const (the surface considered, shown in Fig. 2.14, is purely geometrical and has no physical identity). If all the molecules were moving in the x_1 direction with uniform speed v_1, the number flux J would be simply $J_n = nv_1$ particles per unit area per unit time. Since the molecular velocities vary in both direction and magnitude, this expression must be modified. Consider the molecules in the component velocity range v_1 to $v_1 + dv_1$; the number of such molecules per unit volume is just $n_{v_1}\, dv_1$, and the rate at which they cross the surface is just $(n_{v_1}\, dv_1)v_1$. Then the rate at which molecules in all velocity ranges cross is the sum

$$J_n^+ = \int_0^\infty v_1 n_{v_1}\, dv_1 \tag{2.91}$$

where the integration extends from zero to infinity because we count only those molecules crossing in the positive x_1 direction, with $v_1 \geq 0$ (integration over the entire range must lead to $J_n = 0$, corresponding to the gas being at rest, because the integrand is an odd function). Alternatively (2.91) can be written, by the use of the definition (2.89),

$$J_n^+ = \int_{-\infty}^{\infty} \int_{-\infty}^{\infty} \int_0^\infty f^0 v_1\, dv_1\, dv_2\, dv_3 \tag{2.92}$$

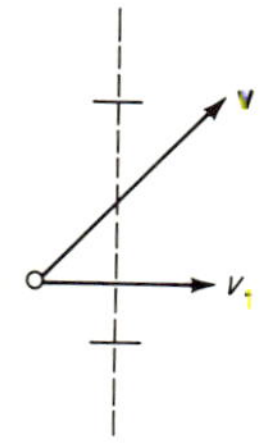

Figure 2.14

where the integral is simply the value of nv_1 averaged over all velocities with $v_1 > 0$. Evaluating the definite integral (2.91) gives

$$J_n^+ = n\sqrt{\frac{kT}{2\pi m}} \tag{2.93}$$

Since just one-half the molecules are traveling in the positive x_1 direction, we can write

$$J_n^+ = \frac{n}{2}\overline{v_1^+}$$

where

$$\overline{v_1^+} = \sqrt{\frac{2kT}{\pi m}} \tag{2.94}$$

is the average component velocity for all molecules traveling in the positive x_1 direction.

From (2.93) the molecular flux of mass is, with $\rho = mn$,

$$J_m^+ = \rho\sqrt{\frac{kT}{2\pi m}} = \rho\sqrt{\frac{RT}{2\pi}} \tag{2.95}$$

For example, in nitrogen at 1 atm and room temperature this has a numerical value of about 140 kg/(m²)(s) or 14 g/(cm²)(s).

The Maxwell distribution in molecular *speed* is found by calculating the number of representative points in velocity space within a spherical shell of thickness dv and radius v. The volume of the shell is $4\pi v^2\,dv$, and so the number of points contained is $f^0(v)4\pi v^2\,dv$; thus the number of points (molecules) per unit speed is $n_v = 4\pi v^2 f^0(v)$, or

$$n_v = 4\pi n\left(\frac{m}{2\pi kT}\right)^{3/2} v^2 e^{-mv^2/2kT} \tag{2.96}$$

a distribution shown in dimensionless form in Fig. 2.15.

The most probable, or modal, molecular speed v_{mp} corresponds by definition to the maximum of (2.96), which occurs at $v = \sqrt{2kT/m}$,

$$v_{mp} = \sqrt{\frac{2kT}{m}} \tag{2.97}$$

The mean molecular speed is the average over the distribution,

$$\bar{v} = \frac{1}{n}\int_0^\infty vn_v\,dv = \frac{1}{n}\iiint_{-\infty}^{\infty} vf^0\,dv_1\,dv_2\,dv_3 \tag{2.98}$$

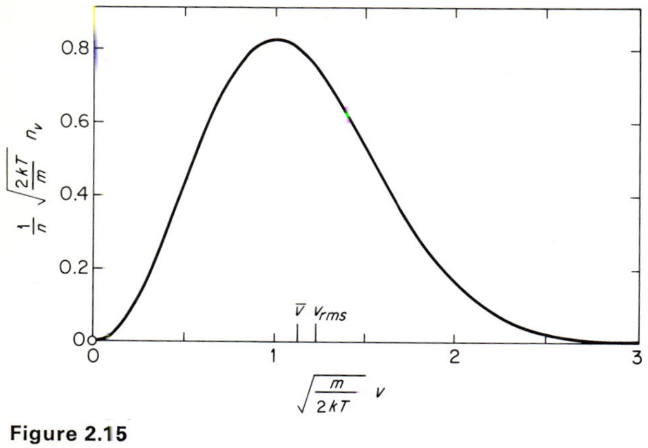

Figure 2.15
Maxwell distribution in speed.

Integration gives

$$\bar{v} = \sqrt{\frac{8kT}{\pi m}} \tag{2.99}$$

The mean-squared speed is the average,

$$\overline{v^2} = \frac{1}{n}\iiint_{-\infty}^{\infty} v^2 f^0 \, dv_1 \, dv_2 \, dv_3 \tag{2.100}$$

and integration gives

$$\overline{v^2} = \frac{3kT}{m}$$

or the root-mean-squared (rms) speed

$$v_{\text{rms}} = \sqrt{\frac{3kT}{m}} \tag{2.101}$$

Note that this is consistent with the equipartition value for the average translational kinetic energy, with $f = 3$,

$$\tfrac{1}{2}m\overline{v^2} = \tfrac{3}{2}kT$$

We have now found the following characteristic molecular speeds, which differ only by a multiplicative numerical constant:

$$\begin{aligned} \overline{v_i^+} &= \sqrt{\frac{2kT}{\pi m}} & v_{mp} &= \sqrt{\frac{2kT}{m}} \\ \bar{v} &= \sqrt{\frac{8kT}{\pi m}} & v_{\mathrm{rms}} &= \sqrt{\frac{3kT}{m}} \end{aligned} \tag{2.102}$$

The Maxwell distribution is not just an academic construction. Experiments have amply confirmed the theoretical predictions; e.g., Eq. (2.90) has been confirmed by doppler-shift measurements of molecular radiation.[1]

Stress and Momentum Flux

Consider a material surface in the sense of continuum mechanics. For the gas at rest, any stationary hypothetical surface within the gas is such a surface. Because the intermolecular forces are negligible in a dilute gas, a stress on the material surface can correspond only to a molecular flux of momentum across it and not to a sum of intermolecular forces. We will calculate the flux of momentum across a surface defined by x_i = const (Fig. 2.16). If all molecules had the same velocity, the momentum flux would be simply $(nv_i)(m\mathbf{v})$, with components mnv_iv_1, mnv_iv_2, and mnv_iv_3, or components denoted in general by mnv_iv_k. Since, however, the molecular velocities are not constant, the momentum flux Π_{ik} is an average over all molecular velocities, $\Pi_{ik} = mn\overline{v_iv_k}$, or

$$\Pi_{ik} = \iiint_{-\infty}^{\infty} mv_iv_kf^0\, dv_1\, dv_2\, dv_3 = \rho\overline{v_iv_k} \tag{2.103}$$

where the limits of integration span the entire molecular velocity range; i.e., we count particles crossing the surface in *both* directions.

[1] Such measurements are given, for example, by *Muntz* [1968]

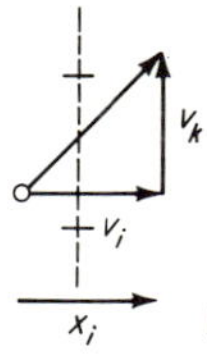

Figure 2.16

For $i \neq k$ this integral is zero because it contains terms of the type

$$\int_{-\infty}^{\infty} v_2 e^{-mv_2^2/2kT}\, dv_2$$

which vanish because the integrand is odd. For $i = k$ (for example Π_{22}) the integral is, after substitution of (2.87),

$$\Pi_{i(i)} = \frac{mn}{\pi^{3/2}} \frac{2kT}{m} \left(\int_{-\infty}^{\infty} e^{-\eta^2}\, d\eta\right)^2 \left(\int_{-\infty}^{\infty} \eta^2 e^{-\eta^2}\, d\eta\right)$$

The definite integrals have the respective values $\sqrt{\pi}$ and $\sqrt{\pi}/2$, giving

$$\Pi_{i(i)} = nkT = mn\frac{k}{m}T = \rho RT = \frac{RT}{v}$$

Since $RT/v = P$ by the perfect-gas law, we have finally

$$\Pi_{ik} = P\delta_{ik} = \begin{bmatrix} P & 0 & 0 \\ 0 & P & 0 \\ 0 & 0 & P \end{bmatrix} \tag{2.104}$$

which is just the negative of the hydrostatic stress tensor (the negative sign in the expression $\sigma_{ik} = -P\delta_{ik}$ is just a consequence of the convention by which tensile stresses are positive). The momentum-flux tensor Π_{ik} is isotropic in consequence of the isotropic molecular velocity distribution.

The idea of local equilibrium with respect to the macroscopic thermodynamic variables in a flowing fluid has already been discussed. It is natural to attempt to extend this idea to the distribution function. If a flowing gas differed from a stationary gas only in respect to the local motion, the distribution function would be a Maxwellian velocity distribution centered on the local continuum velocity **u**,

$$f = n\left(\frac{m}{2\pi kT}\right)^{3/2} \exp\left[-\frac{m}{2kT}(\mathbf{v} - \mathbf{u})^2\right] \tag{2.105}$$

i.e., the center of symmetry of the "cloud" of representative points in Fig. 2.12 is shifted from the origin to a position **u**. In this expression n, T, and **u** may be functions of **x** and t.

It should be apparent that the average molecular velocity $\bar{\mathbf{v}}$ is just the fluid velocity **u**, and this can be confirmed by calculating the average via

(2.105). A material surface now moves with velocity **u**, and the momentum-flux tensor relative to an observer who moves with such a surface is

$$\Pi_{ik} = \iiint_{-\infty}^{\infty} m(v_i - u_i)(v_k - u_k) f \, dv_1 \, dv_2 \, dv_3 \qquad (2.106)$$

It is convenient to write this in terms of the relative or *peculiar* molecular velocity **V**

$$\mathbf{V} \equiv \mathbf{v} - \mathbf{u} \qquad (2.107)$$

in terms of which (2.106) becomes

$$\Pi_{ik} = \iiint_{-\infty}^{\infty} mV_iV_k f \, dV_1 \, dV_2 \, dV_3 = \rho\overline{V_iV_k}$$

but this is formally identical to (2.103). We thus obtain the same momentum-flux tensor as for a gas at rest, $\Pi_{ik} = P\delta_{ik}$. With the momentum-flux tensor identified with the negative of the stress tensor, we conclude that a moving gas with a locally Maxwellian distribution has no viscous stress. Experimentally, however, flowing gases *do* exhibit viscous stresses, though the deviation of the stress from $-P\delta_{ik}$ is usually small. The perturbation of (2.105) which is required to yield a correct distribution function and thus a correct stress tensor is treated by the Boltzmann-Chapman-Enskog kinetic theory and is not pursued here. An excellent introduction to this subject is *Vincenti and Kruger* [1965].

Mean Free Path

A useful though imprecise concept is that of the *mean free path*, the average distance traveled by a molecule between collisions. If the molecules are considered as hard spheres, like billiard balls, then a collision is a clearly defined event. This molecular model is a very rough approximation, as indicated by the force potentials in Fig. 2.17. The collision of two hard-sphere molecules of diameter σ as indicated in Fig. 2.18*a* is kinematically identical to the collision of a sphere of diameter 2σ with a point, the centers of the colliding molecules being respectively at the center of the large sphere and at the point. Then a moving test molecule of effective diameter 2σ and effective molecular speed $\bar{v}_r$ sweeps out volume $\pi\sigma^2\bar{v}_r$ in unit time. The number of (point) molecules lying within this volume is $n(\pi\sigma^2\bar{v}_r)$, and this is just the number of collisions per unit time suffered by the test molecule

$$\text{Number of collisions per unit time} = n\pi\sigma^2\bar{v}_r$$

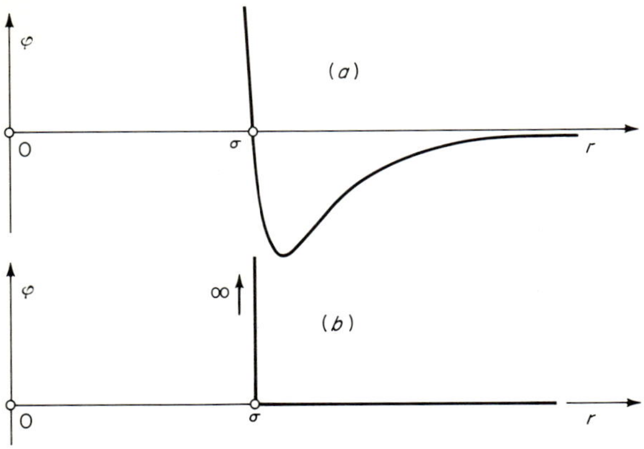

Figure 2.17
Intermolecular potential energy as a function of the separation between molecules: (*a*) realistic Lennard-Jones 6-12 potential; (*b*) hard-sphere potential.

The mean free path Λ is the average distance traveled between collisions, or the distance traveled divided by the number of collisions. If we set the distance traveled per unit time as $\bar{v}$, then

$$\Lambda = \frac{\bar{v}}{n\pi\sigma^2\bar{v}_r} \tag{2.108}$$

The values $\bar{v}$ and $\bar{v}_r$ are not quite the same: $\bar{v}$ is just the average molecular speed, while $\bar{v}_r$ is the average *relative* molecular speed, i.e., the average distance between representative points in velocity space. The calculation of $\bar{v}_r$ (see Prob. 2.14) yields $\bar{v}_r = \sqrt{2}\,\bar{v}$, so that (2.108) becomes

$$\Lambda = \frac{1}{\sqrt{2}\,n\pi\sigma^2} \tag{2.109}$$

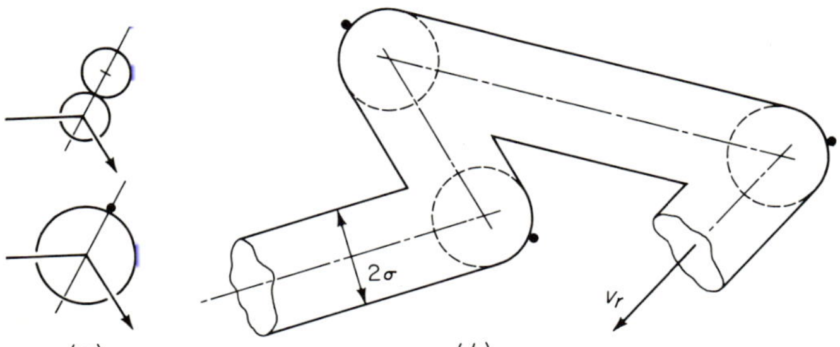

Figure 2.18
(*a*) Equivalent collisions; (*b*) volume swept out by a moving hard-sphere molecule.

Table 2.4 Mean Free Path at 1 Atm†

Gas	Molecular Weight $\tilde{M}$	Temperature T, K	Collision Diameter σ, nm	Mean Free Path Λ, nm	Intermolecular Distance $n^{-1/3}$, nm
He	4.00	100	0.218	64.5	2.39
		300		193.6	3.44
		500		322.6	4.08
Ar	39.9	100	0.364	23.1	2.39
		300		69.4	3.44
		500		115.7	4.08
N_2	28.02	100	0.753	21.8	2.39
		300		65.4	3.44
		500		109.0	4.08

† Adapted from *Hirschfelder et al.* [1964, p. 15].

an equation which is usually credited to Maxwell.[1] For the mean free path in mixtures of perfect gases, see *Chapman and Cowling* [1952, p. 91]. Some values for Λ are given in Table 2.4.

It should be noted that the mean free path Λ is normally much greater than the average distance between molecules $n^{-1/3}$. With the volume occupied by one molecule $\pi\sigma^3/6$, we find from (2.109) that

$$\frac{n^{-1/3}}{\Lambda} \propto \left(\frac{\text{volume occupied by molecules}}{\text{total volume}}\right)^{2/3}$$

The No-slip Condition at a Solid Boundary

A simple application of the mean-free-path concept is the investigation of the relative motion between a flowing gas and a solid boundary, such as a

[1] Found by James Clerk Maxwell (1860). Rudolf Clausius had in 1858 found the nearly identical formula, $\Lambda = (\frac{4}{3}n\pi\sigma^2)^{-1}$, and is credited as the originator of the mean-free-path concept. The true story is more interesting. In 1843 an obscure civil engineer, John James Waterston, published a correct description of the mean free path, equivalent to (2.109) but without the constant of proportionality. This publication, in an obscure journal, was ignored. In 1845, Waterston submitted to the Royal Society of London a paper formulating the *major results of elementary kinetic theory* (as later found by Joule, Clausius, and Maxwell). This bold paper was rejected as, according to one of the reviewers "nothing but nonsense, unfit even for reading before the Society." See *Truesdell* [1968, p. 292].

metal plate. It is widely accepted, with considerable experimental support, that there is *no* relative motion between a flowing fluid and a solid boundary or wall. The postulate of no relative motion is called the *no-slip condition.*

We now assume a slip velocity u_0 at the wall and estimate its magnitude. The assumed velocity distribution is shown in Fig. 2.19. A molecule striking the wall originates, i.e., experienced its last collision, on the average an approximate distance Λ above the wall. Such downward-traveling molecules carry, on the average, an x-component velocity equal to the continuum component velocity in the region of origin, specifically, $u_0 + \Lambda(du/dy)_0$. If the wall were smooth on a molecular scale, the molecule would show a regular reflection and rebound with this component exactly preserved. Now any wall is actually rough on a molecular scale and furthermore holds adsorbed molecules on the surface. It is thus more realistic to expect diffuse reflection, i.e., molecules rebounding with no net x-component velocity, on the average. If we assume, however, that the reflected molecules retain on the average a small fraction β of the incident velocity component, as indicated in the figure, then the upward-traveling molecules may be considered to have an x-component velocity $\beta[u_0 + \Lambda(du/dy)_0]$. Then the fluid velocity u_0 at a plane just above the wall surface is the average of the downward-traveling and upward-traveling molecules, or

$$u_0 \approx \frac{1}{2}\left\{u_0 + \Lambda\left(\frac{du}{dy}\right)_0 + \beta\left[u_0 + \Lambda\left(\frac{du}{dy}\right)_0\right]\right\}$$

which gives

$$u_0 \approx \frac{1+\beta}{1-\beta}\Lambda\left(\frac{du}{dy}\right)_0 \tag{2.110}$$

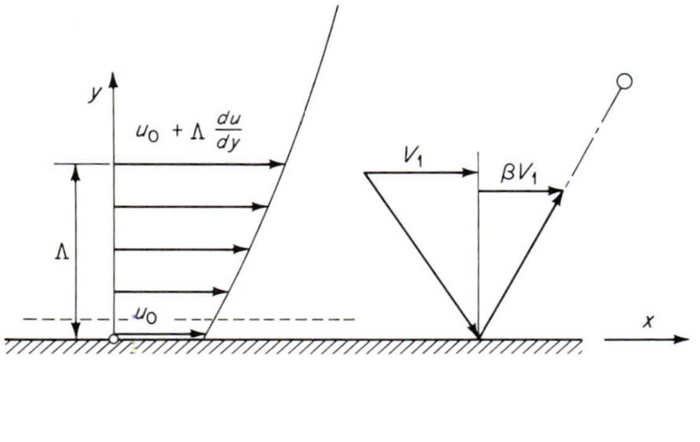

Figure 2.19
Velocity distribution at a wall.

The fraction β is experimentally found to be small, typically 0.2, and the velocity gradient at the wall normally gives a very small change over distances of the order Λ. Except for highly rarefied gases with correspondingly large values for Λ, the slip velocity u_0 is usually altogether negligible. However, at very large shear rates du/dy, as may be found in high-speed gas-lubricated bearings with small clearance, the slip velocity u_0 may be appreciable even at normal densities.

The preceding argument is essentially that of *Lighthill* [1956]. A similar argument applies to *temperature slip* at a boundary; i.e., the gas temperature at the wall may normally be taken as equal to that of the wall.

2.6 Liquids and dense gases

The ideal-gas equations of state, useful and convenient as they are, have a limited range of applicability. For liquids and gases at the high pressures of interest in compressible flows, e.g., in explosions, an alternative description is required.

For certain substances extensive tables of properties are available, and these allow particular compressible-flow problems to be worked out numerically. Of special interest are the properties of water, in both the liquid and vapor phases, as given in the tables of *Keenan et al.* [1969] and in *Meyer et al.* [1967]. Indispensable as such tabulations may be, for our purposes it is desirable to have relatively simple algebraic formulations which will permit analytical prediction, and we concentrate mainly on such descriptions below.

For any given substance, complete state information may be given by a thermal equation of state

$$v = v(P,T)$$

and the dependence of c_p on temperature at a fixed reference pressure P_a,

$$c_p = c_p(P_a,T)$$

We consider particular forms of these in the following paragraphs.

Corresponding States

It is remarkable that a great variety of gases and liquids, the simpler hydrocarbons in particular, can be approximately represented by a single nondimensional equation of state

$$v_r = v_r(P_r,T_r) \tag{2.111}$$

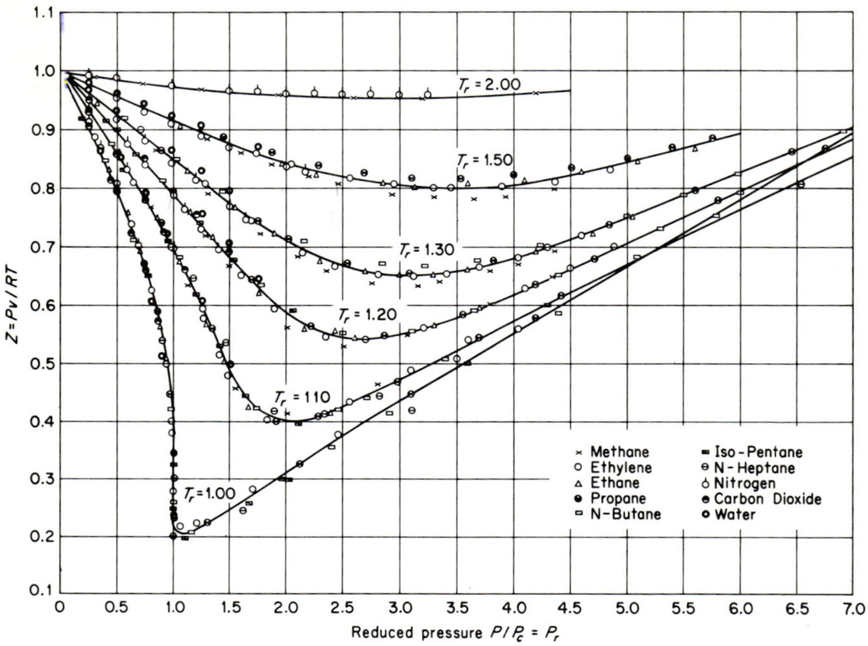

Figure 2.20
Corresponding states correlation of PvT data for various gases. (*From Su* [1946].)

Table 2.5 Critical-point Data

Substance		$\tilde{M}$	P_c, atm	ρ_c, kg/m^3	T_c, K	$\frac{P_c v_c}{RT_c}$
He	Helium	4.003	2.26	69.3	5.19	0.308
Ar	Argon	39.948	47.99	531.	150.72	0.291
Xe	Xenon	131.30	58.	1105.	289.75	0.290
H_2	Hydrogen	1.008	12.797	31.0	33.24	0.304
N_2	Nitrogen	28.013	33.54	311.0	126.2	0.291
O_2	Oxygen	31.999	50.14	430.	154.78	0.292
	Air	28.966	37.25	313.7	132.41	
H_2O	Water	18.015	218.17	317.0	647.29	0.230
CO_2	Carbon dioxide	44.010	72.90	460.	304.20	0.275
CH_4	Methane	16.043	45.8	162.	190.7	0.290
C_3H_8	Propane	44.097	42.1	226.0	370.01	0.277
C_4H_{10}	*n*-Butane	58.124	37.47	225.4	425.17	0.274

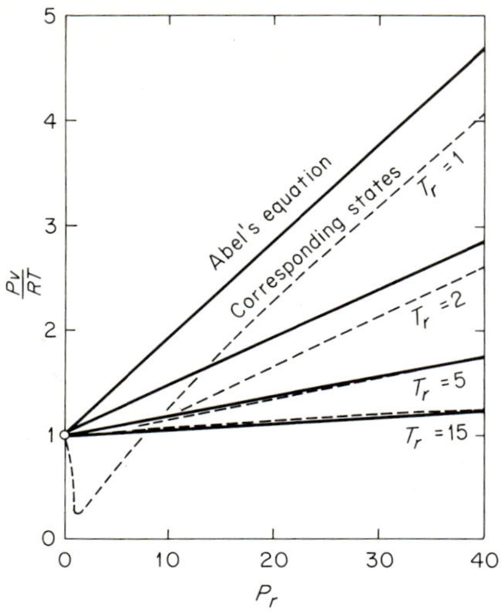

Figure 2.21

where v_r, P_r, and T_r are the volume, pressure, and temperature normalized with respect to their values at the critical point (see Table 2.5). This is often written in the form

$$\frac{Pv}{RT} = f(P_r, T_r) \tag{2.112}$$

by making use of the fact that the critical-point value $RT_c/P_c v_c$ is (approximately) a universal constant. An experimental verification of (2.112) is shown in Fig. 2.20. Further information on corresponding states may be found, for example, in *Hirschfelder et al.* [1964].

The Abel Equation of State

Perhaps the simplest thermal equation for dense gases is

$$P(v - b) = RT \tag{2.113}$$

which is usually called the Abel equation of state (sometimes, the Clausius equation of state). The constant b corresponds to the volume occupied by the molecules themselves, and the pressure goes to infinity when the gas is compressed to this volume. At relatively high temperatures and pressures this equation is reasonably accurate, as shown by the comparison with corresponding states data in Fig. 2.21.

By virtue of Example 2.5 (page 69), we find that fluids described by (2.113) have $e = e(T)$ only, just as the ideal gas, and that $c_p - c_v = R$.

Furthermore, isentropic changes of state are covered by a slight modification of the corresponding perfect-gas equations, provided that γ is constant and that v is replaced by $v - b$; for example, $P(v - b)^\gamma = \text{const}$ describes an isentropic process.

The Tait Equation

This is a particularly convenient equation for liquids at high pressure and may be written

$$\frac{P + B}{B} = \left(\frac{\rho}{\rho_0}\right)^\gamma \tag{2.114}$$

which is a form similar to an isentropic in a perfect gas [compare Eq. (2.86)]. In this expression B is a weak function of the entropy (though in practice usually taken as a constant), and ρ_0 is the liquid density extrapolated to zero pressure, i.e., very nearly the density at 1 atm. The exponent γ is *not* the ratio of specific heats but is indicated by this symbol for convenience because it plays the same role as the ratio of specific heats in a perfect gas undergoing isentropic processes; in particular, it will be possible to use the equations derived for the isentropic flow of a perfect gas merely by replacing P with $P + B$.

The given equation (2.114) is a modification suggested by *Kirkwood* [1942] and bears little resemblance to the equation originally suggested by *Tait* [1889] to fit data for seawater. For a brief discussion of the history of this equation, see *Rowlinson* [1969, chap. 2] and *Hirschfelder et al.* [1964, p. 261].

Table 2.6 Parameters in the Tait Equation

Liquid	B, atm	γ	ρ_0, kg/m^3
Water	3,000	7.15	1,000
Carbon tetrachloride	1,000	9.35	1,600
Mercury	3,000	8.2	13,500
n-Heptane	654	10.6	684
Silicone, 0.65 cks	597	9.1	760

Some values for the Tait equation parameters, appropriate to processes with entropy reasonably close to that at 1 atm and room temperature, are given in Table 2.6.

Hudleston's Equation

Based on a simple intermolecular force model, *Hudleston* [1937] obtained an equation for liquids which is usually written

$$\ln \frac{Pv^{2/3}}{v_0^{1/3} - v^{1/3}} = A + B(v_0^{1/3} - v^{1/3}) \tag{2.115}$$

where $v_0(T)$ is the specific volume at zero pressure and $A = A(T)$ (it would seem preferable, however, to write the equation in such a way that the argument of the logarithm is nondimensional). For a discussion of the accuracy of this equation, see *Bett et al.* [1954].

Other State Equations for Liquids

Several other PvT equations have been proposed for liquids. Many of these, such as that of Tumlirz, are based on a modification of the van der Waals equation. For further discussion, see for example *Reid and Sherwood* [1966] and *Bondi* [1968].

Comments on Liquids in General

Supposing that PvT data are available, either as an equation of state or as numerical data, consider first the specific heats; starting from the Gibbs equation $T\,ds = dh - v\,dP$ one finds for the pressure dependence of c_p

$$\left(\frac{\partial c_p}{\partial P}\right)_T = -T\left(\frac{\partial^2 v}{\partial T^2}\right)_p \tag{2.116}$$

which allows the calculation of c_p at any pressure from PvT data and a knowledge of $c_p(T)$ at some constant pressure (normally, at atmospheric pressure). The right-hand side of (2.116) is quite small for liquids, however, and it is a good approximation to take c_p independent of pressure (for some illustrative data, see *Zemansky and Van Ness* [1966, pp. 210ff]). If c_v is needed, it can then be found by the relation between specific heats,

$$c_p - c_v = -T\left(\frac{\partial P}{\partial v}\right)_T\left(\frac{\partial v}{\partial T}\right)_p^2 \geq 0 \tag{2.117}$$

Now it is sometimes asserted that this difference is quite small, i.e., that the ratio of specific heats γ is only slightly greater than unity. Actually, this assertion is correct *only for water and a few other liquids* (see the values for γ given in Table F.3, Appendix F).

Consider Eq. (2.23), viz.,

$$\left(\frac{\partial P}{\partial v}\right)_s = \gamma\left(\frac{\partial P}{\partial v}\right)_T$$

For the special case of *water* (γ close to unity), it is a consequence of this equation that the pressure-volume relation is not greatly different between an isothermal and an isentropic process. Correspondingly, the temperature change associated with isentropic processes is not very great. Consider the derivative $(\partial T/\partial P)_s$; by calculus,

$$\left(\frac{\partial T}{\partial P}\right)_s\left(\frac{\partial P}{\partial s}\right)_T\left(\frac{\partial s}{\partial T}\right)_p = -1 \qquad \left(\frac{\partial T}{\partial P}\right)_s = -\left(\frac{\partial T}{\partial s}\right)_p\left(\frac{\partial s}{\partial P}\right)_T$$

Now

$$\left(\frac{\partial T}{\partial s}\right)_p = \left(\frac{\partial T}{\partial h}\right)_p\left(\frac{\partial h}{\partial s}\right)_p = \frac{T}{c_p}$$

and with the Maxwell relation $(\partial s/\partial P)_T = -(\partial v/\partial T)_p$ we obtain

$$\left(\frac{\partial T}{\partial P}\right)_s = \frac{T}{c_p}\left(\frac{\partial v}{\partial T}\right)_p \tag{2.118}$$

For example, for liquid water at 20°C, with $\dfrac{1}{v}\left(\dfrac{\partial v}{\partial T}\right)_p \approx 2 \times 10^{-4}\,\text{K}^{-1}$ and $c_p \approx 4{,}200$ J/(kg)(K), this gives $(\partial T/\partial P)_s \approx 1.5 \times 10^{-3}$ K/atm. Isentropic compression from 1 to 1,000 atm thus would produce a temperature rise of only 1.5 K, approximately.

Solids at High Pressures

At pressures which are large compared to the yield stress, solid substances behave essentially as compressible fluids. Such pressures arise, for example, in the collision of a large meteorite with a planet. Laboratory experiments have created pressures sufficiently large to compress solids to one-half their initial volume. For a treatment of state equations, see *Zel'dovich and Raizer* [1967, chap. 11]. The isentropic Pv relation found is similar to (2.114). The behavior of very high speed metal jets has been treated by *Harlow and Pracht* [1966] (see Fig. 2.22). The *Grüneisen equa-*

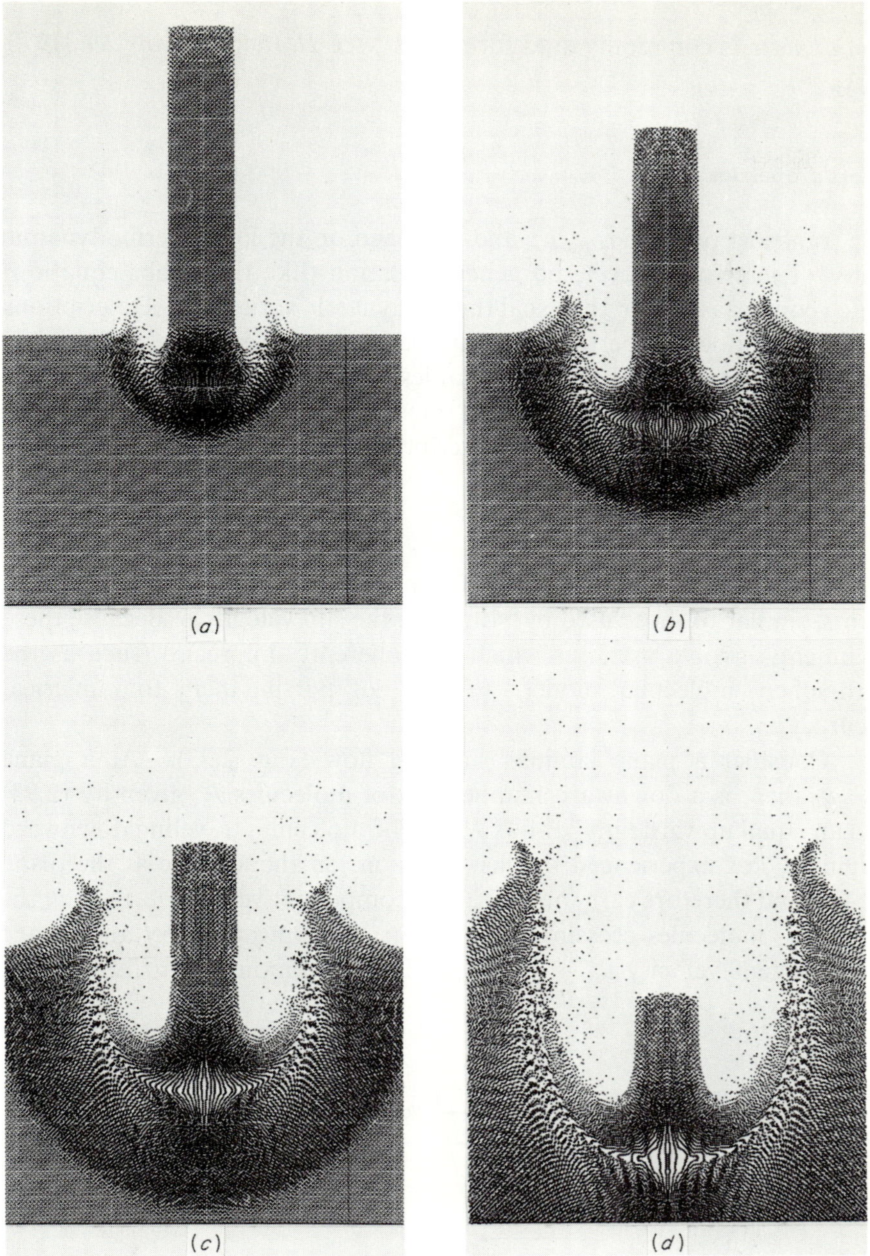

Figure 2.22
Compressible-fluid behavior of a metal. Penetration of an aluminum jet at 15 km/s into an aluminum target. The flow is two-dimensional, with jet thickness 1 mm. Successive frames are at 0.10, 0.20, 0.30, 0.45 μs. (*Courtesy of F. H. Harlow.*)

tion of state is commonly used for metals—see *Harlow and Amsden* [1970, p. 3].

2.7 Transport properties

The transport properties μ, μ_v, and κ depend on the local thermodynamic state. There is, however, no general relation (like the Gibbs equation) which connects them to the usual thermodynamic variables. Descriptions of the transport properties are thus based on observation and kinetic theory, not on classical thermodynamics. In particular, we consider the transport properties in a *dilute gas*, mostly because this is the only case which allows a relatively simple description.

Shear Viscosity μ

We begin by estimating the value of μ in a gas from mean-free-path arguments, similar to those used in estimating the slip velocity. Because these arguments involve premises which are inherently imprecise (such as the hard-sphere molecular model) we will not pursue them to numerical accuracy.

Consider a plane laminar shearing flow (Fig. 2.23). At a plane $y = y_0$ there is a downward number flux of molecules J_n^- given by (2.93) and an equal upward number flux J_n^+. The molecules traveling downward originate, i.e., experienced the last collision, on the average at the plane $y_0 + \Lambda$ and therefore carry an average x-component velocity $u_0 + \Lambda\, du/dy$. Similarly, molecules crossing the plane $y = y_0$ upward carry an average x-component velocity $u_0 - \Lambda\, du/dy$. Then the momentum flux Π_{12} as

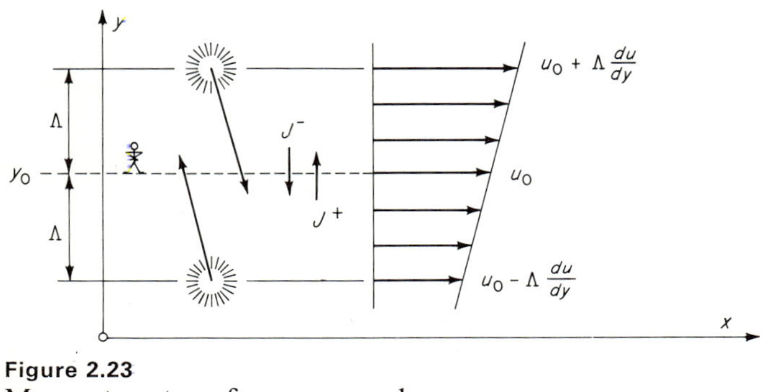

Figure 2.23
Momentum transfer across a plane $y = y_0$.

seen by an observer at the plane, moving with velocity u_0, is the sum of the contributions from downward-traveling and upward-traveling molecules carrying x momentum $+m\Lambda\, du/dy$ and $-m\Lambda\, du/dy$, respectively,

$$\Pi_{12} = -|J_n^-|\left(+m\Lambda \frac{du}{dy}\right) + |J_n^+|\left(-m\Lambda \frac{du}{dy}\right)$$

in which an upward flux is positive. Substituting the value $J_n^+ = J_n^-$ from (2.93) then gives

$$\Pi_{12} = -2\rho\sqrt{\frac{kT}{2\pi m}}\,\Lambda\frac{du}{dy}$$

With the average molecular speed $\bar{v} = \sqrt{8kT/\pi m}$ this becomes

$$\Pi_{12} = -\tfrac{1}{2}\rho\bar{v}\Lambda\frac{du}{dy} \tag{2.119}$$

Comparing this with the stress σ_{12} found in Example 1.2 (page 21),

$$\sigma_{12} = \mu u_{1,2} = \mu\frac{du}{dy}$$

we conclude that (since $\Pi_{12} = -\sigma_{12}$)

$$\mu = \tfrac{1}{2}\rho\bar{v}\Lambda \tag{2.120}$$

which is the desired result. The precise kinetic-theory result for hard-sphere molecules is

$$\mu = \frac{5\pi}{32}\rho\bar{v}\Lambda \tag{2.121}$$

With $5\pi/32 = 0.491$, the agreement with the above approximate result is remarkable but simply fortuitous.

The reader may invent analogies for the momentum exchange involved in this calculation of μ. For example, imagine two trains of flat-cars running on parallel tracks in the same direction but at different speeds. If the two trains carry rival political party picnics (say) and the groups manifest their rivalry by hurling watermelons at each other, there will be an exchange of momentum between trains such that the slower train is speeded up and the faster train is slowed down.

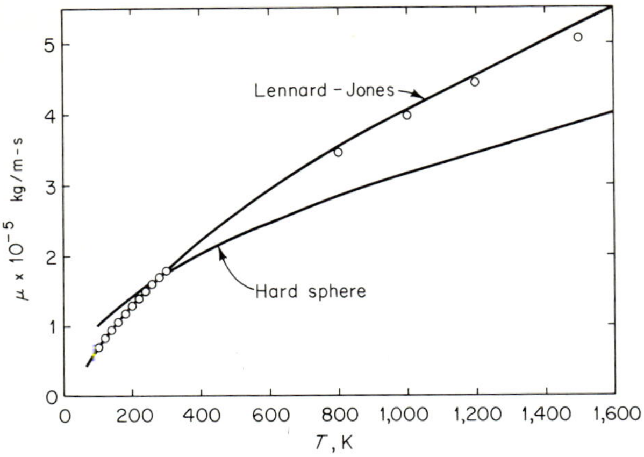

Figure 2.24
Viscosity of nitrogen as a function of temperature. The upper curve is from the Boltzmann theory with a Lennard-Jones intermolecular potential. (*Data from Hirschfelder et al.* [1964, *p.* 562].)

Substitution of Eqs. (2.109) and (2.102) into (2.120) gives for the hard-sphere molecular model

$$\mu = \frac{m}{\pi\sigma^2}\sqrt{\frac{kT}{\pi m}} \tag{2.122}$$

or $\mu \propto \sqrt{T}$. This result is not in very close agreement with experiment, as shown in Fig. 2.24, primarily because *the mean-free-path model is inherently approximate.* What our crude model does predict correctly is that the viscosity of a dilute gas depends only on temperature.[1] Thus we could bring (2.122) into accord with experiment by allowing the molecular diameter σ to decrease with temperature.

It will be convenient to make considerable use of (2.120) as a rough estimate of the shear viscosity μ. Anticipating a result in Chap. 4, the speed of sound c in a dilute gas is

$$c = \sqrt{\frac{\gamma kT}{m}} = \sqrt{\frac{\gamma\pi}{8}}\,\bar{v} \tag{2.123}$$

[1] Thus Maxwell, who found this result, states "If this explanation of gaseous friction be true, the coefficient of friction is independent of the density. Such a consequence of a mathematical theory is very startling . . . ."

i.e., the sound speed is proportional to the characteristic molecular speed. Substituting in (2.120) yields

$$\mu = \sqrt{\frac{2}{\pi\gamma}}\, \rho c \Lambda \tag{2.124}$$

or $\mu = 0.7833\rho c\Lambda/\sqrt{\gamma}$. A convenient rough figure is then

$$\mu \approx \tfrac{2}{3}\rho c \Lambda \tag{2.125}$$

This form of the viscosity dependence is of particular interest because the sound speed c plays a crucial role in compressible flow.

In certain calculations, e.g., in compressible-fluid boundary layers, it is convenient to have relatively simple algebraic formulas for $\mu(T)$ which are more accurate than the hard-sphere equation (2.122). Commonly used is the formula of Sutherland,

$$\mu = \frac{\mu_{\text{hard sphere}}}{1 + T_*/T} = \text{const} \times \frac{T^{1/2}}{1 + T_*/T} \tag{2.126}$$

where T_* is a constant (for values, see, for example, *Hirschfelder et al.* [1964, p. 549]). This equation is based on a simple molecular potential and provides a good approximation to experimental viscosities. Another often used formula is that of Chapman (*Chapman and Rubesin* [1949]),

$$\frac{\mu}{\mu_w} = \frac{T}{T_w} \tag{2.127}$$

where the reference value $\mu_w(T_w)$ is accurately known, e.g., from Sutherland's formula. We will make use of this simple linear expression in Sec. 10.5.

The viscosity mechanism in *liquids* is of course different from that in gases. In contrast to the molecules in a dilute gas, the liquid molecules are constantly subject to intermolecular forces. It is believed that the molecules form ordered clusters, with a locally crystalloid structure, and that resistance to shearing motion is a result of the breakup of such clusters. One analytical approach to liquid viscosity is that of Eyring; a convenient description is given in *Hirschfelder et al.* [1964, p. 624].

The temperature dependence of viscosity in liquids is opposite to that in gases, in that liquid viscosity *decreases* with increasing temperature. The rate of change is greater than in gases; water at its freezing point, for example, has a viscosity 6.3 times greater than at its boiling point.

Thermal Conductivity κ

The viscosity μ has been estimated from simple mean-free-path arguments. For dilute *monatomic* gases the thermal conductivity can be estimated by an analogous method (see, for example, *Fay* [1965, chap. 20]). The result is

$$\kappa = \rho \bar{v} \Lambda R \tag{2.128}$$

where $R = k/m$ is the specific gas constant. The precise Boltzmann kinetic-theory result for hard-sphere molecules is

$$\kappa = \frac{75\pi}{128} \rho \bar{v} \Lambda R \tag{2.129}$$

Note that the (hard-sphere) thermal conductivity has exactly the same dependence as the viscosity in the analogous expressions (2.120) and (2.121).

The comments already made about our viscosity estimates apply also to the above expressions for thermal conductivity. If, however, we form the *Prandtl number*, the nondimensional ratio of viscosity to thermal conductivity, by using (2.121) and (2.129) and $c_p = \gamma R/(\gamma - 1)$ with $\gamma = 5/3$, we find for monatomic gases

$$\mathrm{Pr} = \frac{\mu c_p}{\kappa} = \frac{2}{3} \tag{2.130}$$

a result which is substantiated by experiment (see Table 1.1) and holds for all temperatures.

The behavior of polyatomic gases is similar. It is convenient to express the thermal conductivity via the Prandtl number [assuming that $\mu(T)$ and $c_p(T)$ are known]. A reasonably accurate expression for the Prandtl number is Eucken's formula, which gives

$$\mathrm{Pr} = \frac{4\gamma}{9\gamma - 5} \tag{2.131}$$

With $5/3 \geq \gamma \geq 1$, this gives $1 \geq \mathrm{Pr} \geq 2/3$.

For *liquids* the thermal conductivity is given with reasonable accuracy by the simple formula (see *Hirschfelder et al.* [1964, p. 634])

$$\kappa = \frac{2.80}{\sqrt{\gamma}} \rho c n^{-1/3} R \tag{2.132}$$

which is in the same form as (2.128) and (2.129); c is the speed of sound in the liquid.

Bulk Viscosity μ_v

The bulk viscosity is difficult to measure because the associated stress is

$$\mu_v \nabla \cdot \mathbf{u} = \mu_v \frac{1}{v} \frac{Dv}{Dt}$$

and volume-dilatation rates yielding appreciable stresses are difficult to achieve in a laboratory situation. One technique is the measurement of attenuation in acoustic waves (see Chap. 4).

It can be shown that the bulk viscosity is proportional to the relaxation time τ_r for molecular rotation, provided that the volume dilatation is slow,

$$\tau_r \ll \left(\frac{1}{v} \frac{Dv}{Dt}\right)^{-1}$$

For a brief derivation, see *Zel'dovich and Raizer* [1967, p. 469]. The resulting expression is, for a diatomic gas at low temperature,

$$\mu_v = \tfrac{4}{25}\rho R T \tau_r \tag{2.133}$$

Problems

2.1 Write explicit expressions for the dissipation function Υ for the velocity fields of Examples 1.2 and 1.3 (pages 21 and 24).

2.2 Show that $c_p = T(\partial s/\partial T)_p$.

2.3 For a hypothetical nonconducting incompressible fluid, the energy equation (2.13) reduces to

$$\frac{De}{Dt} = v\Upsilon$$

Calculate the rate of temperature rise for such a fluid, with the pertinent physical properties of water at 20°C, at the representative shear rates in plane laminar shearing (see Example 1.2, page 21),

(*a*) $u_{1,2} = 10^2\ \text{s}^{-1}$
(*b*) $u_{1,2} = 10^4\ \text{s}^{-1}$
(*c*) $u_{1,2} = 10^6\ \text{s}^{-1}$

Answers 2.4×10^{-6} K/s; 2.4×10^{-2} K/s; 2.4×10^{2} K/s

2.4 Given that

$$dh = T\,ds + v\,dP$$

$$c_p \equiv \left(\frac{\partial h}{\partial T}\right)_p$$

$$\left(\frac{\partial s}{\partial P}\right)_T = -\left(\frac{\partial v}{\partial T}\right)_p$$

find, for the variation of c_p with pressure,

$$\left(\frac{\partial c_p}{\partial P}\right)_T = -T\left(\frac{\partial^2 v}{\partial T^2}\right)_p$$

2.5 If the ratio of specific heats γ for an ideal gas varies according to

$$\frac{\gamma}{\gamma - 1} = a_0 + a_1 T$$

where a_0 and a_1 are constants, find an equation relating P and T for an isentropic process analogous to the perfect gas relation

$$\frac{P_2}{P_1} = \left(\frac{T_2}{T_1}\right)^{\gamma/(\gamma-1)}$$

Answer $\quad \dfrac{P_2}{P_1} = e^{a_1(T_2 - T_1)}\left(\dfrac{T_2}{T_1}\right)^{a_0}$

2.6 Find the molecular weight and ratio of specific heats for an equimolar mixture of helium and oxygen at room temperature.

Answer 18.001; 1.500

2.7 Show that

$$\left(\frac{\partial T}{\partial s}\right)_p = -\left(\frac{\partial P}{\partial v}\right)_T\left(\frac{\partial T}{\partial P}\right)_v\left(\frac{\partial T}{\partial P}\right)_s$$

and discuss the slope of an isobar at the critical point on a Ts diagram.

2.8 The identity $(\partial P/\partial v)_s = \gamma(\partial P/\partial v)_T$ is not useful at the critical point because $(\partial P/\partial v)_T \to 0$ and $\gamma \to \infty$. Find the alternative identity,

$$\left(\frac{\partial P}{\partial v}\right)_s = \left(\frac{\partial P}{\partial v}\right)_T - \frac{T}{c_v}\left(\frac{\partial P}{\partial T}\right)_v^2$$

by manipulation of fundamental relations.

2.9 In a certain gas flow which is approximately isentropic, e.g., the flow through a low-pressure fan, the pressure changes only slightly from its initial value P_0.

Show that the corresponding density change is given by

$$\frac{\Delta\rho}{\rho_0} \approx \frac{1}{\gamma}\frac{\Delta P}{P_0}$$

and the flow is nearly "incompressible."

2.10 Evaluate the derivatives $(\partial v/\partial P)_s$, $(\partial^2 v/\partial P^2)_s$, $(\partial^3 v/\partial P^3)_s$ in terms of v, P, and γ for a perfect gas.

2.11 A cryogenic fluid is stored in an insulated tank which is vented to the atmosphere. As heat leaks through the walls at a constant rate $\dot{Q}$ (which can be calculated, with known fluid temperature and known external temperature), there is continuous evaporation of the liquid and loss of vapor at rate $\dot{m}$ through the vent.

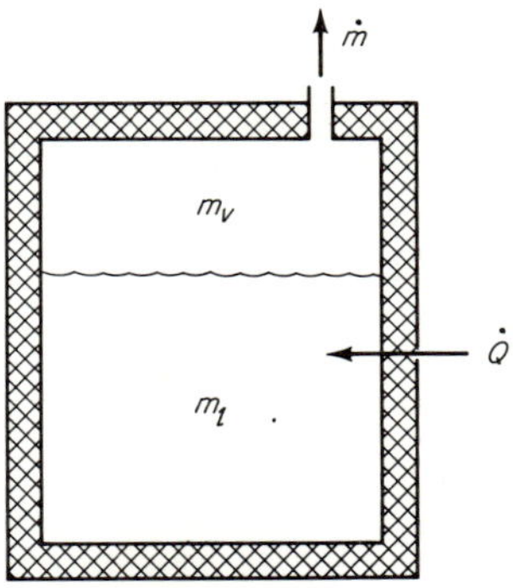

The loss is slow enough, however, so that the efflux velocity is negligible and the pressure of the vapor within the tank is just 1 atm. If the latent heat for evaporation is $L \equiv h_v - h_l$ and the specific volumes of vapor and liquid are v_v and v_l, respectively, find the loss rate $\dot{m}$.

Answer $\qquad \dot{m} = \dfrac{v_v - v_l}{v_v}\dfrac{\dot{Q}}{L}$

2.12 For a gas with an arbitrary distribution function f (not necessarily Maxwellian), determine whether the momentum flux tensor Π_{ik} is symmetric.

2.13 For a gas with a Maxwell distribution of velocities, show that the average relative speed $\bar{v}_r$ is given by $\bar{v}_r = \sqrt{2}\,\bar{v}$.

2.14 Show that the result $\bar{v}_r = \sqrt{2}\,\bar{v}$ holds for any isotropic distribution of molecular velocities.

2.15 For air at 1 atm and 300 K with the velocity distribution of Example 1.3 (page 24), find the value of the velocity gradient $u_{1,1}$ (in units of s^{-1}) at which the normal stress σ_{11} is zero. Discuss this result from the molecular viewpoint.

2.16 A model stratified atmosphere may consist of an *incompressible* fluid, with the density ρ varying with height z according to $\rho = \rho_0 + \rho' z$; that is, the atmosphere consists of horizontal liquid layers of different densities. Find the Brunt-Väisälä frequency N for such an atmosphere and the condition of stability. Assume that the density variation is small, $|\rho - \rho_0| \ll \rho_0$.

Answer $N^2 = -\dfrac{g\rho'}{\rho_0} > 0$

2.17 A certain fluid is compressed isentropically. Will the temperature increase?

three
dimensions, similarity, and magnitude

3.1 Introduction

A given fluid flow is not unique; i.e., exactly similar flows may occur in different fluids at different scales of space and time. For example, the flow about a steel sphere sinking in glycerin may correspond to the flow about a very much smaller spherical pollen grain sinking in air. If an exact correspondence between the two flows can be established, the flow about the steel sphere is said to be a *model* of the flow about the pollen grain; it will of course be essential to establish the meaning for "exact correspondence," and thereby to construct a theory of models.

The theory of models, and related topics which are important in compressible flow, will be approached in this chapter mainly from the point of view of physical dimensions, such as mass, length, and time. An elementary example of dimensional reasoning is the following: if it is asserted that the formula for the area of a circle is $A = 2\pi r$, we reject the assertion simply because the right-hand side does not have the units of area but the units of length. By generalizing this kind of argument, we obtain an apparatus which will help us to answer practical questions such as the following:

1 Can the performance of a model airplane in a wind tunnel be used to predict the performance of a geometrically similar full-sized airplane?

2 Can the electrical performance of a network of coils, resistors, and capacitors be used to predict the acoustic performance of a certain enclosure?

3 Can the problem of solving a partial differential equation of fluid motion be reduced to that of solving an ordinary differential equation?

4 Is it possible to drop certain terms from the differential equations of fluid motion?

5 Can parameters governing the general nature of a flow be discovered?

The answer to each of the above questions is a qualified "yes," and we shall examine the qualifications in this chapter.

The misty beginnings of dimensional reasoning can be found in the writings of Galileo and Newton (for early references, see *Birkhoff* [1960, chap. 4]). The more modern development can be traced through *Stokes*, who made precise use of dimensional arguments in a famous paper on the drag of slowly moving objects [1850]; and to his student J. W. Strutt (Lord Rayleigh). It was *Rayleigh* above all who brought the method to general attention [1915]:

> I have often been impressed by the scanty attention paid even by original workers in physics to the great principle of similitude [i.e., dimensional analysis]. It happens not infrequently that results in the form of "laws" are put forward as novelties on the basis of elaborate experiments, which might have been predicted *a priori* after a few minutes' consideration. However useful verification may be, whether to solve doubts or to exercise students, this seems to be an inversion of the normal order. One reason for the neglect of the principle may be that, at any rate in its applications to particular cases, it does not much interest mathematicians. On the other hand, engineers, who might make much more use of it than they have done, employ a notation which tends to obscure it. I refer to the manner in which gravity is treated. When the question under consideration depends essentially upon gravity, the symbol of gravity (g) makes no appearance, but when gravity does not enter into the question at all, g obtrudes itself conspicuously.

While Rayleigh used *similitude* for what is now called dimensional analysis, the word has acquired a wider (and therefore less definite) meaning, particularly in aeronautical circles. Thus, according to *Hayes and Probstein* [1966, p. 24], "The word similitude usually is used to refer to an equivalence between two physical problems which are different from each other in some fundamental way." The present chapter concerns similitude, in this sense.

Worthy references to the subject are *Bridgman* [1931], *Sedov* [1959], and *Kline* [1965].

3.2 Geometric relation

In the construction of physical models, such as a model airplane built to reduced scale, there is a definite geometrical relation between model and prototype. We define two distinct categories of geometrical relationship.

Geometric Similarity

If *all* of the corresponding lineal dimensions of two bodies *a* and *c* are in constant ratio, the bodies are said to be geometrically similar. For example, all spheres are geometrically similar to each other. Formally, consider two coordinate systems **X** and **x** related by the *mapping* $X_i = \alpha x_i$, where α is a constant. Then the surfaces $f(x_1,x_2,x_3) = 0$ and $f(X_1,X_2,X_3) = 0$ are geometrically similar.

Geometric Affinity

If the corresponding lineal dimensions in *each of the three Cartesian directions* of two bodies *a* and *b* are in constant ratio, the bodies are said to be

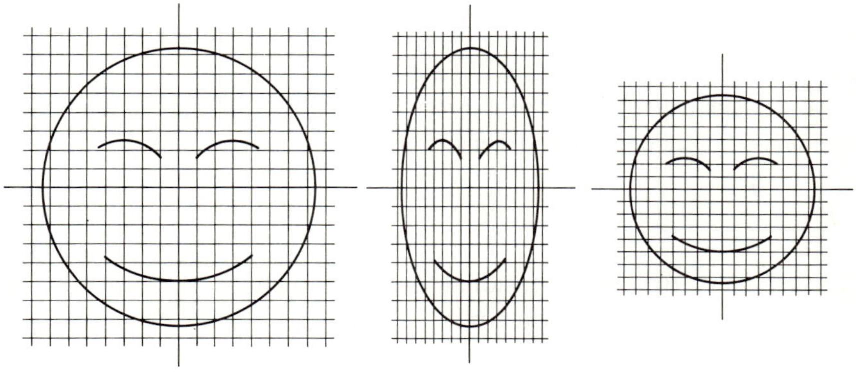

(*a*) (*b*) (*c*)

Figure 3.1
The figures (*a*) and (*b*) are geometrically affine; (*a*) and (*c*) are geometrically similar.

geometrically affine. If for example a figure is drawn on a rubber sheet and the sheet is stretched in one direction only, the resulting figure is affine to the original figure. Formally, if coordinate systems **X** and **x** are related by the mapping $X_i = \alpha_i x_{(i)}$, where the constants α_i have three distinct values, the surfaces $f(x_1,x_2,x_3) = 0$ in the x space and the X space, respectively, are geometrically affine.

3.3 Units and dimensions

The quantity which we call mass can be measured in various units, such as kilograms (abbreviated kg) or pounds mass (abbreviated lb_m). Thus a certain quantity m, for example $m = 0.4536$ kg, has the units of kilograms and the *dimensions* of mass. This is conventionally written as

$$[m] = M$$

where M stands for mass and $[m]$ is read "the dimensions of m."

We will take the *fundamental* or *primary dimensions* to be[1]

Mass M
Length L
Time T
Current I
Temperature θ

This list is neither unique nor irreducible; i.e., the choice of fundamental dimensions is to some extent arbitrary. For example, force F can usually be used in place of mass M, as will be discussed.

All physical quantities are expressible in terms of the fundamental dimensions, e.g., velocity u and density ρ,

$$[u] = \frac{L}{T} \qquad [\rho] = \frac{M}{L^3}$$

A dimensionless or nondimensional quantity such as π, the ratio of circumference to diameter for a circle, is written

$$[\pi] = 1$$

[1] The electrical current I is included for completeness and consistency with the international system of units. An alternative possibility which may be more philosophically satisfying is to use charge Q as the fundamental dimension. In any case, the electrical dimension will not be used in this book.

That is, π has the dimensions of a pure number, unity. Such a convention is algebraically consistent; consider the ratio of two lengths, l_1 and l_2,

$$\left[\frac{l_1}{l_2}\right] = \frac{L}{L} = 1$$

Newton's Second Law

An axiomatic physical law may be used to establish an equivalence between dimensions. Newton's second law connects force and mass:

$$F \propto ma \tag{3.1}$$

The proportionality constant is conventionally written $1/g_c$, so that

$$F = \frac{1}{g_c} ma \tag{3.2}$$

The value of g_c is established ultimately by experiment or, equivalently, defined to be unity, thus fixing the size of the force unit. The numerical value depends on the units used for force, mass, and acceleration; example values are[1]

$$g_c = 32.174 \frac{\text{lb}_m}{\text{lb}_f} \frac{\text{ft}}{\text{s}^2}$$

$$= 1 \frac{\text{kg}}{\text{N}} \frac{\text{m}}{\text{s}^2}$$

By virtue of (3.2) force is always proportional to mass (for any given acceleration), and there is little justification for regarding force and mass as dimensionally independent quantities. Formally, we *define* g_c to be a nondimensional quantity,

$$[g_c] \equiv 1 \tag{3.3}$$

Thus g_c becomes a simple *conversion factor* perfectly analogous to, say, 12 in/ft. Note that conversion factors of this kind are formally and functionally *equal* to unity.[2] For illustration, consider a certain area

$$A = 0.5 \text{ ft}^2$$

[1] That g_c is *numerically* equal to the standard acceleration of gravity in certain unit systems should be regarded as fortuitous. The reader is reminded that g_c is a *universal constant*, unlike the local acceleration of gravity.

[2] For a different conclusion, see *Zemansky and Van Ness* [1966, p. 34]. It is probably safe to assume that controversies about dimensions will go on forever.

Let $\lambda_c = 1 = 12$ in/ft. Multiplying both sides by λ_c^2 then yields the area in converted units, after canceling units,

$$A = 72\ \text{in}^2$$

Similarly we have, formally,

$$g_c = 32.174 \frac{\text{lb}_m}{\text{lb}_f} \frac{\text{ft}}{\text{s}^2} = 1$$

The dimensional statement of (3.2) is now

$$[F] = [ma] = \frac{ML}{T^2} \tag{3.4}$$

Thus, by arbitrarily taking g_c to be dimensionless, the force F can be removed from a list of fundamental dimensions. Alternatively, mass M could be removed while retaining force F. We will follow the former procedure and retain M, as indicated by our list of fundamental dimensions.

With g_c now simply a nondimensional conversion factor of absolute magnitude unity, we will follow the practice of omitting it from all equations (note that this practice has been followed in Chaps. 1 and 2). Then Newton's second law is written

$$F = ma \tag{3.5}$$

with the understanding that a numerical conversion factor g_c may be required in numerical computations, depending on the system of units employed.

As examples, the dimensions of pressure and of viscosity can be worked out in the $MLTI\theta$ fundamental system. Pressure P is force per unit area or

$$[P] = \frac{F}{L^2} = \frac{ML}{T^2}\frac{1}{L^2} = \frac{M}{LT^2}$$

From the linear stress law (1.37) viscosity μ has the dimensions of pressure divided by $D_{mm} = \nabla \cdot \mathbf{u}$ or

$$[\mu] = \left[\frac{P}{\nabla \cdot \mathbf{u}}\right] = \frac{M}{LT^2} T = \frac{M}{LT}$$

Note that bulk viscosity μ_v has the same dimensions as μ.

First Law of Thermodynamics

A similar reduction in the number of dimensions can be achieved via the first law of thermodynamics. Consider a fluid material volume, in particular a certain gas enclosed in a cylinder by a movable piston. If the enclosed gas executes a *cycle*, i.e., starting from some initial state P_0, V_0, undergoes changes in pressure and volume, returning finally to the initial state P_0, V_0, there will be no change in the internal energy of the gas. A traditional statement of the first law is that for such a cycle the net work done by the gas is proportional to the net heat input, or

$$W \propto Q \tag{3.6}$$

Introducing the proportionality constant J, the mechanical equivalent of heat,

$$W = JQ \tag{3.7}$$

The determination of the value of J has been the subject of famous experiments, in particular those of James Joule, who is recognized in the choice of symbol. Example values in different unit systems are

$$J = 778.17 \text{ ft}\cdot\text{lb}_f/\text{Btu} = 4.1868 \text{ J/cal}$$

Just as in the case of g_c, we arbitrarily take the constant J to be nondimensional, $[J] = 1$. Then from (3.7)

$$[W] = [Q] = \frac{ML^2}{T^2} \tag{3.8}$$

where the dimensions of W are those of force times distance. In this way heat is eliminated as a fundamental dimension, and we do not write J explicitly in our equations.

On Further Reduction in the Number of Fundamental Dimensions

By taking the fundamental constants which occur in additional physical laws to be nondimensional and of absolute magnitude unity we can further reduce the number of fundamental dimensions. For example, in the expression

$$\bar{\varepsilon} = {}^3\!/\!_2 kT$$

for the average translational energy of a molecule, the Boltzmann constant k can be taken as nondimensional

$$[k] = \frac{ML^2}{T^2\theta} = 1$$

Thus the dimensions of temperature would become ML^2/T^2, the dimensions of energy. The extensive entropy would be dimensionless in this system, a result which is in harmony with the statistical view of entropy. The size of a "degree" would depend on the unit system: in modern metric units,

$$k = 1.3805 \times 10^{-23}\ \text{J/K} = 1$$

which gives a (rather inconvenient) temperature unit, 1 J = 7.24×10^{24} K.

Similarly, we could take (say) the velocity of light in free space, Planck's constant, and the charge of the electron to be nondimensional constants of magnitude unity. Then *all* quantities would be nondimensional. We would not find this course to be fruitful, however, as will be explained in subsequent sections, and we retain M, L, T, I, θ as the fundamental dimensions except in special circumstances.

Systems of Units

In this book, SI metric units (Système international d'unités, adopted at the 1960 Eleventh General Conference on Weights and Measures) are extensively used. The pertinent units are summarized in Table 3.1. The

Table 3.1 SI Units

Quantity	Unit	Abbreviation	Equivalent
Mass	Kilogram	kg	
Length	Meter	m	
Time	Second	s	
Temperature	Kelvin	K	°C
Force	Newton	N	1 kg·m/s^2
Energy	Joule	J	1 N·m
Power	Watt	W	1 J/s
Frequency	Hertz	Hz	1 s^{-1}

Table 3.2

Exponential	Prefix	Abbreviation
10^{12}	tera	T
10^{9}	giga	G
10^{6}	mega	M
10^{3}	kilo	k
10^{-3}	milli	m
10^{-6}	micro	μ
10^{-9}	nano	n
10^{-12}	pico	p

prefixes commonly used in conjunction with these units are shown in Table 3.2. For example, 3 nm $= 3 \times 10^{-9}$ meter.

We will also have occasion to make use of the common English system of units, in which we distinguish between the pound mass (lb_m) and pound force (lb_f) as already noted.

The use of engineering metric units, in which the units of mass and force can be confused (as in the English system) by the use of the kilogram as a unit for both force and mass, is not encouraged. Factors for conversion between different units are given in Appendix C.

3.4 Dimensionless forms

Physical laws must have a validity independent of the system of units in which they are expressed. This requirement of objectivity is the basic axiom of dimensional analysis and limits the possible forms of a particular physical law.

As an example, consider the time of fall for a stone dropped from a certain height. This is a somewhat artificial example, because the elementary "answer" to this problem is probably already known to the reader; let us see, however, what can be learned from dimensional reasoning alone. The quantities entering the problem may be presumed to be

m = mass of stone	$[m] = M$
h = height	$[h] = L$
g = acceleration of gravity	$[g] = LT^{-2}$
t = total time of fall	$[t] = T$

The preparation of such a list is an act of physical intuition and a potential weak link in the chain of analysis. The omission of g from the list, for example, will certainly lead to incorrect results.

The choice of the above list of variables is equivalent to the statement that there is a particular physical law which can be written in the implicit form

$$f(m,g,h,t) = 0 \tag{3.9}$$

Now this function must be the same, whatever system of units is employed; i.e., it must yield the value of t which is independent of the size of the units of mass, length, and time.[1] A change in the size of the mass unit, however, influences only the magnitude of the quantity m, and this change cannot be compensated by a change in the size of length and time units. Apparently the quantity m cannot enter the function (3.9) at all, because otherwise a change in the mass unit will affect the value of t. Therefore the function must reduce to

$$f(g,h,t) = 0 \tag{3.10}$$

This result, that the time of fall is independent of the mass of the object, is of course a result of Galileo's legendary experiment involving the dropping of cannonballs of different sizes.

Consider now the unit of length. The function will be uninfluenced by the size of this unit only if g and h occur as the combination g/h, so that the length units cancel. Thus

$$f\left(\frac{g}{h}, t\right) = 0 \tag{3.11}$$

Finally, with the dimensional statements $[g/h] = 1/T^2$ and $[t] = T$, these quantities can enter only in the ultimate nondimensional combination gt^2/h, so that

$$f\left(\frac{gt^2}{h}\right) = 0 \tag{3.12}$$

[1] An example of a function which violates this requirement is $h - 16t^2 = 0$, an expression for the free fall of a body from rest which happens to be valid only in the English system of units, with the experiment carried out near sea level on the planet earth.

This has a solution,[1] which we assume is unique,

$$\frac{gt^2}{h} = K \tag{3.13}$$

where K is a universal constant. From elementary mechanics, $K = 2$.

The reduction from four variables in (3.9) to one variable in (3.12) by dimensional reasoning is a great practical advance. Suppose that the physical law corresponding to (3.9) were to be discovered by a series of experiments; though it might soon be realized that mass was not involved, a very large number of experiments would still be required to arrive at (3.13). Starting from (3.13), however, only *one* experiment is required to fix the value of K. Similarly, in the mathematical analysis of complicated problems, it is a great advantage to be able to reduce the number of variables entering the problem in the very beginning.

The essence of dimensional analysis is the formation of nondimensional quantities such as gt^2/h from a list of dimensional quantities such as m, g, h, t, with a resulting functional form such as $f(gt^2/h) = 0$. We have seen that the necessity for this nondimensional formulation is implicit in the basic axiom, the requirement that physical laws be unit-free.

The Pi theorem

In problems more complicated than the one discussed above, several nondimensional quantities (dimensionless groups) may appear in the final dimensionless form. The pi theorem is a general statement predicting the number of dimensionless groups required in any particular problem (the name has no connection with the circumference-to-diameter ratio of a circle but arises from the symbol Π used for dimensionless groups by Buckingham, who first proved the theorem).

Let the list of dimensional quantities which enter a problem be

$$X_1, X_2, \ldots, X_i, \ldots, X_n$$

Some of these quantities may be fixed, and some may vary in the problem under consideration; for example, X_1 might stand for the (constant) thermal conductivity κ, and X_2 might stand for the spatial coordinate x.

[1] Specific examples of a functional statement of this kind are

$$f(x) = x \tan x - x^3 = 0$$
$$f(x) = x^2 - 2x + 1 = 0$$

The second example has the unique solution (the roots of the equation are repeated) $x = 1$.

We will refer to the X's as the dimensional variables. The physical law which describes the problem is represented by

$$f(X_1, X_2, \ldots, X_n) = 0 \tag{3.14}$$

In general only r ($r \leq n$) of the variables have independent dimensions. This means that the dimensions of one variable cannot be formed as a product of powers of the dimensions of the others; e.g., the dimensions L, L/T, and M/LT are independent, while the dimensions L, L/T, and L/T^2 are not independent. The number of variables r with independent dimensions evidently cannot exceed the number of fundamental dimensions f which occur in the problem (e.g., if M, L, T, and θ occur, $f = 4$), or

$$r \leq f \tag{3.15}$$

Let the list of variables be so arranged that the r dimensionally independent quantities occur first in the list

$$X_1, X_2, \ldots, X_r, X_{r+1}, \ldots, X_n$$

Then by our definition of dimensional independence, the dimensions of the last $n - r$ quantities can be written as a product of powers of the first r quantities,

$$[X_i] = [X_1]^{\alpha_1}[X_2]^{\alpha_2} \cdots [X_r]^{\alpha_r} \tag{3.16}$$

Thus we can form the $n - r$ dimensionless groups Π_i,

$$\Pi_i = \frac{X_i}{X_1^{\alpha_1} X_2^{\alpha_2} \cdots X_r^{\alpha_r}} \tag{3.17}$$

The physical law is then in the form

$$\phi(\Pi_1, \Pi_2, \ldots, \Pi_{n-r}) = 0 \tag{3.18}$$

as required by the condition that the law must be unit-free; for a formal proof of the equivalence of (3.14) and (3.18), see *Birkhoff* [1960, chap. 4].

The pi theorem is just the above result, that $n - r$ independent dimensionless variables are implied by a list of n dimensional variables. Let $p = n - r$ be the number of dimensionless variables (number of Π's): from (3.15), $r \leq f$, so that p satisfies

$$\boxed{p \geq n - f} \tag{3.19}$$

In practical problems, the number of dimensionally independent variables r is almost always equal to the number of fundamental dimensions, and the above expression is an equality.

By means of linear algebra, a formal apparatus can be developed for isolating dimensionally independent variables and formulating the dimensionless variables Π_i. Except perhaps in very complicated problems, however, the dimensionless variables can be written down by inspection. Some examples follow.

EXAMPLE 3.1 FALLING SPHERE WITH FLUID RESISTANCE

Following Aristotle, we may suppose that the time of fall for the dropped stone already considered *is* in fact influenced by its mass. Specifically, if the falling stone is sufficiently light in weight, we expect the time of fall to be influenced by fluid resistance. For definiteness, consider a sphere of diameter d; the mass of the sphere will be contained implicitly in our list of variables by including the density ρ_s and the diameter d. The fluid resistance should depend on the viscosity μ and the fluid density ρ_f. Thus the dimensional variables are expected to be

h =	height	$[h] = L$
d =	diameter of sphere	$[d] = L$
g =	acceleration of gravity	$[g] = LT^{-2}$
t =	time of fall	$[t] = T$
ρ_s =	density of sphere	$[\rho_s] = ML^{-3}$
ρ_f =	density of fluid	$[\rho_f] = ML^{-3}$
μ =	viscosity of fluid	$[\mu] = ML^{-1}T^{-1}$

Note that $n = 7, f = 3$. In the previous example we found the dimensionless variable $\Pi_1 = gt^2/h$. In addition, we can form by inspection $\Pi_2 = d/h$ and $\Pi_3 = \rho_s/\rho_f$. Then the only dimensional quantity which has not been incorporated into a dimensionless variable is the viscosity μ. By virtue of the dimensions of μ, a dimensionless variable can be formed by multiplying μ by a length l_0 and a time t_0 and dividing by a mass m_0, forming $\mu l_0 t_0/m_0$. Choosing $l_0 = d, t_0 = \sqrt{h/g}$ (aside from a multiplicative constant, the time of free fall), and $m_0 = \rho_s d^3$ yields $\Pi_4 = \mu/(\rho_s d^2\sqrt{g/h})$. We thus have the nondimensional variables

$$\frac{gt^2}{h} \qquad \frac{d}{h} \qquad \frac{\rho_s}{\rho_f} \qquad \frac{\mu}{\rho_s d^2\sqrt{g/h}}$$

These variables are *independent*, in the sense that none can be formed as a product of the remaining three, and they are four in number, as predicted by the pi-theorem statement (3.19); $4 \geq 7 - 3$. By trial, no additional independent nondimensional variable can be formed.

The above list is not unique, because any product of these variables is also a possible (substitute) variable. For example, taking the square root of the first variable and multiplying the last by the middle two, then inverting, yields the new list,

$$t\sqrt{\frac{g}{h}} \qquad \frac{d}{h} \qquad \frac{\rho_s}{\rho_f} \qquad \frac{\rho_f d\sqrt{gh}}{\mu}$$

The last variable is now in the form of a *Reynolds number*, a dimensionless group which plays a crucial role in fluid mechanics, as will be discussed in Sec. 3.5.

The essential result of the dimensional analysis is that there is a universal law connecting the nondimensional variables. Thus, for example, the dimensionless time required for a sphere falling through glycerin on the moon might be used to predict the time for a spherical pollen grain falling through air on the earth, provided that the last three nondimensional variables have the same values in both experiments.

EXAMPLE 3.2 MIXING OF TWO LIQUIDS

Suppose that it is desired to find the density ρ of a mixture of two pure liquids a and b, with densities ρ_a and ρ_b. Let the *mass* fraction of liquid a be x_a; for example, with a mixture of 2 kg ethyl alcohol (a) and 3 kg water (b), $x_a = 0.40$. The density ρ is presumably completely determined by ρ_a, ρ_b, and x_a. Thus, for some particular pair of liquids,

$$f(\rho,\rho_a,\rho_b,x_a) = 0$$

The mass fraction x_a is already nondimensional. In addition, we can immediately form (say) ρ/ρ_a and ρ_b/ρ_a; thus the nondimensional variables may be taken to be

$$\frac{\rho}{\rho_a} \qquad \frac{\rho_b}{\rho_a} \qquad x_a$$

This is an *exceptional* case (and has been chosen just for that reason) in which the inequality in (3.19) is honored: with $n = 4$, $f = 2$, and $p = 3$ we have $3 \geq 4 - 2$. This particular dimensional analysis does not seem to have any practical value, because ρ_a and ρ_b are already fixed by the choice of liquids and the reduction in number of variables is only an apparent reduction.

Incidentally, experience suggests that a *linear* law

$$\rho = \rho_b + x_a(\rho_a - \rho_b) \tag{3.20}$$

is unlikely to be encountered in practice; e.g., in mixing alcohol and water, total liquid volume is not constant. The nondimensional form is, however,

$$\frac{\rho}{\rho_a} = \frac{\rho_b}{\rho_a} + x_a\left(1 - \frac{\rho_b}{\rho_a}\right) \tag{3.21}$$

In performing a given dimensional analysis, two judgments are necessary at the very beginning. The first is the selection of pertinent variables; in the problem of the falling stone, for example, we obtain additional nondimensional variables by including the effects of fluid resistance. Had we decided that heat conduction also played a role, we would have obtained even more nondimensional variables. The selection of a list of dimensional variables is evidently a matter of experience and luck and is ultimately justified only by experiment.

The second judgment is the selection of the fundamental dimensions appropriate to the problem. In fluid mechanics the dimensions M, L, T, θ usually occur, and this involves some tacit assumptions. For example, omission of the dimension F for force is possible via Newton's second law by setting g_c equal to the pure number unity; this implies that Newton's law plays a role in the motion, a condition which is almost always satisfied. Suppose, however, that we wished to retain F as a fundamental dimension and that Newton's law does play a role; then we must include g_c in our list of dimensional constants. From the pi theorem $p \geq n - f$, and both n and f have each been increased by 1, so that the final number of dimensionless variables p is unchanged.

It can be seen that the selection of fundamental dimensions amounts to an assumption concerning which physical laws are relevant to the problem at hand. In a given problem, the velocity of light c in free space can be included explicitly in our list of dimensional quantities or included implicitly by setting c to unity and thus eliminating either L or T as a fundamental dimension; the procedures are equivalent, as noted above in connection with g_c. If, however, we omit the velocity of light from consideration, we have consciously or unconsciously made a judgment that relativistic effects play a negligible role in the problem. In this way the value of f is in effect increased by 1, and we are rewarded by the appearance of one less dimensionless variable.

As a more immediate example, consider the transfer of heat by convection, as when air passes over a heated metal plate. In many such problems, interconversion of mechanical energy and thermal energy is negligible, i.e., stress-velocity work can be neglected. Then we can treat thermal energy E as a separate fundamental dimension, with units of, for example, calories or Btu's. The number of fundamental dimensions is thereby increased by one, and the utility of dimensional analysis is increased, since one less dimensionless variable will appear (see also *Sedov* [1959, p. 20]).

Thus dimensional analysis is more useful, the greater the effective number of fundamental dimensions f, in that the number of final dimensionless variables $p \geq n - f$ is accordingly reduced. Although it is logically defensible, as explained in Sec. 3.3, to ultimately have a system of units with no dimensions whatever, it is both inconvenient and unnecessary in the problems of interest here to pursue such a course. The M, L, T, θ system will meet our needs.

EXAMPLE 3.3 A UNIVERSAL VISCOSITY LAW?

The Lennard-Jones (6-12) intermolecular potential is shown in Fig. 3.2 and given by the equation

$$\varphi(r) = 4\varepsilon\left[\left(\frac{\sigma}{r}\right)^{12} - \left(\frac{\sigma}{r}\right)^{6}\right]$$

where r is the intermolecular separation. By suitable choice of ε and σ, this formula *approximately* represents the force potential between various molecules.

It is known that the shear viscosity μ of a pure dilute gas depends only on temperature and of course the molecular properties. The dynamics of an intermolecular collision (and therefore ultimately the viscosity) are controlled by the potential, as represented by the parameters ε and σ, the molecular mass m, and the relative molecular speed. Since the average molecular speed is characterized by the temperature, we might expect a functional relation $\mu = \mu(m,\varepsilon,\sigma,T)$. But in this case only the quantity T has dimensions of temperature, and we cannot form a dimensionless variable involving T. This is clearly wrong, but the

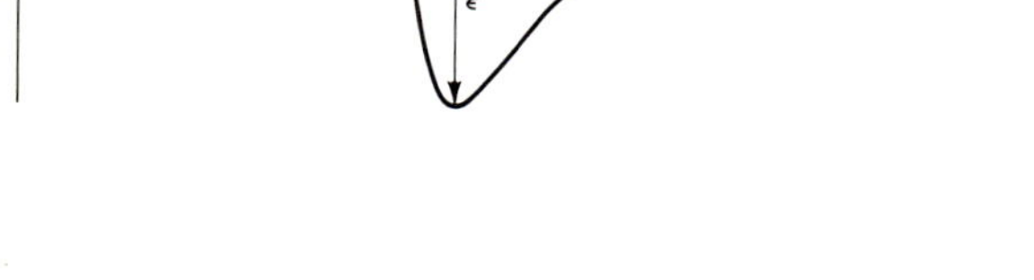

Figure 3.2

difficulty can be remedied by including the Boltzmann constant k in the dimensional variables (note that in the Maxwell distribution T always appears in the product kT). Thus

$$f(\mu, m, \varepsilon, \sigma, kT) = 0 \tag{3.22}$$

represents, at least approximately, a law for the dependence of viscosity.

We can now form the nondimensional variables $\mu\sigma^2/\sqrt{m\varepsilon}$ and kT/ε, with a corresponding universal viscosity law for dilute gases,

$$\frac{\mu\sigma^2}{\sqrt{m\varepsilon}} = F\left(\frac{kT}{\varepsilon}\right) \tag{3.23}$$

Experimental and theoretical data in support of such a law are shown by *Hirschfelder et al.* [1964, p. 560].

Dimensionless Form of a Known Physical Law

When a physical law is already stated in the form of an explicit equation, with given boundary conditions, straightforward reduction will yield an explicit dimensionless form. The problem of choosing relevant dimensional variables does not enter, this judgment having been made at an earlier stage, when the physical law was first derived. The reduction is easy and will yield important results.

In particular, we are interested in the basic differential equations of fluid mechanics, expressing the conservation of mass, momentum, energy, and entropy. Although we are ultimately interested in *compressible* fluids, for simplicity let us first consider the flow of an incompressible fluid. This will allow us to isolate the Reynolds number, a fundamental dynamic parameter for all types of flow.

For an incompressible fluid with constant viscosity and no body force, the continuity equation (1.62) and momentum equation (1.66) become respectively

$$\nabla \cdot \mathbf{u} = 0 \tag{3.24}$$

$$\frac{\partial \mathbf{u}}{\partial t} + \mathbf{u} \cdot \nabla \mathbf{u} = -\frac{1}{\rho} \nabla P + \nu \nabla^2 \mathbf{u} \tag{3.25}$$

where $\nu = \mu/\rho$ is the kinematic viscosity. It is assumed that constant reference values l_0, u_0, t_0 enter the problem in the form of boundary conditions; e.g., in the case of a body moving through the atmosphere, l_0 may be a characteristic dimension of the body (such as the diameter of a sphere, or the length of an airplane), u_0 may be the speed of the body, and t_0 may be the period of oscillation of some part of the body. We can form the nondimensional, or *normalized*, variables

$$\mathbf{X} = \frac{\mathbf{x}}{l_0} \qquad \tau = \frac{t}{t_0} \qquad \mathbf{U} = \frac{\mathbf{u}}{u_0} \qquad \tilde{P} = \frac{P}{\rho {u_0}^2} \tag{3.26}$$

Some variation in these forms is possible; since there is an alternative characteristic time $t_0' = l_0/u_0$, we could form $\tau = t/t_0'$; in aeronautical practice, with some uniform atmospheric pressure P_0 given, it is usual to form the *pressure coefficient* $C_p \equiv (P - P_0)/\frac{1}{2}\rho {u_0}^2$.

The equations of motion can now be rewritten in terms of the nondimensional variables in (3.26). Thus

$$\frac{\partial}{\partial t} = \frac{1}{t_0'}\frac{\partial}{\partial \tau}$$

and in Cartesian coordinates,

$$\frac{\partial}{\partial x_i} = \frac{1}{l_0}\frac{\partial}{\partial X_i}$$

Eqs. (3.24) and (3.25) become

$$\frac{u_0}{l_0}\nabla \cdot \mathbf{U} = 0$$

$$\frac{u_0}{t_0}\frac{\partial \mathbf{U}}{\partial \tau} + \frac{{u_0}^2}{l_0}\mathbf{U} \cdot \nabla \mathbf{U} = -\frac{{u_0}^2}{l_0}\nabla \tilde{P} + \frac{\nu u_0}{{l_0}^2}\nabla^2 \mathbf{U}$$

where ∇ is now the gradient operator in the dimensionless coordinate system, $\nabla = \mathbf{e}_i\, \partial/\partial X_i$. Multiplying these equations through by l_0/u_0 and $l_0/{u_0}^2$, respectively, yields the nondimensional equations of motion

$$\nabla \cdot \mathbf{U} = 0 \tag{3.27}$$

$$\mathrm{St}\frac{\partial \mathbf{U}}{\partial \tau} + \mathbf{U} \cdot \nabla \mathbf{U} = -\nabla \tilde{P} + \frac{1}{\mathrm{Re}}\nabla^2 \mathbf{U} \tag{3.28}$$

where the constant nondimensional coefficients are

$$\text{Strouhal number:} \quad \mathrm{St} \equiv \frac{l_0}{u_0 t_0} \tag{3.29}$$

$$\text{Reynolds number:} \quad \mathrm{Re} \equiv \frac{l_0 u_0}{\nu} \tag{3.30}$$

It is the Reynolds number which is of primary interest. All the dimensionless quantities Re, St, X, τ, $\mathbf{U}$, and $\tilde{P}$ could have been formed, of course, just from a list of the original dimensional quantities. It will, however, be a great advantage to have the explicit nondimensional equations available.

The boundary conditions must also be expressible in nondimensional form. As a simple example, suppose that a sphere of diameter l_0 translates at constant velocity $-\mathbf{u}_0$ through a uniform fluid. It is convenient to consider this as a *streaming flow*, from the viewpoint of an observer moving with the sphere (Galilean transformation): in this reference frame the flow is presumed steady. From the no-slip condition at the surface of the sphere

$$\mathbf{U} = 0 \qquad \text{at } X = \tfrac{1}{2}$$

for a coordinate system with origin at the center of the sphere. Far from the sphere, we have the uniform conditions

$$U = 1 \qquad \text{at } X = \infty$$

$$\tilde{P} = \frac{P_0}{\rho u_0^2} \qquad \text{at } X = \infty$$

In this example the first (unsteady) term in (3.28) is zero.

Now consider the steady flow about two different spheres, designated a and b, in different fluids and with different diameters and flow speeds. The boundary conditions just given apply equally to both flows. The differential equations (3.27) and (3.28) will be exactly the same for both flows *provided* that

$$\text{Re}_a = \text{Re}_b \tag{3.31}$$

That is, the fluid motions a and b have *formally identical descriptions* if the Reynolds number is the same in both cases. It follows that the motions themselves are identical, as measured by the nondimensional variables.[1] Flow a is said to be a *model* of flow b, and vice versa; an equivalent statement is that the two flows are *dynamically similar*.

This result has great practical significance, in that it allows the testing of (say) full-scale airplanes or turbine blades by the use of models, on more

[1] It is of course assumed that the mathematical description is *complete* and *accurate*, conditions which are not easy to achieve in practice. The statement given is based on the idea of physical determinism, an idea on which all of science is dependent. For a discussion, see *Born* [1949].

convenient and economical size scales. In the same vein, analytical solutions can be applied to a variety of physical situations.

In order to construct a valid model experiment, it is not even necessary that the experiment bear any physical resemblance to the phenomenon being modeled. For example, if we could find an electrical system which is described by the nondimensional equations (3.27) and (3.28), and which also satisfies appropriate nondimensional boundary conditions, then we could "model" fluid motions by the use of the electrical system. Such devices are called *analogs*, and are of course of practical importance.

In another aspect of the problem of modeling, the Reynolds number, and some other nondimensional parameters are important in establishing the qualitative nature of a flow. Consider the effect of gradually increasing the Reynolds number in a given flow, e.g., by increasing the free-stream speed u_0. The relative magnitude of the viscous term in (3.25) decreases with increasing Re. With the change in the character of the differential equation, there is a corresponding change in the nature of the motion. In the original experiments of Osborne Reynolds, flow within a pipe was observed to change from *laminar* (smooth, filamentous) to *turbulent* (irregular, chaotic) at a value Re $\approx 10^4$. For a general discussion of turbulence, see *Landau and Lifshitz* [1959, p. 102] or *Schlichting* [1968, p. 431].

The relative importance of various terms in the differential equations and the possible influence of certain nondimensional parameters on the character of the motion will be discussed further below.

3.5 Magnitude estimates

Real physical flows are far too complicated for complete mathematical analysis. It is therefore necessary to consider idealized models of reality, which leave out many real effects. For example, the flow around an airfoil may be treated according to the idealization that the fluid is incompressible and inviscid, endowed only with the physical property of inertia (in Batchelor's phrase[1]).

Such drastic approximations, if judiciously chosen, may be remarkably successful in describing real flows. The object of this section is to form a more or less rational basis for such approximations, i.e., a basis for casting terms out of the complete equations of motion.

[1] *Batchelor* [1967, p. 172].

The hierarchy of the notations of equality as used in this book is as follows:

$\equiv$ is defined to be
$=$ equals
$\approx$ is approximately equal to
$\sim$ is of the order of magnitude of
$\propto$ is proportional to

It will be convenient to make use of nondimensional equations. In forming such equations, it is assumed that we begin with physical laws expressed in dimensionally homogeneous form; i.e., all the additive terms in the physical law have the same dimensions; colloquially, we add "apples and apples, and not apples and oranges" (for a counterexample, however, see *Bridgman* [1931, p. 42]). Then a nondimensional form of the physical law can usually be constructed essentially by multiplying through by certain dimensional constants, as, for example, in deriving Eq. (3.28).

Now if the normalizing constants are artfully selected, the derivatives appearing in the final nondimensional differential equation may have an order of magnitude of roughly unity. In the flow sketched in Fig. 3.3*a*, the velocity gradients have magnitude roughly in the range from zero to u_0/l_0. In Fig. 3.3*b*, the velocity gradients in the region outside of the boundary layer are also roughly in the range from zero to u_0/l_0; the gradient within the boundary layer is approximately u_0/δ, which may be considerably larger than u_0/l_0. Then *provided that* u_0 *and* l_0 *are suitably chosen*, we *estimate* the maximum value of the velocity derivative to be

$$\frac{\partial u_i}{\partial x_k} \sim \frac{u_0}{l_0} \tag{3.32}$$

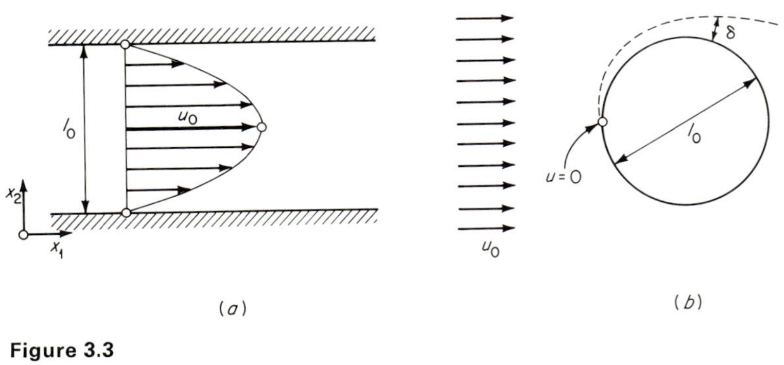

Figure 3.3

where the tilde $\sim$ is read "is of the order of magnitude." Then if u is normalized with respect to u_0 and x with respect to l_0, the nondimensional derivative has

$$\frac{\partial U_i}{\partial X_k} \sim 1 \tag{3.33}$$

A similar result often applies to second derivatives. In Fig. 3.3*a*, u_1 changes from u_0 to zero (at the stagnation point, say) in a distance Δx_1 of order l_0; thus $\partial u_1/\partial x_1$ changes from zero to about u_0/l_0 in a distance of order l_0, so that

$$\frac{\partial^2 u_1}{\partial {x_1}^2} \sim \frac{u_0/l_0 - 0}{l_0} = \frac{u_0}{{l_0}^2}$$

or

$$\frac{\partial^2 U_1}{\partial {X_1}^2} \sim 1$$

The Creeping Limit and Inviscid Limit for Incompressible Flow

For steady incompressible flow, the nondimensional momentum equation (3.28) becomes

$$\mathbf{U} \cdot \nabla \mathbf{U} + \nabla \tilde{P} = \frac{1}{\mathrm{Re}} \nabla^2 \mathbf{U} \tag{3.34}$$

The velocity derivatives are roughly of order unity. Because the pressure is normalized by $\rho {u_0}^2$, the pressure gradient will also be of order unity for flows described by the Bernoulli equation, $P + \frac{1}{2}\rho u^2 = \text{const}$, i.e., for flows not dominated by viscosity or unsteadiness. Thus if $\mathrm{Re} \gg 1$, the last term, of order $1/\mathrm{Re}$, is negligible relative to the others and (3.34) reduces to the *inviscid* equation

$$\mathbf{U} \cdot \nabla \mathbf{U} + \nabla \tilde{P} = 0 \qquad \mathrm{Re} \gg 1 \tag{3.35}$$

Thus, although real fluids are viscous, it will be possible in certain situations (high speed, large characteristic dimension, etc.) to treat them according to the inviscid approximation.

Although gasdynamic flows are strictly not described by (3.34), we can take these results to be indicative. The Reynolds numbers in steady gasdynamic flows, in regions away from boundary layers, are *commonly* in the range 10^5 to 10^8, and the inviscid approximation appears to be well justified.

At the other end of the Reynolds number scale, Re $\ll$ 1, the right-hand (viscous) term becomes much larger than the acceleration $\mathbf{U} \cdot \nabla \mathbf{U}$, which can then be neglected, yielding

$$\nabla \tilde{P} = \frac{1}{\text{Re}} \nabla^2 \mathbf{U} \qquad \text{Re} \ll 1 \tag{3.36}$$

In this case the fluid is in essentially static equilibrium between pressure and viscous forces; the effect of motion appears only in the viscous forces. It should be noted that $\nabla \tilde{P}$ is no longer of order unity, because the Bernoulli equation does not apply. Equation (3.36) is called the equation of *creeping motion* or *Stokes flow*. The very rough nature of the magnitude estimation procedure is reflected in the fact that (3.36) is found in practice to hold fairly well up to Reynolds numbers in the order of 3 or 4, say, rather than the small values required by the condition Re $\ll$ 1.

We should remark that (3.36) is *exact* for flows without acceleration, even if the Reynolds number is large. An example of such a case is *Poiseuille flow*, the laminar flow of incompressible fluid in a constant-area pipe.

A Criterion for Compressibility

It has already been noted, at the beginning of Chap. 1, that although all real fluids are compressible, this property may often be neglected. We now seek to determine the conditions under which compressibility, as manifested in fluid-density changes, must be considered. The continuity equation is

$$\frac{1}{\rho} \frac{D\rho}{Dt} + \nabla \cdot \mathbf{u} = 0 \tag{3.37}$$

We will estimate the relative magnitude of $\nabla \cdot \mathbf{u} = -\dfrac{1}{\rho} \dfrac{D\rho}{Dt}$.

It is noted that we cannot simply put $\nabla \cdot \mathbf{u} \sim u_0/l_0$; in Cartesian coordinates, $\nabla \cdot \mathbf{u} = \partial u_1/\partial x_1 + \partial u_2/\partial x_2 + \partial u_3/\partial x_3$, and *individual* terms are expected to be of order u_0/l_0 but may be of different signs and sum, for example, to zero. In fact, we will take the *criterion* for nonnegligible compressibility to be that $\nabla \cdot \mathbf{u}$ is of order u_0/l_0, for if $\nabla \cdot \mathbf{u} \sim u_0/l_0$, velocity gradients are then necessarily dependent on fluid-density changes. For example, consider a gas confined in a cylinder between two pistons, a distance l_0 apart. If the pistons move toward each other at relative velocity u_0 and we imagine the gas motion to be just one-dimensional, then the velocity gradient is in the direction of motion, with magnitude u_0/l_0, and is associated with compression of the gas, with $\nabla \cdot \mathbf{u} \sim u_0/l_0$. This is the motion described in Example 1.3 (page 24); it may be contrasted with a simple shearing motion described in Example 1.2 (page 21), in which $\nabla \cdot \mathbf{u} = 0$ and the velocity gradient has no associated fluid-density change.

In calculating the value of $\nabla \cdot \mathbf{u}$ we will make use of the continuity, momentum, and energy equations in the forms (1.62), (1.66), and (2.15), viz.,

$$\nabla \cdot \mathbf{u} = -\frac{1}{\rho}\frac{D\rho}{Dt} \tag{3.38}$$

$$\rho \frac{D\mathbf{u}}{Dt} + \nabla P = \rho \mathbf{G} + \mu \nabla^2 \mathbf{u} + (\mu_v + \tfrac{1}{3}\mu)\nabla(\nabla \cdot \mathbf{u}) \tag{3.39}$$

$$\frac{Ds}{Dt} = \frac{v\Upsilon}{T} + \frac{v\kappa}{T}\nabla^2 T \tag{3.40}$$

in which the transport properties have been taken as constants. Considering $P = P(\rho,s)$ we can write

$$\frac{DP}{Dt} = \left(\frac{\partial P}{\partial \rho}\right)_s \frac{D\rho}{Dt} + \left(\frac{\partial P}{\partial s}\right)_\rho \frac{Ds}{Dt} \tag{3.41}$$

The derivative $(\partial P/\partial \rho)_s$ plays an important role in fluid mechanics and will be identified in Chap. 4 as the *square of the sound speed.* With the symbol c for the speed of sound,

$$c^2 \equiv \left(\frac{\partial P}{\partial \rho}\right)_s \tag{3.42}$$

rearranging (3.41) yields

$$\nabla \cdot \mathbf{u} = -\frac{1}{\rho}\frac{D\rho}{Dt} = \frac{1}{\rho c^2}\left[-\frac{\partial P}{\partial t} - \mathbf{u} \cdot \nabla P + \left(\frac{\partial P}{\partial s}\right)_\rho \frac{Ds}{Dt}\right]$$

Substituting Ds/Dt from (3.40) and the value of $\mathbf{u} \cdot \nabla P$ found from (3.39) after solving for ∇P yields

$$\begin{aligned}\nabla \cdot \mathbf{u} = &-\frac{1}{\rho c^2}\frac{\partial P}{\partial t} + \frac{1}{c^2}\frac{D}{Dt}\frac{u^2}{2} - \frac{\mathbf{u} \cdot \mathbf{G}}{c^2} \\ &-\frac{\mu}{\rho c^2}\mathbf{u} \cdot \nabla^2 \mathbf{u} - \frac{\mu_v + \frac{1}{3}\mu}{\rho c^2}\mathbf{u} \cdot \nabla(\nabla \cdot \mathbf{u}) \\ &+\frac{1}{\rho^2 c^2 T}\left(\frac{\partial P}{\partial s}\right)_\rho (\Upsilon + \kappa \nabla^2 T)\end{aligned} \tag{3.43}$$

The coefficient of the final term can be simplified. By definition $1/c^2 = (\partial\rho/\partial P)_s$, and by calculus

$$\left(\frac{\partial \rho}{\partial P}\right)_s \left(\frac{\partial P}{\partial s}\right)_\rho = -\left(\frac{\partial \rho}{\partial s}\right)_p = -\left(\frac{\partial \rho}{\partial T}\right)_p \left(\frac{\partial T}{\partial s}\right)_p$$

With

$$\left(\frac{\partial \rho}{\partial T}\right)_p = -\frac{1}{v^2}\left(\frac{\partial v}{\partial T}\right)_p \qquad \text{and} \qquad \left(\frac{\partial T}{\partial s}\right)_p = \frac{T}{c_p}$$

we have finally

$$\frac{1}{\rho^2 c^2 T}\left(\frac{\partial P}{\partial s}\right)_\rho = \frac{1}{c_p}\left(\frac{\partial v}{\partial T}\right)_p$$

Equation (3.43) becomes finally, after collecting terms,

$$\begin{aligned}\nabla \cdot \mathbf{u} = {} & \frac{1}{2c^2}\mathbf{u} \cdot \nabla \mathbf{u}^2 - \frac{\mathbf{u} \cdot \mathbf{G}}{c^2} + \frac{1}{c^2}\left[\frac{1}{2}\frac{\partial u^2}{\partial t} - \frac{1}{\rho}\frac{\partial P}{\partial t}\right] \\ & - \frac{\mu}{\rho c^2}\left[\mathbf{u} \cdot \nabla^2 \mathbf{u} + \left(\frac{\mu_v}{\mu} + \frac{1}{3}\right)\mathbf{u} \cdot \nabla(\nabla \cdot \mathbf{u})\right] \\ & + \frac{1}{c_p}\left(\frac{\partial v}{LT}\right)_p [\Upsilon + \kappa \nabla^2 T] \end{aligned} \qquad (3.44)$$

In the particular problem under consideration, it is assumed that a characteristic velocity u_0, a characteristic length l_0, and a characteristic time t_0 appear and that velocity changes of order u_0 occur in distance l_0 and time t_0. (In steady flows, however, t_0 will not appear, and in exceptional cases discussed in Sec. 3.6, neither l_0 nor t_0 will appear.) The full list of dimensional constants is now assumed to be

$$u_0, \quad l_0, \quad t_0, \quad \rho_0, \quad c_0, \quad T_0, \quad \mu, \quad \kappa, \quad c_p$$

For example, u_0 may be the free-stream velocity in the flow over an airfoil, l_0 may be the airfoil chord, T_0 may be the temperature difference between the airfoil and the free stream, and the remaining constants may be the free-stream fluid properties. For a steady flow we may formally take t_0 to be infinity. In an unsteady flow, t_0 might be the period of oscillation of a piston and u_0 the velocity amplitude of the piston.

We now form nondimensional variables, such as $\mathbf{U} = \mathbf{u}/u_0$, $\mathbf{X} = \mathbf{x}/l_0$, and so on. There is a variety of choices available, however, for normalizing the pressure; e.g., we could form $P/\rho_0 u_0^2$ or $P/\rho_0 c_0^2$. Noting that the pressure appears only in an *unsteady* term, and anticipating a result of

Chap. 4 that pressure disturbances propagate at the speed of sound c, we can argue from momentum as follows: for an observer traveling with a pressure disturbance, i.e., with a wave, there is a relative fluid mass flux ρc entering the disturbance. If the disturbance produces a velocity change δu, then the associated pressure change δP must be $\delta P \approx \rho c\, \delta u$, that is, the mass flux times the velocity change. We will expand on this rather sketchy argument in Chap. 4 and use it here to suggest that P should be normalized by $\rho_0 c_0 u_0$ (with $\delta u \sim u_0$ in the above). Thus we form the following nondimensional variables:

$$
\begin{aligned}
\mathbf{U} &= \frac{\mathbf{u}}{u_0} & \tilde{\mathbf{G}} &= \frac{\mathbf{G} l_0}{c_0^2} \\
\mathbf{X} &= \frac{\mathbf{x}}{l_0} & \tilde{P} &= \frac{P}{\rho_0 c_0 u_0} \\
\tau &= \frac{t}{t_0} & \tilde{\Upsilon} &= \frac{l_0^2}{\mu u_0^2}\Upsilon \\
\tilde{T} &= \frac{T}{T_0}
\end{aligned}
\tag{3.45}
$$

The nondimensional form of (3.44) is found by substitution to be, with ∇ standing now for the gradient in the dimensionless coordinate system,

1 $$\nabla \cdot \mathbf{U} = \frac{c_0^2}{c^2}\left[\frac{M_0^2}{2}\mathbf{U}\cdot\nabla U^2 - \mathbf{U}\cdot\tilde{\mathbf{G}}\right]$$

2 $$+ \frac{c_0^2}{c^2}\frac{l_0}{c_0 t_0}\left[\frac{M_0}{2}\frac{\partial U^2}{\partial \tau} - \frac{\rho_0}{\rho}\frac{\partial \tilde{P}}{\partial \tau}\right]$$

3 $$- \frac{\rho_0 c_0^2}{\rho c^2}\frac{M_0^2}{\mathrm{Re}_0}\left[\mathbf{U}\cdot\nabla^2\mathbf{U} + \left(\frac{\mu_v}{\mu} + \frac{1}{3}\right)\mathbf{U}\cdot\nabla(\nabla\cdot\mathbf{U})\right]$$

4 $$+ \frac{1}{\mathrm{Re}_0}\frac{T_0}{v_0}\left(\frac{\partial v}{\partial T}\right)_p\left[\frac{u_0^2}{c_p T_0}\tilde{\Upsilon} + \frac{1}{\mathrm{Pr}}\nabla^2\tilde{T}\right] \tag{3.46}$$

where the nondimensional parameters which have been represented by special symbols are

$$\text{Mach number:} \quad M_0 \equiv \frac{u_0}{c_0}$$

$$\text{Reynolds number:} \quad \text{Re}_0 \equiv \frac{\rho_0 u_0 l_0}{\mu} \tag{3.47}$$

$$\text{Prandtl number:} \quad \text{Pr} \equiv \frac{\mu c_p}{\kappa}$$

The criterion for significant compressibility is $\nabla \cdot \mathbf{u} \sim u_0/l_0$ or $\nabla \cdot \mathbf{U} \sim 1$. In general this will be satisfied if any one of the terms on the right-hand side of (3.46) is of order unity (thus, with eight opportunities for nonnegligible compressibility, it is a little surprising that there are *any* cases for which the incompressible approximation is defensible).

The dimensionless variables in (3.46) have been chosen such that each of the derivatives on the right-hand side (including also the dissipation $\tilde{\Upsilon}$) is very roughly of order unity. Then the magnitude of each term is established approximately by its coefficient. The value of c_0/c and ρ_0/ρ, the ratio of the reference to local thermodynamic property, will be of order unity; we are therefore mainly concerned with the values of the dimensionless parameters

$$M_0, \quad \tilde{G}, \quad \frac{l_0}{c_0 t_0}, \quad \text{Re}_0, \quad \frac{\mu_v}{\mu}, \quad \frac{T_0}{v_0}\left(\frac{\partial v}{\partial T}\right)_p, \quad \frac{u_0^2}{c_p T_0}, \quad \text{Pr}$$

We now consider the terms in (3.46) line by line.

The first term in line 1 derives from the convective acceleration and is normally the dominant term in steady flow. Since it is of order M_0^2, the square of the Mach number, it will be important only if the characteristic velocity u_0 is at least of the order of the sound speed c_0, a condition which is fulfilled only for high-speed flows. For example, a light utility airplane with flight speed $u_0 = 150$ mi/h $= 220$ ft/s moving through air with $c_0 = 1{,}100$ ft/s gives $M_0 = 0.2$, $M_0^2 = 0.04$, and compressibility may be neglected. Because of the importance of this term, the *Mach number is often referred to as a measure of the importance of compressibility.* We will consider this further in Chaps. 5 and 6.

The second term in line 1 represents the effect of body force, the most important example of which is gravity. Since U is of order unity, the term has the order of magnitude of $\tilde{G} = Gl_0/c_0^2$; with G equal to the acceleration of gravity g, the term is important if $gl_0/c_0^2 \sim 1$ or $l_0 \sim c_0^2/g$. In the earth's atmosphere c_0^2/g is roughly 10 km [see Example 2.4 (page 64) for a discussion of the scale height of the atmosphere]. It is thus unlikely that this condition should ever be fulfilled; possible exceptions are the case of *very* low frequency sound propagating vertically in the atmosphere and lee waves over mountains.

The terms in line 2 are associated only with unsteady flow. In *acoustic* motions the dimensionless parameter l_0/c_0t_0 is just unity; that is, l_0 is the wavelength λ, t_0^{-1} is the frequency ν, and $\nu\lambda = c_0$. The first term is significant only for wave motions with velocity changes of the order of c_0, which is the case of nonlinear wave propagation considered in Chap. 8. The second term is already of order unity, and all acoustic motions are associated with compressibility, as considered in Chap. 4.

The terms in line 3 form the component of the net viscous force in the direction of motion times the magnitude of the velocity and thus represent dilatation due to direct viscous action. The magnitude, as represented by the coefficient $M_0{}^2/\text{Re}_0$, is small in almost every case; with $\mu \approx \rho_0 c_0 \Lambda_0$ [Eq. (2.125)] we have

$$\text{Re}_0 = \frac{\rho_0 u_0 l_0}{\mu} \approx M_0 \frac{l_0}{\Lambda_0}$$

The ratio Λ_0/l_0 of mean free path to characteristic length is defined to be the *Knudsen number* Kn,

$$\text{Kn}_0 \equiv \frac{\Lambda_0}{l_0} \tag{3.48}$$

As discussed in Sec. 1.1, the continuum model is applicable only if this quantity is small compared to unity. The above estimate of the Reynolds number is rearranged to give

$$\text{Kn}_0 \approx \frac{M_0}{\text{Re}_0} \tag{3.49}$$

With typical Mach numbers of order unity and Reynolds numbers in the range 10^5 to 10^8, this quantity is indeed small. Similarly, the magnitude of the coefficient $M_0{}^2/\text{Re}_0$ is normally small compared to unity.

The first and second terms in line 4 represent respectively fluid-volume increase due to dissipation of mechanical work and volume increase due to accumulation of heat. This line is also normally small; the dimensionless quantity $\dfrac{T_0}{v_0}\left(\dfrac{\partial v}{\partial T}\right)_p$ may be of order unity (gases) or smaller (liquids), while the Reynolds number Re is usually large.

In general the *relative* magnitude of the various terms in (3.46) may be taken as a measure of the corresponding effect, e.g., gravity, in any given problem. There are some problems in which all terms in (3.46) are small, so that the continuity equation reduces to the incompressible limit $\nabla \cdot \mathbf{u} = 0$, but in which the forces associated with density changes cannot be neglected in the momentum equation. For example, consider a fluid confined between horizontal surfaces in a gravity field, with the lower surface heated, e.g., a pan of water heated on a stove. This is called the *Bénard problem.* In this case the gravity-driven buoyant force associated with very small density changes cannot be neglected (it is the sole cause of motion); in Eq. (3.46) the second term in line 1 and the second term in line 4 are small, but of comparable magnitude, and must be represented in the momentum equation.

In summary, we have found that whether a given *fluid* should be considered compressible or incompressible depends on the relative magnitude of $\nabla \cdot \mathbf{u}$; this magnitude in turn depends on the circumstances of the particular flow, as well as the properties of the fluid. Nevertheless, it is a traditional convenience to speak of "compressible fluids" and "incompressible fluids" as though these qualities were inherent in the fluid itself.

Remark on Incompressible Fluids

We have already formally defined an incompressible fluid as satisfying ρ = const for each fluid particle [Example 2.5 (page 69)]. Thus $D\rho/Dt = 0$, and by continuity $\nabla \cdot \mathbf{u} = 0$, consistent with the preceding criterion for compressibility. The above definition does *not* require that the density be the same for each fluid particle; in general, an incompressible fluid need not satisfy $\nabla\rho = 0$ (for an example, see Prob. 2.16, as well as the flow illustrated in Fig. 1.17).

A Criterion for Isentropic Flow

It is a common assumption in compressible-fluid flow that the fluid state changes are isentropic. In practice, this often takes the form of writing $P = P(\rho)$, since s is considered constant. We can test the validity of this assumption by writing the general relation for $P = P(\rho,s)$,

$$\frac{DP}{Dt} = \left(\frac{\partial P}{\partial \rho}\right)_s \frac{D\rho}{Dt} + \left(\frac{\partial P}{\partial s}\right)_\rho \frac{Ds}{Dt} \tag{3.50}$$

It is assumed that compressibility is *not* negligible; i.e., we cannot set $D\rho/Dt = 0$. Then if the second term on the right is small compared to the first term on the right, the pressure varies approximately isentropically with the density; i.e., if

$$\left| c^2 \frac{D\rho}{Dt} \right| \gg \left| \left(\frac{\partial P}{\partial s} \right)_\rho \frac{Ds}{Dt} \right| \tag{3.51}$$

the assumption of isentropic flow can be justified.

This inequality can be translated into estimable terms by methods similar to those used in finding a criterion for compressibility. Omitting the details, the result is that

$$|\nabla \cdot \mathbf{U}| \gg \left| \frac{1}{\mathrm{Re}_0} \frac{T_0}{v_0} \left(\frac{\partial v}{\partial T} \right)_p \left(\frac{u_0{}^2}{c_p T_0} \tilde{\Upsilon} + \frac{1}{\mathrm{Pr}} \nabla^2 \tilde{T} \right) \right| \tag{3.52}$$

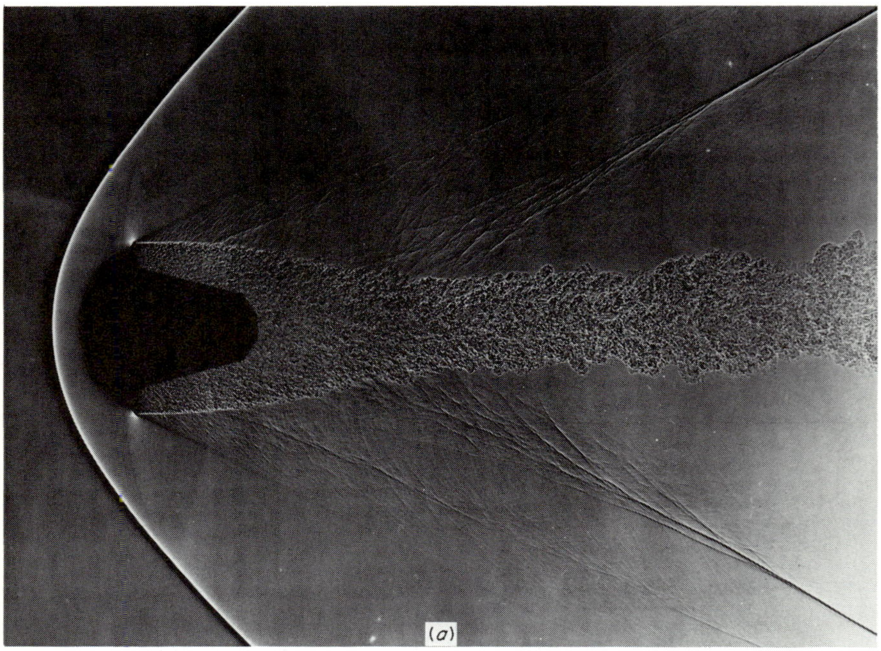

Figure 3.4
Regions of turbulence. (*a*) Turbulent wake of blunt body traveling at supersonic speed. (*Ballistic Research Laboratory*.) (*b*) Pellets emerging from muzzle of shotgun. (*Field Emission Corp*.) (*c*) Turbulent boundary layer and wake; supersonic cone, $M_\infty = 7.5$. (*Courtesy of Z. J. Levensteins, Naval Ordnance Laboratory*.) (*d*) Transition from laminar to turbulent flow in wake of supersonic cone, $M_\infty = 10.2$.

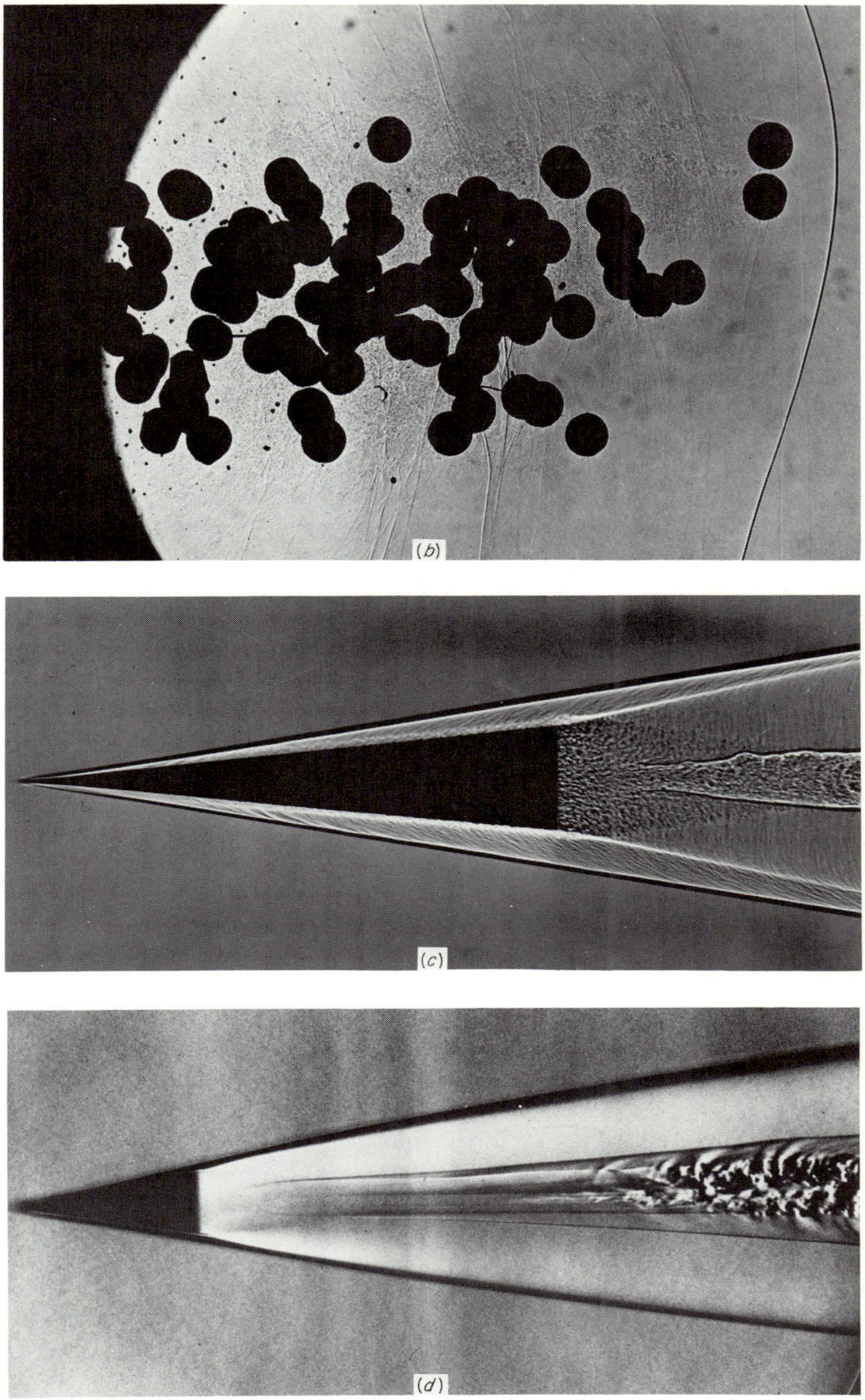

(b)

(c)

(d)

is a criterion for isentropic flow. The physical interpretation is that the overall rate of volume change must be large compared to that due to dissipation and heating. For example, for gas flow in a nozzle with

$$\nabla \cdot \mathbf{U} \sim 1 \qquad \frac{T_0}{v_0}\left(\frac{\partial v}{\partial T}\right)_p \sim 1 \qquad \frac{u_0^2}{c_p T_0} \sim M_0^2 \qquad \text{and} \qquad \text{Pr} \sim 1$$

this becomes

$$1 \gg \frac{M_0^2 + 1}{\text{Re}_0} \tag{3.53}$$

a criterion which is normally well satisfied.

A Remark on Turbulence

At large Reynolds numbers, magnitude estimates indicate that viscous forces are negligible. Within boundary layers and wakes (see Fig. 3.4) the recession of viscous forces is, however, more than compensated by an even more effective mechanism for momentum transfer, viz., turbulence. In the one boundary-layer problem which is explicitly treated in this book (Sec. 10.5) we will assume that the Reynolds number is sufficiently small for the flow to remain laminar.

In the steady *inviscid* flows of interest in gasdynamics (outside the boundary layer) the kinetic energy stored in turbulence is commonly a small fraction (less than 0.1 percent, say) of the free-stream kinetic energy. While the flow is then *turbulent*, it is turbulent only to small degree; the momentum transfer due to turbulence, as well as that due to viscosity, can usually be neglected.

3.6 Self-similar motions

In the conventional (Eulerian) formulation of fluid mechanics the independent variables are three spatial coordinates, for example, x_1, x_2, x_3, and time t. Most problems considered do not, however, contain all four independent variables, as the number is reduced by a regularity of geometry or the condition of steadiness, or both. In certain peculiar problems the number of independent variables is further reduced by the requirement that the physical law be unit-free. The corresponding motion is termed *self-similar* (the descriptiveness of this terminology will be explained below).

Self-similar motions are quite important in gasdynamics. In the simplest cases there is finally only one independent variable, and the equations of motion reduce to one or more *ordinary* differential equations. The following example is for an incompressible fluid and has been chosen primarily for its simplicity.

A stationary fluid of constant density ρ and constant viscosity μ is bounded by an infinite plane surface at $x_2 = 0$ (Fig. 3.5). All of the space $x_2 > 0$ is supposed to be filled with fluid; such a geometry is said to be *semi-infinite*. At time zero the boundary surface impulsively acquires a constant velocity u_0, the velocity vector lying in the plane of the surface (a rough analogy for the boundary motion is the parlor trick of snatching a tablecloth from underneath a table setting). It is desired to find the resulting fluid motion.

Let the boundary motion be in the x_1 direction, and suppose that all the fluid motion is also in this direction, $u_2 = u_3 = 0$. We seek a solution in which the flow properties vary only in the x_2 direction; thus we set $\partial/\partial x_1 = \partial/\partial x_3 = 0$. Then the momentum equation (1.66) in the x_1 direction reduces, with no body force, to

$$\rho \frac{\partial u_1}{\partial t} = \mu \frac{\partial^2 u_1}{\partial {x_2}^2} \tag{3.54}$$

Rewriting this with the kinematic viscosity $\nu = \mu/\rho$ and dropping the subscripts,

$$\frac{\partial u}{\partial t} = \nu \frac{\partial^2 u}{\partial x^2} \tag{3.55}$$

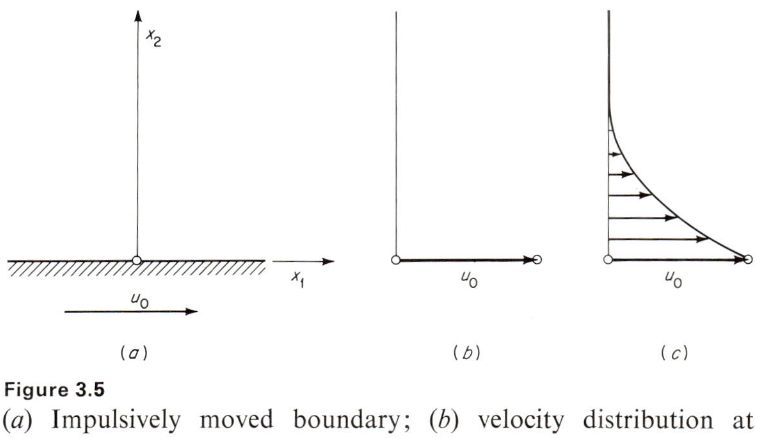

Figure 3.5
(*a*) Impulsively moved boundary; (*b*) velocity distribution at $t = 0$; and (*c*) velocity distribution for $t > 0$.

The boundary conditions for $u(x,t)$ may be taken to be

$$\begin{aligned} u(x,0) &= 0 \\ u(0,t) &= u_0 \\ u(\infty,t) &= 0 \end{aligned} \tag{3.56}$$

Suppose that we require a solution of the form

$$\frac{u}{u_0} = f(\nu,x,t) \tag{3.57}$$

With the dimensions of ν given by L^2/T, the only nondimensional group which can be formed from ν, x, t is $x/\sqrt{\nu t}$ or some power thereof. Thus a formulation which satisfies the requirement that the physical law be unit-free is

$$U = f(\eta) \tag{3.58}$$

where

$$\eta \equiv \frac{x}{\sqrt{\nu_t}} \qquad U \equiv \frac{u}{u_0} \tag{3.59}$$

The quantity η, which is a combination of the old independent variables x and t, is called a *similarity variable*. It now plays the role of a single independent variable for the problem.

The derivative operators appearing in (3.55) are now

$$\left.\frac{\partial}{\partial t}\right)_x = \left(\frac{\partial \eta}{\partial t}\right)_x \frac{d}{d\eta} = -\frac{\eta}{2t}\frac{d}{d\eta}$$

$$\left.\frac{\partial}{\partial x}\right)_t = \left(\frac{\partial \eta}{\partial x}\right)_t \frac{d}{d\eta} = \frac{1}{\sqrt{\nu t}}\frac{d}{d\eta}$$

$$\left.\frac{\partial^2}{\partial x^2}\right)_t = \frac{1}{\nu t}\frac{d^2}{d\eta^2}$$

Substituting yields

$$-\frac{\eta}{2t}\frac{du}{d\eta} = \frac{\nu}{\nu t}\frac{d^2u}{d\eta^2}$$

Dividing through by u_0 gives the nondimensional *ordinary* differential equation

$$\frac{d^2U}{d\eta^2} + \frac{\eta}{2}\frac{dU}{d\eta} = 0 \tag{3.60}$$

The boundary conditions (3.56), rewritten in terms of the dimensionless variables U and η, become respectively

$$\begin{aligned} U(\infty) &= 0 \\ U(0) &= 1 \\ U(\infty) &= 0 \end{aligned} \tag{3.61}$$

The first and third conditions are now redundant, so that we have only two distinct conditions (the proper number for a second-order ordinary differential equation). Formally, the problem is now properly stated in terms of the dimensionless variables.

Equation (3.60) can be rewritten in terms of $U' = dU/d\eta$ as

$$\frac{dU'}{d\eta} + \frac{\eta}{2}U' = 0$$

which has the elementary solution

$$\frac{dU}{d\eta} = Ae^{-\eta^2/4}$$

Integrating this, and making use of the boundary conditions (3.61) gives the solution for the fluid motion,

$$U = 1 - \frac{2}{\sqrt{\pi}}\int_0^{\eta/2} e^{-p^2}\,dp \tag{3.62}$$

where p is a dummy variable for the integration. The value of the integral depends only on the upper limit; it is called the error function, abbreviated erf, and is widely tabulated. Thus (3.62) is conventionally written

$$U = 1 - \operatorname{erf}\left(\frac{\eta}{2}\right) \tag{3.63}$$

The corresponding velocity distribution is shown in Fig. 3.6.

The motion is said to be *self-similar* because the same velocity distribution holds for all time, with an appropriate adjustment of vertical scale; in Fig. 3.6*b*, any of the given velocity distributions can be made to coincide with any other simply by stretching or compressing the vertical scale.

The velocity disturbance is effectively confined to a region of increasing thickness near the displaced boundary. This region is called the *boundary layer* and grows with time according to a simple rule. Consider

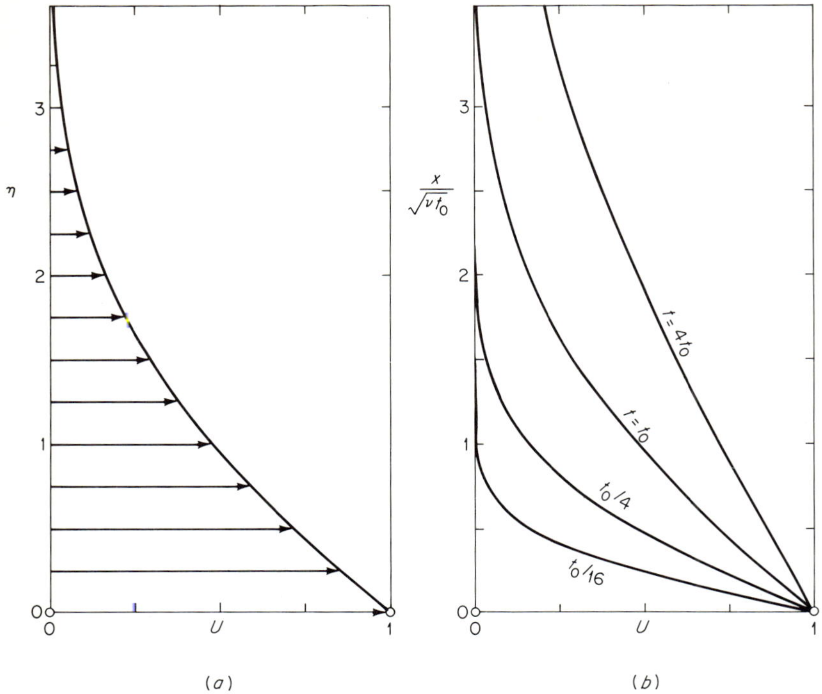

Figure 3.6
Self-similar velocity profiles: (*a*) universal profile plotted against the similarity variable $\eta = x/\sqrt{\nu t}$; (*b*) profiles at various times plotted against the normalized spatial coordinate $x/\sqrt{\nu t_0}$.

some arbitrarily small value for the normalized velocity U, say $U = 0.01$; smaller values of U may arbitrarily be considered to lie in an undisturbed region above the boundary layer, and the edge of the boundary layer itself is signified by this value. Now $U = 0.01$ occurs at $\eta = 3.644$, from Eq. (3.63); thus the edge of the boundary layer occurs at

$$\eta = \frac{x}{\sqrt{\nu t}} = 3.644$$

and the boundary-layer thickness (position of the edge) is given by x^*

$$x^* = 3.644\sqrt{\nu t} \tag{3.64}$$

The numerical factor in this equation will be somewhat changed by a different definition for the edge of the boundary layer. For example, if we choose $U = 0.001$, the boundary-layer thickness is $x^* \approx 4.6\sqrt{\nu t}$. The essential result is that the thickness is proportional to $\sqrt{t}$. This is a typical result for diffusion problems, i.e., for problems governed by (3.55), which is generally called the *diffusion equation* or the *heat equation.*

The problem of the impulsively moved boundary, as considered above and in some of its variations, is often referred to as the *Rayleigh problem* (*Rayleigh* [1911]). This problem was in fact first treated by the ninteenth-century giant *Stokes* [1850].

The simplicity of the preceding analysis flows from the formation of the similarity variable η. It is fair to say that the dimensional arguments leading to the formation of η were somewhat forced, by the instrument of the assumed special form (3.57), which can be justified by its suitability to the boundary conditions. The more general statement $f(u,u_0,\nu,x,t) = 0$ would permit us to form more conventional dimensionless variables, such as a spatial coordinate u_0x/ν. It may be noted that in the analogous heat-transfer problem, in which a fixed temperature is suddenly impressed at the surface of a semi-infinite solid initially at some uniform temperature, this difficulty does not occur. This is because the dependent variable T has a separate dimension, which will prevent its combination with variables such as x. The solution to this heat-transfer problem (which was one of J. B. J. Fourier's original solutions) is formally identical to (3.63).

These cautionary remarks, and those which preceded, may be taken to indicate that the application of dimensional arguments is at least partly art, not completely science.

3.7 Similarity based on coordinate transformation

In the approximate, linearized theory of steady inviscid flow we will find equations of the form

$$\beta^2 \frac{\partial^2\phi}{\partial {x_1}^2} + \frac{\partial^2\phi}{\partial {x_2}^2} = 0 \tag{3.65}$$

where β is a constant. With the affine transformation

$$X_1 = \frac{1}{\beta}x_1 \qquad X_2 = x_2 \tag{3.66}$$

(3.65) reduces to the *Laplace equation*

$$\frac{\partial^2\phi}{\partial {X_1}^2} + \frac{\partial^2\phi}{\partial {X_2}^2} = 0 \tag{3.67}$$

which describes an incompressible flow.

By the affine transformation (3.66) a compressible-flow problem can be reduced to an equivalent incompressible-flow problem, in which the geometry of the flow boundaries is "stretched" according to the transformation. Similarly, two different compressible flows, characterized by different values of β (say β_a and β_b), may be related to one another.

We introduce this form of similitude here only for completeness; a further discussion is postponed to Chap. 5.

Problems

3.1 Find the M, L, T, θ dimensions of the following quantities:

(a) Pressure P
(b) P/ρ
(c) Maxwell distribution function f^0
(d) Momentum flux
(e) Energy flux
(f) Molecular number density n
(g) Thermal conductivity κ
(h) A unit vector $\mathbf{e}$
(i) Avogadro's number $\tilde{N}$
(j) The price of gold

3.2 A spherical gas bubble is surrounded by an infinite sea of liquid. The bubble radius R oscillates periodically with small amplitude at a well-defined frequency Ω; that is, the bubble undergoes spherically symmetric repeated growth and contraction. If the parameters entering the problem are the mean radius R_0, mean pressure P_0, liquid density ρ_l, and ratio of specific heats γ for the gas, find (by dimensional reasoning) an expression for the frequency Ω.

Answer $\Omega^2 = \text{const} \times \dfrac{P}{\rho_l {R_0}^2}$, where the constant depends on γ.

3.3 Find the frequency Ω for Prob. 3.2 from the hydrodynamic equations of motion. The radial oscillation may be assumed sinusoidal and of small amplitude, the liquid may be assumed incompressible, and the gas properties within the bubble may be assumed spatially uniform with $P\rho^{-\gamma} = \text{const}$. Viscous and surface-tension effects may be neglected.

Answer $$\Omega^2 = \frac{3\gamma P_0}{\rho_l {R_0}^2}$$

3.4 An ideal pendulum consists of a mass m at the end of a rigid massless rod of length L. The rod is mounted on a frictionless pivot. If the pendulum oscillates with frequency Ω in a gravitational field g, find an expression for Ω from dimensional reasoning.

3.5 Find the M, L, T, θ dimensions corresponding to the following units:

(a) Radian (angular measure)
(b) Degree (angular measure)
(c) Steradian
(d) Hertz
(e) Radian per second

3.6 Compare the dimensions of specific heat, for example, c_p, and specific entropy.

3.7 Newton's law of universal gravitation can be written

$$F = \gamma \frac{mm'}{r^2}$$

where γ is a universal constant, the dimensions of which are, from the above equation, $[\gamma] = L^3/MT^2$. If a planet orbits a massive central star of mass m_0 at some characteristic distance l_0, for example, the distance of closest approach, find an expression for the period t_0 of the planetary orbital motion. (Compare Kepler's third law.)

3.8 A one-tenth scale model is tested in a wind tunnel under conditions dynamically similar, i.e., at the same Reynolds number, to those for a full-sized prototype. The fluid in question is the same for model and prototype and may be considered incompressible with constant viscosity. If the aerodynamic drag force on the model is measured to be 100 lb_f, predict the drag force on the prototype.

Answer 100 lb_f

3.9 In filming a motion picture, an automobile of length L is to be driven over a vertical cliff at speed V. The normal filming and projection speed is N frames per second. If a *model* car of length l is substituted for economy, what is the proper value (for realism) for the model speed v and filming speed n? Assume that the film will be projected at normal speed and that air resistance can be neglected; i.e., the mechanics are essentially those of a point mass.

Answer $v = V\sqrt{\frac{l}{L}}$ $\quad n = N\sqrt{\frac{L}{l}}$

3.10 In filming a motion picture, a small model of a full-rigged sailing vessel is subjected to a model storm in a model basin. Discuss the problem of similarity, and in particular specify the filming speed n (see Prob. 3.9).

3.11 A racing car accelerates from rest. The equation of motion may be written $F = ma$, where F is the traction force.

(a) Write the equation of motion in a physically appropriate nondimensional form.

(*b*) From the list of variables (F, m, g, x, t) form appropriate dimensionless groups.

(*c*) Suppose $F =$ const. By introducing a similarity variable η, reduce the equation of motion $F = m\ddot{x}$ to an *algebraic* equation and solve.

3.12 In the cooling of a heated body by natural (gravity) convection the constant parameters are α, βg, ν, l_0, ΔT, where

$$\alpha \equiv \frac{\kappa}{\rho c_p} \qquad \beta \equiv \frac{1}{v}\left(\frac{\partial v}{\partial T}\right)_p \qquad \nu \equiv \frac{\mu}{\rho}$$

and l_0 is a characteristic body dimension, and ΔT is a fixed temperature difference. Find the appropriate nondimensional parameters (cf. *Landau and Lifshitz* [1959, p. 212]).

3.13 With the assumptions of steady inviscid flow, the Bernoulli equation (2.25) predicts for the efflux velocity u_e from the tank $u_e = \sqrt{2gH}$. By order-of-magnitude estimates show that the ratio of the convective acceleration $|\mathbf{u} \cdot \nabla \mathbf{u}|$ to the unsteady acceleration $|\partial \mathbf{u}/\partial t|$ is of order $\dfrac{H}{l}\left(\dfrac{D}{d}\right)^2$ and thus that the assumption of steadiness is justified. Find a numerical value for the ratio in some typical situation.

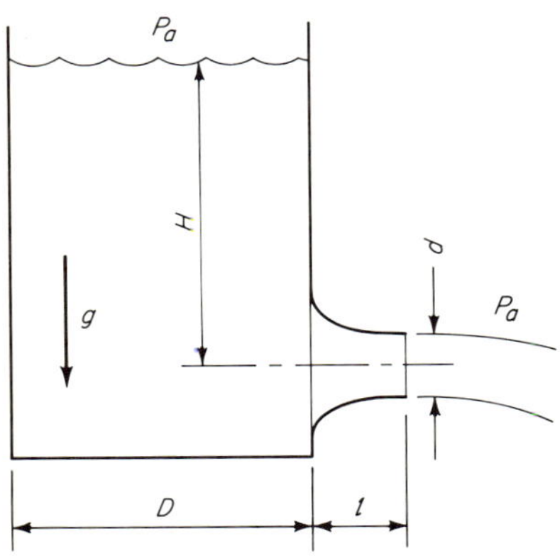

3.14 Suppose that in the flow associated with the impulsively moved plate discussed in Sec. 3.6, the fluid is bounded by a second plate which is stationary and parallel to the impulsively moved plate, a distance L away. Find a condition, in the form of an inequality, such that the solution (3.63) will still apply up to a certain time t_0, at least approximately. *Hint*: Recall the boundary-layer character of the solution.

Answer $\quad t \ll \dfrac{L^2}{10\nu}$

3.15 For the problem of the impulsively moved plate discussed in Sec. 3.6, find the drag on the plate per unit area exerted by the fluid, i.e., just the stress σ_{21}, as a function of time.

Answer $-\dfrac{\mu u_0}{\sqrt{\pi \nu t}}$

3.16 Given two geometrically similar sailboats of different size, which may be expected to heel (tip) more sharply in any given breeze?

four
physical acoustics

4.1 Introduction

The ancient study of acoustics (ἀκουστικός, relating to hearing) has subdivided into specialized fields such as ultrasonics, architectural acoustics, psychoacoustics, physiological acoustics, and so on. In this chapter we consider physical acoustics, which is here taken to mean the study of the *fluid motions associated with the propagation of sound.* (It should be remarked, however, that physical acoustics is sometimes given a more specialized meaning, viz., the study of the physical properties of materials via their interaction with acoustic waves.) Our interest will be centered on the fluid dynamics of sound rather than the mechanical vibrations which may produce it, the transducer system which may record it, or other related topics, however interesting they may be in themselves.

The study of acoustics is motivated first of all by its very simplicity, providing a natural and instructive introduction to more complex problems in compressible flow. Second, acoustic motions provide a physically interesting example of wave motion, with many of the features of electromagnetic, surface, and even quantum waves. Third, acoustics is of interest in connection with noise generation and suppression, underwater detection and communication, interaction of electromagnetic and acoustic waves, and several other problems.

Historical Remarks

The first quantitative observation in acoustics is credited to *Pythagoras* (ca. 580–500 B.C.), who experimented with a monochord, a device which maintains a single string in constant tension over two wooden bridges. Pythagoras found that the pitch of two plucked strings of length l and $l/2$ respectively differed by an octave exactly (the actual observation of the auditor being that the two sounds heard together have maximum *consonance*). As we would now say, the frequency of the sound from the shorter string is just twice that from the longer; the sound frequency is inversely proportional to the length of the string. This simple discovery became the basis not for an exact science but for fantastic speculations on the relation of physics to music, culminating in the famous "harmony of the spheres."

A treatise of *Aristotle's* entitled *Sound and Hearing* contains a passage which conveys a surprisingly accurate physical concept of sound propagation. Yet there is little concrete in these writings, or in those of Euclid on sound, which would now be considered of value. To Aristotle's successor *Theophrastus*, however, we owe what may be called the second quantitative observation in acoustics, viz., that the speed of sound is independent of its frequency (at least in the audible range of frequencies):[1]

> But the high note would not differ in speed [from the low]; for then it would lay hold on the hearing the sooner, so that there would not be concord; thus if there is concord, each goes equally swiftly . . . .

This contradicted the view of Aristotle, who had held in effect that high notes should travel more swiftly than the low.

Not until the time of Galileo (1564–1642), who discovered the laws of vibrating strings, was there an experimental science of acoustics. The fundamental question of how a given sound reached its auditor was a troublesome one; specifically, was the intervening air necessary for transmission? A standard experiment involved the ringing of a bell within an evacuated jar; if the bell could not be heard, it followed that air was necessary for transmission. Unfortunately, the bell had to be mechanically suspended, and early techniques for the production of vacuum were not very effective. The Jesuit Kircher tried the experiment and reported in 1650 that air was not necessary, contradicting the view of Aristotle, and

[1] Quoted in *Truesdell* [1954, vol. 13, p. xx].

von Guericke found the same result. Robert Boyle repeated the experiment more carefully with an improved vacuum pump and concluded (1660) that air was indeed necessary.

The earliest published measurement of the speed of sound is that of Mersenne (1636). Of the many subsequent measurements, most were based on the observed time delay between the muzzle flash of a cannon or musket and its audible report, assuming the velocity of light to be infinite. In this connection, the estimation of the distance of a thunderstorm by the same technique was suggested by William Derham (1708).

Theoretical acoustics begins with *Newton* [1686, bk. 2, props. 41ff], who found for the speed of sound in a gas $c = \sqrt{P/\rho}$, which is nearly right, as we shall see. The further theoretical developments of the eighteenth and nineteenth centuries will not be recounted here, except in a few footnotes. By the time of Lord Rayleigh's monumental *Theory of Sound* (1877) the science of sound, if not complete, was well founded. An excellent account of the history of theoretical acoustics is given by *Truesdell* [1954], and an account on the experimental side is given by *Miller* [1935].

The Nature of Waves

The diverse situations to which the word "wave" is applied have certain features in common. Consider ocean waves coming into a beach. From the point of view of an observer standing on a pier and looking down at the water surface, the motion is complicated and unsteady (though approximately periodic) in time. But much depends on the point of view; for a successful surfboard rider the motion is fairly steady, and a discrete wave has a well-defined identity which changes only slowly as the rider moves with the wave. We take this to be the distinguishing quality of all progressive waves: that for an observer moving with the wave velocity there is a well-defined entity called a wave, the spatial form of which changes only slowly.

Several additional properties often loosely associated with waves are in fact applicable only to special categories of wave motion. These are mentioned below with the specific admonition that they are *restricted*, rather than universal, properties and *will not apply to every case*:

1. There is a well-defined wave speed c. For wave motion of sufficiently small amplitude, the wave speed is a constant; e.g., in air at room temperature, c is about 1,130 ft/s.
2. The material motion has a small amplitude, compared to the motion of the wave. An often-cited example is that of a wind gust traveling across

a field of wheat: the displacement of an individual stalk of wheat is small compared to the distance traveled by the wave disturbance in the same time interval.

3 A superposition principle holds, whereby two or more waves can traverse the same space independently of each other but exhibit additive physical properties, e.g., additive pressure perturbations. This important property, together with properties 1 and 2 above, is restricted to wave motions described by *linear* equations.

4 The material motion is in the same direction as the wave motion. Such waves are said to be *longitudinal.* Acoustic waves are of this type.

5 The wave motion *may* be periodic: waves of length λ and frequency ν satisfy the kinematical relation $c = \nu\lambda$ (Newton, 1686). Musical sounds and those of the human voice are approximately periodic. Many waves of physical interest, however, are not periodic, e.g., the sonic boom or other impulsive disturbance.

4.2 The wave equation

The wave equation is the fundamental equation of acoustics. It is based on two important approximations, viz., that the flow may be treated as *inviscid* and that *convective derivatives are negligible in comparison to unsteady derivatives.* It will prove to be most convenient to test the validity of these assumptions after the theory has been partially developed. For ordinary acoustics—the phenomena of audible sounds—they will turn out to be highly satisfactory approximations.

The fluid medium is assumed to have an undisturbed state of rest, with uniform properties ρ_0, P_0, and $\mathbf{u}_0 = 0$; for example, the air in a hypothetical room is perfectly still before the onset of conversation. The acoustic motions produce deviations from the uniform state which are called *small perturbations*; thus the instantaneous local density ρ is given by

$$\rho = \rho_0 + \hat{\rho} \tag{4.1}$$

where $\hat{\rho}$ is the small perturbation in density. We specifically assume the basic condition $|\hat{\rho}| \ll \rho_0$. In analogy with (4.1) the instantaneous velocity $\mathbf{u}$ can be written

$$\mathbf{u} = 0 + \mathbf{u} \tag{4.2}$$

That is, the velocity is a perturbation from the uniform state $\mathbf{u}_0 = 0$. It will be convenient to postpone a description of the precise sense in which this perturbation is "small" [it should be mentioned, however, that Eq. (3.46) suggests the condition for dropping convective derivatives is that the velocity disturbance be small compared to the speed of sound, that is, that $M^2 \ll 1$].

By the inviscid assumption, viscous dissipation and heat conduction are absent, and there is no generation of entropy. The specific entropy therefore has the uniform value $s = s_0$ everywhere and for all time; the flow is homentropic. This means that the thermodynamic state is fixed by *one* thermodynamic variable, such as the density ρ. In particular $P = P(\rho)$, and the pressure perturbation p is given by the Taylor expansion

$$p \equiv P - P_0 = \left(\frac{\partial P}{\partial \rho}\right)_s (\rho - \rho_0) + \frac{1}{2}\left(\frac{\partial^2 P}{\partial \rho^2}\right)_s (\rho - \rho_0)^2 + \cdots \tag{4.3}$$

Abbreviating the isentropic derivative by the symbol c^2, as in Chap. 3,

$$c_0^2 \equiv \left[\left(\frac{\partial P}{\partial \rho}\right)_s\right]_0 \tag{4.4}$$

and with $\hat{\rho} = \rho - \rho_0$ small the squared and higher-order terms may be neglected, so that the Taylor expansion becomes simply

$$P - P_0 = c_0^2(\rho - \rho_0) \tag{4.5}$$

Note that the thermodynamic property c_0 is known as a function of the uniform state; for example, $c_0 = c_0(P_0, \rho_0)$. As has already been suggested, this property will turn out to be just the speed of sound.

It is convenient and conventional in acoustics to use the relative density change, or *condensation*,[1] $\hat{\rho}/\rho_0$ instead of the density ρ. The usual symbol for this quantity is the lowercase s, which happens to be our symbol for the specific entropy; the uppercase S will therefore be used,

$$S \equiv \frac{\rho - \rho_0}{\rho_0} \tag{4.6}$$

In terms of the condensation, the expression (4.5) for the pressure perturbation is

$$p = P - P_0 = \rho_0 c_0^2 S \tag{4.7}$$

[1] The term condensation does not imply a change of phase but is simply a somewhat archaic expression for compression.

The basic condition of small amplitude, $|\hat{\rho}| \ll \rho_0$, thus takes the equivalent forms

$$|S| \ll 1 \qquad |p| \ll \rho_0 c_0^2 \tag{4.8}$$

By expansions similar to (4.3) it is possible to express *any* thermodynamic property as a linear function of the condensation S.

The continuity equation and the inviscid Euler momentum equation are

$$\frac{\partial \rho}{\partial t} + \mathbf{u} \cdot \nabla \rho + \rho \nabla \cdot \mathbf{u} = 0 \tag{4.9}$$

$$\frac{\partial u}{\partial t} + \mathbf{u} \cdot \nabla \mathbf{u} + \frac{1}{\rho} \nabla P = 0 \tag{4.10}$$

Formally, we replace the coefficients of derivatives by their unperturbed values. In the case of the convective derivative terms, this value is zero, so that these terms disappear. Then

$$\frac{\partial \rho}{\partial t} + \rho_0 \nabla \cdot \mathbf{u} = 0 \tag{4.11}$$

$$\rho_0 \frac{\partial \mathbf{u}}{\partial t} + \nabla P = 0 \tag{4.12}$$

These equations, which do not contain products of unknowns or products of unknowns with derivatives of unknowns [as in Eqs. (4.9) and (4.10)], are by definition *linear*. Inserting (4.6) and (4.7) yields

$$\frac{\partial S}{\partial t} + \nabla \cdot \mathbf{u} = 0 \tag{4.13}$$

$$\frac{\partial \mathbf{u}}{\partial t} + c_0^2 \nabla S = 0 \tag{4.14}$$

The wave equation for the condensation S is obtained by differentiating (4.13) with respect to t, taking the divergence of (4.14), and subtracting. Then with

$$\nabla \cdot \frac{\partial \mathbf{u}}{\partial t} = \frac{\partial}{\partial t} \nabla \cdot \mathbf{u}$$

the velocity terms cancel, and we obtain

$$\frac{\partial^2 S}{\partial t^2} = c_0^2 \nabla^2 S \tag{4.15}$$

which is the *classical wave equation*, first found for fluid motions by Euler (1759) and for the vibrating string by d'Alembert (1747).

Because all thermodynamic properties are expressible as linear functions of the condensation S, direct substitution shows that any such property (P, T, ρ, etc.) will also obey the wave equation (4.15).

By eliminating S from Eqs. (4.13) and (4.14) and inserting the condition that the velocity field be irrotational, it can be found that the vector velocity $\mathbf{u}$ also satisfies the wave equation. It will be more convenient, however, to work with a scalar velocity potential, as explained below.

The acoustic motion is assumed to begin from a state of rest and to be inviscid. It is therefore irrotational, $\boldsymbol{\Omega} = \nabla \times \mathbf{u} = 0$, a result which is formally contained in Eq. (2.47). That $\nabla \times \mathbf{u} = 0$ implies that $\mathbf{u}$ may be expressed as the gradient of a scalar potential ϕ,

$$\mathbf{u} = \nabla\phi \tag{4.16}$$

We call $\phi = \phi(\mathbf{x},t)$ the *velocity potential*.[1] Equation (4.16) defines ϕ and ensures, in a mathematical sense, that the velocity field is irrotational because the curl of the right side is identically zero.

Inserting (4.16) into the momentum equation (4.14) gives

$$\nabla\left(\frac{\partial\phi}{\partial t} + c_0{}^2 S\right) = 0$$

which has an integral

$$\frac{\partial\phi}{\partial t} + c_0{}^2 S = h(t)$$

where $h(t)$ is an arbitrary function of time. This function can always be considered to be incorporated into the velocity potential ϕ, however, without affecting (4.16) in any way, so that there is no loss in generality in taking $h(t)$ to be zero. Then

$$S = -\frac{1}{c_0{}^2}\frac{\partial\phi}{\partial t} \tag{4.17}$$

[1] Some authors write $\mathbf{u} = -\nabla\phi$ in agreement with the convention for force potentials [compare Eq. (1.72)]. Indeed, the negative sign is more commonly used in acoustics than the positive (Rayleigh and some others use the positive sign). In other areas of fluid mechanics, however, use of the positive sign as above is prevalent modern practice. It is adopted here in the interest of consistency.

Thus *both the velocity* **u** *and the condensation* S (and hence *all* thermodynamic properties) are conveniently expressible as derivatives of a single function, the velocity potential.

Taking the time derivative of (4.17) and substituting in (4.13) gives

$$\frac{\partial^2\phi}{\partial t^2} = c_0{}^2\,\nabla^2\phi \tag{4.18}$$

Again, the wave equation. We will apply the preceding results to specific situations in subsequent sections.

The wave equation (4.18) is *linear*, since the coefficient c_0 is constant. This property allows superposition of solutions as follows: if $\phi_1(\mathbf{x},t)$ and $\phi_2(\mathbf{x},t)$ satisfy the wave equation individually, then any linear combination, say $\phi = A\phi_1 + B\phi_2$, also satisfies the wave equation, as can be verified by addition of the equations for $A\phi_1$ and $B\phi_2$ respectively. This important result allows us to build up relatively complicated solutions from elementary solutions of the wave equation.

4.3 The speed of sound

This section is primarily concerned with physical aspects of the speed of sound. Consider a purely one-dimensional problem involving $\phi(x,t)$. The wave equation is

$$\frac{\partial^2\phi}{\partial t^2} - c_0{}^2\frac{\partial^2\phi}{\partial x^2} = 0 \tag{4.19}$$

The same equation can, of course, be written for S, P, ρ, etc. It is a famous result (d'Alembert, 1750) that this equation has a solution of the general form

$$\phi = F(x - c_0t) + G(x + c_0t) \tag{4.20}$$

as can be verified by direct substitution [an alternative and equivalent form is $\phi = F(t - x/c_0) + G(t + x/c_0)$]. The functions F and G represent waves propagating in the $+x$ and $-x$ directions, respectively, with propagation speed c_0.

For simplicity, consider that $G = 0$; then

$$\phi = F(x - c_0t) \tag{4.21}$$

Such a wave is called variously a *simple wave* or *progressive wave*. If the argument of the function F is constant, F is necessarily constant. Thus

$$\phi = \text{const} \qquad \text{if } x = c_0 t + \text{const}$$

Note that, from (4.16) and (4.17), precisely the same statements apply to u and S (it may in fact be more congenial to consider the arguments of this section as applied to some "concrete" physical quantity such as ρ or P rather than to the abstract quantity ϕ). We thus find that for an observer traveling in the $+x$ direction with velocity c_0 there is no change in the fluid conditions; this is exactly the meaning of a wave. As shown in Fig. 4.1, for an observer initially at the arbitrary point a who moves to a' through a distance $\Delta x = c_0 \, \Delta t$, there is no change in amplitude; but this statement applies to all observers moving with velocity c_0, and since the relative position of such observers does not change, the *waveform* does not change with time.[1]

The point here is simply that the quantity c_0 plays the role of a wave speed, the speed of sound. In general, we write (4.4) as

$$c^2 = \left(\frac{\partial P}{\partial \rho}\right)_s \tag{4.22}$$

to emphasize that the sound speed c is a thermodynamic variable.

[1] To be precise, it should be mentioned that the velocity-potential waveform sketched in Fig. 4.1 has a discontinuity in slope (on the left side) and a discontinuity in curvature (on the right side). The corresponding waveforms for velocity and condensation will then be discontinuous in amplitude (left side) and slope (right side). While the derivation given for the wave equation assumes continuous derivatives up to second order, we will later show that discontinuities in amplitude and slope as above do propagate with the speed of sound, so that waveforms like the one shown are valid.

Another viewpoint follows from the differentiation of the simple-wave formula (4.21): we find what is sometimes called the wave equation for simple waves,

$$\frac{\partial \phi}{\partial t} + c_0 \frac{\partial \phi}{\partial x} = 0$$

Note that $\partial/\partial t + c_0 \, \partial/\partial x$ is analogous to the material derivative and that this equation requires again that ϕ be constant for an observer with velocity $+c_0$. The same equation hold for u, P, S, etc. Rewritten in the form $\partial\phi/\partial t = -c_0 \, \partial\phi/\partial x$, this coincides with the intuitive notion that a (reversed) temporal waveform is equivalent to the spatial waveform; e.g., a sinusoidal pressure record $p(t)$ corresponds to a sinusoidal spatial waveform.

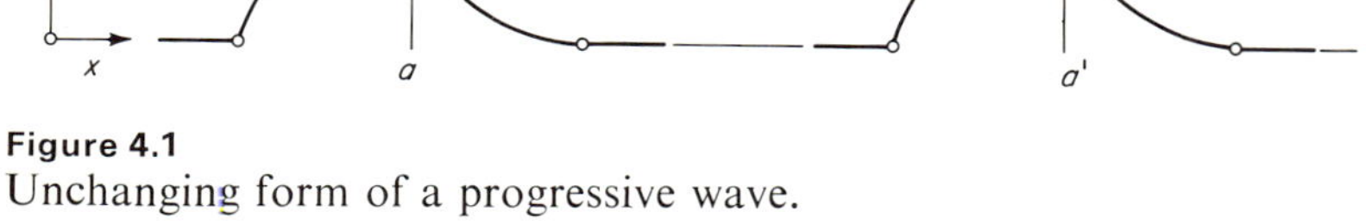

Figure 4.1
Unchanging form of a progressive wave.

The sound speed varies with the isentropic compressibility of the particular medium in question. Substances which are hard to compress, i.e., a large pressure change is required to accomplish a given density change, have a correspondingly large sound speed. For example, iron has a much larger sound speed than air.

Speed of Sound in an Ideal Gas

This is a very important special case. The thermal equation of state for an ideal gas may be written

$$P = R\rho T \tag{4.23}$$

With $\rho = 1/v$, the identity (2.23) can be written in the form

$$\left(\frac{\partial P}{\partial \rho}\right)_s = \gamma \left(\frac{\partial P}{\partial \rho}\right)_T \tag{4.24}$$

From (4.23), $(\partial P/\partial \rho)_T = RT$; thus

$$c^2 = \left(\frac{\partial P}{\partial \rho}\right)_s = \gamma RT$$

Thus, the sound speed is given by the well-known result (Laplace, 1825)

$$c = \sqrt{\gamma RT} = \sqrt{\frac{\gamma P}{\rho}} \tag{4.25}$$

where $\gamma = c_p/c_v$ may be a function of temperature.

It is emphasized again that the sound speed is *not* a constant for a particular fluid; it is a thermodynamic variable and depends on the *state* of the fluid. In the above case of an ideal gas, it varies with the square root of the absolute temperature. It is now usually possible to measure the sound speed with considerable precision; such measurements are used in gases for the determination of γ and for the indirect measurement of temperature.

For air at 0°C, using the properties found in Example 2.10 (page 85), the sound speed is calculated to be

$$c = \sqrt{1.402 \frac{8{,}314}{28.98} 273.15} = 331.5 \text{ m/s}$$

This may be compared with Lenihan's experimental value of 331.45 m/s for dry air at 0°C and 1 atm. Such close agreement between Eq. (4.25) and experiment is not typical, however; a discrepancy in the neighborhood

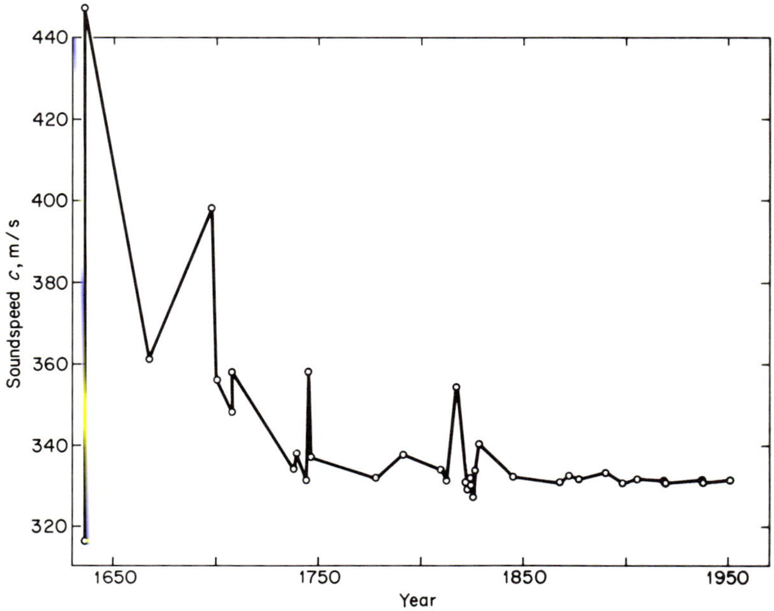

Figure 4.2
Historical measurements of the speed of sound in dry air, corrected to 0°C. (*Data from Lenihan* [1952].)

of 1 percent is normally to be expected at 1 atm, with frequencies in the audible range (see, for example, the tabulation in *Schaaffs* [1967, p. 6]).

The coefficient $\sqrt{\gamma R}$ in Eq. (4.25) gives a large variation of sound speed between chemical species. The value of γ is in the range $1 \leq \gamma \leq 1.667$ and in most cases between 1.20 and 1.40. The specific gas constant $R = \tilde{R}/\tilde{M}$ may vary by two orders of magnitude or more, so that we have roughly

$$c \propto \frac{1}{\sqrt{\tilde{M}}} \tag{4.26}$$

a dependence on molecular weight. Thus hydrogen and helium have relatively large sound speeds. A well-known manifestation is the "chirping" speech of divers who inhale helium.

That the sound speed is given by the isentropic derivative in (4.22) follows from our assumption of inviscid motion, an assumption which will be tested in Sec. 4.6. In general, acoustic entropy-generating processes such as relaxation or viscous dissipation will tend to invalidate (4.22) and result in a frequency-dependent sound speed, a situation which is referred to as *dispersion*. In particular, relaxation processes are important if the relaxation time $\tau \sim 1/\nu$, as will be discussed in Sec. 4.13.

A famous example of a nonisentropic sound speed is Newton's result $c = \sqrt{P/\rho}$, which corresponds exactly to the *isothermal* derivative, though it does not appear that he made this assumption explicitly. Newton however, did make several very special assumptions, e.g., that a fluid particle in acoustic motion follows "the law of the oscillating pendulum," i.e., exhibits simple harmonic motion. In the end he found that his theoretical prediction differed from experiment by about 20 percent, i.e., by a factor equal to $\sqrt{\gamma}$. Newton rationalized this as being due to salt particles and water in the air—"but in this calculation we have made no allowance for the crassitude of the solid particles of the air, by which the sound is propagated instantaneously"—and proceeded to show that a suitable correction precisely aligns theory and experiment. As von Kármán remarks, even very great men sometimes indulge in wishful thinking. We may remark, however, that Newton's calculation preceded by 200 years the invention of the word "entropy" by Rudolf Clausius.

Speed of Sound in General Substances

Where equation-of-state data are available, the sound speed can of course be directly calculated. The identity (4.24) is frequently useful in this connection. Data for the primitive fluids are shown in Fig. 4.3. Extensive data for a variety of fluids are given in *Schaaffs* [1967].

For liquids, the compressibility is sometimes specified in terms of the isentropic bulk modulus K_s

$$K_s \equiv \rho\left(\frac{\partial P}{\partial \rho}\right)_s \tag{4.27}$$

(In accord with the discussion of Sec. 2.6, this differs only slightly, for certain liquids such as water, from the analogous isothermal bulk modulus K_t.) Then the sound speed is given by the formula

$$c^2 = \frac{K_s}{\rho} \tag{4.28}$$

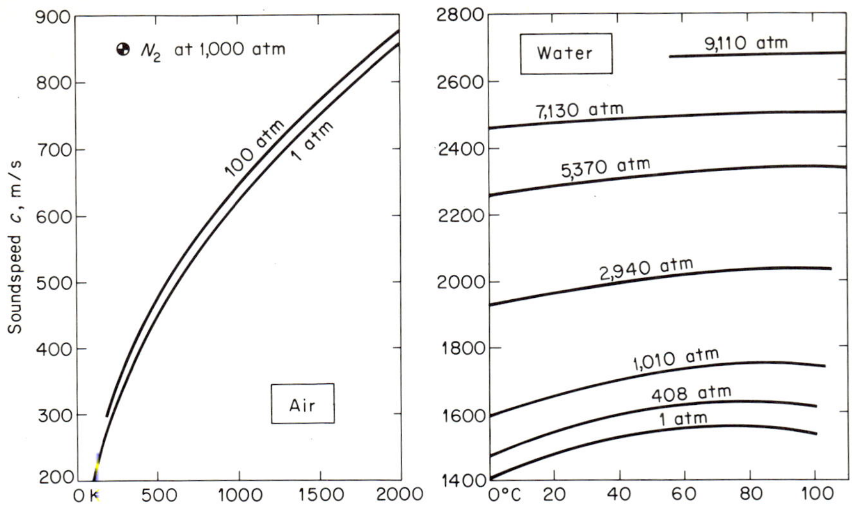

Figure 4.3
Speed of sound vs. temperature for air and water. (*Data for air from Hilsenrath et al.* [1955]; *for water from Smith and Lawson* [1954].)

Elastic solids have two primary modes of wave propagation, longitudinal and lateral (shear). For purely one-dimensional extension and compression, however, only longitudinal waves enter, and these may be treated according to the simple acoustic theory given here.

Table 4.1 Sound Speed in Various Substances
Values Are for 300 K, 1 Atm Unless Noted Otherwise

	Ft/s	Km/s
Hydrogen at low pressure ($\gamma = 1.405$)	4,326	1.32
Helium at low pressure ($\gamma = 1.667$)	3,340	1.02
Air at low pressure ($\gamma = 1.400$)	1,139	0.347
$U^{238}F_6$ gas at low pressure ($\gamma = 1.200$)	303	0.0924
Electron gas at very low pressure ($\gamma = 1.667$)	286,500	87.4
Pure water	4,950	1.51
Iron	16,400	5.00
Mercury	4,760	1.45
Ethyl alcohol	3,940	1.20
Silicone fluid (0.65 cks)	2,860	0.87
Ice	10,500	3.20
Hypothetical incompressible fluid	∞	∞

Molecular Motion and Sound Speed

The motions of the molecules in an ideal gas are described, in a statistical way, by the Maxwell velocity distribution. From this we have calculated a variety of characteristic molecular speeds (mean, rms, most probable), all of which turn out to be some coefficient of order unity times $\sqrt{RT}$. Thus we have molecular speeds which are functionally and numerically very close to the speed of sound. Consider the *mean speed*

$$\bar{v} = \sqrt{\frac{8}{\pi} RT} \tag{4.29}$$

and the *mean component velocity*, i.e., the average component velocity in any specified direction i of all molecules traveling in that direction, which we found to be precisely one-half the above

$$\bar{v}_i^+ = \sqrt{\frac{2}{\pi} RT} \tag{4.30}$$

These values form upper and lower bounds for the sound speed $c = \sqrt{\gamma RT}$.

It is physically reasonable that the "small disturbances" which are acoustic waves should travel with the speed of the molecules; in dilute gases, only the molecules themselves are available to transport energy and momentum.

In dense media, i.e., dense gases, liquids, and solids, the situation is quite different. Each molecule is perpetually within the force field of its neighbors, and disturbances can be propagated by pushing and hauling. The sound speed is thus much greater than the molecular speed.

4.4 One-dimensional wave motion

Plane acoustic waves traveling in one dimension represent the simplest possible case. Practical cases of approximately one-dimensional wave motion include transmission through constant-area ducts and spherical waves far from the source of sound.

The one-dimensional wave equation $\phi_{tt} - c_0^2\phi_{xx} = 0$ has the solution, as we have already noted,

$$\phi = F(x - c_0 t) + G(x + c_0 t) \tag{4.31}$$

for the scalar velocity potential ϕ. From $\mathbf{u} = \nabla\phi$ we find

$$\mathbf{u} = \mathbf{e}(F' + G') \tag{4.32}$$

where $\mathbf{e}$ is the Cartesian unit vector in the x direction and the primes denote the derivative of each function with respect to its peculiar argument.[1] Equation (4.32) indicates fluid motion purely in the x direction, which is also the direction of wave motion, so that the waves are *longitudinal.*

The condensation is given by $S = -\phi_t/c_0^2$, or

$$S = \frac{1}{c_0}(F' - G') \tag{4.33}$$

so that the condensation and velocity have a simple relation. To emphasize this let us define

$$f(x - c_0 t) = +\frac{F'}{c_0}$$

$$g(x + c_0 t) = -\frac{G'}{c_0}$$

Then (4.32) and (4.33) become

$$\begin{aligned} u &= c_0(f - g) \\ S &= f + g \end{aligned} \tag{4.34}$$

where f and g are arbitrary functions of their respective arguments. The acoustic condition $S \ll 1$ is seen to be equivalent to $u \ll c_0$. We now have the following equivalent conditions for the smallness of acoustic perturbations:

$$\begin{gathered} \left| S = \frac{\rho - \rho_0}{\rho_0} \right| \ll 1 \\ \left| \frac{u}{c_0} \right| \ll 1 \\ |p| \ll \rho_0 c_0^2 \end{gathered} \tag{4.35}$$

For air or other ideal gases, $\rho_0 c_0^2 = \gamma P_0$, so that the pressure perturbations are "small" if they are much smaller than the undisturbed pressure, which is usually about 1 atm. For water, on the other hand, $\rho_0 c_0^2$ is about 20,000 atm, so that *positive* pressure perturbations of several

[1] For example, let $\eta \equiv x - c_0 t$ and $F = F(\eta)$. Then $F' \equiv dF/d\eta$. Note also that $(\partial/\partial x)_t F[\eta(x,t)] = (dF/d\eta)(\partial\eta/\partial x)_t = F'(\partial\eta/\partial x)_t$ and in this case $(\partial\eta/\partial x)_t = +1$. Similar expressions apply to $(\partial F/\partial t)_x$ and to higher-order derivatives.

hundred atmospheres may still be in the acoustic range. Negative pressure perturbations in liquids, with magnitude equal to or greater than P_0, will result in *cavitation*; i.e., the liquid will not sustain negative pressures and evaporates (for equilibrium fluid states, cavitation takes place when the liquid pressure falls to the vapor pressure). This phenomenon occurs, for example, in some underwater explosions, and is utilized in certain ultrasonic cleaning devices.

Simple One-dimensional Waves

We make the final simplification that waves travel in one direction only. These *simple* waves may be random, periodic, or discrete.

If we visualize a speaking tube (essentially a long pipe open at both ends) of the kind found on old ships or in Victorian houses, simple waves correspond to the case in which there is a speaker at one end and an auditor at the other: waves travel only in the direction from speaker to auditor.

To obtain the simple wave case we set either f or g to zero in Eq. (4.34), depending on whether "positive" or "negative" waves are wanted. There is no physical distinction between the two cases, since the choice of the positive x direction is arbitrary. Let $g = 0$; then

$$\begin{aligned} S &= f(x - c_0 t) \\ u &= c_0 f(x - c_0 t) \end{aligned} \tag{4.36}$$

For waves traveling in the $+x$ direction, $u = c_0 S$.

As already noted, we have the elementary and essential result that all variables (u, S, P, ρ, etc.) are constant for an observer who travels with the wave, i.e., with velocity $dx/dt = +c_0$. In addition, S and u are interdependent; in particular, since they have the same sign, *in simple waves fluid with* $S > 0$ *(compressed) travels in the same direction as the waves while fluid with* $S < 0$ *(rarefied) travels in the opposite direction.* This generalization can be physically visualized as the motion produced by an accelerating piston, in which the fluid ahead of the piston is compressed ($S > 0$) and waves propagate in the direction of motion.

It is convenient to graphically represent the acoustic motions in the space of the *independent* variables, x and t. Such a picture is called a *wave diagram* or xt diagram. As an illustration, consider a bug starting from the origin and crawling along the x axis (one-dimensional motion!), as sketched in Fig. 4.4. We assume that the bug's displacement is given by $X_b = \frac{1}{2}at^2$ and plot this function in the xt plane, with t chosen as

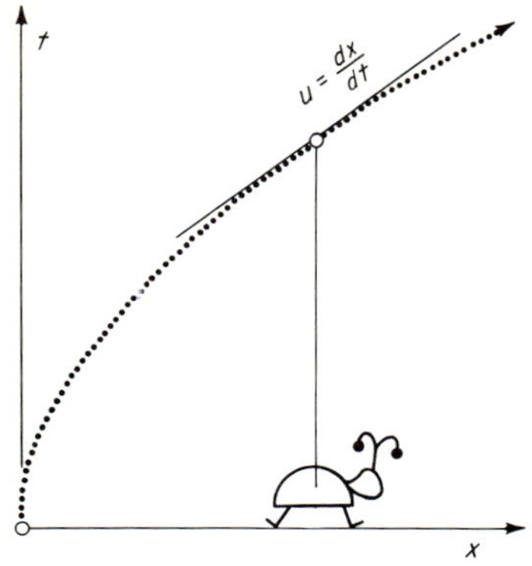

Figure 4.4

ordinate by convention. The velocity dx_b/dt appears on the diagram as the inverse slope of the plotted path line.

In unsteady acoustic flow problems we are interested in fluid *particle paths* ($dx/dt = u$) and wave paths, which are called *characteristic lines* ($dx/dt = \pm c_0$). Because $|u| \ll c_0$, we expect the particle paths to be nearly vertical on the wave diagram.

Consider the *displacement* ξ of a particular fluid particle. If x_0 is the initial undisturbed position of the particle, the position at any time is $x_0 + \xi(x_0,t)$. The displacement of the particle is given by

$$d\xi = u\,dt \tag{4.37}$$

where from (4.36)

$$u = c_0 f(x_0 + \xi - c_0 t) \tag{4.38}$$

It is customary and convenient to neglect ξ in the argument of f; this is possible because $\xi \sim ut \ll c_0 t$. As long as f is a slowly-varying function of its argument, the approximation is admissible. A somewhat awkward nonlinear form is thus avoided in (4.37), which integrates to

$$\xi = c_0 \int_0^t f(x_0 - c_0 \tau)\,d\tau \tag{4.39}$$

in which t has been replaced in the integrand by the running variable τ, the upper limit t is the "present" time at which ξ is evaluated, and x_0 is constant for the integration.

It is not difficult to show that this gives the result that $\xi = \xi(x_0 - c_0 t)$;

that is, the function $\xi(x_0,t)$ itself satisfies the wave equation. Let $\eta = x_0 - c_0\tau$; then the integral (4.39) becomes

$$\xi = -\int_{x_0}^{x_0 - c_0 t} f(\eta)\, d\eta$$

This is rewritten formally

$$\xi = -\int_{0}^{x_0 - c_0 t} f(\eta)\, d\eta + \int_{0}^{x_0} f(\eta)\, d\eta$$

Let $x_0 > 0$ and the first (wave) disturbance be on $x_0 - c_0\tau = 0$. Then $f(\eta) = 0$ for $\eta > 0$ and the second integral vanishes. Thus $\xi = \xi(x_0 - c_0 t)$, as was to be shown.

A particular example of a simple progressive wave may be helpful. We take the case of the sinusoidal wave.

EXAMPLE 4.1 PARTICLE DISPLACEMENT

Let the condensation be given by

$$S = f(x - c_0 t) = -S_0 \sin \frac{2\pi}{\lambda}(x - c_0 t) \qquad x - c_0 t < 0 \tag{4.40}$$

where S_0 is the amplitude. The coefficient $2\pi/\lambda$ is called the *wave number* and given the symbol k. This equation completely specifies the wave, as all other quantities can be specified in terms of the condensation S. Thus

$$u = c_0 S \qquad \rho = \rho_0(1 + S) \qquad p = \rho_0 c_0{}^2 S$$

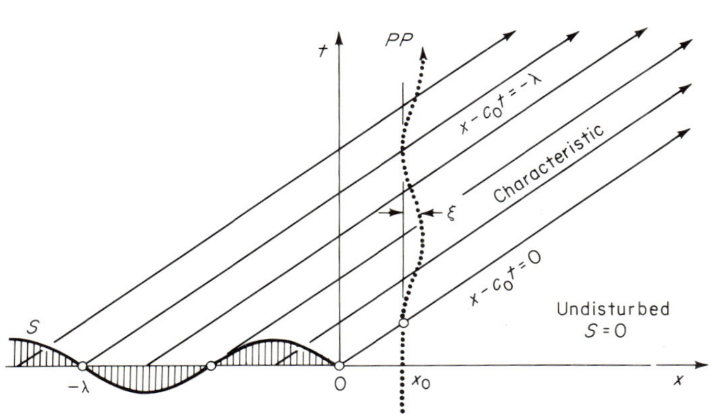

Figure 4.5
Wave diagram for simple sinusoidal wave.

The wave diagram is shown in Fig. 4.5. It is assumed that the first wave arrives at the origin at $t = 0$. The displacement of a particle originally at x_0 is easily calculated from (4.39)

$$\xi = -S_0 c_0 \int_{t_0}^{t} \sin \frac{2\pi}{\lambda} (x_0 - c_0 \tau)\, d\tau$$

with the lower limit t_0 taken as the time of first arrival of the wave, $t_0 = x_0/c_0$. (Note that $\int_0^{t_0} u\, d\tau = 0$.) Evaluation gives

$$\xi = \frac{S_0 \lambda}{2\pi} \left[1 - \cos \frac{2\pi}{\lambda} (x_0 - c_0 t) \right] \tag{4.41}$$

The corresponding particle path *PP* is shown on the wave diagram. As might have been anticipated, the particle-displacement *amplitude* is proportional to $S_0 \lambda$.

For waveforms other than sinusoidal we find similar results. In Fig. 4.6 the temporal displacement is plotted with the temporal (condensation) waveform. It is *not* necessary that the waves in question be periodic.

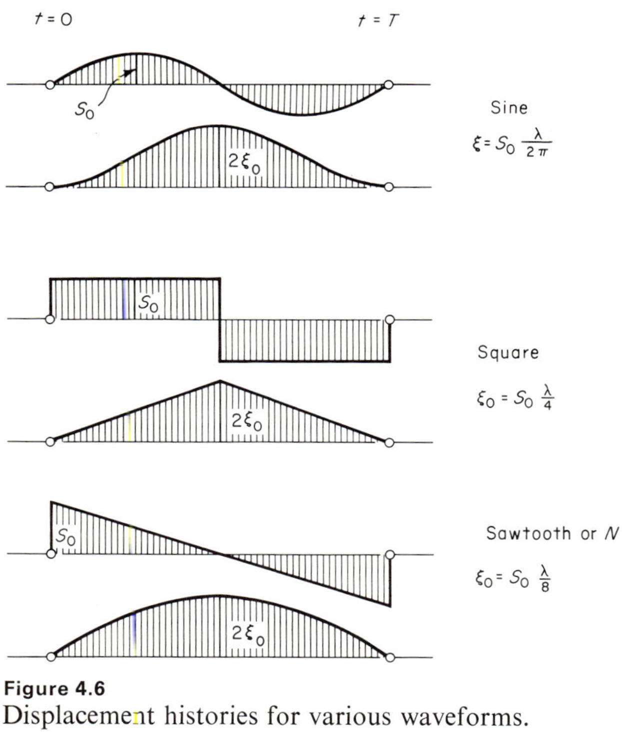

Figure 4.6
Displacement histories for various waveforms.

The wavelength λ and period T are related via $\lambda = c_0 T$. Note that the displacement amplitude $2\xi_0$ is a peak-to-peak value.

A formula which simplifies certain displacement calculations is given in Prob. 4.36.

4.5 Spherically symmetric wave motion

This simple case is of physical interest to the extent that a source of sound can be approximated as a point or as a perfect sphere.

The wave equation (4.18) written in spherical coordinates, with spherical symmetry (note that ϕ is still the velocity potential, *not* a spherical coordinate!) is

$$\frac{\partial^2 \phi}{\partial t^2} - \frac{c_0{}^2}{r} \frac{\partial^2}{\partial r^2} r\phi = 0 \tag{4.42}$$

where the second term is just $c_0{}^2 \nabla^2 \phi$. The equation can be rewritten as

$$\frac{\partial^2}{\partial t^2} r\phi - c_0{}^2 \frac{\partial^2}{\partial r^2} r\phi = 0 \tag{4.43}$$

which is precisely the one-dimensional wave equation in the dependent variable $r\phi$. We thus have the one-dimensional solution for $r\phi$

$$r\phi = F(r - c_0 t) + G(r + c_0 t)$$

or

$$\phi = \frac{1}{r} [F(r - c_0 t) + G(r + c_0 t)] \tag{4.44}$$

There are two families of waves, as before. In this case the F waves travel outward from the origin ($r = c_0 t + \text{const}$) while the G waves travel inward toward the origin. It is normally the divergent outward-traveling waves which are of physical interest, as in the case of sound radiated from a source.

From (4.16) and (4.17) we have for the condensation and velocity corresponding to the above velocity potential

$$S = \frac{1}{c_0 r} (F' - G') \tag{4.45}$$

$$\mathbf{u} = \mathbf{e}_r \left[\frac{1}{r} (F' + G') - \frac{1}{r^2} (F + G) \right] \tag{4.46}$$

the velocity equation again indicating longitudinal fluid motion. Let us pass immediately to the physically interesting case of simple diverging waves with the above equations rewritten with $f \equiv F'/c_0$

$$S = \frac{1}{r} f(r - c_0 t) \tag{4.47}$$

$$\frac{u}{c_0} = \frac{1}{r} f(r - c_0 t) - \frac{1}{c_0 r^2} F(r - c_0 t) \tag{4.48}$$

The condensation waves (and therefore the pressure waves, density waves, etc.) propagate outward *without change of form but attenuate as* $1/r$. The velocity does not show such a simple relation to the condensation as in the one-dimensional case: the second term introduces a certain *distortion* or change of form to the velocity waves. The second term on the right side of (4.48) attenuates as $1/r^2$, however, and at large distances from the origin is negligible;[1] in this situation we have just the one-dimensional result but with a $1/r$ attenuation coefficient.

These mathematical results are verified by ordinary experience. If a tenor sings "Celeste Aïda" in a very large room, the listener hears the same music whether he is 20 or 200 ft away; there is only a difference in amplitude corresponding to $1/r$. Acoustic distortion is not present, or at least not detected.[2]

Somewhat strangely, the treatment of *cylindrically* symmetric waves is more complicated than that for spherically symmetric waves (for cylindrical waves, see *Lamb* [1932, art. 302]).

[1] For periodic waves "large distances" means $r \gg \lambda$ (see Prob. 4.11).

[2] Distortion can be produced by either *viscous* or *nonlinear* effects or both acting together. These effects will be discussed. The effects produced by boundary reflections are of course important in practical architectural acoustics.

Figure 4.7
Spherical propagation.

4.6 The acoustic approximations

We now turn our attention to the question of the validity and limits of the acoustic approximations which led to the wave equation.

As a model consider a progressive simple wave of length λ and amplitude S_0. Then there are variations in the density, pressure, and velocity, over corresponding variations in distance and time:

$$\delta\rho \sim \rho_0 S_0 \qquad \delta P \sim \rho_0 c_0^2 S_0 \qquad \delta u \sim c_0 S_0$$

$$\delta x \sim \lambda \qquad \delta t \sim \frac{\lambda}{c_0}$$

The order of magnitude of the various terms in the full one-dimensional equations for continuity and momentum is then as indicated below

$$\underset{\left(\frac{\rho_0 c_0 S_0}{\lambda}\right)}{\frac{\partial\rho}{\partial t}} + \underset{\left(\frac{\rho_0 c_0 S_0^2}{\lambda}\right)}{u\frac{\partial\rho}{\partial x}} + \underset{\left(\frac{\rho_0 c_0 S_0}{\lambda}\right)}{\rho\frac{\partial u}{\partial x}} = 0 \tag{4.49}$$

$$\underset{\left(\frac{c_0^2 S_0}{\lambda}\right)}{\frac{\partial u}{\partial t}} + \underset{\left(\frac{c_0^2 S_0^2}{\lambda}\right)}{u\frac{\partial u}{\partial x}} + \underset{\left(\frac{c_0^2 S_0}{\lambda}\right)}{\frac{1}{\rho}\frac{\partial P}{\partial x}} = \underset{\left(\frac{\mu'}{\rho_0}\frac{c_0 S_0}{\lambda}\right)}{\frac{4}{3}\frac{\mu'}{\rho}\frac{\partial^2 u}{\partial x^2}} \tag{4.50}$$

where the viscosity has been treated as constant and for conciseness indicated by the abbreviation μ'

$$\tfrac{4}{3}\mu' \equiv \tfrac{4}{3}\mu + \mu_v \tag{4.51}$$

The nonlinear convective terms $u\rho_x$ and uu_x are evidently negligible for $S_0 \ll 1$, the already anticipated small-condensation condition.

Ordinary audible sound has condensation amplitudes of order 10^{-6}, roughly. The condition $S \ll 1$ is thus very well satisfied. There are, however, problems in *gasdynamics*, such as blast-wave propagation, for which S is relatively large and a nonlinear theory of waves is required. This is called the *theory of waves of finite amplitude.* We will consider such problems in Chap. 8.

The relative magnitude of the viscous-force term in the momentum equation (4.50) depends partly on the peculiar properties of the fluid medium. As an important and convenient case we take the *ideal gas* for which $\mu' \sim \mu \approx \rho c\Lambda$ [see Eq. (2.125)]. Now the relative magnitudes of the terms in the momentum equation (4.50) become

$$\begin{array}{ccccccc} \dfrac{\partial u}{\partial t} & + & u\dfrac{\partial u}{\partial x} & + & \dfrac{1}{\rho}\dfrac{\partial P}{\partial x} & = & \dfrac{4}{3}\dfrac{\mu'}{\rho}\dfrac{\partial^2 u}{\partial x^2} \\ (1) & & (S_0) & & (1) & & (\Lambda/\lambda) \end{array} \tag{4.52}$$

The viscous term is of the order of the ratio of the mean free path to the wavelength of sound (or, equivalently, the ratio of sound frequency to molecular collision frequency). This is a form of *Knudsen number*. For air at standard conditions, with $\Lambda = 6.5 \times 10^{-8}$ m, a sound frequency of $\nu \approx 5 \times 10^9\ \mathrm{s}^{-1}$ is required to make Λ/λ unity. Thus, either very high sound frequency or extreme gas rarefaction (or some combination) is required to give the viscous term magnitude of order unity. Under standard conditions with $v = 1{,}000\ \mathrm{s}^{-1}$ in air we find $\Lambda/\lambda \approx 2 \times 10^{-7}$ or apparently utterly negligible.

When the viscous force is, under extreme conditions, of the same order as the pressure force, we have $\Lambda \sim \lambda$ and the *continuum* model has already broken down because we have a "macroscopic" characteristic length λ which comes close to molecular dimensions: our continuum has become very "grainy." In this case the Boltzmann kinetic theory is required for the study of acoustics.

Even when the viscous force is very small, however, as in the above example, it has an effect on the *attenuation* of sound and must be considered, together with the additional entropy-generating phenomena of heat conduction and relaxation. Although ordinary sound propagation is very close to reversible (sufficiently close for one to be tempted to call it the best approximation in nature to a *macroscopic* reversible process), the dissipative effects make themselves felt in propagation over very long distances or within resonant "hard" enclosures.

A standard (and fallacious) argument, originally due to Laplace, holds that the state changes of a fluid particle in acoustic motion are isentropic because the compression and rarefaction occur so rapidly that there is insufficient time for heat transfer. This argument implies that the isentropic model should improve with increasing acoustic frequency. The heat transfer, however, can be shown to be proportional to Λ/λ, that is, like the viscous term in Eq. (4.52), hence proportional to frequency (see Prob. 4.28). It is only at relatively *low* frequency that gradients are small enough to render dissipation negligible.

4.7 Transport of energy and momentum

It is remarkable that all modes of wave propagation transport momentum and energy with a characteristic wave speed; ordinary acoustics is an important and instructive case.

In the absence of heat conduction and viscous stress the *instantaneous* energy flux or rate of work transfer is the vector

$$P\mathbf{u} = P_0\mathbf{u} + p\mathbf{u} \tag{4.53}$$

We will calculate the average value of this quantity for simple one-dimensional waves. Let $\mathbf{e}$ be the unit vector in the direction of propagation. Then $\mathbf{u} = \mathbf{e}c_0S$, and $p = \rho_0 c_0{}^2 S$, so that the time average of (4.53) becomes

$$\mathbf{\Phi}_e = \frac{\mathbf{e}}{T}\int_0^T (P_0 + \rho_0 c_0{}^2 S)c_0 S\,dt \tag{4.54}$$

which is the *average energy flux* or vector *sound intensity*, a vector in the direction of wave propagation. The first integral in (4.54) is simply P_0 times the net displacement of a fluid particle over the time interval T. It is usually true that T can be chosen such that the net displacement is zero; in the case of *periodic* waves (the only one for which the average energy flux is of interest, actually) this is accomplished by letting T correspond to an integral number of periods or be very large (in which case the first integral becomes negligibly small compared to the second). For definiteness, let $T =$ one period; then

$$\mathbf{\Phi}_e = \mathbf{e}\rho_0 c_0{}^3 \overline{S^2} \tag{4.55}$$

where $\overline{S^2} \equiv \int_0^T S^2\,dt/T$ is the mean-squared condensation. As an important example, consider sinusoidal waves with $S = S_0 \sin \omega t$ at a point in space; with $\omega T = 2\pi$,

$$\overline{S^2} = \frac{1}{T}\int_0^T S_0{}^2 \sin^2 \omega t\,dt = \frac{S_0{}^2}{2}$$

Note that with $S \ll 1$, the energy flux Φ_e is a quantity of second-order smallness.

Let us find the corresponding momentum flux Φ_m. For a fixed surface in space with unit normal $\mathbf{n}$, the instantaneous momentum flux is $(\rho\mathbf{u}\cdot\mathbf{n})\mathbf{u}$. We consider again simple one-dimensional waves, with the propagation unit vector $\mathbf{e}$ aligned with the surface normal $\mathbf{n}$; that is, $\mathbf{e} = \mathbf{n}$. Then the time-average momentum flux is

$$\boldsymbol{\Phi}_m = \frac{\mathbf{e}}{T}\int_0^T \rho_0(1 + S)c_0{}^2S^2\,dt \tag{4.56}$$

and retaining quantities up to second-order smallness,

$$\boldsymbol{\Phi}_m = \mathbf{e}\rho_0 c_0{}^2\overline{S^2} \tag{4.57}$$

We thus have the simple relation

$$\boldsymbol{\Phi}_e = c_0\boldsymbol{\Phi}_m \tag{4.58}$$

which is common to various forms of wave propagation. The quantity $\rho_0 c_0{}^2\overline{S^2}$ is called the acoustic *radiation pressure*. It has the physical significance of a potential force transfer by a kind of "action at a distance" via acoustic radiation. Thus it is possible, in principle, for a vociferous debater to drive his opponent back by sufficiently strong argument.

The physical intensity of sound is conventionally measured on the *decibel scale*.[1] The weakest audible sound or *threshold of hearing* (at 1,000 Hz) has an intensity, for an average person, of about

$$\Phi_{e(\mathrm{ref})} = 10^{-12}\ \mathrm{W/m^2}$$

This remarkably small flux is taken as a reference value for the decibel scale. Weber's law of sensation suggests that hearing mechanisms respond essentially to the logarithm of the intensity:[2] response $\sim \log_{10}\Phi_e$; consistent with this, the decibel scale is defined by

$$\Delta \equiv 10\log_{10}\frac{\Phi_e}{\Phi_{e(\mathrm{ref})}} \tag{4.59}$$

which is the *sound intensity level in decibels* (dB). Thus the weakest perceptible sound has a level $\Delta = 0$ dB. The level of sound at which immediate damage[3] to the ear is almost certain, and above which it does not discriminate, is roughly at $\Delta = 140$ dB. Over an enormous range of energies—some 14 orders of magnitude—the ear responds. Confirming

[1] A decibel is formally one-tenth of a *bel*, named in honor of Alexander Graham Bell, who worked on problems of acoustics and hearing. Historically, however, the decibel unit originated as a measure of attenuation in telephony in the early 1900s (see *Huntley* [1970]).

[2] More precisely, the smallest change δS in a stimulus S leading to a perceived change is such that $\delta S/S$ is a constant; thus $\delta E = K\,\delta S/S$ integrates to $E = K \log S$ and some constant difference ΔE corresponds to the smallest perceptible change.

[3] Repeated exposure to $\Delta = 90$ dB approximately is considered likely to produce progressive damage, particularly at high frequencies. Levels considerably above this have been measured at "rock" concerts, where instruments are amplified.

Table 4.2 Energy and Momentum Constants

Sinusoidal Waves in Air: 300 K, 1 Atm, 100 Hz					
Δ, dB	Φe, W/m^2	Φm, atm	S_{rms}	P_{rms}, atm	ξ_0, m
0	10^{-12}	2.84×10^{-20}	1.43×10^{-10}	1.99×10^{-10}	1.12×10^{-11}
140	10^2	2.84×10^{-6}	1.43×10^{-3}	1.99×10^{-3}	1.12×10^{-4}

Acoustic Constants			
	Air		Water
	0°C, 1 atm	27°C, 1 atm	0°C, 1 atm
ρ, kg/m^3	1.293	1.177	1,000
c, m/s	331.4	347.6	1,500
ρc, kg/(m^2)(s)	429	408	1.50×10^6
ρc^2, N/m^2	1.42×10^5	1.42×10^5	2.25×10^9
ρc^3, W/m^2	4.70×10^7	4.92×10^7	3.37×10^{12}

Weber's law, it is found that the smallest perceptible change in sound level is very close to 1 dB over most of the range.

To illustrate the extreme smallness of acoustic phenomena, Table 4.2 lists various amplitudes at the extremes of the audible range. For convenience, acoustic constants in air and water are also listed.

Rayleigh has calculated that the minimum acoustic energy flux $\Phi_{e(\text{ref})}$ perceptible to the ear is slightly less than the minimum *light* intensity perceptible to the eye. At this level of sound and 1,000 Hz the fluid-particle amplitude of motion is about 1 Å, and smaller for higher frequencies. This is less than the atomic diameter of the smallest atom: the acoustic motion is a very faint perturbation of statistical order superimposed on molecular chaos.

The human voice has a total acoustic power, in ordinary conversation, of about 10^{-6} W. A symphony orchestra in crescendo passage produces 70 W.† The power radiated from the exhaust of a large rocket may be kilowatts.

† In the concert hall (Bell Telephone Labs measurement, quoted by C. F. Officer, *Introduction to the Theory of Sound Transmission*, McGraw-Hill, New York, 1958). The acoustic output of a high-fidelity reproduction system with a 70-W amplifier is of course much smaller.

The decibel scale is also applied to the *sound pressure level*, as follows:

$$\text{Pressure level in decibels} \equiv 20 \log_{10} \frac{p_{\text{rms}}}{p_{\text{ref}}} \tag{4.60}$$

If we suppose for a moment that the value of p_{ref} corresponds exactly to the value of $\Phi_{e(\text{ref})}$, then assuming both the measurement and the reference values are for the same fluid, at the same conditions

$$\frac{p_{\text{rms}}}{p_{\text{ref}}} = \frac{\rho_0 c_0{}^2 S_{\text{rms}}}{\rho_0 c_0{}^2 S_{\text{ref,rms}}} = \sqrt{\frac{\overline{S^2}}{(\overline{S^2})_{\text{ref}}}} = \sqrt{\frac{\Phi_e}{\Phi_{e(\text{ref})}}}$$

and (4.60) is precisely equivalent to (4.59); the decibel scales for intensity and pressure would then be identical and, in fact, redundant. In practice, reference and test may not involve the same fluids or the same conditions of pressure and temperature, and so the scales become distinct. In addition there are *two* different values of p_{ref}, each one widely used. In ordinary acoustics $p_{\text{ref}} = 2.2 \times 10^{-5}\ \text{N/m}^2$, corresponding *roughly* to $\Phi_{e(\text{ref})}$, is normally used. In geophysics and underwater acoustics the larger value $p_{\text{ref}} = 0.1\ \text{N/m}^2$ is taken and corresponds *roughly* to the level of background noise in the sea.

There are a variety of units related to the decibel scale and used in connection with physiological acoustics. In the ancient field of sound, there seems to have been little reluctance, from the time of Pythagoras, to propose a new unit or a new standard, leading to such esoteric units as the phon, sone, sabin, bel, pulsatance, rayl, mel, noys, pnoys, bar, and neper.

Balance of Acoustic Energy

A more general description of acoustic energy transport is possible via the energy equation (1.60). The energy per unit mass of fluid at any instant is $e + u^2/2$. We calculate the internal energy e by the Taylor expansion of $e(v,s)$:

$$e - e_0 = \left(\frac{\partial e}{\partial v}\right)_s (v - v_0) + \frac{1}{2}\left(\frac{\partial^2 e}{\partial v^2}\right)_s (v - v_0)^2 + \cdots$$

With $(\partial e/\partial v)_s = -P$ and $v - v_0 = (\partial v/\partial P)_s p + \cdots$, this yields for the specific fluid energy

$$e + \frac{u^2}{2} = e_0 - P_0(v - v_0) + \frac{p^2}{2\rho_0{}^2 c_0{}^2} + \frac{u^2}{2} + O(p^3)$$

Retaining only terms through second order, the energy equation (1.60) becomes, for inviscid motion,

$$\rho \frac{D}{Dt}\left[e_0 - P_0(v - v_0) + \frac{p^2}{2\rho_0{}^2c_0{}^2} + \frac{u^2}{2}\right] = -\nabla\cdot(P\mathbf{u})$$

Writing $P = P_0 + p$ and with $\rho\, Dv/Dt = \nabla\cdot\mathbf{u}$ from continuity, this becomes

$$\rho \frac{D}{Dt}\left(\frac{p^2}{2\rho_0{}^2c_0{}^2} + \frac{u^2}{2}\right) = -\nabla\cdot(p\mathbf{u})$$

Finally, with the convective derivative dropped in the acoustic approximation $D/Dt \approx \partial/\partial t$, this yields the acoustic-energy equation

$$\frac{\partial}{\partial t}\left(\frac{p^2}{2\rho_0c_0{}^2} + \frac{\rho_0u^2}{2}\right) + \nabla\cdot(p\mathbf{u}) = 0 \tag{4.61}$$

This has a simple interpretation. Consider the continuity equation written in the analogous form

$$\frac{\partial\rho}{\partial t} + \nabla\cdot(\rho\mathbf{u}) = 0 \tag{4.62}$$

where ρ is the mass per unit volume and $\rho\mathbf{u}$ is the flux of mass; this equation states that the accumulation of mass in a unit volume is necessarily balanced by a decrement in the mass flux. By analogy, the quantity

$$E \equiv \frac{p^2}{2\rho_0c_0{}^2} + \frac{\rho_0u^2}{2} \tag{4.63}$$

represents the instantaneous acoustic energy per unit volume and $p\mathbf{u}$ represents the instantaneous energy flux.

For simple waves which are nearly one-dimensional, e.g., spherical waves far from the origin, the perturbation pressure p and the velocity $\mathbf{u}$ have a simple relation[1]

$$p = \rho_0c_0{}^2S \qquad \mathbf{u} = \mathbf{e}c_0S \tag{4.64}$$

[1] It should be emphasized that this relation is more general than it may first appear. It will be shown in Sec. 4.10 that it holds for simple waves in a duct with arbitrary area variation, provided that

$$\frac{\lambda}{A}\frac{dA}{dx} \ll 1$$

For spherical or cylindrical waves, this condition corresponds roughly to $kr \gg 4\pi$ (see also *Malecki* [1969, sec. 1.6]).

where $\mathbf{e}$ is a unit vector in the direction of propagation. Then the acoustic-energy density is

$$E = \rho_0 c_0{}^2 \frac{S^2}{2} + \rho_0 c_0{}^2 \frac{S^2}{2} = \rho_0 c_0{}^2 S^2 \tag{4.65}$$

Note that the internal-energy and kinetic-energy contributions are equal; there is equipartition of energy. The energy flux becomes

$$p\mathbf{u} = \mathbf{e}\rho_0 c_0{}^3 S^2 = \mathbf{e}c_0 E \tag{4.66}$$

which agrees with the (time-averaged) equation (4.55). Now the energy equation (4.61) takes the elegant form

$$\frac{\partial E}{\partial t} + \nabla \cdot (E\mathbf{c}_0) = 0 \tag{4.67}$$

where $\mathbf{e}c_0$ has been written as $\mathbf{c}_0$ to show the analogy with the continuity equation (4.62). This verifies that the acoustic energy propagates with the wave speed c_0.

It is characteristic of wave motion that energy propagates with a well-defined velocity (though there is some confusion with respect to *dispersive* waves). This is not necessarily true in other forms of energy flow; in Fourier heat conduction, for example, there is no identifiable velocity of energy propagation.

4.8 Weak discontinuities

A discontinuity in amplitude (pressure, density, velocity) may propagate as a wave. If the discontinuity is sufficiently weak, the acoustic approximations will apply, and we consider this particular case. Physical examples of such a discontinuity are the initial front of a sonic boom and the waves produced by the impact of two plane surfaces, such as a hammer on an anvil. Weak discontinuities of this kind are actually degenerate shock waves (a comprehensive treatment of shock waves will be postponed to Chap. 7).

Let a discontinuity wave be moving in the positive direction with velocity c *with respect to the fluid* ahead of the wave. For generality, the fluid ahead of the wave is given velocity u_1, not necessarily small compared to c (though we can satisfy this condition, if needed, by a Galilean transformation). It is not required that the conditions ahead of and behind the wave be spatially uniform (see Fig. 4.8).

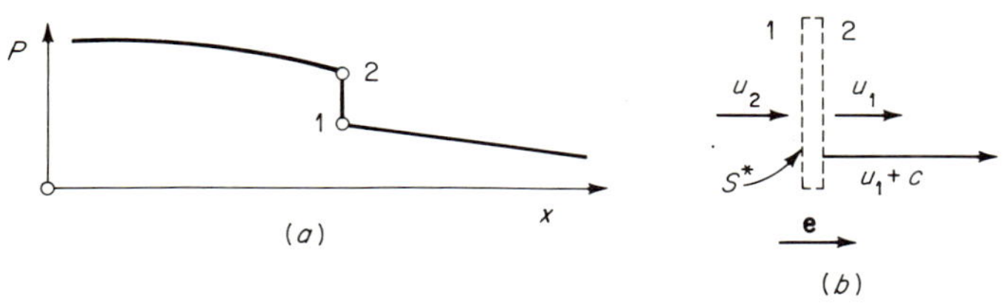

Figure 4.8
(*a*) Discontinuity in pressure; (*b*) control volume.

A control surface S^* of unit area on the faces 1 and 2 just encloses the discontinuity and moves with it at velocity $\mathbf{b} = \mathbf{e}(u_1 + c)$ such that it overtakes fluid ahead with relative velocity $\mathbf{e}c$. The control volume may be taken arbitrarily thin, as it encloses an idealized discontinuity of zero thickness. Making use of this, we can neglect the storage of mass and momentum within the volume, and the integral statements for balance of mass (1.79) and momentum (1.80) become respectively

$$\int_{S^*(t)} \rho(\mathbf{u} - \mathbf{b}) \cdot \mathbf{n}\, dA = 0$$

$$\int_{S^*(t)} \rho\mathbf{u}(\mathbf{u} - \mathbf{b}) \cdot \mathbf{n}\, dA = \int_{S^*(t)} \mathbf{T}\, dA$$

These reduce, for the propagation direction, to

$$\rho_1 c = \rho_2(u_1 + c - u_2) \tag{4.68}$$

$$-\rho_1 u_1 c + \rho_2 u_2(u_1 + c - u_2) = P_2 - P_1 \tag{4.69}$$

where the traction $\mathbf{T}$ has been set equal to $-P\mathbf{n}$. Substituting (4.68) into (4.69) gives

$$\rho_1 c(u_2 - u_1) = P_2 - P_1 \tag{4.70}$$

and (4.68) can be rearranged to give

$$c^2 = \frac{\rho_2}{\rho_1}\frac{P_2 - P_1}{\rho_2 - \rho_1} \tag{4.71}$$

To simplify these fundamental relations further, we introduce the acoustic approximation that $(\rho_2 - \rho_1)/\rho_1 \ll 1$, equivalent to $S \ll 1$. Then, with

$\delta P \equiv P_2 - P_1$ and $\delta\rho \equiv \rho_2 - \rho_1$ (the small delta being intended to suggest small but finite changes) the above equations become

$$\delta P = \rho c \, \delta u \tag{4.72}$$

$$c^2 = \frac{\delta P}{\delta \rho} \tag{4.73}$$

The relation between pressure and velocity changes (4.72) is precisely equivalent to the relations (4.64) for simple waves.

The derivation of (4.73) is one of the conventional derivations for the speed of sound; by arguing that state changes are isentropic, $\delta P/\delta\rho \to (\partial P/\partial\rho)_s$ and one obtains our Eq. (4.22). Since we have considered a *discontinuity*, however, it seems that velocity and temperature gradients must actually be *infinite*, which suggests difficulties for a model which neglects dissipation! We shall not attempt the resolution of this seeming paradox here but postpone it until the discussion of shock waves in Chap. 7; it turns out that (4.73) is in fact equivalent to the isentropic derivative provided that the acoustic condition $\delta\rho/\rho \ll 1$ is satisfied. From (4.72) and (4.73) we have the equivalent conditions $\delta P \ll c^2\rho$ and $\delta u \ll c$.

If the above derivation is repeated for a wave running in the *negative* direction, we obtain the same equation but with a minus sign on the right side. Thus for one-dimensional problems in general we obtain

$$\delta P = \pm \rho c \, \delta u \tag{4.74}$$

Table 4.3 Characteristic Acoustic Impedance

Unless Otherwise Noted, Values are for 15°C, 1 Atm, Approximately

Substance	$\mathscr{R} = \rho c$, kg/(m²)(s)
Hydrogen	110
Air	430
Water	1.5×10^6
Iron	4.7×10^7
Mercury	1.9×10^7
Ice (0°C)	2.9×10^5
ρc rubber	1.55×10^6
Glass	1.44×10^6
Platinum	8.4×10^7
Hypothetical incompressible fluid	∞

where the delta operator means the quantity ahead of the wave minus the quantity behind the wave. This simple relation embodies the "theory" of weak discontinuities.

The thermodynamic property ρc is called the *characteristic acoustic impedance* and is of course peculiar to the propagation medium. For brevity, and corresponding to electrical usage, we will adopt the symbol $\mathscr{R}$ in place of ρc (this quantity has no relation to the specific gas constant). The acoustic impedance is a measure of the stiffness of a material, in the sense that it is a proportionality constant between impressed velocity and pressure (the usual elastic moduli correspond however to ρc^2). Note that the dimensions are those of mass flux, corresponding to the mass flux seen by an observer riding with the wave.

EXAMPLE 4.2 CROSSING OF TWO WEAK DISCONTINUITIES

Let two plane surfaces of discontinuity be parallel and traveling toward each other, in the $+x$ and $-x$ directions, say. Given the strength of each discontinuity (as measured by the pressure jump δP, for example) it is desired to find the fluid conditions after the surfaces "collide" and pass each other.

The wave diagram is shown in Fig. 4.9. The quantities $p_a = P_a - P_0$ and $p_b = P_b - P_0$ are known, and from (4.74)

$$p_a = \mathscr{R}\hat{u}_a \qquad p_b = -\mathscr{R}\hat{u}_b$$

where $\hat{u} \equiv u - u_0$. Applying (4.74) again yields

$$p_c = p_a - \mathscr{R}(\hat{u}_c - \hat{u}_a) = p_b + \mathscr{R}(\hat{u}_c - \hat{u}_b)$$

Solving gives simply the algebraic sums

$$\hat{u}_c = \hat{u}_a + \hat{u}_b \tag{4.75}$$

$$p_c = p_a + p_b \tag{4.76}$$

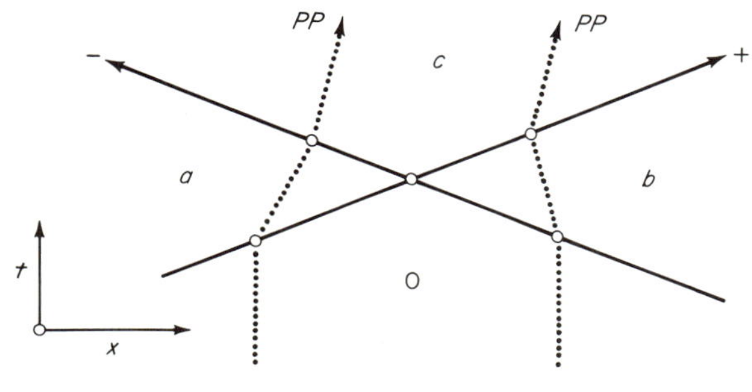

Figure 4.9
Wave diagram for crossing discontinuities.

The fluid conditions in fields a and b are not necessarily uniform; i.e., there may be spatial and temporal changes in the velocity and other properties. The above calculation is then valid only in the immediate neighborhood of the point of wave intersection. Note also that these quantities are given by a simple superposition, just as in the case of continuous waves.

4.9 Reflection and transmission at a boundary

All compressible media have boundaries, and all waves must eventually strike such boundaries, except in highly abnormal circumstances. In addition to the wave equation, we must consider the interaction with the boundaries.

For a boundary surface separating two media, 1 and 2, the *inviscid* matching conditions which must be satisfied are

$$P_1 = P_2 \tag{4.77}$$

$$u_{n1} = u_{n2} \tag{4.78}$$

where u_n is the component of velocity normal to the surface. In the case of solids the absolute "pressure" may become negative, corresponding to a positive tensile stress. The boundary represents a material discontinuity separating media with different properties. It is not necessary that the two media be chemically dissimilar; e.g., we may have a surface separating hot helium from cold helium.[1] In the language of gasdynamics, such a boundary is a *contact surface.*

Weak Discontinuity at Normal Incidence

We consider first the simplest possible one-dimensional case, that of a plane acoustic discontinuity normally incident on a plane contact surface. The situation is sketched in Fig. 4.10*b* in physical space, before and after the incident wave i strikes the contact surface (CS). The matching conditions (4.77) and (4.78) can be satisfied by the postulation of a transmitted wave t and a reflected wave r. Let the conditions ahead of the incident wave be P_0 and u_0. We adopt the notation, as before, $p = P - P_0$ and $\hat{u} = u - u_0$, where p and $\hat{u}$ are small in the sense of the acoustic approximations.

[1] Such a surface is inevitably smeared out by the action of heat transfer, diffusion, and turbulent mixing. To treat it as a discontinuity is an idealization.

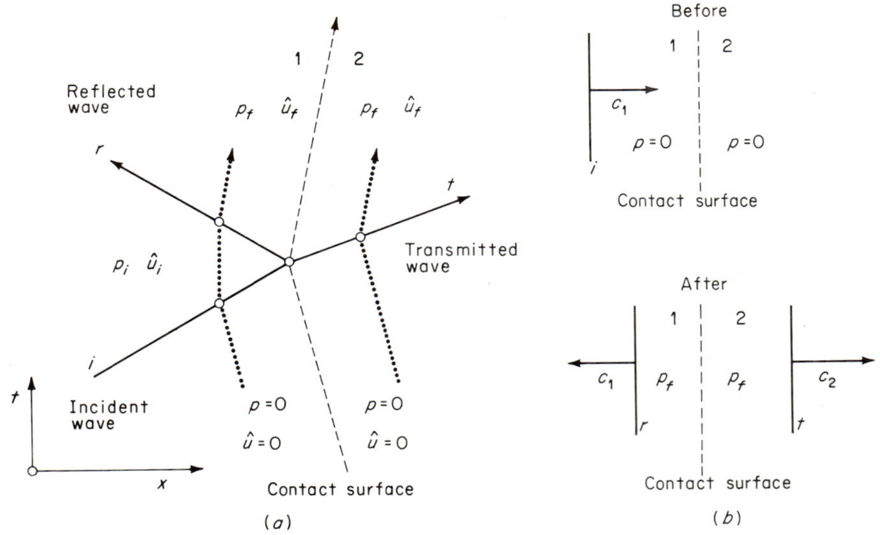

Figure 4.10
Weak discontinuity normally incident on a contact surface.

To find the conditions after intersection it is only necessary to apply the elementary equation (4.74). Let the strength of the incident wave be given as p_i, $p_i = \mathscr{R}_1\hat{u}_i$. Specification of p_i is in the nature of an initial condition: if $p_i > 0$, we have a *compression wave*; if $p_i < 0$, a *rarefaction wave*. The matching conditions require that $p_f = p_{f_1} = p_{f_2}$ and $\hat{u}_f = \hat{u}_{f_1} = \hat{u}_{f_2}$, and (4.74) gives

$$p_f = p_i - \mathscr{R}_1(\hat{u}_f - \hat{u}_i) = \mathscr{R}_2\hat{u}_f$$

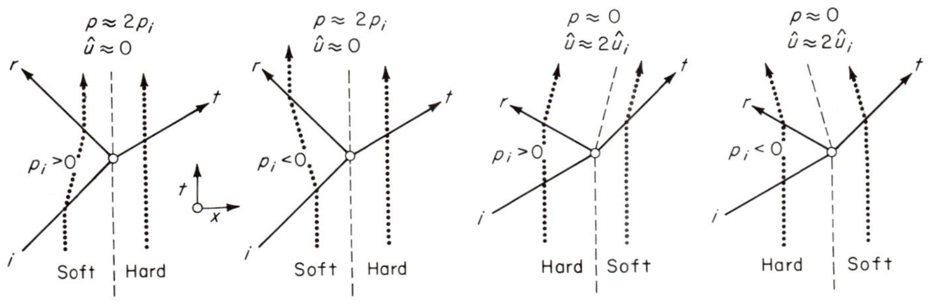

Figure 4.11
Mismatch in acoustic impedance.

Note that conditions ahead of and behind the wave are respectively taken at earlier and later times on the wave diagram. With $p_i = \mathscr{R}_1 \hat{u}_i$, this gives

$$\hat{u}_f = \frac{2\mathscr{R}_1}{\mathscr{R}_1 + \mathscr{R}_2} \hat{u}_i \tag{4.79}$$

$$p_f = \frac{2\mathscr{R}_2}{\mathscr{R}_1 + \mathscr{R}_2} p_i \tag{4.80}$$

Figure 4.12
Experimental wave diagram (smear-camera shadowgraph) showing acoustic discontinuity in water incident on (opaque) test specimen. (*Courtesy of N. L. Coleburn, Naval Ordnance Laboratory.*)

If $\mathscr{R}_1 = \mathscr{R}_2$, the discontinuity vanishes for purposes of wave transmission (even though the materials may be dissimilar) because the reflected wave disappears. This is the motivation for using "ρc" rubber in sonar housings so that the acoustic impedance of the rubber matches that of seawater. This situation is referred to as *impedance matching*. If $\mathscr{R}_1$ and $\mathscr{R}_2$ are greatly different, we have *impedance mismatch*; the four possible cases of mismatch are illustrated in Fig. 4.11.

An experimental wave diagram showing the reflection of a weak discontinuity is shown in Fig. 4.12.

EXAMPLE 4.3 NORMAL IMPACT

A solid plate (substance 1) of thickness L impacts normally against a semi-infinite slab (substance 2) with relative velocity U. The impact takes place in a vacuum. It is desired to find the subsequent motion of substances 1 and 2.

A trial wave diagram is shown in Fig. 4.13*b*. After impact the plate and slab are assumed to be in contact, with $p_{11} = p_{21}$, $u_{11} = u_{21}$. Applying these matching conditions and (4.74) yields

$$p_{11} = -\mathscr{R}_1(u_{11} - U) = \mathscr{R}_2 u_{11}$$

hence

$$u_{11} = \frac{\mathscr{R}_1}{\mathscr{R}_1 + \mathscr{R}_2} U \qquad p_{11} = \frac{\mathscr{R}_1 \mathscr{R}_2}{\mathscr{R}_1 + \mathscr{R}_2} U$$

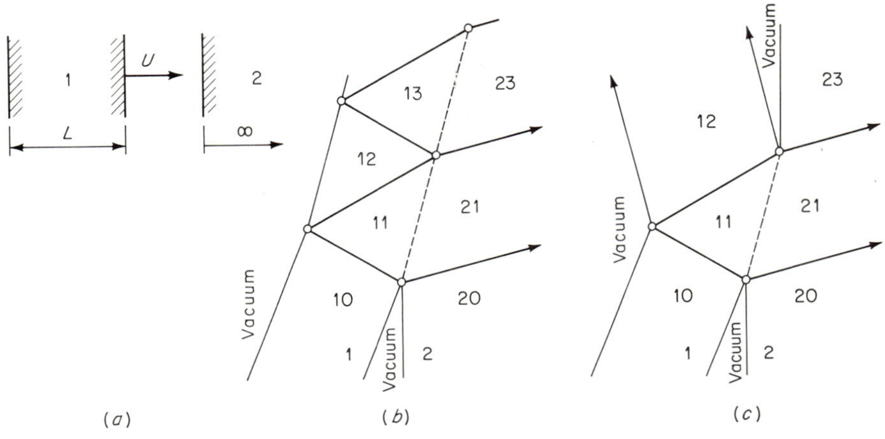

Figure 4.13
(*a*) Physical space; (*b*) trial wave diagram; (*c*) wave diagram for $\mathscr{R}_2 > \mathscr{R}_1$.

Table 4.4

Field	p	$\hat{u}$
10 20	0 0	U 0
11, 21	$\dfrac{\mathscr{R}_1\mathscr{R}_2}{\mathscr{R}_1 + \mathscr{R}_2} U$	$\dfrac{\mathscr{R}_1}{\mathscr{R}_1 + \mathscr{R}_2} U$
12	0	$\dfrac{\mathscr{R}_1 - \mathscr{R}_2}{\mathscr{R}_1 + \mathscr{R}_2} U$
13, 23	$\dfrac{\mathscr{R}_1\mathscr{R}_2(\mathscr{R}_1 - \mathscr{R}_2)}{(\mathscr{R}_1 + \mathscr{R}_2)^2} U$	$\dfrac{\mathscr{R}_1(\mathscr{R}_1 - \mathscr{R}_2)}{(\mathscr{R}_1 + \mathscr{R}_2)^2} U$

With the additional condition that $p_{12} = 0$ (since the left-hand side of the plate is exposed to vacuum) the calculation can be continued by repeated application of (4.74) and the matching conditions, under the assumption that the plate and slab remain in contact. The values that result are shown in Table 4.4. If $\mathscr{R}_1 > \mathscr{R}_2$, the calculation can be similarly extended to later fields. If, however, $\mathscr{R}_2 > \mathscr{R}_1$, the pressure $p_{13} = p_{23}$ is predicted negative, which is impossible (in the absence of adhesive on the colliding surfaces!). Let us take the case $\mathscr{R}_2 > \mathscr{R}_1$; evidently, the assumption that the plates are in contact at fields 13 and 23 must be abandoned, together with the tabulated values for p and u. If the plates are *not* in contact, the appropriate field pressures are zero (see Fig. 4.13*c*). Then with $p_{23} = 0$ we find

$$u_{23} = \frac{p_{23} - p_{21}}{\mathscr{R}_2} + u_{21} = 0$$

so that the slab is brought to rest, and the plate rebounds with velocity u_{12} as shown in the wave diagram.

It can be verified that this solution satisfies conservation of initial momentum and energy.

Arbitrary Wave at Any Angle of Incidence

We now consider the more general problem of plane waves of arbitrary form incident at any angle θ_i on a plane material boundary. The physical plane shown in Fig. 4.14 is the plane defined by the propagation unit

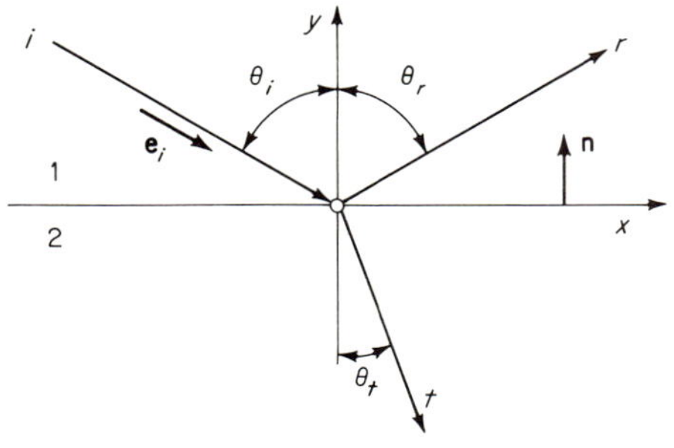

Figure 4.14
Physical plane, showing propagation directions of incident, reflected, and transmitted waves.

vector $\mathbf{e}_i$ of the incident wave and the boundary unit normal $\mathbf{n}$. The propagation directions of the incident wave i, reflected wave r, and transmitted wave t, assumed to be plane waves, are as shown at the (positive) angles θ_i, θ_r, and θ_t, respectively.

Again the incident wave may be considered in the nature of a specified initial condition. It is desired to find the resulting transmitted and reflected waves.

It will be more convenient to use wave coordinates of the form $\eta = t - l/c$ instead of $l - ct$ [note that $f(t - l/c)$ and $f(l - ct)$ both represent simple waves propagating in the positive direction of l]. By geometry, the distances from the origin along the propagation directions i, r, and t are respectively

$$\begin{aligned} l_i &= x \sin \theta_i - y \cos \theta_i \\ l_r &= x \sin \theta_r + y \cos \theta_r \\ l_t &= x \sin \theta_t - y \cos \theta_t \end{aligned} \tag{4.81}$$

which are positive in the propagation direction away from the origin. The respective plane-wave coordinates are then

$$\begin{aligned} \eta_i &= t - \frac{x \sin \theta_i - y \cos \theta_i}{c_1} \\ \eta_r &= t - \frac{x \sin \theta_r + y \cos \theta_r}{c_1} \\ \eta_t &= t - \frac{x \sin \theta_t - y \cos \theta_r}{c_2} \end{aligned} \tag{4.82}$$

By *superposition*, the velocity potentials in the two media, 1 and 2, are

$$\begin{matrix} \phi_i(\eta_i) + \phi_r(\eta_r) & y > 0 \\ \phi_t(\eta_t) & y < 0 \end{matrix} \tag{4.83}$$

where no sum is intended by repeated indices. We now make use of the matching conditions (4.77) and (4.78). The velocities are given by $\mathbf{u} = \nabla\phi$: hence $u_n = \mathbf{n}\cdot\nabla\phi$. Then in medium 1 the y-component velocity u_n^+ is

$$\begin{aligned} u_n^+ &= \mathbf{n}\cdot(\nabla\phi_i + \nabla\phi_r) \\ &= \mathbf{n}\cdot\left[\mathbf{e}_i\frac{d\phi_i}{d\eta_i}\left(\frac{\partial\eta_i}{\partial l_i}\right)_t + \mathbf{e}_r\frac{d\phi_r}{d\eta_r}\left(\frac{\partial\eta_r}{\partial l_r}\right)_t\right] \\ &= \phi_i'\frac{\cos\theta_i}{c_1} - \phi_r'\frac{\cos\theta_r}{c_1} \end{aligned}$$

where the primes denote differentiation with respect to the appropriate wave coordinates (η_i and η_r, respectively). Similarly, the normal component velocity in medium 2 is

$$u_n^- = \phi_t'\frac{\cos\theta_t}{c_2}$$

The pressure is given by $p = \rho c^2 S = -\rho\,\partial\phi/\partial t$. Then

$$\begin{aligned} p^+ &= -\rho_1\phi_i' - \rho_1\phi_r' \\ p^- &= -\rho_2\phi_t' \end{aligned}$$

Then matching conditions $u_n^+ = u_n^-$ and $p^+ = p^-$ applied at $y = 0$ (which is the *approximate* position of the boundary) are, respectively, with the arguments written out in full,

$$\phi_i'\left(t - \frac{x\sin\theta_i}{c_1}\right)\frac{\cos\theta_i}{c_1} - \phi_r'\left(t - \frac{x\sin\theta_r}{c_1}\right)\frac{\cos\theta_r}{c_1} = \phi_t'\left(t - \frac{x\sin\theta_t}{c_2}\right)\frac{\cos\theta_t}{c_2} \tag{4.84}$$

$$\phi_i'\left(t - \frac{x\sin\theta_i}{c_1}\right)\rho_1 + \phi_r'\left(t - \frac{x\sin\theta_r}{c_1}\right)\rho_1 = \phi_t'\left(t - \frac{x\sin\theta_t}{c_2}\right)\rho_2 \tag{4.85}$$

The reflected and transmitted waves may be expected to reproduce the incident wave but with different direction and amplitude. Then under this assumption

$$\phi_r = C_r\phi_i \qquad \phi_t = C_t\phi_i \tag{4.86}$$

where the constants C_r and C_t are respectively the *reflection coefficient* and *transmission coefficient*. Then the matching conditions (4.84) and (4.85) are each of the form

$$A_i f\left(t - \frac{x \sin\theta_i}{c_1}\right) + A_r f\left(t - \frac{x \sin\theta_r}{c_1}\right) = A_t f\left(t - \frac{x \sin\theta_t}{c_2}\right)$$

These relations must hold for *all* x and t, and the arguments of the functions must therefore be equal to each other, which gives

$$\frac{\sin\theta_i}{c_1} = \frac{\sin\theta_r}{c_1} = \frac{\sin\theta_t}{c_2} \tag{4.87}$$

This statement contains the *law of reflection*,

$$\theta_i = \theta_r \tag{4.88}$$

and *Snell's law of refraction*

$$\frac{\sin\theta_i}{c_1} = \frac{\sin\theta_t}{c_2} \tag{4.89}$$

That the three arguments appearing in (4.84) and (4.85) are equal to each other is equivalent to the requirement that the point of intersection of the three wavefronts I, R, T move along the axis (Fig. 4.15). This can be verified by simple geometry, with each wavefront moving normal to itself at the appropriate wave speed c.

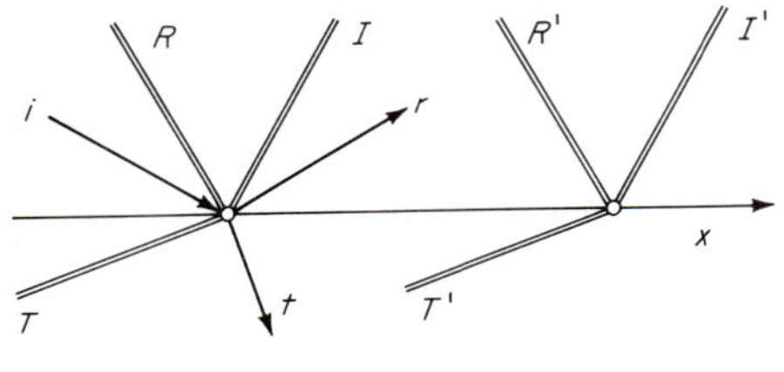

Figure 4.15

Substitution of (4.86) into (4.84) and (4.85) gives

$$\begin{aligned} C_r &= \frac{\mathscr{R}_2 \cos \theta_i - \mathscr{R}_1 \cos \theta_t}{\mathscr{R}_2 \cos \theta_i + \mathscr{R}_1 \cos \theta_t} \\ C_t &= \frac{2\rho_1 c_2 \cos \theta_i}{\mathscr{R}_2 \cos \theta_i + \mathscr{R}_1 \cos \theta_t} \end{aligned} \tag{4.90}$$

in which θ_t is known in terms of θ_i from (4.89).

The reflection coefficient has interesting behavior for certain values of the angle of incidence. If $\theta_t = \pi/2$, $C_r = 1$ and we have *total reflection* of the incident wave; from (4.89) this occurs at a value

$$\sin \theta_i = \frac{c_1}{c_2} \tag{4.91}$$

Thus, total reflection is possible only if $c_2 > c_1$. Note also that in cases of extreme impedance mismatch the reflection coefficient has unit magnitude: for $\mathscr{R}_2 \gg \mathscr{R}_1$, $C_r \to 1$; for $\mathscr{R}_2 \ll \mathscr{R}_1$, $C_r \to -1$. We find *zero reflection* or $C_r = 0$ when, from (4.90),

$$\sin \theta_i = \sqrt{\frac{(\mathscr{R}_2/\mathscr{R}_1)^2 - 1}{(\mathscr{R}_2/\mathscr{R}_1)^2 - (c_2/c_1)^2}} \tag{4.92}$$

which is real and less than unity only if $\mathscr{R}_2/\mathscr{R}_1 > 1 > c_2/c_1$ or $\mathscr{R}_2/\mathscr{R}_1 < 1 < c_2/c_1$.

There is a precise correspondence between extreme impedance mismatch at a material boundary and the classical boundary conditions for organ pipes. Specifically, the fixed closed end of an organ pipe corresponds to a boundary with $C_r = +1$, while the constant-pressure open end of an organ pipe corresponds to a free boundary with $C_r = -1$ (see Prob. 4.17).

It is the reflection and transmission of *energy* at a boundary which is of greatest physical interest. Let us verify that the results already obtained in terms of the velocity potential satisfy conservation of energy. The energy flux for a simple wave is, from Eq. (4.55),

$$\Phi_e = \rho c^3 \overline{S^2}$$

From (4.17) the condensation $S = -\dfrac{1}{c^2}\dfrac{\partial \phi}{\partial t}$. Then for ϕ in the form $\phi(t - l/c)$ we have $S = -\phi'/c^2$, so that the energy flux becomes

$$\Phi_e = \frac{\rho}{c}\,\overline{(\phi')^2} \tag{4.93}$$

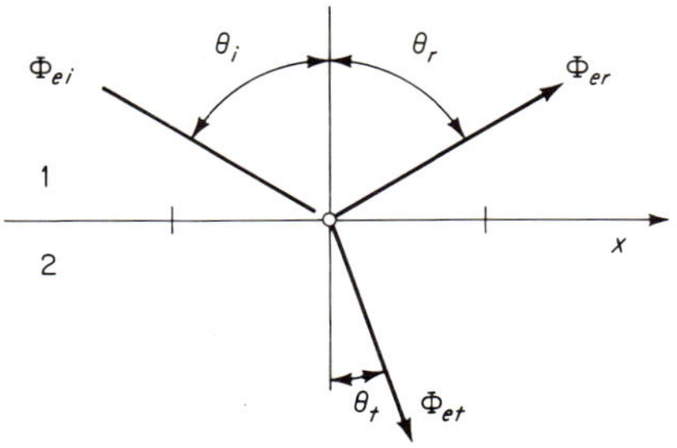

Figure 4.16

With Eq. (4.86), the incident, reflected, and transmitted energy fluxes become respectively

$$\Phi_{ei} = \frac{\rho_1}{c_1}\overline{(\phi_i')^2}$$

$$\Phi_{er} = C_r^2\Phi_{ei} \tag{4.94}$$

$$\Phi_{et} = \frac{\rho_2 c_1}{\rho_1 c_2} C_t^2\Phi_{ei}$$

We can now verify conservation of energy. Consider a unit area of the boundary surface (Fig. 4.16). Since there can be no energy storage within the surface itself, the normal component of the incident energy flux must be equal to the sum of the normal components of the reflected and transmitted fluxes, or

$$\Phi_{ei}\cos\theta_i = \Phi_{er}\cos\theta_i + \Phi_{et}\cos\theta_t$$

With Eq. (4.94) this gives

$$1 - C_r^2 = \frac{\rho_2 c_1}{\rho_1 c_2}\frac{\cos\theta_t}{\cos\theta_i} C_t^2 \tag{4.95}$$

By Eq. (4.90) this is just an identity, so that energy is conserved. The fraction of the incident energy which is reflected is C_r^2, and the fraction transmitted is $1 - C_r^2$ (see Fig. 4.17).

It is apparent from (4.90) that in the case of extreme impedance mismatch at normal incidence, very little energy is transmitted; that is, C_r^2 is very close to unity. In the familiar case of waves from air incident on a water surface with $\mathscr{R}_2/\mathscr{R}_1 \gg 1$ we find $1 - C_r = 5.74 \times 10^{-4}$, giving a fractional energy transmission $1 - C_r^2$ of 1.15×10^{-3}, which corresponds to an apparent attenuation at the surface of about 30 dB. It is very difficult for a speaker above water to be heard by an underwater

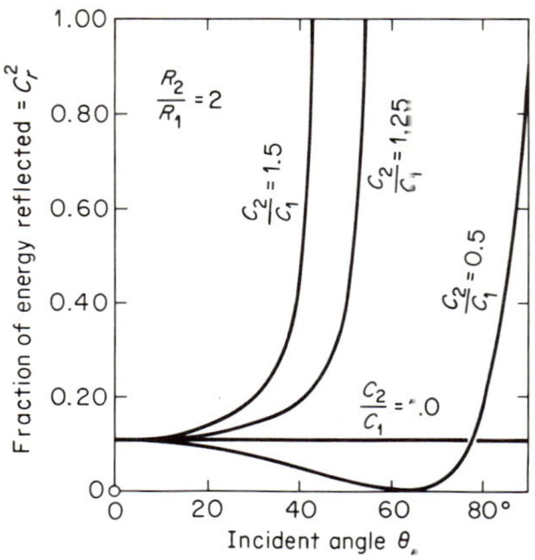

Figure 4.17
Energy reflection from a material boundary for the particular case $\mathscr{R}_2/\mathscr{R}_1 = 2$.

auditor (and vice versa, since C_r^2 is symmetric with respect to interchange of $\mathscr{R}_2$ and $\mathscr{R}_1$, for normal incidence). At an angle of only $\theta_i \approx 14°$ we have total reflection and zero transmission for the same case, for an apparent attenuation of ∞ dB.

Additional boundary-reflection problems, including the case of an absorptive boundary, are treated by *Morse and Ingard* [1968]. Instructive photographs of reflection/refraction and related phenomena can be found in *Bergmann* [1954, chap. 3, sec. d].

The phenomena of zero reflection, total reflection, and refraction are of course familiar in *optics*. It is not surprising that such analogies should appear between acoustics and optics, both of which are governed by the wave equation. The details of the matching conditions at a material boundary, however, are somewhat different. In Sec. 4.11 we will consider geometrical ray theory, the phenomenon of *continuous* refraction, which is formally analogous between acoustics and optics.

4.10 Propagation of sound in a duct

Acoustic propagation for geometries other than one-dimensional, cylindrical, or spherical is of practical interest. Specifically, consider inviscid motion in a duct with a cross-sectional area which may vary slowly along its length. With the transverse dimension, e.g., the diameter of a circular duct, small compared to the wavelength of sound, it can be assumed that the motion is quasi-one-dimensional; i.e., the spatial variation of properties is only along the length of the duct and not over the cross section. Further, the duct walls are assumed to be rigid and stationary.[1]

The continuity equation can be found with the help of the control volume shown in Fig. 4.18. The differential equation of continuity, integrated over the volume, is

$$\int_{V*} \left[\frac{\partial \rho}{\partial t} + \nabla \cdot (\rho \mathbf{u})\right] dV = 0$$

Applying the divergence theorem to the second term and integrating yields, with $A(x)$ the cross-sectional area of the duct,

$$A\,\Delta x \frac{\partial \rho}{\partial t} + (\rho u A)_{x+\Delta x} - (\rho u A)_x = 0$$

Dividing by Δx and taking the limit as $\Delta x \to 0$ gives the continuity equation

$$A \frac{\partial \rho}{\partial t} + \frac{\partial}{\partial x} \rho u A = 0 \tag{4.96}$$

where $A(x)$ is some known function.

The momentum equation is simply the x component of (1.67), or

$$\frac{\partial u}{\partial t} + u \frac{\partial u}{\partial x} + \frac{1}{\rho}\frac{\partial P}{\partial x} = 0 \tag{4.97}$$

Note that variation in area appears only in the continuity equation.

[1] If the duct walls are *elastic*, it is well known that compressibility of the fluid is augmented by the ability of the duct itself to expand and contract. This leads to a finite wave speed, even for an incompressible fluid; if the duct has radius R, thickness t, and elastic modulus E, $c^2 = Et/(2\rho R)$, a formula due to Thomas Young (1808). For example, the wave speed in the human main aortic artery (the large artery from the heart, $R \approx 1$ cm) is in the range 5 to 10 m/s. Wave propagation in elastic tubes is treated by the formalism of one-dimensional acoustics, and is sometimes called *water-hammer theory*. A recent treatment of blood flow is given by *Attinger* [1966].

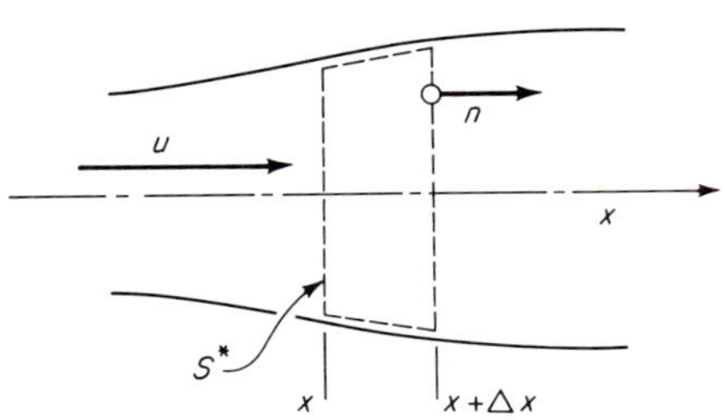

Figure 4.18

Under the acoustic assumptions introduced in Sec. 4.2, these equations are linearized to give

$$A\frac{\partial \rho}{\partial t} + \rho_0 \frac{\partial}{\partial x} uA = 0 \tag{4.98}$$

$$\frac{\partial u}{\partial t} + \frac{1}{\rho_0}\frac{\partial P}{\partial x} = 0 \tag{4.99}$$

Note, however, that we have retained the small convective term $\rho_0 u \, dA/dx$ in (4.98); if this term were dropped, the result would be just the (constant-area) one-dimensional wave equation.

With $S = (\rho - \rho_0)/\rho_0$ and $P - P_0 = \rho_0 c_0{}^2 S$ as before, Eqs. (4.98) and (4.99) are combined to yield the wave equation for a variable-area duct

$$A\frac{\partial^2 S}{\partial t^2} - c_0{}^2 \frac{\partial}{\partial x}\left(A\frac{\partial S}{\partial x}\right) = 0 \tag{4.100}$$

We can of course obtain the same equation for the pressure, velocity, etc.

Progressive or simple waves, which propagate along the tube in only one direction, are of primary interest. A plausible choice for a solution of this type is of the form

$$S = a(x)f(x - c_0 t) \tag{4.101}$$

With this form for S, Eq. (4.100) reduces to

$$\frac{f'}{a}\frac{d}{dx}Aa^2 + f\frac{d}{dx}\left(A\frac{da}{dx}\right) = 0 \tag{4.102}$$

With f an arbitrary function, this equation has exact solutions only for two particular cases of area variation, viz.,

$$A = \text{const} \rightarrow a = \text{const}$$

$$A \propto x^2 \qquad \rightarrow a = \frac{\text{const}}{x}$$

corresponding to one-dimensional (constant-area duct) and spherical (conical duct) propagation.

We can however obtain an important approximate solution to (4.102). The coefficient of the first term is f', which is of order f/λ, while the coefficient of the second term is just f; thus, for sufficiently short wavelength λ (high frequencies) we may expect the first term to dominate. Accordingly, the second term is dropped, and (4.102) becomes simply

$$\frac{d}{dx} A a^2 = 0 \tag{4.103}$$

with the solution

$$a(x) = \frac{\text{const}}{\sqrt{A}} \tag{4.104}$$

(which is exact for one-dimensional and spherical waves). Making use of this solution in estimating the order of magnitude of the terms in (4.102), it is found that the last term can indeed be neglected if

$$\frac{1}{\lambda} \gg \frac{1}{A} \frac{dA}{dx} \tag{4.105}$$

Furthermore, the additional relation

$$u = c_0 S \tag{4.106}$$

is found from (4.98) and (4.99) to hold under the same condition.

Thus, for sufficiently high frequencies, the propagation of simple waves in a variable-area duct is equivalent to one-dimensional propagation, with the addition of an amplitude factor $a \propto A^{-1/2}$.

Finally, these results are consistent with acoustic energy propagation. For periodic waves, the time average of the energy equation (4.67) is just

$$\nabla \cdot \bar{E} \mathbf{c}_0 = 0 \tag{4.107}$$

where $\bar{E}$ is the (average) energy density. For the duct problem, this corresponds to

$$A \bar{E} c_0 = \text{const} \tag{4.108}$$

which states that the flow of acoustic energy is the same at all sections of the duct. With $u = c_0 S$, Eq. (4.65) is applicable, and this becomes

$$A \rho_0 c_0{}^3 \overline{S^2} = \text{const} \tag{4.109}$$

For any given waveform, the amplitude is thus proportional to $A^{-1/2}$, which is the result already found.

The simple results found in this section can be usefully applied to the problem of geometrical acoustics, as discussed in the next section.

4.11 Geometrical acoustics

On any moderately large scale of distance (in the order of hundreds of meters, say) the atmosphere and the oceans are distinctly nonuniform. That is, the physical properties such as density, temperature, and in particular the sound speed vary significantly. We therefore reconsider the problem of small-amplitude (acoustic) motions, in a continuous medium which is no longer required to be spatially homogeneous. Specifically, we assume a simple fluid medium which is in static equilibrium in the undisturbed state, with no other restrictions on the spatial variation of thermodynamic properties.

Interest in nonuniform media is motivated by several peculiar phenomena. Instrumented dynamite explosions in Alaska showed clearly defined zones of complete silence, even relatively near the explosion source. Underwater sonar transmissions are possible over very great, even trans-oceanic, distances, with far less amplitude attenuation than would be expected from spatial (spreading) effects alone. Sonic booms show regions of anomalous high (and low) intensity. In general, the harmony or cacophony of man, machine, and beast is subject to the idiosyncrasies of the atmosphere.

The appropriate modification of the wave equation can now be derived. The equations of inviscid motion are linearized, as in Sec. 4.2, to yield

$$\frac{\partial \rho}{\partial t} + \rho_0 \nabla \cdot \mathbf{u} = 0 \tag{4.110}$$

$$\rho_0 \frac{\partial \mathbf{u}}{\partial t} + \nabla P = \rho_0 \mathbf{g} \tag{4.111}$$

where ρ_0 is now spatially dependent, $\rho_0 = \rho_0(\mathbf{x})$, and $\mathbf{g}$ is the gravitational body force. The pressure P is the sum of the hydrostatic pressure $P_0(\mathbf{x})$ and the acoustic perturbation $p(\mathbf{x},t)$; with $\nabla P_0 = \rho_0 \mathbf{g}$, (4.111) reduces to[1]

$$\rho_0 \frac{\partial \mathbf{u}}{\partial t} + \nabla p = 0 \tag{4.112}$$

[1] To be precise, it should be mentioned that if the body-force term $\rho\mathbf{g}$ is not linearized, a term $(\rho - \rho_0)\mathbf{g}$ survives on the right side of (4.112). Consistent with the discussion of Sec. 3.5, however, this term is significant only for sound with wavelength $\lambda \sim c^2/g$ (about 10 km in the atmosphere and much greater in the ocean). If, however, the body force were sufficiently large (in a centrifuge, perhaps), the term in question should be retained.

where

$$p(\mathbf{x},t) \equiv P(\mathbf{x},t) - P_0(\mathbf{x}) \tag{4.113}$$

Because the acoustic motions are isentropic (but not homentropic, in general), we can write as before

$$p = c_0^2(\mathbf{x})[\rho(\mathbf{x},t) - \rho_0(\mathbf{x})] \tag{4.114}$$

In terms of the perturbation pressure p and the velocity, the equations of motion (4.110) and (4.112) can now be written

$$\frac{1}{\rho_0 c_0^2}\frac{\partial p}{\partial t} + \nabla \cdot \mathbf{u} = 0$$

$$\frac{\partial u}{\partial t} + \frac{1}{\rho_0}\nabla p = 0$$

Combining these equations gives

$$\frac{1}{c_0^2}\frac{\partial^2 p}{\partial t^2} - \nabla^2 p + \frac{1}{\rho_0}\nabla\rho_0 \cdot \nabla p = 0 \tag{4.115}$$

The estimated magnitudes of the second and third terms are p/λ^2 and $p/L\lambda$, respectively, where L is the distance over which ρ_0 changes by its own magnitude (in the atmosphere, L is typically of the order of the scale height $c^2/g \approx 10$ km). The last term is then negligible in normal circumstances, and (4.115) becomes finally

$$\frac{\partial^2 p}{\partial t^2} - c^2\,\nabla^2 p = 0 \tag{4.116}$$

where $c = c(\mathbf{x})$ and the subscript zero has been dropped. This is just the classical wave equation, but with a variable sound speed.

The approximations used in obtaining the wave equation are based on the assumption that the properties of the undisturbed atmosphere have small spatial gradients. By making further use of these approximations, the wave equation (4.116) can also be obtained for the condensation S and the velocity potential [note that taking the curl of Eq. (4.112) does not yield $\partial\Omega/\partial t = 0$, so that setting $u = \nabla\phi$ involves an immediate approximation, which is justifiable for $\nabla\rho_0$ small]. For the present purpose, however, it is sufficient to have a wave equation in any variable.

The Eikonal and the Geometry of Rays

We seek now equations for the trajectories (rays) of wavefronts. Assume a harmonic solution (or sum of such solutions) to the wave equation

$$\phi = \beta(\mathbf{x})e^{\pm i\omega t} \tag{4.117}$$

Complex numbers are used only as a computational convenience: we obtain "real" waves by retaining, in the end, only the real part of a complex solution (alternatively, we may retain only the imaginary part). Substituting the above in the wave equation [(4.116) written in terms of ϕ] gives a differential equation for the space variable β

$$k^2\beta + \nabla^2\beta = 0 \tag{4.118}$$

where

$$k \equiv \frac{\omega}{c} \tag{4.119}$$

is the *wave number*, in general a function of position through the sound speed c. We write a solution for (4.118) in the form

$$\beta = B(\mathbf{x}) \exp [i\Sigma(\mathbf{x})] \tag{4.120}$$

where the function $\Sigma(\mathbf{x})$ is called the *eikonal*[1] and $B(\mathbf{x})$ is the amplitude. The overall form (4.117) becomes

$$\phi(\mathbf{x},t) = Be^{i(\Sigma \pm \omega t)} \tag{4.121}$$

which may be compared to a known solution for one-dimensional wave motion in a homogeneous medium,

$$\phi = Be^{i(kx \pm \omega t)}$$

In the one-dimensional case a surface instantaneously at constant x is called a *wavefront*; similarly, a surface $\Sigma(\mathbf{x}) = \text{const}$ represents a wavefront in the three-dimensional case at some instant. Whereas in the one-dimensional situation the wave amplitude is constant over the entire wavefront, this is no longer true in the three-dimensional case represented by (4.120) unless the coefficient $B(\mathbf{x})$ happens to be constant.

[1] Greek εἰκών, image. We again encounter the problem of ambiguous notation in that the usual symbol for the eikonal is S, which we have already used for the condensation, not to mention the total entropy.

In the one-dimensional case the *phase* is given by $kx \pm \omega t$ (equivalently, $x \pm ct$). If this is held constant, we have $x = \text{const} \mp ct$, which is the motion of a wavefront. By analogy

$$\Sigma(\mathbf{x}) \pm \omega t = \text{const}$$

represents a moving wavefront of fixed identity, or moving surface of constant phase, in three dimensions (the simplest example is the expanding spherical wavefront $kr - \omega t = \text{const}$, where the wave number k is constant). Then $d[\Sigma(\mathbf{x}) \pm \omega t] = 0$, and with $d\Sigma = \nabla\Sigma \cdot d\mathbf{l}$,

$$\nabla\Sigma \cdot d\mathbf{l} \pm \omega\, dt = 0$$

Let us choose $d\mathbf{l}$ to be in the direction of $\nabla\Sigma$, that is, in the direction of a *ray* (Fig. 4.19). Then $|\nabla\Sigma|\, dl = \omega\, dt$, and the motion of the wavefront is given by

$$\frac{dl}{dt} = \frac{\omega}{|\nabla\Sigma|}$$

Now it is intuitive that the wavefront should advance with the speed c, the local speed of sound. If we impose this condition, the above equation becomes $dl/dt = c$, or with $\omega/c = k$

$$|\nabla\Sigma| = k \tag{4.122}$$

which is the *eikonal equation*. With $|\nabla\Sigma|^2 = \nabla\Sigma \cdot \nabla\Sigma$, this can be written in the alternative form

$$\left(\frac{\partial\Sigma}{\partial x_1}\right)^2 + \left(\frac{\partial\Sigma}{\partial x_2}\right)^2 + \left(\frac{\partial\Sigma}{\partial x_3}\right)^2 = k^2 \tag{4.123}$$

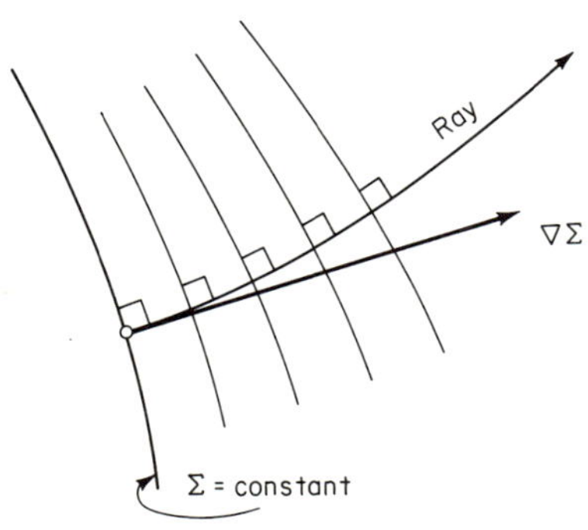

Figure 4.19

We shall now see that (4.122) does not hold in general but only in the case of the approximations appropriate to *geometrical acoustics.* Substituting (4.120) into the differential equation (4.118) gives two equations, for the real and imaginary parts, respectively,

$$\nabla^2 B + [k^2 - (\nabla\Sigma)^2]B = 0 \qquad (4.124)$$

$$B\,\nabla^2\Sigma + 2\nabla B \cdot \nabla\Sigma = 0 \qquad (4.125)$$

The approximation (4.122) consists in neglecting the first term in (4.124). To put this approximation on a rational basis, we rewrite the equation as

$$\frac{\nabla^2 B}{k^2 B} + 1 - \frac{(\nabla\Sigma)^2}{k^2} = 0$$

and the first term is clearly negligible if

$$|\nabla^2 B| \ll |k^2 B| \qquad (4.126)$$

in which case (4.124) reduces to just the eikonal equation. Condition (4.126), together with the requirement that the properties of the undisturbed medium vary slowly in space, are the necessary conditions for geometrical acoustics.

There are three circumstances in which (4.126) will fail: (1) near a *source*, where the amplitude may vary rapidly; (2) at a *caustic*, which will be discussed later; (3) near an obstacle which is not large compared to the wavelength, i.e., where *diffraction* is important. The latter case follows from the estimate that the amplitude will fall from a uniform value B_0 to near zero in a distance of the order d (Fig. 4.20). Then $\nabla^2 B \sim B_0/d^2$, and (4.126) is satisfied only if $d \gg \lambda$.

We will not pursue the solution of (4.124) and (4.125) but focus on the problem of finding the *rays*, as governed by the eikonal equation. Let

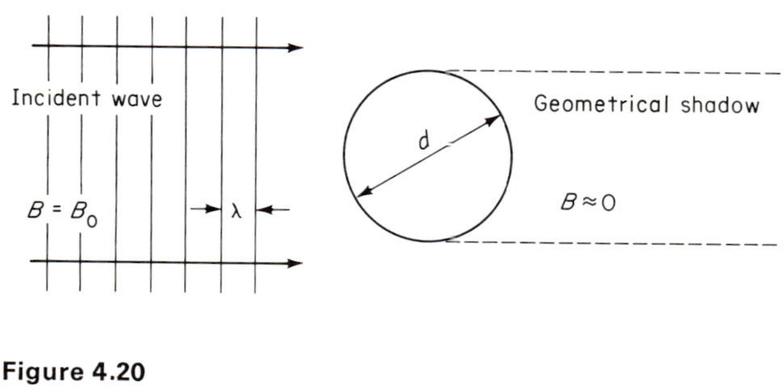

Figure 4.20

$\mathbf{e}$ be a unit vector along the ray, i.e., in the direction of $\nabla\Sigma$. Then with the wave-number vector $\mathbf{k}$ defined by $\mathbf{k} = \mathbf{e}k$, (4.122) is

$$\nabla\Sigma = \mathbf{k} \tag{4.127}$$

It should be mentioned that this is the mathematical equivalent of the famous Huygens wavefront construction by means of "wavelets" (we will later give an example of the Huygens construction for the slightly more general case in which motion is allowed in the undisturbed fluid). Since $\nabla \times \nabla\Sigma \equiv 0$, we can write

$$\nabla \times \mathbf{k} = 0 \tag{4.128}$$

This eliminates the eikonal Σ from further consideration here. We interpret this result by the following manipulation: by vector identity

$$\nabla(\mathbf{e} \cdot \mathbf{k}) = \nabla k = (\mathbf{e} \cdot \nabla)\mathbf{k} + (\mathbf{k} \cdot \nabla)\mathbf{e} + \mathbf{e} \times (\nabla \times \mathbf{k}) + \mathbf{k} \times (\nabla \times \mathbf{e})$$

The third term on the right is zero from (4.128), and the second and fourth terms cancel by the further identity $(\mathbf{e} \cdot \nabla)\mathbf{e} = -\mathbf{e} \times (\nabla \times \mathbf{e})$, after the scalar k is factored out. Then the above equation becomes just

$$(\mathbf{e} \cdot \nabla)\mathbf{k} = \nabla k \tag{4.129}$$

which has a simple interpretation. Let dl be a differential length along the ray; then (4.129) is simply, with $\mathbf{e} \cdot \nabla \equiv d/dl$,

$$\frac{d\mathbf{k}}{dl} = \nabla k \tag{4.129a}$$

Since the frequency ω is a constant, this has other equivalent forms, such as

$$\frac{d}{dl}\frac{\mathbf{e}}{c} = \nabla \frac{1}{c} \tag{4.129b}$$

In the notation of optics, the *index of refraction* n is $n \equiv c_0/c$. In optics, c_0 is conventionally the speed of light in vacuum; in acoustics, there is no equivalent universal standard, but c_0 can be taken as any constant reference sound speed, and the above can be written

$$\frac{d\mathbf{n}}{dl} = \nabla n \tag{4.129c}$$

where $\mathbf{n} \equiv \mathbf{e}n$. The path of any given ray is determined by (4.129) provided that $c(\mathbf{x})$ is known or, equivalently, that $n(\mathbf{x})$ is known. Some examples follow.

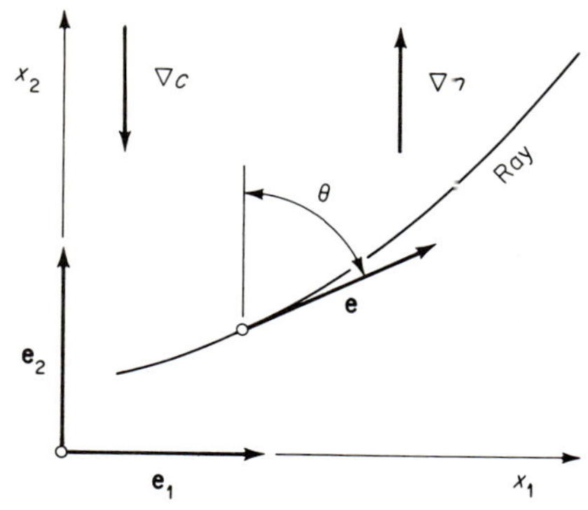

Figure 4.21

Suppose in all of the discussion that follows that the sound speed c varies only in one direction; i.e., the vector ∇n has constant direction.

In the degenerate but important case of a *homogeneous* medium with n = const, $d\mathbf{n}/dl = 0$ or $\mathbf{e}$ = const on each ray, and all rays are *straight lines.*

The vectors $\mathbf{e}$ and ∇n in general define a plane. A given ray is then confined to this (constant) plane because the change in $\mathbf{e}$ is confined to the plane by virtue of (4.129*c*). Let the plane of the ray be the x_1x_2 plane, as shown in Fig. 4.21. From (4.129*c*),

$$\frac{d\mathbf{n}}{dl} = \nabla n = \mathbf{e}_2 \frac{dn}{dx_2}$$

Then

$$\frac{d}{dl}(\mathbf{n}\cdot\mathbf{e}_1) = \mathbf{e}_1\cdot\frac{d\mathbf{n}}{dl} = \mathbf{e}_1\cdot\mathbf{e}_2\frac{dn}{dx_2} = 0$$

Therefore $\mathbf{n}\cdot\mathbf{e}_1 = n\sin\theta$ is constant along any given ray,

$$n\sin\theta = \text{const} \tag{4.130}$$

which can be recognized as the continuous form of *Snell's law* [compare Eq. (4.89)]. Media in which ∇n is unidirectional, and therefore in which (4.130) holds, are sometimes referred to as *stratified* or *layered.* For a spherically symmetric atmosphere with $n = n(r)$, the appropriate form of Snell's law is (see Prob. 4.29)

$$nr\sin\theta = \text{const} \tag{4.131}$$

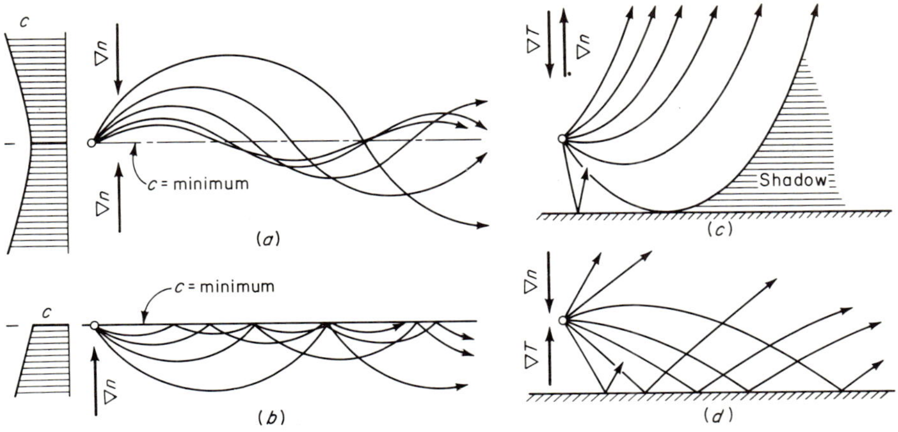

Figure 4.22
(*a,b*) Ray trapping in the sea; (*c*) ray deflection in a normal atmosphere; (*d*) ray deflection in an atmosphere with a temperature inversion, similar to the situation shown in (*b*).

where θ is the angle between the spherical position vector $\mathbf{r}$ and the ray direction.

It is a consequence of (4.129) that rays are bent in the direction of ∇n. It may be useful to think of ∇n as a "force" acting on the ray vector.

Some manifestations are shown in Fig. 4.22. In the ocean at moderate latitudes there is a distinct minimum in the sound speed at a depth of roughly 1 km. The layers in the vicinity of this minimum behave as a waveguide and form the so-called SOFAR channel (sound fixing and ranging), which serves to propagate sound with relatively little attenuation. A similar situation occurs when there is a sound-speed minimum at the surface, where rays are almost completely reflected because of the impedance mismatch (Fig. 4.22*a* and *b*). In the earth's atmosphere $c^2 \propto T$, so that ∇n has a direction opposite to ∇T. Some possible effects are shown in Fig. 4.22*c* and *d*. The situation in which the air is coolest near the ground is referred to as a temperature inversion, as found, for example, on clear winter nights or over a cold lake [see Example 2.4 (page 64)]. Then ray trapping will occur, and the sound intensity will fall off, on the average, like $1/r$ rather than the $1/r^2$ characteristic of spherical propagation. In addition, the stable and quiescent atmosphere associated with inversions will not give rise to turbulent scattering of sound, so that conditions are favorable for clear transmission of sound over considerable distances.

Another statement of the ray equation (4.129), and possibly a more elegant one, is *Fermat's principle* of least time, which is stated in terms of variational calculus

$$\delta \int \frac{dl}{c} = 0 \tag{4.132}$$

where the integral is taken between fixed points in physical space and measures the time of travel of the wave. In words, the ray path between fixed points is such that the time is an extremum. A very simple application of this principle is given in Prob. 4.23. We will not pursue this formulation of geometrical radiation theory here; for a more complete discussion, see, for example, *Born and Wolf* [1965] or *Pearson* [1966].

Sound Amplitude in Geometrical Acoustics

A *ray tube* is a (hypothetical) duct, the walls of which are formed by rays as sketched in Fig. 4.23 (the ray tube is a direct analog of the *stream tube* formed by the streamlines in steady flow). The assumptions of Sec. 4.10 for propagation of sound in a duct are in general very well satisfied by this hypothetical duct, and we can take advantage of the results of that section.

The acoustical energy flowing along each ray tube is constant, and Eq. (4.109) is applicable, viz.,

$$A\rho c^3 \overline{S^2} = \text{const} \tag{4.133}$$

so that the sound amplitude can be determined along a ray, provided that the variation of A is known; i.e., the amplitude S_0 is given by $S_0 \propto 1/\sqrt{\rho c^3 A}$.

It is necessary to find how the cross-sectional area A varies along a ray in each particular problem. One possibility is to make use of the kinematical relation

$$\frac{1}{A}\frac{dA}{dl} = \nabla \cdot \mathbf{e} \tag{4.134}$$

(which can be found with the help of the divergence theorem). Another possibility is to find $A(l)$ by geometrical construction of the rays. A

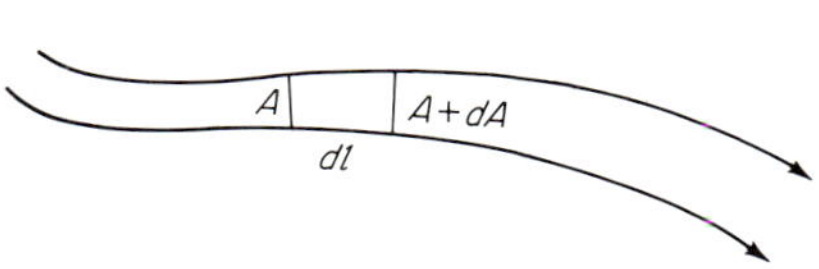

Figure 4.23
Representation of a ray tube.

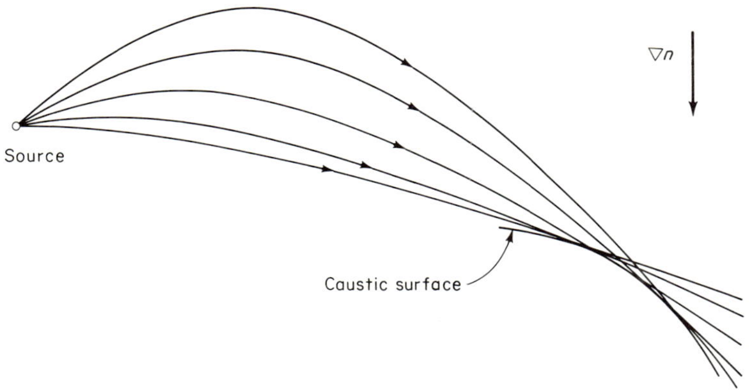

Figure 4.24
Caustic.

simple illustration of the task of finding A will be given in Example 4.5 (page 213).

In many cases the refracted rays from a localized source of sound form an envelope called a caustic surface or *caustic*[1] (Fig. 4.24). Geometrical acoustics breaks down near a caustic because the amplitude varies rapidly [note that Eq. (4.133) predicts *infinite* amplitude at a caustic, with $A \to 0$]. The appearance of caustics as regions of anomalously large sound intensity is of obvious practical interest; in the case of sonic booms, the appearance of a caustic is referred to as a *superboom.*

EXAMPLE 4.4 RAY PATH IN A STRATIFIED ATMOSPHERE WITH LINEAR SOUND-SPEED VARIATION

Let the sound speed vary linearly with height according to

$$c = c_0 + c'x_2$$

where c_0 and c' are constants. This can be rewritten $c = c'X_2$, where $X_2 = x_2 + c_0/c'$. Then Snell's law (4.130) is just

$$\frac{c_0 \sin \theta}{c'X_2} = \sin \theta_0 \tag{4.135}$$

[1] Presumably from the corresponding effect in optics, as produced by a magnifying glass focusing the sun's rays (burning glass), since caustic means burning.

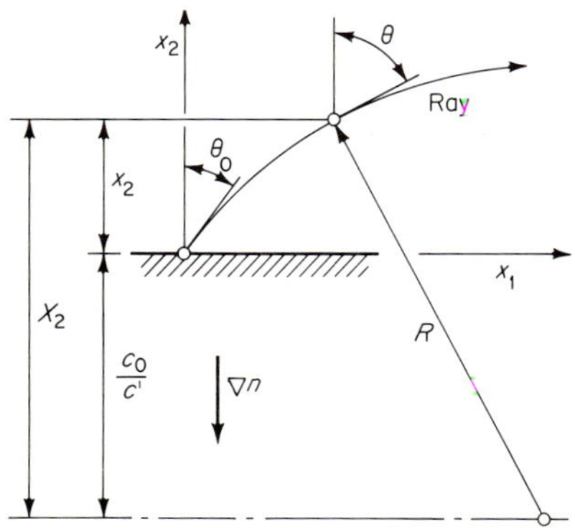

Figure 4.25

where θ_0 is the ray angle where $x_2 = 0$, as shown in Fig. 4.25. Rearranging,

$$\sin\theta = \frac{c' \sin\theta_0}{c_0} X_2 = \frac{X_2}{R}$$

which is just the equation for a *circle*, with radius $R = c_0/(c' \sin\theta_0)$.

An alternative approach to the same conclusion is to integrate Snell's law: Eq. (4.135) can be written

$$\frac{c_0}{c' X_2} \frac{dx_1}{\sqrt{dx_1{}^2 + dX_2{}^2}} = \sin\theta_0$$

which rearranges to

$$dx_1 = \frac{X_2\, dX_2}{\sqrt{\left(\frac{c_0}{c' \sin\theta_0}\right)^2 - X_2{}^2}}$$

On integration, this gives

$$(x_1 - \text{const})^2 + X_2{}^2 = \left(\frac{c_0}{c' \sin\theta_0}\right)^2$$

which is again the equation for a circle.

A similar analysis for the case in which the *square* of the sound speed varies linearly (as with a linear temperature variation in the atmosphere)

$$c^2 = c_0{}^2 + kx_2$$

gives rays in the form of *cycloids*.

EXAMPLE 4.5 AREA VARIATION ALONG A RAY

The simplest possible stratified atmosphere has a linear sound-speed variation, leading to circular rays, as found in Example 4.4. For this case, consider adjacent rays from a point source or line source, as shown in Fig. 4.26. The rays initially diverge from each other at the (arbitrarily small) angle ε. It is desired to find the variation of the separation h with distance from the source.

The upper ray is a circle centered at A', and the lower ray is a circle centered at A. Elementary analytic geometry yields the result

$$h = \frac{\varepsilon x_1}{\sin \theta_0}$$

If the source is a *line* source (perpendicular to the paper), the geometry is two-dimensional and $h(x_1)$ is just the area variation. For the physically more interesting case of a *point* source we have cylindrical symmetry (the rays lie in planes passing through the X_2 axis) and $x_1 \leftrightarrow r$; for this case there is an additional divergence of the rays, proportional to r and $A \propto r^2/\sin \theta_0$. This can be written

$$A = \frac{\delta\Omega}{\sin \theta_0} r^2$$

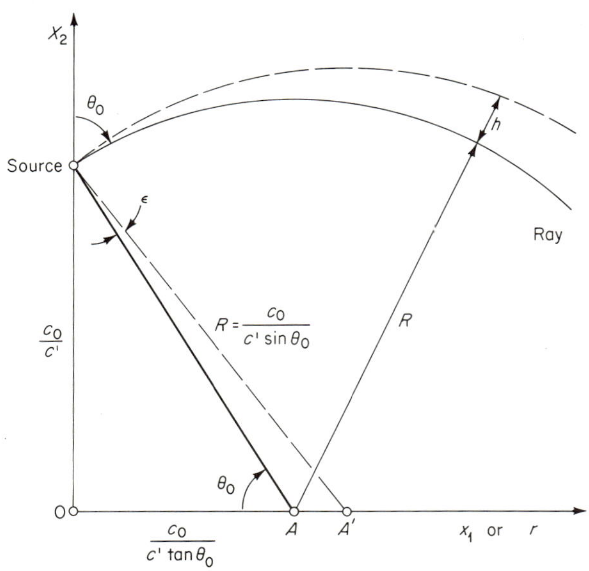

Figure 4.26
Geometry for area variation.

where $\delta\Omega$ is the initial solid angle formed by the ray tube at the source. Thus, if a spherically symmetric source radiates total acoustic power W, a fraction $\delta\Omega/4\pi$ enters the ray tube and we find for the energy flux along a ray

$$\rho c^3 \overline{S^2} = \frac{W \sin\theta_0}{4\pi r^2}$$

More General Formulations of Geometrical Acoustics

In the derivation of the eikonal equation it was assumed that the sound was periodic. One consequence is that the rays are stationary in space. In many cases of nonperiodic sound, as, for example, in sonic-boom theory, it may however be a good approximation to treat the rays as stationary (for a mathematical treatment of nonperiodic sound, see *Friedlander* [1958]).

The presence of winds in the atmosphere (and, to a lesser extent, of currents in the sea) is a significant effect which has not been considered. In particular, the presence of *wind shear* will produce refractive effects somewhat comparable to those produced by sound-speed gradients. The construction of Huygens is shown for the case of wind shear in Fig. 4.27*a*. In this construction, successive wavefronts are the envelopes of the spherical wavelets, of radius $c\,\delta t$, which are convected by the wind a

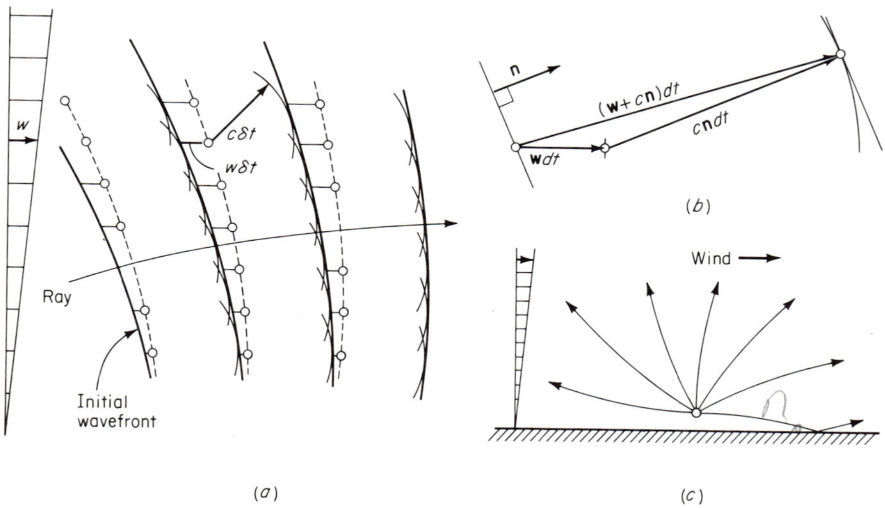

Figure 4.27
Refraction due to wind shear: (*a*) Huygens' construction; (*b*) ray direction and propagation velocity; (*c*) ray refraction.

distance $\mathbf{w}\,\delta t$, where $\mathbf{w}$ is the wind velocity. In this case rays are not in general normal to the wavefronts, as indicated by the construction of successive wavefronts, for a differential time increment, in Fig. 4.27*b*. The propagation velocity is now

$$\mathbf{c} = \mathbf{w} + c\mathbf{n} \tag{4.136}$$

where $\mathbf{n}$ is the unit vector normal to the wavefront.

The distinct weakness of sound as observed by an auditor upwind of a source, as compared to that observed by a downwind auditor, is due to just this refractive effect. Useful treatments of geometrical acoustics in moving media are those of *Hayes* [1968] and *Ryshov and Shefter* [1962].

Remark on Geometrical Optics

The formalism of geometrical radiation theory as given in this section is perfectly applicable to geometrical optics. This is pertinent here because it happens that the optical flow-visualization techniques important in

Table 4.5 Refractive Constants for Gases†
ρ_0 Taken at 0°C, 1 Atm; $\lambda = 589.3$ nm

Gas	$\beta \times 10^5$
Air	29.26
Ammonia	37.6
Argon	28.1
Carbon dioxide	45.1
Carbon monoxide	34.0
Chlorine	77.3
Chloroform	145
Ethyl ether	153
Helium	3.6
Hydrogen	13.2
Methane	44.3
Nitrogen	29.7
Nitric oxide	29.7
Nitrous oxide	51.6
Oxygen	27.2
Sulphur dioxide	68.6

† Data from *Smithsonian Physical Tables*, 1954, table 554.

gasdynamics, viz., *shadowgraph*, *schlieren*, and *interferometry*, are based on the continuous refraction of light by gases.

To make optical use of the results, such as Snell's law, given in this section, it is only necessary to know how the *optical* index of refraction of gas varies with density. Following *Liepmann and Roshko* [1957, chap. 6], we write for this relation

$$n = 1 + \beta \frac{\rho}{\rho_0} \tag{4.137}$$

where $n \equiv c_0/c$ (with c_0 the velocity of light in vacuum and c the velocity of light at density ρ), β is a constant nearly independent of wavelength, and ρ_0 the density at standard conditions.

4.12 Some basic radiation fields

We return to the homogeneous atmosphere and consider some simple solutions for the three-dimensional wave equation. The corresponding wave motions are *simple* in the sense that only progressive waves are present; this requires that the medium be infinite in extent, in order to avoid waves reflected from boundaries. In practice this situation is approximated by large bodies of fluid such as the atmosphere, lakes, and oceans; and for certain experiments by an *anechoic chamber*, i.e., a room

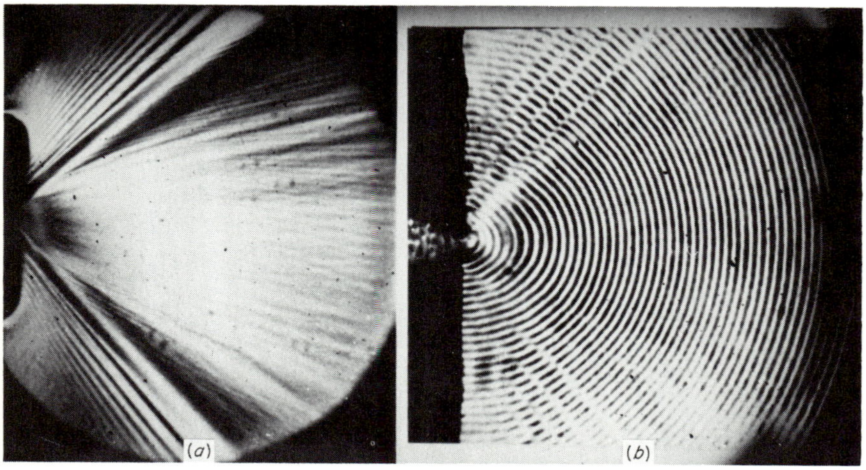

Figure 4.28
Sound field from a transducer: (*a*) radiation pattern; (*b*) wavefronts. (*Courtesy of B. D. Cook, University of Houston*).

with almost perfectly absorbing boundaries. Photographs of simple radiation are shown in Fig. 4.28.

Formally, any choice of the velocity potential $\phi(\mathbf{x},t)$ which satisfies the wave equation $\phi_{tt} - c^2 \nabla^2\phi = 0$ represents a possible wave motion, with all variables of interest derivable from ϕ by the following equations, which are repeated for convenience:

$$S = -\frac{1}{c^2}\frac{\partial\phi}{\partial t} \qquad p = \rho c^2 S \qquad \mathbf{u} = \nabla\phi \tag{4.138}$$

The subscript zero, for example, c_0, has been dropped for convenience, but it is to be understood that all coefficients are evaluated at the undisturbed state.

The Simple Point Source

The case of spherically symmetric motion is the most elementary. Let the velocity potential be

$$\phi = \frac{A}{r} e^{i(\omega t - kr)} \tag{4.139}$$

where A is an amplitude constant (which can be complex). As noted before, only the real part of this function and its derivatives is to be taken, that is, $\mathscr{Re}(\phi)$. This function represents spherically symmetric waves moving outward from the origin [note that the exponent can be rewritten as $-(i\omega/c)(r - ct)$].

The condensation calculated from (4.138) is

$$S = -\frac{Ai\omega}{c^2 r} e^{i(\omega t - kr)} \tag{4.140}$$

in which the coefficient $Ai\omega/c^2r$ is the condensation amplitude. Clearly as $r \to 0$, there is some value of r at which the fundamental acoustic assumption $S \ll 1$ will fail. This anomalous behavior near the source is common to all mathematical models for three-dimensional radiation fields. To the extent that velocity potentials such as (4.139) describe physical sounds, they are of interest, even though the behavior near the source is such that the acoustic approximations are invalid; indeed, near the source, the velocity potential may not represent *any* possible fluid motion. In order for (4.139) to describe an acoustic motion we require that the coefficient in (4.140) be small or, with $k \equiv \omega/c$,

$$r \gg \left|\frac{Ak}{c}\right| \tag{4.141}$$

The velocity field calculated from (4.139) is $\mathbf{u} = \mathbf{e}_r u$, where

$$u = -A\left(\frac{1}{r^2} + \frac{ik}{r}\right)e^{i(\omega t - kr)} \tag{4.142}$$

That is, the motion is purely radial. We can now show that (4.139) represents the sound field due to a *pulsing sphere*, where the surface radius of the sphere is given by

$$R = R_0 - ae^{i\omega t} \tag{4.143}$$

where $a \ll R_0$, with surface velocity at $r = R$ of

$$\dot{R} = -ia\omega e^{i\omega t} \tag{4.144}$$

It is required to show that this boundary velocity matches the fluid velocity given by (4.142), which is, at the point $r = R$,

$$-\frac{Ae^{-ikR}}{R^2}(1 + ikR)e^{i\omega t} = u(R,t) \tag{4.145}$$

Under the approximation that $R \approx R_0$ we have $u = \dot{R}$ at $r = R$ if the coefficients in (4.144) and (4.145) match, or

$$\mathscr{I}m[A(1 + ikR_0)e^{-ikR_0}] = R_0{}^2 a\omega \tag{4.146}$$

which relates the amplitudes a and A. (It can be verified that the substitution $R \approx R_0$ is valid if $a \ll R_0$ and $a \ll \lambda$; the latter condition gives a velocity field which has negligible change over distance a.)

The simple source is the first member, the *monopole*, in a hierarchy of sources which includes the dipole, quadrupole, and so on.

The Dipole

An acoustic dipole can be considered as two simple sources quite close together. These two sources are 180° out of phase; if we think of them as pulsating spheres, then one sphere expands while the other contracts: one puffs while the other huffs.

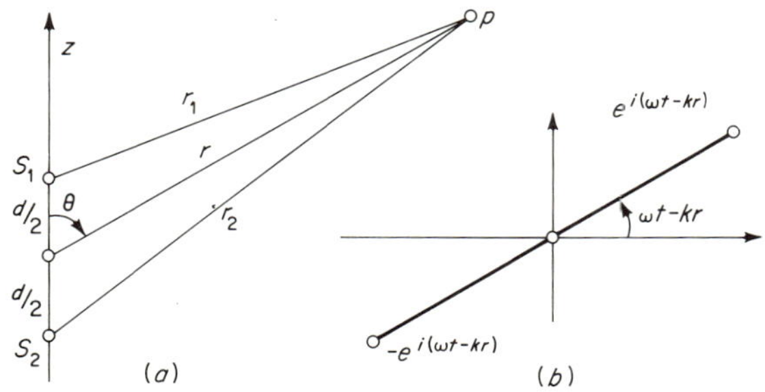

Figure 4.29
Geometry for dipole: (*a*) physical plane; (*b*) complex plane.

Let the two sources S_1 and S_2 be separated by a distance d and centered about the origin on the z axis; r and θ are spherical coordinates (Fig. 4.29). The two sources have scalar velocity potentials ϕ_1 and ϕ_2, respectively, which, evaluated at the general point $p(r,\theta)$, are

$$\phi_1 = \frac{A}{r_1} e^{i(\omega t - kr_1)} \qquad \phi_2 = -\frac{A}{r_2} e^{i(\omega t - kr_2)} \tag{4.147}$$

The dipole source is just the superposition $\phi_1 + \phi_2$. From the figure, with $r \gg d$,

$$r_1 \approx r - \frac{d}{2}\cos\theta \qquad r_2 \approx r + \frac{d}{2}\cos\theta \tag{4.148}$$

which gives

$$\frac{1}{r_1} \approx \frac{1}{r}\left(1 + \frac{d}{2r}\cos\theta\right) \qquad \frac{1}{r_2} \approx \frac{1}{r}\left(1 - \frac{d}{2r}\cos\theta\right) \tag{4.149}$$

With the additional condition $kd \ll 1$, that is, the separation d is small compared to the wavelength $\lambda = 2\pi/k$,

$$\exp\left(\pm ik\frac{d}{2}\cos\theta\right) \approx 1 \pm ik\frac{d}{2}\cos\theta \tag{4.150}$$

Substituting (4.149) and (4.150) into (4.147) and forming the superposition $\phi = \phi_1 + \phi_2$, we have $\phi(r,\theta,t)$

$$\phi = \frac{Ad\cos\theta}{r}\left(\frac{1}{r} + ik\right)e^{i(\omega t - kr)} \tag{4.151}$$

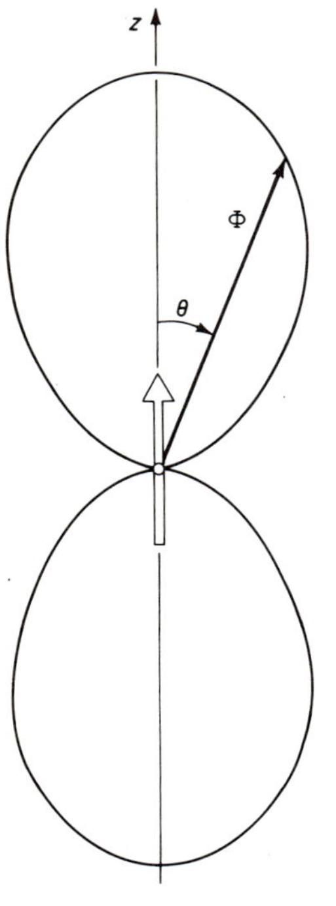

Figure 4.30
Angular distribution of intensity for a dipole, in the far field.

This is the velocity potential for the dipole. In spite of the approximations used in the above (didactic) derivation, it can be verified that this potential satisfies the wave equation *exactly*, except at the origin $r = 0$.

In the *far field* of the dipole, where $kr \gg 1$, the first term $1/r$ drops out; even in this case the dipole does not have spherical symmetry but cylindrical symmetry. Calculating the pressure and velocity far fields from (4.151) gives

$$p = \frac{Ak^2d}{r}\rho c(\cos\theta)e^{i(\omega t - kr)} \tag{4.152}$$

$$\mathbf{u} = \mathbf{e}_r\frac{Ak^2d}{r}(\cos\theta)e^{i(\omega t - kr)} \tag{4.153}$$

Even though the motion is not spherically symmetric, it is purely radial. The energy flux is simply $\overline{pu}$, and from the above we find

$$\mathbf{\Phi}_e = \frac{|A|^2k^4d^2}{r^2}\rho c(\cos^2\theta)\mathbf{e}_r \tag{4.154}$$

as shown in Fig. 4.30. This gives a total radiated power, through a sphere enclosing the source, of $(4\pi\rho\omega^4/3c^3)|A|^2d^2$.

A working approximation for the acoustic dipole is supposed to be an ordinary tuning fork. It should be remarked, however, that simple sources such as dipoles are primarily of interest for representing physical sound sources by means of *distributions* of such simple sources, in analogy with the methods of incompressible potential flow.

The Quadrupole

We adopt the convention that the dipole is represented by a vector pointing from the "negative" to "positive" monopole source, i.e., from S_2 to S_1 in Fig. 4.29. Arrays of two dipoles constitute *quadrupoles* and are illustrated in Fig. 4.31 for the cases where the quadrupoles are located in the x_1x_2 plane. The situation is analogous to the Cartesian representation of the system of stresses in a continuum. There are nine possible quadrupoles in three-dimensional space, and they can be represented as a matrix analogous to the stress tensor

$$Q_{ik} = \begin{bmatrix} Q_{11} & Q_{12} & Q_{13} \\ Q_{21} & Q_{22} & Q_{23} \\ Q_{31} & Q_{32} & Q_{33} \end{bmatrix}$$

Quadrupoles of the type Q_{11} are called *axial quadrupoles* and those of the type Q_{12} *tesseral quadrupoles*.

This correspondence between stress and quadrupole configuration is more than fortuitous; it forms the basis for the theory of aerodynamically generated sound, e.g., by a jet of gas (see Fig. 4.32). The now classical reference on aerodynamically generated sound is *Lighthill* [1952, 1962]: a

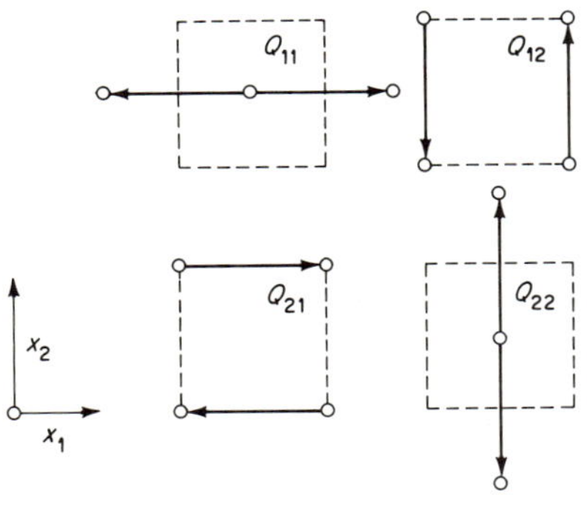

Figure 4.31
Quadrupoles in the x_1x_2 plane.

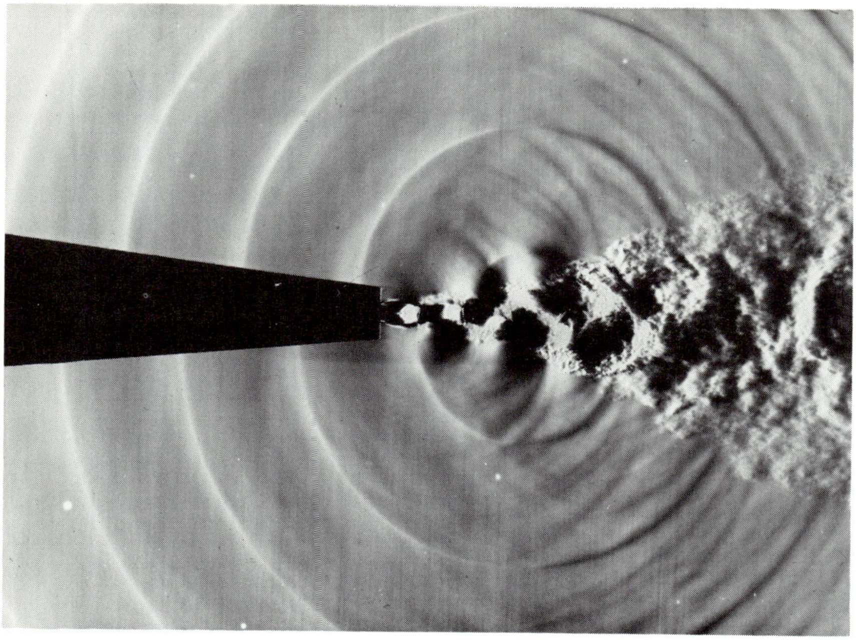

Figure 4.32
Sound field produced by an unstable axisymmetric supersonic jet. Nozzle diameter = 3.7 mm; pressure ratio = 3.7. (*Courtesy of M. G. Davies, University of Liverpool.*)

recent survey is given by *Sears* [1969]. For more information on various sound sources, see *Morse and Ingard* [1968].

EXAMPLE 4.6 RADIATION FROM AN OPEN ORGAN PIPE

Certain simple wind instruments, the flute, recorder, and open-pipe organ in particular, have two ends open to the atmosphere. It is natural to inquire whether the sound is radiated primarily from the mouthpiece end or from the end opposite: where does the sound come from?

This question was investigated by *Coltman* [1969], who employed a double-ended organ pipe specially constructed for the purpose (Fig. 4.33). The pipe is symmetric about its midplane but blown from one end only. According to the simple model for the internal wave motion, the open ends are at constant pressure (pressure nodes): the idealized standing-wave pattern shown in Fig. 4.33*b* corresponds to the second mode satisfying this condition, with

$L \approx \lambda$. This is the mode which was excited for the experiment under discussion (that the standing-wave model is indeed an idealization is illustrated by Coltman's data: $L = 0.652$ m and with $\nu = 469$ Hz, $t = 24.5°$C, $\lambda = 0.748$ m; that is, $\lambda \approx L$ only very roughly). In this mode, air is leaving the mouthpiece at one end while entering the mouthpiece at the other end: the acoustic currents at the two apertures are 180° out of phase.

In Coltman's experiment (in an anechoic chamber) the acoustic pressure p_{rms} was measured on a horizontal circle of radius R, centered on the midpoint of the organ pipe (Fig. 4.34). These measurements can be compared to the pressure due to two monopoles S_1 and S_2, of equal amplitude and 180° out of phase (if theory and experiment agree, the sound is radiated from both ends in equal amount). The velocity potentials are thus

$$\phi_1 = \frac{A}{r_1} e^{i(\omega t - kr_1)}$$

$$\phi_2 = -\frac{A}{r_2} e^{i(\omega t - kr_2)}$$

These sources together thus correspond to an "extended" dipole. The pressure is just the superposition

$$p = -\rho \frac{\partial \phi}{\partial t} = -\rho \frac{\partial}{\partial t}(\phi_1 + \phi_2)$$

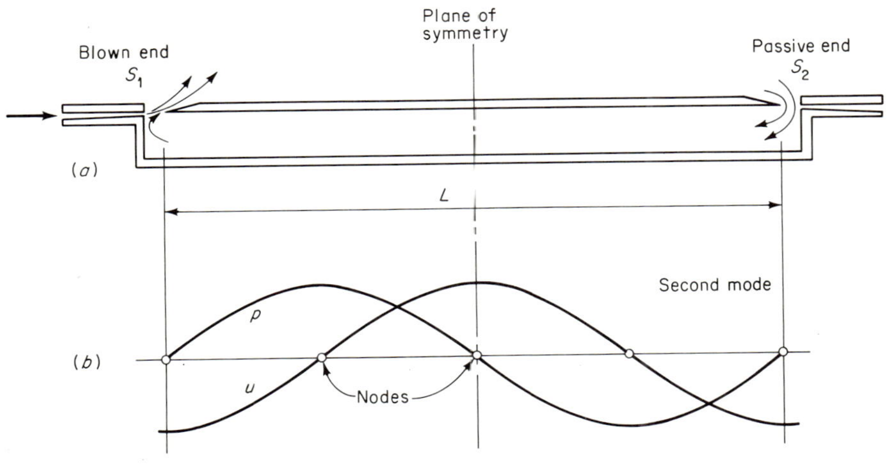

Figure 4.33
Double-ended organ pipe: (*a*) pipe geometry; (*b*) idealized standing-wave pattern within the pipe.

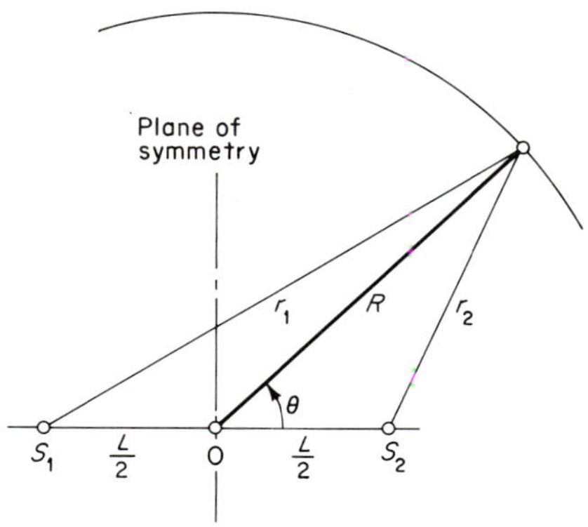

Figure 4.34
Geometry for pressure measurements and monopole calculation.

Performing the differentiation and taking the real part yields (assuming A to be real)

$$p = \frac{\rho A \omega^2}{c} (M \sin \omega t - N \cos \omega t)$$

where

$$M \equiv \frac{\cos (2\pi r_1/\lambda)}{r_1/\lambda} - \frac{\cos (2\pi r_2/\lambda)}{r_2/\lambda}$$

$$N \equiv \frac{\sin (2\pi r_1/\lambda)}{r_1/\lambda} - \frac{\sin (2\pi r_2/\lambda)}{r_2/\lambda}$$

The rms pressure is thus

$$p_{\text{rms}} = \frac{\rho A \omega^2}{\sqrt{2}\, c} \sqrt{M^2 + N^2}$$

The values of r_1 and r_2 are related to the polar angle θ by the cosine law, viz.,

$$r_1{}^2 = R^2 + \frac{L^2}{4} + LR \cos \theta$$

$$r_2{}^2 = R^2 + \frac{L^2}{4} - LR \cos \theta$$

so that, for fixed R, p_{rms} can be found as a function of θ. The comparison of

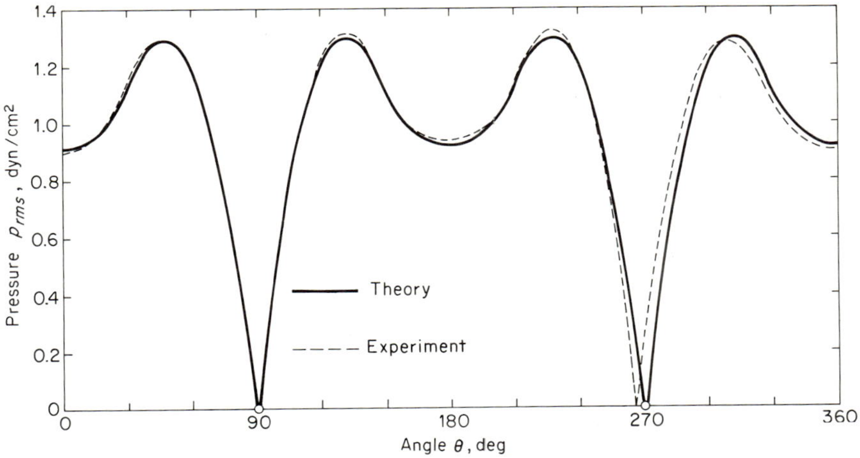

Figure 4.35
Comparison of theory and experiment. The experimental data are the result of filtering out all harmonics other than the second.

this calculation with experiment is shown in Fig. 4.35. The experimental conditions were (in air)

$$
\begin{aligned}
L &= 0.6523 \text{ m}\\
\nu &= 469 \text{ Hz}\\
t &= 24.5°\text{C}\\
R &= 0.695 \text{ m}
\end{aligned}
$$

The conclusion is that sound is emitted equally from each end. The amplitude coefficient A is of course empirically adjusted to produce the agreement shown.

It is interesting to note that according to the simple model given here, the open ends are pressure nodes at zero (acoustic) pressure; viewed as monopole sources, however, these same open ends have infinite acoustic pressure (at the source itself). Thus, we have in a sense set *zero* equal to *infinity*.

3 Attenuation of sound

The wave equation is an approximate model based on isentropic processes. In the range of audible sounds, this model is highly satisfactory for most purposes but has the disadvantage of preserving acoustic energy for all time. If this condition were in fact realized, life would be unbearable: things are noisy enough as they are.

The major source of attenuation in normal architectural spaces is boundary absorption due to viscosity and thermal conduction in the vicinity of porous surfaces. We will not discuss this problem, but consider instead propagation in unbounded (or partially bounded) space.

Spatial attenuation results from a variety of effects, often acting simultaneously. Primary is simple spherical spreading (which may, however, be modified by refraction in a nonhomogeneous atmosphere, as we have noted). The attenuation of spherical waves in a homogeneous atmosphere due to simple spreading is easy to calculate. From (4.47) we have

$$S = \frac{1}{r} f(r - ct)$$

which gives an energy flux Φ_e

$$\Phi_e = \frac{1}{r^2} \Phi_0 \tag{4.155}$$

where Φ_0 is a constant. Between a distance $r = r_1$ and $r = 2r_1$ there is an attenuation corresponding to

$$\frac{\Phi_2}{\Phi_1} = \left(\frac{r_1}{r_2}\right)^2 = \frac{1}{4}$$

or an attenuation

$$\begin{aligned} \Delta_1 - \Delta_2 &= 10\left(\log_{10} \frac{\Phi_1}{\Phi_{\text{ref}}} - \log_{10} \frac{\Phi_2}{\Phi_{\text{ref}}}\right) \\ &= 10 \log_{10} 4 \approx 6 \text{ dB} \end{aligned}$$

Thus, for example, a source which produces 100 dB at $r = 10$ ft gives 94 dB at 20 ft, 88 dB at 40 ft, and $100 - 6n$ dB at 10×2^n ft.†

Additional agents of spatial attenuation are bulk dissipation due to viscosity and thermal conductivity, turbulent scattering[1] (via velocity and density gradients on a small scale), particulate scattering (dust, fog, etc.), dissipation in the neighborhood of particulate boundaries, dissipation via

† Note that there is nothing to prevent the decibel level being negative; a level of $\Delta = -17$ dB, for example, represents "real" but subaudible sound.

[1] An observable manifestation appears to be the diffuse rumble of distant thunder, as opposed to the sharp crack of a nearby lightning strike.

a variety of molecular relaxation processes, and other effects such as thermal radiation. We will discuss only the first of these.

Bulk Attenuation Due to Viscous and Thermal Dissipation

Consider periodic progressive plane waves, which may show a slight diminution of amplitude with distance. The following calculation of this attenuation is based on associating the loss in energy flux with entropy production in the fluid.

According to the idealization implicit in the wave equation, the passage of waves through a certain material volume V is an isentropic process and produces no net change in the state of the fluid. Actually, the small dissipation which does occur increases the internal energy of the fluid and causes it to expand slightly. Thus, the balance of energy may be written as the average over time (see Fig. 4.36)

$$A(\Phi_1 - \Phi_2) = \overline{\int_V \rho \frac{De}{Dt}\, dV} + \overline{\int_V P\rho \frac{Dv}{Dt}\, dV}$$

The first term on the right is the increase in internal energy, and the second term is the net work done by expansion (in the isentropic idealization both these terms are zero). The equivalent local statement is

$$-\frac{d\Phi}{dx} = \rho \overline{\frac{De}{Dt}} + P\rho \overline{\frac{Dv}{Dt}} \tag{4.156}$$

With the Gibbs equation, this becomes simply

$$-\frac{d\Phi}{dx} = \rho T \overline{\frac{Ds}{Dt}} \tag{4.157}$$

which describes the dissipative "heating" of the fluid due to loss of energy flux.

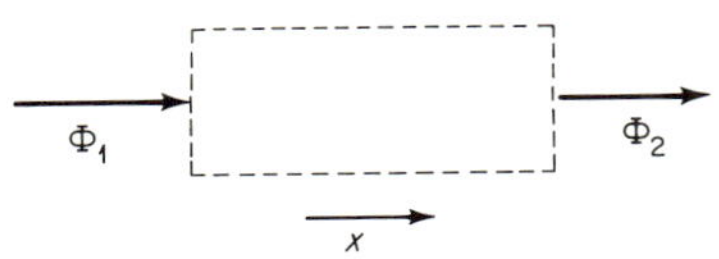

Figure 4.36

The rate of entropy increase is given by Eq. (2.15); then (4.157) becomes

$$\overline{\rho T \frac{Ds}{Dt}} = \bar{\Upsilon} + \overline{\frac{\kappa}{T}(\nabla T)^2} + \overline{T\nabla \cdot \frac{\kappa \nabla T}{T}} \tag{4.158}$$

The last term vanishes when averaged over time, and (4.157) becomes, retaining terms up to second order in the wave amplitude,[1]

$$-\frac{d\Phi}{dx} = \bar{\Upsilon} + \frac{\kappa}{T}\overline{(\nabla T)^2} \tag{4.159}$$

which is just ρT times the (average) entropy-production rate. The dissipation function Υ given by (2.9), for one-dimensional motion in the x direction, is

$$\Upsilon = (\tfrac{4}{3}\mu + \mu_v)\left(\frac{\partial u}{\partial x}\right)^2 \tag{4.160}$$

By Taylor expansion of $T(\rho,s)$ the local temperature is given in terms of the condensation by

$$T = T_0 + \left(\frac{\partial T}{\partial \rho}\right)_s \rho S = T_0 - \left(\frac{\partial T}{\partial v}\right)_s vS \tag{4.161}$$

With $\nabla T = \mathbf{e}\, \partial T/\partial x$ and making use of (4.161) and the identity

$$\left(\frac{\partial T}{\partial v}\right)_s^2 = (\gamma - 1)\frac{c^2}{v^2}\frac{T}{c_p} \tag{4.162}$$

the second term on the right side of (4.159) becomes

$$\frac{\kappa}{T}\overline{\left(\frac{\partial T}{\partial x}\right)^2} = \frac{(\gamma - 1)\kappa c^2}{c_p}\overline{\left(\frac{\partial S}{\partial x}\right)^2} \tag{4.163}$$

[1] Suppose the waves are one-dimensional and sinusoidal, with condensation amplitude S_0; then direct evaluation of the right-hand side of (4.158) gives terms, respectively, of order $S_0{}^2$, $S_0{}^2$, and $S_0{}^3$. Thus, the last term is dropped, since $S_0 \ll 1$.

For a progressive wave the condensation and velocity are related by $u = cS$, and (4.159) becomes, with (4.160) and (4.163),

$$-\frac{d\Phi}{dx} = \mu c^2\left(\frac{4}{3} + \frac{\mu_v}{\mu} + \frac{\gamma - 1}{\Pr}\right)\overline{\left(\frac{\partial S}{\partial x}\right)^2} \tag{4.164}$$

The coefficient combines viscous and thermal transport properties. Following Lighthill and Kirchhoff, we define the *diffusivity* δ

$$\delta \equiv \frac{\mu}{\rho}\left(\frac{4}{3} + \frac{\mu_v}{\mu} + \frac{\gamma - 1}{\Pr}\right) \tag{4.165}$$

The time average in (4.164) is proportional to wave amplitude squared, as is the energy flux Φ itself, suggesting the conventional attenuation relation

$$\frac{1}{\Phi}\frac{d\Phi}{dx} = -2\alpha \tag{4.166}$$

which has the integrated form[1]

$$\Phi = \Phi_0 e^{-2\alpha x} \tag{4.167}$$

With $\Phi = \rho c^3 \overline{S^2}$, the *attenuation coefficient* 2α is given by (4.164) and (4.166) as

$$2\alpha = \frac{\delta}{c}\frac{1}{\overline{S^2}}\overline{\left(\frac{\partial S}{\partial x}\right)^2} \tag{4.168}$$

This is independent of amplitude (for any given waveform) but proportional to the frequency squared, through the spatial derivative. The most important case is the sinusoidal wave,

$$S = S_0 \sin k(x - ct) \tag{4.169}$$

[1] The attenuation coefficient has been written as 2α, rather than α, by convention (unfortunately, not universal). Then the wave *amplitude* will be proportional to $e^{-\alpha x}$, for example, $S = e^{-\alpha x} f(x - ct)$.

Table 4.6 Viscous and Thermal Terms in the Diffusivity δ
Values at 20°C and 1 Atm, Approximately

Substance	$\frac{4}{3} + \frac{\mu_v}{\mu}$	$\frac{\gamma - 1}{\text{Pr}}$	$\frac{\mu}{\rho} \times 10^5$, m^2/s	$\delta \times 10^5$, m^2/s
Air	1.93	0.563	1.57	3.91
Helium	1.33	0.953	12.2	27.9
Water	4.4	0.00093	0.101	0.44
Mercury	2.5	5.4	0.012	0.092
Glycerin	1.7	0.000015	118	200
Alcohol	5.8	0.011	0.152	0.88

Calculating the averages according to the general formula

$$\bar{F} \equiv \frac{1}{T}\int_0^T F(t)\,dt \tag{4.170}$$

where $T = 1/\nu$ is the period, we find the attenuation coefficient

$$2\alpha = \frac{\omega^2\delta}{c^3} \tag{4.171}$$

This result is due to Stokes and Kirchhoff.

The coefficient $\omega^2\delta/c^3$, which has the dimensions of reciprocal length, is small except at extremely high frequencies. Consider an ideal gas, for which $\delta \sim c\Lambda$ from (2.125); we find $\omega^2\delta/c^3 \sim (\Lambda/\lambda)(1/\lambda)$. Now the percentage diminution of energy flux in one wavelength is $\omega^2\delta\lambda/c^3 \sim \Lambda/\lambda$, which is exceedingly small for ordinary sound.

The terms $\frac{4}{3} + \mu_v/\mu$ and $(\gamma - 1)/\text{Pr}$ represent the relative importance of viscous and thermal attenuation.[1] Approximate values for these two factors are given in Table 4.6.

For ideal gases, $c^2 = \gamma P/\rho$, and (4.171) can be written

$$2\alpha = \frac{\omega^2\delta}{c^3} = \frac{\omega^2(\rho\delta)}{\gamma P c} \tag{4.172}$$

[1] The presence of viscous attenuation is perhaps consistent with intuitive notions of loss through "dissipative heating." Thermal attenuation can be thought of as resulting from the leakage of heat from the relatively hot compression regions into the relatively cold rarefaction regions, thus smearing out the waveform.

where

$$\rho\delta = \mu\left(\frac{4}{3} + \frac{\mu_v}{\mu} + \frac{\gamma - 1}{\mathrm{Pr}}\right)$$

is a function only of the temperature in the undisturbed state. At fixed temperature, therefore, the attenuation depends on ω^2/P and increases with *decreasing* pressure; one consequence is that most attenuation measurements are carried out at low pressure. Experimental data for gases often show the nondimensional quantities $\alpha c/\omega$ versus $\mu\omega/P$; according to the above theory, such a plot should be just a straight line (see Fig. 4.37). The data shown substantiate the continuum theory very well, up to a frequency of the order of the reciprocal molecular relaxation time, $\nu \sim \tau^{-1}$, corresponding to $\mu\omega/P \approx 0.1$ (at 1 atm $\mu\omega/P = 0.1$ corresponds to a frequency $\nu = 9.02 \times 10^7$ Hz). Such data provide in fact one of the few means for inferring a value for the bulk viscosity μ_v; from Fig. 4.37, taking the values of μ, Pr, and so on from other sources, one obtains $\mu_v = 0.62\mu$.

The presence of minute quantities of foreign gases may result in attenuation coefficients far from the predictions of the Stokes-Kirchhoff

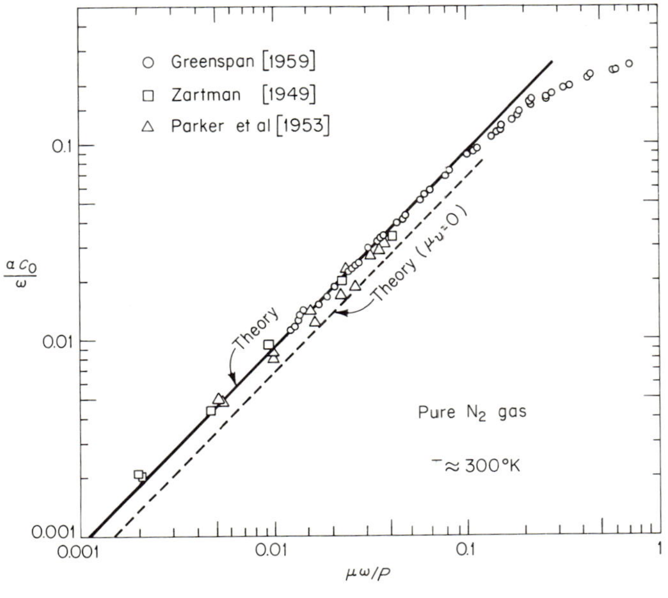

Figure 4.37
Attenuation data for pure nitrogen at $T_0 \approx 300$ K.

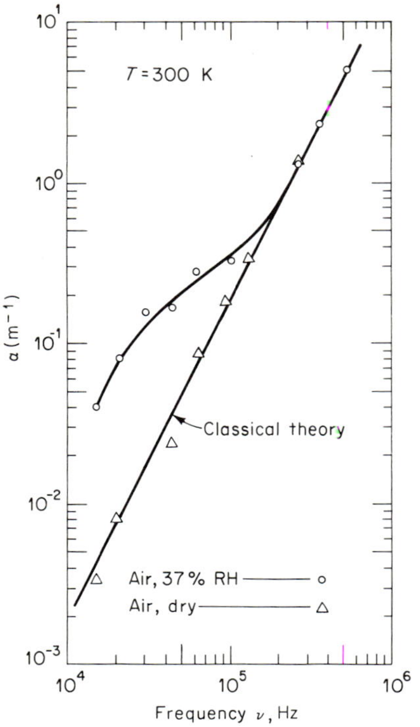

Figure 4.38
Attenuation data of *Sivian* [1947].

theory. In particular, the presence of water vapor and carbon dioxide in atmospheric air leads to very large values for the attenuation coefficient, even in the audible range of frequencies (see Fig. 4.38; additional data are given by *Piercy* [1969]). With respect to water vapor, the value of α depends strongly on the relative humidity, with a distinct maximum at a relatively low humidity (see, for example, the data of *Harris* [1966]). For a brief discussion of the relaxation mechanism, see *Lighthill* [1956].

Since the attenuation coefficient varies with the frequency squared, the frequency spectrum of a general (Fourier sum) periodic wave changes with distance, the higher-frequency components being progressively filtered out. This effect is heard in the rumbling sound of high-altitude jet aircraft or rockets by a listener on the ground. It is also responsible for the increased "purity" of music heard from a distance.

At very low or *infrasonic* frequencies the bulk attenuation is apparently very small. The explosion of the volcano Krakatoa in 1883 was *heard* almost 3,000 mi away in Africa and detected by various barometers during four circuits of the globe (*Miller* [1935]). The Soviet nuclear blast of 1962 behaved similarly. Waves of this kind are, to be sure, not simple acoustic waves in a homogeneous atmosphere but acoustic-gravity waves in a distinctly nonuniform medium.

Attenuation due to Dissipation in the Boundary Layer of a Duct

The acoustic waves transmitted in ducts or pipes are essentially plane waves, provided that the duct diameter and sound wavelength are sufficiently large. In this case there is dissipation in the boundary layer near the walls and a corresponding attenuation. We will not reproduce the calculation here (see, for example, *Morse and Ingard* [1961]) but merely quote the results. By satisfying the no-slip conditions at a stationary and isothermal wall, with an acoustic simple wave for the interior flow one obtains velocity and thermal boundary layers of thickness δ_v and δ_t, respectively,

$$\delta_v^2 = \frac{2\mu}{\rho\omega} \qquad \delta_t^2 = \frac{2\kappa}{\rho\omega c_p} \tag{4.173}$$

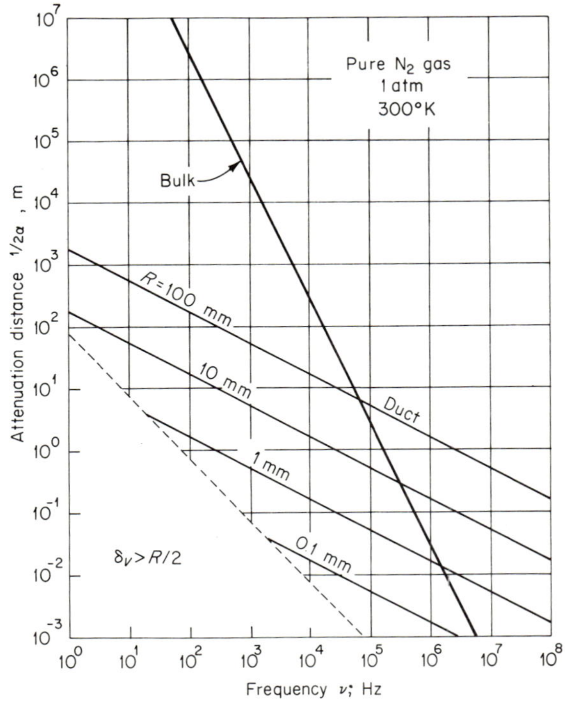

Figure 4.39 Attenuation distance vs. frequency.

[The symbol δ used here has no connection with the diffusivity defined in Eq. (4.165).] Gases have a Prandtl number Pr $\sim$ 1, and the boundary layers have about the same thickness, $\delta \sim \sqrt{\lambda\Lambda}$. For the given results to be valid, it is required that the boundary-layer thickness δ be small compared to the duct radius R.

The attenuation is given by

$$\Phi_e = \Phi_0 e^{-2\alpha x} \tag{4.174}$$

where for a circular duct of radius R

$$2\alpha = \frac{\omega}{Rc}\,[\delta_v + (\gamma - 1)\delta_t] \tag{4.175}$$

found by Kirchhoff (1868). This coefficient varies as the square root of the frequency; in consequence, boundary-layer attenuation is more important than bulk attenuation at low frequencies (Fig. 4.39). The above results are well substantiated by experiments.[1]

Problems

4.1 Given that the condensation S satisfies the wave equation; which of the quantities P, ρ, T, and s satisfy the wave equation?

4.2 Show formally that a function $F(t - x/c_0)$ satisfies the one-dimensional wave equation.

4.3 For a hypothetical substance which has an equation of state $P = a^2\rho$, where a is a constant, find an expression for the sound speed.

Answer $c^2 = a^2$

4.4 Assume that the dense plasma at the center of the sun is made up of equal numbers of protons, deuterons, and He^3 nuclei, the density is $\rho = 129$ g/cm^3, and temperature $T = 16 \times 10^6$ K. If the gas is ideal, find the pressure in atmospheres and sound speed in kilometers per second.

Answer 8.47×10^{10} atm; 333 km/s

4.5 The equation of state for a Clausius gas is $P(v - b) = RT$, where b is a constant representing the molecular volume and R is the gas constant. Find an expres-

[1] The first was perhaps that of J. B. Biot, who experimented with iron water pipes in Paris in 1808, finding that the human voice could be clearly heard over a distance of 951 m. For more recent data, see *Mason* [1928].

sion for the sound speed in such a gas in terms of γ, R, T, v, and b. If possible, interpret the result physically.

Answer $$c^2 = \gamma RT\left(\frac{v}{v - b}\right)^2$$

4.6 For acoustic motion in an ideal gas with the local value of condensation S, find the corresponding value of the relative sound-speed change, $(c - c_0)/c_0$.

Answer $$\frac{c - c_0}{c_0} = \frac{\gamma - 1}{2} S$$

4.7 A two-phase mixture of gas and liquid (small air bubbles dispersed in water, for example) may be treated as a continuum for the transmission of sound of long wavelengths. The liquid behaves as a heat reservoir, and pressure changes are approximately isothermal. Let r_v be the ratio of gas volume to liquid volume and r_m the ratio of gas mass to liquid mass. Show that

$$\rho = \rho_l \frac{1 + r_m}{1 + r_v}$$

$$Pr_v = \rho_l r_m RT = \text{const}$$

If liquid compressibility is neglected, show that the sound speed is given approximately by

$$c^2 \approx \frac{(1 + r_v)^2}{r_v} \frac{P}{\rho_l}$$

Find the value of r_v corresponding to minimum sound speed. Obtain a numerical value for an air-water mixture at 1 atm and room temperature.

Answer $r_v = 1$; 20.1 m/s

4.8 Calculate the particle displacement amplitude for one-dimensional sinusoidal progressive waves in air at 300 K at the following intensity levels and frequencies:

(*a*) 20 dB, 10 kHz
(*b*) 60 dB, 50 Hz

These levels are roughly the lowest at which the sound can be clearly heard at the particular frequency.

Answer 1.11×10^{-11} m; 2.23×10^{-7} m

4.9 A loudspeaker advertisement claims a total (peak-to-peak) diaphragm displacement of $^5/_8$ in. for bass notes. Assuming one-dimensional sinusoidal motion at frequency $\nu = 35$ Hz, calculate the one-dimensional energy flux and find the corresponding decibel level at the speaker.

Answer 148 dB

4.10 Making use of the calculation of the particle displacement ξ given in Example 4.1 (page 173), make a second-order calculation of ξ by iteration. Show that the second-order correction is negligible at acoustic amplitudes.

4.11 For acoustic radiation with spherical symmetry, the velocity potential is $\phi = (1/r)F(r - ct)$. If F is sinusoidal, with wavelength λ, find a criterion for the negligibility of the $1/r^2$ term in the expression for particle velocity; i.e., find the condition on r such that this term is negligible.

Answer $kr \gg 1$

4.12 If a semi-infinite slab (substance 1) impacts against a slab of thickness L (substance 2) with relative velocity U, find the resulting motion. The impact takes place in a vacuum. Assume that $\mathscr{R}_1 > \mathscr{R}_2$.

4.13 At large distance r from the source, the amplitude of cylindrically symmetric sound varies as r^n. Find the value of n and the intensity attenuation in decibels for a doubling of the distance.

Answer $n = -\frac{1}{2}$; 3.0 dB

4.14 Ideal acoustic wave motion is *reversible* in a literal as well as a thermodynamic sense. Show that the case of reversed zero reflection is *also* zero reflection.

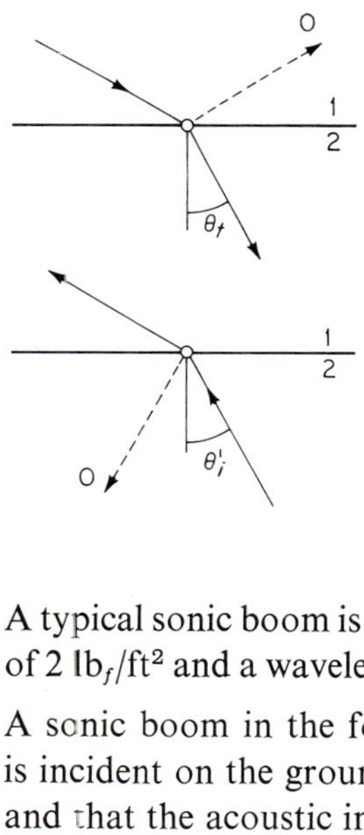

4.15 A typical sonic boom is a solitary N wave with a peak-to-peak pressure amplitude of 2 lb_f/ft^2 and a wavelength of 50 ft. Find the maximum particle displacement.

4.16 A sonic boom in the form of an N wave with amplitude p_0 and wavelength λ is incident on the ground surface at angle θ_i. Assume that the surface is plane and that the acoustic impedance of the ground is effectively infinite. Find $p(t)$ and $\mathbf{u}(t)$ at some fixed point in the atmosphere adjacent to the surface and sketch the pattern of incident and reflected wavefronts.

4.17 Consider a medium of thickness L bounded on one side by a "hard" material ($\mathscr{R} \to \infty$) and on the other side by a "soft" material ($\mathscr{R} \to 0$). Find explicit expressions for possible *standing* one-dimensional sinusoidal waves in the medium and find the characteristic frequencies. This is the analog of an organ pipe closed at one end.

Answer $\phi = A \sin \omega t \cos kx$

$$\omega = \frac{(2n+1)\pi c}{2L} \qquad k = \frac{(2n+1)\pi}{2L} \qquad n = 0, 1, 2, 3, \ldots$$

4.18 By integrating the eikonal equation (4.122) around a closed path, show that the optical (acoustic) path for all rays connecting two different wave surfaces is the same, where the optical path is defined as $\int n\, dl$. Note that this result can be expressed in the form that successive wavefronts are everywhere separated by the same increment of time Δt.

4.19 An acoustic ray leaves a source S at the small angle ε to the horizontal. There is an atmospheric inversion, with the vertical temperature gradient ∇T. Find the distance L at which the ray intersects the (horizontal) source plane, i.e., where the ray strikes the ground. Find the particular value for the case $|\nabla T| = 5$ K per 1,000 ft, with surface temperature $T_0 = 273$ K and $\varepsilon = 0.1$. Use appropriate approximations.

Answer 4.14 mi

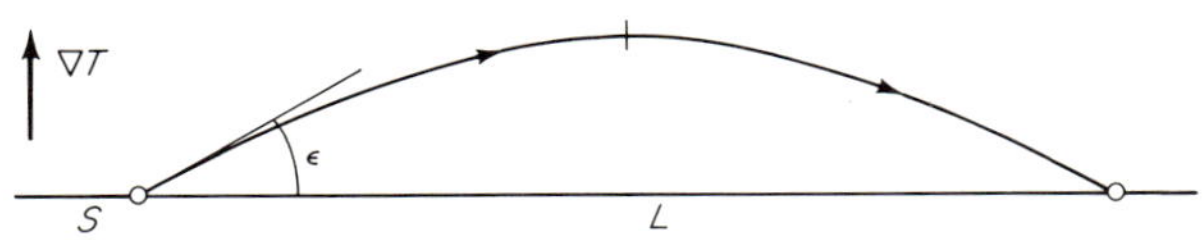

4.20 A liquid with density $\rho = 1$ g/cm^3 and sound speed $c = 10^5$ cm/s flows steadily in a constant-area pipe with velocity $u_1 = 30$ cm/s and pressure $P_1 = 1$ atm. A quick-closing gate valve in the line is suddently closed. Assuming that the vapor pressure of the liquid is zero, find a simple model for the flow conditions on the *downstream* side of the valve.

4.21 Calculate the deflection angle ε for a light ray which traverses a gas of thickness 1 cm, with a density gradient normal to the ray such that the density changes from ρ_0 to $2\rho_0$ in a distance of 1 cm. The density gradient is normal to the initial ray direction, and the index of refraction is given by (4.137).

Answer 0.017° in air

4.22 Make an estimate of the *horizontal atmospheric refraction* δ, that is, the angle below the horizon of a star which *appears* to be exactly on the horizon. The atmosphere may be assumed spherically symmetric, and the density varies with height z according to

$$\rho = \rho_0 e^{-z/H}$$

where H is approximately 10 km.

Answer 0.6°

4.23 Consider reflection and refraction of a ray at the boundary between two uniform media. Apply Fermat's principle of least time between fixed points and find the laws of reflection and refraction, i.e., Eqs. (4.88) and (4.89). Note that the rays are straight lines, so that the path time can be evaluated algebraically.

4.24 Consider the spherically symmetric acoustic-velocity potential $\phi = A/r$, where A is a constant. Find the corresponding pressure and velocity fields and the physical significance of these results.

4.25 The ear canal is an essentially cylindrical channel leading to the human eardrum. The average length is about 2.7 cm. Assuming that this cavity behaves like a semiopen organ pipe, calculate the resonant frequency at which human hearing presumably is most acute.

Answer 3,210 Hz at 300 K

4.26 An *A natural* (440 Hz) is blown on the flute with the distance between the mouthpiece (embouchure) and first open finger hole of 13 in. These two openings are approximately at constant pressure. Calculate the "theoretical" frequency, i.e., the frequency of the lowest-order standing wave, or *fundamental*, for 300 K.

Answer 525 Hz

4.27 The following derivation for the sound speed is due to G. Riccati (1777). It is known that the particle displacement ξ satisfies the wave equation

$$\frac{\partial^2 \xi}{\partial t^2} = c^2 \frac{\partial^2 \xi}{\partial x^2}$$

Canceling $\partial^2 \xi$ and multiplying through by ∂x^2, we obtain

$$\left(\frac{\partial x}{\partial t}\right)^2 = c^2$$

so that the propagation velocity is just c. Is there any grain of truth in this derivation?

4.28 For progressive one-dimensional acoustic waves, estimate the ratio of the instantaneous heat flux $-\kappa\, \partial T/\partial x$ to the instantaneous mechanical-energy transfer rate Pu. Let the waves be characterized by an amplitude S_0 and wavelength λ. The fluid is a gas, with a Prandtl number of order unity.

4.29 Show that for refraction in a spherically symmetric atmosphere

$$nr \sin \theta = \text{const}$$

where θ is the angle between the position vector $\mathbf{r}$ and the ray direction.

4.30 Discuss the acoustical analog of a lens. In particular, if the lens is formed from a bubble of hot gas embedded in a cooler uniform gas, what form is required in order that the lens be converging?

4.31 A poorly heated opera house has a stratified atmosphere with a temperature difference of 10°C between floor and ceiling. The height and length of the house are on the order of 100 m. Is geometrical acoustics relevant to the problem of sound propagation, or can the rays be approximated as straight lines?

4.32 A frequently used illustration of Newtonian mechanics involves the impact of a single moving sphere against a row of stationary spheres. Consider the one-dimensional analog of this experiment; i.e., the spheres are replaced by parallel flat plates of identical thickness and material. Find the corresponding wave diagram.

4.33 Are two noisy children twice as noisy as one? As an analytical model, consider two coincident but independent point sources (different frequencies). Is the total decibel level equal to the sum of the decibel levels of the individual sources; i.e., is $\Delta_{tot} = \Delta_1 + \Delta_2$? It may be convenient to assume that the sources are sinusoidal.

Answer $\Delta_{tot} \approx \Delta_{one} + 3$ dB

4.34 The (acoustic) pressure wave from a spherical explosion is sometimes represented by a function

$$p = \begin{cases} \dfrac{B}{r} \exp\left(-\dfrac{t - r/c}{t_0}\right) & t \geq \dfrac{r}{c} \\ 0 & t < \dfrac{r}{c} \end{cases}$$

(*a*) Does this represent a possible solution for the wave equation?
(*b*) Find the corresponding velocity field; in particular, show that there is a finite residual velocity $Bt_0/\rho r^2$ when $t \to \infty$.
(*c*) Draw the corresponding wave diagram.
(*d*) Suggest a modification for the pressure wave, such that the residual velocity is zero.

4.35 Find the pressure amplitude of the reflected step wave shown, for one-dimensional motion. First find a general formula for this situation; then obtain a numerical value. The amplitude p_i may be considered as given.

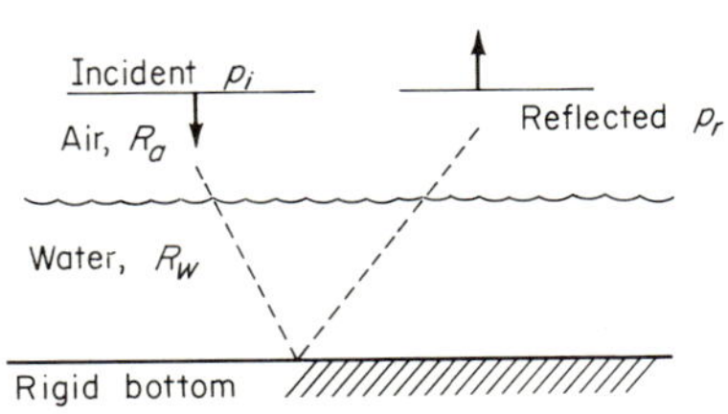

4.36 The displacement of a particle in a one-dimensional simple wave is given by Eq. (4.39), which may be written

$$\xi = c_0 \int_{t_0}^{t} f(x_0 - c_0\tau)d\tau + c_0 \int_{0}^{t_0} f(x_0 - c_0\tau)d\tau$$

where $S = f(x - c_0 t)$ is the condensation. Let $t_0 = x_0/c_0$ be the time of arrival for the first signal, or equivalently, the time at which we take ξ to be zero (then the second integral necessarily vanishes). Show formally that the first integral reduces to

$$\xi = c_0 \int_{0}^{t'} f(-c_0\tau)d\tau$$

where $t' = t - t_0$ is the "local" time, measured from the arrival of the first signal. (This formula is useful in simplifying displacement calculations).

4.37 A canonical equation of state is given in the form $a = a(T, v)$. Find an expression for the sound speed squared (c^2) in terms of v and the derivatives of this function with respect to its independent variables.

five
the nature of steady compressible flow

5.1 Mach number

In Sec. 3.5 it was suggested that the Mach number M is a rational measure of the importance of density changes in a steady flow. In this chapter we expand on this idea.

The Mach number M is defined as the ratio of the local flow speed to the local speed of sound

$$M \equiv \frac{u}{c} \tag{5.1}$$

It is thus a (scalar) dimensionless quantity. In general both u and c are functions of time and position, so that the Mach number is not merely the flow speed made dimensionless by dividing by a constant; thus, we cannot write $M \propto u$. It is almost always true, however, that M increases monotonically with u.

Consider steady one-dimensional flow in a pipe at three different Mach numbers. We imagine that the wall of the pipe is tapped with a small hammer, introducing sound pulses which travel in the fluid, approximately as shown in Fig. 5.1. In every case the sound pulses travel at speed c *with respect to the fluid.* The absolute motion of the sound pulses is thus given by

$$v_{\text{pulse}} = u \pm c$$

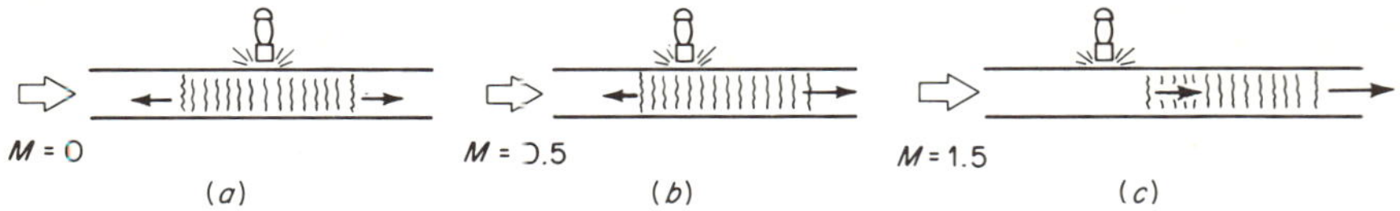

Figure 5.1
Convection of a pressure disturbance by a flowing fluid.

In the case of *supersonic* flow, $u > c$, both waves are swept downstream, $v > 0$. *In supersonic flow, disturbances cannot propagate upstream.* This means that the upstream flow remains unaffected by changes in conditions at a given point. Suppose, for example, that the pipe suddenly develops a leak; this cannot affect any of the upstream flow conditions; the upstream flow remains "ignorant" of the leak, although the downstream flow is certainly altered. In the *subsonic* case, however, both upstream and downstream flows are affected.

The idea of the limited range of influence of disturbances in supersonic flow is more generally illustrated by the situation of a small disturbance in a moving three-dimensional medium. Consider a small source of sound, a bumblebee, say, which moves with respect to the fluid in such a way as to remain stationary in our "absolute" reference frame; e.g., the bumblebee flies into the wind just fast enough to remain stationary with respect to the ground. We consider respectively stationary, subsonic, and supersonic fluid motion, though the last case may be a bit hard for the bumblebee. Suppose that sound pulses are initiated at fixed intervals. Again it is necessary to recognize that the sound propagation is with respect to the fluid medium: each sound pulse will thus have a *spherical* wavefront. The three cases are illustrated in Fig. 5.2. The first case is not very interesting; the sound eventually extends to infinity in every direction. This is also true for the second case, the motion of the fluid producing only a distortion of the wavefront array. In the third case ($M > 1$) things are quite different. All the disturbances are confined to a conical region, the *Mach cone*. A stationary observer situated outside of the Mach cone is not given any indication of the presence of the sound source and is said to be in the *zone of silence.* Just as in the one-dimensional case, disturbances cannot propagate upstream; in addition, however, their lateral penetration is also limited.

If the viewpoint is changed by allowing the observer to move with the fluid (Galilean transformation), the geometry of Fig. 5.2 is unchanged. The source now moves relative to the observer with a flight speed which is respectively zero, subsonic, and supersonic. In the subsonic case, a *Doppler effect* now appears, the sound decreasing in frequency as the source moves past the observer. In the supersonic case, the Mach cone now moves past the observer at the source flight speed; no sound appears until the Mach cone reaches his ear. This effect is familiar in the passage of supersonic aircraft.

The cone angle μ, called the *Mach angle*, is given from the construction of the figure by

$$\sin \mu = \frac{1}{M} \tag{5.2}$$

In photographs of steady supersonic flows, taken by either the schlieren or shadowgraph method, Mach cones are visible as waves emanating from points of local disturbance such as a rough spot on a smooth surface or the tip of a needle probe (see Fig. 5.3).

In two-dimensional flow it is appropriate to consider the local disturbance in Mach's construction as a *line* disturbance rather than a point disturbance. This line is perpendicular to the flow plane, and disturbances appear as cylindrical wavefronts; the envelope is then strictly a *Mach wedge*. In any case, the geometry of Fig. 5.2 is unchanged; we will refer to the projection of the Mach cone or Mach wedge into the flow plane as *Mach waves* or *Mach lines*.

The construction shown in Fig. 5.2, which has appeared in hundreds of books, was first given by Ernst Mach (1838–1916) in a paper on supersonic projectiles presented in 1887. It is largely on the basis of this paper, together with the development of the optical interferometer (which was, however, due chiefly to his son Ludwig) that Mach's name has become

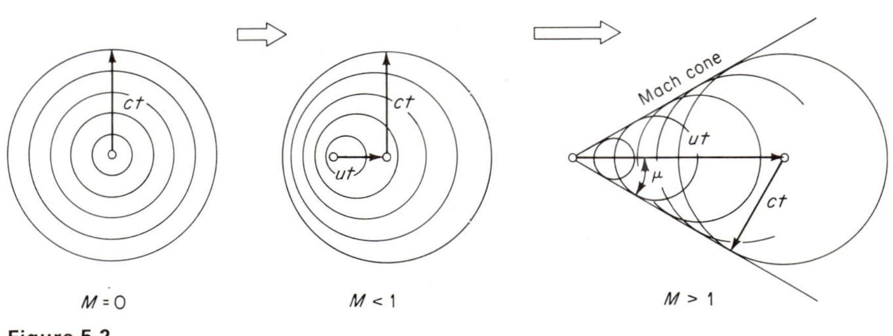

Figure 5.2
Mach's construction for the convection of a disturbance in a three-dimensional medium.

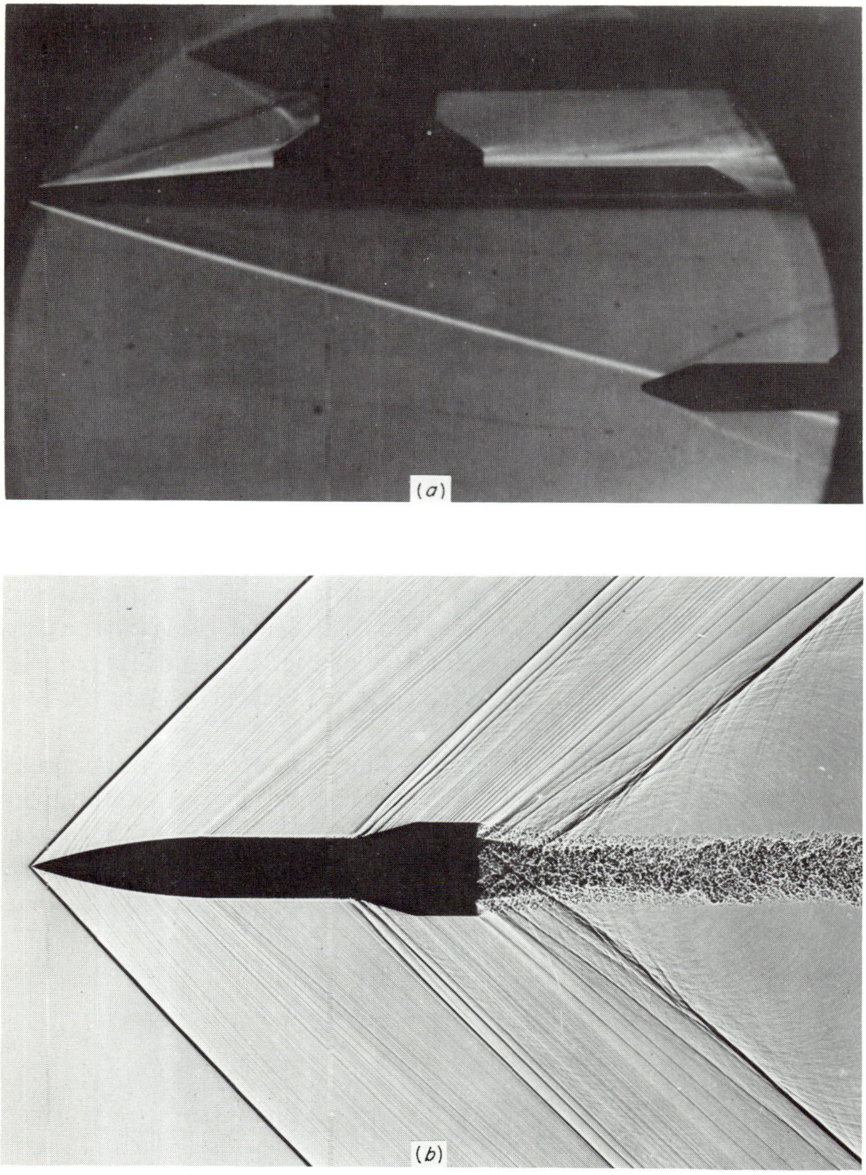

Figure 5.3
(*a*) Mach waves from the tip of a probe, $M_\infty = 4.5$. (*Courtesy of P. K. Pierpont, NASA.*) (*b*) Mach waves from axisymmetric projectile in free flight, $M_\infty = 1.4$. (*Courtesy of Z. J. Levensteins, Naval Ordnance Laboratory.*)

associated with compressible flow. The term *Mach number* was suggested by Jakob Ackeret in 1929. It was not immediately accepted, the Russians at one time preferring Bairstow number and the French having proposed (1947) Moisson number.

Flows are conventionally classified according to the Mach number (of the free stream) as follows:

$M < 1$, **Subsonic flow**: Low-speed flows, up to about $M = 0.3$, can be fairly accurately treated as incompressible. Up to about $M = 0.75$ (depending on the situation), the effects of density changes are accounted for by applying a *compressibility correction* to the results of incompressible theory. The equations are relatives of the heat equation.

$M \approx 1$, **Transonic flow**: A special mathematical theory is required.

$M > 1$, **Supersonic flow**: Above Mach number unity, new phenomena, including shock waves, appear. The equations of motion are more complicated relatives of the wave equation.

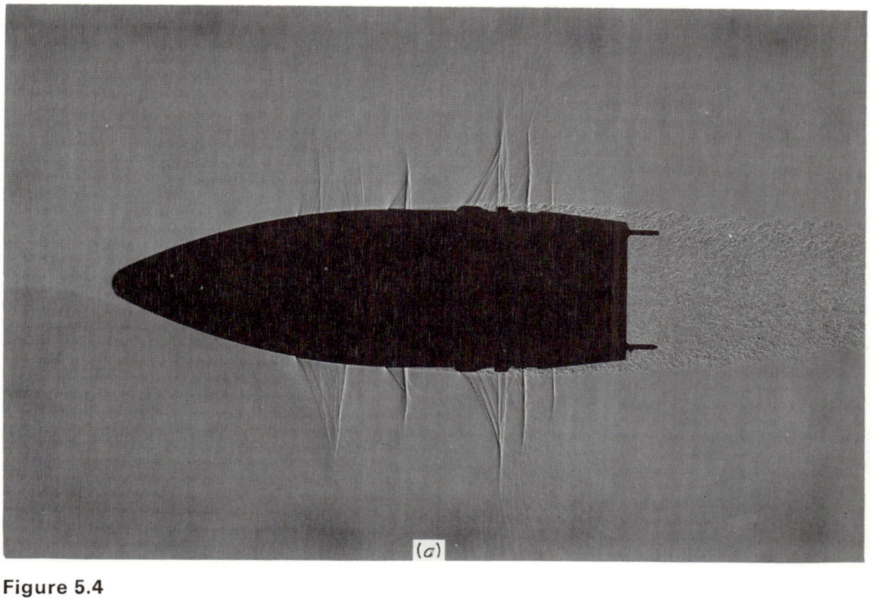

Figure 5.4
Shadowgraphs of a 30-mm projectile in free flight through air: (*a*) $M_\infty = 0.89$. (*Ballistic Research Laboratory.*)

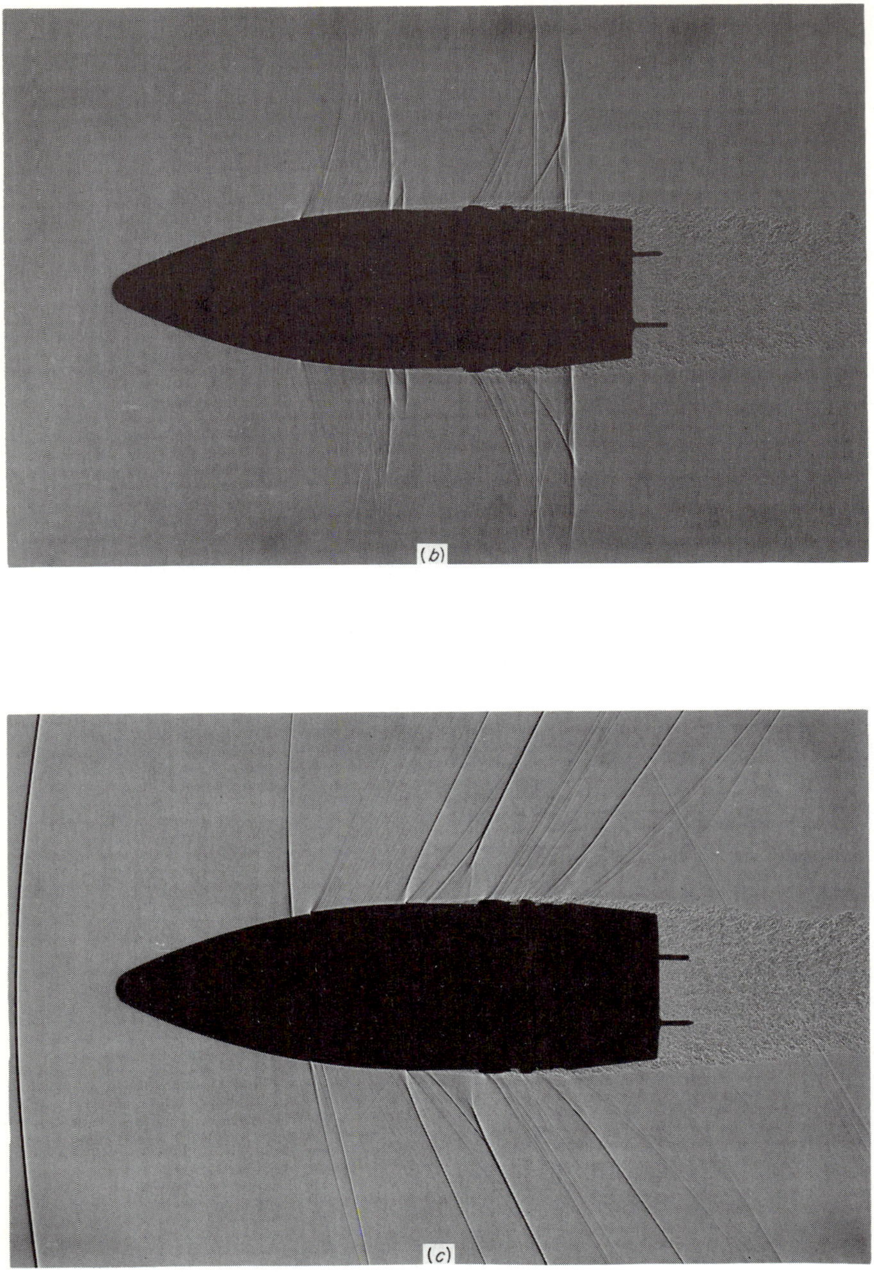

Figure 5.4 ***(Continued)***

Shadowgraphs of a 30-mm projectile in free flight through air: (*b*) $M_\infty = 0.91$; (*c*) $M_\infty = 1.06$; (*d*) $M_\infty = 1.42$; (*e*) $M_\infty = 1.99$.

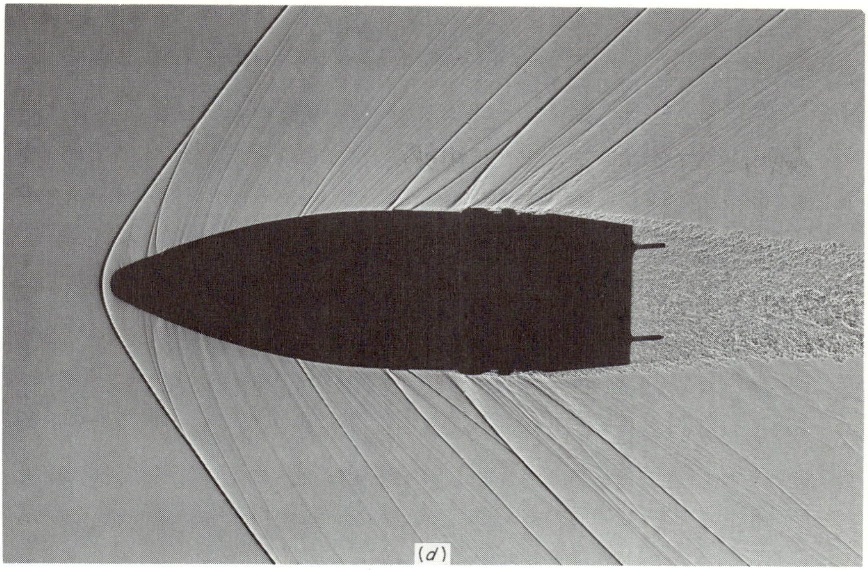

(d)

(e)

$M \gg 1$, Hypersonic flow: At sufficiently high speeds, essentially new phenomena, such as gas ionization and substantial thickening of viscous layers, come into play. The inviscid equations of motion are inherently nonlinear. This regime has been of interest primarily in connection with atmospheric reentry.

Shadowgraphs of streaming flow at various free-stream Mach numbers are shown in Fig. 5.4.

As we have seen, the distinction between supersonic and subsonic flow is in the limited range of influence of pressure disturbances (supersonic) as opposed to the *unlimited* range of influence (subsonic). This physical (or geometrical) difference is clearly manifested also in the equations of motion, as discussed in Secs. 5.3 and 5.4.

Microscopic Interpretation

The distinction between subsonic and supersonic flows can be roughly interpreted in microscopic terms. Specifically, the absence of upstream-traveling disturbances in supersonic flow is reflected in the fact that relatively few *molecules* are traveling in the upstream direction.

Consider the uniform (streaming) flow of an ideal gas at Mach number M in the x_1 direction. The Maxwell distribution (2.90) applies to a reference frame moving with the fluid; only those molecules with component velocity $v_1 \leq -u = -Mc$ will be traveling counter to the flow direction, i.e., will be traveling in the upstream direction. The fraction F of molecules so traveling is simply the (positive) value

$$F = \frac{1}{n}\int_{-\infty}^{-u} n_{v_1}\,dv_1$$

which is equivalent to

$$F = \frac{1}{n}\int_{u}^{\infty} n_{v_1}\,dv_1$$

Substituting n_{v_1} from (2.90) and carrying out the calculation yields

$$F = \tfrac{1}{2}\left[1 - \operatorname{erf}\left(\sqrt{\frac{\gamma}{2}}\,M\right)\right] \tag{5.3}$$

as illustrated in Fig. 5.5. For the case illustrated ($\gamma = 5/3$), 90 percent of the molecules are traveling in the downstream direction at Mach number unity.

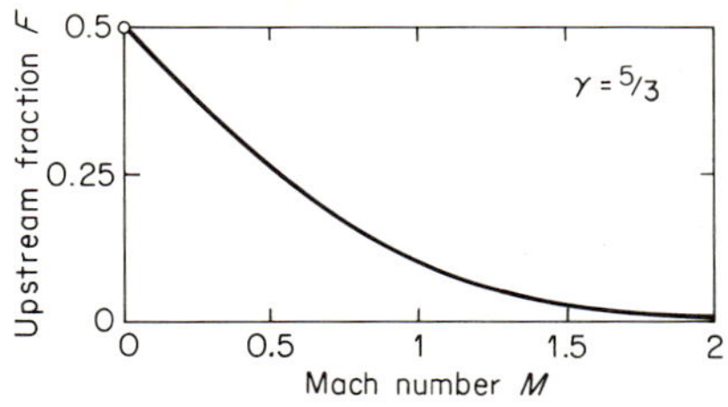

Figure 5.5
Fraction of molecules traveling in the upstream direction for a stationary observer ($\gamma = 5/3$).

What the preceding calculation does *not* show, of course, is the discontinuity in the nature of the flow at $M = 1$.

5.2 The inviscid energy equation

For steady inviscid flow the energy equation reduces to (1.75), viz.,

$$h + \frac{u^2}{2} + \Psi = \text{const}$$

Consistent with the discussion of Sec. 3.5, the body force (with potential Ψ) is almost always negligible, and this equation becomes

$$h + \frac{u^2}{2} = h_0 \tag{5.4}$$

The constant h_0 is called the *stagnation enthalpy*.

Stagnation Properties

The energy equation (5.4) in general describes an isentropic flow (an exception is adiabatic flow with friction, treated in Sec. 6.4). With the entropy fixed, the thermodynamic state at any point on a streamline is determined by any single thermodynamic variable and in particular by the enthalpy h.

In most practical cases, the entropy and stagnation enthalpy are the same for all streamlines. This will be true if the flow originates from a uniform region, as in the two cases shown in Fig. 5.6. Such flows are homentropic and *homenergic*. The *stagnation state*, corresponding to $u = 0$ in Eq. (5.4), is characterized by two constants, h_0 and s. Then the fluid at rest has a unique thermodynamic state; e.g., the stagnation temperature is given via a state equation by $T_0 = T(h_0,s)$. In general, any property evaluated at the stagnation state is referred to as a stagnation

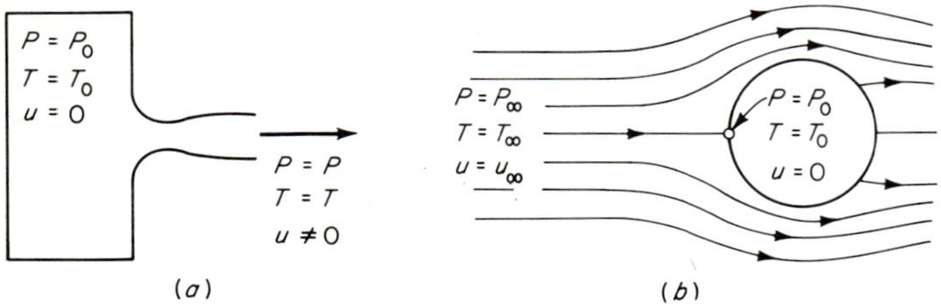

Figure 5.6
Flows characterized by a uniform stagnation state: (*a*) reservoir; (*b*) stagnation point.

property and will be indicated by subscript zero as P_0, ρ_0, T_0, and so on. In some places the word "total" is used instead of "stagnation," e.g., the *total temperature.*

The stagnation condition in a vessel as in Fig. 5.6*a* is sometimes called a *reservoir state.* The special point on a blunt body where the streaming flow stagnates, shown in Fig. 5.6*b*, is called a *stagnation point.*

Even if a region with zero flow speed cannot be found in a particular flow, the stagnation state is still "real" because the entire flow is characterized by the stagnation properties, for example, h_0 and s. At any given point in the flow, stagnation properties such as pressure and temperature can be measured by means of a stagnation probe, indicated schematically in Fig. 5.7. To the extent that the flow leading to the probe can be brought to rest by an isentropic process, a direct measurement is possible.

The concept of stagnation properties is useful even for certain unsteady nonhomentropic flows. The stagnation properties at a given time and a given point in space are those which would be obtained after a hypothetical steady isentropic deceleration to stagnation.

Static Properties

The local values of pressure, temperature, etc. in the flowing stream are conventionally (and confusingly!) referred to as the *static values.* The

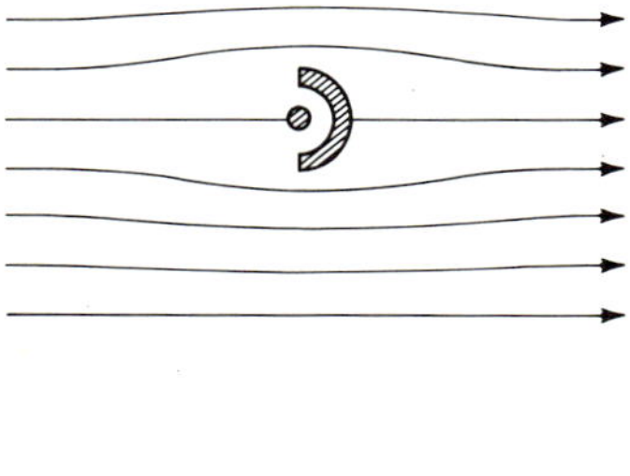

Figure 5.7
Stagnation probe.

implication is that such values can be measured, ideally, by means of a device which is static *with respect to the fluid.*

EXAMPLE 5.1 STAGNATION OF A PERFECT GAS

Suppose that a subsonic airplane flies at 600 mi/h through air at $P_\infty = 0.5$ atm and $t_\infty = 0°\text{F}$ (these are the static properties of the free stream). Find the fluid properties at the stagnation points in the flow.

The situation is conceptually as shown in Fig. 5.6*b*. From (5.4), with $\Delta h = c_p\,\Delta T$,

$$T_0 - T_\infty = \frac{u_\infty{}^2}{2c_p}$$

Inserting numbers, with the properties of air from Example 2.10 (page 85), gives $T_0 - T_\infty = 64.8°\text{R}$ or $t_0 = 64.8°\text{F}$. The stagnation pressure is found from the isentropic relation (2.86),

$$\frac{P_0}{P_\infty} = \left(\frac{T_0}{T_\infty}\right)^{\gamma/(\gamma-1)} \rightarrow P_0 = 0.79 \text{ atm}$$

Relation between Velocity and Mach Number

Let us determine whether M always increases with u in isentropic flow. The velocity change along a streamline can be written from (2.24) as

$$u\,du + v\,dP = 0 \tag{5.5}$$

where du and dP are the change in velocity and pressure along a differential length of the streamline. As the pressure changes, there will be a corresponding change in the sound speed, given by

$$dP = \left(\frac{\partial P}{\partial c^2}\right)_s dc^2 \tag{5.6}$$

It will be helpful to write the above thermodynamic derivative in another form; the sound speed is given by

$$c^2 = \left(\frac{\partial P}{\partial \rho}\right)_s = -v^2\left(\frac{\partial P}{\partial v}\right)_s = \frac{-v^2}{(\partial v/\partial P)_s}$$

Taking the derivative with respect to P, we find

$$\left(\frac{\partial c^2}{\partial P}\right)_s = 2v\left[\frac{c^4}{2v^3}\left(\frac{\partial^2 v}{\partial P^2}\right)_s - 1\right] \tag{5.7}$$

The first term within the brackets is a quantity which will persistently reappear in the following chapters, and we denote it by Γ,

$$\Gamma \equiv \frac{c^4}{2v^3}\left(\frac{\partial^2 v}{\partial P^2}\right)_s \tag{5.8}$$

This nondimensional thermodynamic property is of sufficient importance to justify calling it the *fundamental gasdynamic derivative.*

Substituting (5.6) and (5.7) into (5.5) now yields

$$u\,du + \frac{c\,dc}{\Gamma - 1} = 0 \tag{5.9}$$

The Mach number is defined by $M = u/c$, and logarithmic differentiation gives

$$\frac{dM}{M} = \frac{du}{u} - \frac{dc}{c} \tag{5.10}$$

Combining this with (5.9) yields the result

$$\frac{du}{u} = \frac{dM/M}{1 + (\Gamma - 1)M^2} \tag{5.11}$$

Thus, the Mach number will always increase monotonically with fluid velocity, provided that $\Gamma \geq 1$. This condition is satisfied for normal fluids, as discussed below.

The Fundamental Derivative

We have already defined this quantity in Eq. (5.8), viz.,

$$\Gamma \equiv \frac{c^4}{2v^3}\left(\frac{\partial^2 v}{\partial P^2}\right)_s$$

Manipulation of derivatives shows that the following forms are equivalent:

$$\Gamma = \frac{v^3}{2c^2}\left(\frac{\partial^2 P}{\partial v^2}\right)_s = \frac{1}{c}\left(\frac{\partial \mathscr{R}}{\partial \rho}\right)_s \tag{5.12}$$

where $\mathscr{R} = \rho c$ is the acoustic impedance.

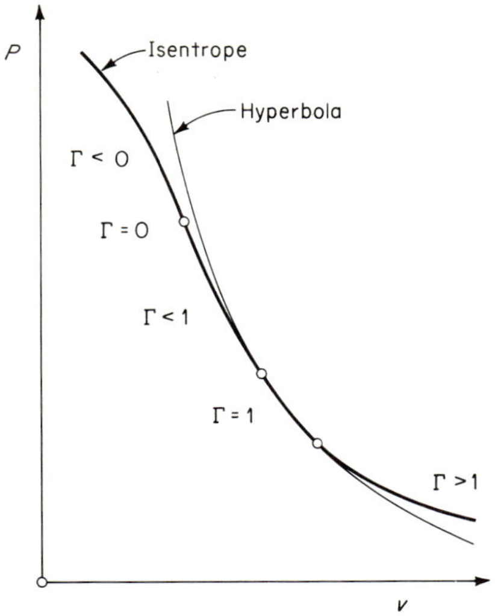

Figure 5.8
Possible behavior of the fundamental derivative.

For a perfect gas, differentiation of the isentropic relation $Pv^{\gamma} =$ const yields the constant value

$$\Gamma = \frac{\gamma + 1}{2} \qquad \text{perfect gas} \tag{5.13}$$

With $\gamma \geq 1$, $\Gamma \geq 1$ and of order unity. For a liquid satisfying the Tait equation (2.114) the same expression is obtained (γ is not the ratio of specific heats but still satisfies $\gamma \geq 1$); e.g., water has $\gamma \approx 7$, giving $\Gamma \approx 4$. Values of Γ for various liquids are given in Table F.4, Appendix F. Roughly, we can take Γ to be of order unity for fluids in general.

There is, however, no fundamental reason for us to expect Γ to be greater than unity or even universally positive. Indeed, in bizarre cases (such as the neighborhood of the critical point) available data indicate that isentropes can show an inflection point in the Pv plane, corresponding to Γ passing through zero (see Fig. 5.8). The case $\Gamma = 1$ corresponds geometrically to an isentrope in the form of a rectangular hyperbola, $P = K_1/v + K_2$, as can be found from (5.8). This case might be called the *acoustic* equation of state, because pressure is a linear function of density.

The case $\Gamma = 0$ corresponds to a linear relation $P(v)$. The equations of motion for such a fluid tend to be considerably simplified, as in the celebrated Kármán-Tsien theory for two-dimensional flows. The physical behavior of such a fluid is peculiar, however; e.g., shocks do not form (see

Sec. 8.5). If, as a convenient artifice, we set $\Gamma = 0$ for a fluid which is really a perfect gas, we assert implicitly that the ratio of specific heats $\gamma = -1$!

Compressible Form of the Bernoulli Equation

Isentropic flow of an *incompressible* fluid satisfies the Bernoulli equation (2.25), viz.,

$$P_0 - P = \tfrac{1}{2}\rho u^2 \tag{5.14}$$

where ρ and P_0 are constants.

A simple argument based on this equation shows that compressibility effects can be neglected for small values of the Mach number squared. For steady flow from stagnation conditions, $\Delta P/\Delta\rho \approx c^2$; with $\Delta P \sim \tfrac{1}{2}\rho u^2$ from (5.14) this yields

$$\frac{\Delta\rho}{\rho} \sim \tfrac{1}{2}M^2$$

Hence, density changes are negligible for small values of M^2. Let us now proceed more carefully and see how (5.14) is modified by compressibility.

A more general statement of (5.14) is the energy equation (5.4), which can be written

$$\frac{h_0 - h}{c^2} = \frac{M^2}{2} \tag{5.15}$$

By expanding $h - h_0$ and $c^2 - c_0{}^2$ in a Taylor series in $P - P_0$, a form like (5.14) can be obtained. Thus, with $p = P - P_0$,

$$h - h_0 = \left(\frac{\partial h}{\partial P}\right)_s p + \frac{1}{2}\left(\frac{\partial^2 h}{\partial P^2}\right)_s p^2 + \frac{1}{6}\left(\frac{\partial^3 h}{\partial P^3}\right)_s p^3 + \cdots$$

$$c^2 - c_0{}^2 = \left(\frac{\partial c^2}{\partial P}\right)_s p + \frac{1}{2}\left(\frac{\partial^2 c^2}{\partial P^2}\right)_s p^2 + \frac{1}{6}\left(\frac{\partial^3 c^2}{\partial P^3}\right)_s p^3 + \cdots$$

From Eq. (2.18), $(\partial h/\partial P)_s = v$; then $(\partial^2 h/\partial P^2)_s = (\partial v/\partial P)_s = -v^2/c^2$, and so on. From Eq. (5.7), $(\partial c^2/\partial P)_s = 2v(\Gamma - 1)$, and the higher derivatives can be found. Substituting the Taylor expansions into (5.15), one finds after considerable algebra the result

$$P_0 - P = \tfrac{1}{2}\rho_0 u^2[1 - \tfrac{1}{4}M^2 + \tfrac{1}{24}(4\Gamma - 3)M^4 \cdots] \tag{5.16}$$

which is the modified Bernoulli equation.

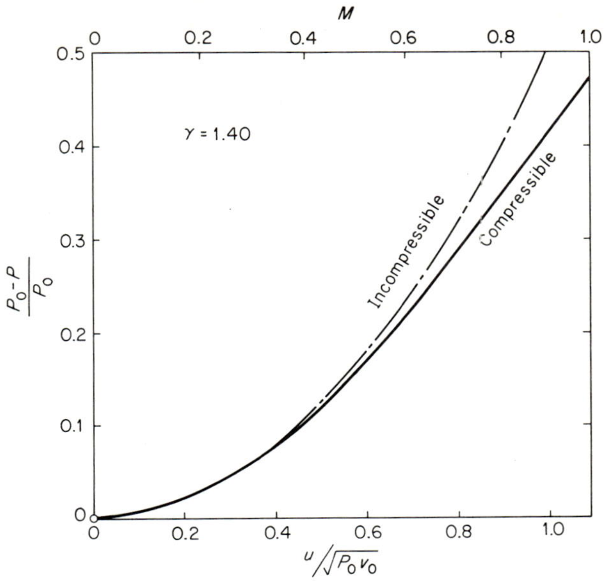

Figure 5.9
Pressure drop vs. velocity for a perfect gas with $\gamma = 1.40$.

Consistent with the discussion of Sec. 3.5, compressibility effects vary as the square of the Mach number. Up to some arbitrary "small" Mach number, for example, $M = 0.3$, compressibility may be neglected entirely. A comparison between the compressible and incompressible calculations is shown in Fig. 5.9.

5.3 Potential flow

In the treatment of acoustics it was found useful to introduce a velocity potential ϕ. We will now find the equation for the velocity potential in a general inviscid motion. Under the restrictions described in Sec. 2.4, we assume the motion to be irrotational and set

$$\mathbf{u} = \nabla\phi \tag{5.17}$$

The inviscid momentum equation (1.67) is

$$\frac{\partial \mathbf{u}}{\partial t} + (\mathbf{u} \cdot \nabla)\mathbf{u} + \frac{1}{\rho}\nabla P = 0 \tag{5.18}$$

The scalar product of the last term with **u** may be written

$$\frac{1}{\rho}\mathbf{u}\cdot\nabla P = \frac{1}{\rho}\frac{DP}{Dt} - \frac{1}{\rho}\frac{\partial P}{\partial t}$$

For isentropic flow, $dP = c^2\,d\rho$; taking advantage of this fact and using the continuity equation (1.62), this becomes

$$\frac{1}{\rho}\mathbf{u}\cdot\nabla P = -c^2\nabla\cdot\mathbf{u} - \frac{1}{\rho}\frac{\partial P}{\partial t}$$

The scalar product of (5.18) with **u** is now

$$\frac{1}{2}\frac{\partial u^2}{\partial t} + \frac{1}{2}\mathbf{u}\cdot\nabla u^2 - c^2\nabla\cdot\mathbf{u} = \frac{1}{\rho}\frac{\partial P}{\partial t} \tag{5.19}$$

[in obtaining this result, the identity $(\mathbf{u}\cdot\nabla)\mathbf{u} = \nabla(u^2/2) - \mathbf{u}\times\mathbf{\Omega}$ has been used]. This is *almost* an equation in the velocity potential only, i.e., containing only terms involving **u**; to express the right-hand side in terms of the velocity potential, we take the time derivative of (5.18), which yields, after insertion of the velocity potential,

$$\frac{\partial^2}{\partial t^2}\nabla\phi + \frac{1}{2}\frac{\partial}{\partial t}\nabla(\nabla\phi)^2 + \frac{\partial}{\partial t}\left(\frac{1}{\rho}\nabla P\right) = 0 \tag{5.20}$$

Now if $f(P)$ is a function only of P, it is an identity that

$$\frac{\partial}{\partial t}f(P)\,\nabla P = \nabla\left(f(P)\frac{\partial P}{\partial t}\right) \tag{5.21}$$

Applying this identity to the last term, (5.20) becomes

$$\nabla\left[\frac{\partial^2\phi}{\partial t^2} + \frac{1}{2}\frac{\partial}{\partial t}(\nabla\phi)^2 + \frac{1}{\rho}\frac{\partial P}{\partial t}\right] = 0$$

The quantity in square brackets is (at most) an arbitrary function of time; this function can be incorporated into ϕ, however, so that

$$\frac{\partial^2\phi}{\partial t^2} + \frac{1}{2}\frac{\partial}{\partial t}(\nabla\phi)^2 + \frac{1}{\rho}\frac{\partial P}{\partial t} = 0 \tag{5.22}$$

This can in turn be integrated with respect to time to give

$$\frac{\partial\phi}{\partial t} + \tfrac{1}{2}(\nabla\phi)^2 + \int\frac{dP}{\rho} = \text{const} \tag{5.23}$$

which is a variant of the (compressible) Bernoulli equation (2.24).

Substituting (5.22) into (5.19) and expressing $\mathbf{u}$ in terms of the velocity potential finally gives

$$\frac{\partial^2\phi}{\partial t^2} + \frac{\partial}{\partial t}(\nabla\phi)^2 + \tfrac{1}{2}\nabla\phi \cdot \nabla(\nabla\phi)^2 - c^2\,\nabla^2\phi = 0 \tag{5.24}$$

where the sound speed c is known in terms of the velocity potential via (5.23).

There are two celebrated special cases of (5.24). For incompressible fluids $c \to \infty$, and we obtain the *Laplace equation*, $\nabla^2\phi = 0$. For small-amplitude motions the terms in $(\nabla\phi)^2$ may be neglected, and we obtain the *wave equation* $\phi_{tt} - c^2\,\nabla^2\phi = 0$.

For steady flow, (5.24) can be written in indicial notation, with $\mathbf{u}\cdot\nabla = u_k\,\partial/\partial x_k$ and $\nabla\phi = \mathbf{e}_i\,\partial\phi/\partial x_i$,

$$u_i u_k \phi_{,ik} - c^2\phi_{,ii} = 0 \tag{5.25}$$

where the coefficient of the first term has been written as $u_i u_k$ rather than $\phi_{,i}\phi_{,k}$ to emphasize the comparison with the coefficient c^2. According to the theory of partial differential equations, (5.25) has distinctly different behavior depending on whether $u < c$ or $u > c$, that is, $M < 1$ or $M > 1$. If $u < c$, the equation is said to be *elliptic*; Laplace's equation is an example of this type, which in general describes diffusive processes. If $u > c$, the equation is said to be *hyperbolic*; the wave equation is an example of this type, which is in general associated with waves and shock-type discontinuities. We shall distinguish more specifically between the subsonic and supersonic behavior of (5.25) in the following section.

.4 Linearized descriptions of potential flow

There are very few known solutions for the nonlinear potential-flow equation (5.25). The most important approximate technique for dealing with this equation is that of linearization, as briefly described in this section.

A thin airfoil is an example of a *slender body* which disturbs only slightly a uniform streaming flow (Fig. 5.10). This disturbance is then a small perturbation on a uniform state, analogous to the acoustic motions considered in Chap. 4. We begin with the full potential equation (5.25) written out for two dimensions

$$u_1{}^2\phi_{,11} + 2u_1u_2\phi_{,12} + u_2{}^2\phi_{,22} - c^2(\phi_{,11} + \phi_{,22}) = 0 \tag{5.26}$$

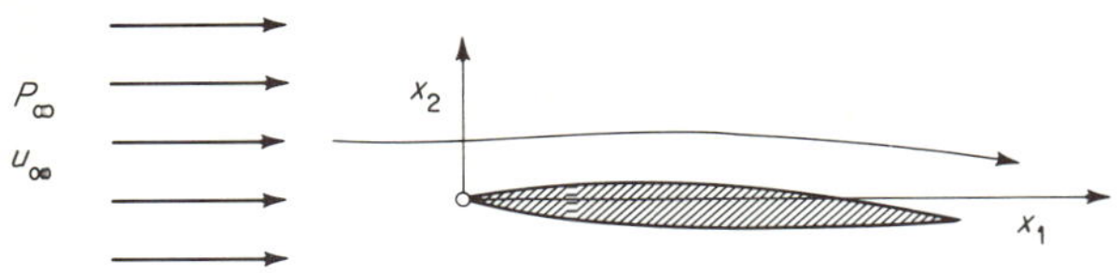

Figure 5.10
Two-dimensional slender body in streaming flow.

Without justifying the steps in detail,[1] the linearization consists (as in the acoustic case) in replacing the coefficients by their undisturbed values; in particular, $u_1 \to u_\infty$, $u_2 \to 0$, and $c \to c_\infty$, so that the equation becomes

$$(u_\infty{}^2 - c_\infty{}^2)\phi_{,11} - c_\infty{}^2\phi_{,22} = 0$$

or

$$(1 - M_\infty{}^2)\phi_{,11} + \phi_{,22} = 0 \tag{5.27}$$

which is the linearized equation of plane potential flow. Depending on whether the free stream is subsonic ($M_\infty < 1$) or supersonic ($M_\infty > 1$), the coefficient $1 - M_\infty{}^2$ is either positive or negative; the two cases are entirely distinct, and will be discussed separately below.

We now introduce the perturbation quantities $\hat{\mathbf{u}} = \mathbf{u} - \mathbf{u}_\infty$ and $p = P - P_\infty$. The velocity $\mathbf{u}$ can be written

$$\mathbf{u} = \mathbf{e}_1(u_\infty + \hat{u}_1) + \boldsymbol{\epsilon}_2\hat{u}_2 \tag{5.28}$$

The perturbation velocity potential $\hat{\phi}$ is defined by

$$\mathbf{u} = \mathbf{u}_\infty + \nabla\hat{\phi} \tag{5.29}$$

Because $\mathbf{u}_\infty$ is just a constant, this potential also satisfies the basic equation (5.27). Henceforth we shall drop any special notation for perturbation quantities and regard all variables as describing just the perturbations; thus $\hat{\phi} \to \phi$, $\hat{u}_1 \to u_1$, $\hat{u}_2 \to u_2$, etc.

From (5.23), for steady flow,

$$\int_{P_\infty}^{P} \frac{dP}{\rho} + \tfrac{1}{2}u^2 = \tfrac{1}{2}u_\infty{}^2$$

With (5.28), and neglecting terms of order u_2/u_∞, this gives

$$p = P - P_\infty = -\rho_\infty u_\infty u_1$$

[1] For a fuller discussion, see, for example, *Liepmann and Roshko* [1957, chap. 8].

or

$$p = -\rho_\infty u_\infty \phi_{,1} \tag{5.30}$$

The analogy between the small perturbations represented here and those represented by the wave equation is quantitative. In fact, Eq. (5.27) *is* the one-dimensional wave equation (4.19) expressed in a moving reference frame, for which unsteady terms are presumed to vanish (see Prob. 5.21).

Subsonic Case ($M_\infty < 1$)

Equation (5.27) becomes, with $M_\infty < 1$,

$$\beta^2 \frac{\partial^2 \phi}{\partial x_1{}^2} + \frac{\partial^2 \phi}{\partial x_2{}^2} = 0 \tag{5.31}$$

where the constant coefficient β^2 is

$$\beta^2 \equiv 1 - M_\infty{}^2 \qquad 1 > \beta > 0 \tag{5.32}$$

and the positive root is always assumed.

The potential equation (5.31) is virtually the classical Laplace equation, which describes incompressible potential flow, and becomes the Laplace equation via a simple coordinate transformation, as noted below. This affine, or coordinate-stretching, transformation was described in Sec. 3.7 and is the basis for the several *similarity rules* of compressible flow. These rules have considerable practical importance in aerodynamics. A comprehensive treatment requires more space than is available here, and only the basic outline will be sketched. For a more complete discussion see *Shapiro* [1953, chap. 10] or *Liepmann and Roshko* [1957, chap. 10].

With the affine transformation

$$X_1 = x_1 \qquad X_2 = \beta x_2 \tag{5.33}$$

and with

$$\Phi(X_1,X_2) = \lambda\phi(x_1,x_2) \tag{5.34}$$

where λ is a constant, the potential equation (5.31) reduces to the Laplace equation,

$$\frac{\partial^2 \Phi}{\partial X_1{}^2} + \frac{\partial^2 \Phi}{\partial X_2{}^2} = 0 \tag{5.35}$$

An incompressible flow in the X_1X_2 space may thus be compared with a compressible flow in the x_1x_2 space.

Let the perturbation velocities be u_1, u_2 in the x space and U_1, U_2 in the X space; thus,

$$u_i = \frac{\partial \phi}{\partial x_i} \qquad U_i = \frac{\partial \Phi}{\partial X_i} \tag{5.36}$$

The boundary conditions far from the body are, consistent with the transformation (5.33),

$$\begin{aligned} u_1 &= u_2 = 0 \qquad x = \infty \\ U_1 &= U_2 = 0 \qquad X = \infty \end{aligned} \tag{5.37}$$

and the velocity potentials $\phi(x)$ and $\Phi(X)$ thus have a common boundary condition at infinity.

There is more than one choice of stretching for the slender-body shape which will lead to the necessary common boundary conditions on the body, i.e., that the flow follow the body surface in both cases. One possibility is to let the body surface be given by the *same* function in both coordinate systems, e.g.,

$$\begin{aligned} (x_2)_s &= T(x_1) \\ (X_2)_s &= T(X_1) \end{aligned} \tag{5.38}$$

where the subscript s denotes the surface coordinate. The physical meaning of such a choice is that we compare the *same profile* in compressible and incompressible flow. The flow must follow the local surface; this requires that the slope of the surface equal that of the streamline, or

$$\begin{aligned} \frac{dT(x_1)}{dx_1} &= \frac{u_2}{u_1 + u_\infty} \approx \frac{1}{u_\infty}\frac{\partial \phi}{\partial x_2} \\ \frac{dT(X_1)}{dX_1} &= \frac{U_2}{U_1 + U_\infty} \approx \frac{1}{U_\infty}\frac{\partial \Phi}{\partial X_2} \end{aligned} \tag{5.39}$$

Thus

$$\frac{1}{u_\infty}\frac{\partial \phi}{\partial x_2} = \frac{1}{U_\infty}\frac{\partial \Phi}{\partial X_2} \tag{5.40}$$

With (5.33) and (5.34) this is an identity if we take $U_\infty = u_\infty$ and $\lambda = \beta$ (note, however, that other choices are possible). Then

$$\Phi = \beta\phi \tag{5.41}$$

The perturbation velocity components in the two spaces are now related, at corresponding points $X_1 = x_1$, $X_2 = x_2$, by

$$\begin{aligned} U_1 &= \frac{\partial \Phi}{\partial X_1} = \beta \frac{\partial \phi}{\partial X_1} = \beta \frac{\partial \phi}{\partial x_1} = \beta u_1 \\ U_2 &= \frac{\partial \Phi}{\partial X_2} = \frac{\partial \phi}{\partial X_2} = \frac{\beta}{\beta} \frac{\partial \phi}{\partial x_2} = u_2 \end{aligned} \tag{5.42}$$

Note that with the same profile and $u_\infty = U_\infty$, the relation $U_2 = u_2$ could have been predicted in advance from the flow-following condition. From Eq. (5.30) the pressure perturbations p and $\hat{P}$ are related, at corresponding points, by

$$\hat{P} = \beta p \tag{5.43}$$

By an analogous procedure, we could compare two compressible flows characterized by different values of β, that is, by different free-stream Mach numbers.

Steady subsonic flow has been found to be qualitatively like incompressible flow. The supersonic case will be found to be qualitatively different.

Supersonic Case ($M_\infty > 1$)

Equation (5.27) becomes, with $M_\infty > 1$,

$$\frac{\partial^2 \phi}{\partial {x_2}^2} - \beta^2 \frac{\partial^2 \phi}{\partial {x_1}^2} = 0 \tag{5.44}$$

where the constant coefficient β^2 is now

$$\beta^2 = {M_\infty}^2 - 1 \qquad \beta > 0 \tag{5.45}$$

and the positive root is always assumed. The geometrical relation between β and the Mach angle is indicated in Fig. 5.11.

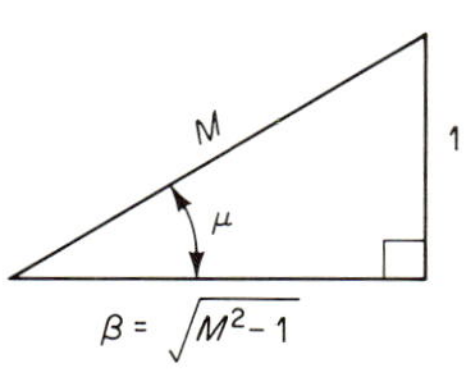

Figure 5.11

The potential equation (5.44) is *formally analogous to the one-dimensional wave equation* (4.19), $\phi_{tt} - c^2\phi_{xx} = 0$. Taking advantage of this analogy, we can immediately write down the solution

$$\phi = F(x_1 - \beta x_2) + G(x_1 + \beta x_2) \tag{5.46}$$

which may be compared to (4.20). Consider, for example, simple F waves with $G = 0$. The argument of F, and therefore F and its derivatives, is constant along the *characteristics* $x_1 - \beta x_2 = \text{const}$. (The notion of characteristic lines was introduced in connection with acoustics in Sec. 4.4.) In the present case, a characteristic line has slope $dx_2/dx_1 = 1/\beta = \tan \mu_\infty$ and therefore is a *Mach line*, or a line lying on the surface of the Mach cone (in two-dimensional flow, the Mach cone becomes a Mach wedge, as already noted in Sec. 5.1).

An illustrative solution of the form (5.46) is shown in Fig. 5.12. All flow properties are constant along each characteristic (Mach line) shown. Taking advantage of the fact that disturbances cannot propagate upstream in supersonic flow, the waves above the airfoil are necessarily simple F waves and those below simple G waves. The properties along any given characteristic can be related to the local slope of the surface, where it is intersected by the characteristic, as follows: let the upper surface be given by

$$(x_2)_s = T(x_1) \tag{5.47}$$

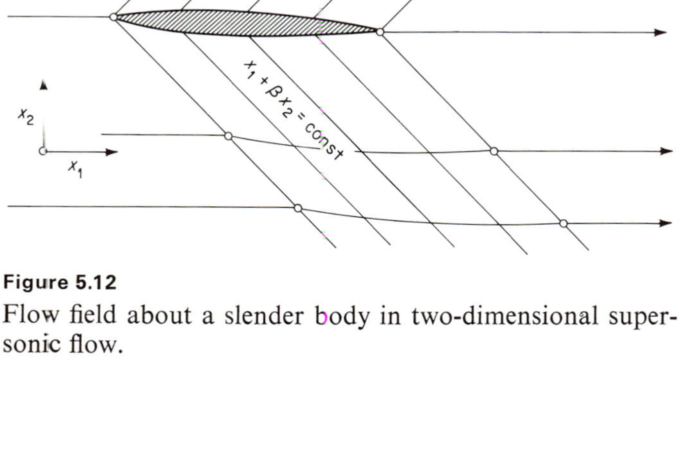

Figure 5.12

Flow field about a slender body in two-dimensional supersonic flow.

as before. From the flow-following condition, the slope of a streamline matches the local surface slope, so that

$$\frac{u_2}{u_\infty} = T' \tag{5.48}$$

Along the characteristic $x_1 - \beta x_2 = \text{const}$, the flow variables are given by (5.29) and (5.30) as

$$\begin{aligned} u_1 &= \frac{\partial \phi}{\partial x_1} = F'(x_1 - \beta x_2) \\ u_2 &= \frac{\partial \phi}{\partial x_2} = -\beta F'(x_1 - \beta x_2) \\ p &= -\rho_\infty u_\infty \frac{\partial \phi}{\partial x_1} = -\rho_\infty u_\infty F'(x_1 - \beta x_2) \end{aligned} \tag{5.49}$$

Combining (5.48) and (5.49) yields the results

$$\frac{u_1}{u_\infty} = -\frac{T'}{\beta} \qquad \frac{u_2}{u_\infty} = T' \qquad \frac{p}{\rho_\infty {u_\infty}^2} = \frac{T'}{\beta} \tag{5.50}$$

All these properties may thus be discontinuous along the leading characteristic. If the lower surface is given by $x_2 = T_l(x_1)$, we obtain similar expressions for the flow field below the airfoil:

$$\frac{u_1}{u_\infty} = \frac{T_l'}{\beta} \qquad \frac{u_2}{u_\infty} = T_l' \qquad \frac{p}{\rho_\infty {u_\infty}^2} = -\frac{T_l'}{\beta} \tag{5.51}$$

The *streamlines* in the flow depicted in Fig. 5.12 may be compared to the *path lines* in the wave diagram of Fig. 4.5. The streamlines and path lines run generally in different directions in the respective figures only

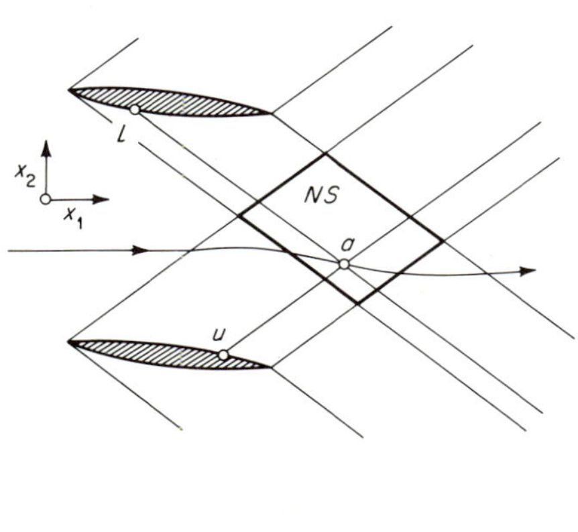

Figure 5.13
Nonsimple region (NS) formed by the intersection of F waves and G waves.

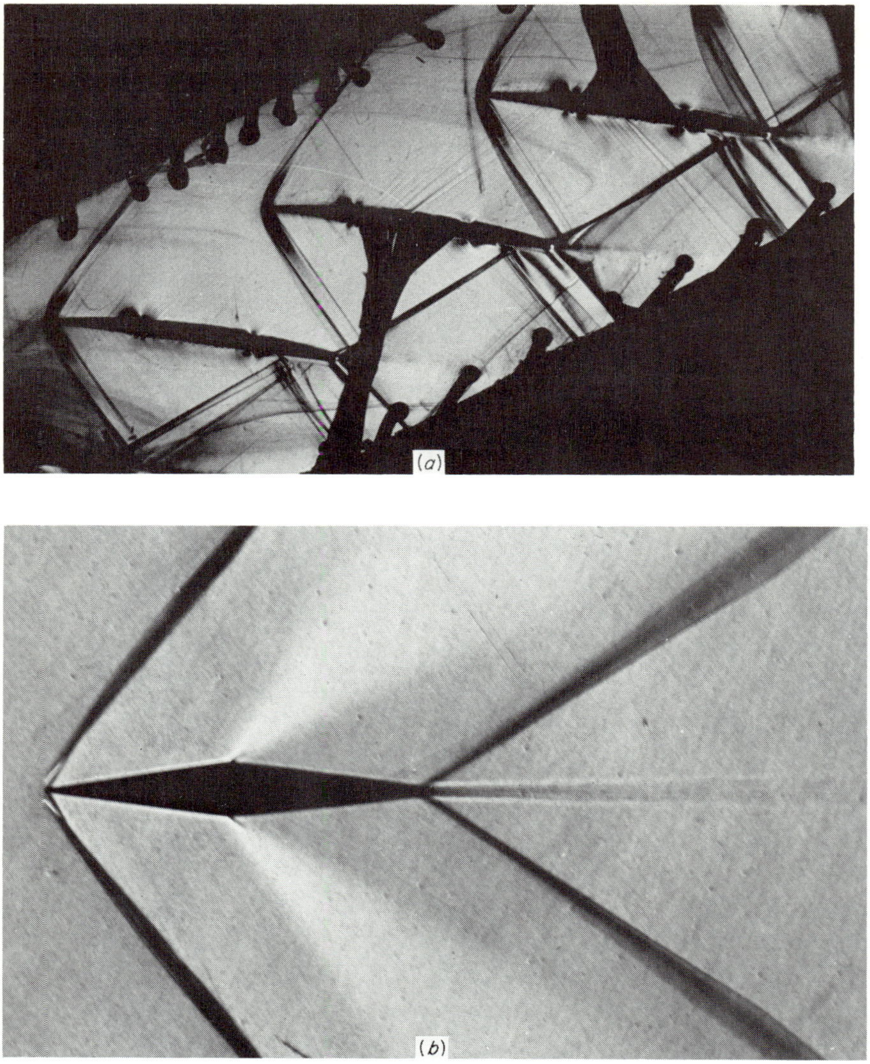

Figure 5.14
(*a*) Supersonic flow over slender airfoils in a turbine cascade $M_\infty = 1.30$. (*Courtesy of H. Starken, DFVLR Institut für Luftstrahlantriebe.*) (*b*) Supersonic flow over a double-wedge airfoil, $M_\infty = 1.50$. The airfoil is sufficiently thick for nonlinear effects to be important in this case. (*Courtesy of R. North, National Physical Laboratory; Crown Copyright reserved.*)

by virtue of the different conventions for choice of axes in the two problems. If, for example, we choose the supersonic streaming flow to be in the x_2 direction, the graphical analogy between the two flows becomes precise.

For *nonsimple waves*, the flow is given by the sum $F + G$ of Eq. (5.46) or, correspondingly, by the sum of solutions (5.50) and (5.51). In particular, for the flow shown in Fig. 5.13 (as might be seen in a turbine cascade, for example), the streamline slope at the point a is the algebraic sum of the surface slopes at u and l; that is, we have the superposition of (5.50) and (5.51)

$$\frac{u_2}{u_\infty} = T_u' + T_l'$$

A photograph of flow over a slender airfoil is shown in Fig. 5.14.

EXAMPLE 5.2 PRESSURE DISTRIBUTION ON AN AIRFOIL

Let the upper and lower surfaces of a slender airfoil be formed by the respective parabolas

$$\frac{T_u}{L} = \frac{x_{2u}}{L} = A\left[\frac{x_1}{L} - \left(\frac{x_1}{L}\right)^2\right] - \alpha\frac{x_1}{L}$$

$$\frac{T_l}{L} = \frac{x_{2l}}{L} = -B\left[\frac{x_1}{L} - \left(\frac{x_1}{L}\right)^2\right] - \alpha\frac{x_1}{L}$$

where α is the *angle of attack* and L is the *chord*, as shown in Fig. 5.15. By differentiation, obtain

$$x_{2u}' = A\left(1 - 2\frac{x_1}{L}\right) - \alpha$$

$$x_{2l}' = -B\left(1 - 2\frac{x_1}{L}\right) - \alpha$$

The pressure distributions, from (5.50) and (5.51) are $p_u/\rho_\infty u_\infty^2 = x_{2u}'/\beta$ and $p_l/\rho_\infty u_\infty^2 = -x_{2l}'/\beta$, so that

$$\frac{p_u}{\rho_\infty u_\infty^2} = \frac{1}{\beta}\left[A\left(1 - 2\frac{x_1}{L}\right) - \alpha\right]$$

$$\frac{p_l}{\rho_\infty u_\infty^2} = \frac{1}{\beta}\left[B\left(1 - 2\frac{x_1}{L}\right) + \alpha\right]$$

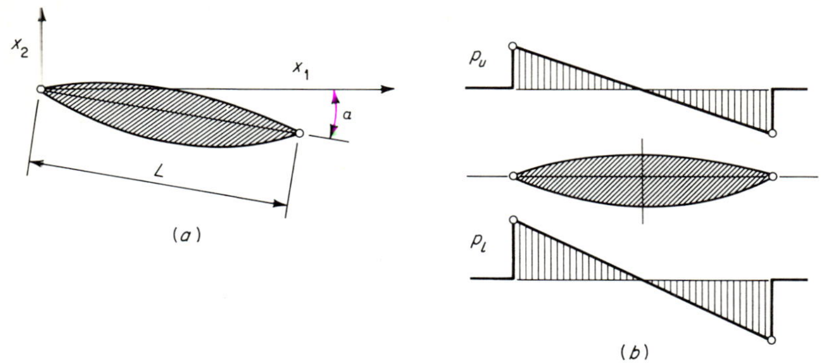

Figure 5.15
(*a*) Airfoil at angle of attack (thickness exaggerated); (*b*) pressure distribution at zero angle of attack.

These results, for $\alpha = 0$, are shown in Fig. 5.15*b*. It may be remarked that the pressure "signature" from the lower surface is essentially that of a sonic boom.

There is no net lifting force on the airfoil unless $\alpha \neq 0$; the lifting force (in the x_2 direction) per unit span is, by integration, making use of (5.50) and (5.51),

$$F_L = \int_0^L (p_l - p_u)\, dx_1 = \frac{2\rho_\infty {u_\infty}^2 L}{\beta}\,\alpha \tag{5.52}$$

There is however a *drag* force at any angle of attack. The reader can verify that the drag force (in the x_1 direction) per unit span is

$$F_D = \int_0^L (p_u T_u' - p_l T_l')\, dx_1$$

$$= \frac{\rho_\infty {u_\infty}^2 L}{\beta}\,[\tfrac{1}{3}(A^2 + B^2) + \alpha^2] \tag{5.53}$$

The second term, proportional to α^2, is called the *drag due to lift*, since the lift is associated with α by (5.52).

The linearized treatment of supersonic flow is agreeably simple and predicts with reasonable accuracy the flow in the close neighborhood of a slender body. Far from the body, however, this simple theory predicts a disturbance undiminished in amplitude from that near the body. This is not in accord with observations, and will be remedied by the nonlinear theory given in Chap. 9.

5.5 Isentropic flow of a perfect gas

Over a considerable range of pressure and temperature, many fluids of practical importance (such as atmospheric air!) are adequately described as perfect gases. The algebraic simplicity of the perfect gas relations will lead to simple results from the integral energy statement $h + u^2/2 = \text{const}$ and from the integral mass statement $\rho u A = \text{const}$.

Energy

For the perfect gas, $h = c_p T + \text{const}$, and the energy equation (5.4) becomes

$$c_p T + \frac{u^2}{2} = c_p T_0 \tag{5.54}$$

where T_0 is the stagnation temperature. Thus, as a particle of fluid accelerates along a streamline (as in flow from a reservoir, say), the temperature of the particle drops. Most routine compressible-flow calculations are based on this simple equation.

With $c^2 = \gamma R T$ and $c_p = \gamma R/(\gamma - 1)$, Eq. (5.54) becomes

$$c^2 + \frac{\gamma - 1}{2} u^2 = c_0{}^2 \tag{5.55}$$

where c_0 is the stagnation sound speed. Alternatively, this relation can be written in terms of the sound speed at $M = 1$, the so-called *sonic* condition. By definition $u = c$ at this condition, which will be denoted by a subscript star; thus, with $u = c = c_*$, (5.55) gives

$$c_*^2 = \frac{2}{\gamma + 1} c_0{}^2 \tag{5.56}$$

Then (5.55) can be rewritten in terms of this new constant as

$$c^2 + \frac{\gamma - 1}{2} u^2 = \frac{\gamma + 1}{2} c_*^2 \tag{5.57}$$

Equation (5.55) can be rewritten once again by dividing through by c^2 to obtain

$$\frac{T_0}{T} = 1 + \frac{\gamma - 1}{2} M^2 \tag{5.58}$$

Table 5.1 Thermodynamic Propert es at $M = 1$
Isentropic Flow, Perfect Gas

$\gamma \Rightarrow$	$9/7 = 1.286$	$7/5 = 1.400$	$5/3 = 1.667$
T_*/T_0	0.8750	0.8333	0.7500
P_*/P_0	0.5483	0.5283	0.4871
ρ_*/ρ_0	0.6267	0.6339	0.6495

For isentropic flow, simple relations connect the temperature and other variables; in particular,

$$\frac{T_0}{T} = \left(\frac{c_0}{c}\right)^2 = \left(\frac{P_0}{P}\right)^{(\gamma-1)/\gamma} = \left(\frac{\rho_0}{\rho}\right)^{\gamma-1} \tag{5.59}$$

Note that T, c, P, and ρ all decrease monotonically with increasing Mach number. All thermodynamic quantities are now fixed by the single quantity M; numerical values are given in the *isentropic-flow tables* (Table D.1, Appendix D).

Setting $M = 1$ in (5.58) or using (5.56) directly gives the following values for thermodynamic quantities at the sonic condition:

$$\frac{T_*}{T_0} = \frac{2}{\gamma + 1} \qquad \frac{P_*}{P_0} = \left(\frac{2}{\gamma + 1}\right)^{\gamma/(\gamma-1)} \qquad \frac{\rho_*}{\rho_0} = \left(\frac{2}{\gamma + 1}\right)^{1/(\gamma-1)} \tag{5.60}$$

For representative values of the ratio of specific heats γ, this gives the numerical values shown in Table 5.1. The value of P_*/P_0 is of particular interest for nozzle flows; if P_a is the atmospheric pressure at the nozzle discharge, the relative values of P_a/P_0 and P_*/P_0 will determine the nature of the flow, as will be discussed in Chap. 6.

A simple physical application of the isentropic-flow equations is a nozzle exhausting into a vacuum (Fig. 5.16). As the pressure drops with distance along the axis of the nozzle, the Mach number gradually increases

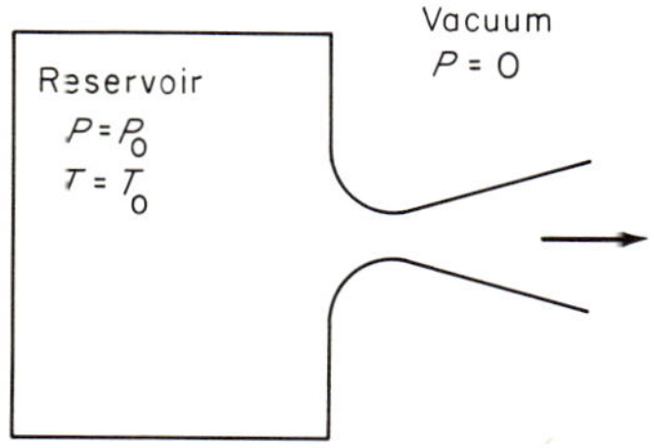

Figure 5.16

from a value of zero in the reservoir. In the limit, as the pressure approaches zero outside the nozzle, the Mach number goes to infinity, from (5.58) and (5.59). The flow speed, however, remains finite ($M \to \infty$ because $c \to 0$) and from (5.55) is just

$$u_{\max} = \sqrt{\frac{2}{\gamma - 1}}\, c_0 \tag{5.61}$$

the *limit* or *maximum speed.* This is the maximum speed which can be attained in any (inviscid) steady flow. In order to calculate the detailed flow, to find the streamlines, say, it would be necessary to solve the equation of motion (5.25). The flow through the nozzle itself can be treated, however, by the approximate methods to be given in Chap. 6.

The Normalized Mach Number M_*

It is sometimes convenient to express the isentropic-flow relations in terms of a variant of the Mach number, $M_* \equiv u/c_*$ (the notation here, which is conventional, is not quite consistent; that is, M_* is *not* the Mach number at the sonic condition). Since c_* is a constant for any given stagnation conditions, M_* is a true normalized speed

$$M_* \equiv \frac{u}{c_*} = M \frac{c}{c_0} \frac{c_0}{c_*}$$

Equations (5.58) to (5.60) then give, for a perfect gas,

$$M_* = \frac{M}{\left[\dfrac{2}{\gamma + 1} + \dfrac{\gamma - 1}{\gamma + 1} M^2\right]^{1/2}} \tag{5.62}$$

which gives the corresponding values

M	M_*
0	0
1	1
∞	$\sqrt{\dfrac{\gamma + 1}{\gamma - 1}}$

The last set of values corresponds to the maximum speed.

Mass

A tubular surface containing a set of streamlines forms a hypothetical duct for the flow and is called a *stream tube* (formally, if C is a closed curve in the region of the flow, the streamlines passing through C generate the surface of the stream tube). Consider a stream tube of cross-sectional area δA, where δA is small enough so that all properties may be considered uniform at any given cross section (see Fig. 5.17). The condition that mass is conserved is just

$$\rho u \, \delta A = \text{const} \tag{5.63}$$

The mass flow per unit area, or *mass flux*, is just ρu.

For any fluid, the mass flux can be written, with $u_* = c_*$,

$$\frac{\rho u}{\rho_* u_*} = \frac{\rho}{\rho_0}\frac{\rho_0}{\rho_*}\frac{u}{c}\frac{c}{c_0}\frac{c_0}{c_*} = \frac{\rho_0}{\rho_*}\frac{c_0}{c_*}M\frac{\rho}{\rho_0}\frac{c}{c_0}$$

For a perfect gas, this is just a function of Mach number; substituting from (5.58) to (5.60), we obtain

$$\frac{\rho u}{\rho_* u_*} = \frac{M}{\left(\dfrac{2}{\gamma+1} + \dfrac{\gamma-1}{\gamma+1}M^2\right)^{(\gamma+1)/[2(\gamma-1)]}} \tag{5.64}$$

By differentiation, this function has a maximum at $M = 1$; *the mass flux is a maximum at the sonic condition* (as will be verified in Chap. 6, this result holds for all normal fluids, not just for the perfect gas). By virtue of (5.63), the area δA of a stream tube is inversely proportional to the mass flux: it follows that δA *will be a minimum at the sonic condition.* We should remark that the converse of this statement does not necessarily follow: a stream tube can have an area minimum (as in a garden nozzle or venturi tube) without having sonic flow.

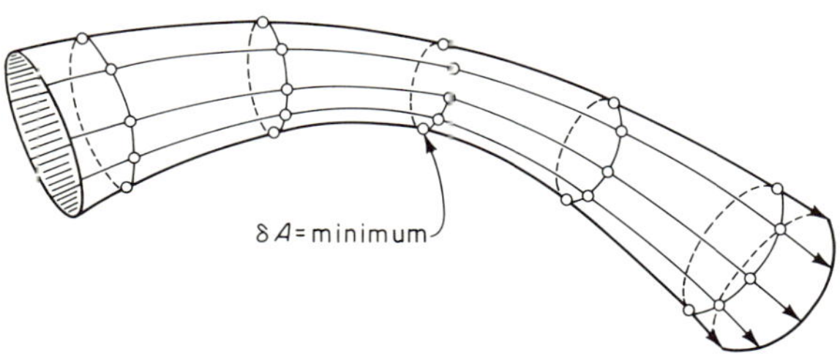

Figure 5.17
Stream tube with an area minimum.

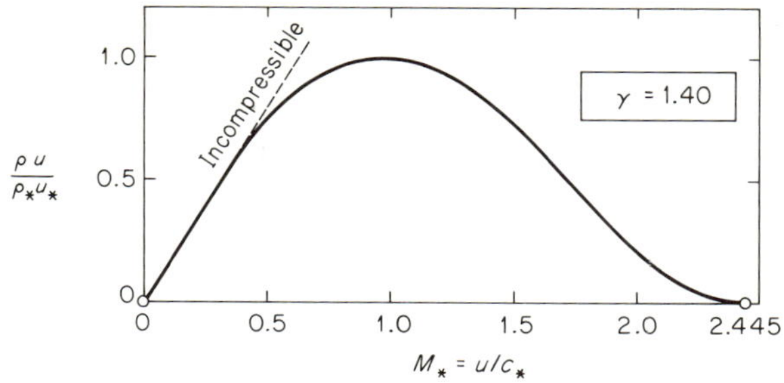

Figure 5.18
Mass flux as a function of M_*, for a perfect gas with $\gamma = 1.40$.

In terms of M_*, the normalized mass flux becomes, from (5.62),

$$\frac{\rho u}{\rho_* u_*} = M_* \left(\frac{\gamma + 1}{2} - \frac{\gamma - 1}{2} M_*^2 \right)^{1/(\gamma - 1)} \tag{5.65}$$

which is shown in Fig. 5.18. This illustrates the maximum at $M = M_* = 1$. For low Mach numbers, the curve rises as a straight line, corresponding to the linear increase in mass flux with speed for incompressible flow. At $M = \infty$ $[M_* = \sqrt{(\gamma + 1)/(\gamma - 1)}]$ the mass flux is zero, so that infinite stream-tube area is required at the maximum speed because the density is zero.

Problems

5.1 Find the Mach angle μ corresponding to free-stream Mach numbers of 1, 1.2, 2, and 4.

5.2 Consider the Mach construction shown in Fig. 5.1. Explain why a *Doppler effect* appears in the case of a moving disturbance in a stationary medium (as when a tooting car passes a pedestrian at high speed) but does not appear in the case of a stationary source in a moving medium. The observer in each case is considered to be stationary.

5.3 Draw the Mach construction for a moving disturbance which accelerates through the speed of sound. This construction is helpful in understanding the peculiar form of the sonic boom associated with transition from subsonic to supersonic flight, i.e., breaking the sound barrier.

5.4 In a certain flow both the velocity and the velocity gradient are purely in the x direction [as in Example 1.3 (page 24)]. If the Mach number of the flow increases linearly with x, sketch several of the Mach lines (point disturbances may be inserted wherever desired to generate the Mach lines).

5.5 A point source of sound moves through a stationary homogeneous atmosphere at velocity u_0. The constant source frequency is ν_0, and the sound speed is c_0. Find the apparent frequency ν for an observer on the ground, $\nu = \nu(\theta, M_0)$, where θ is the source position at the time of sound emission, i.e., earlier than the time of reaching the observer, and $M_0 = u_0/c_0$. Determine whether propaganda broadcast from a supersonic source ($M_0 > 1$) would have to be played backward.

Answer $$\nu = \frac{\nu_0}{1 - M_0 \cos\theta}$$

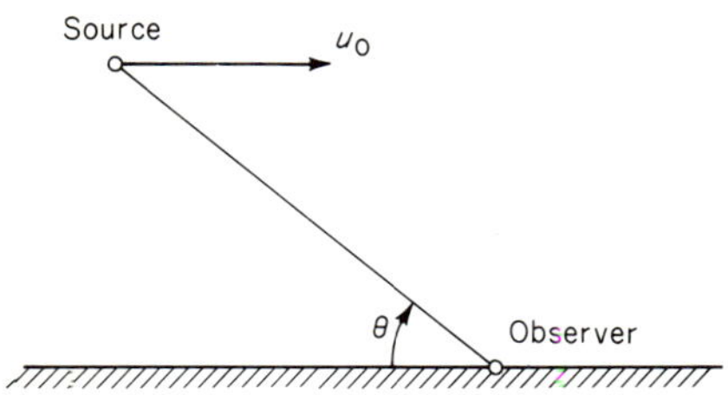

5.6 A slow isentropic flow satisfies the condition

$$\frac{P_0 - P}{P_0} \ll 1$$

If the fluid is a perfect gas, find a relation between the pressure drop $P_0 - P$ and the flow speed u. Compare this with the incompressible form of Bernoulli's equation.

5.7 For a lookout on the bow of a ship, stagnation conditions may be approximately realized at the tip of the lookout's nose. If the ship is moving at 20 knots, estimate the stagnation temperature rise $T_0 - T_\infty$. In general, at what fluid speeds would the heating due to isentropic compression become significant?

Answer $T_0 - T_\infty = 0.095$ K

5.8 By differentiation show that

$$\left(\frac{\partial^2 P}{\partial v^2}\right)_s = \frac{c^6}{v^6}\left(\frac{\partial^2 v}{\partial P^2}\right)_s$$

5.9 For steady isentropic flow of an arbitrary fluid, show that the condensation $S \equiv (\rho - \rho_0)/\rho_0$ is given by

$$S = -\tfrac{1}{2}M^2$$

provided that $|S| \ll 1$.

5.10 For the steady flow of liquid water, at what flow speed (in meters per second, say) may compressibility effects become important?

5.11 Air flows at a speed of 1,200 ft/s and a static pressure of 1 atm. The air is isentropically brought to rest in a steady-flow process. Find the Mach number and stagnation pressure if the static temperature is

(*a*) 1000°F

(*b*) −100°F

Answer (*a*) 0.641, 1.318 atm; (*b*) 1.291, 2.735 atm

5.12 A supersonic airplane is designed to travel at $M_\infty = 4$ at an altitude of 50,000 ft, where $t_\infty = -67.6$°F and $P_\infty = 242.2\ \text{lb}_f/\text{ft}^2$. Find the adiabatic stagnation temperature in degrees Fahrenheit.

Answer 1187°F

5.13 If the static pressure P and stagnation pressure P_0 are known for a certain flow, such that $P_0/P = 2$, what is the corresponding flow Mach number? Assume that the fluid in question is a perfect gas with $\gamma = 1.40$.

5.14 It is desired to find the Mach number of a certain flow. The static pressure P is accurately measured, but the measured stagnation pressure $(P_0)_m$ differs from the true (isentropic) stagnation pressure. It can be assumed (more or less realistically) that the deceleration process yielding $(P_0)_m$ is adiabatic and satisfies the inviscid energy equation $h + u^2/2 = \text{const}$ but is not now isentropic; i.e., the specific entropy of the fluid increases. If the ratio $(P_0)_m/P = 2$, what can be said about the flow Mach number? Assume a perfect gas with $\gamma = 1.40$. Compare with the result from the previous problem.

Answer $M > 1.047$

5.15 Find the approximate form of the potential equation (5.25) appropriate to steady *transonic* flow, $M_\infty \approx 1$.

5.16 The *lift coefficient* C_L and *drag coefficient* C_D are respectively defined by

$$C_L \equiv \frac{F_L}{\tfrac{1}{2}\rho_\infty u_\infty^2 L} \qquad C_D \equiv \frac{F_D}{\tfrac{1}{2}\rho_\infty u_\infty^2 L}$$

where the notation is the same as that of Example 5.2 (page 265); note that F_L and F_D are respectively the lift and drag per unit span. Find C_L and C_D for an airfoil in the form of a flat plate of zero thickness, at angle of attack α, using the linear theory. The flow is supersonic.

Answer $\dfrac{4\alpha}{\beta} \quad \dfrac{4\alpha^2}{\beta}$

5.17 Give a detailed derivation of Eq. (5.30), $p = -\rho_\infty u_\infty u_1$.

5.18 In a horizontal streaming flow, a flat plate moves vertically upward with velocity u_v, $u_v \ll u_\infty$. Given that the free-stream flow is supersonic, find the overpressure $p = P - P_\infty$ on the upper face of the plate in the form

$$p = \rho_\infty u_\infty u_v f(M_\infty)$$

from the linear theory. Also find the same result from the integral momentum statement applied to an appropriate control volume, making use of the fact that the disturbed field is bounded by a Mach wave from the leading edge, with $\mu = \mu_\infty$. Discuss the case $M_\infty \approx 1$.

Answer $p = \dfrac{\rho_\infty u_\infty u_v}{\beta}$

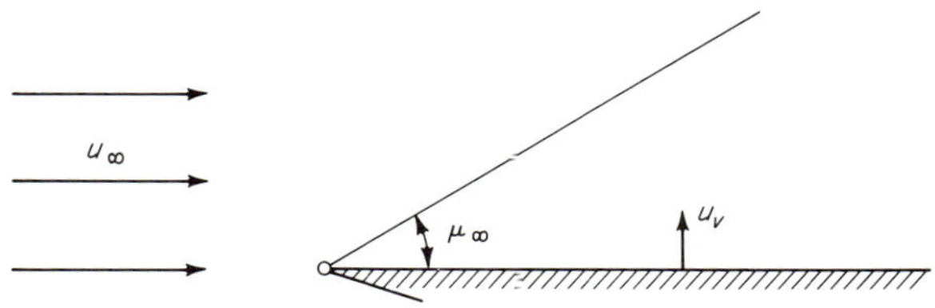

5.19 (*Busemann biplane*). Find the drag force on the special symmetric airfoil shown, at zero angle of attack. Use the linearized theory for supersonic airfoils and note that M_∞ has a special value, such that the Mach wave geometry is as shown.

Answer Zero

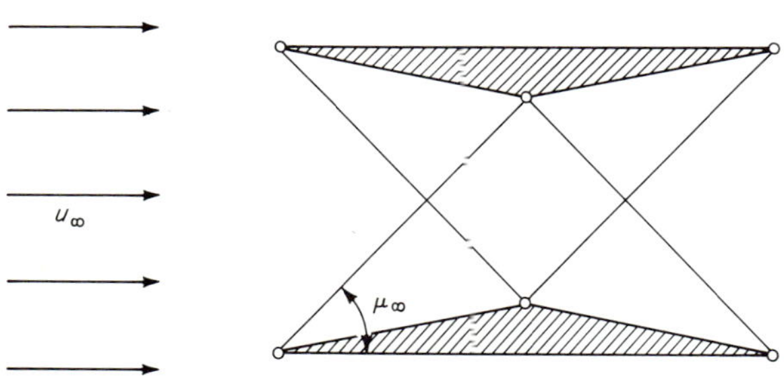

5.20 (*Potential vortex in compressible flow*). For a potential vortex with uniform entropy and stagnation enthalpy, find the radial variation of Mach number $M(r)$ for a perfect gas. It may be assumed that the flow is sonic at some radius $r = r_*$. Sketch the Mach lines for the flow.

5.21 The wave equation in two dimensions can be written

$$\phi_{,11} + \phi_{,22} = \frac{1}{c_\infty{}^2} \phi_{tt}$$

Write this equation for an observer who moves along the x_1 axis at velocity $-u_\infty$, where $u_\infty > 0$. Assuming that the flow is steady in the observer's reference frame, show that the wave equation thus reduces to (5.27), the linearized potential equation for steady flow.

5.22 A Clausius gas has the thermal equation of state

$$P(v - b) = RT$$

where b and R are constants.

(*a*) Show that the ratio of specific heats γ satisfies $\gamma = \gamma(T)$.

(*b*) Find an expression for the fundamental derivative Γ, in terms of γ and other quantities. Note that this expression reduces to that for a perfect gas.

six
one-dimensional steady flow

6.1 Introduction

Compressible flow can be treated at various levels of complexity. In this chapter we take the most elementary level possible, and the results, by virtue of their simplicity, are among the most instructive.

The basic approximations are that the flow is *steady* and *one-dimensional* or, more precisely, steady and quasi-one-dimensional. By this we mean that the flow can be treated according to a one-dimensional model even though the "real" flow is in fact three-dimensional. Consider the distribution of fluid velocities in a pipe of variable cross-sectional area, as shown in Fig. 6.1. The "real" velocity distribution has variable velocity over the cross section and is not necessarily symmetric; the idealized one-dimensional distribution has a single average value for the velocity component. The real flow also has in general a radial component of velocity such that the flow at the pipe wall is parallel to the wall (in the absence of separation). This component is accounted for only by way of the continuity equation, ρuA = const.

Other flow variables of interest, such as pressure and density, will also be treated according to the one-dimensional model. That is, in any given cross section all flow variables have constant values.

Thus all properties are assumed to vary only along the axis of the duct, which is taken to be the x direction. It is not essential that the duct

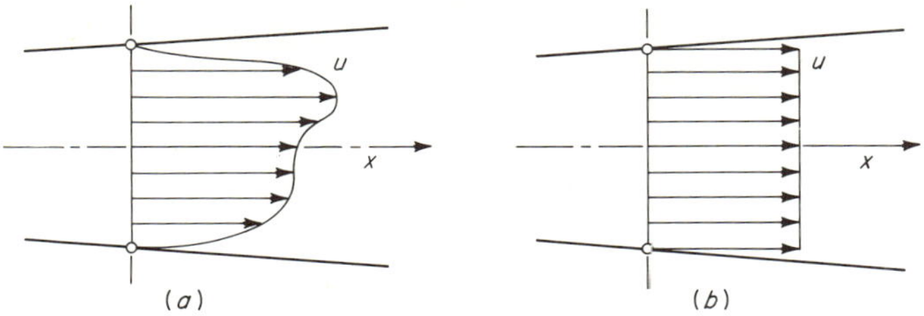

Figure 6.1
Distribution of axial velocity component: (*a*) real; (*b*) idealized.

be straight, as we can bend the coordinate x axis around in such a way as to follow the duct.

The relative success of the one-dimensional model will of course depend on the geometry of the duct or pipe. If h is some lineal measure of the cross section, such as the radius or half-width, and R is the radius of curvature of the wall (so that $R \to \infty$ for a straight duct), we require

$$\frac{dh}{dx} \ll 1 \qquad \frac{h}{R} \ll 1 \tag{6.1}$$

These conditions can of course be met by a *stream tube*, which can be taken arbitrarily small. Thus, if the streamlines are somehow known for a three-dimensional flow, the methods of this chapter are perfectly applicable to the resulting stream tubes. The isentropic-flow model presented in the following section is usually better suited to this case than to flow in a pipe because the assumption of negligible dissipation is likely to be closer to realization.

The quasi-one-dimensional model originated in the celebrated *Hydrodynamics* of Daniel Bernoulli (1738). He discussed limitations of this model, e.g., violations of the conditions (6.1), and suggested that "a noticeable error can arise from defects of this sort only very rarely." In view of modern experiments, such errors are not rare but somewhat common. We pursue the one-dimensional treatment more for its simplicity than for its accuracy.

.2 Isentropic flow

The flow near the walls of a duct is subject to viscous and thermal effects, and entropy normally accumulates in the fluid. For sufficiently short ducts, however, the isentropic model is a satisfactory approximation. We will later investigate the effects of friction and obtain a rough criterion for shortness.

The equations of continuity and momentum were given in Sec. 4.10 and are, respectively, for steady flow

$$\frac{1}{\rho}\frac{d\rho}{dx} + \frac{1}{u}\frac{du}{dx} + \frac{1}{A}\frac{dA}{dx} = 0 \tag{6.2}$$

$$\rho u \frac{du}{dx} + \frac{dP}{dx} = 0 \tag{6.3}$$

Equation (6.2) is just the differentiated form of $\rho u A =$ const, and the cross-sectional area $A(x)$ is a known function.

With

$$\frac{dP}{dx} = \left(\frac{\partial P}{\partial \rho}\right)_s \frac{d\rho}{dx} = c^2 \frac{d\rho}{dx}$$

combining (6.2) and (6.3) yields

$$\frac{1}{u}\frac{du}{dx} = \frac{1}{M^2 - 1}\frac{1}{A}\frac{dA}{dx} \tag{6.4}$$

where $M \equiv u/c$, as before. This relation has the following consequences: an acceleration, corresponding to a positive value of du, will be associated with a positive value of $dA/(M^2 - 1)$, a deceleration with a negative value of $dA/(M^2 - 1)$. Table 6.1 summarizes the various cases. The sign of

Table 6.1

Geometry		$M < 1$	$M > 1$
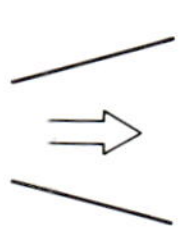	$dA > 0$ Diverging channel	$du < 0$ $dP > 0$ Subsonic diffuser	$du > 0$ $dP < 0$
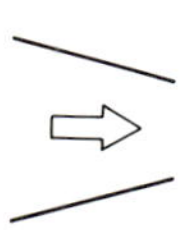	$dA < 0$ Converging channel	$du > 0$ $dP < 0$	$du < 0$ $dP > 0$ Supersonic diffuser

dP is always opposite to that of du from Eq. (6.3); one can also see that the density change will have the same sign as the pressure change because $(\partial P/\partial \rho)_s$ is necessarily positive. The behavior of subsonic and supersonic flows is thus opposite in many respects.

Transonic Flow at a Throat

In order for a flow to proceed continuously from subsonic to supersonic, it must at some point pass through $M = 1$. This is logically trivial but physically nontrivial; let us consider how it might be achieved.

From (6.4), the acceleration can remain finite where $M = 1$ only if $dA/dx = 0$, so that the relation is of the form zero/zero. Thus is appears that transonic flow might be achieved at a *throat*, i.e., an area minimum, in the duct. To confirm this, we substitute (5.11) into (6.4) to obtain an equation explicitly in the Mach number

$$\frac{1}{M}\frac{dM}{dx} = \frac{1 + (\Gamma - 1)M^2}{M^2 - 1}\frac{1}{A}\frac{dA}{dx} \tag{6.5}$$

For the case $dA/dx = 0$ and $M = 1$, we can apply l'Hospital's rule to obtain the elegant relation

$$\left(\frac{dM}{dx}\right)^2 = \frac{\Gamma}{2A}\frac{d^2A}{dx^2} \tag{6.6}$$

For a physical solution to exist, the right-hand side must be positive; that is, Γ must have the same sign as d^2A/dx^2. There are then two roots, $dM/dx > 0$ and $dM/dx < 0$, corresponding respectively to subsonic →

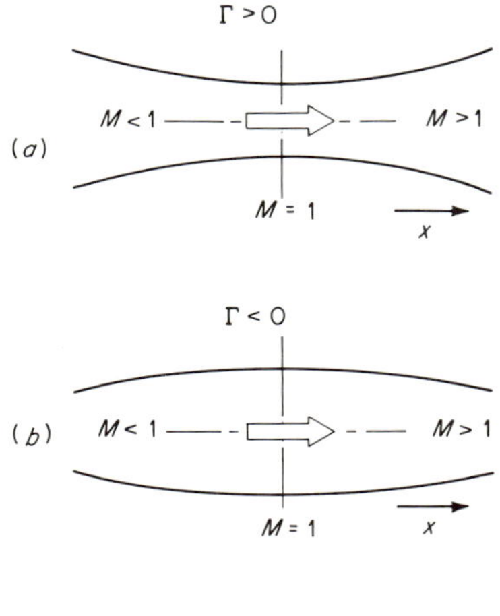

Figure 6.2
Transition from subsonic to supersonic flow.

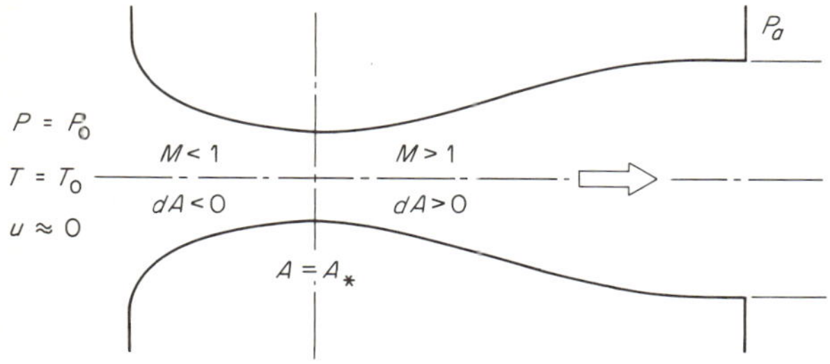

Figure 6.3
Supersonic nozzle, with transition to supersonic flow at the throat.

supersonic transition and to supersonic → subsonic transition. The possible cases for $dM/dx > 0$ are shown in Fig. 6.2. As discussed in Sec. 5.2, the fundamental derivative Γ is positive for almost all fluids and for the perfect gas in particular. The case shown in Fig. 6.2*a* is therefore the one of practical interest; the academic case shown in Fig. 6.2*b* will not be considered further.

Thus, for normal fluids *the Mach number can pass through unity only at an area minimum or throat* (this conclusion was already reached, for a perfect gas, in Sec. 5.5). The corresponding geometrical configuration is illustrated again in Fig. 6.3. Such a configuration is called variously a supersonic nozzle, converging-diverging nozzle, or Laval nozzle.[1] The

[1] Carl Gustav Patrik de Laval (1845–1913) was a Swedish inventor who developed such a nozzle as part of a high-speed turbine for driving a cream separator (1883).

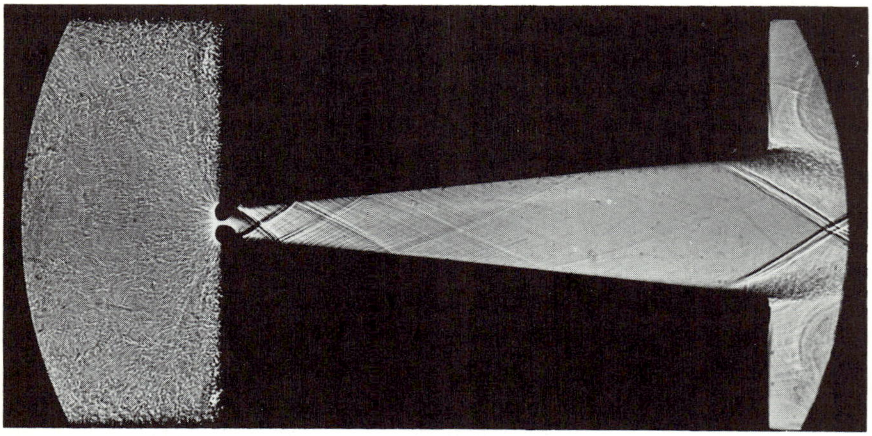

Figure 6.4
Supersonic flow in the diverging section of a supersonic nozzle. (*Courtesy of H. O. Amann, Ernst Mach Institut.*)

fluid is accelerated from rest in the converging section ($dA < 0$), passes through $M = 1$ at the throat ($dA = 0$), and continues to accelerate in the diverging section ($dA > 0$). This is the normal operating mode for such a nozzle (see Fig. 6.4). The case of *reversed* flow, in which supersonic → subsonic transition occurs at the throat, is also allowed by the equations, however, and is approximately realized in some practical situations.

It should be noted that while subsonic → supersonic transition can occur only at a throat, it will not necessarily occur there. From Eqs. (6.4) if $M \neq 1$ at a throat, the acceleration simply vanishes; this means that a subsonic flow in the converging section will remain subsonic in the diverging section, as in a *venturi tube*; similarly, a flow which is supersonic in the converging section will remain supersonic in the diverging section.

As noted earlier, the above arguments are applicable also to stream tubes in a three-dimensional flow.

6.3 Isentropic flow of perfect gas in a duct

All the relations developed in Sec. 5.5 are applicable. For convenience we repeat the main relations:

$$\frac{T_0}{T} = 1 + \frac{\gamma - 1}{2} M^2 \tag{6.7}$$

$$\frac{T_0}{T} = \left(\frac{c_0}{c}\right)^2 = \left(\frac{P_0}{P}\right)^{(\gamma-1)/\gamma} = \left(\frac{\rho_0}{\rho}\right)^{\gamma-1} \tag{6.8}$$

The continuity condition is written $\rho u A = \rho_* u_* A_*$, where the *star subscript refers to sonic conditions* (for steady isentropic flow, sonic conditions can occur only at a *throat*; in other flows, sonic conditions may be achieved differently; see Sec. 6.4). Then Eq. (5.64) gives

$$\frac{A}{A_*} = \frac{1}{M}\left(\frac{2}{\gamma + 1} + \frac{\gamma - 1}{\gamma + 1} M^2\right)^{\gamma + 1/(2(\gamma - 1))} \tag{6.9}$$

Numerical values for these functions of the Mach number are tabulated in the isentropic-flow table, Table D.1.

A plot of the continuity equation (6.9) is shown in Fig. 6.5. The curve may be thought of as the distorted profile of a supersonic nozzle, in which Mach number increases with distance. In particular flows, however, the sonic area A_* may not appear as a physically realized area; it will

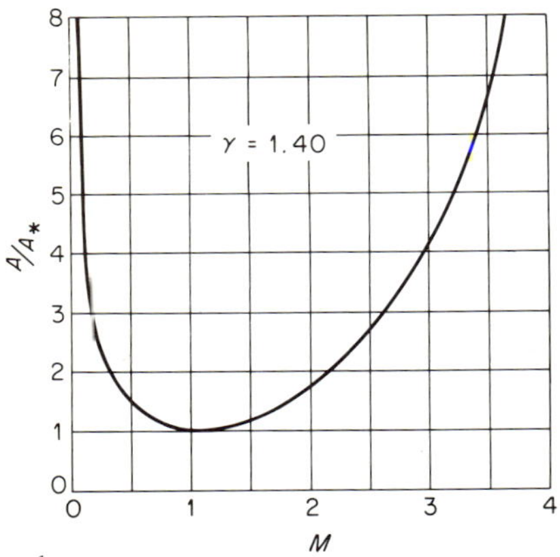

Figure 6.5
Area vs. Mach number for a perfect gas, $\gamma = 1.40$.

not be realized if the flow is everywhere subsonic, or everywhere supersonic, as will be illustrated shortly.

Equations (6.7) to (6.9) allow the calculation of all flow variables when $A(x)$ and entry/exit conditions are given for the pipe. As an important specific case we will discuss a nozzle supplied by a reservoir with known conditions P_0, T_0 and discharging into a stationary atmosphere of known pressure P_a.

Discharge Conditions

The outflow of fluid from a cylindrical pipe into a large volume in which the fluid is essentially at rest is usually characterized by *separation*. That is, the flow is unable to follow the sharp break in the pipe wall at the discharge section and separates from it, forming a *contact surface* between the discharge fluid and the atmospheric fluid (see Fig. 6.6). This is distinct from the classical potential-flow solution, in which the flow adheres to the walls, producing a more or less radial streamline pattern. Across the contact surface, the inviscid matching conditions (4.77) and (4.78) for pressure and velocity are applicable.

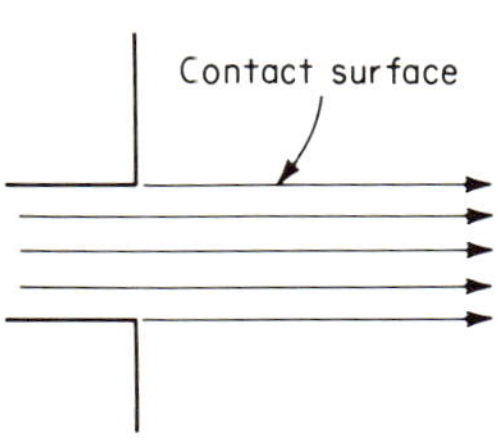

Figure 6.6
Separated discharge.

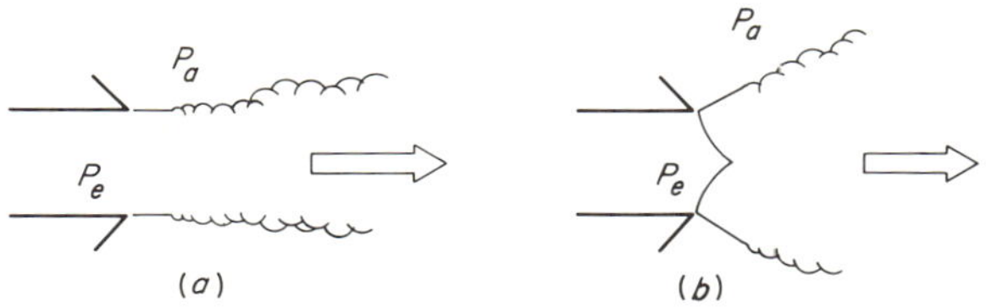

Figure 6.7
Discharge into a stationary atmosphere: (*a*) subsonic, $P_e = P_a$; (*b*) supersonic overpressure, $P_e > P_a$.

The fluid discharge, e.g., at the exit plane of a nozzle, may be either subsonic or supersonic (see Figs. 6.7 and 6.8), and these two cases are considered separately below.

Subsonic Discharge

The matching condition across the contact surface requires that the exit pressure P_e be equal to the ambient atmosphere pressure P_a (although the atmosphere actually acquires some local motion as a result of momentum transfer from the jet, the corresponding reduction in ambient pressure is usually small). We thus have the simple boundary condition $P_e \approx P_a$.

Supersonic Discharge

The pressure-matching condition still holds across the contact surface, so that the pressure on the boundary of the jet just beyond the exit section is

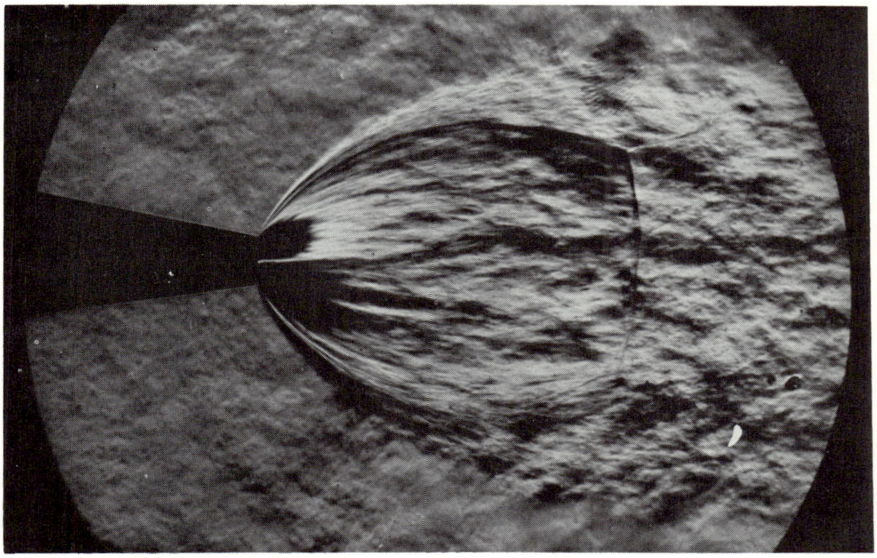

Figure 6.8
External expansion of the jet from a choked converging nozzle, $P_e/P_a \approx 105$. (*Courtesy of E. S. Love, NASA*).

approximately ambient. Because the outflow is supersonic, however, *this pressure does not propagate upstream* and the exit pressure P_e may be greater than, less than, or equal to the ambient pressure P_a. If $P_e = P_a$, the discharge is like the subsonic discharge shown in Fig. 6.7*a*. If $P_e > P_a$, the outflow expands immediately upon leaving the pipe, such that the boundary pressure is reduced to P_a (this expansion flow is distinctly two-dimensional and will be treated in Chap. 9; see also Figs. 6.7*b* and 6.8). The case $P_e < P_a$ can occur only if a *shock wave* stands in the exit section, such that the boundary pressure rises discontinuously to P_a.

Nozzle Operation

We begin with the simple *converging nozzle*. For some fixed value of the reservoir pressure P_0, consider the effect of varying the discharge ambient pressure P_a over the levels *a*, *b*, *c*, *d*, as shown in Fig. 6.9. In case *a*, $P_a = P_0$, and there is no flow. In case *b*, $P_0 > P_a > P_*$, the flow is everywhere subsonic, and $P_e = P_a$ at the exit section; e.g., if $P_a/P_0 = 0.65$ and $\gamma = 1.4$, we find from the isentropic-flow tables that the Mach number $M = 0.81$ in the minimum-area section of the nozzle. In case *c* the ambient pressure is just P_*, sonic conditions ($M = 1$, etc.) are achieved

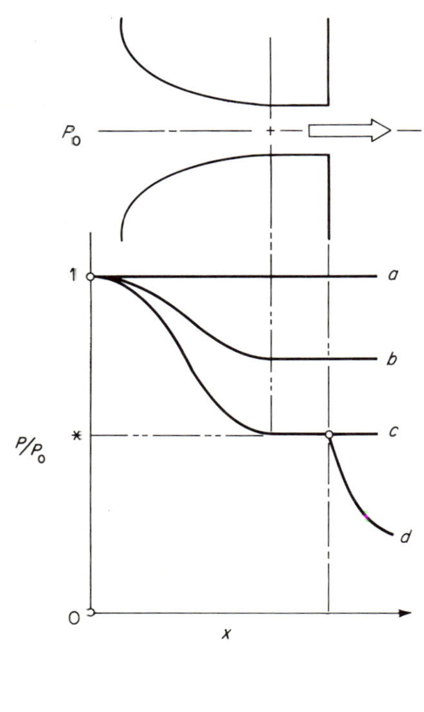

Figure 6.9
Pressure distributions in a converging nozzle.

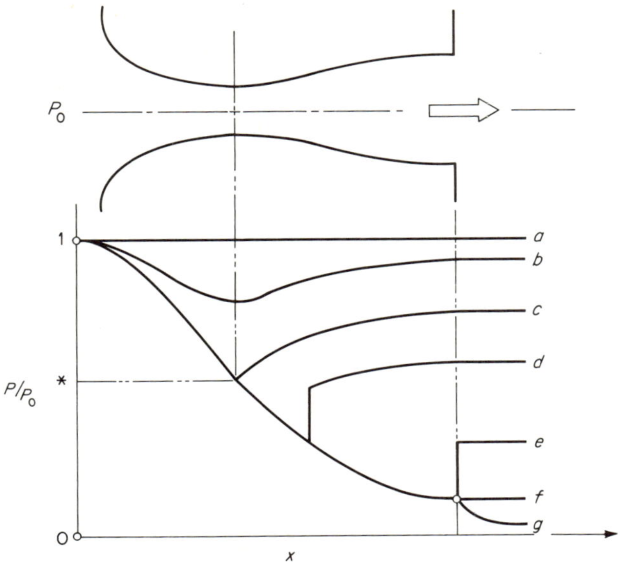

Figure 6.10
Pressure distributions in a supersonic nozzle.

in the minimum-area section, and $P_e = P_a$ at the exit section. *Lowering the pressure P_a further will not produce any additional change in the internal flow*, as the maximum Mach number in the minimum-area section is unity, with a corresponding minimum pressure P_*. Then in case *d*, with $P_a < P_*$, the internal flow is exactly the same as in case *c*, and the expansion from P_* to P_a takes place outside the nozzle.

A nozzle flow which has reached sonic conditions at the minimum-area section and which expands to a lower pressure outside the nozzle is said to be *choked*. The absence of any response to a further lowering of the ambient pressure P_a can be rationalized by observing that the signals announcing the lowering of pressure cannot propagate upstream if $M \geq 1$. Put differently, one cannot shout through a choked nozzle from the downstream side.

The case of a *supersonic*, or converging-diverging, nozzle is slightly more complicated. The nozzle profile and various possible pressure distributions are sketched in Fig. 6.10. Again, consider the effect of varying the ambient atmosphere pressure over the levels *a* to *g*, as shown. In case *a*, $P_a = P_0$, and there is no flow. In case *b*, the nozzle flow is totally subsonic, and the nozzle throat functions as a venturi. At *c*, the throat conditions are exactly sonic, but the flow in the diverging section must be

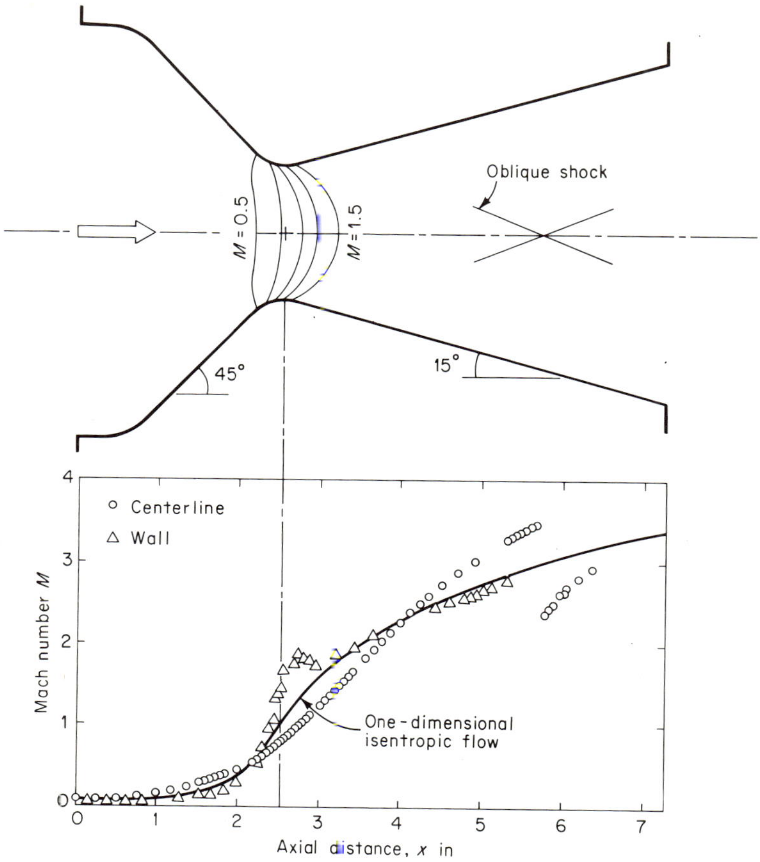

Figure 6.11
Comparison of the experimental Mach number distribution in a nozzle with the one-dimensional isentropic flow model, Eq. (6.9). Gas is air with $P_0 = 151$ psia, $T_0 = 530°R$; conical 45°/15° nozzle, 1.60-in throat diameter. Lines of constant Mach number in the upper figure are respectively $M = 0.5, 0.8, 1.0, 1.2, 1.5$. (*Data from Back and Cuffel* [1966].)

subsonic in order to achieve a pressure rise so that the discharge pressure matches the ambient pressure. If the atmospheric pressure is lowered to f, the supersonic flow in the diverging section again leads to exact pressure matching; this is the special case of *design operation*. Lowering the outside pressure to g does not affect the internal flow (one could say that the flow was choked), and expansion takes place outside the nozzle.

Between the special ambient pressures corresponding to c and f, there is some adjustment of the internal flow not accounted for in Eqs.

(6.7) to (6.9). Specifically, in case d a shock wave can occur inside the nozzle, as indicated by the (idealized) pressure discontinuity in the sketch. Transition from supersonic flow to subsonic flow occurs across this shock. Such cases can be calculated, approximately, only with the help of the shock conditions to be developed in Chap. 7.

A comparison of the one-dimensional theory with experiment is shown in Fig. 6.11. The geometry of this particular nozzle is quite angular and violates the slowly-varying assumptions (6.1); in spite of this and the presence of "real" effects such as friction, the one-dimensional model works fairly well.

EXAMPLE 6.1 CONVERGING NOZZLE

A large tank contains air ($\gamma = 1.4$) at 600°R and 1 atm. The tank is discharged through a simple converging nozzle with throat area $A_t = 1.0\ \text{in}^2$. Find the velocity in the throat and the mass-flow rate if the outside atmosphere has pressure (*a*) 0.6 atm, (*b*) 0.2 atm, (*c*) zero.

The numerical values can be worked out from formulas (6.7) to (6.9) or, more conveniently, can be found directly in the isentropic-flow tables.

In case a the "atmospheric" pressure is greater than the sonic value ($P_* = 0.528$ atm), and the flow is totally subsonic. The exit pressure is just atmospheric; then with $P/P_0 = 0.6$, tables give

$$M_t = 0.886 \qquad \frac{c_t}{c_0} = 0.930 \qquad \frac{\rho_t}{\rho_0} = 0.694$$

Then the velocity is

$$u_t = M_t c_t = (0.886)(0.930c_0) = 0.824c_0$$

The mass flow is

$$\dot{m} = \rho_t u_t A_t = 0.572\rho_0 c_0 A_t$$

For case b the flow is choked, and throat conditions are exactly sonic, giving

$$M_t = 1.00 \qquad \frac{c_t}{c_0} = 0.913 \qquad \frac{\rho_t}{\rho_0} = 0.634$$

$$u_t = 0.913c_0 \qquad \dot{m} = 0.579\rho_0 c_0 A_t$$

These values are exactly the same in case c, although the *outside* conditions are greatly different. Working out the stagnation density and sound speed from the ideal-gas laws

$$\rho_0 = \frac{P_0}{RT_0} = \frac{(14.7)(144)}{(53.34)(600)} = 0.0661 \text{ lb}_m/\text{ft}^3$$

$$c_0 = \sqrt{\gamma RT_0} = \sqrt{1.4 \frac{49{,}700}{28.97} 600} = 1{,}200 \text{ ft/s}$$

The answers are:

Case	P_a, atm	u_t, ft/s	$\dot{m}$, lb$_m$/s
a	0.6	990	0.316
b	0.2	1,097	0.319
c	0	1,097	0.319

EXAMPLE 6.2 CONVERGING NOZZLE

Conditions are exactly the same as in Example 6.1, except that the reservoir pressure is reduced to 0.1 atm with corresponding atmospheric pressures (*a*) 0.06, (*b*) 0.02, (*c*) zero.

The calculations are the same as above, except that the stagnation density is reduced to one-tenth of its former value. Thus:

Case	P_a, atm	u_t, ft/s	$\dot{m}$, lb$_m$/s
a	0.06	990	0.0316
b	0.02	1,097	0.0319
c	0	1,097	0.0319

EXAMPLE 6.3 CONNECTED RESERVOIRS

A large reservoir discharges gas to the atmosphere via a secondary reservoir, as shown in Fig. 6.12. The nozzles shown are simple converging nozzles with a common minimum area, $A_1 = A_3$. For essentially steady flow (as would be the case, for example, if the primary reservoir were sufficiently large) find the mass throughput $\dot{m}$ and the reservoir pressure P_2. It is given that the atmospheric pressure is small compared to P_0.

For the given large pressure difference, $P_0 \gg P_a$, it would appear that at least one of the nozzle flows is choked; we will show that just the second nozzle is choked.

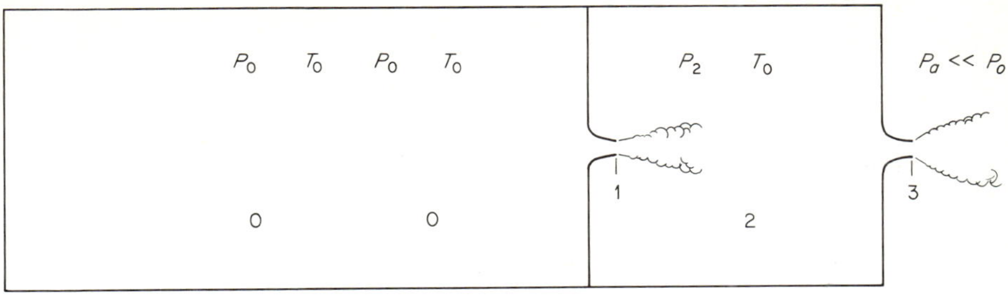

Figure 6.12

Applying the steady-flow energy equation (1.85) to a control surface enclosing the secondary reservoir and passing through nozzle sections 1 and 3 gives

$$h_1 + \frac{u_1^2}{2} = h_3 + \frac{u_3^2}{2} = h_0 \tag{6.10}$$

where h_0 is the (constant) stagnation enthalpy. This equation holds provided that there is negligible heat transfer at the control surface; it does *not* imply that the flow between 1 and 3 is isentropic. In fact, we now specifically introduce the assumption $s_3 > s_1$ because the stagnation of the jet in the secondary reservoir is an irreversible process. Let $J = \rho u$ be the mass flux common to 1 and 3; then (6.10) can be written

$$h + \tfrac{1}{2}J^2v^2 = h_0 = \text{const}$$

which defines the parabola shown in Fig. 6.13; points 1 and 3 lie on this curve. It will be shown in Sec. 6.4 that the entropy increases along such a curve in the direction shown. With $s_3 > s_1$, point 3 then necessarily lies below point 1, as shown. Equation (5.58) holds along the curve, in consequence of (6.10), for a

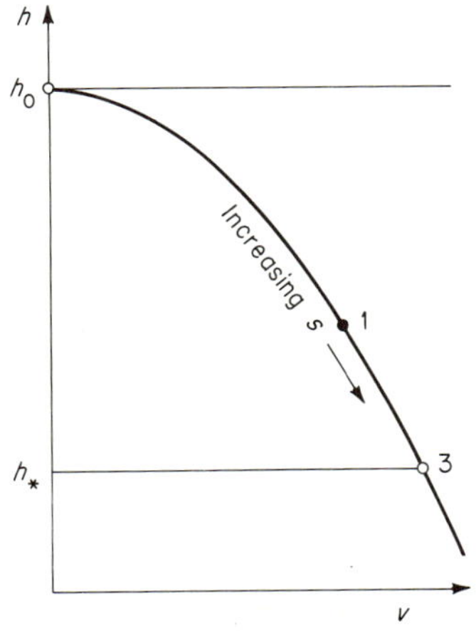

Figure 6.13

perfect gas; it follows that there is a unique sonic point $h_*/h_0 = 2/(\gamma + 1)$, and points above this represent subsonic flow. Thus if one nozzle flow is choked, it can only be the flow at section 3; the flow at section 1 is then subsonic (the arguments of Sec. 6.4 will show that this conclusion holds for a general fluid, not just for a perfect gas).

We now introduce two assumptions about the flow: (1) flow through the *nozzles* is isentropic; and (2) the jet issuing from section 1 stagnates at constant pressure; that is, $P_1 = P_2$ (note that this is just the matching condition on the jet contact surface). For steady flow the mass flux $J_1 = J_3$, or

$$\rho_1 u_1 = (\rho_* c_*)_3$$

Rewriting this gives

$$\frac{\rho_1}{\rho_0}\frac{u_1}{c_1}\frac{c_1}{c_0}\rho_0 c_0 = \frac{\rho_*}{\rho_2}\frac{c_*}{c_2}\rho_2 c_2$$

Into this equation we can substitute the relations

$$\frac{\rho_1}{\rho_0} = \left(\frac{P_2}{P_0}\right)^{1/\gamma} \qquad \frac{\rho_2}{\rho_0} = \frac{P_2}{P_0}$$

$$\frac{c_1}{c_0} = \left(\frac{P_2}{P_0}\right)^{(\gamma-1)/2\gamma} \qquad \frac{c_2}{c_0} = 1$$

$$\frac{\rho_*}{\rho_2} = \left(\frac{2}{\gamma + 1}\right)^{1/(\gamma-1)} \qquad \frac{c_*}{c_2} = \left(\frac{2}{\gamma + 1}\right)^{1/2}$$

$$M_1 = \sqrt{\frac{2}{\gamma - 1}\left[\left(\frac{P_0}{P_2}\right)^{(\gamma-1)/\gamma} - 1\right]}$$

to obtain the quadratic equation

$$X^2 - X - \frac{\gamma - 1}{2}\left(\frac{2}{\gamma + 1}\right)^{(\gamma+1)/(\gamma-1)} = 0$$

where $X \equiv (P_0/P_2)^{(\gamma-1)/\gamma}$. This has the solution

$$X = \frac{1}{2} \pm \sqrt{\frac{1}{4} + \frac{\gamma - 1}{2}\left(\frac{2}{\gamma + 1}\right)^{(\gamma+1)/(\gamma-1)}}$$

For $\gamma = 1.40$, this gives $P_2/P_0 = 0.807$, with a corresponding Mach number $M_1 = 0.562$. Isentropic-flow relations [as in Example 6.1 (page 287)] give

$$\dot{m} = 0.467 \rho_0 c_0 A$$

From Example 6.1, this is just 0.807 times the mass flow through a choked nozzle with the same reservoir conditions.

EXAMPLE 6.4 THRUST OF A SIMPLE ROCKET

A conventional rocket develops thrust by discharging gas through a supersonic nozzle. Let us calculate the thrust force (see Fig. 6.14).

Suppose the rocket is operating in a vacuum. From Example 1.7 (page 43), the equivalent net external force, or *thrust force*, is, with $\dot{m}$ the (positive) rate of mass discharge,

$$(F_t)_{\text{net}} = -\dot{m}\mathbf{v}_e - P_e A_e \mathbf{n} \tag{6.11}$$

This is the force which tends to accelerate the rocket; in a static firing it is just the negative of the force required to restrain the rocket. With $v_e \to u_e$ and $\dot{m} = \rho_e u_e A_e$, (6.11) becomes

$$|F_t| = A_e(P_e + \rho_e u_e^2) \tag{6.12}$$

For a perfect gas, with $c^2 = \gamma P/\rho$, this becomes

$$|F_t| = A_e P_e(1 + \gamma M_e^2) \tag{6.13}$$

which, with (6.7) and (6.8), gives

$$|F_t| = P_0 A_e \frac{1 + \gamma M_e^2}{\left(1 + \dfrac{\gamma - 1}{2} M_e^2\right)^{\gamma/(\gamma-1)}} \tag{6.14}$$

where M_e is known in terms of $A_e/A_t = A_e/A_*$.

In ordinary rockets, $\gamma M_e^2 \gg 1$, and the first term in (6.14) is negligible, corresponding to

$$|F_t| \approx \dot{m} u_e \tag{6.15}$$

The net *specific impulse* is then just the exit velocity,[1] $I_s = u_e$. With $u_e = M_e c_e = M_e c_0(c_e/c_0)$ we find

$$I_s = u_e = c_0 \frac{M_e}{\left(1 + \dfrac{\gamma - 1}{2} M_e^2\right)^{1/2}} \tag{6.16}$$

[1] If we *define* the thrust to be $\dot{m}u_e$, then $I_s = u_e$ exactly. If we define the thrust to be the equivalent external force, then $I_s \approx u_e$ if $\dot{m}u_e$ is large compared to the external force $P_e A_e$.

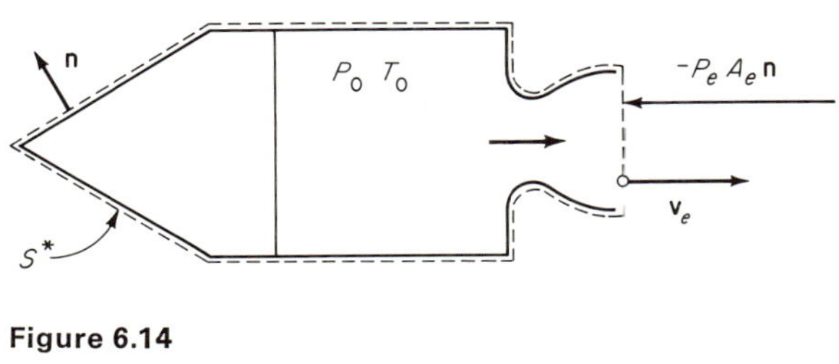

Figure 6.14

which is of the form $c_0 f(M_e)$. This function depends on the nozzle geometry via A_e/A_t; the dependence on the ratio of specific heats γ can be neglected because the range of variation in γ is small. Thus, for fixed nozzle geometry, approximately

$$I_s = u_e \propto c_0 \tag{6.17}$$

where $c_0 = \sqrt{\gamma R T_0}$. Then for fixed reservoir temperature T_0, we have roughly

$$I_s \propto \frac{1}{\sqrt{\tilde{M}}} \tag{6.18}$$

That is, low-molecular-weight propellants are to be preferred. For example, in a nuclear rocket hydrogen is the usual propellant gas.

Separation in Supersonic Nozzles

When the discharge ambient pressure falls between the levels c and f shown in Fig. 6.10 the flow in "real" supersonic nozzles will usually separate from the nozzle walls, as indicated in Fig. 6.15. This behavior is in contrast to that required by the ideal model, in which there is merely a pressure rise across a normal shock, with the flow continuing to fill the nozzle channel.

Shadowgraph pictures of separated nozzle flow are shown in Fig.

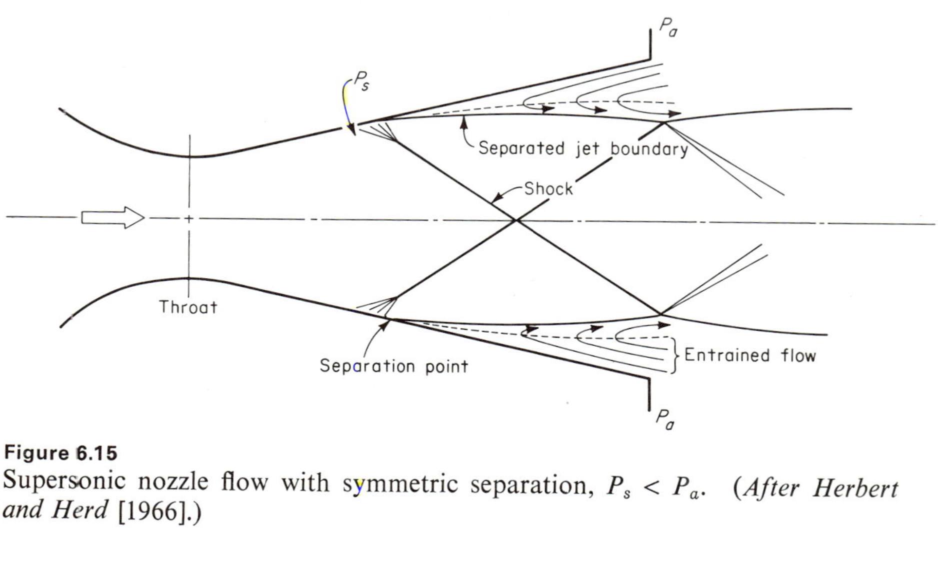

Figure 6.15
Supersonic nozzle flow with symmetric separation, $P_s < P_a$. (*After Herbert and Herd* [1966].)

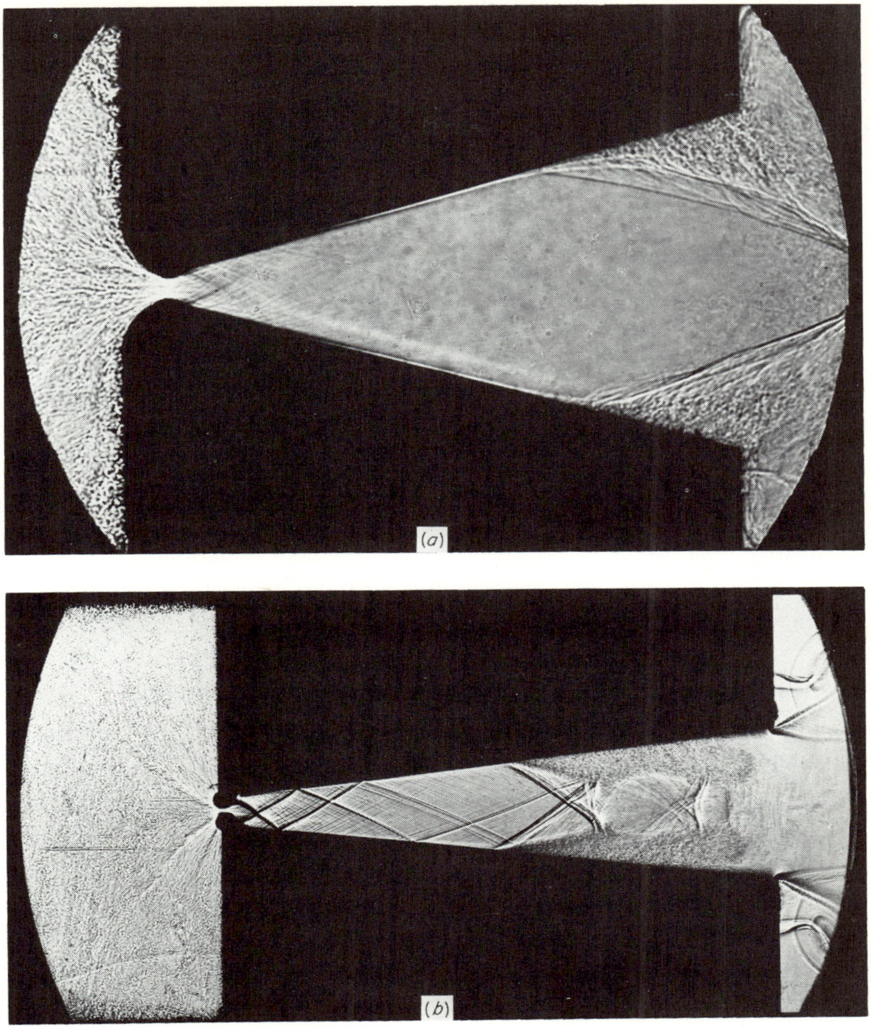

Figure 6.16
Shadowgraphs showing flow separation in supersonic nozzles. (*Courtesy of H. O. Amann, Ernst Mach Institut.*)

6.16. It should be remarked that the separation is not always symmetric; i.e., the flow may separate from one side and remain attached on the other.

A useful empirical rule is that the pressure P_s just ahead of the separation point is about four-tenths of the discharge ambient pressure, $P_s \sim 0.4P_a$. For further information, see *Herbert and Herd* [1966] and *Lawrence and Weynand* [1968].

6.4 Flow with friction in a constant-area pipe

If a high-speed flow travels through a pipe of sufficient length, the effects of viscosity and associated entropy change cannot be neglected. We suppose that the pipe has constant area, that the one-dimensional model is applicable, and that the walls of the pipe are perfectly insulated (adiabatic wall). This conventional and idealized model gives some insight into the effect of entropy change.

Since the mass throughput is constant, we have

$$J = \rho u = \text{const} \tag{6.19}$$

With no work or heat transfer at the walls, the steady-flow energy equation applied to the control volume shown in Fig. 6.17 gives[1]

$$h_1 + \frac{u_1^2}{2} = h_2 + \frac{u_2^2}{2}$$

Thus, at any position X,

$$h + \frac{u^2}{2} = h_0 = \text{const} \tag{6.20}$$

This is formally identical to the energy equation (5.4); it does not imply isentropic flow here but rather *adiabatic* flow (in the sense that there is no heat transfer at the walls). Thus, we cannot now consider all thermodynamic quantities to be known in terms of h because the entropy is not fixed.

Combining (6.19) and (6.20) gives

$$h + \tfrac{1}{2}J^2v^2 = h_0 = \text{const} \tag{6.21}$$

where J is the mass flux and v the specific volume. This relation defines families of curves (the particular curve depending on the choice of the

[1] To verify that the work-transfer term $\int \mathbf{T} \cdot \mathbf{u}\, dA$ in (1.81) is zero at the walls, consider that the surface S^* lies just inside the *solid* wall, where $u \equiv 0$. A less general argument takes advantage of the no-slip condition at the wall, so that $u = 0$ in the *fluid*. This argument, while correct, tends to cloud the quasi-one-dimensional model.

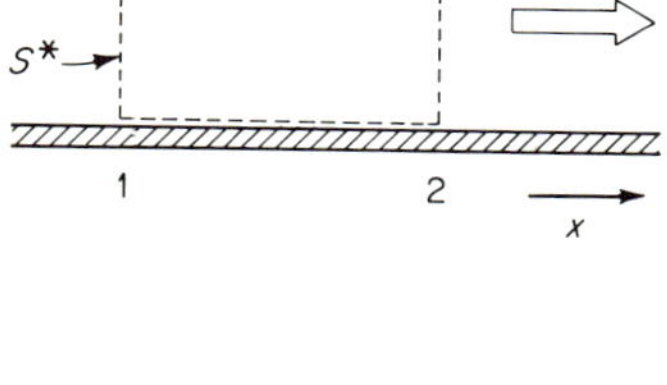

Figure 6.17

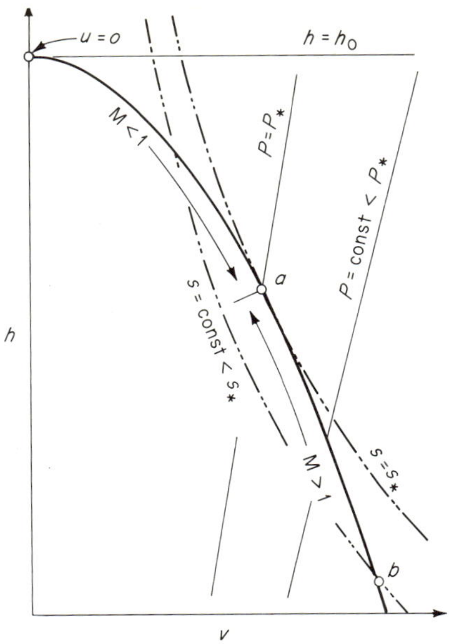

Figure 6.18
Fanno line in the hv plane.

parameters J and h_0) in the plane of any two thermodynamic variables; in the hv plane the curves are simple parabolas (see Fig. 6.18). Such curves are in general called *Fanno lines.*

With zero entropy flux at the boundaries of the control volume shown in Fig. 6.17, the entropy balance (1.82) reduces to

$$s_2 \geq s_1$$

That is, there can only be a creation of entropy. Expressed as a derivative, this is

$$\frac{ds}{dx} \geq 0 \tag{6.22}$$

The equal sign corresponds to isentropic flow, and the inequality corresponds to creation of entropy by friction, i.e., via the dissipation function Υ.

Let us say that a flow enters the pipe with given values of v, J, and h_0, thereby fixing the Fanno curve. The initial point on the curve is fixed by the value of v. With increasing x this point (which represents the state of the fluid) moves along the Fanno curve. Equation (6.22) requires, however, that it move in such a way that the entropy increases, corresponding to the arrows in Fig. 6.18. It will be shown that the point a, at which the entropy becomes stationary, is just a sonic point, where $M = 1$.

By definition, an isentrope is just tangent to the Fanno curve at point a, so that

$$\left(\frac{\partial h}{\partial v}\right)_s = \left(\frac{dh}{dv}\right)_{\text{Fanno}} \tag{6.23}$$

We can construct the identity

$$\left(\frac{\partial h}{\partial v}\right)_s = \left(\frac{\partial h}{\partial P}\right)_s\left(\frac{\partial P}{\partial v}\right)_s = v\left(-\frac{c^2}{v^2}\right) = -\frac{c^2}{v} \tag{6.24}$$

By differentiation $(dh/dv)_{\text{Fanno}} = -J^2v = -u^2/v$, and (6.23) becomes

$$u^2 = c^2$$

at the point of maximum entropy, point a, which is thus a sonic point.

We now assume that the isentropes are concave upward, as shown in the figure.[1] Then $dh/dv > (\partial h/\partial v)_s$ for $h > h_a$ and $dh/dv < (\partial h/\partial v)_s$ for $h < h_a$, the latter condition occurring, for example, at point b. Then it follows, by the same manipulations as above, that

$$\begin{aligned} u^2 &< c^2 \qquad M < 1 \qquad \text{for } h > h_a \\ u^2 &> c^2 \qquad M > 1 \qquad \text{for } h < h_a \end{aligned}$$

The two distinct branches of the Fanno line thus represent subsonic and supersonic flows. In the absence of discontinuities there is no possibility of passing from one branch to another, for this would require crossing point a and the condition $ds/dx > 0$ would be violated. Thus, an initially supersonic flow remains supersonic but moves in the direction of $M = 1$; an initially subsonic flow remains subsonic but also moves in the direction of $M = 1$. *The effect of friction is to drive the Mach number toward unity.* This has an interesting consequence. An initially subsonic flow can be accelerated only to point a; thus, for specified entrance conditions, there is a maximum possible pipe length $L_{\max}$. This phenomenon is usually referred to as *frictional choking* and will be illustrated later. A similar comment applies to initially supersonic flow; the difference is that discontinuous jump to the subsonic branch of the Fanno curve (via a shock wave) is possible.

[1] This condition is somewhat stronger than actually needed but is taken for convenience. By differentiation of (6.24) and the use of (5.7)

$$\left(\frac{\partial^2 h}{\partial v^2}\right)_s = \frac{c^2}{v^2}(2\Gamma - 1)$$

so that the isentropes are concave upward if $\Gamma > \frac{1}{2}$.

Finally, it will be shown that the pressure increases with x in a supersonic flow and decreases with x in a subsonic flow. To do this, it is only necessary to show that dv/dx and dP/dx have opposite signs. From (6.21)

$$dh + J^2 v\, dv = 0$$

and with the Gibbs equation this becomes

$$T\,ds + v\,dP + J^2 v\,dv = 0 \tag{6.25}$$

with $v = v(P,s)$

$$dv = \left(\frac{\partial v}{\partial P}\right)_s dP + \left(\frac{\partial v}{\partial s}\right)_p ds$$

Now $(\partial P/\partial v)_s = -\rho^2 c^2 = -J^2/M^2$, and this becomes

$$T\,ds = T\left(\frac{\partial s}{\partial v}\right)_p dv + \frac{M^2}{J^2}\left(\frac{\partial s}{\partial v}\right)_p T\,dP$$

Substituting this into (6.25) gives

$$\left[\frac{M^2}{J^2}\left(\frac{\partial s}{\partial v}\right)_p T + v\right] dP + \left[T\left(\frac{\partial s}{\partial v}\right)_p + J^2 v\right] dv = 0 \tag{6.26}$$

Now $(\partial s/\partial v)_p = c_p(\partial T/\partial v)_p/T$, which is positive except in extraordinary cases (such as that of liquid water below 4°C). We conclude that the coefficients are ordinarily positive and that dP and dv have opposite signs.

We summarize the above results in a table.

	ds/dx	du/dx	dv/dx	dP/dx
$M < 1$	+	+	+	−
$M > 1$	+	−	−	+

Friction-factor Formulation

The conventional engineering representation of the wall friction gives the shear stress Σ_w at the wall

$$\Sigma_w = \frac{f\rho u^2}{2} \tag{6.27}$$

where f is a dimensionless numerical factor, the *Fanning friction factor*.[1] By dimensional arguments, it depends upon the Mach and Reynolds numbers and the roughness of the wall

$$f = f(M,\mathrm{Re},\varepsilon) \tag{6.28}$$

where ε is a dimensionless measure of the surface roughness. The friction factor is analogous to a drag coefficient.

The tangential friction force per unit length is $\Sigma_w \pi D$, opposing the motion (where πD is the circumference of a circular pipe). The friction force per unit area of cross section is thus per unit length $-\Sigma_w \pi D/(\pi D^2/4) = -2f\rho u^2/D$. Adding this term to the right-hand side of (6.3) yields the momentum equation

$$\rho u \frac{du}{dx} = -\frac{dP}{dx} - \frac{2}{D} f \rho u^2 \tag{6.29}$$

If the pipe is noncircular, the same equation holds, provided that D is the *hydraulic diameter*, defined as 4 times the cross-sectional area divided by the perimeter (the value of the friction factor f will in general vary with the shape of the cross section). Equation (6.29) can be written

$$\frac{1}{u^2}\frac{du^2}{dx} + \frac{4f}{D} = -\frac{2}{\rho u^2}\frac{dP}{dx} \tag{6.30}$$

In addition, we have available the energy and continuity equations (6.20) and (6.19), viz.,

$$h + \frac{u^2}{2} = h_0 = \text{const}$$

$$\rho u = J = \text{const}$$

For specified entrance conditions these give $h(u)$ and $\rho(u)$, and hence any thermodynamic quantity in terms of u, allowing (6.30) to be written as an equation in one unknown.

We take now the specific case of a *perfect gas*, for which the energy equation (6.20) reduces to

$$c^2 + \frac{\gamma - 1}{2} u^2 = c_0{}^2 = \text{const} \tag{6.31}$$

[1] The reader is cautioned that the Darcy factor used in hydraulics in the same connection is numerically exactly 4 times as large.

We now rewrite (6.30) in terms of the Mach number M. By definition, $M^2 = u^2/c^2$, and logarithmic differentiation gives

$$\frac{dM^2}{M^2} = \frac{du^2}{u^2} - \frac{dc^2}{c^2}$$

where for brevity the dx is omitted; i.e., we write dM^2 instead of dM^2/dx. Inserting (6.31) gives

$$\frac{dM^2}{M^2} = \left(1 + \frac{\gamma - 1}{2} M^2\right) \frac{du^2}{u^2} \tag{6.32}$$

which allows replacement of the first term in (6.30). For the last term we begin with $P = \rho c^2/\gamma$ and with the use of (6.31) find

$$dP = -\frac{\gamma - 1}{2\gamma} \rho \, du^2 + \frac{c^2}{\gamma} d\rho$$

and with (6.19) in the form $d\rho/\rho = -du/u$ this gives

$$\frac{2dP}{\rho u^2} = -\frac{1 + (\gamma - 1)M^2}{\gamma M^2} \frac{du^2}{u^2} \tag{6.33}$$

Inserting (6.32) and (6.33) into the momentum equation (6.30) gives finally

$$\frac{1 - M^2}{\gamma M^4\left(1 + \frac{\gamma - 1}{2} M^2\right)} \frac{dM^2}{dx} = \frac{4f}{D} \tag{6.34}$$

which is integrable if $f(x)$ is known. With $f > 0$, we have confirmed the previous general conclusion that the Mach number goes toward unity.

Equation (6.34) can be integrated from an arbitrary initial section $x = 0$, $M = M$, to a frictionally choked final section $x = L_{\text{max}}$, $M = 1$ to give

$$\frac{4\bar{f}L_{\text{max}}}{D} = \frac{1 - M^2}{\gamma M^2} + \frac{\gamma + 1}{2\gamma} \ln \frac{(\gamma + 1)M^2}{2\left(1 + \frac{\gamma - 1}{2} M^2\right)} \tag{6.35}$$

where $\bar{f}$ is the friction coefficient averaged over the length, $\bar{f} \equiv \frac{1}{L}\int_0^L f\,dx$. The average friction factor commonly has a value of roughly 2×10^{-3}.

EXAMPLE 6.5 ELONGATED CONVERGING NOZZLE

A duct 100 in long with a 1-in diameter is supplied via a converging nozzle from an air reservoir with $P_r = 10$ atm, $T_r = 300$ K, and exhausts into a region at 1 atm (Fig. 6.19). Find the exit velocity and the pressure and Mach number at the entrance; compare the mass flow with that through a short converging nozzle with the same reservoir conditions. The friction factor is given as $f = 0.0025$.

Assume the flow is adiabatic and choked. Then $L = L_{\max}$ and $4fL_{\max}/D = 1.00$. From (6.35), $M_1 = 0.51$. Under the reasonable assumption that the flow in the converging entrance section is isentropic, $P_1 = 8.37$ atm from tables. The mass flow is

$$\dot{m} = \rho_1 u_1 A = \rho_0 c_0 A \frac{\rho_1}{\rho_0} \frac{u_1}{c_1} \frac{c_1}{c_0}$$

$$= (0.51)(0.881)(0.975)\rho_0 c_0 A = 0.438\rho_0 c_0 A$$

For choked flow through a short converging nozzle the corresponding mass flow is

$$\dot{m}_0 = (1.00)(0.634)(0.913)\rho_0 c_0 A = 0.531\rho_0 c_0 A$$

With $M_2 = 1$ at section 2, Eq. (6.31) gives

$$\frac{\gamma + 1}{2} u_2^2 = c_0^2$$

$$u_2 = 1{,}040 \text{ ft/s}$$

This is precisely the exit velocity from the corresponding short converging nozzle, but the density is somewhat lower.

To illustrate clearly that the flow is choked, let us calculate the exit pressure P_2

$$\frac{P_2}{P_0} = \frac{\rho_2 c_2^2}{\rho_0 c_0^2} = \frac{\rho_2 c_2}{\rho_0 c_0} \frac{c_2}{c_0}$$

From the mass-flow calculation, the quotient $\rho_2 c_2/\rho_0 c_0 = 0.438$; from the condition $u_2 = c_2$, $c_2/c_0 = 1{,}040/1{,}139$. Thus we find $P_2 = 4.00$ atm, which is

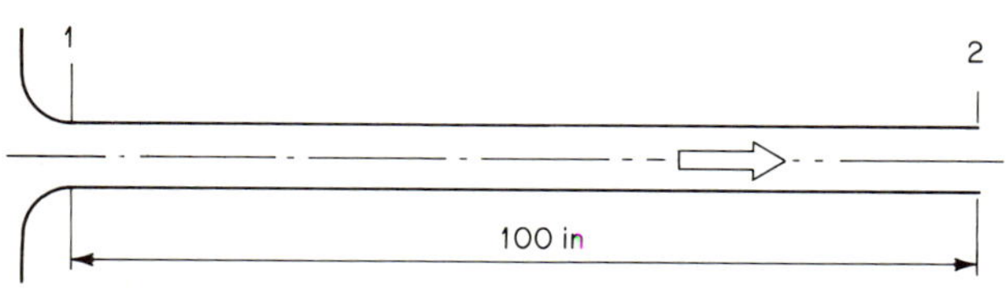

Figure 6.19
Duct supplied by a converging nozzle.

above the ambient pressure in the discharge region. The exit pressure P_2 is in fact independent of the ambient pressure P_a as long as $P_2 > P_a$.

For more detailed treatment of quasi-one-dimensional flow, the reader is referred to *Shapiro* [1953] and to *Crocco* [1958].

Problems

6.1 A supersonic nozzle with exit area of 1 in² discharges nitrogen at $M = 2$, $P = 1$ atm, $T_0 = 600°R$. For steady isentropic flow, determine:

(*a*) The throat area in square inches
(*b*) The reservoir pressure in atmospheres
(*c*) The static-discharge temperature in degrees Rankine
(*d*) The mass throughput in pounds mass per second

Answer (*a*) 0.593 in²; (*b*) 7.83 atm; (*c*) 333.4°R; (*d*) 1.454 lb_m/s

6.2 A supersonic nozzle has exit area $2\frac{1}{2}$ times the throat area. For steady isentropic flow ($\gamma = 1.40$) discharging into an atmosphere with pressure P_a, find the Mach number at the throat and at the exit plane for

(*a*) $\dfrac{P_a}{P_0} = 0.0600$

(*b*) $\dfrac{P_a}{P_0} = 0.9725$

Answer (*a*) 1, 2.443; (*b*) 0.602, 0.200

6.3 If air is expanded isentropically from room temperature (540°R) through a supersonic nozzle to a Mach number of 4, the static temperature is only 129°R. Perhaps the low-temperature gas could be used as the heat sink in a refrigeration device. Are there any serious difficulties in such a proposal?

6.4 A hypersonic wind tunnel using helium has a test-section Mach number of 15. Flow is approximately isentropic and steady. For reservoir conditions $P_0 = 50$ atm, $T_0 = 800$ K, what is the test-section pressure and temperature? Will helium condensation be a problem?

Answer 10.5 K; 9.93×10^{-4} atm

6.5 Consider the purely radial three-dimensional steady flow of a perfect gas, with spherical symmetry. Find an implicit relation $M(r)$ for the Mach number, where r is the distance from the origin. Is there a maximum or minimum value for r? If so, what is the physical significance?

6.6 Liquid water is approximately described by the equation of state

$$P + B = B\left(\frac{\rho}{\rho_0}\right)^{\gamma}$$

where $B = 3{,}000$ atm, $\gamma = 7.15$, and ρ_0 is the reference density. For steady flow from a reservoir at pressure P_0 show that the Mach number is given by

$$M^2 = \frac{2}{\gamma - 1}\left[\left(\frac{P_0 + B}{P + B}\right)^{(\gamma-1)/\gamma} - 1\right]$$

(*Hint:* One possibility is to integrate the one-dimensional momentum equation.) Find the critical reservoir pressure $(P_0)_c$ in atmospheres required to give Mach number unity at 1 atm.

Answer $P_0 = 12{,}400$ atm

6.7 Derive the identity

$$\left(\frac{\partial h}{\partial v}\right)_s = -\frac{c^2}{v}$$

6.8 A zero-power (reversible) supersonic wind tunnel is proposed with the geometry shown. The cross-sectional area of the test section is 4 times the throat area. Find the test-section Mach number and mention any practical difficulties which might be encountered.

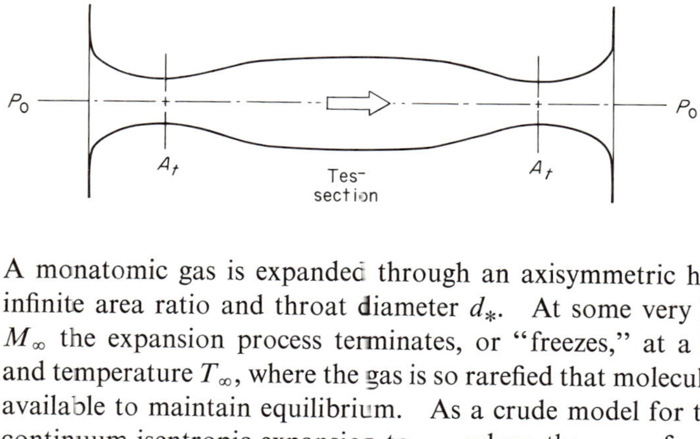

6.9 A monatomic gas is expanded through an axisymmetric hypersonic nozzle of infinite area ratio and throat diameter d_*. At some very high Mach number M_∞ the expansion process terminates, or "freezes," at a limiting density ρ_∞ and temperature T_∞, where the gas is so rarefied that molecular collisions are not available to maintain equilibrium. As a crude model for this process, assume continuum isentropic expansion to ρ_∞, where the mean free path Λ is equal to the local nozzle diameter d. Note that Λ is inversely proportional to density and may be conveniently written $\Lambda = kd_*\rho_0/\rho$, where k is a dimensionless constant and ρ_0 is stagnation density. Find an *approximate* expression for the freezing-density ratio ρ_∞/ρ_0 in terms of k.

Answer $\dfrac{\rho_\infty}{\rho_0} \approx 3.1k^2$

6.10 Consider the flow described in Example 6.3 (page 288). By relaxing the restriction that $A_1 = A_3$, find the area ratio A_3/A_1 which just allows sonic conditions at section 1 as well as section 3.

Answer $\left(\frac{\gamma + 1}{2}\right)^{\gamma/(\gamma - 1)}$

6.11 Steam enters a long perfectly insulated pipe of constant area at 200 psia and 900°F with velocity 10 ft/s. Frictional resistance results in a pressure drop, to an exit pressure of 100 psia. Find the temperature and velocity at the discharge section. Assume steady flow.

Answer 893.5°F; 20.02 ft/s

6.12 A long rectangular duct has a cross section of height h and width w. If $h \ll w$, find the ratio L/h such that the exit flow from the duct is choked and the entrance flow has Mach number $M = 0.1$. Assume laminar flow, with the friction factor f given by $f = 16/\mathrm{Re}$, and $\mathrm{Re} = 160$.

6.13 Given a very slight constriction in an initially uniform two-dimensional channel flow as shown. Find the value of P_2/P_1 for a perfect gas ($\gamma = 1.40$) from the ideal theories for:

(*a*) Linearized supersonic flow
(*b*) Quasi-one-dimensional flow

Answer (*a*) 1.13; (*b*) 1.15

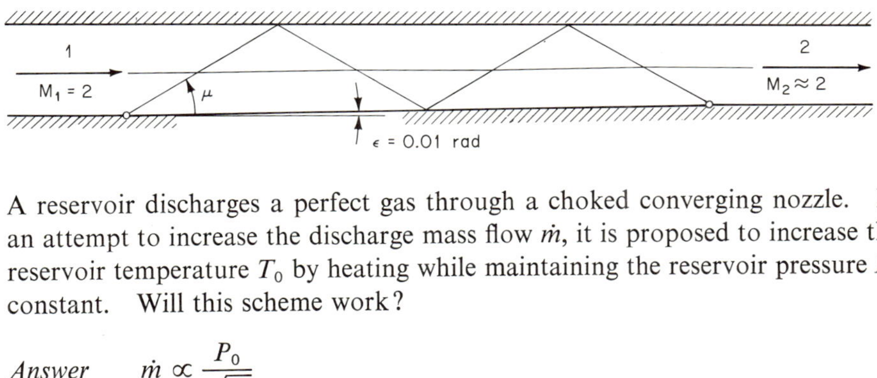

6.14 A reservoir discharges a perfect gas through a choked converging nozzle. In an attempt to increase the discharge mass flow $\dot{m}$, it is proposed to increase the reservoir temperature T_0 by heating while maintaining the reservoir pressure P_0 constant. Will this scheme work?

Answer $\dot{m} \propto \frac{P_0}{\sqrt{T_0}}$

6.15 Helium with stagnation temperature $T_0 = 300$ K is discharged from a 5-mm-diameter converging nozzle at $M = 1$, $P = 10^{-3}$ atm. Find the mass flow rate in kilograms per second. Assume steady flow. First find a general formula for this situation; then substitute numbers.

seven
shock waves and related discontinuities

7.1 Introduction

A shock wave is a relatively thin region of rapid state variation across which there is a flow of matter. Because the region of variation is thin, it can almost always be idealized as a *surface of discontinuity* in space. This surface propagates into the fluid and is not necessarily stationary. In

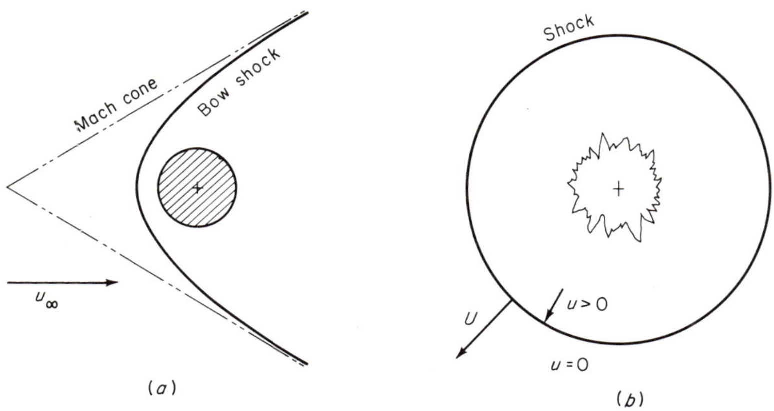

Figure 7.1
(*a*) Bow shock in supersonic streaming flow about a blunt body; (*b*) blast wave.

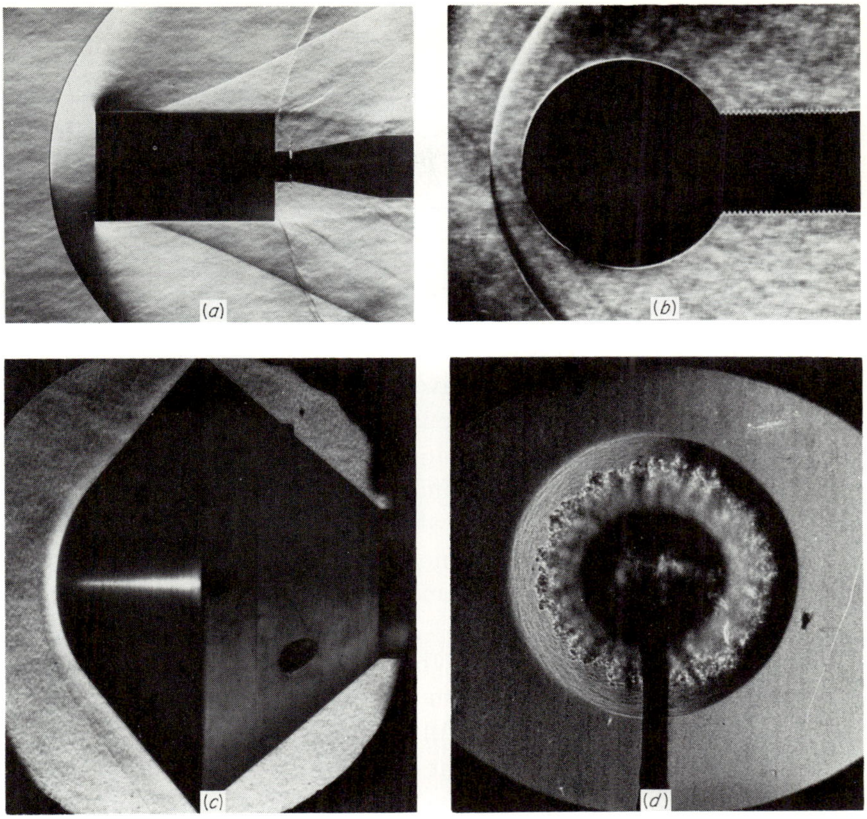

Figure 7.2
(*a*) Bow shock on longitudinal cylinder, $M_\infty = 1.99$. (*Naval Supersonic Laboratory, M.I.T.*) (*b*) Bow shock on transverse cylinder in hypersonic flow, $M_\infty = 19.6$ (*Courtesy of H. T. Nagamatsu, General Electric Co.*) (*c*) Bow shock on cone hemisphere in hypersonic flow. The air behind the shock front is incandescent, with $T_0 \approx 4000$ K. (*d*) Spherical shock front in water from an exploding glass sphere. (*Courtesy of I. I. Glass, University of Toronto.*)

general, all fluid properties—pressure, velocity, density, etc.—are discontinuous across the surface. Alternative names for the surface are *shock*, *shock wave*, and *shock front*.

The treatment of shock waves as *discontinuities*, or surfaces of zero thickness, is an idealization of inviscid gasdynamics. Physically, shocks are found to have a finite and measurable thickness, commonly of the order of 10^{-6} m. Discussion of the internal structure of the shock is postponed to Sec. 7.10; in the following, shocks are treated as true discontinuities.

Figure 7.3
Traffic shock.

Common examples of shock waves are shown in Figs. 7.1 and 7.2. When a blunt projectile travels through the atmosphere at supersonic speeds, a *bow wave* travels ahead of and with the projectile; this wave is roughly a hyperboloid of revolution, and the *Mach cone* forms an envelope such that the wave appears conical when viewed from a distance. An atmospheric explosion produces a more or less spherical shock front, which advances into stationary and undisturbed fluid. Such shocks are often referred to as *blast waves*.

A more picturesque example is the *traffic shock*. On certain freeways, particularly under foggy (or smoggy) conditions, extensive chain-reaction collisions may occur; a fanciful version is shown in Fig. 7.3. The shock front is identified with the point of collision, which moves counter to the flow of traffic with increasing time. Individual motorcars are somewhat comparable to the molecules of a gas in which a shock wave occurs.

Further examples of shock waves observable in ordinary experience are a thunderclap, sonic boom, and the hydraulic jump (the latter will be treated in Chap. 11).

7.2 Shock conditions

The flow of matter across the shock front must satisfy the conditions of balance for mass, momentum, energy, and entropy. Application of these conditions in the form of the integral balance statements of Sec. 1.6 will result in a set of simple equations, called the *shock conditions*, which relate the flow variables on one side of the shock front to those on the other.

Let $S^*(t)$ be a moving control surface which encloses a portion of the shock front and moves with the local shock front at velocity $\mathbf{b}$; the outward unit normal from the control surface is $\mathbf{n}$ (see Fig. 7.4*a*). Since the shock front is treated as a discontinuity, we can make the control volume

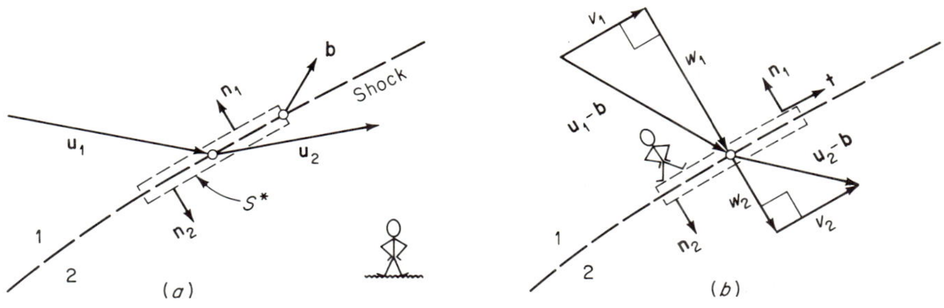

Figure 7.4
Control surface enclosing the shock front: velocities as seen by (*a*) stationary observer and (*b*) observer on the control volume.

arbitrarily *thin* and still enclose the shock; since the *local* conditions are desired, we also wish to make the surface parallel to the shock front small. Formally, these conditions can be met by letting the side dimension of the control surface be L and thickness be εL, where ε is an arbitrarily small number; then as $L \to 0$, both the "edge" surface area and the volume can be neglected.

The absolute fluid velocity is $\mathbf{u}$. Since properties differ from one side of the shock to the other, we label the two sides of the shock 1 and 2; thus, for example, $\mathbf{u}_1$ is the velocity on side 1. By convention, side 1 is the side at which fluid *enters* the control volume relative to an observer moving with the volume; similarly, side 2 is the side from which fluid leaves the volume.[1] Formally, with $\mathbf{u} - \mathbf{b}$ the fluid velocity relative to the control volume,

$$(\mathbf{u} - \mathbf{b}) \cdot \mathbf{n} < 0 \qquad \text{side 1; inflow}$$
$$(\mathbf{u} - \mathbf{b}) \cdot \mathbf{n} > 0 \qquad \text{side 2; outflow}$$

It will be convenient to define the (positive) components of the relative velocity

$$\begin{aligned} w_1 &\equiv -(\mathbf{u}_1 - \mathbf{b}) \cdot \mathbf{n}_1 \\ w_2 &\equiv (\mathbf{u}_2 - \mathbf{b}) \cdot \mathbf{n}_2 \end{aligned} \tag{7.1}$$

as shown in Fig. 7.4*b*.

In applying the integral balance statements of Sec. 1.6, the condition that the control volume be thin allows us to neglect both the volume

[1] The use of subscripts 1 and 2 to distinguish the two sides of the shock front does *not* have any connection with the index notation introduced in Chap. 1. The notation adopted seems to be the most practical one, and the ambiguity should not lead to any confusion.

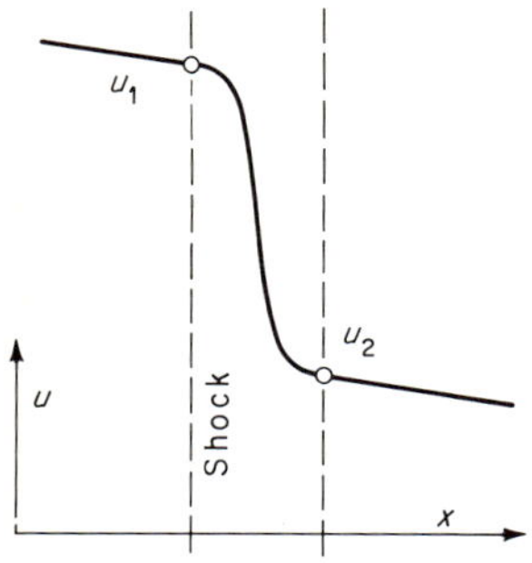

Figure 7.5

integrals (because there can be no storage of mass, momentum, etc. in essentially zero volume) and the contribution of the edges to the surface integrals. Then the balance statements (1.79) to (1.82) for mass, momentum, energy, and entropy become respectively, per unit area of shock front surface,

$$\rho_2 w_2 - \rho_1 w_1 = 0 \tag{7.2}$$

$$\rho_2 \mathbf{u}_2 w_2 - \rho_1 \mathbf{u}_1 w_1 = -P_2 \mathbf{n}_2 - P_1 \mathbf{n}_1 \tag{7.3}$$

$$\rho_2\left(e_2 + \frac{u_2^{\,2}}{2}\right)w_2 - \rho_1\left(e_1 + \frac{u_1^{\,2}}{2}\right)w_1 = -P_2 \mathbf{n}_2 \cdot \mathbf{u}_2 - P_1 \mathbf{n}_1 \cdot \mathbf{u}_1 \tag{7.4}$$

$$\rho_2 s_2 w_2 - \rho_1 s_1 w_1 \geq 0 \tag{7.5}$$

In the last three statements it has been assumed that the viscous stress and heat flux are negligible at the control surface. Consider, for example, the viscous stress: the change in velocity from $\mathbf{u}_1$ to $\mathbf{u}_2$ across the shock front occurs physically in a very thin region (as we have noted), in which the velocity gradients must be very large (see Fig. 7.5). Thus, viscous stresses can hardly be negligible *within* the physical shock front. We assume, however, that it is possible to select the control surface such that the velocity gradients are sufficiently small to justify the inviscid approximation and still satisfy the condition of small thickness. These conditions are normally satisfied in practice; exceptions are shocks embedded in hypersonic boundary layers and cases where a slow chemical or nuclear reaction takes place within the shock.[1] Similar comments apply to the assumption of negligible heat flux.

[1] Another point of view is that the shock conditions are the *exact* conditions at a true discontinuity in an *inviscid* flow.

We now proceed to simplify each of the balance statements, in the order in which they have been given.

Mass: Equation (7.2) is already in simplified form and just states that the mass flux across the shock front is the same on both sides.

Momentum: Adding $\mathbf{b}(\rho_1 w_1 - \rho_2 w_2) = 0$ to Eq. (7.3) gives, with $\mathbf{n}_2 = -\mathbf{n}_1$,

$$\rho_2 w_2(\mathbf{u}_2 - \mathbf{b}) - \rho_1 w_1(\mathbf{u}_1 - \mathbf{b}) = \mathbf{n}_1(P_2 - P_1) \tag{7.6}$$

This is a vector equation and implies three component equations. Let a *basis* be defined by the unit vectors $\mathbf{n}_1$ and $\mathbf{t}$ (in the plane of $\mathbf{n}_1$ and $\mathbf{u}_1 - \mathbf{b}$) and a third unit vector $\mathbf{p}$ perpendicular to this plane. Taking the scalar product of (7.6) with the unit vectors $\mathbf{n}_1$, $\mathbf{t}$, and $\mathbf{p}$ yields, respectively,

$$P_1 + \rho_1 w_1^2 = P_2 + \rho_2 w_2^2 \tag{7.7}$$

$$v_1 = v_2 \tag{7.8}$$

$$v_{2p} = 0 \tag{7.9}$$

This last result simply states that the relative velocity vector $\mathbf{u}_2 - \mathbf{b}$ is in the plane of $\mathbf{n}$ and $\mathbf{u}_1 - \mathbf{b}$. The bending of the velocity vector is thus analogous to the refraction of a (light or sound) ray at a material boundary. Equation (7.8) states that the tangential component of velocity is invariant across the shock. These points are all expressed by combining (7.6) and (7.7) and making use of the continuity equation (7.2) to form

$$\mathbf{u}_2 - \mathbf{u}_1 = \mathbf{n}_1(w_1 - w_2) \tag{7.10}$$

In words, *the velocity change across a shock is normal to the shock front*, with magnitude $w_1 - w_2$ (we will later show that $w_1 - w_2$ is almost always positive).

The momentum equation (7.7) has a striking but *spurious* resemblance to the incompressible Bernoulli equation, $P + \frac{1}{2}\rho u^2 = \text{const.}$ Aside from the fact that the latter equation describes *incompressible* fluids, these two relations are distinct in other ways: (1) the momentum equation (7.7) is a matching condition across a discontinuity whereas Bernoulli's equation describes a continuous flow; (2) the term ρw^2 in (7.7) represents a momentum flux whereas the term $\frac{1}{2}\rho u^2$ in Bernoulli's equation represents the kinetic energy per unit volume.

Energy: Adding $P_2w_2 - P_1w_1$ to both sides of (7.4) gives, with $\rho w = \rho_2 w_2 = \rho_1 w_1$,

$$\rho w\left[h_2 + \frac{u_2^2}{2} - \left(h_1 + \frac{u_1^2}{2}\right)\right] = (P_2 - P_1)\mathbf{b}\cdot\mathbf{n}_1$$

With (7.2) and (7.6) this becomes

$$h_2 + \frac{u_2^2}{2} - \left(h_1 + \frac{u_1^2}{2}\right) = \mathbf{b}\cdot(\mathbf{u}_2 - \mathbf{u}_1)$$

An important consequence of this equation is that for *stationary* shocks ($b = 0$) the *stagnation enthalpy* $h_0 = h + u^2/2$ *is invariant across the shock.* Making use of $(\mathbf{u} - \mathbf{b})^2 = w^2 + v^2$ and $v_1 = v_2$, this simplifies to

$$h_1 + \frac{w_1^2}{2} = h_2 + \frac{w_2^2}{2} \tag{7.11}$$

which contains only the normal component of velocity and is identical in form to the adiabatic steady-flow energy equation (5.4).

Entropy: With (7.2), the entropy equation (7.5) becomes just

$$s_2 \geq s_1 \tag{7.12}$$

There is an increase in entropy across the shock and a corresponding production within the shock front due to thermal and mechanical dissipation and relaxation processes.

This completes the reduction of the balance or conservation statements to the elementary shock conditions. For convenience, the main results are rewritten below. We adopt the notational convention that the jump in any quantity across the shock is represented by enclosing it in special square brackets; thus, for example, the jump in pressure is $P_2 - P_1 \equiv [P]$. In this format the main shock conditions (7.2), (7.7), (7.8), (7.11), and (7.12) are respectively

$$\boxed{\begin{aligned} [\rho w] &= 0 \\ [P + \rho w^2] &= 0 \\ [v] &= 0 \\ [h + 1/2w^2] &= 0 \\ [s] &\geq 0 \end{aligned}} \tag{7.13}$$

As we will show, the shock conditions represent real discontinuities which are consistent with physical principles; i.e., the relations (7.13) are satisfied simultaneously by known fluids. Moreover, shock waves are physically observed in a variety of experimental situations. Yet the recognition of this peculiar phenomenon was slow historically.

Historical Remarks

George Gabriel Stokes suggested the possibility of shock waves in a note, "On a Difficulty in the Theory of Sound," published in the *Philosophical Magazine* in 1848. Stokes had observed that a certain waveform becomes infinitely steep:

> Of course, after the instant at which the expression (A) becomes infinite, some motion or other will go on, and we might wish to know what the nature of the motion was. Perhaps the most natural supposition to make for trial is, that a surface of discontinuity is formed, in passing across which there is an abrupt change of density and velocity. The existence of such a surface will presently be shown to be possible . . . .

Almost 30 years later, a former student of Stokes, Lord Rayleigh, was preparing to publish his *Theory of Sound* and offered the following objection:

> 4 Carlton Gardens, S.W.
> June 2/77
>
> Dear Prof. Stokes,
>
> In consequence of our conversation the other evening I have been looking at your paper "On a difficulty in the theory of sound," *Phil. Mag.* Nov. 1848. The latter half of the paper appears to me to be liable to an objection, as to which (if you have time to look at the matter) I should be glad to hear your opinion . . . . [There follows a "proof."] It would appear therefore that on the hypotheses made, no discontinuous change is possible.
>
> I have put the matter very shortly, but I dare say what I have said will be intelligible to you.

Stokes, who had some reputation for diffidence, replied at once:

Cambridge
5th June, 1877

Dear Lord Rayleigh,

Thank you for pointing out the objection to the queer kind of motion I contemplated in the paper you refer to. Sir W. Thomson [Lord Kelvin] pointed the same out to me many years ago, and I should have mentioned it if I had had occasion to write anything bearing on the subject, or if, without that, my paper had attracted attention. It seemed, however, hardly worth while to write a criticism on a passage in a paper which was buried among other scientific antiquities . . . .

P.S. You will observe I wrote somewhat doubtfully about the possibility of the queer motion.[1]

For Rayleigh this apparently settled the matter. The following passage appears in his *Theory of Sound* [1894; vol. 2, p. 40]:

> but it would be improper to pass over in silence an error on the subject of discontinuous motion into which Riemann and other writers have fallen. It has been held that a state of motion is possible in which the fluid is divided into two parts by a surface of discontinuity.

There follows a proof of the impossibility of such motions based on a "violation" of the conservation of energy. Even Horace Lamb's celebrated *Hydrodynamics* [1932, p. 484] treads this same ground. A source of confusion was the entropy condition; the recognition that the entropy should increase, i.e., that dissipative terms could not be neglected *within* the shock front, was, however, pointed out by Zemplén (1905) and acknowledged by Rayleigh himself (1910). In addition, Mach's paper of 1887 contains a schlieren photograph which (in retrospect) clearly shows a shock front.

Algebraic shock conditions, incorporating energy conservation, are due to Rankine (1870) and to Hugoniot (1887, 1889), and their names have

[1] This correspondence is taken from Truesdell's preface to Stokes' collected *Mathematical and Physical Papers*, vol. 1, Johnson Reprint, New York, 1966.

come to be associated with the shock conditions generally.[1] Stokes' claim to recognition for his discovery has been diverted by circumstance.[2]

Generality of the Shock Conditions and Equivalence of Different Shocks

The shock conditions (7.13) involve only the components of the *relative* velocities $\mathbf{u}_1 - \mathbf{b}$ and $\mathbf{u}_2 - \mathbf{b}$ and are independent of the rate of change in velocity or other quantities. It follows that:

1 *The shock conditions hold for any observer.*

2 *The shock conditions hold whether or not the shock-front velocity and fluid properties ahead of the shock are constant.*

The only motion of the shock front which is physically identifiable is the motion normal to itself. The tangential component of $\mathbf{b}$ has no physical or geometric significance; this is reflected in that the tangential velocity component v enters the shock conditions (7.13) only in a very simple way, i.e., as an invariant across the shock. Thus, the choice of the shock-front velocity $\mathbf{b}$ is arbitrary to the extent that it can always be altered locally in such a way as to keep its normal component unchanged. In particular, we can choose $\mathbf{b}$ in such a way as to make $v_1 = v_2 = 0$; the relative velocities 1 and 2 are then both normal to the shock front, and the shock is by definition a *normal shock* (see Fig. 7.6). These observations are summarized by the following statement:

3 *Every shock can be reduced to an equivalent normal shock, for which* $v = 0$.

Formally, the reduction of a given shock to an equivalent normal shock is accomplished by adding a tangential motion $v\mathbf{t}$ to the shock-front

[1] According to Truesdell, the modern theory of shocks should be credited to Hugoniot, Hadamard, Duhem, and Zemplén.

[2] In a remarkably similar situation, Stokes originally derived the theoretical statement for what is now called Poiseuille flow but discounted it as being contradicted by contemporary experiments.

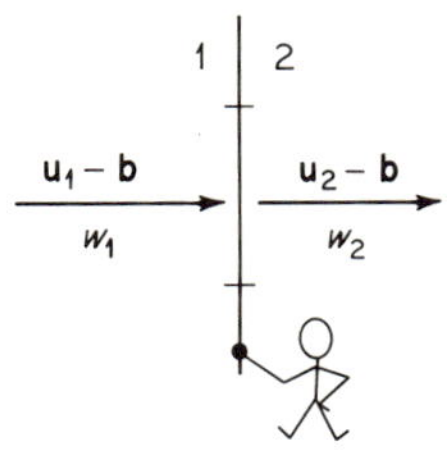

Figure 7.6
Normal shock.

velocity $\mathbf{b}$ (in effect, defining a new shock-front velocity $\mathbf{b}' = \mathbf{b} + v\mathbf{t}$). The relative velocities at the given shock are, making use of Fig. 7.4b,

$$\begin{aligned} \mathbf{u}_1 - \mathbf{b} &= -w_1\mathbf{n}_1 + v\mathbf{t} \\ \mathbf{u}_2 - \mathbf{b} &= -w_2\mathbf{n}_1 + v\mathbf{t} \end{aligned} \tag{7.14}$$

Adding the tangential motion $v\mathbf{t}$ to $\mathbf{b}$ gives the relative velocities for the equivalent normal shock

$$\begin{aligned} \mathbf{u}_1 - (\mathbf{b} + v\mathbf{t}) &= -w_1\mathbf{n}_1 \\ \mathbf{u}_2 - (\mathbf{b} + v\mathbf{t}) &= -w_2\mathbf{n}_1 \end{aligned} \tag{7.15}$$

This is illustrated by the following example.

EXAMPLE 7.1 EQUIVALENT NORMAL SHOCK FOR A BOW WAVE

Consider the bow wave on a supersonic projectile (Fig. 7.7). For simplicity, suppose that the projectile is traveling with constant velocity $-\mathbf{u}_\infty$ into a uniform medium. Then the bow wave will have a stationary form. For an observer on the projectile, $\mathbf{b} = 0$ and $\mathbf{u}_1 = +\mathbf{u}_\infty$; for an observer on the ground, $\mathbf{b} = -\mathbf{u}_\infty$ and $\mathbf{u}_1 = 0$. The reduction to the same equivalent normal shock, as described above, is illustrated on the right-hand side of each figure.

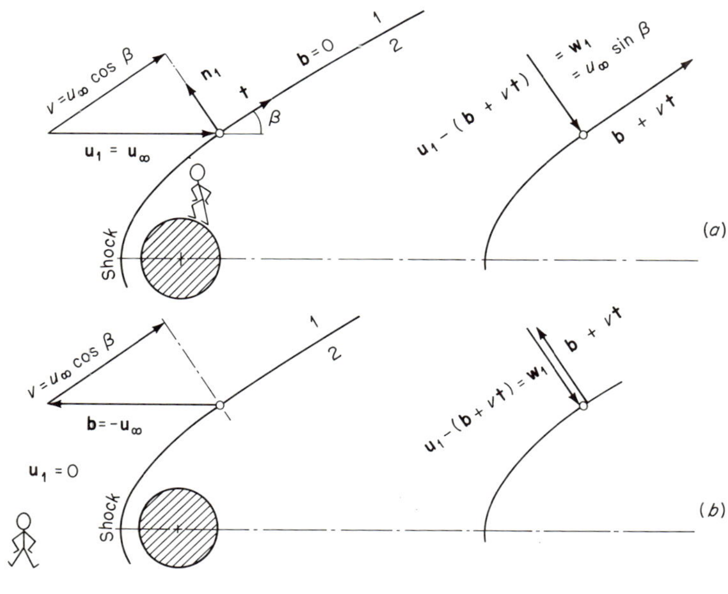

Figure 7.7
Bow wave: observer on (a) projectile and (b) ground.

All shocks can thus be treated as normal shocks; there is no physical distinction between a given shock and its normal-shock equivalent. The shock conditions (7.13), omitting the relation $[v] = 0$, are in fact often called the *normal-shock conditions.*

7.3 The properties of shock waves

For convenience the principal shock conditions are rewritten from (7.13), for continuity, momentum, energy, and entropy, respectively,

$$
\begin{aligned}
[\rho w] &= 0 \\
[P + \rho w^2] &= 0 \\
[h + 1/2w^2] &= 0 \\
[s] &\geq 0
\end{aligned}
\tag{7.16}
$$

where, for example, $[\rho w] = 0$ denotes $\rho_2 w_2 - \rho_1 w_1 = 0$. The solution of these equations may be thought of in the following physical situation: all the conditions ahead of the shock (side 1, or the upstream side) are known, and the shock-front velocity **b** is known. This would be the situation, for example, if a shock front advanced into stationary fluid at known velocity. We now inquire whether the conditions behind the shock (side 2, or the downstream side) are uniquely fixed by the shock conditions (see Fig. 7.8).

It is assumed that the fluids ahead of and behind the shock are essentially the same fluid and satisfy the same equation of state (this assumption precludes certain chemical reactions, e.g., the decomposition of a solid explosive; such cases are treated in Sec. 7.8). Then the thermodynamic state of the fluid is fixed by two variables, say P and v. The known quantities are w_1, P_1, v_1, and the unknowns are w_2, P_2, v_2; to determine the unknowns three equalities and one inequality are available in (7.16).

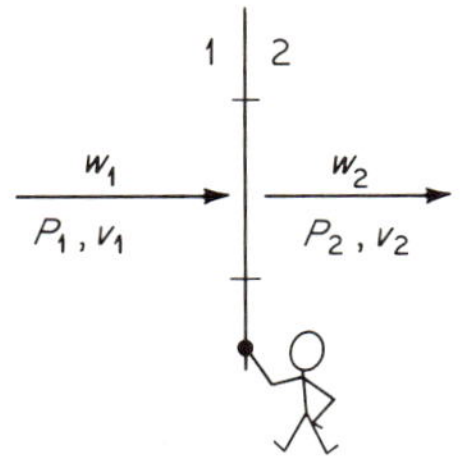

Figure 7.8

Combining the continuity and momentum equations in (7.16) yields

$$w_1 w_2 = \frac{[P]}{[\rho]} \tag{7.17}$$

Multiplying through by $\rho_1\rho_2$ gives the alternative form

$$J^2 = -\frac{[P]}{[v]} \tag{7.18}$$

where $J = \rho_1 w_1 = \rho_2 w_2$ is the *mass flux across the shock front*. We now define the *shock Mach number* M_{1n},

$$M_{1n} \equiv \frac{w_1}{c_1} \tag{7.19}$$

(We will later have occasion to use the symbol M_s for this same quantity.) Further manipulation of the continuity and momentum conditions then leads to the nondimensional relation

$$\frac{[P]}{\rho_1 {c_1}^2} = -M_{1n}\frac{[w]}{c_1} = -{M_{1n}}^2\frac{[v]}{v_1} \tag{7.20}$$

This gives in turn the simple relation

$$[w]^2 = -[P][v] \tag{7.21}$$

It will be convenient to denote the nondimensional pressure jump by the symbol Π

$$\Pi \equiv \frac{[P]}{\rho_1 {c_1}^2} \tag{7.22}$$

By definition, the numerical value of Π is a measure of the *strength* of the shock. The two extreme cases

$\Pi \ll 1$ weak shock
$\Pi \gg 1$ strong shock

will be considered separately in Sec. 7.6. For a perfect gas $c^2 = \gamma P/\rho$, and the strength of the shock is just $[P]/\gamma P_1$: thus, for example, a shock in a gas is weak if the pressure jump $[P]$ is small compared to the pressure ahead of the shock. For liquid water, however, $\rho_1 {c_1}^2 \approx 2.2 \times 10^4$ atm, and a shock with a pressure jump of several hundred atmospheres may still be considered weak.

We now introduce the energy condition in (7.16) for the first time; making use of (7.17), this can be put in the form

$$h_2 - h_1 = \tfrac{1}{2}(P_2 - P_1)(v_2 + v_1) \tag{7.23}$$

This is called the *Rankine-Hugoniot equation* and has the useful property that it contains only *thermodynamic* quantities. If we consider the upstream conditions P_1, v_1 to be known and fixed and the function $h_2 = h_2(P_2, v_2)$ to be known from a state equation, then (7.23) reduces to a function $P_2 = P_2(v_2)$. This function is called the *shock adiabat* and will be discussed in detail.

As an important example, consider the *perfect gas*. With $h = c_pT + \text{const} = [\gamma/(\gamma - 1)]Pv + \text{const}$, (7.23) reduces to

$$\frac{P_2}{P_1} = \frac{\dfrac{\gamma + 1}{\gamma - 1} - \dfrac{v_2}{v_1}}{\dfrac{\gamma + 1}{\gamma - 1}\dfrac{v_2}{v_1} - 1} \tag{7.24}$$

This is the equation for a shock-adiabat curve for a perfect gas. It may be compared to the equation for an isentropic curve

$$\frac{P_2}{P_1} = \left(\frac{v_1}{v_2}\right)^{\gamma} \tag{7.25}$$

These equations are both of the form $P_2(v_2)$, with P_1 and v_1 as constant parameters, and represent possible end states for a shock process (a process which is adiabatic by virtue of the absence of any *net* heat transfer to a fluid particle) and for an isentropic process (*reversible* adiabatic).

In general, shocks are classified according to the sign of the pressure jump $[P]$:

$[P] > 0$	compression shocks
$[P] < 0$	rarefaction shocks

Points lying above $P_2/P_1 = 1$ on the shock adiabat in Fig. 7.9 thus represent compression shocks. For a perfect gas it can be found directly from the Gibbs equation (2.17*a*) that $(\partial s/\partial P)_v = c_v/P$; it follows that points above the isentrope in Fig. 7.9 have $s > s_1$. Thus for compression shocks (where the shock adiabat lies above the isentrope), $s_2 > s_1$; similarly, rarefaction shocks have $s_2 < s_1$. Since the entropy condition in (7.16) is just that $[s] > 0$, it follows that *only compression shocks are possible*, at least for the case of a perfect gas.

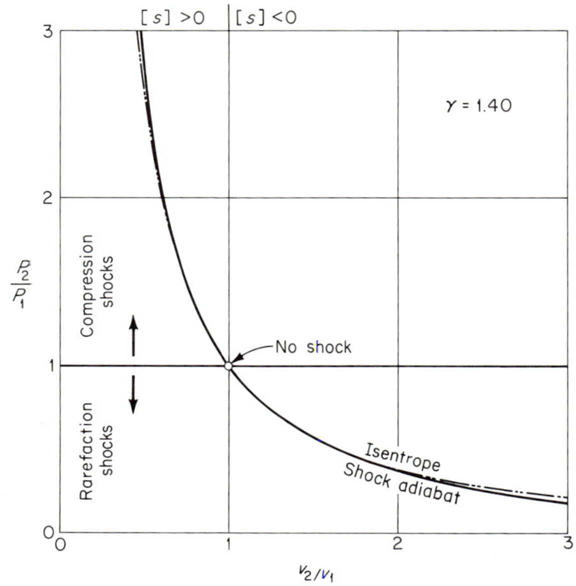

Figure 7.9
Shock adiabat and isentrope for a perfect gas with $\gamma = 1.40$.

In the neighborhood of the no-shock point (where downstream conditions are exactly the same as upstream) the isentropic and shock-adiabat curves are coincident: the first and second derivatives of the functions are equal, as can be verified by differentiation. The entropy is thus *stationary* in the vicinity of the no-shock point.

The above conclusions are true for almost all fluids and not just for the case of a perfect gas. To establish this we begin by calculating the entropy change, in terms of the pressure change, for a weak shock. The Rankine-Hugoniot equation (7.23) can be rewritten

$$[h] = v_1[P] + \tfrac{1}{2}[v][P] \tag{7.26}$$

To convert this into a relation between $[P]$ and $[s]$ we expand $h(s,P)$ and $v(s,P)$ in Taylor series.[1] Then using the identities $T = (\partial h/\partial s)_p$ and

[1] Symbolically, $f(x,y)$ is expanded about (x,y) as follows:

$$f(x + \Delta x, y + \Delta y) =$$
$$f(x,y) + \left(\Delta x \frac{\partial}{\partial x} + \Delta y \frac{\partial}{\partial y}\right) f(x,y) + \frac{1}{2!}\left(\Delta x \frac{\partial}{\partial x} + \Delta y \frac{\partial}{\partial y}\right)^2 f(x,y) + \cdots$$
$$+ \frac{1}{n!}\left(\Delta x \frac{\partial}{\partial x} + \Delta y \frac{\partial}{\partial y}\right)^n f(x,y) + \cdots$$

In the shock notation, Δx is, for example, replaced by $[x]$.

$v = (\partial h/\partial P)_s$ and retaining terms up to third order, (7.26) gives the important result

$$[s] = \frac{1}{12T_1}\left(\frac{\partial^2 v}{\partial P^2}\right)_s [P]^3 + O([P]^4) \tag{7.27a}$$

Because higher-order terms are omitted, this is applicable only to weak shocks. Rewritten in nondimensional form,

$$\frac{T_1[s]}{c_1^2} = \frac{1}{6}\Gamma_1\Pi^3 + O(\Pi^4) \tag{7.27b}$$

where we have already defined the fundamental derivative

$$\Gamma \equiv \frac{c^4}{2v^3}\left(\frac{\partial^2 v}{\partial P^2}\right)_s$$

in Sec. 5.2 and $\Pi \equiv [P]/\rho_1 c_1^2$.

The entropy condition requires that $[s] > 0$; thus from (7.27) the pressure jump $[P]$ necessarily has the same sign as the fundamental derivative Γ. It was explained in Sec. 5.2 that except in bizarre cases $\Gamma > 0$; we will henceforth assume Γ to be positive. It follows that $[P] > 0$, and *only compression shocks are possible for normal fluids.* Correspondingly, $[w] < 0$ and $[\rho] > 0$ from (7.20).

A further consequence of (7.27) is that the entropy becomes *stationary* as $[P] \to 0$, because $[s] \propto [P]^3$. We have already defined a weak shock as one for which Π is small compared to unity. Since the (dimensionless) entropy jump is given by (7.27*b*) as a coefficient of order unity times Π^3, the entropy jump for a weak shock is negligibly small. It will be a considerable practical advantage to treat such shocks as isentropic.

A shock-adiabat curve *A* and an isentrope *I* for a general fluid are shown in Fig. 7.10. The isentrope is concave upward, consistent with $\Gamma > 0$. The upstream (unshocked) state is represented by point 1, and a typical downstream state is represented by point 2, on the shock adiabat.

From the condition $[s] > 0$ we know that the entropy associated with points on *A* is greater than s_1. For typical fluids the adiabat will then lie above the isentrope as shown; manipulation of derivatives gives

$$\left(\frac{\partial s}{\partial P}\right)_v = \frac{c_p v^2}{c^2 T(\partial v/\partial T)_p} \tag{7.28}$$

which is typically positive (a well-known exception is water below 4°C).

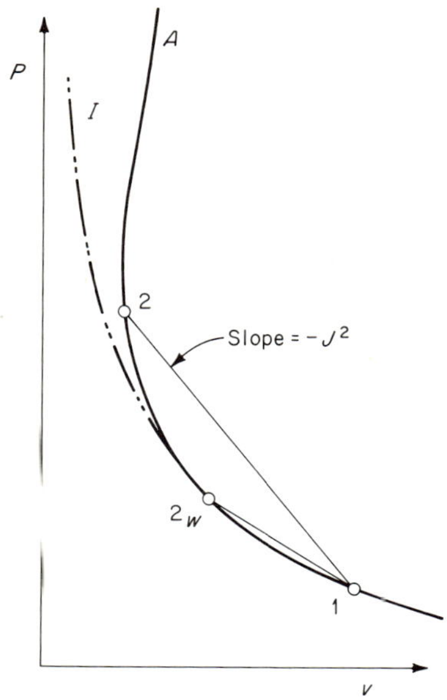

Figure 7.10
Shock adiabat.

Certain *slopes* on this diagram have physical significance. A tangent to the isentropic curve has slope

$$\left(\frac{\partial P}{\partial v}\right)_s = -\frac{c^2}{v^2} = -\rho^2 c^2 \tag{7.29}$$

and a chord connecting 1 and 2 has slope $-J^2$ from Eq. (7.18). With ρ_1 fixed, the strength of the shock increases as $J = \rho_1 w_1$ increases; correspondingly, the magnitude of the slope of the chord $1 \to 2$ increases and the point 2 slides upward on the shock adiabat. Consider now a weak shock $1 \to 2_w$, for which the isentrope and adiabat are essentially coincident; from the geometry of the figure

$$\left(\frac{\partial P}{\partial v}\right)_{s(2)} \leq -J^2 \leq \left(\frac{\partial P}{\partial v}\right)_{s(1)}$$

Making use of (7.29) and with $J = \rho_1 w_1 = \rho_2 w_2$, this becomes

$$\rho_2^2 c_2^2 \geq (\rho_2^2 w_2^2 = \rho_1^2 w_1^2) \geq \rho_1^2 c_1^2 \tag{7.30}$$

Hence

$$\begin{aligned} w_1 &\geq c_1 \\ w_2 &\leq c_2 \end{aligned} \tag{7.31}$$

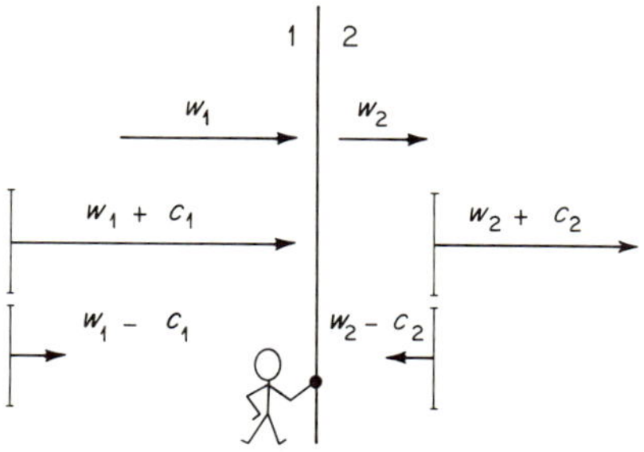

Figure 7.11

Thus the *relative normal upstream flow is supersonic*, and the *relative normal downstream flow is subsonic*. These inequalities hold for strong shocks as well, but the proof (which also depends on the assumption $\Gamma > 0$) is omitted; see, for example, *Landau and Lifshitz* [1959, chap. 9] or *Hayes* [1958].

As the strength of a shock is progressively diminished, the inequalities in (7.31) approach equalities: $w_1 \to c_1$ and $w_2 \to c_2$. In addition $[w] \to 0$ by virtue of (7.21). Thus, in the limit, the two expressions

$$w_1 w_2 = \frac{[P]}{[\rho]} \qquad c^2 = \left(\frac{\partial P}{\partial \rho}\right)_s$$

become identical. *In the limit of vanishing strength, shock waves become acoustic discontinuities, propagating with speed c relative to the fluid.* Such discontinuities were discussed in Sec. 4.8; weak shocks will be treated more generally in Sec. 7.6.

Consider acoustic wavefronts parallel to a shock front, as shown in Fig. 7.11. In general, there are two families of acoustic waves on each side of the shock, traveling in opposite directions with respect to the fluid. For an observer on the shock front, the acoustic waves have directions of travel as shown, in consequence of (7.31). In particular, on the upstream, or supersonic, side both families of waves travel toward the shock front. One result of this situation is that there will typically be no acoustic prewarning of a blast wave or sonic boom propagating in the atmosphere.

Here is a summary of the principal properties of shock discontinuities:

1 Only compression shocks, $[P] > 0$, are possible, assuming that $(\partial^2 v/\partial P^2)_s > 0$. Correspondingly, $[\rho] > 0$ and $[w] < 0$.

2 The upstream flow is supersonic with respect to the shock ($M_{1n} \geq 1$) and the downstream flow subsonic ($M_{2n} \leq 1$). In consequence, disturbances downstream of the shock can have no effect on the upstream flow.

3 For weak shocks the entropy change is proportional to the third power of the pressure change. $[s] \propto [P]^3$. Shocks thus become isentropic in the limit of vanishing strength.

EXAMPLE 7.2 SHOCK-INDUCED MOTION

A shock advances into stationary fluid. What is the direction of travel of the fluid immediately behind the shock front?

With $\mathbf{u}_1 = 0$, Eq. (7.10) is

$$\mathbf{u}_2 = -\mathbf{n}_1[w]$$

where $\mathbf{n}_1$ is the unit normal to the shock front, pointing in the propagation direction. Since $[w] < 0$, $\mathbf{u}_2$ is in the propagation direction, with magnitude $[w]$. Note that this is consistent with the acoustic result for simple waves, that compressed fluid travels in the same direction as the wave.

EXAMPLE 7.3 OVERTAKING SHOCK

The parallel shock fronts A and B shown in Fig. 7.12 are traveling in the same direction and separate the fluid into uniform regions 1, 2, and 3. Will shock front B overtake A?

From the standpoint of an observer who is stationary with respect to the fluid in region 2, shock front A travels to the left subsonically and B travels to the left supersonically. Therefore B will overtake A. Note that even if B were just an acoustic discontinuity, it would still overtake.

7.4 Normal shocks

It has already been emphasized that all shock waves can be considered to be normal shocks. The designation of certain discontinuities as normal shocks and others as oblique shocks is therefore artificial. Nevertheless, it is conventional and convenient to make such a distinction, particularly when the shock front is stationary; several shocks which are naturally considered normal are shown in Fig. 7.13.

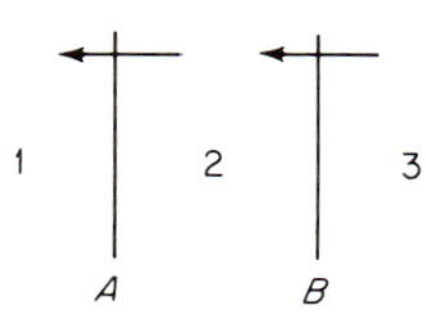

Figure 7.12

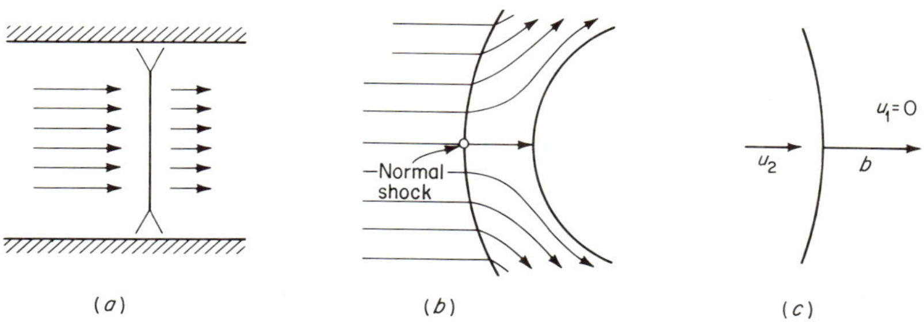

Figure 7.13
Normal shocks: (*a*) in a duct; (*b*) portion of a bow shock; and (*c*) spnerical shock advancing into stationary fluid.

Normal-shock Relations for a Perfect Gas

It is convenient to express the jump in all quantities in terms of the shock Mach number M_{1n}. The resulting relations will of course be applicable to all shocks; it is conventional to call them normal-shock relations only because they are expressed in terms of the normal Mach number.

The Rankine-Hugoniot equation (7.24) for a perfect gas is rearranged to give

$$\frac{[v]}{v_1} = -\frac{2[P]/P_1}{2\gamma + (\gamma + 1)[P]/P_1}$$

Substituting this into (7.20) then gives the desired relations

$$\frac{[P]}{P_1} = \frac{2\gamma}{\gamma + 1}(M_{1n}{}^2 - 1) \tag{7.32}$$

$$\frac{[w]}{c_1} = -\frac{2}{\gamma + 1}\left(M_{1n} - \frac{1}{M_{1n}}\right) \tag{7.33}$$

$$\frac{[v]}{v_1} = -\frac{2}{\gamma + 1}\left(1 - \frac{1}{M_{1n}{}^2}\right) \tag{7.34}$$

These relations are complete in that they fix the relative downstream velocity and thermodynamic state. It is, however, convenient to find additional relations. The downstream Mach number is by definition $M_{2n} = w_2/c_2$; this can be rewritten

$$M_{2n} = \frac{w_1 + [w]}{c_1}\frac{c_1}{c_2} = \left(M_{1n} + \frac{[w]}{c_1}\right)\frac{c_1}{c_2}$$

Thus, with $c^2 \propto T$,

$$M_{2n}{}^2 = \left(M_{1n} + \frac{[w]}{c_1}\right)^2 \frac{T_1}{T_2} = \left(M_{1n} + \frac{[w]}{c_1}\right)^2 \frac{P_1 v_1}{P_2 v_2}$$

Using (7.32) and (7.34) to find $P_2 v_2 / P_1 v_1$, this becomes

$$M_{2n}{}^2 = \frac{M_{1n}{}^2 + \dfrac{2}{\gamma - 1}}{\dfrac{2\gamma}{\gamma - 1} M_{1n}{}^2 - 1} \tag{7.35}$$

The jump conditions (7.32) to (7.35) are tabulated for $\gamma = 1.40$ in the *shock tables*, Table D.2, in the form of various property ratios (P_2/P_1, ρ_2/ρ_1, etc.) for the single argument M_{1n}, and shown in Fig. 7.14.

A practical and experimental measure of the entropy jump across a standing normal shock is the decrease in stagnation pressure. That is, if the fluid downstream of a shock is brought to rest in a steady isentropic flow, the stagnation pressure realized is less than the corresponding upstream stagnation pressure. This is illustrated by the steady flow between

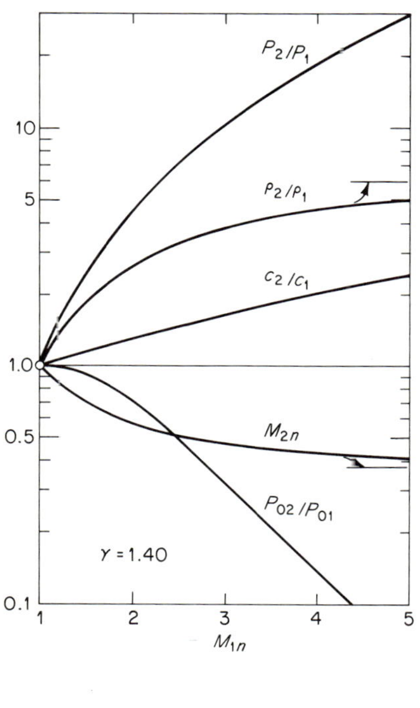

Figure 7.14
Downstream conditions as a function of the shock Mach number for a perfect gas with $\gamma = 1.40$.

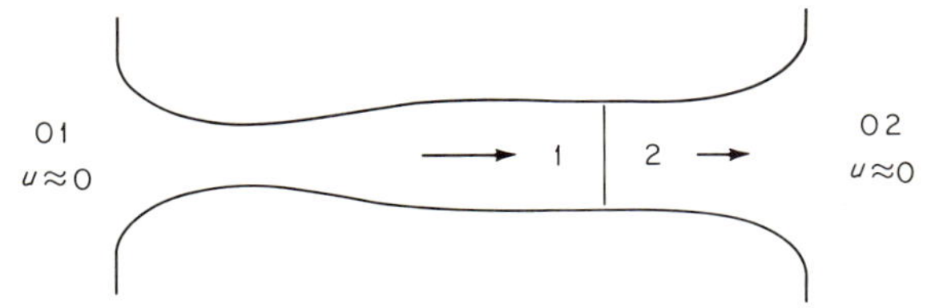

Figure 7.15

reservoirs shown in Fig. 7.15 (the flow is considered to be isentropic everywhere *except* across the shock). In general, the entropy change is given by Eq. (2.81), viz.,

$$\Delta s = c_p \ln \frac{T_2}{T_1} - R \ln \frac{P_2}{P_1} \tag{7.36}$$

With $s_{02} - s_{01} = s_2 - s_1$ and $T_{02} = T_{01}$ (because the stagnation enthalpy is unchanged across the shock), applying this equation between the stagnation states gives

$$[s] = -R \ln \frac{P_{02}}{P_{01}}$$

or

$$\frac{P_{02}}{P_{01}} = \exp\left(-\frac{[s]}{R}\right) \tag{7.37}$$

With $[s] > 0$, $P_{02} < P_{01}$, a result which also holds for a general fluid.[1] Applying the shock conditions (7.32) and (7.34) to (7.36), we can calculate the entropy jump in terms of the shock Mach number,

$$\exp \frac{[s]}{R} = \left(\frac{2\gamma}{\gamma + 1} M_{1n}^2 - \frac{\gamma - 1}{\gamma + 1}\right)^{1/(\gamma-1)} \left[\frac{2}{(\gamma + 1)M_{1n}^2} + \frac{\gamma - 1}{\gamma + 1}\right]^{\gamma/(\gamma-1)} \tag{7.38}$$

It is an objective of normal aerodynamic design to minimize the decrease in stagnation pressure and the corresponding decrease in stagnation density. The presence of a strong shock at the inlet of a (supersonic) jet engine, for example, will result in lower combustion pressures and a loss of thrust.

[1] In general, $(\partial P/\partial s)_h = -\rho T < 0$. With equal stagnation enthalpies, $h_{02} = h_{01}$, it follows that $P_{02} < P_{01}$.

EXAMPLE 7.4 TEMPERATURE AND PRESSURE

A supersonic airplane moves at 2,000 mi/h through air at $P_\infty = 0.5$ atm and $T_\infty = 440°\text{R}$. If a normal shock stands in front of a blunt surface on the airplane, what are the temperature and pressure immediately behind the shock?

The free-stream sound speed is calculated

$$c_\infty = \sqrt{\frac{(1.4)(8{,}314)(440)}{28.98}\frac{5}{9}} = 313 \text{ m/s} = 1{,}028 \text{ ft/s}$$

with corresponding Mach number

$$M_\infty = \frac{88}{60}\frac{2{,}000}{1{,}028} = 2.85 = M_{1n}$$

From the shock tables (Appendix D)

$$\frac{P_2}{P_\infty} = 9.310 \qquad \frac{T_2}{T_\infty} = 2.507$$

Thus, $T_2 = 1{,}100°\text{R}$, and $P_2 = 4.65$ atm. In general, the design of supersonic aircraft is intended to *avoid* the appearance of normal shocks.

EXAMPLE 7.5 PLUGGED CHANNEL

Air flows steadily in a constant-area pipe with $T_1 = 540°\text{R}$, $P_1 = 1$, and velocity $u_1 = 200$ ft/s. At some point in the pipe a valve is suddenly closed and a shock wave propagates upstream (Fig. 7.16). What are the shock velocity and the temperature and pressure behind it?

With $c_1 = 1{,}139$ ft/s, the velocity jump is

$$-\frac{[w]}{c_1} = \frac{200}{1{,}139} = 0.1757$$

The shock tables (Appendix D) then give $M_{1n} = 1.111$; hence $P_2 = 1.274$ atm, $T_2 = (1.072)(540) = 576°\text{R}$. Also,

$$w_1 = u_1 - b = (1.111)(1{,}139) = 1{,}265 \text{ ft/s}$$

Then

$$b = u_1 - w_1 = 200 - 1{,}265 = -1{,}065 \text{ ft/s}$$

is the (absolute) velocity with which the shock propagates.

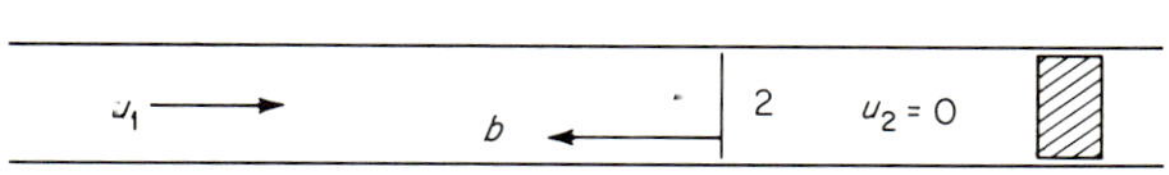

Figure 7.16

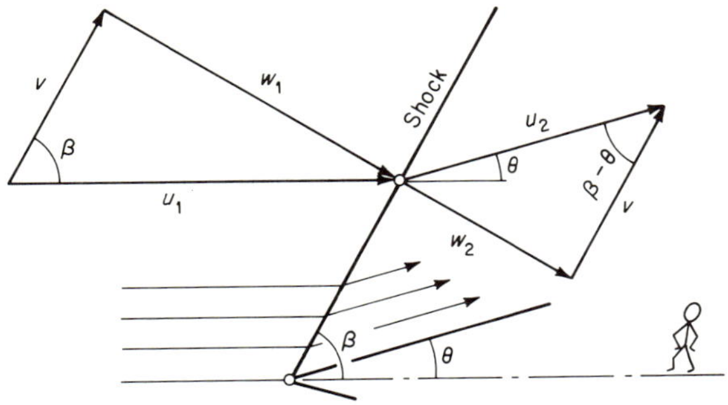

Figure 7.17
Oblique-shock geometry.

7.5 Oblique shocks

An oblique shock is defined by the property that the relative fluid velocities are apparently oblique to the shock front. This is in contrast to the normal shock, for which the relative velocities are apparently normal to the shock front. It is natural to consider a shock to be oblique, for example, in the streaming flow over a wedge shown in the lower part of Fig. 7.17 (see also the photograph in Fig. 7.18).

The flow is considered from the point of view of an observer for whom the shock front is stationary ($b = 0$). The angles shown in Fig. 7.17 are the *shock angle* β and the *turning angle* θ. The normal-shock conditions apply to the components $w_1 = u_1 \sin \beta$ and w_2. The tangential component $v = u_1 \cos \beta$ is, as we have found, invariant across the shock. Then if the upstream conditions and the shock angle β are known, the downstream conditions are fully known.

The fundamental requirement for supersonic upstream normal flow is now

$$M_1 \sin \beta \geq 1 \tag{7.39}$$

where $M_1 \equiv u_1/c_1$. The downstream normal flow is of course subsonic, but the resultant velocity u_2 may be supersonic

$$M_2 \gtreqless 1 \tag{7.40}$$

where $M_2 \equiv u_2/c_2$.

From the geometry of Fig. 7.17, $\tan \beta = w_1/v$, and $\tan (\beta - \theta) = w_2/v$. Subtracting the second statement from the first and applying trigonometric identities yields

$$-\frac{[w]}{v} = \frac{\tan \theta}{\cos^2 \beta(1 + \tan \beta \tan \theta)}$$

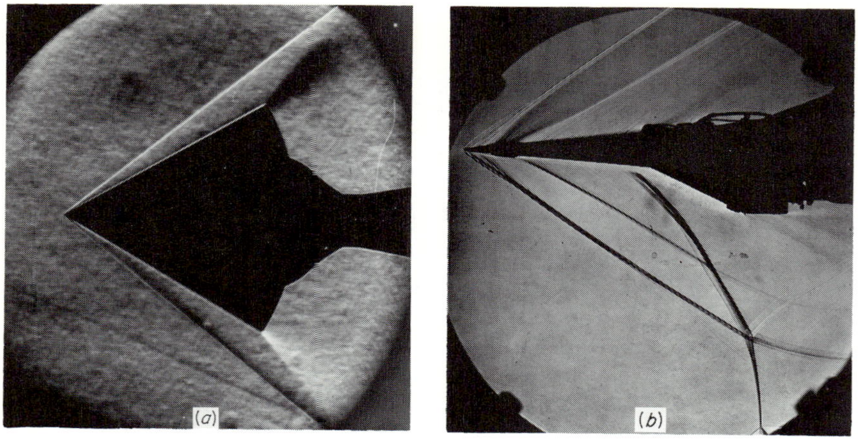

Figure 7.18
(*a*) Oblique shock attached to a wedge in hypersonic flow, $M_\infty \approx 10$. (*Courtesy of H. T. Nagamatsu, General Electric Co.*) (*b*) Oblique shocks from a two-dimensional body, $M_\infty = 2.02$. (*Courtesy of J. J. Ginoux, von Kármán Institute.*)

Multiplying through by $v/c_1 = M_1 \cos \beta$ then gives

$$-\frac{[w]}{c_1} = \frac{M_1 \tan \theta}{\cos \beta + \sin \beta \tan \theta} \tag{7.41}$$

With (7.20) and (7.22) and with $M_{1n} = M_1 \sin \beta$, this can also be written

$$\frac{\Pi}{M_1{}^2} = \frac{\tan \theta}{\cot \beta + \tan \theta} \tag{7.42}$$

This shows that $\tan \theta$ (and therefore θ) is of the order of the strength of the shock provided that β is not close to $\pi/2$, that is, the shock is not close to being a normal shock, and provided that M_1 is not too large.

A photograph of an oblique shock on a wedge is shown in Fig. 7.18.

Oblique-shock Relations for a Perfect Gas

With $[w]/c_1$ given by (7.33) and $M_{1n} = M_1 \sin \beta$, Eq. (7.41) yields on rearrangement

$$\tan \theta = \frac{2(\cot \beta)(M_1{}^2 \sin^2 \beta - 1)}{(\gamma + 1)M_1{}^2 - 2(M_1{}^2 \sin^2 \beta - 1)} \tag{7.43}$$

For given upstream Mach number M_1, this gives $\theta(\beta)$ or, implicitly, $\beta(\theta)$. This is presented graphically in the *oblique-shock chart* and numerically in the *oblique-shock tables* of Appendix D. There are two degenerate cases for which $\theta = 0$: $\beta = \pi/2$ (normal shock) and $\beta = \sin^{-1}(1/M_1)$ (Mach wave).

In general, the function $\beta(\theta)$ is double-valued; i.e., for any specified turning angle θ there are two possible shock angles. This is conveniently described by the *shock-polar relation*, as derived in the following.

With a stationary shock front, the energy condition (7.11) is just

$$h_1 + \frac{u_1^2}{2} = h_2 + \frac{u_2^2}{2}$$

For a perfect gas $h_2 - h_1 = (c_2^2 - c_1^2)/(\gamma - 1)$, and this becomes

$$u_1^2 + \frac{2}{\gamma - 1} c_1^2 = u_2^2 + \frac{2}{\gamma - 1} c_2^2 = \text{const} \tag{7.44}$$

The constant will be written in the following way: if the flow were assumed to pass continuously from state 1 to state 2 such that $u^2 + [2/(\gamma - 1)]c^2 =$ const, there would necessarily be a sonic point where $u = c = c_*$, and substitution gives for the constant just $(\gamma + 1)[c_*^2/(\gamma - 1)]$. The sonic condition thus defined is algebraically consistent with that of Sec. 5.5 but of course does not occur physically within the shock. Then with $w^2 + v^2 = u^2$, (7.44) becomes

$$w_1^2 + \frac{2}{\gamma - 1} c_1^2 = w_2^2 + \frac{2}{\gamma - 1} c_2^2 = \frac{\gamma + 1}{\gamma - 1} c_*^2 - v^2 \tag{7.45}$$

For a perfect gas, $c_2^2 - c_1^2 = \gamma(P_2/\rho_2 - P_1/\rho_1)$. Making use of this, together with (7.2) and (7.17), Eq. (7.45) reduces after considerable algebra to

$$w_1 w_2 = c_*^2 - \frac{\gamma - 1}{\gamma + 1} v^2 \tag{7.46}$$

This is known as the *Prandtl relation*. Note that c_* is known in terms of the upstream conditions, i.e.,

$$c_*^2 = \frac{2}{\gamma + 1} c_1^2 + \frac{\gamma - 1}{\gamma + 1} u_1^2$$

The oblique-shock geometry is redrawn in Fig. 7.19. Let the Cartesian components of the downstream velocity $\mathbf{u}_2$ be u_{2_x} and u_{2_y}, where

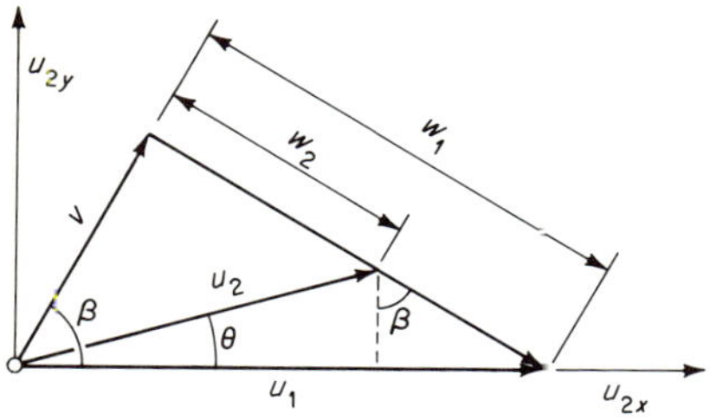

Figure 7.19
Oblique-shock geometry with Cartesian velocity components.

the x axis is aligned with the upstream velocity (we have reverted to x and y notation rather than using indicial notation to avoid confusion with the subscripts labeling upstream and downstream). From the geometry of the figure,

$$\frac{v}{w_1} = \frac{u_{2_y}}{u_1 - u_{2_x}}$$

$$\frac{v}{u_1} = \frac{u_{2_y}}{w_1 - w_2}$$

$$u_1{}^2 = w_1{}^2 + v^2$$

Combining these relations with (7.45) gives the equation of the *shock polar*

$$U_{2_y}{}^2 = \frac{(U_1 - U_{2_x})^2(U_1U_{2_x} - 1)}{\dfrac{2}{\gamma + 1}U_1{}^2 - U_1U_{2_x} + 1} \tag{7.47}$$

where the velocities are normalized with respect to c_*, $U \equiv u/c_*$ (in the notation of Chap. 5, $U = M_*$). With upstream conditions fixed, this relation is of the form $U_{2_y} = f(U_{2_x})$, with $U_1 = M_1^*$ as parameter. This is drawn for particular values of γ and M_1^* in Fig. 7.20.

The resulting figure is a classical cubic curve.[1] Every point on the curve represents a possible terminus for the downstream velocity vector $\mathbf{U}_2 = \mathbf{M}_2^*$. Beginning on the U_{2_x} axis, the labeled points are *mw*, corresponding to a *Mach wave*, a degenerate oblique shock with $M_2^* = M_1^*$; *w*, a typical *weak solution* for a given turning angle θ; the *sonic* downstream condition $M_2^* = M_2 = 1$; *m*, corresponding to the *maximum turning angle*; *s*, a typical *strong solution* for a given turning angle θ; *ns*, corresponding to a normal shock.

[1] The curve has been variously called the *folium of Descartes*, *Cartesian leaf*, and *strophoid*. We will have occasion to refer to a polar velocity diagram of this type as a *hodograph* (ὁδός, way, + γράφειν, write; a way of writing).

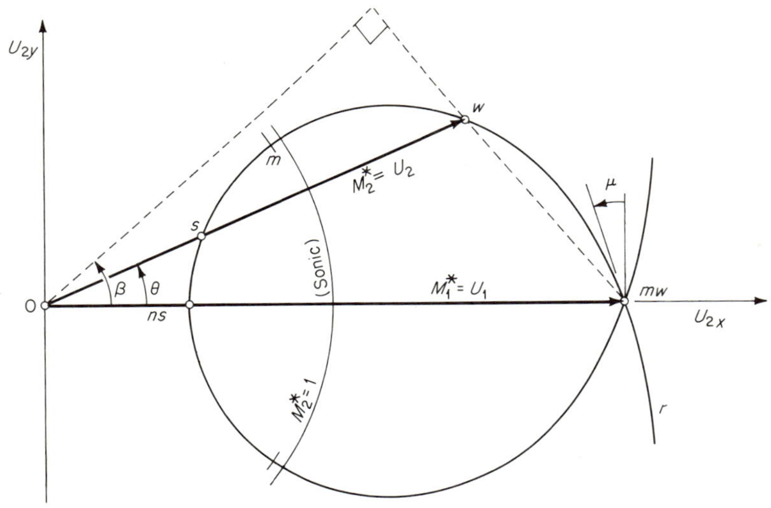

Figure 7.20
Oblique-shock polar diagram for the particular case $U_1 = M_1^* = 2$, $\gamma = 1.40$.

For any given turning angle θ, there are two possible distinct solutions, which are conventionally called respectively the *weak solution* and the *strong solution.* It should be made clear that this nomenclature has no direct connection to the weak shock ($\Pi \ll 1$) and strong shock ($\Pi \gg 1$) as already defined in Sec. 7.3. In the context of the shock polar we use the nomenclature weak and strong because the strength of the shock is always greater for the strong solution, $[P]_s > [P]_w$, as follows from the fact that $[w]$ is exactly the distance between *mw* and the terminus of the vector $\mathbf{U}_2^*$.

The shock angle β is constructed as shown, making use of the condition that the tangential component v is common to both M_1^* and M_2^* (viewed as vectors). The lower portion of the shock polar corresponds to rightward turning and is of course symmetric with the upper portion.

Just as in the case of the normal shock, an expansion or *rarefaction shock* (branch r) results in a decrease of fluid entropy and is impossible. In the vicinity of *mw*, however, the entropy is again stationary and some *infinitesimal* excursion onto the branch r is conceivable; this does, in fact, have physical significance and will be developed in Chap. 9.

It is fair to ask whether the weak or the strong solution actually occurs when the turning angle is physically prescribed. A rough general answer is that the *weak* solution, corresponding to minimum entropy

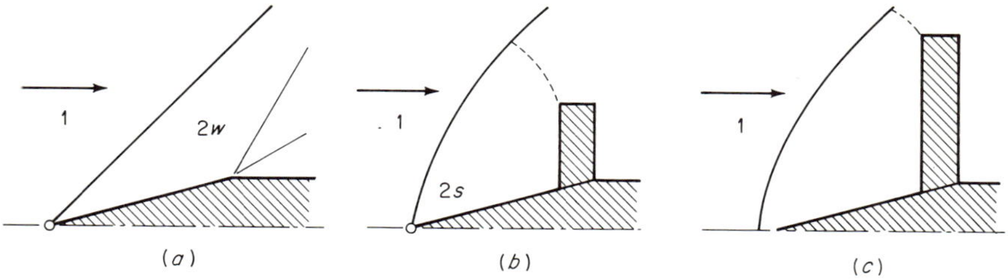

Figure 7.21
Oblique shocks for streaming flow over a wedge: (*a*) weak solution; (*b*) strong solution at the point of the wedge; and (*c*) detached shock.

production, usually occurs when permitted by downstream boundary conditions. In Fig. 7.21*a*, the weak solution holds for streaming flow over a wedge followed by a flat plate. If the flow is dammed, as in Fig. 7.21*b*, the pressure behind the shock is greater and the strong solution obtains, at least at the point of the wedge, where the turning angle is prescribed. If the dam is sufficiently enlarged, the shock detaches, as in Fig. 7.21*c*.

The extreme cases of the weak and strong solutions are respectively the *Mach wave* and the *normal shock*.

EXAMPLE 7.6 SHOCK ATTACHED TO A WEDGE

A uniform supersonic free stream ($M_1 = 3.00$, $\gamma = 1.40$) flows over an infinite wedge with semiangle of 10°. Find the shock angle and the downstream conditions.

The weak-shock solution has a shock angle $\beta = 27.38°$ from the oblique-shock tables (Table D.3). From the same table we also find

$$\frac{P_2}{P_1} = 2.055$$

$$M_2 = 2.505$$

An alternative procedure is to make use of the normal-shock tables, with the shock Mach number

$$M_{1n} = M_1 \sin\beta = 3 \sin 27.38° = 1.380$$

The postulated *uniform* downstream conditions provide a solution for the entire flow field, and this is approximately realized physically. The presence of a boundary layer on the surface of the wedge alters the above conditions somewhat.

EXAMPLE 7.7 DETACHED SHOCK

Conditions are the same as in the preceding problem except that the free-stream Mach number $M_1 = 1.30$. What is now the shock configuration?

For $M_1 = 1.30$, the *maximum* turning angle is about 6.7°. No attached oblique shock can yield the necessary flow inclination on the surface of the wedge. Under these conditions, the shock is generally *detached*, as in the case of flow around a blunt body. The downstream flow field is now necessarily nonuniform; its reconstruction requires a detailed solution of the equations of motion.

EXAMPLE 7.8 REFLECTED SHOCK

A plane shock A with shock Mach number $M_{1n} = b/c_1$ advances into stationary air. The shock is intercepted by a rigid wall, as shown in Fig. 7.22*a*. Find the flow in the neighborhood of the point where the shock front A intersects the wall surface.

A possible flow field is shown in Fig. 7.22*b*. If the shock were sufficiently weak, the reflection would be regular and could be treated according to the linear theory of Sec. 4.9, giving just a reflected shock of the same strength as the incident shock. For a shock which is not necessarily weak, one possibility is to consider the flow from the point of view of an observer who moves with the point of intersection. Such an observer sees stationary oblique shocks, with $u_1 = b/(\sin\beta)$. Conditions in field 2 can be found from the known $M_1 = u_1/c_1$ and given β. In particular, the turning angle θ and Mach number M_2 can be found. Then the flow in field 3 can be found from the condition that the flow must follow the wall, so that the required turning angle is the same as across shock front A. The above procedure requires only that the flow in field 2 be supersonic.

Further oblique-shock calculations are given in Examples 7.9 and 7.10 (pages 343ff).

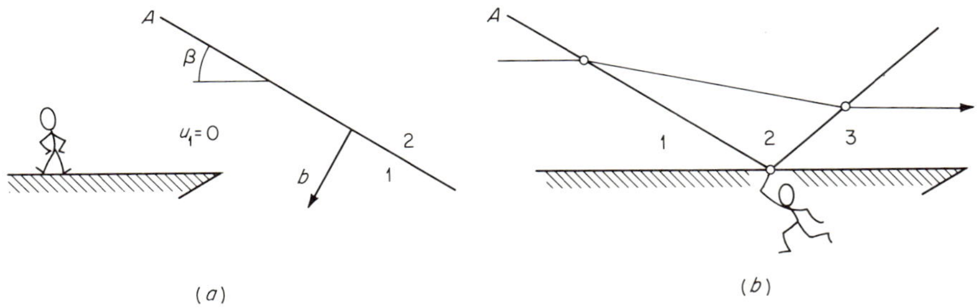

Figure 7.22

Figure 7.23
Oblique-shock reflection at a wall with attached turbulent boundary layer. Upstream Mach number = 2.5, turning angle = 9.5°. (*Courtesy of J. E. Green, Royal Aircraft Establishment.*)

Remark on the Reflection of an Oblique Shock from a Wall

The reflection of an oblique shock (as shown in Fig. 7.22*b*) can usually be *calculated* from the condition that the flow behind the reflection follows the wall. At least in steady flows, however, the flow near the wall is in the form of a *boundary layer* (see Fig. 7.23). This complicates the reflection somewhat; e.g., the flow adjacent to the wall cannot possibly be supersonic! In the most significant form of departure from the simple calculation, the flow actually separates from the wall at the reflection point. For a discussion of this problem, see *Schlichting* [1968, chap. 13, Sec. e]. For a more complete discussion of the general problem of shock reflection, see *Shapiro* [1953, sec. 16.7].

7.6 Weak-shock and strong-shock approximations

We have already (in Sec. 7.3) defined a weak shock as satisfying $\Pi \ll 1$ and a strong shock as satisfying $\Pi \gg 1$, where $\Pi \equiv [P]/\rho_1 c_1^2$ is the dimensionless pressure jump or shock strength. Each of these extreme cases leads to useful simplifying approximations. At the low end of the strength scale, the weak-shock approximations are useful in slender-body aerodynamics and a variety of nearly acoustic phenomena. At the high end, the strong-shock approximations are useful in the theory of blast waves and blunt reentry bodies.

Weak Shocks

As we shall soon verify, the shock Mach number is very close to unity. Then with (7.20), a weak shock satisfies the following mutually equivalent conditions:

$$\left.\begin{aligned} \Pi &= \frac{[P]}{\rho_1 c_1^2} \\ &-\frac{[w]}{c_1} \\ &-\frac{[v]}{v_1} \\ M_{1n} &- 1 \end{aligned}\right\} \ll 1 \tag{7.48}$$

By virtue of (7.27), $[s] \propto \Pi^3$ and is therefore extremely small if (7.48) is satisfied. In practice, depending on the degree of approximation which is acceptable, certain shocks may be treated as weak even though the requirement (7.48) is somewhat strained, for example, $\Pi = \frac{1}{2}$.

In Appendix A, the jump in various quantities is expressed as a power series in Π; this is done by expressing the desired quantity in a Taylor series about the upstream state and making use of the shock conditions (7.20) and (7.27). Typical and important results are

$$\begin{aligned} -\frac{[w]}{c_1} &= \Pi - \frac{\Gamma_1}{2}\Pi^2 \\ &\quad -\frac{1}{4}\left[\frac{c_1^2\Gamma_1}{3v_1 T_1}\left(\frac{\partial T}{\partial P}\right)_s + \frac{\Gamma_1^2}{2} + \frac{c_1^6}{3v_1^4}\left(\frac{\partial^3 v}{\partial P^3}\right)_s\right]\Pi^3 \cdots \end{aligned} \tag{7.49}$$

$$\begin{aligned} M_{1n} &= 1 - \frac{\Gamma_1}{2}\frac{[w]}{c_1} \\ &\quad +\frac{1}{4}\left[\frac{c_1^2\Gamma_1}{3v_1 T_1}\left(\frac{\partial T}{\partial P}\right)_s + \frac{5\Gamma_1^2}{2} + \frac{c_1^6}{3v_1^4}\left(\frac{\partial^3 v}{\partial P^3}\right)_s\right]\left(\frac{[w]}{c_1}\right)^2 \cdots \end{aligned} \tag{7.50}$$

In every case, retaining only the first term corresponds to the approximation appropriate to an *acoustic discontinuity*, which propagates at the speed of sound [compare Eq. (4.72)]. Retaining only the first two terms corresponds to the *isentropic* approximation for a weak shock. Retaining the first three terms accounts for the (small) entropy jump across the shock.

Weak Oblique Shocks

It is somewhat more convenient to take $[w]/c_1$ (rather than Π) as a measure of the shock strength. From the construction of Fig. 7.20, it is clear that a weak oblique shock, with $[w]$ small, is very close to a Mach wave; i.e., the point w is adjacent to point mw. We begin by finding the (small) difference between the Mach angle μ and the shock angle β.

For conciseness, Eq. (7.50) is rewritten as

$$M_{1n} = 1 - \tfrac{1}{2}\Gamma_1 \frac{[w]}{c_1} + \tfrac{1}{4}\Psi_1 \left(\frac{[w]}{c_1}\right)^2 \cdots \tag{7.51}$$

where Ψ_1 is an abbreviation for the coefficient of the last term. From the oblique shock geometry (Fig. 7.17)

$$\sin\beta = \frac{w_1}{u_1} = \frac{M_{1n}}{M_1}$$

With M_{1n} given by (7.51) and $1/M_1 = \sin\mu_1$, this becomes

$$\sin\beta = \sin\mu_1 - \frac{1}{2}\frac{\Gamma_1}{M_1}\frac{[w]}{c_1} + \frac{1}{4}\frac{\Psi_1}{M_1}\left(\frac{[w]}{c_1}\right)^2 \cdots \tag{7.52}$$

By trigonometric manipulation this can be reduced to

$$\beta - \mu_1 = -\frac{\Gamma_1}{2\sqrt{M_1^2 - 1}}\frac{[w]}{c_1} + \frac{1}{4\sqrt{M_1^2 - 1}}\left[\Psi_1 + \frac{\Gamma_1^2}{2(M_1^2 - 1)}\right]\left(\frac{[w]}{c_1}\right)^2 \cdots \tag{7.53}$$

A similar result can now be obtained for the turning angle θ. Equation (7.41) rearranges to

$$\tan\theta = \frac{-([w]/c_1)\cos\beta}{M_1 + ([w]/c_1)\sin\beta}$$

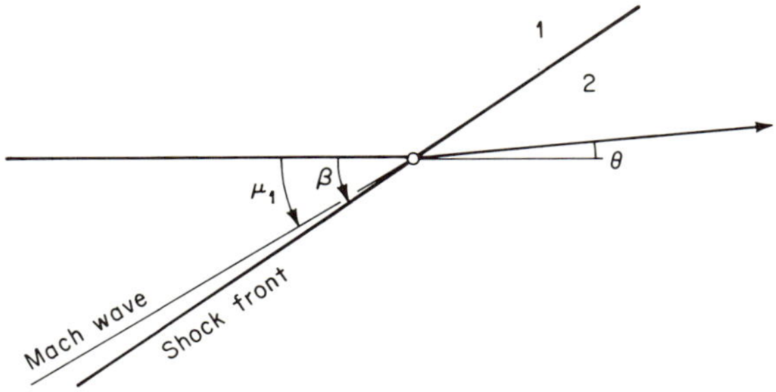

Figure 7.24
Weak oblique shock.

With $\cos\beta = \sqrt{1 - \sin^2\beta}$ and $\sin\beta$ given by (7.52) this yields, after some algebra,

$$\theta = -\frac{\sqrt{M_1^2 - 1}}{M_1^2}\frac{[w]}{c_1} + \frac{\sqrt{M_1^2 - 1}}{M_1^2}\left[\frac{1}{M_1^2} - \frac{\Gamma_1}{2(M_1^2 - 1)}\right]\left(\frac{[w]}{c_1}\right)^2 \cdots \tag{7.54}$$

Thus, $\beta - \mu_1$ and θ *vary as the shock strength* and are of the same order (see Fig. 7.24).

It is convenient to take the turning angle θ as an alternative measure of the strength of a weak oblique shock. For example, with (7.54) and (7.20), the entropy-jump expression (7.27) can be rewritten as

$$\frac{T_1[s]}{c_1^2} = \frac{\Gamma_1}{6}\frac{M_1^6}{(M_1^2 - 1)^{3/2}}\theta^3 \cdots \tag{7.55}$$

This result guarantees that turning through a finite angle may be accomplished via a sequence of oblique shocks, with a very small net entropy change. Let $\Delta\theta$ be the fixed total turning angle accomplished through n oblique shocks (Fig. 7.25), and let $[\theta]$ be the turning angle across each shock. Then the total turning angle and total entropy change

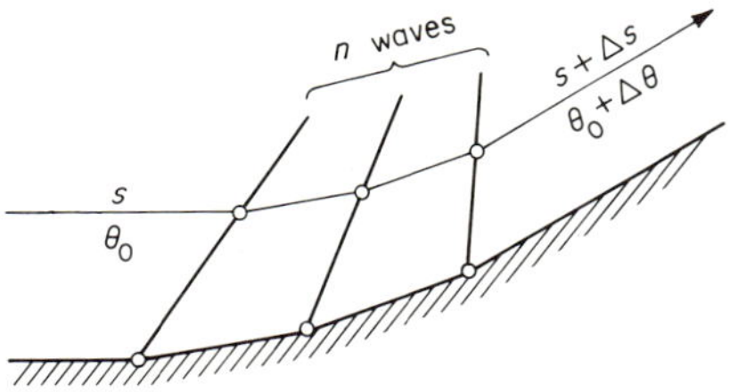

Figure 7.25

are, respectively, $\Delta\theta = n[\theta]$ and $\Delta s = \sum [s]_n$. With $[s]_n = K_n[\theta]^3$, where K_n is the coefficient in (7.55), we have

$$\Delta s = \sum_n [s]_n = \sum_n K_n[\theta]^3 = \sum_n K_n\left(\frac{\Delta\theta}{n}\right)^3$$

or

$$\Delta s = \bar{K}\frac{(\Delta\theta)^3}{n^2} \tag{7.56}$$

where $\bar{K}$ is the average coefficient. By increasing the number of shocks, and thereby decreasing the strength of each individual shock, Δs can be made arbitrarily small. We will make extensive use of this result in Chap. 9.

Strong Shocks

Since fluid specific volume cannot become negative, $-[v] < v_1$ and it follows algebraically from (7.20) that

$$M_{1n}^2 > \Pi \tag{7.57}$$

By definition, a strong shock has $\Pi \gg 1$; thus, a strong shock satisfies the following mutually equivalent conditions

$$\left.\begin{array}{r} \Pi = \dfrac{[P]}{\rho_1 c_1^2} \\ -M_{1n}\dfrac{[w]}{c_1} \\ M_{1n}^2 \end{array}\right\} \gg 1 \tag{7.58}$$

It appears that $P < \rho c^2$ for all known fluids.[1] This being true, or even nearly true, the condition $\Pi \gg 1$ is equivalent to $P_2 \gg P_1$.

With P_1 negligible, (7.20) can be written

$$P_2 = -\rho w[w] \tag{7.59}$$

[1] In general this statement is equivalent to

$$\left(\frac{\partial P}{\partial v}\right)_s < -\frac{P}{v}$$

For a perfect gas, this corresponds to $\gamma > 1$.

That is, the downstream pressure results from momentum transfer only. Similarly, if h represents in some sense an *absolute* enthalpy, for example, $h = c_pT$, then h_1 is negligible compared to h_2 and the Rankine-Hugoniot relation (7.23) can be written

$$h_2 = \tfrac{1}{2}P_2(v_2 + v_1) \tag{7.60}$$

To develop further relations it is necessary to introduce a specific equation of state. For the *perfect gas*, Eqs. (7.32) to (7.34) become, respectively,

$$P_2 = \frac{2}{\gamma + 1}\rho_1 w_1^2 \tag{7.61}$$

$$[w] = -\frac{2}{\gamma + 1}w_1 \tag{7.62}$$

$$[v] = -\frac{2}{\gamma + 1}v_1 \tag{7.63}$$

An alternative statement of (7.63) is $\rho_2/\rho_1 = (\gamma + 1)/(\gamma - 1)$; the density ratio approaches this limiting value as the shock strength increases. Similarly, by (7.35) the downstream Mach number approaches the limiting value

$$M_{2n} = \sqrt{\frac{\gamma - 1}{2\gamma}} \tag{7.64}$$

7.7 Contact surfaces

A contact surface separates two fluids of different properties, e.g., water and air, hot air and cold air, or nitrogen and helium. Like the shock wave, the contact surface may be *idealized* as a surface of discontinuity, not necessarily stationary; unlike the shock wave, there is no flow of matter across the contact surface.

Alternative and specialized names for the contact surface are *material boundary*, *entropy discontinuity*, *tangential discontinuity*, *vortex sheet*, and *slip surface*.

Familiar examples include the wavy surface of a body of water and meteorological fronts[1] separating warm from cold air. This latter

[1] It has been suggested that this name, coined at the time of the First World War, was taken by analogy with the line separating opposing armies, as in the Western Front.

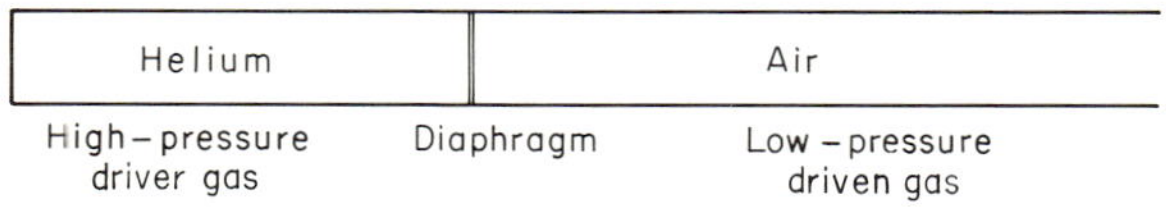

Figure 7.26
Shock tube.

example serves to emphasize that the treatment of a contact surface as a discontinuity may be a considerable idealization, as turbulent mixing and diffusion may result in a "surface" of very great thickness. This is in contradistinction to the case of a shock wave, in which, as a practical matter, the thickness of the shock front can almost always be neglected.

A particularly simple example of a contact surface occurs in the *shock tube*. A possible arrangement is shown in Fig. 7.26. When the pressure of the helium driver gas is sufficiently raised, either by heating or from an external supply, the relatively thin *diaphragm* ruptures. The high-pressure helium gas then expands to the right, compressing the low-density air, and separated from it by a moving contact surface. The details of this motion will be discussed in the next chapter.

A further example is the boundary of a fluid jet, as in the case of water flowing from a tap or the exhaust of a rocket. Such flows are often unstable, with the result that treatment of the contact surface as a discontinuity is questionable, except relatively near the point of discharge (Figs. 7.27 and 7.28).

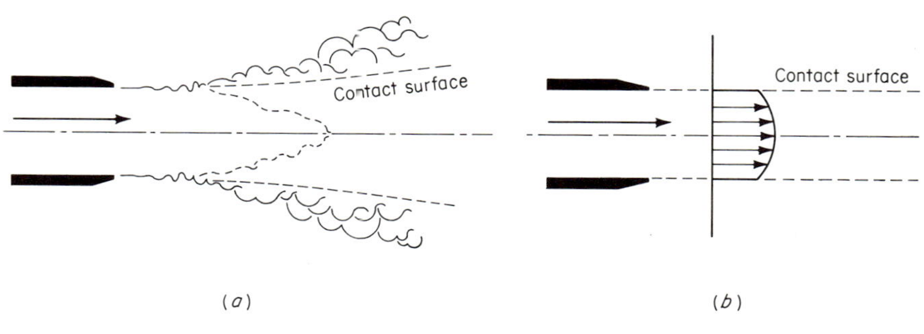

Figure 7.27
Fluid jet: (*a*) unstable contact surface with unsteady turbulent mixing; (*b*) idealized velocity distribution with the contact surface a surface of infinite vorticity, or *vortex sheet*.

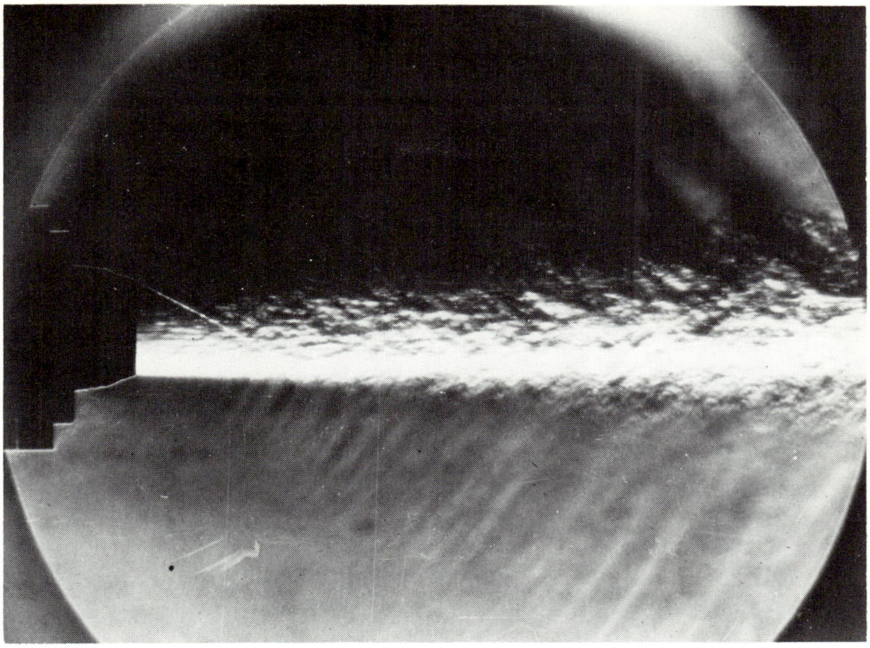

Figure 7.28
Breakup of the contact surface bounding a supersonic jet, $M_e = 2.22$. (*Courtesy of J. M. Eggers, NASA.*)

Matching Conditions for a Surface of Discontinuity

If turbulent mixing and viscous-thermal diffusion are neglected, the contact surface can be idealized as a surface of discontinuity. The balance statements (1.79) to (1.82) are applicable to the control volume shown in Fig. 7.29. The separate media are labeled 1 and 2, but these designations are

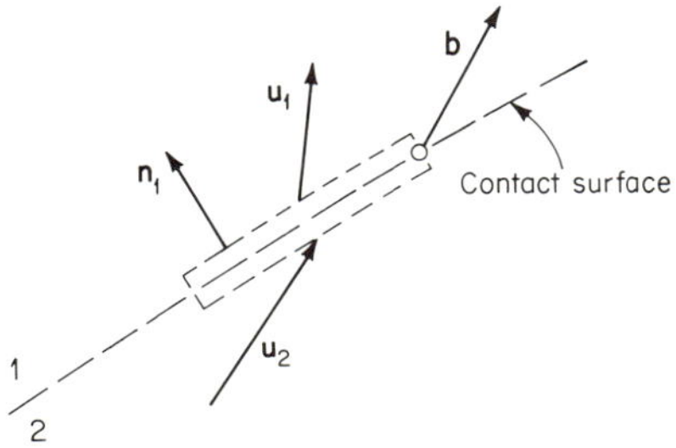

Figure 7.29

arbitrary and interchangeable. By definition, there is no mass flux across the surface; hence $(\mathbf{u}_1 - \mathbf{b}) \cdot \mathbf{n}_1 = (\mathbf{u}_2 - \mathbf{b}) \cdot \mathbf{n}_1 = 0$, or

$$(\mathbf{u}_2 - \mathbf{u}_1) \cdot \mathbf{n}_1 = 0 \tag{7.65}$$

This is the fundamental condition. *The velocity change is purely tangential to the surface* (as opposed to the case of a shock, in which it is purely normal). That is, there may be relative tangential motion, or *slippage*, between the two media, a situation sketched in Fig. 7.27*b*. Application of the balance statements (1.80) to (1.82) then gives, with no mass flux across the surface,

$$\begin{aligned} \mathbf{T}_1 &= -\mathbf{T}_2 \\ \mathbf{T}_1 \cdot (\mathbf{u}_2 - \mathbf{u}_1) &= \mathbf{n}_1 \cdot (\mathbf{q}_2 - \mathbf{q}_1) \\ \left(\frac{\mathbf{q}_2}{T_2} - \frac{\mathbf{q}_1}{T_1}\right) \cdot \mathbf{n}_1 &\leq 0 \end{aligned} \tag{7.66}$$

Physically these may be described respectively as equilibration of forces in the absence of accelerated mass, equality of the net energy input as heat to the net energy output as work, and a net outflow of entropy.

Inviscid Approximation

Consistent with our treatment of shock waves, both sides of the control surface may be treated as inviscid, with $\mathbf{T} = -P\mathbf{n}$ and $\mathbf{q} = 0$. Then the conditions (7.66) reduce to simply

$$P_1 = P_2 \tag{7.67}$$

Extensive use is made of this approximation [together with the condition $(\mathbf{u}_2 - \mathbf{u}_1) \cdot \mathbf{n}_1 = 0$] in ordinary gasdynamics.

Matching Conditions in the Absence of Slip

For ordinary viscous fluids it is found by experiment and from elementary kinetic theory that there is negligible velocity and temperature slip (*no-slip* conditions) at such a boundary, and the conditions (7.65) and (7.66) reduce to

$$\begin{aligned} \mathbf{u}_1 &= \mathbf{u}_2 \\ \mathbf{T}_1 &= -\mathbf{T}_2 \\ T_1 &= T_2 \\ \mathbf{q}_1 &= \mathbf{q}_2 \end{aligned} \tag{7.68}$$

There may still be a discontinuity in other quantities such as stress and density. If both media are gases, however, interdiffusion eliminates even the density discontinuity. Consistent with the treatment of the boundary region as a discontinuity, we will not pursue these matters here; to do so is to inquire into the *structure* of the contact surface, somewhat analogous to the structure of a shock.

EXAMPLE 7.9 CONTACT SURFACE PRODUCED BY INTERSECTING SHOCKS

An initially uniform supersonic flow (field 1) encounters the duct geometry shown in Fig. 7.30. It is desired to find the resulting flow field.

Let the local angle of inclination of a streamline be θ. Then θ_2 and θ_3 are just the angles of the adjacent duct boundaries. Thus $\theta_2 - \theta_1$ and $\theta_3 - \theta_1$ are the known turning angles across the respective shocks, and conditions in fields 2 and 3 are given by the oblique-shock relations after choosing between the weak and strong solutions.

If the weak solutions are tenable, the flow in fields 2 and 3 will normally be supersonic and shocks will run downstream from the point of intersection *a*. Then conditions in fields 4 and 5 are determined by the matching conditions across the contact surface

$$P_4 = P_5 \qquad \theta_4 = \theta_5$$

corresponding to (7.67) and (7.65), respectively.

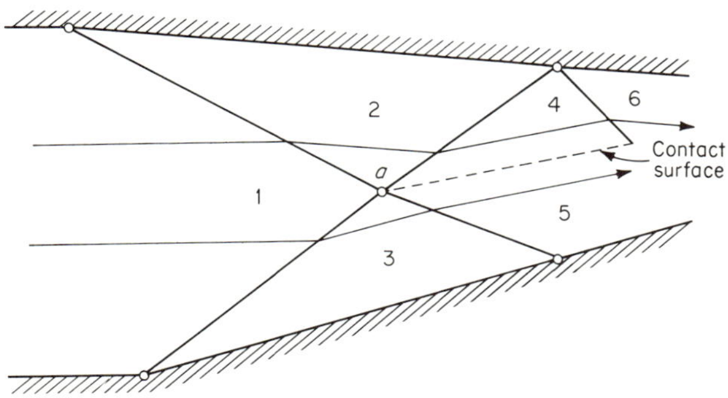

Figure 7.30
Shock intersections resulting from boundary geometry.

At the upper wall the oblique shock is shown reflected from the wall [compare Example 7.8 (page 333)]. The flow angle θ_6 is just the wall angle, and conditions in field 6 can then be found (it has been assumed in this discussion that the prescribed turning angle does not exceed the maximum possible turning angle for the given upstream Mach number; if this angle is exceeded, the situation is more complicated).

Numerical values have been worked out for the case of a perfect gas, $\gamma = 1.40$, $M_1 = 2.50$, $\theta_1 = 0$, $\theta_2 = -5°$, $\theta_3 = +15°$.

With the given initial Mach number and given turning angles (5 and 15°) the oblique-shock tables (with linear interpolation) give the values for fields 2 and 3 shown in the table below. Conditions in fields 4 and 5 can be determined by *trial* with the help of the matching conditions. The author took as a first guess $\theta_4 = \theta_5 = +10°$, giving $[\theta]_{2\to 4} = 15°$ and $[\theta]_{3\to 5} = -5°$. The resulting pressure ratios are, respectively, $P_4/P_2 = 2.343$ and $P_5/P_3 = 1.301$, giving

$$P_4 = 2.343(1.382P_1) = 3.24P_1$$

$$P_5 = 1.301(2.470P_1) = 3.21P_1$$

Improbably enough, the pressures were matched (very nearly) on the first trial. As a second trial, take $[\theta]_{2\to 4} = 14.9°$ and $[\theta]_{3\to 5} = -5.1°$, giving

$$P_4 = 2.332(1.382P_1) = 3.22P_1$$

$$P_5 = 1.308(2.470P_1) = 3.23P_1$$

so that this represents a solution to the nearest tenth of a degree. The resulting values are given in Table 7.1. With the Mach number M_4 now known and

Table 7.1

<table>
<tr><th>Field</th><th>θ, Degrees</th><th>P/P₁</th><th>M</th><th>Shock Angle, Degrees</th></tr>
<tr><td>1</td><td>0</td><td>1</td><td>2.500</td><td rowspan="6">$\beta_{12} = 27.44$
$\beta_{13} = 36.94$
$\beta_{24} = 39.8$
$\beta_{35} = 36.9$
$\beta_{46} = 56.0$</td></tr>
<tr><td>2</td><td>−5</td><td>1.382</td><td>2.292</td></tr>
<tr><td>3</td><td>15</td><td>2.470</td><td>1.873</td></tr>
<tr><td>4</td><td>9.9</td><td>3.22</td><td>1.696</td></tr>
<tr><td>5</td><td>9.9</td><td>3.23</td><td>1.707</td></tr>
<tr><td>6</td><td>−5</td><td>6.87</td><td>1.120</td></tr>
</table>

turning angle $[\theta]_{4\to6}$ prescribed, the reflected oblique shock is determined, giving the values shown for field 6.

The contact surface between fields 4 and 5 carries a very weak discontinuity in the tangential velocity. Because the stagnation enthalpy h_0 is the same everywhere in the flow, Eq. (5.58) is applicable and $c_0/c = \sqrt{T_0/T}$ can be found as a function of Mach number (it is most convenient just to use the values tabulated in the isentropic-flow tables for this function). Thus we find $c_4/c_0 = 0.7968$ and $c_5/c_0 = 0.7948$, giving

$$u_4 = M_4 c_4 = (1.696)(0.7968)c_0 = 1.351c_0$$

$$u_5 = M_5 c_5 = (1.707)(0.7948)c_0 = 1.357c_0$$

where c_0 is the stagnation sound speed.

Figure 7.31
Oblique-shock interactions in air, with formation of a contact surface. The upstream Mach number $M_\infty = 2.48$. Upper surface inclined at 16.1° to the free stream, lower surface at 15.5°. (*Courtesy of I. I. Glass, University of Toronto.*)

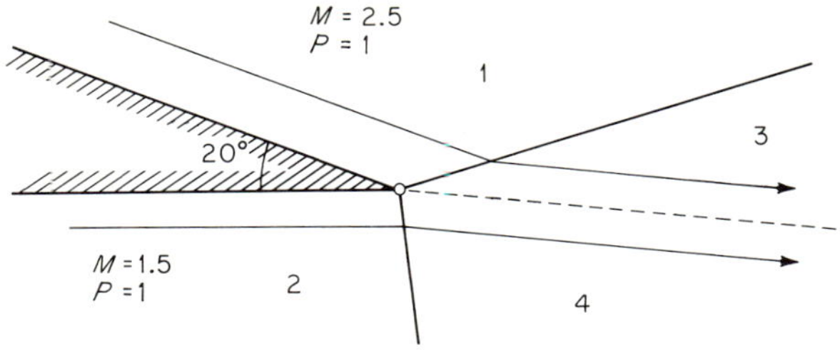

Figure 7.32
Confluence.

In general, two solutions that satisfy the matching conditions are possible, and the "weaker," i.e., lower-pressure, solution has been found above. We will illustrate both types of solution in the following example.

A photograph of oblique-shock intersection is shown in Fig. 7.31.

EXAMPLE 7.10 CONFLUENCE INTERACTION

Two streams come together as shown in Fig. 7.32. Assuming that both fluids are perfect gases with $\gamma = 1.40$, find the downstream flow.

The discontinuity in flow angle θ can, in this case, be resolved by two shocks, as shown in the figure. To illustrate the two possible conditions $\theta_3 = \theta_4$ and $P_3 = P_4$, the downstream P, θ states for both shocks have been plotted in Fig. 7.33, from the data in Table D.3 (Appendix D). These curves are called *shock polars in the pressure-flow angle plane.*

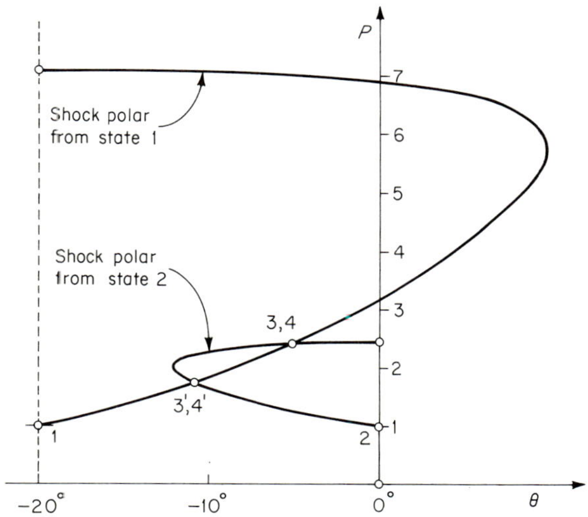

Figure 7.33
Intersecting shock polars, from upstream states 1 and 2.

The solutions are represented by the two points of intersection. For example, the stronger solution is, numerically,

$$\theta_3 = \theta_4 = -5.3^\circ \qquad P_3 = P_4 = 2.43$$

$$M_3 = 1.887 \qquad M_4 = 0.719$$

$$\beta_{1\to3} = 36.64^\circ \qquad \beta_{2\to4} = 83.59^\circ$$

Which of the two solutions actually occurs depends on the downstream conditions, as we have already noted.

In the case of the stronger solution given above, there is a strong discontinuity in the tangential velocity; we find

$$u_3 = 1.887(0.7645)c_{01} = 1.445c_{01}$$

$$u_4 = 0.719(0.9520)c_{02} = 0.684c_{02}$$

7.8 Detonation waves

The derivation of the shock conditions (7.16) has not precluded the possibility of chemical reaction or phase change within the shock. We consider here the case in which an exothermic chemical reaction or combustion takes place within the discontinuity; the incoming fluid (reactants) and outgoing fluid (products) are thus chemically different substances.

Relative to the complexity of the subject of combustion-type discontinuities, our treatment will be brief. A recent survey article, with further references, is given by *Edwards* [1969].

Example reactions are

$$H_2(gas) + \tfrac{1}{2}O_2(gas) \to H_2O(gas)$$

$$\text{TNT}(solid) \to \text{gaseous products}$$

The reactants are usually far from thermodynamic stability, and the reaction may go almost to completion. A *detonation wave* is thus characterized by an exothermic reaction originating in reactants which are not in stable equilibrium. The wave is physically pictured as made up of two parts, an ordinary shock, which raises the reactants to high temperature and pressure, followed by a somewhat thicker reaction zone, in which the chemical reaction goes more or less to completion (see Fig. 7.34). The overall thickness of the detonation front may be typically of the order of 1 cm.

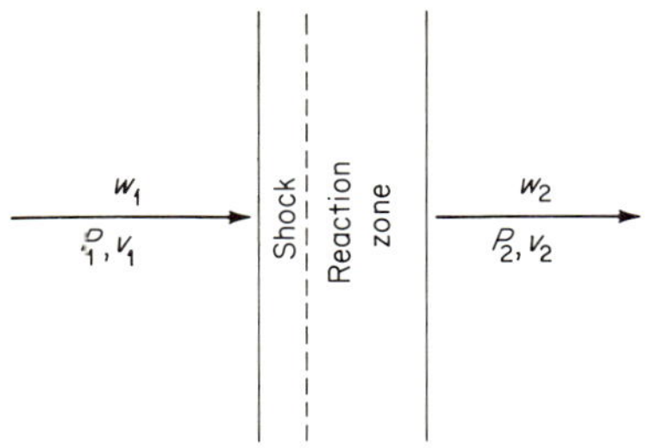

Figure 7.34

The equations already developed in Secs. 7.2 and 7.3 are applicable. In particular, use will be made of the following:

$$[h + \tfrac{1}{2}w^2] = 0 \tag{7.16}$$

$$J^2 = -\frac{[P]}{[v]} \tag{7.18}$$

$$h_2 - h_1 = \tfrac{1}{2}(P_2 - P_1)(v_2 + v_1) \tag{7.23}$$

We again make use of the Hugoniot diagram, Fig. 7.35. The dotted curve is the shock adiabat for the chemically unchanged reactants, i.e., a shock adiabat in the ordinary sense. The solid curve, of principal interest, is the *detonation adiabat* representing all possible end states for the combustion process. Unlike the shock adiabat, it does not pass through the

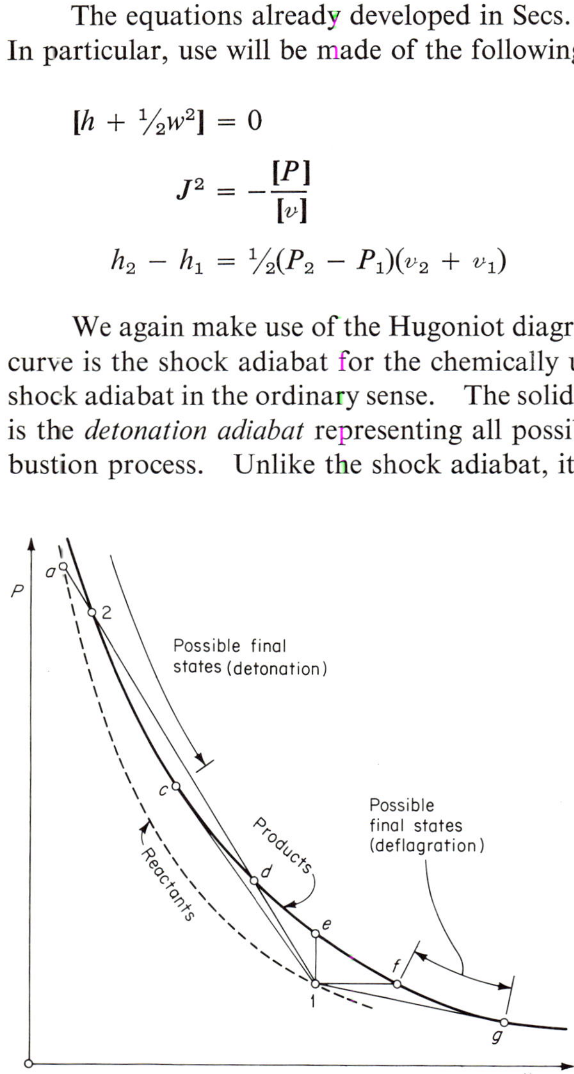

Figure 7.35
Detonation adiabat.

point 1 since $h_2(P_2,v_2)$ is a different *function* than $h_1(P_1,v_1)$, because of the chemical difference between the two substances (in the simplest case, both products and reactants are perfect gases with $h = [\gamma/(\gamma - 1)]Pv +$ const, but the constants are not the same, differing by the heat of reaction). A possible detonation process is represented by the chord $1 \rightarrow a \rightarrow 2$. The process $1 \rightarrow a$ is an ordinary shock in the reactants and is followed by expansion, during chemical reaction, along $a \rightarrow 2$. If this shock-reaction sequence is the correct one, we see that point d is an impossible final state, as it would require a rarefaction shock along $2 \rightarrow d$. Thus no shock-reaction sequence is possible with an end point below point c, the *Chapman-Jouguet point*. Processes $1 \rightarrow c$ are called Chapman-Jouguet detonations and are the processes which most often occur physically (this case will be treated in detail later).

We now discuss some of the general properties of the detonation.[1] The upstream *thermodynamic* quantities h_1, P_1, v_1 are regarded as fixed, and possible downstream states are on the detonation adiabat *above* point c. As the slope of the chord $1 \rightarrow 2$ varies, the mass flux varies according to (7.18) and w_1 must be regarded as variable.

A tangent to the dotted line at point 1 has slope $(\partial P/\partial v)_s = -\rho_1{}^2c_1{}^2$, just as in the case of a shock. The slope of the chord $1 \rightarrow 2$ is $[P]/[v] = -J^2$ from (7.18): thus, from the figure (assuming that the shock and detonation adiabats do not cross and are concave upward),

$$-\rho_1{}^2c_1{}^2 > -J^2 = -\rho_1{}^2w_1{}^2$$

or

$$w_1 > c_1 \tag{7.69}$$

The *detonation velocity w_1 is always supersonic.*[2]

From (7.18), considering differentiation to correspond to infinitesimal changes along the detonation adiabat,

$$dP_2 + J^2\,dv_2 = (v_1 - v_2)\,dJ^2 \tag{7.70}$$

The energy-conservation condition (7.16) is rewritten as

$$h_1 + \tfrac{1}{2}J^2v_1{}^2 = h_2 + \tfrac{1}{2}J^2v_2{}^2$$

[1] This treatment follows closely that given in *Landau and Lifshitz* [1959, chap. 14].

[2] Since the detonation wave normally propagates into a stationary medium, it is conventional to call $w_1 = |b|$ the detonation velocity.

and differentiated

$$dh_2 + J^2 v_2 \, dv_2 = \tfrac{1}{2}(v_1{}^2 - v_2{}^2) \, dJ^2$$

From the Gibbs equation,

$$dh_2 = T_2 \, ds_2 + v_2 \, dP_2$$

Substituting this into the above and using (7.70) gives

$$T_2 \, ds_2 = \tfrac{1}{2}(v_1 - v_2)^2 \, dJ^2$$

Thus, since $(v_1 - v_2)^2$ and T_2 are necessarily positive,

$$\frac{ds_2}{dJ^2} > 0 \tag{7.71}$$

From this we can draw some conclusions about the Chapman-Jouguet point c. Since J^2 increases in either direction from c, so also does the entropy s_2: *the Chapman-Jouguet point is a point of minimum entropy for the detonation products.* Furthermore, the entropy is necessarily stationary in the vicinity of c since, with $dJ/dv_2 = 0$ at this point,

$$\left(\frac{ds_2}{dv_2}\right)_c = \left(\frac{ds_2}{dJ}\right)_c \left(\frac{dJ}{dv_2}\right)_c = 0$$

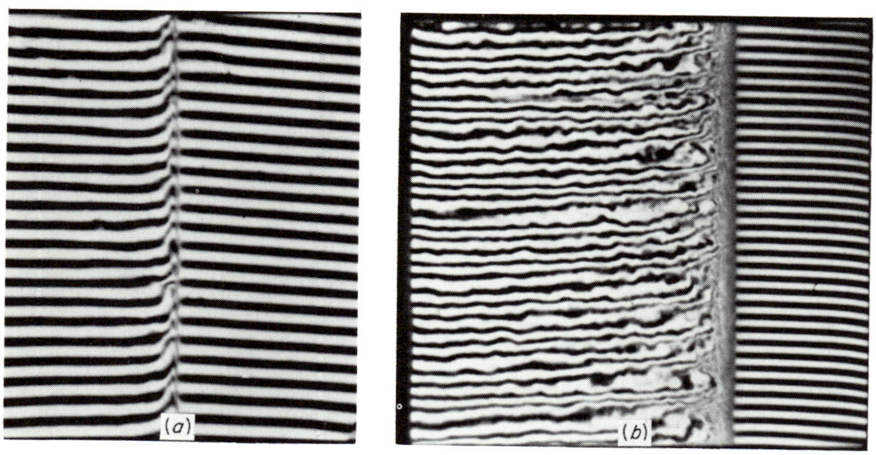

Figure 7.36
Interferograms of detonations in $2H_2 + O_2 + 2CO$ mixtures: (*a*) laminar detonation, $P_1 = 0.0263$ atm, detonation velocity 2.7 km/s; (*b*) turbulent detonation, $P_1 = 0.300$ atm, detonation velocity 2.2 km/s. Flow is from right to left. (*Courtesy of D. R. White, General Electric Co.*)

A tangent at c thus has slope $(\partial P_2/\partial v_2)_s = -\rho_2{}^2c_2{}^2 = -J^2$. Thus

$$J^2 = +\rho_2{}^2w_2{}^2 = +\rho_2{}^2c_2{}^2$$

or

$$w_2 = c_2 \tag{7.72}$$

In a Chapman-Jouguet detonation, the normal velocity of the products with respect to the detonation front is exactly sonic.

Similar arguments show that the outflow for points above c is subsonic and for points below c is supersonic.[1]

An interferogram of detonation in a gas is shown in Fig. 7.36.

Other Combustion Modes

Every point on the detonation adiabat represents a possible end state for a combustion (or decomposition, etc.) process which is without external friction and is adiabatic, and these conditions are fulfilled reasonably well in a number of combustion situations. By convention, detonation refers to points above e; above c the process is called strong detonation, and between c and e weak detonation, although the latter process has never been observed. In all detonation processes the incoming flow is supersonic. There is no physical process corresponding to points between e and f because the mass flux is imaginary. Point g is another Chapman-Jouguet point. Between f and g are weak deflagrations with incoming flow subsonic and outgoing flow subsonic. Below g are strong deflagration solutions, but these are not realizable in practice; this case may be shown to be unstable. It should be noted that in the case of subsonic inflow (deflagration, below f) the speed with which a flame can propagate is controlled by heat conduction and mass diffusion, in other words, by the transport and thermodynamic properties of the substance. The various combustion modes of physical interest are summarized below

Strong detonation, above c:	$w_1 > c_1$	$w_2 < c_2$
Chapman-Jouguet detonation, point c:	$w_1 > c_1$	$w_2 = c_2$
Weak deflagration, between f and g (slow burning):	$w_1 < c_1$	$w_2 < c_2$

[1] *Some* process with end state below c is conceivable, even if the shock-reaction sequence is not.

Of course, many combustion processes *cannot* be treated as normal discontinuities. For example, turbulent flames and the flame in a tube of small diameter fall in this category. While such phenomena are of great practical interest, they will not be discussed here.

In cases where a combustion front propagates as an oblique discontinuity, the normal conditions may be applied to the normal component of the flow, just as in the case of an oblique shock.

Chapman-Jouguet Detonation

This is the detonation usually encountered in practice. It can be shown, for example, that combustion initiated at the closed end of a rigid tube should result in formation of a Chapman-Jouguet wave (*Landau and Lifshitz* [1959, chap. 14]). As we have seen, the entropy production is a minimum for this case; it has been suggested that there is a *principle of minimum entropy production* applicable to irreversible processes.

Since $v_2 < v_1$, we have

$$c_2 = w_2 < w_1 \tag{7.73}$$

and from before

$$w_1 > c_1$$

Except for the sonic (or supersonic, in the case of a strong detonation) outflow, the direction of change of the flow variables across a detonation is the same as for a normal shock. In particular, the detonation products move toward an observer who is stationary with respect to the reactants; i.e., the products chase the detonation front.

Chapman-Jouguet Detonation in a Perfect Gas

In the case where products and reactants can be treated as perfect gases and the Chapman-Jouguet condition holds, especially simple results are obtained. We take account of the heat of reaction by writing

$$h_1 = h^{01} + c_{p1}T_1$$

$$h_2 = h^{02} + c_{p2}T_2$$

$$h^{01} - h^{02} = \Delta h^0$$

where Δh^0 is the heat of reaction (a positive number for exothermic reactions) extrapolated to absolute zero temperature. While c_{p1} and c_{p2} are constants, they will in general have different values.

For a perfect gas, the first three jump conditions (7.16) can now be written, in respective order,

$$\frac{\gamma_2 P_2}{v_2} M_{2n}{}^2 = \frac{\gamma_1 P_1}{v_1} M_{1n}{}^2$$

$$P_2(1 + \gamma_2 M_{2n}{}^2) = P_1(1 + \gamma_1 M_{1n}{}^2)$$

$$\frac{\gamma_2}{\gamma_2 - 1} P_2 v_2\left(1 + \frac{\gamma_2 - 1}{2} M_{2n}{}^2\right) = \frac{\gamma_1}{\gamma_1 - 1} P_1 v_1\left(1 + \frac{\gamma_1 - 1}{2} M_{1n}{}^2\right) + \Delta h^0$$

Setting $M_{2n} = 1$ (the Chapman-Jouguet condition) and solving for M_{1n} yields

$$M_{1n} = \sqrt{\mathscr{H} + \frac{(\gamma_1 + \gamma_2)(\gamma_2 - 1)}{2\gamma_1(\gamma_1 - 1)}} \pm \sqrt{\mathscr{H} + \frac{(\gamma_2 - \gamma_1)(\gamma_2 + 1)}{2\gamma_1(\gamma_1 - 1)}} \tag{7.74}$$

where the nondimensional heat of reaction $\mathscr{H}$,

$$\mathscr{H} \equiv \frac{(\gamma_2 - 1)(\gamma_2 + 1)\,\Delta h^0}{2\gamma_1 R_1 T_1} \tag{7.75}$$

is of order 10 in a typical case.

The greater value ($M_{1n} > 1$) corresponds to the Jouguet point c for *detonation*, and the lesser value ($M_{1n} < 1$) corresponds to the Jouguet point g for *deflagration*.[1] If, as a further approximation, we take $\gamma_1 = \gamma_2$, (7.74) reduces to

$$M_{1n} = \sqrt{\mathscr{H} + 1} \pm \sqrt{\mathscr{H}} \tag{7.76}$$

Note that if $\mathscr{H} \to 0$, then $M_{1n} \to 1$, corresponding to a shock of vanishing strength; this behavior is consistent, because we have already assumed $M_{2n} \equiv 1$.

In terms of M_{1n}, the pressure ratio can be obtained directly from $[P + \rho w^2] = 0$,

$$\frac{P_2}{P_1} = \frac{\gamma_1 M_{1n}{}^2 + 1}{\gamma_2 + 1} \tag{7.77}$$

[1] Chapman-Jouguet *deflagrations* are almost never observed because (subsonic) deflagration velocities are determined by transport properties such as thermal conductivity and by chemical kinetics.

The density ratio can be found from the condition that the outflow is sonic, $w_2 = \sqrt{\gamma_2 P_2/\rho_2}$. This, combined with the continuity equation (7.2) and Eq. (7.17), gives

$$\frac{\rho_2}{\rho_1} = \frac{(\gamma_2 + 1)P_2/P_1 - 1}{\gamma_2 P_2/P_1} \tag{7.78}$$

With (7.77) this becomes

$$\frac{\rho_2}{\rho_1} = \frac{\gamma_1(\gamma_2 + 1)M_{1n}^2}{\gamma_2(1 + \gamma_1 M_{1n}^2)} \tag{7.79}$$

The upstream and downstream gas constants are in the ratio of the number of moles (found from the chemical reaction), $R_2/R_1 = \mathcal{N}_2/\mathcal{N}_1$. The temperature ratio is then

$$\frac{T_2}{T_1} = \frac{P_2}{P_1}\frac{\rho_1}{\rho_2}\frac{\mathcal{N}_1}{\mathcal{N}_2} \tag{7.80}$$

Chapman-Jouguet Detonation in a Solid or Liquid Explosive

Because the temperatures and pressures generated by a condensed explosive are very great indeed, it is very difficult to measure downstream properties and to determine equations of state for the products. It is not certain, furthermore, that the Chapman-Jouguet condition is satisfied in every case, though it is usually assumed. The one quantity which can be accurately measured is the detonation velocity w_1. Some reported values, at various loading densities ρ_1, are given in Table 7.2. In the (rather unrealistic)

Table 7.2†

Explosive	ρ_1, g/cm³	P_2, kbar	w_1, m/s
TNT (solid)	1.630	225	6,948
TNT (solid)	1.051	115	5,188
TNT (liquid)	1.450	182	6,590
Nitromethane (liquid)	1.159	148	6,378
PETN	1.670	310	7,975
9404	1.844	390	8,811
Composition B	1.714	293	7,990
Dynamite	0.54–1.3		1,200–7,000

† Data from *Davis et al.* [1965]. Dynamite data from *Anon.* [1966].

event that the detonation products can be treated as perfect gases, all the downstream properties may be simply expressed in terms of the detonation velocity w_1 with the help of the Chapman-Jouguet conditions. From (7.78) with $P_2/P_1 \to \infty$,

$$\frac{\rho_2}{\rho_1} = \frac{\gamma_2 + 1}{\gamma_2} \tag{7.81}$$

and the momentum and continuity conditions in (7.16) give

$$P_2 = \frac{1}{\gamma_2 + 1} \rho_1 {w_1}^2 \tag{7.82}$$

Note that this differs from the result (7.61) for very strong shocks. For the values listed in Table 7.2, the required values of the adiabatic exponent γ_2 for consistency with Eq. (7.82) are in the range of 2 to 3, in contradiction of equilibrium thermodynamics.

Weak Deflagration, or Slow Combustion

It is a matter of everyday experience that "ordinary" flames have very nearly constant pressure and low gas speeds. To treat such phenomena as normal adiabatic discontinuities is an approximation: it neglects three-dimensional effects, radiation from the flame, etc. We will give such a treatment, however, with the understanding that the results are indicative and appropriate only to certain large-scale combustion processes and certain laminar flames (see Fig. 7.37).

The processes of interest correspond to the neighborhood of point f, where $P_2 \approx P_1$ and J^2 is very small. Because the flow velocities are small (of the order of a few meters per second), the balance conditions (7.16) reduce to

$$\begin{aligned} P_1 &\approx P_2 \\ h_1 &\approx h_2 \end{aligned} \tag{7.83}$$

The densities, however, are not the same. Using the same notation for enthalpy used previously, and again assuming calorically and thermally perfect gases,

$$\begin{gathered} c_{p1}T_1 + h^{01} = c_{p2}T_2 + h^{02} \\ T_2 = \frac{c_{p1}}{c_{p2}} T_1 + \frac{\Delta h^0}{c_{p2}} \end{gathered} \tag{7.84}$$

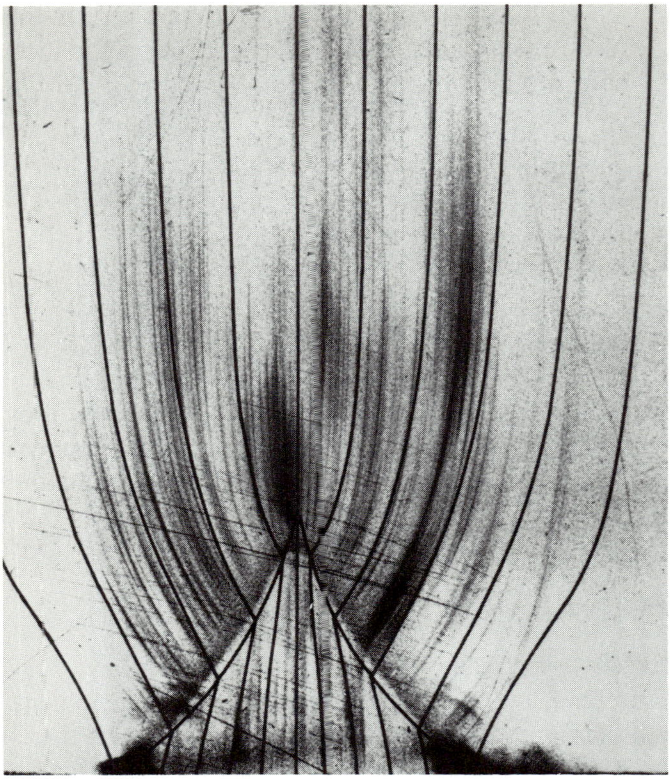

Figure 7.37
Bunsen burner flame, flow-field streamlines computed with discontinuous flame-front model, superimposed on experimental streaklines. (*Courtesy of M. S. Uberoi, University of Colorado.*)

and since pressure is constant

$$\frac{\rho_2}{\rho_1} = \frac{T_1 \mathcal{N}_1}{T_2 \mathcal{N}_2}$$

EXAMPLE 7.11 DETONATION IN A PERFECT GAS

Estimate the speed with which a detonation wave travels into a stationary $5N_2 + 2H_2 + O_2$ mixture at $P_1 = 1$ atm and $T_1 = 291$ K. Find also the temperature and pressure behind the wave.

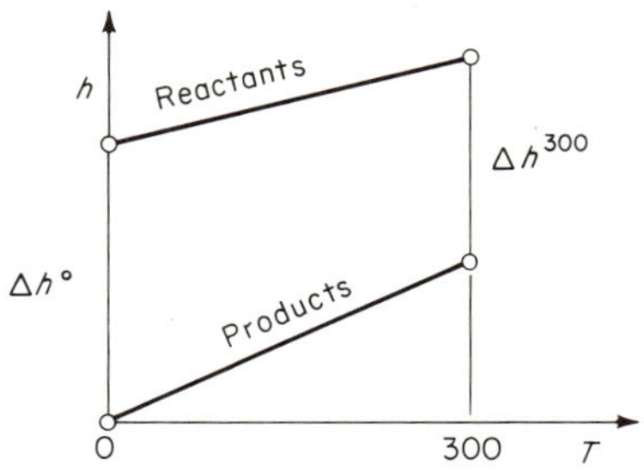

Figure 7.38

This is a diluted stoichiometric mixture of hydrogen and oxygen

$$5N_2 + 2H_2 + O_2 \rightarrow 2H_2O + 5N_2$$

$$m = 140 + 4 + 32 = 176 \text{ g mixture}$$

$$\mathcal{N}_1 = 8 \text{ mol reactants} \qquad \mathcal{N}_2 = 7 \text{ mol products}$$

$$\tilde{M}_1 = {}^{176}\!/_8 = 22 \qquad \tilde{M}_2 = {}^{176}\!/_7 = 25.1$$

The standard heat of reaction at 300 K is

$$\Delta H^{300} = 57.1 \text{ kcal/g mol } H_2O$$

and this is to be converted to Δh^0, which is on a mass (rather than mole) basis. We assume

$$\gamma_1 = {}^7\!/_5 = 1.400$$

$$\gamma_2 = {}^9\!/_7 = 1.286$$

i.e., as an approximation the classical equipartition values for *diatomic* gases at low and high temperatures, respectively, have been taken. Then (see Fig. 7.38)

$$\Delta h^0 + c_{p1}T_{300} = c_{p2}T_{300} + \Delta h^{300}$$

$$\Delta h^0 = \Delta h^{300} + (c_{p2} - c_{p1})T_{300}$$

$$= \frac{2\Delta H^{300}}{m} + \left(\frac{\gamma_2}{\gamma_2 - 1}\frac{\tilde{R}}{\tilde{M}_2} - \frac{\gamma_1}{\gamma_1 - 1}\frac{\tilde{R}}{\tilde{M}_1}\right)T_{300}$$

$$= \frac{2\Delta H^{300}}{m} + \frac{\tilde{R}T_{300}}{m}\left(\frac{\gamma_2}{\gamma_2 - 1}\mathcal{N}_2 - \frac{\gamma_1}{\gamma_1 - 1}\mathcal{N}_1\right)$$

and with $\tilde{R} = 1.987 \times 10^{-3}$ kcal/(g mol)(K), this gives

$$\Delta h^0 = \frac{114.2 + 2.1}{m} = \frac{116.3}{m} \qquad \text{kcal/g}$$

so that the dimensionless heat of reaction is

$$\mathscr{H} = \frac{(\gamma_2 - 1)(\gamma_2 + 1)}{2\gamma_1 R_1 T_1}\,\Delta h^0 = \frac{\dfrac{32}{49}\,\dfrac{116.3}{m}}{\dfrac{14}{5}\,\dfrac{\tilde{R}\mathscr{N}_1}{m}\,291} = 5.86$$

The initial sound speed is

$$c_1 = \sqrt{\frac{1.4(8{,}317)(291)}{22}} = 392 \text{ m/s}$$

Substituting the value of $\mathscr{H}$ into (7.74), we find for the *detonation Mach number*

$$M_{1n} = \sqrt{5.86 + 0.69} + \sqrt{5.86 - 0.23} = 4.93$$

so that the detonation speed is

$$w_1 = 4.93(392) = 1{,}930 \text{ m/s}$$

The pressure is obtained from (7.77)

$$P_2 = 15.3 \text{ atm}$$

and the temperature from (7.79) and (7.80)

$$T_2 = 2920 \text{ K}$$

These results can be compared with experimental results and an "exact" calculation which accounts for variations in specific heat with temperature *and* the dissociation of H_2O into H^+ and OH^- (thus effectively lowering the heat of reaction).

	u_1, m/s	P_2, atm	T_2, K
This calculation	1,930	15.3	2,920
"Exact" theory	1,850	14.4	2,685
Experiment	1,822		

Details of the numerical "exact" calculation may be found in *Taylor and Tankin* [1958].

The *deflagration* velocity at the other Jouguet point (maximum velocity for slow combustion) is approximately

$$M_{1n} = 0.19 \rightarrow u_1 = 74 \text{ m/s}$$

from (7.74).

7.9 Condensation discontinuities

During the acceleration of a gas through a supersonic nozzle, the entropy is nearly constant while pressure and temperature decrease. Even for modest exit Mach numbers, the pressure and temperature may be very low; e.g., for isentropic flow of air from a reservoir where $P_0 = 1$ atm, $T_0 = 300$ K to a point where $M = 3$, we find the local values

$$P = 0.027 \text{ atm}$$

$$T = 107 \text{ K}$$

Apparently, if the expansion process is continued to sufficiently high Mach numbers, the gas (perhaps more appropriately called a *vapor* here) eventually reaches *saturation*, and further expansion would result in condensation if equilibrium obtained. In the above example the oxygen component of air reaches saturation at about 50 K, corresponding to a Mach number of about 5.

If the air flowing through a nozzle is moist, the water-vapor component will reach saturation conditions long before air liquefaction becomes a problem. The gases in supersonic wind tunnels are in fact carefully dried in order to prevent moisture condensation.

Another case of interest is that of the *steam nozzle*, involving only one component. We will discuss this case, for the qualitative features are typical (see Fig. 7.39). Here the flowing vapor reaches saturation temperature and pressure at point s. Condensation does not, however, begin at this point; the flow continues as supersaturated vapor until point c, where condensation appears to be suddenly initiated. The distance $x_c - x_s$ corresponds to the time required for spontaneous nucleation to take place; this distance is relatively large because the nucleation is a "slow" nonequilibrium process.[1]

Once condensation processes are initiated at c, condensation proceeds fairly rapidly, so that equilibrium is attained at d resulting in the "condensation shock" cd. This discontinuity is, however, fairly thick compared to a normal shock.

[1] Strictly, the saturation vapor pressure depends on the liquid-drop size. The reluctance of a vapor to condense, even when the pressure is well below the saturation pressure, is the basis for the well-known *Wilson cloud chamber*. If a closed volume of moisture-saturated but dust-free air is adiabatically suddenly expanded as much as 25 per cent in volume (corresponding to a path like $s \rightarrow c$), no condensation is observed for some time. If an extremely energetic particle, as from cosmic radiation, enters the chamber, this particle "knocks some sense into the heads of the water molecules," and the well-known droplet track is formed. The more modern and sensitive device, the hydrogen bubble chamber, operates on nonequilibrium evaporation instead of condensation.

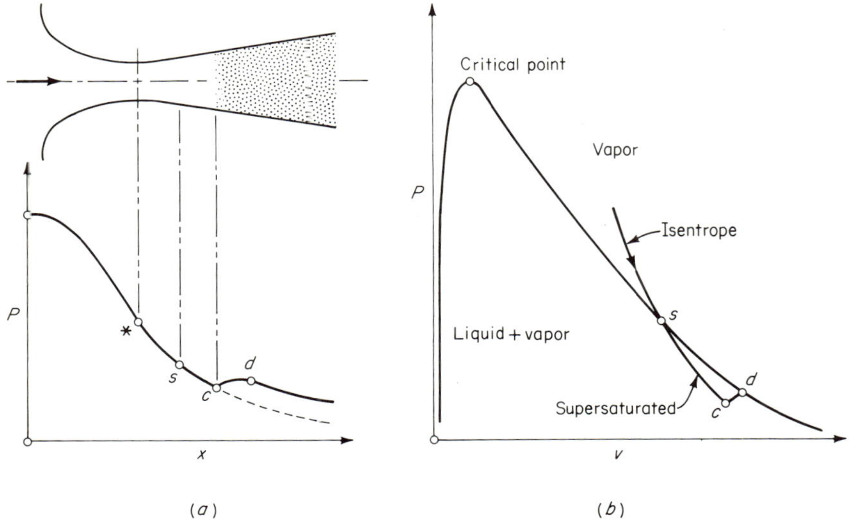

Figure 7.39
Condensation shock: (*a*) nozzle and pressure distribution; (*b*) pressure-volume diagram.

It should be noted that this phenomenon is quite distinct from an ordinary shock wave. The pressure rise is, in general, quite modest, and in the above example the downstream flow is still supersonic.

There is close similarity between this *condensation discontinuity* and the *combustion waves* considered in the preceding section. In each case there is a release of energy—from "burning" or condensation—and in each case the "reaction" is exothermic. Then the discussion for the Hugoniot diagram with combustion is applicable also to this case. There is, however, no Chapman-Jouguet condition for condensation.

The jump conditions can be computed with relative algebraic simplicity for the following special but practically important case: air (or other gas) flows with small concentration of water vapor (or other vapor) with specific humidity ω_1, the mass of water vapor per unit mass of gas. The condensate (which may be ice) forms from an upstream condition 1 of considerable supersaturation, and almost all the vapor condenses. Then, approximately,

$$h_1 - h_2 = c_{pa}(T_1 - T_2) + \omega_1 L \tag{7.85}$$

where L is the (positive) latent heat of evaporation. This is formally identical to the enthalpy difference in combustion, with $\omega_1 L = \Delta h^0$ and the added simplification that $\gamma_1 = \gamma_2$, $R_1 = R_2$. The resulting jump conditions may be found in the article by *Stever* [1958].

7.10 Continuum shock structure

The thickness of shock waves and the detailed variation of properties across them, i.e., their structure, are of theoretical and philosophical interest. An elementary discussion of this rather involved problem is offered here.

As has been suggested in the treatment of shock conditions, the thickness of shock waves may be neglected for most practical purposes. Yet if the shock were a true discontinuity, the velocity and temperature gradients would necessarily be infinite, suggesting infinite viscous stress and heat flux; this paradox is avoided by the shock, or *shock layer*, of finite thickness. In this connection the shock interior and the boundary layer are analogous, both being thin regions of large viscous stress and high heat flux bounded by extensive regions which may (usually) be considered inviscid. In both cases the Reynolds number based on thickness, i.e., the thickness of the shock or of the boundary layer, is small.

Continuum theory, Boltzmann kinetic theory, and experiments all lead to shock thicknesses amounting to a few mean free paths in the downstream gas, for shocks of moderate strength.[1] This result suggests difficulties for a continuum model of shock structure. Nevertheless, the Navier-Stokes equations predict structures which agree surprisingly well with experiment, particularly for monatomic gases, i.e., without relaxation effects, up to moderate shock Mach numbers. For weak shocks, $M_{1n} - 1 \ll 1$, the shock thickness is, as will be shown, quite large, and there is no difficulty in justifying the continuum theory.

[1] A common order-of-magnitude argument gives a somewhat different result: if the viscous and acceleration terms in the one-dimensional momentum equation are taken to be of the same order,

$$\rho u \frac{du}{dx} \sim \mu \frac{d^2u}{dx^2}$$

If the shock thickness is represented by Δ, the derivatives are estimated as $du/dx \approx [u]/\Delta$ and $d^2u/dx^2 \approx [u]/\Delta^2$ and substitution gives, with $\mu \approx \rho c \Lambda$ and $u/c = M \sim 1$,

$$\Delta \sim \Lambda$$

which underestimates the shock thickness by roughly one order of magnitude. This argument neglects the fact that d^2u/dx^2 necessarily *changes sign* across the shock. A more nearly correct estimate for shock thickness is given in the following pages.

Continuum Equations for a Stationary Normal Shock

Let the direction normal to the shock front be the x direction. Assuming that the shock thickness Δ is small compared to the radius of curvature R of the shock front (a condition usually fulfilled in practice), the flow is one-dimensional. Assuming the shock to be stationary means that the *structure* has become stationary for some Galilean observer and that $\partial/\partial t$ terms may be omitted. In addition, the fluid is treated as though it were in equilibrium (more precisely, near enough to equilibrium so that the departure is adequately treated by the linear transport relations), a somewhat bold assumption in the circumstances. Thermal radiation and diffusion are neglected. With this rather formidable list of assumptions, the equations are the ordinary continuum statements for balance of mass, momentum (Navier-Stokes), and energy, respectively,

$$(\rho u)_x = 0 \tag{7.86}$$

$$\rho u u_x + P_x - (\tfrac{4}{3}\mu' u_x)_x = 0 \tag{7.87}$$

$$\rho u\left(h + \frac{u^2}{2}\right)_x - (\tfrac{4}{3}\mu' u u_x)_x - (\kappa T_x)_x = 0 \tag{7.88}$$

where for conciseness d/dx is indicated by subscript x and μ' is an abbreviation for the full viscosity coefficient, that is, $\tfrac{4}{3}\mu' \equiv \tfrac{4}{3}\mu + \mu_v$. These equations can immediately be integrated with respect to x, giving

$$J = \rho u \tag{7.89}$$

$$K = P + \rho u^2 - \tfrac{4}{3}\mu' u_x \tag{7.90}$$

$$L = h + \frac{u^2}{2} - \frac{4}{3}\frac{\mu' u_x}{\rho} - \frac{\kappa T_x}{\rho u} \tag{7.91}$$

where J, K, and L are the *constants* for mass flux, momentum flux, and specific energy, respectively. Note that the transport coefficients μ' and κ have not been assumed constant. With flow to the right, the upstream flow condition 1 is at $x = -\infty$, and the downstream flow condition 2 at $x = +\infty$, giving the uniform boundary conditions

$$\left.\begin{aligned} u &= u_1 \\ P &= P_1 \\ \mu' &= \mu_1' \\ &\cdots\cdots \end{aligned}\right\} x = -\infty \qquad \left.\begin{aligned} u &= u_2 \\ P &= P_2 \\ \mu' &= \mu_2' \\ &\cdots\cdots \end{aligned}\right\} x = +\infty \tag{7.92}$$

With the gradients, such as u_x, vanishing at $x = \pm\infty$, note that (7.89) to (7.91) are just the *shock conditions* (7.16).

Weak-shock Thickness Estimate

The momentum integral (7.90) contains only one derivative, u_x. This forms the basis for a simple thickness estimate. Solving for the derivative term gives, with $K = P_1 + \rho_1 u_1^2$,

$$-\tfrac{4}{3}\mu' u_x = P_1 - P + \rho_1 u_1(u_1 - u) \tag{7.93}$$

We evaluate this derivative at the midpoint of the velocity distribution, where $u = u_1 + [u]/2$, and define a shock thickness Δ_m by the slope u_x at this point, $\Delta_m \equiv [u]/u_x$, as indicated in Fig. 7.40 (note, however, that u_x is not necessarily an extremum at this point). Then the only unknown on the right-hand side of (7.93) is the pressure P, which we now proceed to calculate.

From continuity $u = J\mathscr{v}$, and this gives at the midpoint

$$\mathscr{v} = \mathscr{v}_1 + \frac{[u]}{2J} \tag{7.94}$$

(note that $[u]$ is interchangeable with $[w]$ in the shock conditions). Then a Taylor expansion of $P(\mathscr{v},s)$ gives

$$P - P_1 = \left(\frac{\partial P}{\partial \mathscr{v}}\right)_s (\mathscr{v} - \mathscr{v}_1) + \frac{1}{2}\left(\frac{\partial^2 P}{\partial \mathscr{v}^2}\right)_s (\mathscr{v} - \mathscr{v}_1)^2 + \cdots$$

$$+ \left(\frac{\partial P}{\partial s}\right)(s - s_1) + \cdots \tag{7.95}$$

If terms only through second order in $\mathscr{v} - \mathscr{v}_1$ are retained (only for economy of calculation), the estimate will be restricted to weak shocks.

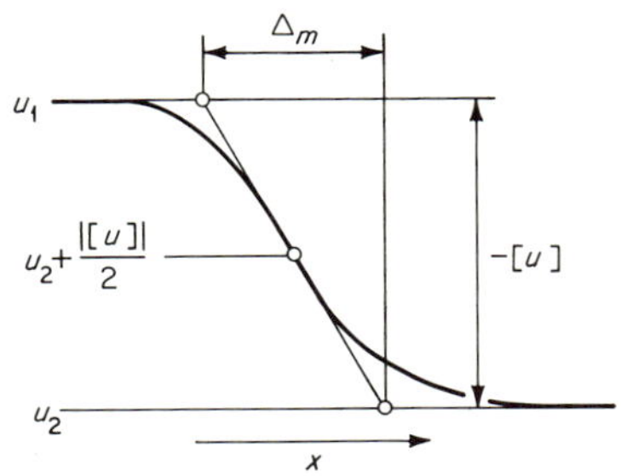

Figure 7.40

The term in $s - s_1$ may then be dropped, as it is at least of order $(v - v_1)^3$. Then the Taylor expansion gives, with the substitution of (7.94) and $J = \rho_1 u_1$,

$$P - P_1 = -\rho_1 c_1^2 \frac{[u]}{2u_1} + \rho_1 c_1^2 \Gamma_1 \frac{[u]^2}{4u_1^2} \cdots \tag{7.96}$$

where $\Gamma \equiv \dfrac{c^4}{2v^3}\left(\dfrac{\partial^2 v}{\partial P^2}\right)_s$ as defined in Eq. (5.8). Substituting this in (7.93) gives

$$-\frac{8}{3}\frac{\mu'}{\rho_1 c_1 \Delta_m} = \frac{1}{M_{1n}} - M_{1n} - \frac{\Gamma_1 [u]}{2M_{1n}^2 c_1} \cdots \tag{7.97}$$

where $M_{1n} \equiv u_1/c_1$ is the shock Mach number. From the weak-shock condition (7.50), $-[u]/c_1 = 2(M_{1n} - 1)/\Gamma_1 \cdots$, and substituting this gives finally

$$\frac{8}{3}\frac{\mu'}{\rho_1 c_1 \Delta_m} \approx M_{1n} - 1 \tag{7.98}$$

which provides the desired estimate[1] for the shock thickness Δ_m.

For *dilute gases* we have roughly $\mu' \approx \rho c \Lambda \approx \rho_1 c_1 \Lambda_1$, which gives

$$\frac{\Delta_m}{\Lambda} \approx \frac{8}{3(M_{1n} - 1)} \tag{7.99}$$

A remarkable verification of this formula is given by Taylor's solution for shock structure, as shown below.

Taylor's Solution for a Weak Shock

For a *perfect gas* with $h = c_p T + \text{const}$ and $P = R\rho T$, Eqs. (7.89) to (7.91) reduce to a pair of equations in u and T and the transport coefficients (which have a known dependence on T)

$$\frac{4}{3}\frac{\mu' u_x}{J} = \frac{RT}{u} + u - \frac{K}{J} \tag{7.100}$$

$$\frac{\kappa T_x}{J} = \frac{RT}{\gamma - 1} - \frac{u^2}{2} - \frac{K}{J} u - L \tag{7.101}$$

[1] An estimate for shock thickness which is not confined to weak shocks is obtained by equating the entropy generation via dissipation to the (known) entropy jump $[s]$ (see Prob. 7.17).

Eliminating T gives the single differential equation in u (and the transport coefficients)

$$-\frac{4\kappa}{3RJ^2}(\mu' u u_x)_x + \left[\frac{4}{3}\frac{\mu' u}{(\gamma - 1)J} + \frac{\kappa}{RJ}\left(2u - \frac{K}{J}\right)\right]u_x$$

$$= \frac{\gamma + 1}{2(\gamma - 1)}u^2 - \gamma\frac{K}{J}u + L \quad (7.102)$$

All gradients vanish at the upstream and downstream conditions; hence, the left-hand side (and with it the quadratic on the right side) must vanish when $u = u_1$ and when $u = u_2$. The quadratic then necessarily has the *roots* u_1 and u_2, giving the final form

$$-\frac{4\kappa}{3RJ^2}(\mu' u u_x)_x + \left[\frac{4}{3}\frac{\mu' u}{(\gamma - 1)J} + \frac{\kappa}{RJ}\left(2u - \frac{K}{J}\right)\right]u_x$$

$$= -\frac{\gamma + 1}{2(\gamma - 1)}(u_1 - u)(u - u_2) \quad (7.103)$$

This equation is of course not restricted to weak shocks.

The solution of G. I. Taylor (1910) is based on neglecting the first term in (7.103). The ratio of magnitudes of the first and second terms is, with $u \sim c$,

$$\frac{(\mu' u u_x)_x/J}{u u_x} \sim \frac{\mu'[u]/\rho\,\Delta^2}{u[u]/\Delta} \sim \frac{\Lambda}{\Delta}$$

which by (7.99) is small for $M_{1n} - 1 \ll 1$. As a further approximation the entire coefficient $[\cdot]$ of the second term is taken to be constant; with $u \approx c \approx u_1$ this coefficient becomes $\delta/(\gamma - 1)$, where

$$\delta \equiv \frac{4}{3}\frac{\mu'}{\rho} + (\gamma - 1)\frac{\kappa}{\rho c_p} \quad (7.104)$$

is the diffusivity found in Chap. 4 (originally by Kirchhoff) for the problem of acoustic attenuation. Then (7.103) becomes simply

$$u_x = -\frac{\gamma + 1}{2\delta}(u_1 - u)(u - u_2) \quad (7.105)$$

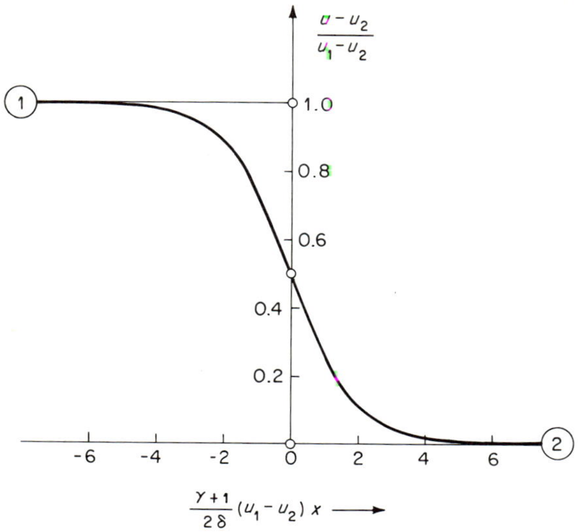

Figure 7.41
Taylor shock structure.

in which the right-hand side is negative or zero. With the boundary conditions (7.92) this has a solution

$$\frac{\gamma + 1}{2\delta}(u_1 - u_2)x = \ln\frac{u_1 - u}{u - u_2} \tag{7.106}$$

as shown in Fig. 7.41.

The shock thickness is somewhat arbitrary. If the edges of the shock are reckoned to be at $(u - u_2)/(u_1 - u_2) = 1 - \varepsilon$ and $(u - u_2)/(u_1 - u_2) = \varepsilon$, then (7.106) gives

$$\frac{\gamma + 1}{2\delta}(u_1 - u_2)\,\Delta = 2\ln\frac{1 - \varepsilon}{\varepsilon} \equiv A \tag{7.107}$$

Table 7.3

A	Basis
4	Max. slope
5.89	$\varepsilon = 0.05$
9.19	$\varepsilon = 0.01$
13.8	$\varepsilon = 0.001$

Depending on the basis selected, the values of the numerical factor A shown in Table 7.3 are obtained. The coefficient δ may be rewritten

$$\delta = \frac{\mu}{\rho}\left(\frac{4}{3} + \frac{\mu_v}{\mu} + \frac{\gamma - 1}{\mathrm{Pr}}\right) \tag{7.108}$$

For hard-sphere molecules, $\mu = 0.7833\rho c\Lambda/\sqrt{\gamma}$ (as given in Chap. 2). The Prandtl number is given approximately by Eucken's formula as $4\gamma/(9\gamma - 5)$. The bulk viscosity varies considerably and according to no simple formula; for monatomic gases it is zero, while for nitrogen it is approximately 0.8μ. Putting these values in gives

$$\delta = 1.414c\Lambda \qquad \text{monatomic gas}$$

$$\delta = 1.785c\Lambda \qquad \text{nitrogen}$$

and we will use $\delta \approx {}^3\!/_2 c\Lambda$. Then using the weak-shock condition (7.50), Eq. (7.107) becomes

$$\frac{\Delta}{\Lambda} \approx \frac{3}{4}\frac{A}{M_{1n} - 1} \tag{7.109}$$

in agreement with (7.99). For the value $A = 4$ obtained from the maximum slope (see Fig. 7.40) we have

$$\frac{\Delta_m}{\Lambda} \approx \frac{3}{M_{1n} - 1} = \frac{12\gamma}{\gamma + 1}\frac{P_1}{[P]} \tag{7.110}$$

As a numerical example, consider an atmospheric shock with $[P] = 10^{-3}$ atm (equivalent to a fairly strong sonic boom). With $\Lambda = 0.65 \times 10^{-7}$ m, we obtain $\Delta_m \approx 0.5$ mm.

For very weak shocks, the thickness predicted above may be enormous, since $\Delta \to \infty$ as $M \to 1$. Such shocks may never be physically realized, however, as the assumption of stationary flow may not be satisfied. A further discussion of viscous effects in waves of finite amplitude appears in Chap. 8.

In Fig. 7.42 is shown a comparison of the Taylor theory with experiment and with a more laborious calculation from the full Navier-Stokes equation. It should be remarked that this case stretches rather severely the Taylor assumption $M_{1n} - 1 \ll 1$ yet shows good agreement. The

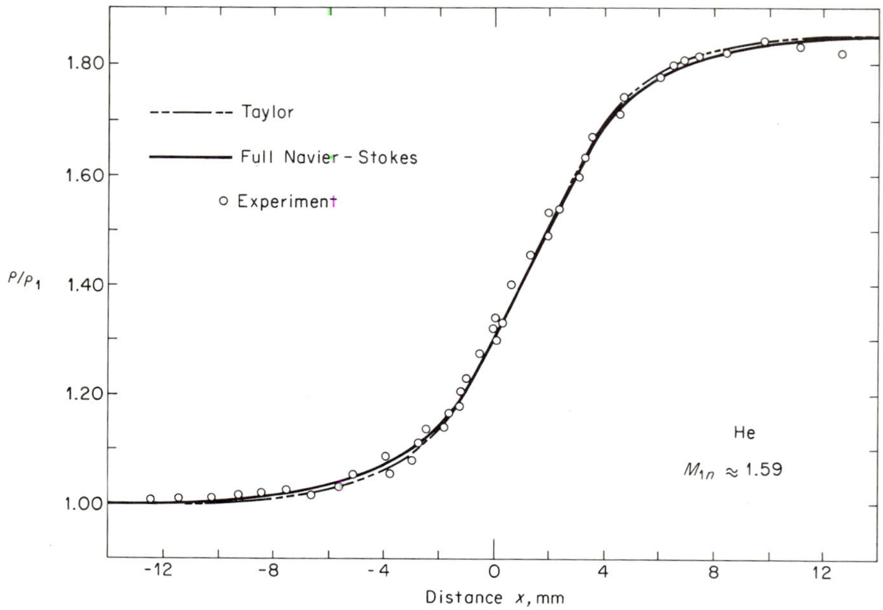

Figure 7.42
Comparison of the simple Taylor theory with the data and Navier-Stokes result given by *Muntz and Harnett* [1969] for a shock in helium, $M_{1n} \approx 1.59$.

density distribution here was calculated by the author from (7.106) written in the form (with ρu = const)

$$\frac{\gamma + 1}{2} \frac{c_1}{\delta} \frac{|[w]|}{c_1} = \ln \frac{\dfrac{\rho}{\rho_1} - 1}{1 - \dfrac{\rho}{\rho_1} \dfrac{\rho_1}{\rho_2}} \tag{7.111}$$

The diffusivity δ, however, was calculated at the density and temperature intermediate between the states 1 and 2, as suggested by *Lighthill* [1956]. This leads to a mean free path, for example, of $\overline{\Lambda} = 1.09$ mm for the conditions given by *Muntz and Harnett* [1969].

Problems

7.1 Consider the possibility of a shocklike discontinuity in temperature within a rigid heat-conducting solid with constant properties. For simplicity, let the problem be one-dimensional with a temperature distribution $T(x,t)$. Are such temperature discontinuities possible in principle? Are they observed in practice?

7.2 A spherical blast wave from an explosion advances into stationary air, $P_1 = 1$ atm, $T_1 = 300$ K, with a relative Mach number $M_{1n} = 2$. Find the pressure behind the wave and the absolute air velocity.

Answer 4.50 atm; 434 m/s chasing shock

7.3 A shock advances into stationary perfect gas. Provided that the shock is sufficiently strong, can the gas be set into supersonic motion with respect to a stationary observer?

7.4 Derive Eq. (7.21), viz., $[w]^2 = -[P][v]$.

7.5 A moving shock advances into stationary air which has $P_1 = 1$ atm, $T_1 = 300$ K. What is the shock speed V_s which gives a pressure $P_2 = 2$ atm behind the shock? What is the absolute air velocity behind the shock?

Answer 473 m/s; 182 m/s chasing shock

7.6 The Hugoniot equation for a perfect gas can be written

$$\frac{P_2}{P_1} = \frac{\dfrac{\gamma + 1}{\gamma - 1}\dfrac{\rho_2}{\rho_1} - 1}{\dfrac{\gamma + 1}{\gamma - 1} - \dfrac{\rho_2}{\rho_1}}$$

Considering the upstream state 1 to be fixed, the derivative $dP_2/d\rho_2$ can be calculated.

(*a*) Evaluate this derivative at $P_2 = P_1$ and comment on the significance of the result.

(*b*) Across the shock-type discontinuity in a strong supersonic bang (sonic boom) the density change is 0.1 percent. Find the corresponding percentage change in pressure.

Answer 0.14 percent

7.7 From the geometry of the shock adiabat, show that for a very weak shock

$$M_{1n} - 1 \approx 1 - M_{2n}$$

Hint: Start from the condition that the slope of a chord is approximately the average of the curve slopes at the chord end points.

7.8 For a hypothetical substance which has an equation of state $P = a^2\rho$, where a is a constant, find the normal-shock conditions M_{2n}, P_2/P_1, ρ_2/ρ_1 as functions of M_{1n}. Also find an expression for the entropy change $[s]$ across a weak shock in terms of a^2, T_1, and M_1.

7.9 Assume that a stream of people moving in a narrow passage, i.e., just wide enough for single-file motion, may be treated as a continuum. Let n be the number of persons per unit length and u be velocity. Show that a possible shock velocity is

$$V_s = \frac{n_2u_2 - n_1u_1}{n_2 - n_1}$$

Estimate this velocity for panic conditions, where the downstream state is stationary, $u_2 = 0$. Also estimate the downstream compression force. Assume $n_1 = 0.25 \text{ ft}^{-1}$, $n_2 = 1 \text{ ft}^{-1}$, $u_1 = 20$ ft/s, and the mass per person is 150 lb_m.

7.10 A compression shock in a simple spring is sketched. Here ρ is the mass per unit length and the "equation of state" is $F = k^2(\rho - \rho_0)$, where F is the compressive force and k and ρ_0 are constants. Find the Prandtl relation $w_1w_2 = f(k)$ and the "speed of sound."

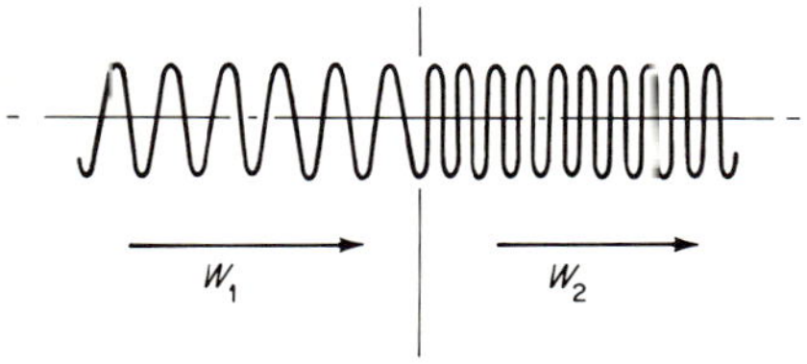

7.11 Nitrogen at upstream state $P_1 = 1$ atm, $T_1 = 300$ K enters a shock. Estimate the shock pressure ratio P_2/P_1 required to produce a 5 percent mole fraction of dissociated nitrogen. *Hint*: Use the data given in Fig. 2.10.

7.12 For a Chapman-Jouguet detonation in a perfect gas, the pressure ratio P_2/P_1 is given by Eq. (7.77). Assuming that $\gamma_1 = \gamma_2 = 1.40$, compare this pressure ratio with that for a normal shock with the same Mach number M_{1n}.

7.13 In deriving the oblique-shock conditions, the component of velocity parallel to the shock surface is invariant across the shock; that is, $v_1 = v_2$. Will such a condition hold for an oblique detonation?

7.14 A supersonic wedge moves through a stationary atmosphere ($u_1 = 0$) at $M = 4$. Find the absolute flow direction in region 2. The atmosphere is a perfect gas with $\gamma = 1.40$.

Answer Inclined at 59.97° to the line of flight

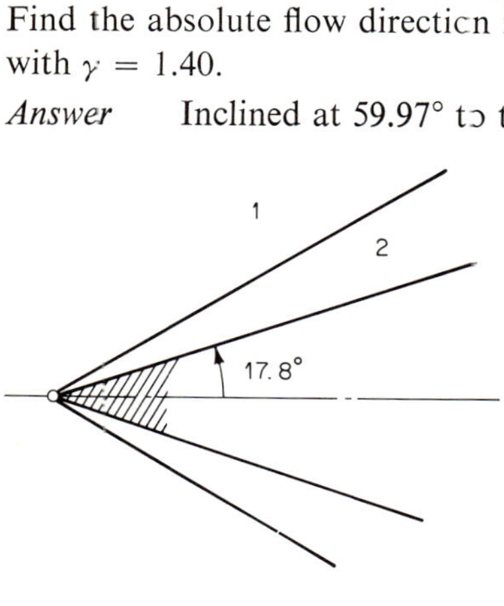

7.15 A plane shock A with shock Mach number $M_s = 1.5$ advances into stationary air at $P_1 = 1$ atm and $T_1 = 300$ K. The shock is intercepted by a rigid wall, as discussed in Example 7.8 (page 333). Find the flow conditions in the neighborhood of the point of intersection between the shock and the wall.

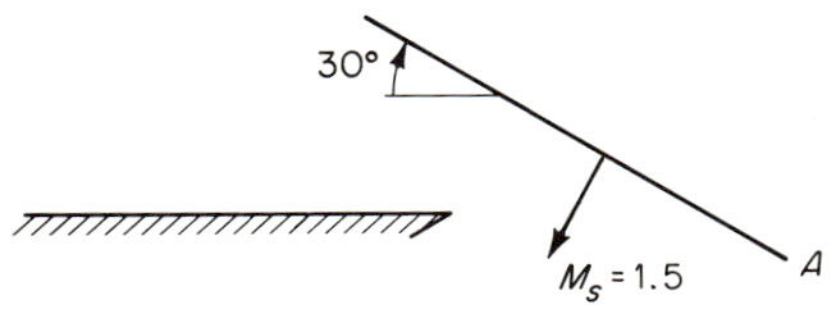

7.16 Show that Eq. (2.7) for the rate of entropy increase can be written, for a steady one-dimensional flow,

$$\rho u \frac{ds}{dx} = \frac{4}{3}\frac{\mu}{T}\left(\frac{du}{dx}\right)^2 + \frac{1}{T}\frac{d}{dx}\left(\kappa \frac{dT}{dx}\right)$$

Consider the application of this equation across a stationary normal shock. If the velocity distribution $u(x)$ and temperature distribution $T(x)$ are essentially S-shaped (see, for example, Fig. 7.41), show that it is plausible that the entropy distribution $s(x)$ have a maximum *within* the shock, such that $s_{\max} > s_2$.

7.17 Assume that the temperature and velocity distributions across a stationary normal shock can be represented as straight lines, i.e.,

$$\frac{dT}{dx} = \frac{[T]}{\Delta} \qquad \frac{du}{dx} = \frac{[u]}{\Delta}$$

where $0 \leq x \leq \Delta$ and Δ is the shock thickness. Also assume that the viscosity and thermal conductivity increase linearly with temperature,

$$\frac{4}{3}\frac{\mu}{T} = m \qquad \frac{\kappa}{T} = k$$

where m and k are constants. Making use of the entropy formula given in Prob. 7.16, show that the shock thickness Δ is given by

$$\frac{\Delta}{\mu_1/\rho_1 c_1} = \frac{1}{\Pr}\frac{1}{M_1}\frac{\frac{4}{3}\Pr\frac{[u]^2}{c_p T_1} + \frac{[T]}{T_1}\ln\frac{T_2}{T_1}}{[s]/c_p}$$

where Pr is the Prandtl number.

7.18 A perfect gas ($\gamma = 1.40$) is accelerated in steady flow from $P_0 = 7.83$ atm to $P_2 = 1$ atm, then brought to rest by a moving shock as shown. Find the pressure P_3 of the stagnant gas behind the shock.

Answer 8.73 atm

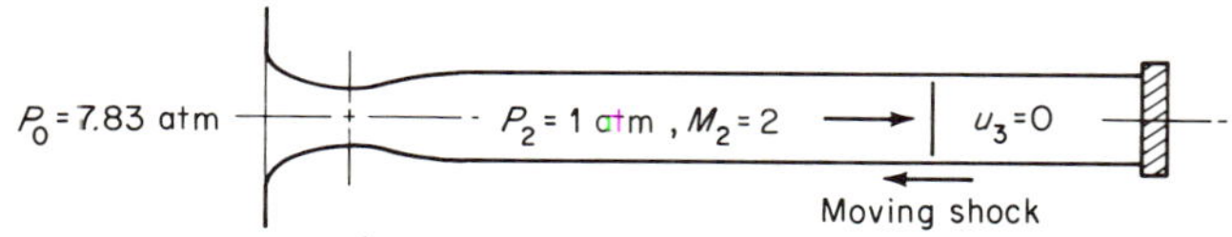

7.19 Find the minimum value of A_3/A_1 which will allow the steady flow shown, for a perfect gas with $\gamma = 1.40$. Assume that the flow is isentropic, except across the shock.

Answer 1.385

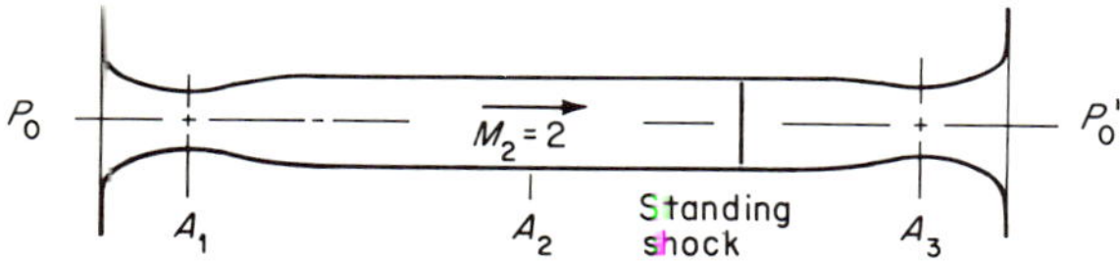

7.20 Given M_1 and the turning angle θ, $\theta \ll 1$, find the shock angle β for an arbitrary fluid. *Hint*: Use the weak-shock relations, retaining only terms up to the first power in the shock strength in each case.

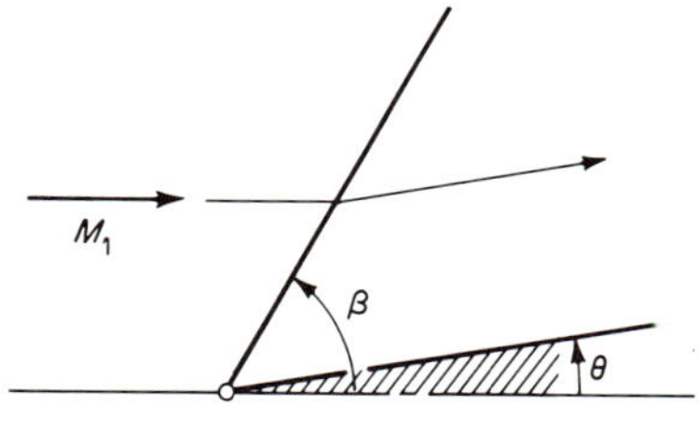

7.21 A relatively weak oblique shock in (liquid) water has an upstream Mach number $M_1 = 4$ and a turning angle $[\theta] = 1°$. Find the corresponding value of the pressure jump $[P]$. The upstream conditions can be taken to be room temperature and 1 atm.

Answer 1,600 atm

7.22 Find the important properties of a shock in a negative-Γ fluid. Draw the Hugoniot diagram.

7.23 The position of a certain shock front is given by $y = ax + bt$ ($y \geq 0$), where a and b are positive constants. The fluid ahead of the shock ($y > ax + bt$) is at rest with known state 1. Find:

(*a*) The shock Mach number
(*b*) The pressure behind the shock from weak-shock theory
(*c*) The pressure behind the shock from linearized steady-flow theory
(*d*) A possible physical situation leading to this shock-front geometry

7.24 Find the temperature jump $[T]$ across a shock in liquid water with a pressure jump $[P] = 100$ atm. The upstream state is $P_1 = 1$ atm, $T_1 = 300$ K. First find a general formula for this situation; then obtain a numerical value.

7.25 The geometry of an oblique shock is essentially fixed by continuity. Let $r \equiv \rho_2/\rho_1$ be the density ratio across an oblique shock in an *arbitrary* fluid.

(*a*) Show that $r = \tan \beta / \tan (\beta - \theta)$.
(*b*) For any given value of r, show that the maximum turning angle θ_m and corresponding shock angle β_m are given by

$$\tan \theta_m = \frac{r - 1}{2\sqrt{r}}$$

$$\tan \beta_m = \sqrt{r}$$

7.26 Show that the following relation holds for weak shocks:

$$-\frac{[w^2]}{2c_1^2} = \Pi - \Pi^2 + \Gamma_1 \Pi^3 \cdots$$

eight

one-dimensional unsteady flow

8.1 Introduction

The unsteady motions considered in the chapter on acoustics were restricted to small amplitudes. This restriction is now removed, and we consider large-amplitude motions in one dimension. This subject is variously called *waves of finite amplitude* or *nonlinear wave propagation*. The pertinent mathematical technique is the *method of characteristics*. All these labels relate to the idea of waves, which plays an important role in what is to follow.

Physical situations to which the theory is applicable usually involve explosions or other mechanisms for rapid energy transfer. Devices of interest include shock tubes, various ballistic devices, and engines making use of wave processes.

A convenient description of the motion of interest is given by the *piston problem*. In Fig. 8.1, the tube contains a gas which is initially

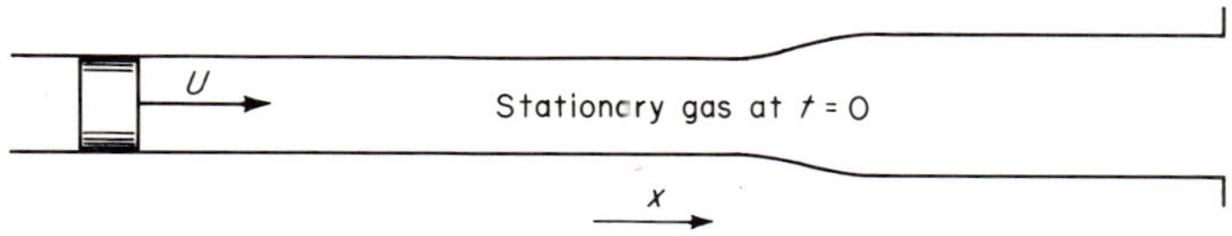

Figure 8.1
Piston problem.

uniform and stagnant. If a piston motion $U(t)$ is initiated at $t = 0$, what is the resulting gas motion $u(x,t)$ and distribution of gas $\rho(x,t)$? A general attack for problems of this kind is outlined below.

8.2 Characteristic equations for homentropic flow

We introduce two major assumptions: (1) the flow is *quasi-one-dimensional*, as discussed in Sec. 6.1, and all variables of interest depend only on x and t; (2) the flow is *homentropic*, as discussed in Sec. 2.3.

The flow may be considered to be confined to a duct with known cross-sectional area $A(x)$. The continuity equation was found in Sec. 4.10 and can be written

$$\frac{\partial \rho}{\partial t} + u\frac{\partial \rho}{\partial x} + \rho\frac{\partial u}{\partial x} = -\frac{\rho u}{A}\frac{dA}{dx} \tag{8.1}$$

The momentum equation is just the x component of (1.67), or

$$\frac{\partial u}{\partial t} + u\frac{\partial u}{\partial x} + \frac{1}{\rho}\frac{\partial P}{\partial x} = G \tag{8.2}$$

where G is the component of body force in the x direction (strictly, to be consistent with the one-dimensional model, the body force should be purely in the x direction).

The area term $A(x)$, which appears only in the continuity equation, allows us to treat plane waves with $A = \text{const}$, cylindrical waves with $A \propto x$, or spherical waves with $A \propto x^2$. These cases correspond respectively to a constant-area duct, a wedge-shaped duct, and a conical duct.

With the homentropic assumption, there is only one independent thermodynamic variable. If P is chosen as the independent variable, we can write $\rho = \rho(P)$ and $d\rho = (\partial\rho/\partial P)_s\, dP = 1/c^2\, dP$; then (8.1) becomes

$$\frac{1}{\rho c}\frac{\partial P}{\partial t} + \frac{u}{\rho c}\frac{\partial P}{\partial x} + c\frac{\partial u}{\partial x} = -\frac{cuA'}{A} \tag{8.3}$$

where $A' \equiv dA/dx$.

The equations of motion are *nonlinear*; i.e., they contain products of unknowns, e.g., the term $u\,\partial u/\partial x$ in (8.2). Nonlinear partial differential equations are difficult, since the extensive techniques for linear equations are not applicable and superposition of solutions is impossible. The only methods available for nonlinear equations are (1) direct numerical solution, (2) analog experiment on another more convenient physical system, (3) formulation as a self-similar problem (not usually possible), and (4) the method of characteristics. We pursue the latter, which has a direct connection to physical waves.

To put the equations of motion (8.2) and (8.3) in characteristic form, we add (8.3) to (8.2) and subtract (8.3) from (8.2) to obtain the *two* equations

$$\left(\frac{\partial u}{\partial t} \pm \frac{1}{\rho c}\frac{\partial P}{\partial t}\right) + (u \pm c)\left(\frac{\partial u}{\partial x} \pm \frac{1}{\rho c}\frac{\partial P}{\partial x}\right) = G \mp \frac{cuA'}{A} \tag{8.4}$$

It appears doubtful that this is any simplification, but it is, as we proceed to show. Define a new thermodynamic function $F = F(P,s)$,

$$F \equiv \int_{P_0}^{P} \frac{dP}{\rho c} \tag{8.5}$$

where P_0 is any fixed reference state and the integration is carried out at constant entropy. The function F will have meaning only for homentropic flows. The pressure derivatives appearing in (8.4) can be written in terms of F,

$$\frac{\partial F}{\partial t} = \frac{dF}{dP}\frac{\partial P}{\partial t} = \frac{1}{\rho c}\frac{\partial P}{\partial t}$$

$$\frac{\partial F}{\partial x} = \frac{dF}{dP}\frac{\partial P}{\partial x} = \frac{1}{\rho c}\frac{\partial P}{\partial x}$$

and (8.4) can be rewritten as

$$\left[\frac{\partial}{\partial t} + (u \pm c)\frac{\partial}{\partial x}\right](u \pm F) = G \mp \frac{cuA'}{A} \tag{8.6}$$

In this form, *each equation contains only one derivative operator.* By definition, an equation is in characteristic form if it contains only one derivative operator; Eqs. (8.6) are thus the equations of motion in characteristic form.

The derivative operators appearing in (8.6) are, respectively,

$$\begin{aligned}\frac{\partial}{\partial t} + (u + c)\frac{\partial}{\partial x} &\equiv \frac{D^+}{Dt} \\ \frac{\partial}{\partial t} + (u - c)\frac{\partial}{\partial x} &\equiv \frac{D^-}{Dt}\end{aligned} \tag{8.7}$$

and the two equations, written in this notation, are finally

$$\boxed{\begin{aligned}\frac{D^+}{Dt}(u + F) &= G - \frac{cuA'}{A} \\ \frac{D^-}{Dt}(u - F) &= G + \frac{cuA'}{A}\end{aligned}} \tag{8.8}$$

The remainder of this chapter is mainly devoted to the implications of these equations.

Characteristic Lines

The derivative operators D^+/Dt and D^-/Dt have a simple interpretation. They may be compared to the *material* derivative D/Dt, which is, in one dimension,

$$\frac{D}{Dt} \equiv \frac{\partial}{\partial t} + u\frac{\partial}{\partial x}$$

This gives the time rate of change (of the quantity operated on) for an observer who moves with a fluid particle, i.e., with velocity u. In general, if V_0 is the velocity of any moving observer, then $\partial/\partial t + V_0\, \partial/\partial x$ is the time rate of change as viewed by that observer. In a one-dimensional flow $u + c$ and $u - c$ are the velocities of *positive* (tending to travel in the positive x direction) and *negative* (tending to travel in the negative x direction) waves respectively; for conciseness we will call these C^+ sound waves and C^- sound waves. Then we have the following interpretation for the derivative operators:

$$\frac{D^+}{Dt} = \frac{\partial}{\partial t} + (u + c)\frac{\partial}{\partial x} =$$ time rate of change for observer traveling at velocity $u + c$, that is, with a positive C^+ sound wave

$$\frac{D^-}{Dt} = \frac{\partial}{\partial t} + (u - c)\frac{\partial}{\partial x}$$ = time rate of change for observer traveling at velocity $u - c$, that is, with a negative C^- sound wave

As an illustration, recall the case of simple acoustic waves. With $|u| \ll c$ and c approximately constant, the derivative operator D^+/Dt becomes

$$\frac{D^+}{Dt} = \frac{\partial}{\partial t} + c_0\frac{\partial}{\partial x}$$

The result that a property such as the condensation S does not vary on the wave, e.g., a simple C^+ wave, is expressed

$$\frac{D^+S}{Dt} = 0$$

which is formally identical to (4.36).

The paths of C^+ and C^- sound waves appear as lines on the xt diagram, as shown in Fig. 8.2. Such lines are called *characteristic lines* or simply *characteristics*. The differential equations (8.8) give the rate of change with respect to time along a characteristic. In the acoustic case, the characteristics C^+ and C^- were just straight lines with $dx/dt = \pm c_0$ (see, for example, Fig. 4.5); in addition, the particle paths were essentially vertical lines. In the more general case considered here, the characteristics and particle paths are curves, and these curves are not normally known in advance.

Alternative Forms for the Thermodynamic Function F

This quantity was defined in Eq. (8.5), which in differential form is

$$dF \equiv \frac{dP}{\rho c} \tag{8.9}$$

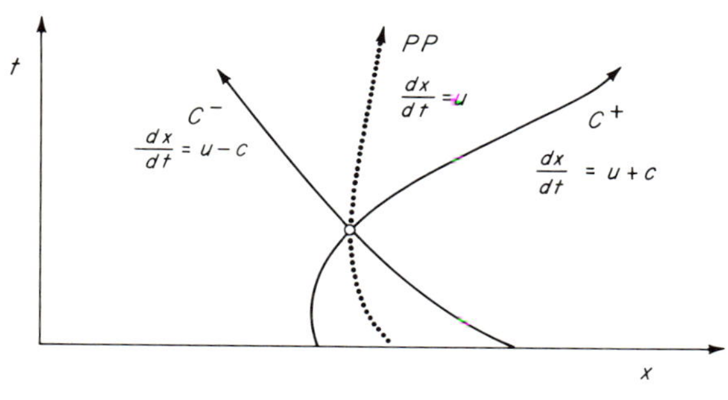

Figure 8.2
Characteristic lines.

With entropy fixed, we can write $dP = c^2\,d\rho$ and, with the help of (5.7), $dP = (\rho c\,dc)/(\Gamma - 1)$. Then (8.9) has the alternative forms

$$dF = \frac{dP}{\rho c} = c\,\frac{d\rho}{\rho} = \frac{dc}{\Gamma - 1} \tag{8.10}$$

with corresponding integral forms

$$F = \int_{P_0}^{P} \frac{dP}{\rho c} = \int_{\rho_0}^{\rho} c\,\frac{d\rho}{\rho} = \int_{c_0}^{c} \frac{dc}{\Gamma - 1} \tag{8.11}$$

where P_0, ρ_0, c_0 refer to the same reference state.

For a *perfect gas*, $\Gamma = (\gamma + 1)/2$ and is constant, so that (8.11) gives $F = [2/(\gamma - 1)](c - c_0)$. It is convenient to take the reference state $c_0 = 0$, so that

$$F = \frac{2}{\gamma - 1}\,c \tag{8.12}$$

Riemann Invariants

For constant-area flow (plane waves) with negligible body force,[1] the equations of motion (8.8) are

$$\begin{aligned} \frac{D^+}{Dt}(u + F) &= 0 \\ \frac{D^-}{Dt}(u - F) &= 0 \end{aligned} \tag{8.13}$$

The quantities $u + F$ and $u - F$ thus do not vary along their respective characteristics and are called the *Riemann invariants* after Georg Friedrich Bernhard Riemann, who developed the theory in a classical paper.[2]

[1] The body force G is important only in unusual situations such as the atmosphere of a star, or in an accelerated frame where the inertial force is treated as a body force, or in magnetogasdynamics. In the latter case, however, the one-dimensional wave-propagation model is likely to fail.

[2] *Riemann* [1859]. This paper, which forms the foundation of much of gasdynamics, represents a small corner in the space of Riemann's interests and was completed only a few years before his death at the age of thirty-nine. (Riemann died of consumption, following long years of poverty as an underpaid instructor.)

These invariants will be labeled[1]

$$J^+ \equiv u + F \qquad J^- \equiv u - F \tag{8.14}$$

Then the integrals of (8.13) are of the form J = const on a particular characteristic; e.g., if we label a certain characteristic C_i^+, we write $u + F = J_i^+$ = const on C_i^+. In general, the value of J^+ will vary from one C^+ characteristic to another.

The result that the Riemann invariant is constant along a certain characteristic is a generalization of, and analogous to, the one-dimensional acoustic result that certain quantities are constant along characteristics [see, for example, Eq. (4.20)]. In addition, the acoustical result $\delta P = \mp \rho c \delta u$ now holds along C^+ and C^- characteristics, respectively.

For a *perfect gas*, the Riemann invariants are just, from (8.12),

$$J^+ = u + \frac{2}{\gamma - 1} c \qquad J^- = u - \frac{2}{\gamma - 1} c \tag{8.15}$$

8.3 Boundary conditions and integration

In a typical problem of physical interest, the state of the fluid is given at some instant, and it is desired to calculate the future motion. Given the present situation, what happens in the future? This is an *initial-value problem.*

Consider first the plane-wave case represented by (8.13). Suppose that complete initial data are given at $t = 0$, as represented in Fig. 8.3 [the initial data might be given, for example, as $u(x)$ and $P(x)$ with a specified entropy constant]. It is desired to find the value of u and some representative thermodynamic quantity, say P, at a later time at some general point d. From (8.13) $J_d^+ = J_a^+$ and $J_d^- = J_b^-$, or

$$u_d + F_d = J_a^+ \qquad u_d - F_d = J_b^-$$

[1] In some books P and Q are used in place of the symbols given here. The symbols originally used by Riemann are r and s. The symbol J here has no connection with the mass flux earlier represented by the same symbol.

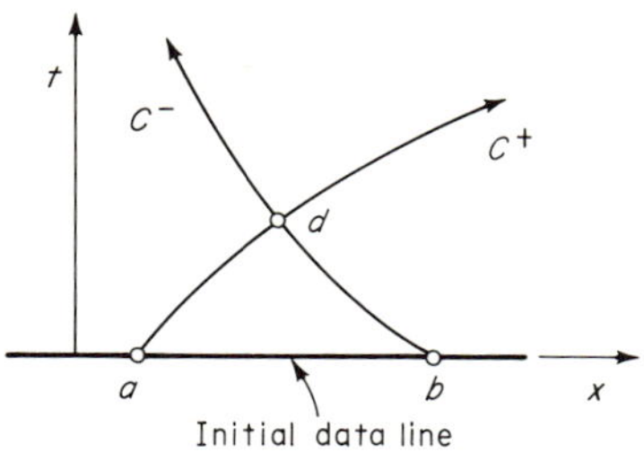

Figure 8.3

where J_a^+ and J_b^- are known initial values. Solving for the unknowns u_d and F_d gives

$$u_d = \tfrac{1}{2}(J_a^+ + J_b^-) \qquad F_d = \tfrac{1}{2}(J_a^+ - J_b^-) \tag{8.16}$$

Then, since F is a thermodynamic quantity, say $F(P)$, all conditions are known at the point d in terms of given initial data at a and b. Similarly, conditions at any point on the interior of the triangle abd are given by initial conditions on the line ab.

Physically, *characteristic lines are identified with sound waves*, which may be thought of as infinitesimal signals of pressure disturbance. In the absence of electromagnetic radiation or overtaking shock waves, no signal travels faster than the sound wave. This implies the existence of a *limited region of influence from a given physical disturbance*, as was shown graphically in Fig. 5.1. A more precise description is possible via the characteristic equations, which imply that characteristics serve as carriers of information. Suppose that initial data (or a physical disturbance) are confined to a certain range of values for x, $x_b > x > x_a$, as shown in Fig.

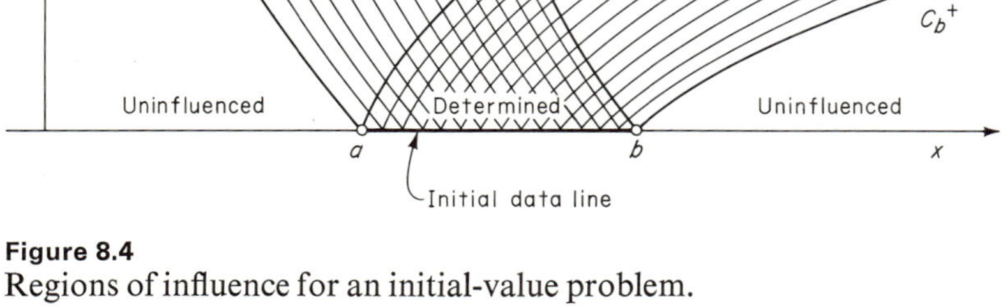

Figure 8.4
Regions of influence for an initial-value problem.

8.4. The future flow is completely determined by the initial data in the triangular region *abd* as before. Outside this region the flow field is only partly influenced, or completely uninfluenced, depending on whether characteristics from the initial data line reach the region in question; thus, we arrive at the various regions labeled in the figure. It is interesting that there can be an uninfluenced region at a physical location overlapping the initial data location.

The boundary conditions can of course be given in other ways; i.e., information can be given along a variety of lines in the *xt* plane. In general, the existence of characteristics places restrictions on the boundary conditions which may be imposed; e.g., it is not possible to arbitrarily set the values of u and P along the curve *ad* in Fig. 8.4 because it is a characteristic line with $u + F = J_a^+$ and u and P are not independent. A general rule is: along an arbitrary data curve as many unknowns must be specified as there are characteristics running into the region to be calculated, e.g., along the curve *ab* two unknowns must be specified. In the remainder of this chapter, various physical boundary conditions will be discussed by example.

Because the path of characteristics is not known in advance (except in the acoustic case), the integration may have to be performed numerically. In particular, in the nonhomogeneous case we have a variation of the quantities $J^+ = u + F$ and $J^- = u - F$ along their respective characteristics, as given by (8.8), viz.,

$$\frac{D^+}{Dt}(u + F) = G^+(x,u,F)$$
$$\frac{D^-}{Dt}(u - F) = G^-(x,u,F) \tag{8.17}$$

where for conciseness the right side has been written $G^\pm \equiv G \mp cuA'/A$. The integration along characteristics is best illustrated by an incremental numerical calculation. Let conditions be known at time t; conditions are to be calculated at time $t + \Delta t$, where Δt is small. Points a and b in Fig. 8.5 are neighboring points on the initial-data line, and the approximate location of point d is the intersection of the characteristics C_a^+ and C_b^-, taken to be straight lines with known initial slopes $(u + c)_a$ and $(u - c)_b$, respectively. Then (8.17) gives, approximately,

$$(u + F)_d = (u + F)_a + \overline{G}^+ \Delta t$$
$$(u - F)_d = (u - F)_b + \overline{G}^- \Delta t \tag{8.18}$$

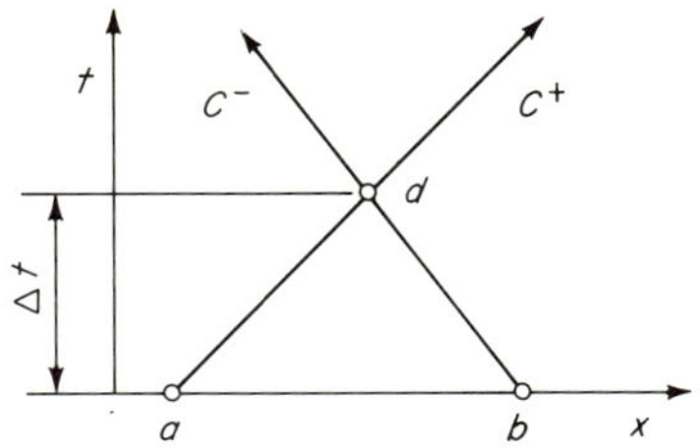

Figure 8.5

where $\overline{G}$ is an average value between the end points (to a slightly poorer approximation, it can be taken as just the initial value). These equations can then be solved for u_d and F_d, with a result similar to (8.16).

For the remainder of the chapter, we will be concerned only with the homogeneous case (8.13).

8.4 Simple waves

Physically, simple waves arise from the propagation of disturbances from one direction only into a region of uniform flow. For example, suppose there is a constant steady flow in a pipe; at some instant a perturbation (not necessarily small) is initiated at some point a, as by the closing of a valve (Fig. 8.6). Sound waves traveling downstream carry the disturbance into the region of uniform flow.

The corresponding *simple region* on a wave diagram is, formally, any region bounded by a uniform region, as illustrated in Fig. 8.7. The characteristics in the simple region have distinctive properties, as we proceed to show. Consider the characteristics C^-, all of which pass through the uniform region. Along *each* of these characteristics $J^- = u_0 - F_0 =$ const, with known value. Then *everywhere*

$$u - F = J_0^- = \text{const} \tag{8.19}$$

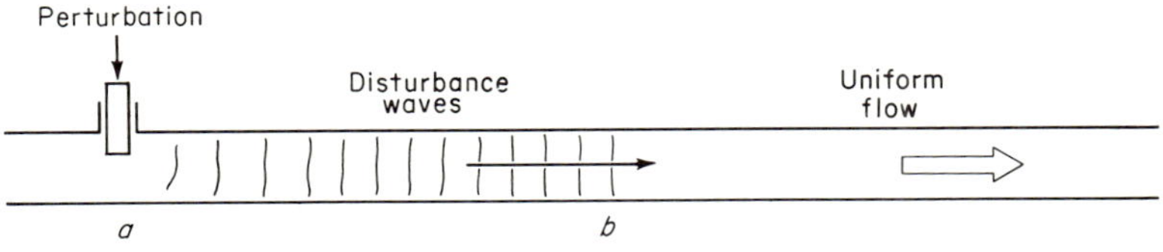

Figure 8.6
Physical illustration of simple waves.

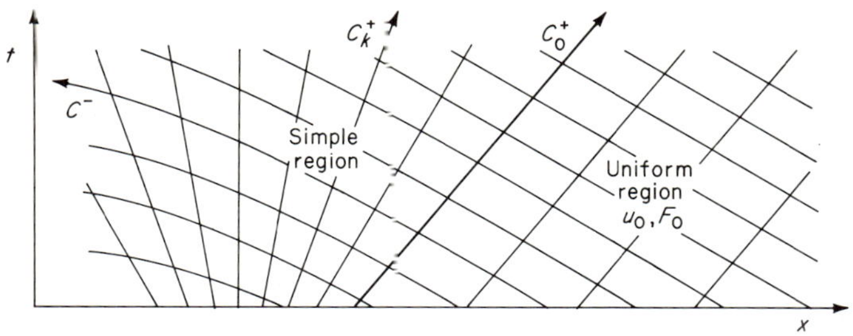

Figure 8.7
Simple region on a wave diagram.

Now consider a typical characteristic C_k^+ in the simple region; along this characteristic

$$u + F = J_k^+$$

This value will, in general, vary from one C^+ characteristic to another. Substituting from (8.19), we obtain

$$\begin{aligned} u_k &= \tfrac{1}{2}(J_k^+ + J_0^-) \\ F_k &= \tfrac{1}{2}(J_k^+ - J_0^-) \end{aligned} \tag{8.20}$$

Since J_0^- is just a constant, *the value of u and the value of F are uniform along each C^+ characteristic* (though these values will vary from one C^+ characteristic to another). Because F is uniform along the characteristic, all the thermodynamic properties (P, ρ, T, etc.) will be uniform; in particular, the sound speed c will be uniform. Therefore the value of $u + c$ will be uniform along each C^+ characteristic, and *the C^+ characteristics are straight lines.* This requires, for example, that each of the sound waves in the disturbance wave of Fig. 8.6 travel at constant velocity (though different waves travel at different velocities).

Exactly analogous results will of course be obtained for a disturbance traveling in the opposite direction, i.e., for simple C^- waves.

8.5 Shock formation

In most unsteady flows shock waves will form sooner or later. In this section we consider the formation of a shock front from the distortion of an initially continuous waveform which is advancing into uniform stationary fluid. Such a waveform, or *pressure pulse*, may be formed, for example, by the forward movement of a piston (see Fig. 8.1) which is initially and finally at rest. The resulting wave is a *simple* wave.

This is a classical problem considered long ago by Stokes and Riemann, among others. We will first work out the details for the important special case of a *perfect gas* and later for a general fluid.

Shock Formation in a Perfect Gas

Let the initial wave be traveling in the $+x$ direction (a C^+ wave) into stationary gas with properties c_0, P_0, as represented in Fig. 8.8. The particular form of the wave is not important to the arguments which follow. It is convenient to represent the wave as a sound-speed distribution, but the distribution of other thermodynamic quantities is similar; by the isentropic relations,

$$\frac{T}{T_0} = \left(\frac{c}{c_0}\right)^2 \qquad \frac{P}{P_0} = \left(\frac{c}{c_0}\right)^{2\gamma/(\gamma-1)} \qquad \frac{\rho}{\rho_0} = \left(\frac{c}{c_0}\right)^{2/(\gamma-1)}$$

giving the alternative distributions shown in Fig. 8.9.

With C^- characteristics coming from the uniform stationary region ahead of the wave, $J^- = J_0^- =$ const everywhere, or

$$u - \frac{2}{\gamma - 1} c = -\frac{2}{\gamma - 1} c_0$$

which gives

$$u = \frac{2}{\gamma - 1}(c - c_0) \tag{8.21}$$

Thus, the fluid within regions of compression ($c > c_0$) travels in the $+x$ direction, the direction of travel of the wave, just as in the acoustic case.

Any portion of a simple C^+ wave advances with constant wave velocity $u + c$; in particular, from (8.21)

$$u + c = \frac{\gamma + 1}{\gamma - 1} c - \frac{2}{\gamma - 1} c_0 \tag{8.22}$$

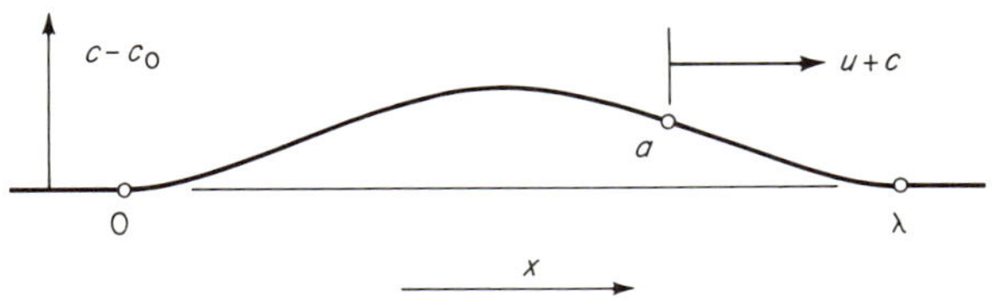

Figure 8.8
Initial wave.

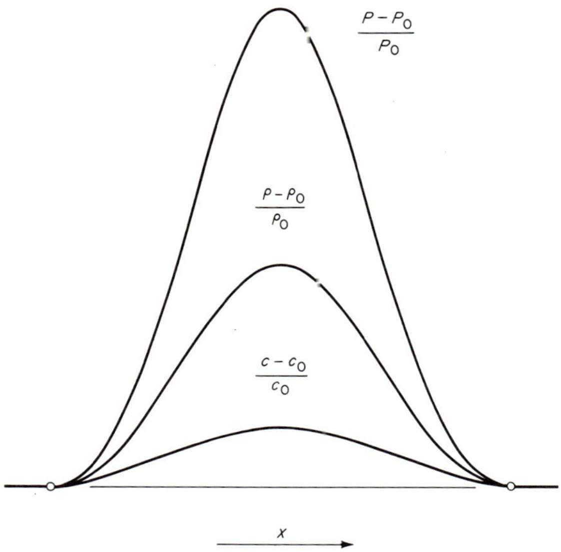

Figure 8.9

It is this monotonic increase of wave speed with wave amplitude c which will give rise to the peculiar distortion of the wave. Consider a general point, such as a on Fig. 8.8; at a time $t_0 + \Delta t$ this point will have advanced a distance $(u + c)\,\Delta t$. Relative to the head and tail ($x = \lambda, 0$) of the wave, which travel at constant velocity c_0, the advance is $(u + c - c_0)\,\Delta t$ which is, from (8.22),

$$\frac{\gamma + 1}{\gamma - 1}(c - c_0)\,\Delta t$$

Thus, the future waveform can be sketched for any value of Δt; this has been done for three different values of Δt in Fig. 8.10.

At time t_1 the waveform shows distortion, in the form of a steepening of the compressive part $Q\lambda$ of the wave and a spreading out of the expansive part OQ. This kind of distortion is characteristic of *nonlinear* waves, as opposed to the linear acoustic case, in which distortion is negligible and is formally excluded by the assumptions built into the wave equation. Because points on the wave with the same amplitude (such as a and a' in the figure) move with the same velocity, their separation distance remains constant in time: the wave distortion thus may be visualized as the shearing of a stack of thin laminae, as shown in Fig. 8.11.

At time t_2 in Fig. 8.10 a portion s of the leading edge of the wavefront has just become vertical, so that the velocity and sound-speed (also pres-

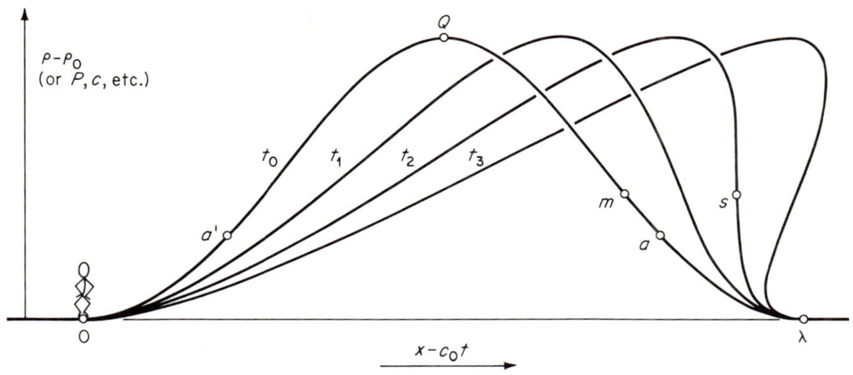

Figure 8.10
Progressive distortion and breaking of the waveform.

sure, density, etc.) gradients become infinite. This time is identified as the *instant of shock formation*; for if the calculation is formally carried further, to time t_3 as shown, the sound speed and other variables becomes triple-valued, a situation which is physically absurd. The progressive distortion of the wave to time t_3 is analogous to the breaking of an ocean wave as it approaches a sloping beach: the difference is that the height $h(x)$ of the ocean wave may *physically* become triple-valued, at least briefly, whereas the amplitude of the gasdynamic wave may not.

The wave diagram corresponding to Fig. 8.10 is shown in Fig. 8.12. The point s of shock formation is the point at which two or more C^+ characteristics first intersect, for each C^+ characteristic carries a unique value of velocity u and sound speed c; an intersection then dictates multiple values of u and c. A shock wave is a physically allowable mechanism for multiple values (two values) in this case. Calculations beyond the point s require the use of the shock conditions. Since the flow becomes non-homentropic (due to the entropy rise of fluid particles crossing the shock front), some modification of the method of characteristics may be required, as will be discussed.

Figure 8.11
Waveform generated by shearing laminae.

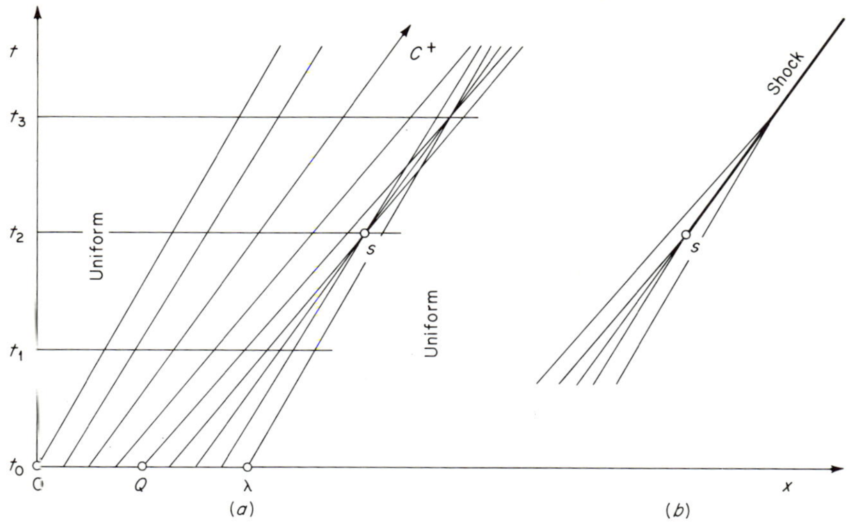

Figure 8.12
(*a*) Wave diagram; (*b*) insertion of a shock discontinuity from the point *s*, where characteristics first intersect.

The time $t_2 - t_0 = t_*$ in which a shock first forms is easily calculated. Let the wave velocities of adjacent points on the wave be v and $v + dv$, where $v \equiv u + c$, as shown in Fig. 8.13. Then the wave becomes vertical when the point behind just overtakes the point ahead (which of course requires that dv be negative). With dv the relative velocity, the distance dx between the points is just closed in time t_* according to $-dv\, t_* = dx$, or

$$t_* = \frac{-1}{dv/dx} \tag{8.23}$$

Then a shock necessarily forms first from a point where dv/dx is a minimum. From (8.22) we have

$$\frac{dv}{dx} = \frac{\gamma + 1}{\gamma - 1}\frac{dc}{dx}$$

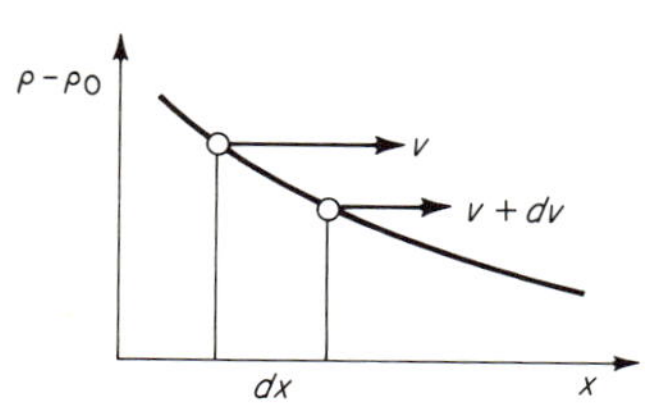

Figure 8.13

This is a minimum at the steepest point in the wave, e.g., at point *m* in Fig. 8.10. The corresponding distance traveled by the wave is $x_* = vt_*$. For *small-amplitude* waves with $c \approx c_0$ this gives

$$x_* \approx -\frac{\gamma - 1}{\gamma + 1} \frac{c_0}{(dc/dx)_{\min}} \tag{8.24}$$

For typical *acoustic* waves this gives a distance of the order of kilometers. Such a calculation should not be taken too seriously in this case, because the viscous terms in the equations of motion may be of the same order of magnitude as the nonlinear terms [see Eq. (4.52)]; the viscous effects in large-amplitude waves will be discussed in Sec. 8.11.

Shock Formation in a General Fluid

The behavior just cited for simple waves in a perfect gas is typical of ordinary fluids, as we proceed to show.

The initial-value problem considered is the same as before, with a pulse propagating into stationary fluid. Let the initial pulse be represented by the *pressure* distribution shown superimposed on the wave diagram of Fig. 8.14. Again, the particular form of the pulse is not important to the arguments that follow. For an arbitrary fluid, it is not necessarily true that the sound speed varies monotonically with pressure, so that the sound-speed distribution may not be similar to the pressure distribution shown.

This is again a simple wave, and the C^+ characteristics are straight lines carrying constant values of pressure, sound speed, and velocity. In general, the C^+ characteristics will not be parallel (as shown in the figure)

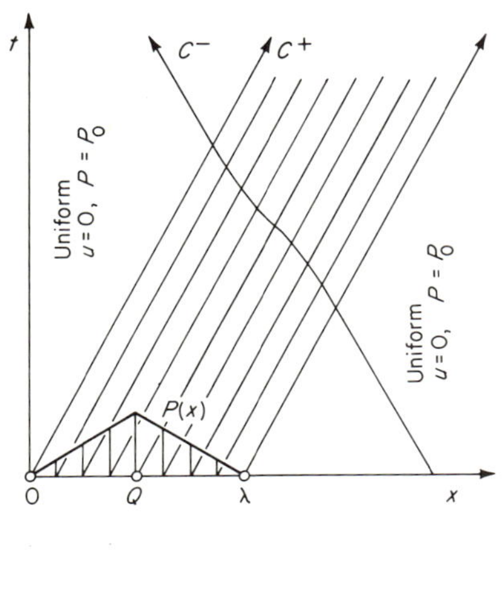

Figure 8.14
Wave diagram with superimposed initial pressure pulse.

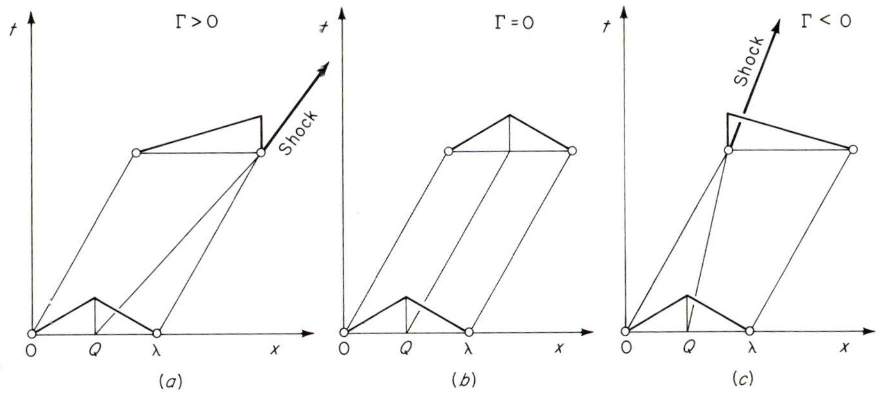

Figure 8.15
Behavior of the wave for various values of the fundamental derivative.

but will either converge or diverge; where they *intersect* a shock will form, and this is the basis for our investigation. Consider the characteristic C^-; if the wave speed $u + c$ increases along this characteristic (moving along C^- in the sense of increasing time), the characteristics C^+ will be progressively bent over to the right and converge, forming a shock, whereas if $u + c$ decreases, they will diverge. Along C^-, from Eq. (8.13), $du - dF = 0$; then

$$d(u + c) = du + dc = dF + dc$$

along C^-. With dc given by (8.10), this becomes

$$d(u + c) = \Gamma\, dF = \frac{\Gamma}{\rho c}\, dP \tag{8.25}$$

where Γ is the fundamental derivative. The sign of $d(u + c)$ is thus just the sign of $\Gamma\, dP$. Now dP is positive in the compressive part of the wave λQ and negative in the expansive part QO. Thus if $\Gamma > 0$, the compressive part of the wave will steepen to form a shock (as in the case of the perfect gas), and if $\Gamma < 0$, the expansive part will steepen to form a shock, which will necessarily be an expansion (rarefaction) shock. Finally, if $\Gamma = 0$, the characteristics will be parallel and the wave will have stationary form, i.e., undistorted. The three possible cases are shown in Fig. 8.15.

These dynamic results are a remarkable confirmation of the thermodynamic condition (7.27*b*),

$$\frac{T_1[s]}{c_1^{\,2}} = \frac{\Gamma_1}{6}\,\Pi^3 + \cdots$$

Table 8.1 Thermodynamic Variables and Shock Formation

Compression shocks	$\Gamma > 1$	$\left(\frac{\partial c}{\partial P}\right)_s > 0$	Normal case, e.g., perfect gas
	$\Gamma = 1$	$\left(\frac{\partial c}{\partial P}\right)_s = 0$	Acoustic equation of state; P is a linear function of ρ; $c =$ const
	$0 < \Gamma < 1$	$\left(\frac{\partial c}{\partial P}\right)_s < 0$	Possible in some liquids and low-temperature gases
No shocks	$\Gamma = 0$	$\left(\frac{\partial c}{\partial P}\right)_s < 0$	Stationary waves; P a linear function of v
Rarefaction shocks	$\Gamma < 0$	$\left(\frac{\partial c}{\partial P}\right)_s < 0$	Possible near a critical point

found for weak shocks in Chap. 7. Thus, either compression shocks ($\Pi > 0$) or rarefaction shocks ($\Pi < 0$) are allowable, depending on whether Γ is positive or negative. It is the conclusion of Eq. (8.25) that shocks will form dynamically according to the same selection rule.

For completeness, the possible cases are listed in Table 8.1. All but the first case are somewhat esoteric. The case $\Gamma = 0$, leading to undistorted (stationary) waves, is superficially like the acoustic case; however, it follows not from the constancy of c but from the tendency of increases in u to just balance decreases in c so that $u + c$ remains constant. In *every* case it follows from $du - dP/\rho c = 0$ that fluid in regions of positive condensation ($P > P_0$) travels in the same direction as the wave.

Shock Location

Suppose that the density waveform is calculated beyond the point at which a shock first forms, as shown in Figs. 8.10 and 8.11. The required shock discontinuity can be fitted to such a waveform by making use of mass conservation, as shown in Fig. 8.16.

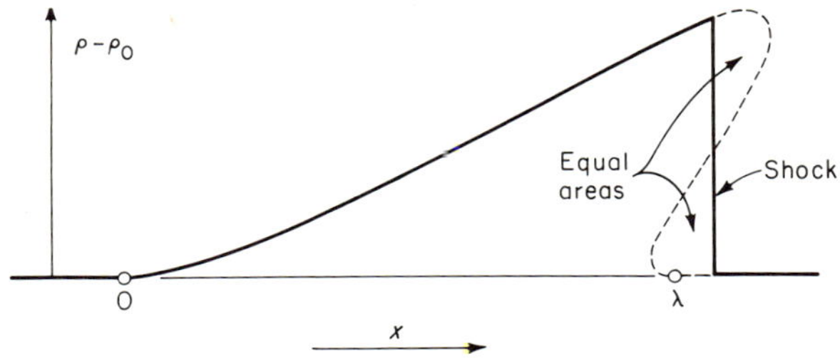

Figure 8.16
Shock fitted to a density waveform.

The characteristics calculation conserves mass (in a bizarre way, however, after the waveform becomes triple-valued) so that $\int_{-\infty}^{\infty} \rho \, dx =$ const. Then the shock can be fitted such that this condition remains satisfied, as shown in the figure. In general, this technique is applicable only to weak shocks, because the characteristics calculation for regions ahead of the shock is in error according to the strength of the shock, as discussed in Sec. 8.7.

8.6 Piston-induced motion of a perfect gas

As a further application of the theory of simple waves we take up the motion of an initially stationary gas confined within a cylinder closed by a piston. The cases of piston *withdrawal* and piston *advance* will be treated separately.

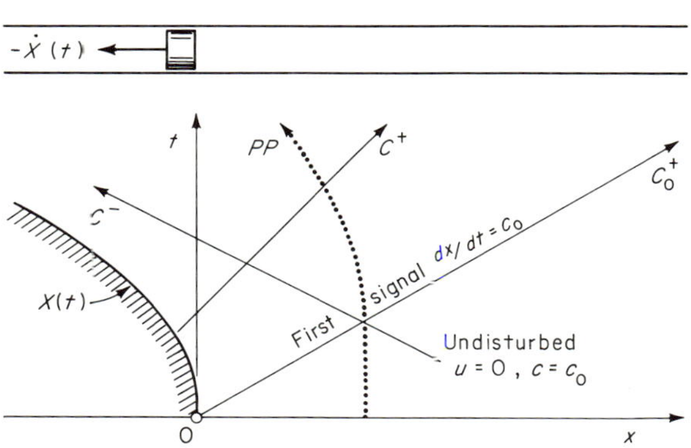

Figure 8.17
Wave diagram for continuous piston withdrawal.

Continuous Piston Withdrawal

At time zero a tube contains gas at uniform state $u = 0$, $c = c_0$. Piston withdrawal is initiated at this time, with finite acceleration (Fig. 8.17). The tube is assumed to be sufficiently long to ensure that no reflected disturbances from the right-hand end appear during the time of interest; i.e., the wave motion remains simple.[1]

This problem can be solved analytically. Let the piston position be given by the known function $X(t)$, with $\dot{X} < 0$. At $t = 0$ a sound wave leaves the piston and travels into undisturbed fluid; this is the *first signal*, and below this characteristic is necessarily an *undisturbed region*. We then have simple waves with the characteristics C^- originating in this region. We assume, subject to future verification, that every C^- characteristic reaches the piston path line $X(t)$. Then with $J^- = \text{const}$ everywhere, with (8.15),

$$u - \frac{2}{\gamma - 1} c = -\frac{2}{\gamma - 1} c_0 \qquad \text{everywhere} \tag{8.26}$$

This equation reduces the number of unknowns to one; the specified piston motion then determines the problem. A typical (straight) characteristic C^+ carries constant c and constant u; the value of u is thus necessarily just the piston velocity $\dot{X}$ where C^+ intersects $X(t)$, assuming that the gas does not separate from the piston. Then on C^+

$$u = \dot{X} \qquad c = c_0 + \frac{\gamma - 1}{2} \dot{X} \tag{8.27}$$

the latter relation following from (8.26). Note that $c \leq c_0$ because $\dot{X} < 0$; then $P \leq P_0$ from the isentropic relation, as one would intuitively expect.

The (inverse) slope of the C^+ characteristic is $dx/dt = u + c$, which is from (8.26) and (8.27)

$$u + c = c_0 + \frac{\gamma + 1}{2} \dot{X} \tag{8.28}$$

For sufficiently large piston speeds this goes to zero (so that C^+ is vertical) when $u = -c$, corresponding to a sonic flow to the left, with conditions

[1] Physical examples of piston-withdrawal (rarefaction) flows include the internal flows associated with the popping of a champagne cork and the discharge of a cannon.

necessarily *constant* in time at the corresponding value of x. This sonic value of u, from (8.28) set to zero, is $u = \dot{X} = -[2/(\gamma + 1)]c_0$ or

$$|u|_{\text{sonic}} = \frac{2}{\gamma + 1} c_0 \tag{8.29}$$

This may be compared with the sonic flow speed for *steady* isentropic acceleration, which is, from (5.56),

$$|u|_{\text{sonic}} = \sqrt{\frac{2}{\gamma + 1}}\, c_0 \tag{8.30}$$

At very large piston speed, the sound speed given by (8.27) goes to *zero* (with the corresponding values of pressure, density, and temperature also formally zero from the isentropic relations). This occurs at $\dot{X} = -[2/(\gamma - 1)]c_0$ or

$$|u|_{\text{escape}} = \frac{2}{\gamma - 1} c_0 \tag{8.31}$$

which is the *escape speed*, as will be explained. With $c = 0$, the slopes of the particle path and characteristics C^+ and C^- ($dx/dt = u$, $u + c$, and $u - c$, respectively) are necessarily identical, so that these three lines are congruent (see the characteristic labeled $C^\pm$ in Fig. 8.18). Even though the piston may accelerate to still higher speed, the gas cannot physically expand to lower sound speed, so that the gas must *separate* from the

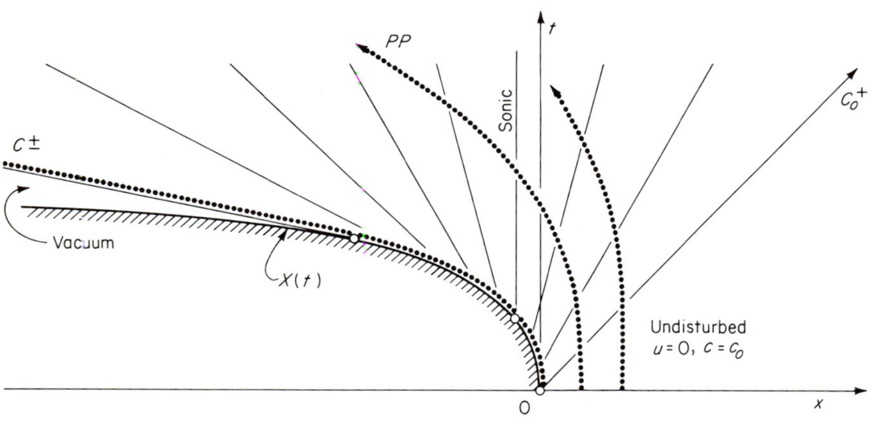

Figure 8.18
Wave diagram illustrating separation of the gas from the piston at the escape speed.

piston, as illustrated in the wave diagram. The rather large flow speed given by (8.31), on the order of a few kilometers per second, is called the escape speed because it is just the limiting speed of gas which is allowed to expand suddenly into a vacuum. It may be compared to the limiting speed of a perfect gas expanded into a vacuum in *steady* isentropic flow, which is, from Eq. (5.61),

$$|u|_{\max} = \sqrt{\frac{2}{\gamma - 1}}\, c_0 \tag{8.32}$$

where c_0 is the stagnation sound speed.

The physical attainment of the escape speed (8.31) is doubtful because the requisite piston speed would be difficult to attain and because the *equilibrium* isentropic expansion of a gas would eventually result in condensation of liquid or solid. Even if condensation did not set in, by virtue of the very rapid expansion, the gas would eventually become so dilute ($\Lambda \to \infty$ as $\rho \to 0$) that the continuum theory of gasdynamics would no longer be tenable.

The problem of putting the preceding results for piston withdrawal (which are that u and c have known values along the straight C^+ characteristics of known slope) into analytical form is an exercise in elementary analytic geometry which we will not pursue.

Impulsive Piston Withdrawal

The simplest possible piston withdrawal is the impulsive motion

$$\dot{X} = \begin{cases} 0 & t < 0 \\ \text{const} < 0 & t > 0 \end{cases}$$

That is, the piston instantaneously acquires some constant velocity $\dot{X} < 0$. As before, the gas is initially at rest with uniform conditions $u = 0$, $c = c_0$, as shown in Fig. 8.19.

In this problem the origin is a singular point where the velocity takes on all possible values between zero and $\dot{X}$; this can be accomplished only by the intersection of C^+ characteristics at the origin, giving multiple values for u and c at that point. This is superficially like a shock discontinuity; the difference is that this discontinuity does not persist in time but immediately decays, following the rule that *rarefaction waves spread out*.

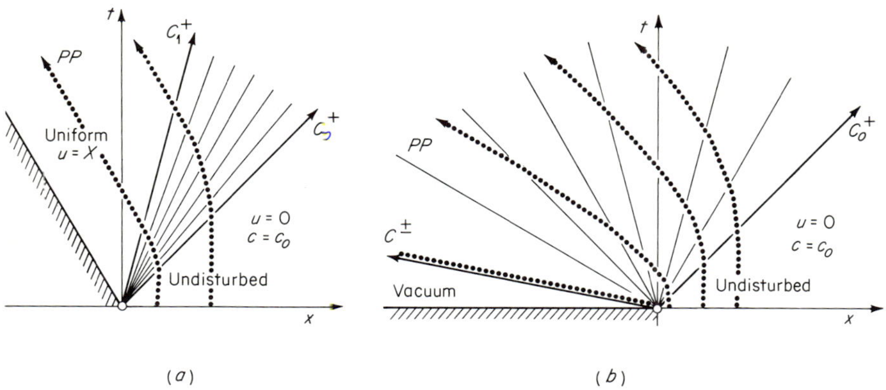

Figure 8.19
Wave diagram for impulsive piston withdrawal: (*a*) finite piston speed; (*b*) infinite piston speed.

The radiation of C^+ characteristics from the origin may be thought of as the limit for increasing piston acceleration to constant velocity. In the finite-acceleration case considered previously, the C^+ characteristics spread out from the curved piston path. As the curvature (acceleration) of the path is increased, these characteristics crowd together, until finally in the limit they radiate from a single point, i.e., the origin.

The region of radiating characteristics is called the *fan* or *expansion fan* or *centered rarefaction*. The xt field is then divided into three regions, viz., the undisturbed region, the fan, and the uniform region where $u = \dot{X}$. Consider the fan region of Fig. 8.19*a*; because the wave is simple, with $J^- =$ const everywhere, we have, from Eq. (8.27),

$$c = c_0 + \frac{\gamma - 1}{2} u$$

and from geometry the C^+ characteristics are straight lines given by

$$\frac{x}{t} = u + c = c_0 + \frac{\gamma + 1}{2} u \qquad 0 \geq u \geq \dot{X}$$

Solving this for u and c in terms of x/t gives the dimensionless forms (with the conditions in the undisturbed region listed also, for completeness)

$$\left.\begin{aligned} \frac{u}{c_0} &= 0 \\ \frac{c}{c_0} &= 1 \end{aligned}\right\} \quad \infty > \frac{x}{c_0 t} \geq 1 \tag{8.33}$$

$$\left.\begin{aligned} \frac{u}{c_0} &= -\frac{2}{\gamma+1}\left(1-\frac{x}{c_0 t}\right) \\ \frac{c}{c_0} &= 1-\frac{\gamma-1}{\gamma+1}\left(1-\frac{x}{c_0 t}\right) \end{aligned}\right\} \quad 1 \geq \frac{x}{c_0 t} \geq 1 + \frac{\gamma+1}{2}\frac{\dot{X}}{c_0} \tag{8.34}$$

so that in this case we have simple explicit formulas for u and c in terms of the independent variables.

The expansion process is terminated on the C^+ characteristic where u reaches its final value $\dot{X}$. The properties in the uniform region are then found to be

$$\left.\begin{aligned} \frac{u}{c_0} &= \frac{\dot{X}}{c_0} \\ \frac{c}{c_0} &= 1+\frac{\gamma-1}{2}\frac{\dot{X}}{c_0} \end{aligned}\right\} \quad 1 + \frac{\gamma+1}{2}\frac{\dot{X}}{c_0} \geq \frac{x}{c_0 t} \geq \frac{\dot{X}}{c_0} \tag{8.35}$$

It is a remarkable feature of some unsteady flows that such regions of (temporarily) *steady* flow exist.

Just as before, the sound speed (and therefore the pressure and density) is zero at the escape speed, $u = -2c_0/(\gamma - 1)$. For piston speeds greater than this, the above formulas are invalid, as they would imply negative pressures; the expansion process is then terminated at $c = 0$ (Fig. 8.19*b*). We may conveniently call this the case of *infinite piston speed* [although the flow field is the same for all piston speeds greater than $2c_0/(\gamma - 1)$]. Physically, such a flow is approximated by sudden expansion into a vacuum, as when a barrier at $x = 0$ is instantaneously removed, e.g., by the rapid rupture of a diaphragm. The gas then escapes into the vacuum region.

The distribution of pressure, density, and sound speed, determined from (8.34) and the isentropic relations, is shown in Fig. 8.20 for the case of expansion to zero pressure (infinite piston speed or expansion into a vacuum) and expansion to finite pressure. These figures illustrate a property of characteristics which has not been mentioned, that *discontinuities in the derivatives of the independent variables*, for example in $\partial c/\partial x$, *may occur on characteristics.* In the problem at hand, the derivatives of u and c are discontinuous across C_0^+ and C_1^+ (Fig. 8.19*a*). Certain characteristics thus represent second-order surfaces of discontinuity in physical space and are somewhat analogous to shock waves. (A further discussion of these points appears in Sec. 9.4.)

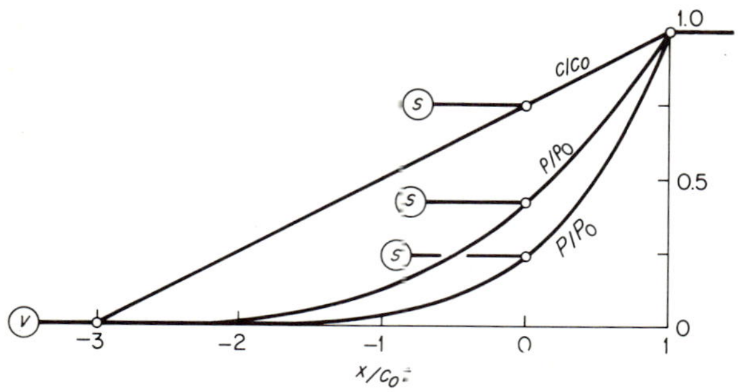

Figure 8.20
Variation of properties across a centered rarefaction terminating at vacuum v and sonic s conditions (calculated for $\gamma = 5/3$).

The *particle paths* (*PP*) in the centered rarefaction can be found from (8.34). They have the (inverse) slope

$$u = \frac{dx}{dt} = -\frac{2}{\gamma + 1} c_0 + \frac{2}{\gamma + 1} \frac{x}{t}$$

This is a first-order ordinary differential equation for the fluid-particle position $x(t)$. Standard solution techniques give

$$x = -\frac{2}{\gamma - 1} c_0 t + A t^{2/(\gamma+1)}$$

Let x_0 be the initial position of the fluid particle. Then the expansion starts at $t_0 = x_0/c_0$ (along C_0^+ in Fig. 8.19); for $t < t_0$ we are in the undisturbed region, and the above solution is inapplicable; that is, $x = x_0 = \text{const}$ in the undisturbed region. With $u = 0$ at $t = t_0$, evaluation of the constant A gives

$$x = -\frac{2}{\gamma - 1} c_0 t + \frac{\gamma + 1}{\gamma - 1} x_0 \left(\frac{c_0 t}{x_0}\right)^{2/(\gamma+1)} \tag{8.36}$$

One final comment: we have found that the dependent variables u/c_0 and c/c_0 depend only on the *similarity variable* $x/c_0 t$. This variable can be deduced from dimensional arguments, as discussed in Sec. 3.6. The self-similar quality of the motion is manifested in the wave diagram of Fig. 8.19, which is indifferent to changes in scale; i.e., we see exactly the same figure after photographic enlargement (the reader can verify that an equivalent statement cannot be made about Fig. 8.18, for example).

Continuous Piston Advance

This is just the case $\dot{X} > 0$ (see Fig. 8.21). The equations already developed for continuous withdrawal are applicable; an important difference, however, is the inevitable formation of one or more shocks, leading to possible invalidation of the homentropic model.

A C^+ characteristic coming from the piston path at a point where the piston velocity is $\dot{X}$ has (inverse) slope given by (8.27)

$$u + c = c_0 + \frac{\gamma + 1}{2}\dot{X}$$

This value increases monotonically in moving up the path of the accelerating piston (physically the sound speed increases with the motion-induced compression), so that characteristics are progressively bent over and eventually converge to form a shock.

A shock first forms at point *a*, where two C^+ characteristics intersect (this point does not necessarily lie on the first-signal characteristic, however). In the region to the left of the C^- characteristic shown we have simple homentropic waves; i.e., the method of characteristics already given is still applicable. In the region to the right of this, however, the shock conditions appear, and our simple calculation scheme will require modification. If the shock is *weak*, it can be fitted into the calculation with very little difficulty, as will be shown in Sec. 8.7; if it is strong, a new calculation procedure is in general required, as will be discussed in Sec. 8.10.

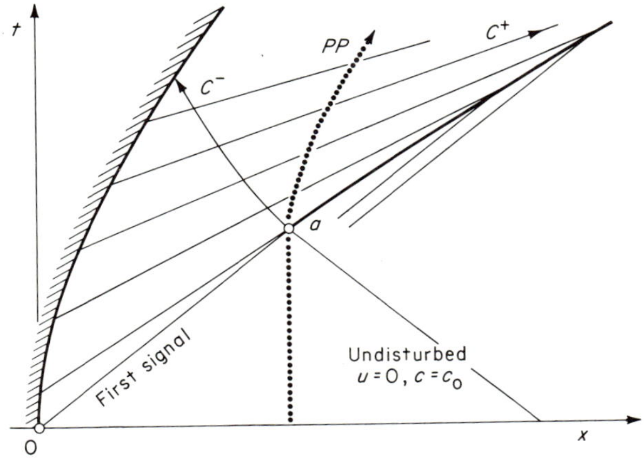

Figure 8.21
Wave diagram for continuous piston advance.

EXAMPLE 8.1 TIME OF SHOCK FORMATION

Find the time t_* at which a shock first forms under a given continuous piston advance $X(t)$ (see Fig. 8.22).

This is similar to the calculation leading to Eq. (8.23), except that the boundary conditions are differently put. From some arbitrary point t on the piston path characteristics will intersect in a time

$$t_* - t = \frac{\delta x}{d(u + c)}$$

[compare Eq. (8.23)]. It is now required to find the differential distance δx. From (8.28),

$$u + c = c_0 + \frac{\gamma + 1}{2} \dot{X}$$

From geometry, $\delta x + dx = (u + c)\, dt$ and $dx = \dot{X}\, dt$; then

$$\delta x = (c_0 + \frac{\gamma - 1}{2} \dot{X})\, dt$$

and the time interval is

$$t_* - t = \frac{\delta x}{d(u + c)} = \frac{(c_0 + \frac{\gamma - 1}{2} \dot{X})\, dt}{\frac{\gamma + 1}{2} \ddot{X}\, dt}$$

where $d(u + c)$ was found from the expression above for $u + c$. This rearranges to

$$t_* = t + \frac{1}{\ddot{X}} \left(\frac{2}{\gamma + 1} c_0 + \frac{\gamma - 1}{\gamma + 1} \dot{X} \right)$$

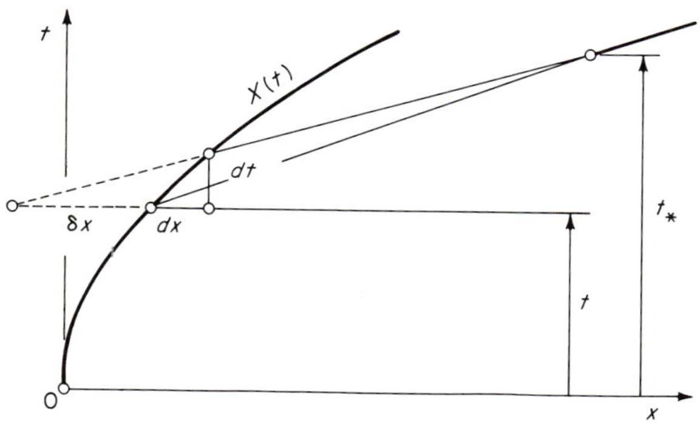

Figure 8.22

This is a minimum where $dt_*/dt = 0$. Performing the differentiation and setting to zero gives

$$(\ddot{X})^2 = \frac{1}{\gamma}\left(c_0 + \frac{\gamma - 1}{2}\dot{X}\right)\dddot{X}$$

which is an implicit relation for $t^*_{\min}$. For the special case of a power-law advance, $X = At^n$, this gives

$$\left(\frac{n-1}{n-2} - \frac{\gamma-1}{2\gamma}\right)nAt^{n-1} = \frac{c_0}{\gamma}$$

For $n = 2$ (parabolic piston path) this in turn gives $t = 0$; hence,

$$t^*_{\min} = \frac{c_0}{(\gamma + 1)A}$$

The corresponding position x_* at which the shock first forms is $X(t) + (u + c) \times (t_* - t)$, which works out to be, for the parabolic case,

$$x^*_{\min} = \frac{{c_0}^2}{(\gamma + 1)A}$$

(Strictly, it should be remarked that t_* is a linear increasing function of t in the parabolic case; thus, there is a sharp and obvious minimum at $t = 0$, and the minimum condition is actually satisfied at a discontinuity.)

Impulsive Piston Advance

If the piston instantaneously acquires velocity $\dot{X} > 0$, a shock must form at the origin of motion (see Fig. 8.23). Just as in the withdrawal motions already discussed, the impulsive case may be considered as the limit of

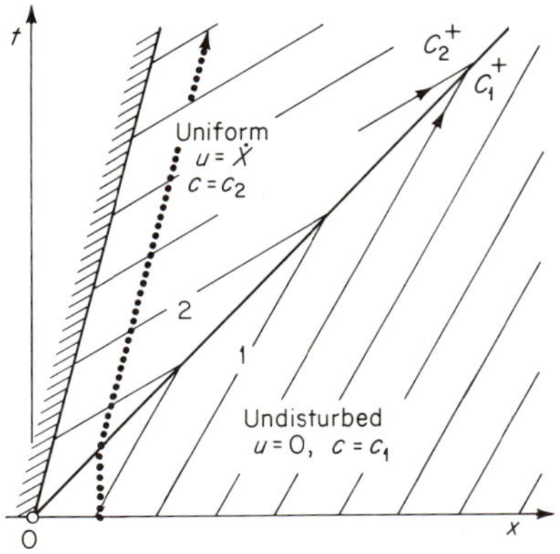

Figure 8.23
Wave diagram for impulsive piston advance.

continuous piston motion with increasing acceleration; the point of shock formation moves closer to the origin and in the limit occurs there.

Let the undisturbed region be denoted by subscript 1 and the region behind the shock by subscript 2 so that the notation is consistent with that of the shock conditions. With $[w] = -\dot{X}$, (7.33) gives

$$\frac{\dot{X}}{c_1} = \frac{2}{\gamma + 1}\left(M_{1n} - \frac{1}{M_{1n}}\right) \tag{8.37}$$

This gives the shock Mach number M_{1n} in terms of $\dot{X}/c_1$, and all quantities of interest (P_2/P_1, T_2/T_1, etc.) can then be calculated.

The rightward-traveling shock shown in Fig. 8.23 is called a C^+ shock because it travels at a speed intermediate between that of the upstream and downstream C^+ waves, as shown in Chap. 7 (in the limit of vanishing strength, the shock path becomes of course a characteristic). The motion shown in Fig. 8.23 is divided into two uniform regions by the shock path. The flow is then *piecewise homentropic*. The motion is also self-similar, depending only on x/c_0t, like impulsive withdrawal.

Each of the motions in this section and in the previous one was discussed in terms of C^+ waves. The results for simple C^- waves (e.g., with the piston at the *right-hand* end of the tube in Fig. 8.17 and x still positive to the right—the piston is withdrawn for piston velocities $\dot{X} > 0$) are entirely analogous; it is only necessary to change the sign of u (and $\dot{X}$) in each of the solutions. This reflects the fact that simple C^+ waves and simple C^- waves differ only in direction of travel, i.e., the sense in which x is chosen positive.

8.7 Flow with weak shocks

For weak shocks, we have already found that the entropy jump varies as the third power of the pressure jump, $[s] \sim [P]^3$. Shocks with small pressure jump are then very nearly isentropic. As an approximation, the flow field containing weak shocks may be treated as homentropic, and the basic assumption for the elementary method of characteristics remains satisfied. It will be shown below that the characteristic crossing a weak shock has negligible change in its invariant; with this result, calculations proceed as before, the shock path being found according to a simple rule given below.

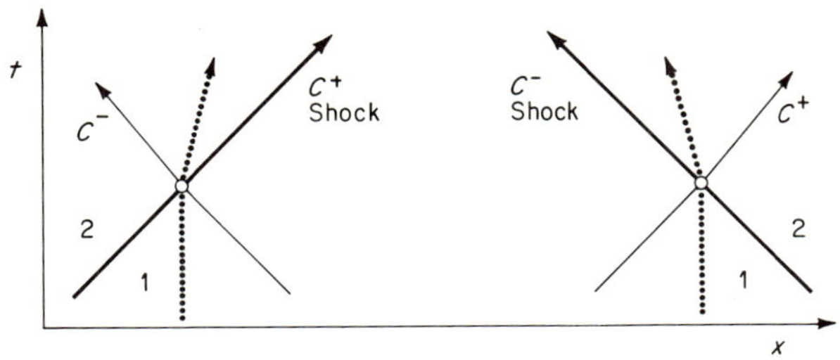

Figure 8.24

The shock may be either a C^+ shock or a C^- shock (see Fig. 8.24). To show that the invariant (J^+ or J^-) carried by the crossing characteristic has in effect negligible change, we will calculate the velocity jump $u_2 - u_1$ from simple wave theory, i.e., assuming that J does not show any change, and from the shock conditions and then compare results.

Consider a C^+ shock; if $J^- = u - \int dP/\rho c$ is unchanged across the shock, we have $J_1^- = J_2^-$, or

$$u_2 - u_1 = \int_{P_1}^{P_2} \frac{dP}{\rho c}$$

With

$$\left(\frac{\partial v}{\partial P}\right)_s = \frac{dv}{d\rho}\left(\frac{\partial \rho}{\partial P}\right)_s = -\frac{1}{\rho^2 c^2}$$

this can be written

$$u_2 - u_1 = \int_{P_1}^{P_2} \sqrt{-\left(\frac{\partial v}{\partial P}\right)_s}\, dP \tag{8.38}$$

Expanding the derivative by a Taylor series about the upstream state 1 gives

$$\left(\frac{\partial v}{\partial P}\right)_s = \left(\frac{\partial v}{\partial P}\right)_s + \left(\frac{\partial^2 v}{\partial P^2}\right)_s (P - P_1) + \frac{1}{2}\left(\frac{\partial^3 v}{\partial P^3}\right)_s (P - P_1)^2 + \cdots$$

where all the derivatives on the right-hand side are evaluated at state 1. Taking the square root (by the binomial theorem) and integrating (8.38) yields

$$\frac{u_2 - u_1}{c_1} = \Pi - \frac{\Gamma_1}{2}\Pi^2 - \frac{1}{6}\left[\Gamma_1{}^2 + \frac{c_1{}^6}{2v_1{}^4}\left(\frac{\partial^3 v}{\partial P^3}\right)_s\right]\Pi^3 \cdots \tag{8.39}$$

where Γ is the fundamental derivative defined in (5.8) and $\Pi \equiv [P]/\rho_1 c_1^2$ is the shock strength, as before. The corresponding calculation across a shock (including the effect of entropy change) was given by Eq. (7.49). With $u_2 - u_1 = -[w]$, subtracting these two results gives

$$\frac{[u]_{\text{shock conditions}} - [u]_{J=K}}{c_1} = \frac{\Gamma_1}{24}\left[\Gamma_1 - \frac{2c_1^2}{v_1 T_1}\left(\frac{\partial T}{\partial P}\right)_s\right]\Pi^3 \cdots \tag{8.40}$$

This is of third order in the shock strength and therefore *of the same order as the entropy jump*. For a *perfect gas*, with

$$\left(\frac{\partial T}{\partial P}\right)_s = \frac{\gamma - 1}{\gamma}\frac{T}{P} \qquad \text{and} \qquad \Gamma = \frac{\gamma + 1}{2}$$

this becomes

$$\frac{[u]_{\text{shock conditions}} - [u]_{J=K}}{c_1} = \frac{(\gamma + 1)(5 - 3\gamma)}{96}\Pi^3 \cdots \tag{8.41}$$

In the case of a perfect gas, another way of arriving at the same conclusion is to calculate the jump in the invariant directly from the weak-shock conditions given in Appendix A. This gives

$$\frac{[J^-]}{c_1} = \frac{\left[u - \dfrac{2}{\gamma - 1}c\right]}{c_1} = -\frac{3(\gamma + 1)^2}{96}\Pi^3 \cdots \tag{8.42}$$

The remarkable smallness of the (effective) jump in the Riemann invariant for weak shocks is of course a great practical advantage, permitting us to retain a fairly simple method of characteristics, in which shocks are treated as isentropic discontinuities.

The shock path can be found from the condition that the *shock velocity is approximately the average of the upstream and downstream wave velocities*. For example, for a C^+ shock we can write for the shock velocity V

$$V = \tfrac{1}{2}[(u_1 + c_1) + (u_2 + c_2)] + c_1 E \tag{8.43}$$

The error term $c_1 E$, which will be shown to be small, is introduced to guarantee equality. With $u_1 = V - w_1$ and $u_2 = V - w_2$ this is solved for the error term $c_1 E$ and put in the form

$$c_1 E = c_1(M_1 - 1) - \tfrac{1}{2}([c] - [w])$$

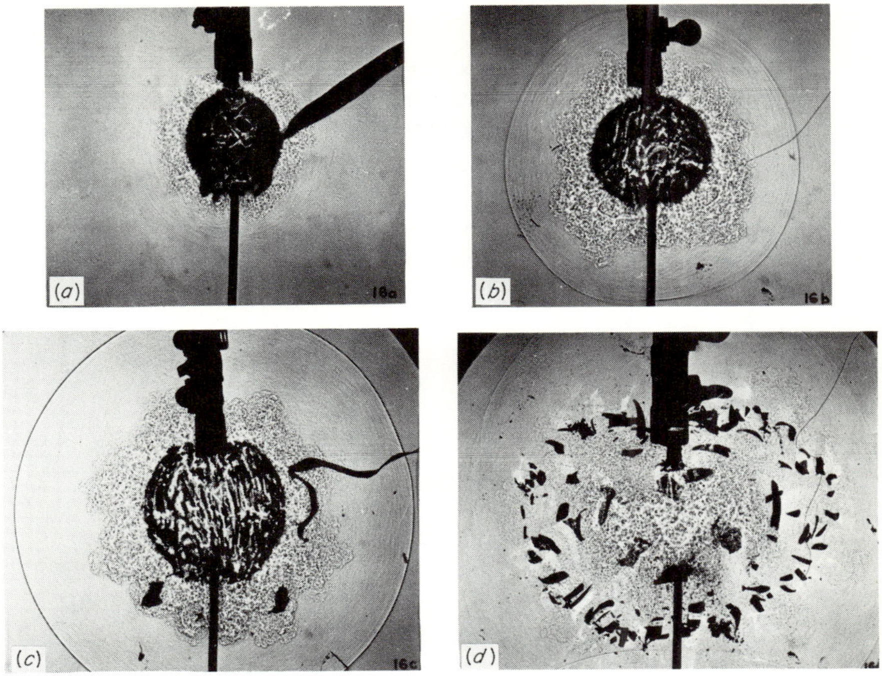

Figure 8.25
Shock formation in water resulting from the rupture of a glass sphere pressurized to 20 atm: (*a*) 200 μs after rupture; (*b*) 255 μs; (*c*) 375 μs; (*d*) 850 μs. (*Courtesy of I. I. Glass, University of Toronto.*)

With the expansions given in Appendix A, this gives

$$E = \tfrac{1}{4}\left\{\Gamma_1\left[1 - \tfrac{3}{2}\Gamma_1 + \frac{c_1^{\,2}}{3v_1T_1}\left(\frac{\partial T}{\partial P}\right)_s\right] - \frac{c_1^{\,6}}{6v_1^{\,4}}\left(\frac{\partial^3 v}{\partial P^3}\right)_s\right\}\Pi^2 \cdots \tag{8.44}$$

which is of *second order* in the shock strength Π; that is, the approximation is not so good as that which neglects the jump in the invariant. For a perfect gas, this reduces to

$$E = \frac{(\gamma + 1)^2}{32}\,\Pi^2 \cdots \tag{8.45}$$

Formation of a weak shock in water is shown in Fig. 8.25.

EXAMPLE 8.2 APPLICATION OF THE WEAK-SHOCK APPROXIMATIONS

A tube contains stagnant perfect gas ($\gamma = 1.40$) with sound speed $c_1 = 1{,}100$ ft/s. A piston closing the end of the tube is moved impulsively inward at a speed of 412 ft/s. Find the velocity of the resulting shock and the pressure behind it by (1) the method of characteristics with weak-shock approximations and (2) exact shock conditions.

For variety, let the piston move in the negative x direction, generating a C^- shock (see Fig. 8.26). The shock is relatively weak with $-[w]/c_1 \sim [P]/P_1 < 1$. Taking $[J^+] = 0$ gives $u_2 + [2/(\gamma - 1)]c_2 = [2/(\gamma - 1)]c_1$ or, with $u_2 = \dot{X} = -412$ ft/s,

$$c_2 = c_1 - \frac{\gamma - 1}{2}\dot{X} = 1{,}182 \text{ ft/s}$$

The shock velocity is (approximately) the average of C^- velocities,

$$V = \tfrac{1}{2}[(u_2 - c_2) + (u_- - c_1)] = -1{,}347 \text{ ft/s}$$

The pressure is found from the isentropic relation

$$P_2 = P_1\left(\frac{c_2}{c_1}\right)^{2\gamma/(\gamma-1)} = 1.653P_1$$

The corresponding exact values, from the shock tables, are $V = -1.25(1{,}100) = -1{,}375$ ft/s and $P_2 = 1.656P_1$.

The shock strength is $\Pi = 0.656/1.4 \approx 0.47$. This is hardly small compared to unity, yet the weak-shock approximations are relatively successful.

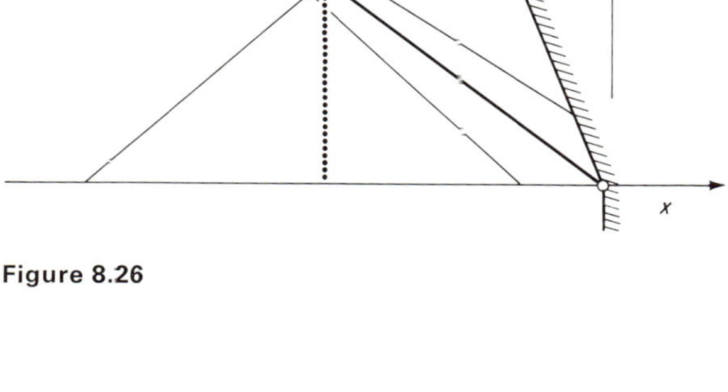

Figure 8.26

EXAMPLE 8.3 ATTENUATION OF A WEAK-SHOCK PULSE

The result that the velocity of a weak shock is approximately the average of the upstream and downstream wave velocities is the basis for a simple calculation of shock attenuation.

The local velocity of a wave traveling in the $+x$ direction is $u + c$; by (8.25), this velocity increases with pressure according to

$$\frac{d}{dP}(u + c) = \frac{\Gamma}{\rho c}$$

which can be written

$$\frac{d}{d\rho}(u + c) = \frac{\Gamma}{\rho c}\left(\frac{\partial P}{\partial \rho}\right)_s = \frac{c}{\rho}\Gamma$$

Let a uniform upstream state be denoted by subscript zero, for example, c_0. Then by Taylor expansion, with $u_0 = 0$,

$$u + c = c_0 + c_0\Gamma_0 S + O(S^2) \tag{8.46}$$

where $S \equiv (\rho - \rho_0)/\rho_0$ is the condensation, or relative density change, as already used in acoustics. In the acoustical approximation, only the zero-order term is retained in this expansion; i.e., the wave velocity is $u + c = c_0 = \text{const}$. Here we will retain the first-order term and thus account for nonlinear effects. Now let S be the amplitude of a wave immediately behind the shock front (Fig. 8.27). The velocity of the shock front itself is approximately, by (8.43), $V = \frac{1}{2}[c_0 + (u + c)_S]$; with (8.46), this is simply

$$V = c_0 + \tfrac{1}{2}c_0\Gamma_0 S + O(S^2) \tag{8.47}$$

The asymptotic form of a deforming plane wave, or shock pulse, is triangular, as shown in the figure (see also Figs. 8.10 and 8.11). From conservation of mass, the area under the wave is constant (this can also be expressed by equality of areas $bb'd$ and $cc'd$, as discussed in Sec. 8.5; see Fig. 8.16). Then

$$LS = \text{const} \equiv L^0 S^0 \tag{8.48}$$

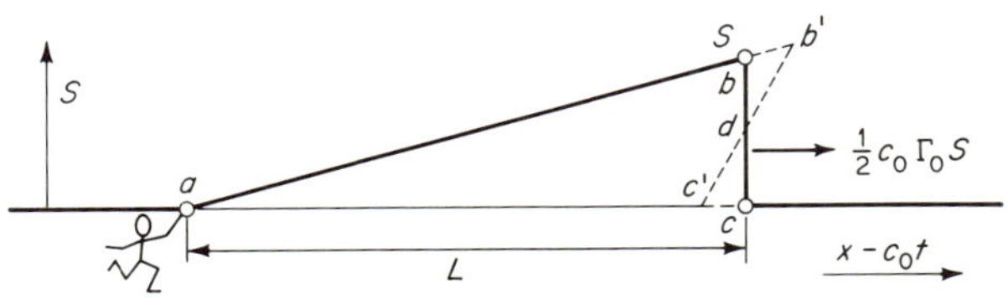

Figure 8.27
Asymptotic form of a simple plane wave.

where L is the length of the wave. From (8.47) and (8.48)

$$\frac{dL}{dt} = \tfrac{1}{2}c_0\Gamma_0 S = \tfrac{1}{2}c_0\Gamma_0\frac{L^0S^0}{L}$$

Integrating,

$$L^2 - (L^0)^2 = c_0\Gamma_0 L^0 S^0 t$$

Since S is small, the entire pulse is traveling at velocity c_0, very nearly. We can therefore put the propagation distance $X = c_0 t$, and the above becomes, at large distances,

$$L \propto x^{1/2} \tag{8.49}$$

Since LS = const, the shock strength (wave amplitude) varies according to

$$S \propto x^{-1/2} \tag{8.50}$$

This same variation applies also, of course, to the amplitude of velocity, pressure, etc.

It is remarkable that a theory which retains terms only to the first power in S predicts *any* (nonzero) attenuation. The entropy jump $[s]$ varies as the third power in S, and dissipation is thus left out of the theory—yet the wave attenuates. The *energy* carried by the wave is proportional to LS^2 and therefore varies as $x^{-1/2}$ (the above solution thus does not strictly satisfy conservation of energy, since it does not account for the residual internal energy left after the passage of the wave).

The physically more interesting case of a *cylindrical-geometry* wave can be treated similarly. In a heuristic derivation (following *Landau and Lifshitz* [1959, chap. 10], we will make use of three results: (1) that the amplitude of cylindrical waves far from the source varies as $r^{-1/2}$ in the acoustic approximation in the absence of shocks, (2) that (8.46) will still apply because the wave motion is locally nearly one-dimensional, (3) that (8.47) will apply, from the weak-shock conditions.

The wave geometry is shown in Fig. 8.28. In the absence of a shock, the wave amplitude is given by $S^0\sqrt{r^0/r}$, where S^0 is the amplitude at some given position r^0. From (8.46), the excess velocity of the point b' is then

$$(u + c) - c_0 = c_0\Gamma_0 S^0\sqrt{\frac{r^0}{r}}$$

Then the distance by which this point has advanced between r^0 and r is

$$G = \int_{t^0}^{t} c_0\Gamma_0 S^0\sqrt{\frac{r^0}{r}}\,dt$$

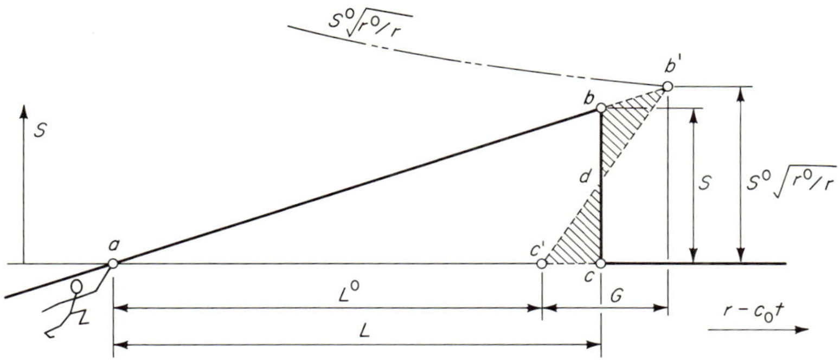

Figure 8.28
Asymptotic form of a simple cylindrical wave.

With $dt = dx/c_0$, this integrates to

$$G = 2\Gamma_0 S^0\sqrt{r^0}(\sqrt{r} - \sqrt{r^0}) \tag{8.51}$$

From the geometry of the figure,

$$\frac{S}{L} = \frac{S^0\sqrt{r^0/r}}{L^0 + G} \tag{8.52}$$

It is no longer true in the present case that the pulse area *abc* is constant. It is, however, still true that the shaded areas are equal, i.e., that area *dbb'* equals area *dcc'*, a result which follows from (8.46) and (8.47), the condition that the shock front advance half as fast as *b'*. Then it follows that the areas *abc* and *ab'c'* are equal; this can be written

$$LS = L^0 S^0\sqrt{\frac{r^0}{r}} \tag{8.53}$$

With (8.52) and (8.51), this gives

$$L^2 = (L^0)^2 + 2\Gamma_0 S^0\sqrt{r^0}(\sqrt{r} - \sqrt{r^0})L^0$$

Thus, for large r,

$$L \propto r^{1/4} \tag{8.54}$$

From (8.53) the wave amplitude S then varies according to

$$S \propto r^{-3/4} \tag{8.55}$$

This important result predicts, for example, that the strength of a sonic boom in a *uniform* atmosphere will vary as $y^{-3/4}$, where y is the distance from the line of flight.

It should be mentioned that shock pulses in cylindrical or spherical geometries almost always include a compression region followed by a rarefaction region, such that $\int S\, dr \approx 0$. This follows from the requirement that the fluid following the pulse must be at rest in a realistic case (see Prob. 4.34). In consequence of this, the asymptotic form of such pulses is an *N wave.*

The problem sketched in this example has been extensively treated. Classical references are *Landau* [1945] and *Whitham* [1956]. A collection of papers has been published under the title *Sonic Boom Theory*, American Institute of Aeronautics and Astronautics, 1969.

8.8 Wave interactions

Sooner or later simple waves and shocks reach boundaries, resulting in *reflection* or *interaction* processes at the boundary. In this section we discuss a variety of such processes.

Reflection of a Characteristic at a Rigid Boundary

As an example of the physical process of interest, the reflection of a centered rarefaction from a stationary wall is shown in Fig. 8.29. It is convenient to consider the C^- characteristics shown as *reflections* of the C^+ characteristics incident on the wall.

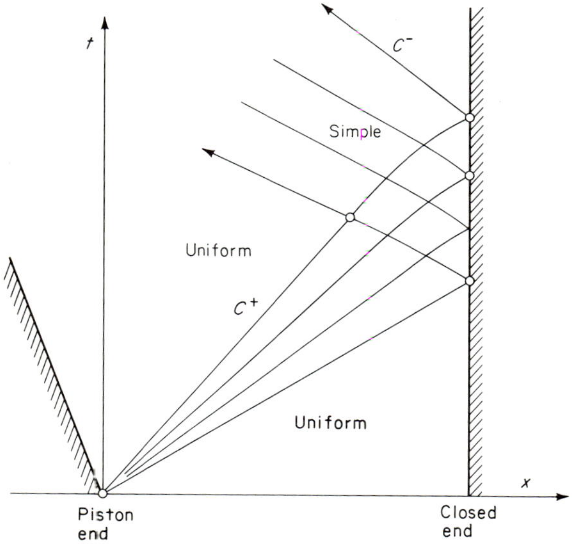

Figure 8.29
Reflection of a centered rarefaction at a closed end.

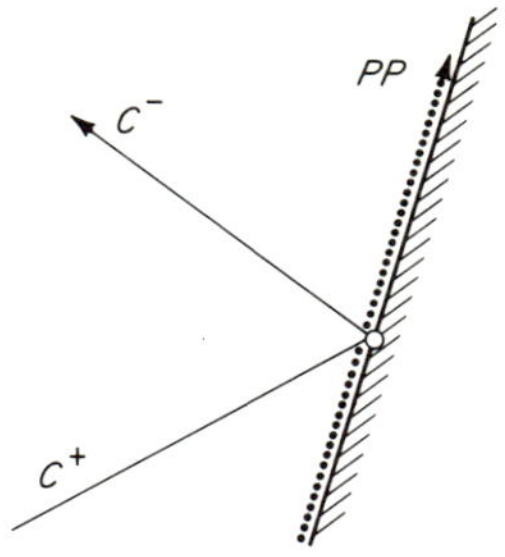

Figure 8.30

The incident C^+ characteristic carries a value J_i^+, presumed known; it is desired to find the value of J_r^- on the reflected characteristic C^-. For generality, let the reflecting boundary have some velocity u_b, as in the case of a piston (see Fig. 8.30). At the boundary the fluid velocity is necessarily u_b, so that

$$J_i^+ = u_b + F_b$$

$$J_r^- = u_b - F_b$$

where F_b is the value of the thermodynamic property F at the wall. Eliminating F_b,

$$J_r^- = -J_i^+ + 2u_b \tag{8.56}$$

The corresponding result for reflection from a left-hand boundary, with incident characteristic C^-, is

$$J_r^+ = -J_i^- + 2u_b \tag{8.57}$$

For a *stationary* boundary ($u_b = 0$), *characteristics simply reflect with a change in sign.*

Reflection at an Open End with Outflow

Consider unsteady flow from a tube into a stationary atmosphere. In steady flow the discharge flow has precisely the ambient pressure P_a of the outside atmosphere *provided that the discharge is subsonic*; the boundary condition is thus simply $P = P_a$. This condition is normally applied also to *unsteady flow*. This is an approximation, since fluctuation in the discharge flow locally affects the ambient pressure; in practice, however, this boundary condition seems to work quite well. For homentropic flow,

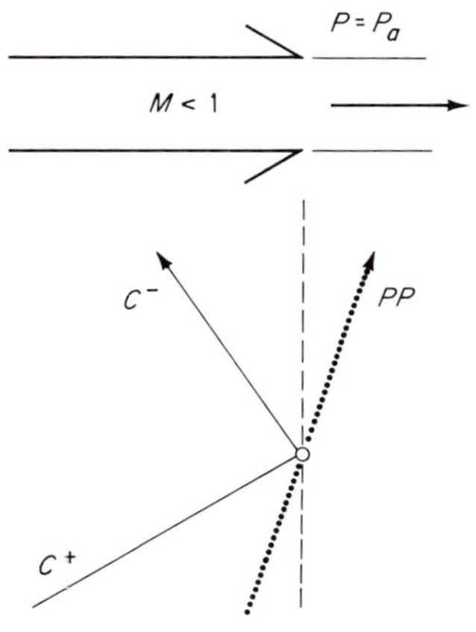

Figure 8.31
Subsonic outflow.

let the constant value of F corresponding to ambient pressure be $F_a = F(P_a)$. Again with J_i^+ known we wish to calculate J_r^- (see Fig. 8.31):

$$J_i^+ = u + F_a$$

$$J_r^- = u - F_a$$

Eliminating u gives

$$J_r^- = J_i^+ - 2F_a \tag{8.58}$$

The corresponding result for an open end on the left-hand side is

$$J_r^+ = J_i^- + 2F_a \tag{8.59}$$

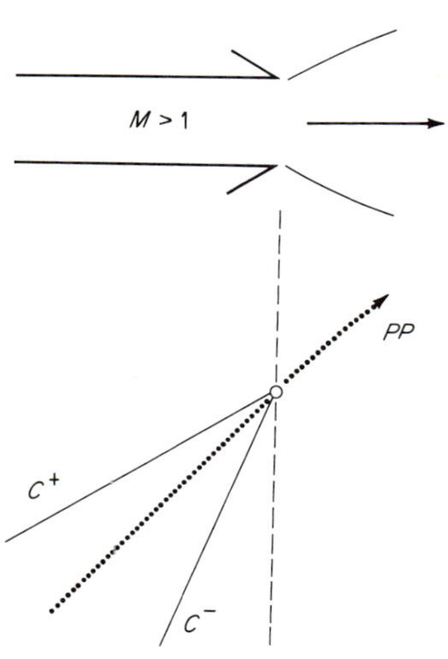

Figure 8.32
Supersonic outflow.

For *supersonic* outflow, both families of characteristics are incident on the open end from within the tube, as shown in Fig. 8.32, and there is no reflected characteristic; i.e., no sound wave can travel upstream in a supersonic flow. In this case conditions in the exit plane where C^+ and C^- intersect are predetermined by the values of J^+ and J^-. The exit pressure is in general different from atmospheric.

The case of a *choked* outflow ($M = 1$ at the exit plane) may occur in practice but does not have the physical significance it has for steady flow. For an illustration of this case, see Example 8.7 (page 421).

Reflection and Transmission at a Contact Surface

For an ideal (inviscid) contact surface separating substances 1 and 2, the matching conditions given in Sec. 7.7 are simply $P_1 = P_2$ and $u_1 = u_2$ (see Fig. 8.33). The known quantities are J_1^+ and J_2^-. Writing

$$J_1^+ = u + F_1(P)$$

$$J_2^- = u - F_2(P)$$

and eliminating u gives

$$F_1(P) + F_2(P) = J_1^+ - J_2^- \tag{8.60}$$

so that P, and consequently F_1 and F_2 and finally u, can be found. Then the transmitted invariants J_1^- and J_2^+ can be found from their definitions. Since the path of the contact surface is not known in advance, it may be constructed step by step.

The simplest case is that of a contact surface separating perfect gases

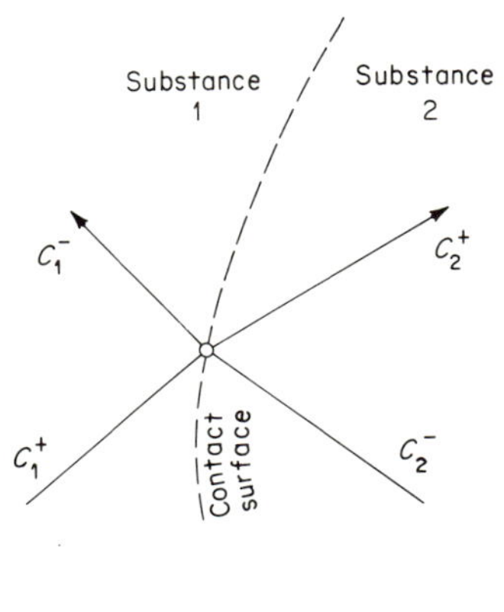

Figure 8.33

at different temperatures but of the same chemical species, e.g., separating hot and cold helium. The state equation can be written in the form

$$\frac{P}{P_0} = e^{(s_0 - s)/R}\left(\frac{T}{T_0}\right)^{\gamma/(\gamma-1)} \tag{8.61}$$

where $P_0(s_0,T_0)$ represents a fixed reference state. Then setting $P_2 = P_1$ gives for the ratio of sound speeds α at the contact surface

$$\alpha \equiv \frac{c_2}{c_1} = \exp\left(\frac{s_2 - s_1}{R}\frac{\gamma - 1}{2\gamma}\right) \tag{8.62}$$

which is necessarily a *constant* for the (piecewise) homentropic flow; this quantity is a dimensionless measure of the entropy change across the contact surface. With $c^2 = \gamma P v$ and $P_1 = P_2$ we also find that

$$\alpha = \frac{\rho_1 c_1}{\rho_2 c_2}$$

is just the constant ratio of acoustic impedances (introduced in Sec. 4.8). The condition that $u_1 = u_2$ can be written in the form

$$J_1^+ + J_1^- = J_2^+ + J_2^-$$

while the condition $\alpha c_1 = c_2$ (which we found from $P_1 = P_2$) can be written

$$\alpha(J_1^+ - J_1^-) = J_2^+ - J_2^-$$

Solving for the unknown transmitted invariants gives

$$\begin{aligned} J_2^+ &= \frac{2\alpha J_1^+ - (\alpha - 1)J_2^-}{\alpha + 1} \\ J_1^- &= \frac{(\alpha - 1)J_1^+ + 2J_2^-}{\alpha + 1} \end{aligned} \tag{8.63}$$

The reader can verify that this gives proper results for the degenerate cases $\alpha = 0, 1, \infty$, corresponding to various degrees of relative "hardness" of the two gases.

EXAMPLE 8.4 NUMERICAL METHOD OF CHARACTERISTICS

A tube of length L contains air at rest. The left end is closed by a movable piston which is withdrawn impulsively at one-half the stagnant-gas sound speed. The right end is permanently closed. Find the flow conditions in the xt region,

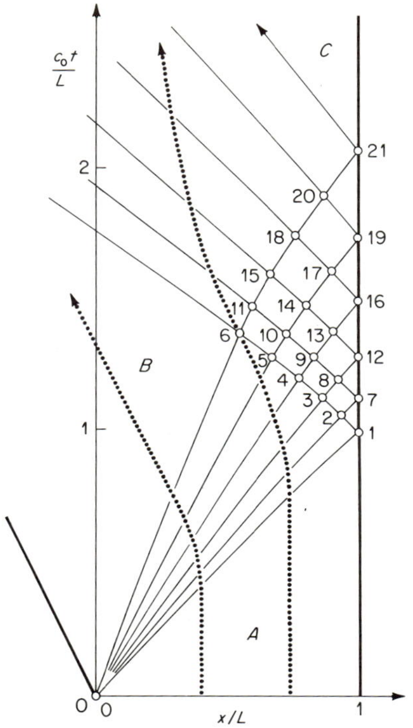

Figure 8.34
Wave diagram.

where the rarefaction fan reflects from the permanently closed end (see Fig. 8.34).

In a problem of this kind, it is useful to *normalize* the variables

$$X = \frac{x}{L} \qquad \tau = \frac{c_0 t}{L} \qquad \frac{2}{\gamma - 1} = 5$$

$$U = \frac{u}{c_0} \qquad C = \frac{c}{c_0}$$

Then the invariants are given by

$$J^+ = U + 5C \qquad U = \tfrac{1}{2}(J^+ + J^-)$$

$$J^- = U - 5C \qquad C = \tfrac{1}{10}(J^+ - J^-)$$

Conditions in the uniform regions are as follows:

A: $U = 0 \qquad C = 1$
B: $U = -0.5$ (given) $\qquad J^- = U - 5C = -5$
$C = 1 - \tfrac{1}{5}(0.5) = 0.9$
C: $U = 0 \qquad J^+ = U + 5C = 4 \qquad C = 0.8$

Table 8.2

	U	C	J^+	J^-	$U-C$	$U+C$
1	0	1	5.0	−5.0	−1	1
2	−0.1	0.98	4.8	−5.0	−1.08	0.88
3	−0.2	0.96	4.6	−5.0	−1.16	0.76
4	−0.3	0.94	4.4	−5.0	−1.24	0.64
5	−0.4	0.92	4.2	−5.0	−1.32	0.52
6	−0.5	0.90	4.0	−5.0	−1.40	0.40
7	0	0.96	4.8	−4.8	−0.96	0.96
8	−0.1	0.94	4.6	−4.8	−1.04	0.84
9	−0.2	0.92	4.4	−4.8	−1.12	0.72
10	−0.3	0.90	4.2	−4.8	−1.20	0.60
11	−0.4	0.88	4.0	−4.8	−1.28	0.48
12	0	0.92	4.6	−4.6	−0.92	0.92
13	−0.1	0.90	4.4	−4.6	−1.00	0.80
14	−0.2	0.88	4.2	−4.6	−1.08	0.68
15	−0.3	0.86	4.0	−4.6	−1.16	0.56
16	0	0.88	4.4	−4.4	−0.88	0.88
17	−0.1	0.86	4.2	−4.4	−0.96	0.76
18	−0.2	0.84	4.0	−4.4	−1.04	0.64
19	0	0.84	4.2	−4.2	−0.84	0.84
20	−0.1	0.82	4.0	−4.2	−0.92	0.72
21	0	0.80	4.0	−4.0	−0.80	0.80

By choosing the characteristics in the fan to have simple values of U (0, −0.1, −0.2, −0.3, −0.4, −0.5, respectively) the numerical values J^+ and J^- are made simple throughout and the necessary arithmetic can be done mentally. The table is calculated first; then the wave diagram is constructed on a drawing board, with the characteristics assigned average slopes between connected points. For example, the C^- characteristic joining points 1 and 2 has $\Delta X/\Delta\tau = (-1 + 1.08)/2$. One particle path is shown passing through point 6. This has been obtained graphically, such that the path line crosses each characteristic with appropriate slope corresponding to the known value of U on that characteristic. Let the distance of this path from the closed end be l. Then

$$\frac{l_C}{l_A} = \frac{V_C}{V_A} = \frac{\rho_A}{\rho_C} = \left(\frac{c_A}{c_C}\right)^{2/(\gamma-1)} = 3.05$$

An *analytical* solution for the above problem is given by *Landau and Lifshitz* [1959, chap. 10].

Shock Interactions

Consider the intersection of a shock with another surface of discontinuity. The second surface may be a shock of opposite family, a shock of the same family, a contact surface, an open end, or a closed end. In every case there is a pressure and/or velocity matching condition to be satisfied, and the various calculations proceed along the lines of those already given in Sec. 4.9 for acoustical discontinuities incident on a contact surface. Examples are shown in Fig. 8.35 (see also Fig. 4.11).

The various interactions are all in the form of initial-value problems, in which conditions are known up to the time of wave intersection, e.g., in interactions *a* to *f* conditions are known respectively in fields 1-2, 1-2, 1-2-3, 1-2-3, 1-2-3, and 1-2-3. Calculation of the unknown fields depends on a matching of pressures and velocities; for this purpose it is often helpful to put the shock conditions and simple-wave (centered rarefaction) equations explicitly in terms of pressure and velocity, which we proceed to do for the special case of the *perfect gas*.

Combining the shock equations (7.32) and (7.33), with $[w]^2 = [u]^2$, yields

$$\frac{[P]}{P_0} = \frac{\gamma(\gamma + 1)}{4}\left(\frac{[u]}{c_0}\right)^2\left[1 + \sqrt{1 + \left(\frac{4}{\gamma + 1}\frac{c_0}{[u]}\right)^2}\,\right] \tag{8.64}$$

where subscript zero indicates the *upstream* condition. Application of this formula to the case illustrated in Fig. 8.35*a*, for example, gives the pressure P_3 directly (with P_2, c_2, and $[u] = -u_2$ known in advance). The corresponding formula across a centered rarefaction, e.g., between fields 2 and 3 of the case shown in Fig. 8.35*b*, is based on the invariance of J^+ or J^-, as the case may be,

$$u \pm \frac{2}{\gamma - 1}c = u_0 \pm \frac{2}{\gamma - 1}c_0$$

Solving for c/c_0 and setting $P/P_0 = (c/c_0)^{2\gamma/(\gamma-1)}$ gives

$$\frac{P}{P_0} = \left(1 \mp \frac{\gamma - 1}{2}\frac{u - u_0}{c_0}\right)^{2\gamma/(\gamma-1)} \tag{8.65}$$

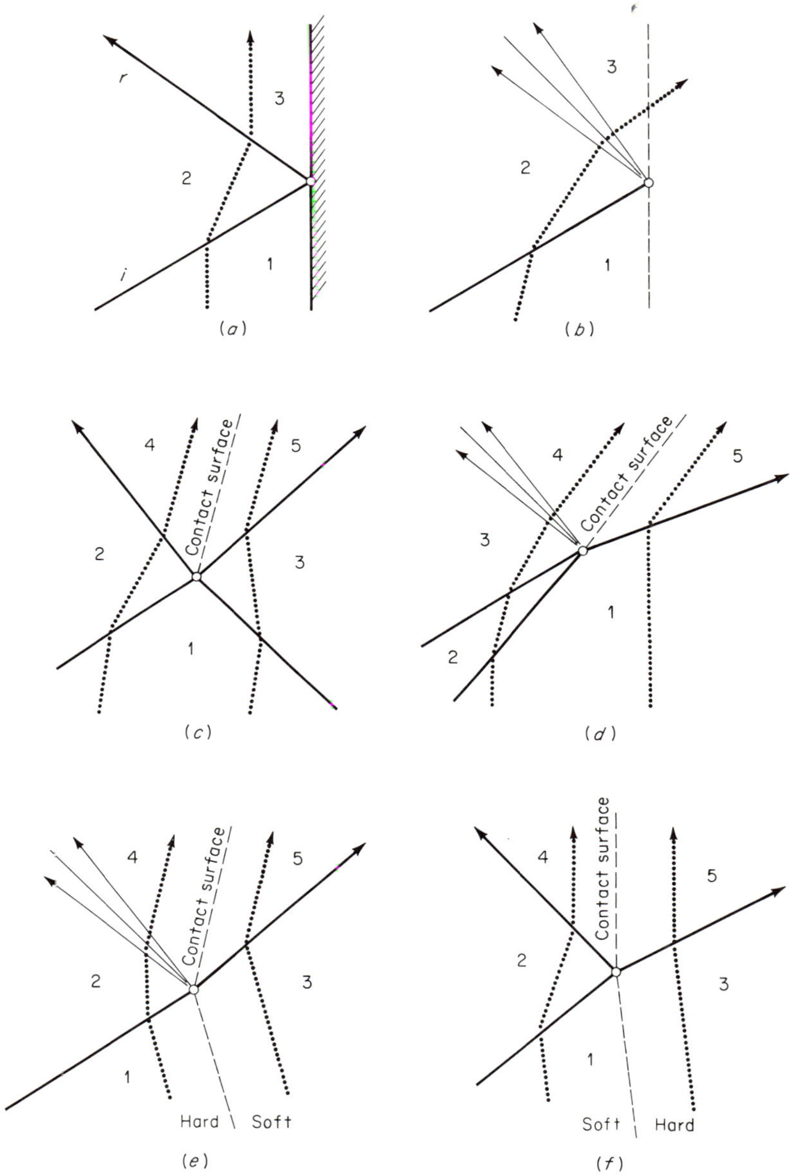

Figure 8.35
Wave diagram for shock interactions: (*a*) shock reflection from a rigid wall or closed end; (*b*) shock incident on open end at constant pressure; (*c*) intersection of shocks of opposite family; (*d*) intersection of shocks of common family; (*e*) shock incident on contact surface; and (*f*) shock incident on contact surface.

As an example application, consider the case shown in Fig. 8.35*b*. For subsonic outflow, $P_3 = P_a$ is known; and with $u_0 = u_2$, $c_0 = c_2$, $P_0 = P_2$ known, this formula gives u_3 directly.[1]

EXAMPLE 8.5 SHOCK REFLECTION

A shock is reflected from a rigid wall, as shown in Fig. 8.35*a*. Find the pressure ratio of the reflected shock *r* as a function of the pressure ratio of the incident shock *i*. The fluid medium is a perfect gas.

Because the gas adjacent to the wall is necessarily at rest, $u_1 = u_3 = 0$. It follows that the change in velocity across the reflected shock has the same magnitude as the change in velocity across the incident shock, $[w]_r = [w]_i$. From (7.21), $[w]^2 = -[P][v]$; thus,

$$[P]_r[v]_r = [P]_i[v]_i \tag{8.66}$$

Eliminating M_{1n} between (7.32) and (7.34) yields for the jump in specific volume

$$\frac{[v]_i}{v_1} = \frac{-\dfrac{[P]_i}{\gamma P_1}}{\dfrac{\gamma + 1}{2}\dfrac{[P]_i}{\gamma P_1} + 1}$$

and a corresponding relation holds for $[v]_r$. Substituting this into (8.66) and making use of the identities

$$\frac{P_2}{P_1} = 1 + \frac{[P]_i}{P_1} \qquad \frac{v_2}{v_1} = 1 + \frac{[v]_i}{v_1}$$

we obtain after some algebra

$$\frac{(\xi_r - 1)^2}{(\gamma + 1)\xi_r + (\gamma - 1)} = \frac{(\xi_i - 1)^2}{\xi_i[(\gamma - 1)\xi_i + (\gamma + 1)]} \tag{8.67}$$

where the pressure ratios are $\xi_r \equiv P_3/P_2$ and $\xi_i \equiv P_2/P_1$. This quadratic has the relevant root

$$\xi_r = \frac{(3\gamma - 1)\xi_i - (\gamma - 1)}{(\gamma - 1)\xi_i + (\gamma + 1)} \tag{8.68}$$

[1] If the outflow is already supersonic in field 1, the incident shock serves only to further accelerate this flow and no C^- characteristics can propagate back upstream. If the outflow is initially subsonic and the incident shock is sufficiently strong, the reflected expansion fan will have a vertical characteristic, stationary in the tube exit plane, corresponding to sonic outflow, with $P_e > P_a$ [see Example 8.7 (page 421)].

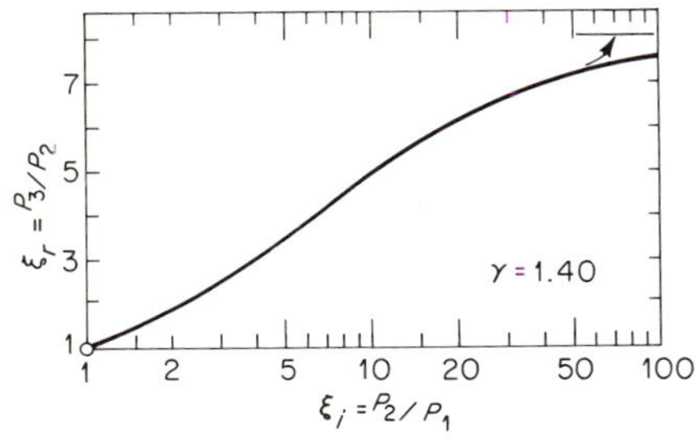

Figure 8.36
Pressure ratio of shock reflected as a function of the pressure ratio of the incident shock (perfect gas, $\gamma = 1.40$).

This relation is plotted in Fig. 8.36. The reader can verify that ξ_r has the asymptotic value $\xi_r = (3\gamma - 1)/(\gamma - 1)$ as $\xi_i \rightarrow \infty$ and that $\xi_r < \xi_i$ for all $\xi_i > 1$.

Additional relations of this kind are given by *Polachek and Seeger* [1958].

EXAMPLE 8.6 SHOCK INTERSECTION

Air at $P_1 = 1$ atm and $T_1 = 300$ K travels with velocity $u_1 = 500$ ft/s in a rigid pipe of constant area. The flow is brought to rest by a leftward-traveling shock A. At the same time, a rightward-traveling shock B with pressure ratio 5 overtakes the flow. Find the conditions in the neighborhood of the shock collision point (see Fig. 8.37).

The sound speed c_1 is found to be $c_1 = \sqrt{\gamma R T_1} = 1{,}139$ ft/s. Then with the given information and the shock tables, the values listed in Table 8.3 are found for fields 2 and 3. To find conditions in fields 4 and 5 we match velocities and pressure. One way to proceed is to assume a final pressure and calculate u_4 and u_5; if they do not match, the process is repeated until they do. [An alternative is to apply Eq. (8.64) directly, with $P_4 = P_5$ and $u_4 = u_5$.] A graphical plot may be helpful. For example, if we guess $P_4 = P_5 = 10$ atm, we find from the shock tables

$$u_5 = +1{,}254 \text{ ft/s}$$
$$u_4 = +1{,}822 \text{ ft/s}$$

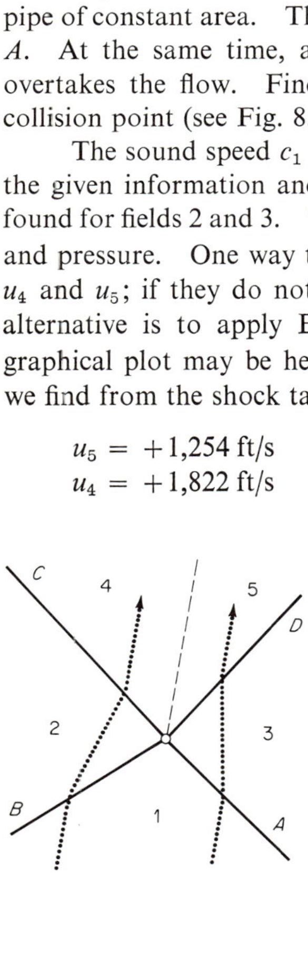

Figure 8.37

Table 8.3

Field	u, ft/s	P, atm	c, ft/s	M_s
1	500	1.00	1,139	$M_A = 1.297$
2	2,047	5.00	1,517	$M_B = 2.105$
3	0	1.80	1,240	$M_C = 1.227$
4	1,527	7.95		$M_D = 1.982$
5	1,527	7.95		

This guess was too high. The final result is

$$P_4 = P_5 = 7.95 \text{ atm}$$
$$u_4 = u_5 = +1{,}527 \text{ ft/s}$$

An entropy discontinuity has been developed. To demonstrate this, we can, for example, calculate the final densities and find

$$\rho_5 = 3.92\rho_1$$
$$\rho_4 = 3.99\rho_1$$

EXAMPLE 8.7 SHOCK INCIDENT ON AN OPEN END

A shock travels down a tube containing perfect gas ($\gamma = 1.40$) and approaches an open end. The fluid ahead of the shock is at rest. Find the flow conditions within the tube after the shock reaches the exit plane, for three different shock Mach numbers M_{1n},

Case a: $M_{1n} = 1.25$
Case b: $M_{1n} = 1.50$
Case c: $M_{2n} = 2.50$

These three cases have been calculated and are illustrated in Fig. 8.38.

Case a. The shock tables give directly that $u_2 = 0.375c_1$, $c_2 = 1.077c_1$, and $P_2 = 1.656P_1$. Assuming that the outflow is subsonic, $P_3 = P_1$; then

$$\frac{c_3}{c_2} = \left(\frac{P_3}{P_2}\right)^{(\gamma-1)/2\gamma} = \left(\frac{1}{1.656}\right)^{1/7} = 0.9305$$

This gives $c_3 = 1.002c_1$. From the constancy of the Riemann invariant between fields 2 and 3, find

$$\frac{u_3}{c_1} = \frac{u_2}{c_1} + 5\left(\frac{c_2}{c_1} - \frac{c_3}{c_1}\right) = 0.375 + 0.375 = 0.750$$

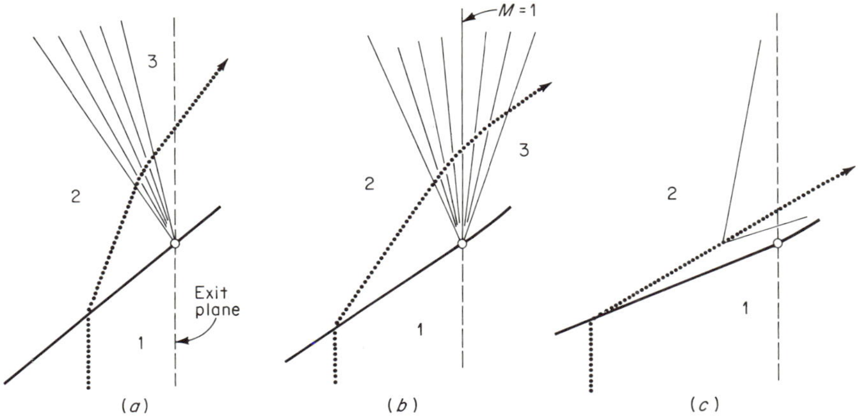

Figure 8.38
Shock incident on an open end.

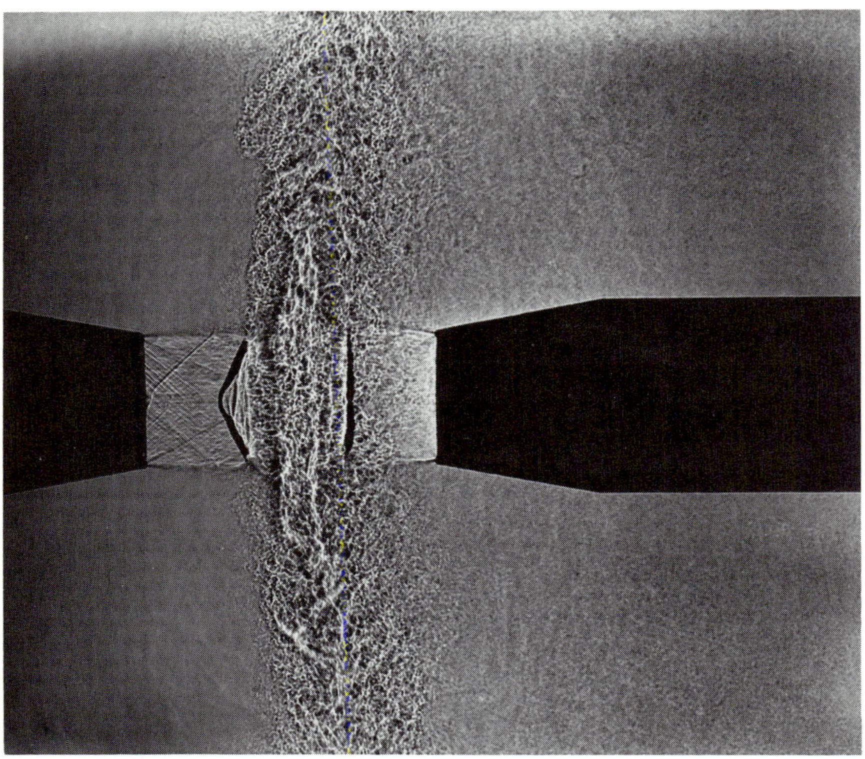

Figure 8.39
The choked jet from the right results from the passage of a shock out of the constant-area tube (unsteady flow).

That the velocity jump is the same across the shock and the reflected rarefaction fan is a consequence of the weakness of the incident shock. The flow in field 3, the outflow, is indeed subsonic since

$$M_3 = \frac{u_3}{c_3} = \frac{u_3}{c_1}\frac{c_1}{c_3} = 0.748$$

Case b. Under the same assumptions as above, and proceeding in exactly the same way, obtain

$$\frac{u_3}{c_1} = 1.389 \qquad \frac{c_3}{c_1} = 1.010$$

The flow is now supersonic in field 3, which in fact lies *outside* the tube [this can be verified by calculating the slope of the final characteristic C^- in the fan: $(u_3 - c_3)/c_1 = +0.373$, as drawn in Fig. 8.38*b*]. That part of the calculation which falls outside the tube is of course invalid. The result which is valid is that the exit flow is sonic, corresponding to the vertical characteristic standing in the mouth of the tube. The pressure P_e along this characteristic is found to be $P_e = 1.639P_1$, with expansion to pressure P_1 taking place outside the tube. For a photograph of such an expansion, see Fig. 8.39.

Case c. The shock itself is sufficiently strong to accelerate the gas to supersonic speed. From the shock tables

$$\frac{u_2}{c_1} = 1.750 \qquad \frac{c_2}{c_1} = 1.462$$

Thus $M_2 = 1.197$, and both sets of characteristics travel toward the exit plane. There can be no reflected wave, and the outflow is supersonic at state 2.

8.9 Elementary devices

The Simple Shock Tube

The shock tube is a device for producing high-temperature gas flows with relatively high Mach numbers and short time duration. It has a variety of applications in high-temperature research for physics, chemistry, and aeronautics.

In its simplest form the shock tube is a pipe closed at the ends and divided into two interior regions by a *diaphragm*. The diaphragm, which is purposely of limited strength, separates high-pressure *driver gas* from low-pressure *test gas* (driven gas). Under sufficient pressure loading the diaphragm ruptures; as an idealization it may be considered to disappear entirely at the time of rupture.

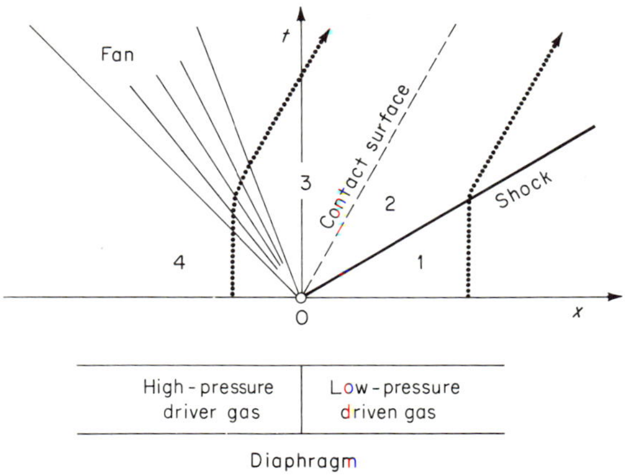

Figure 8.40
Shock-tube wave diagram.

The shock tube was first developed by Vieille (1899), who used variously steel foil, glass, paper, and collodion as diaphragm materials and obtained shock speeds up to 1,980 ft/s in air, i.e., well above the sound speed. The modern revival is due to Payman and Shepherd (1940).

The ideal wave diagram for flow in the neighborhood of the diaphragm is shown in Fig. 8.40. The pressure discontinuity that exists after rupture propagates into the test gas 1 as a strong shock and into the driver gas 4 as a centered rarefaction. The shock compresses the gas considerably, resulting in high temperature and velocity in region 2, the *test region*. Calculation of flow conditions depends upon matching pressure and velocity in regions 2 and 3. It may be noted that the flow indicated in the figure represents a *similarity solution* for the inviscid equations; conditions depend only upon x/t.

In order to achieve strong shocks, the initial pressure ratio P_4/P_1 across the diaphragm may be as great as 10^6 (such large pressure ratios are obtained by making P_1 very small). For a perfect gas, we can find the shock Mach number (and consequently all quantities in the test region) in terms of the diaphragm pressure ratio P_4/P_1. From (8.65) we have for the pressure ratio of the centered rarefaction

$$\frac{P_3}{P_4} = \left(1 - \frac{\gamma_4 - 1}{2}\frac{u_3}{c_4}\right)^{2\gamma_4/(\gamma_4 - 1)}$$

The velocity u_2 imparted by the shock is, from (7.33),

$$u_2 = \frac{2}{\gamma_1 + 1} c_1\left(M_s - \frac{1}{M_s}\right)$$

where M_s is the shock Mach number. This velocity is equal to u_3. Combining these two equations,

$$\frac{P_3}{P_4} = \left[1 - \frac{\gamma_4 - 1}{\gamma_1 + 1}\frac{c_1}{c_4}\left(M_s - \frac{1}{M_s}\right)\right]^{2\gamma_4/(\gamma_4 - 1)} \tag{8.69}$$

From (7.32) the pressure ratio across the shock is

$$\frac{P_2}{P_1} = \frac{2\gamma_1 M_s^2 - (\gamma_1 - 1)}{\gamma_1 + 1} \tag{8.70}$$

Taking advantage of pressure matching ($P_2 = P_3$) we can write for the diaphragm pressure ratio

$$\frac{P_4}{P_1} = \frac{P_4}{P_3}\frac{P_3}{P_1} = \frac{P_4}{P_3}\frac{P_2}{P_1}$$

or, with (8.69) and (8.70),

$$\frac{P_4}{P_1} = \frac{2\gamma_1 M_s^2 - (\gamma_1 - 1)}{\gamma_1 + 1}\left[1 - \frac{\gamma_4 - 1}{\gamma_1 + 1}\frac{c_1}{c_4}\left(M_s - \frac{1}{M_s}\right)\right]^{-2\gamma_4/(\gamma_4 - 1)} \tag{8.71}$$

which gives the shock Mach number M_s implicitly in terms of the diaphragm pressure ratio.

To gain some insight into the problem of arranging a shock tube for attainment of high test temperatures and velocities (corresponding to large values of M_s) consider that $P_4/P_1 \to \infty$. Then if M_s remains finite, the quantity in square brackets, that is, P_3/P_4, approaches zero, so that

$$M_s - \frac{1}{M_s} \to \frac{\gamma_1 + 1}{\gamma_4 - 1}\frac{c_4}{c_1}$$

or

$$M_s \approx \frac{\gamma_1 + 1}{\gamma_4 - 1}\frac{c_4}{c_1} \tag{8.72}$$

Thus, given that the pressure ratio P_4/P_1 is large, it is desirable that the driver gas 4 have high sound speed and, if possible, a low specific-heat

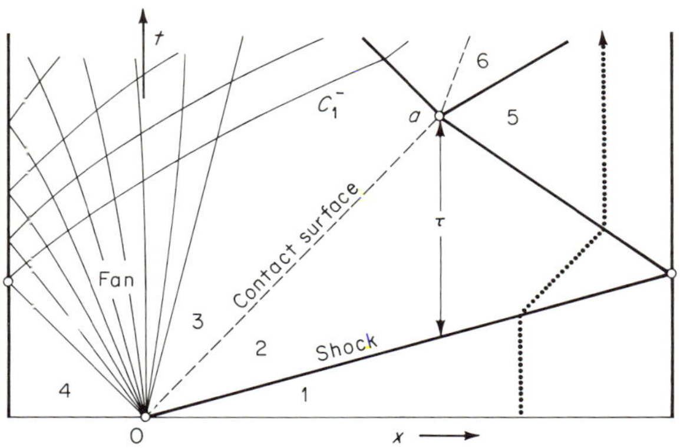

Figure 8.41
Wave diagram for shock tube of finite length.

ratio. One concludes that from the point of view of wishing strong shocks, heated hydrogen is the ideal driver gas.

After a certain time, waves reflected from the closed ends of the shock tube intersect the contact surface. In the case shown in Fig. 8.41, the reflected shock reaches the contact surface ahead of the first rarefaction characteristic C_1^-. The maximum time interval τ for which uniform high-temperature conditions 2 prevail at a fixed position is referred to as the *test time* and is of the order of 1 ms. This brevity represents one of the inherent limitations of shock tubes (another limitation is the relatively low density available).

One might think that the test time could be increased without limit simply by increasing the length of the tube. In practice this is not possible because the viscous boundary layer behind the shock builds up in thickness (Fig. 8.42) with increasing time, eventually resulting in boundary-layer closure. Note also that the flow area of the inviscid core, i.e., outside the boundary layer, changes with time; in effect the inviscid flow is confined

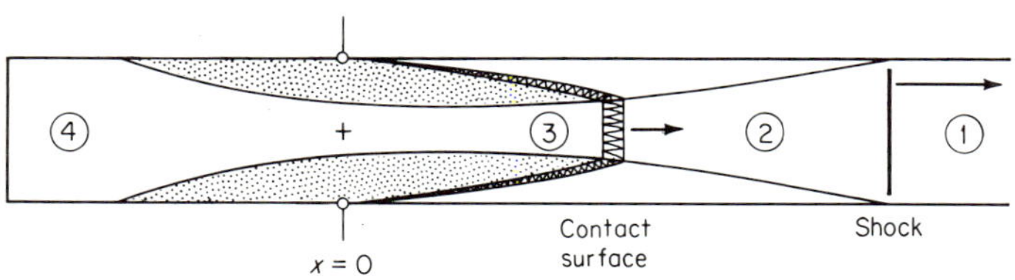

Figure 8.42
Boundary-layer growth behind the shock. (*After Gaydon and Hurle* [1963].)

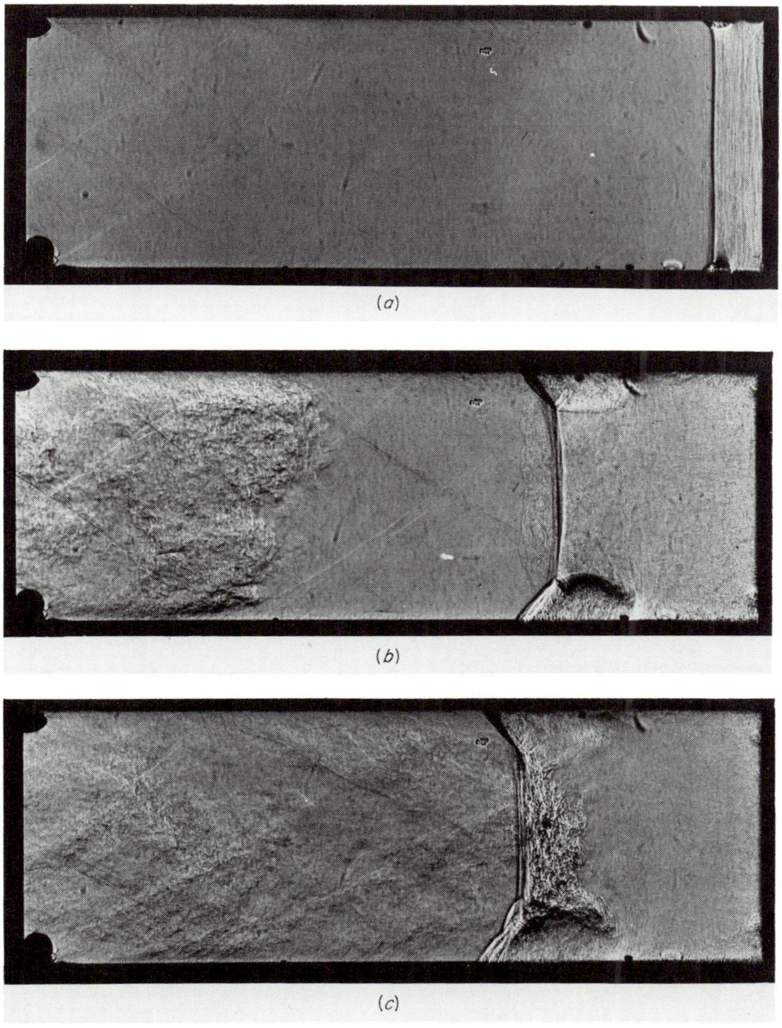

Figure 8.43
Interaction of the shock reflected from the end wall with the boundary layer and contact surface. Shock Mach number $M_s = 6$; driver gas hydrogen, driven gas air. The time in microseconds after shock reflection is: (*a*) 20 μs; (*b*) 120 μs; (*c*) 145 μs; (*d*) 195 μs; (*e*) 245 μs; (*f*) 345 μs. [(*d*)–(*f*) on following page.] (*Courtesy of R. North, National Physical Laboratory; Crown Copyright reserved.*)

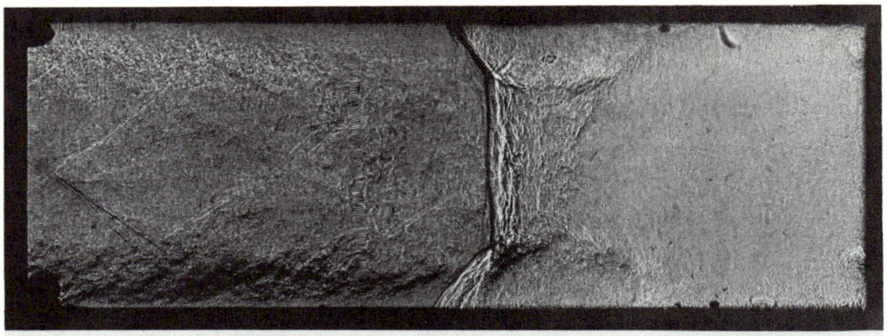

(d)

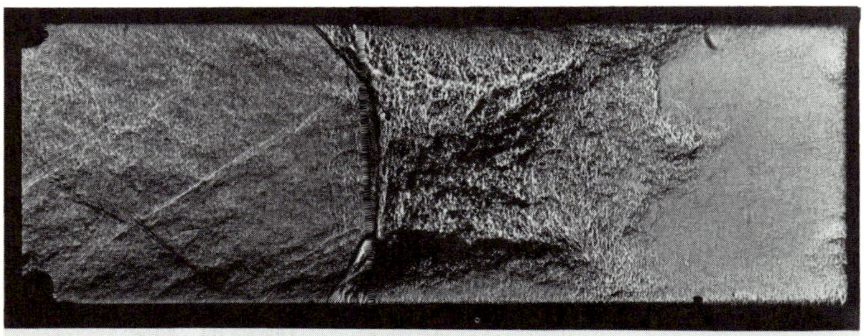

(e)

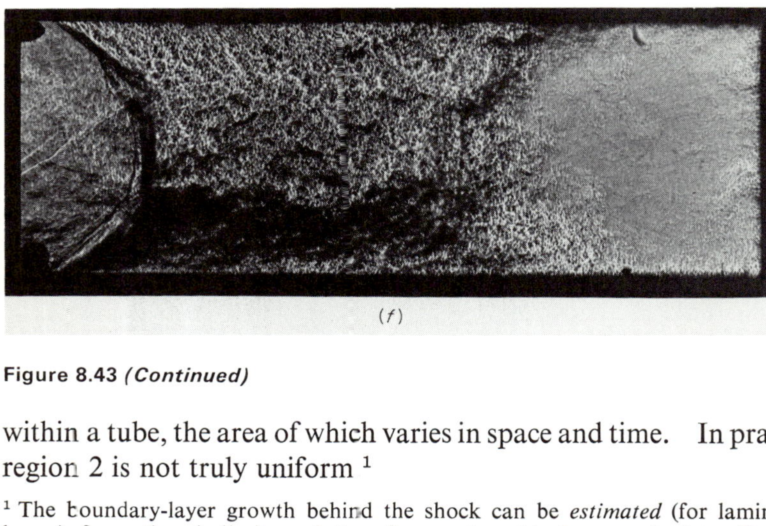

(f)

Figure 8.43 ***(Continued)***

within a tube, the area of which varies in space and time. In practice, then, region 2 is not truly uniform [1]

[1] The boundary-layer growth behind the shock can be *estimated* (for laminar boundary layers) from the similarity solution for an impulsively moved plate (Sec. 3.6). The boundary-layer thickness δ is in that case $\delta \sim \sqrt{\nu t}$, where ν is kinematic viscosity. Applying this to the above problem, from the standpoint of an observer stationary in the fluid of region 2

$$\delta_2 \sim \sqrt{\frac{\nu^2 \Delta x}{M_s c_1 - u_2}}$$

where Δx is the distance behind the shock front.

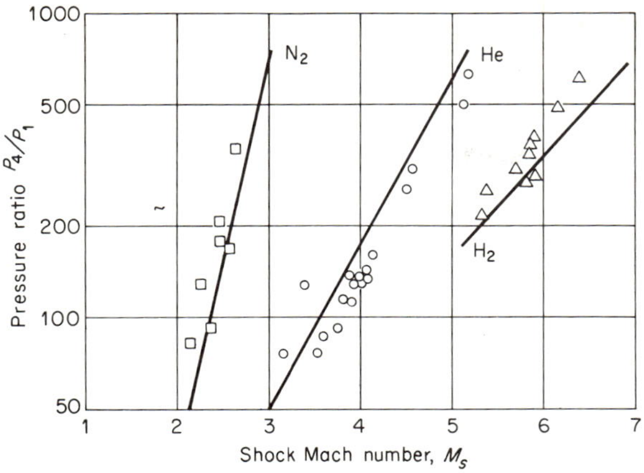

Figure 8.44
Diaphragm pressure ratio vs. shock Mach number, comparison of simple theory with experimental points from the National Physical Laboratory 6-in shock tunnel. The three different gases shown are driver gases (near room temperature); driven gas is nitrogen. (*Redrawn from Pennelegion et al.* [1965].)

Further departure from ideality results from real-gas effects, e.g., dissociation, at high temperatures. In practice, the actual test time τ may be less than half the theoretical (inviscid) time.

If the desired test requires only high temperature, region 5, where the gas is at rest, represents an alternative test region. Under special circumstances, the shock interaction at *a* may produce *no* reflected wave (separating 5 and 6) so that uniform conditions in 5 can persist for a relatively long time. Another possibility, exploited in the *shock tunnel*, is to utilize region 5 as a high-pressure reservoir for expansion through a supersonic nozzle mounted in the tube end. A photograph of the sequence of events following shock reflection from the end wall is shown in Fig. 8.43.

A comparison between the prediction of Eq. (8.71) for the shock Mach number and the performance of a particular shock tube is shown in Fig. 8.44.

The initial motion in an ideal shock tube is the resolution of a physically unsupportable discontinuity in pressure, which is present at the instant of diaphragm annihilation. Because this problem was first considered by *Riemann* [1859] in his classical paper, it is often called *Riemann's problem*. It is instructive to compare the resolution of this discontinuity with that of a *temperature* discontinuity in a heat-conducting solid (this

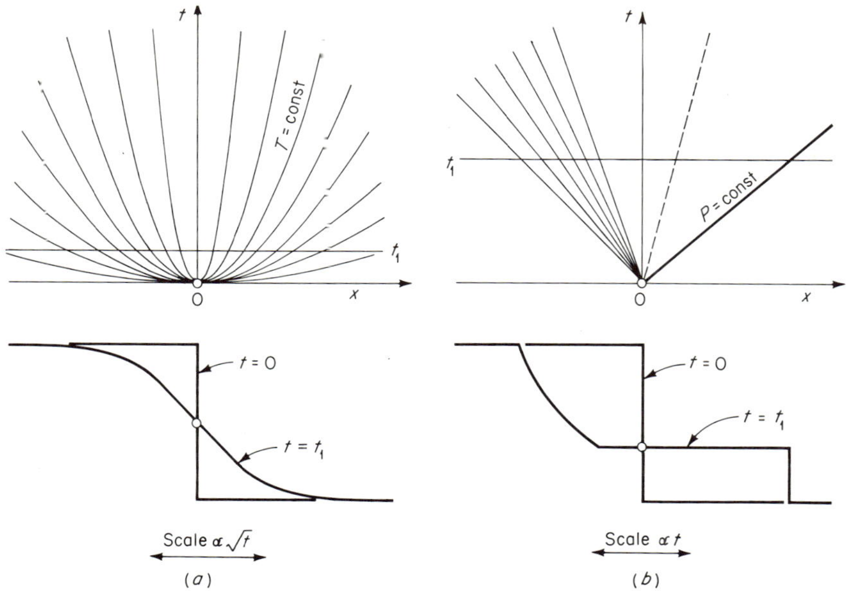

Figure 8.45
Resolution of a discontinuity: (*a*) initially discontinuous temperature distribution in a heat-conducting solid; (*b*) initially discontinuous pressure distribution in a gas.

problem was considered formally in Chap. 3), as shown in Fig. 8.45. The most striking dissimilarity is that no discontinuity can persist after the initial instant in the heat-conduction case, whereas it persists indefinitely in the gas. Further, the heat-conduction "characteristics" $T =$ const have no well-defined velocity (all values from zero to infinity occur), and the spreading is diffusive rather than wavelike, whereas the gas motion is essentially characterized by waves.

Slow Motion: The Lagrange Ballistics Problem

Consider a gas which is confined to a cylinder by a slowly moving piston. Such a problem is at the other end of the speed scale from the high-pressure shock tube. Examples are the internal flows in cannons and in the cylinders of combustion engines.

We will interpret "slow" motion as involving fluid speeds small compared to the sound speed (in this respect the flow is like acoustic motion; in this case, however, we will *not* assume small density changes).

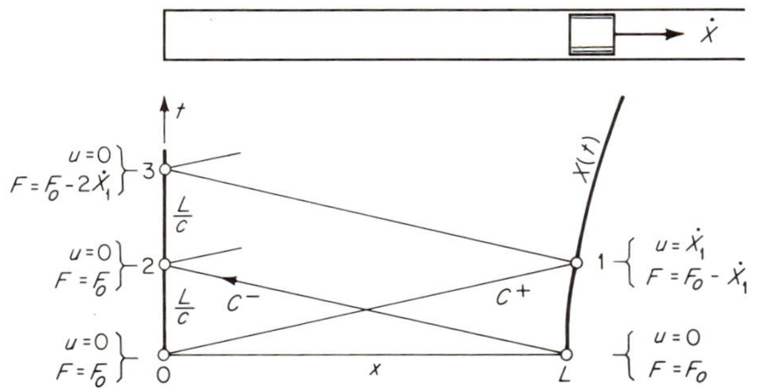

Figure 8.46
Wave diagram for slow motion in a cylinder.

In particular, let the piston velocity $\dot{X}$ be small compared to c (see Fig. 8.46).

The equations of motion are (8.1) and (8.2), with constant area and no body force. The order of magnitude of each term can be estimated with the help of the "exact" method of characteristics and will allow us to neglect certain terms.

With the piston starting from rest, the initial values are $u = 0$ and $F(P_0) = F_0$. Tracing the invariants $u \pm F$ along their respective characteristics then gives the values shown at points 1, 2, and 3 in Fig. 8.46. From $dF = dP/\rho c = c\, d\rho/\rho$ we have

$$P - P_0 \approx \rho c(F - F_0)$$

$$\rho - \rho_0 \approx \frac{\rho}{c}(F - F_0)$$

With u of order $\dot{X}$, we obtain the following estimates for the magnitude of the derivatives[1] in the continuity equation:

$$\frac{\partial \rho}{\partial t} + u\frac{\partial \rho}{\partial x} + \rho\frac{\partial u}{\partial x} = 0 \tag{8.73}$$

$$-\frac{\rho}{c}\frac{\dot{X}}{L/c} \qquad -\dot{X}\frac{\rho}{c}\frac{\dot{X}}{L} \qquad \rho\frac{\dot{X}}{L}$$

$$(1) \qquad (\dot{X}/c) \qquad (1)$$

[1] The derivatives are estimated between points 1, 2, and 3 and *not* across the first-signal characteristic C^- (where derivatives are discontinuous).

with the relative magnitude listed in the last row The convective term $u\rho_x$ is thus negligible. Similarly, the relative magnitudes in the momentum equation are

$$\frac{\partial u}{\partial t} + u\frac{\partial u}{\partial x} + \frac{1}{\rho}\frac{\partial P}{\partial x} = 0 \qquad (8.74)$$

$$(1) \qquad (\dot{X}/c) \qquad (1)$$

Just as in the acoustic case, the convective terms are negligible.

The end-to-end density difference is of order $\rho_2 - \rho_1 \approx \rho\dot{X}/c$; then the relative density difference is

$$\frac{\rho_2 - \rho_1}{\rho} \approx \frac{\dot{X}}{c}$$

which is negligible. The density can thus be treated as spatially uniform,[1] $\rho = \rho(t)$. Then the continuity equation (8.73) becomes

$$\frac{1}{\rho}\frac{d\rho}{dt} + \frac{\partial u}{\partial x} = 0 \qquad (8.75)$$

or $u_x = f(t)$, which integrates to $u = xf(t) + \text{const}$. With the boundary conditions $u(0,t) = 0$ and $u(X,t) = \dot{X}$ this gives the simple result

$$u = \frac{x}{X}\dot{X} \qquad (8.76)$$

That is, the fluid column 'stretches" linearly; e.g., at the midpoint $x = X/2$, the velocity is just $u = \dot{X}/2$. Then an integral of (8.75) is simply

$$\rho X = \rho_0 L = \text{const} \qquad (8.77)$$

i.e., total mass is conserved.

We return to the momentum equation. Since $x/X = \text{const}$ for any given fluid particle, (8.76) gives

$$\frac{Du}{Dt} = \frac{x}{X}\ddot{X}$$

[1] A similar estimate gives a relative pressure difference

$$\frac{P_2 - P_1}{P} \approx \frac{\rho c^2}{P}\frac{\dot{X}}{c}$$

For *gases*, this is of order $\dot{X}/c$, hence, negligible. The pressure variation is, however, of practical interest, and we will not neglect it but calculate it in the following.

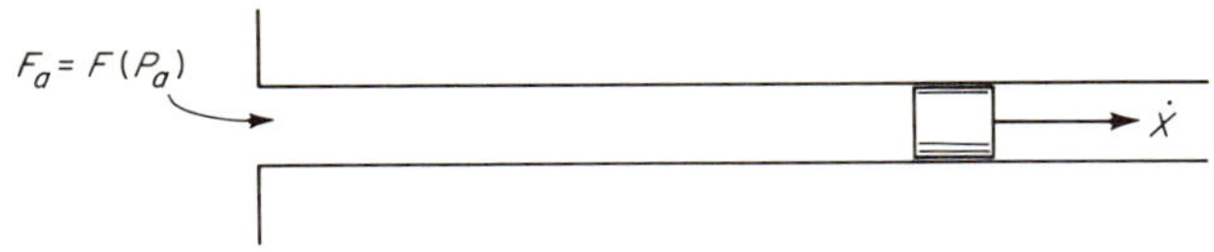

Figure 8.47

and the full momentum equation (8.74) becomes, retaining the convective term for convenience,

$$\frac{\partial P}{\partial x} = -\rho \frac{x}{X} \ddot{X}$$

With (8.77) this becomes

$$\frac{\partial P}{\partial x} = -\frac{\rho_0 L \ddot{X}}{X^2} x$$

which integrates to

$$P = P(x = 0) - \frac{\rho_0 L \ddot{X}}{2X^2} x^2 \tag{8.78}$$

The same line of reasoning can be applied to the case where the left end of the tube is open and at constant atmospheric pressure,[1] as shown in Fig. 8.47. The principal results in this case are $u(t) = \dot{X}$ (the column of fluid moves like a solid body attached to the piston) and $P = P_a =$ const.

The preceding results are approximate and incorrect in detail. It is implicit that the piston motion be smooth, since perturbations in the motion will propagate along characteristics as waves. The main condition that $\dot{X} \ll c$ ensures that several wave transits of the cylinder will occur in a time interval over which there is significant piston motion; i.e., the characteristics will appear nearly horizontal in a wave diagram such as Fig. 8.46.

8.10 Characteristic equations for isentropic flow

For isentropic flow the entropy of each fluid particle does not change with time, $Ds/Dt = 0$, but the entropy of different fluid particles may be different, $\nabla s \neq 0$. Physically this situation arises if the fluid initially has

[1] In the case of *outflow* at the open end, the jet-separation condition gives $P = P_a =$ const; for quasi-steady *inflow* Bernoulli's equation gives $(P_a - P)/P_a \sim \rho u^2/P$, which for a perfect gas is of order M^2 (very small).

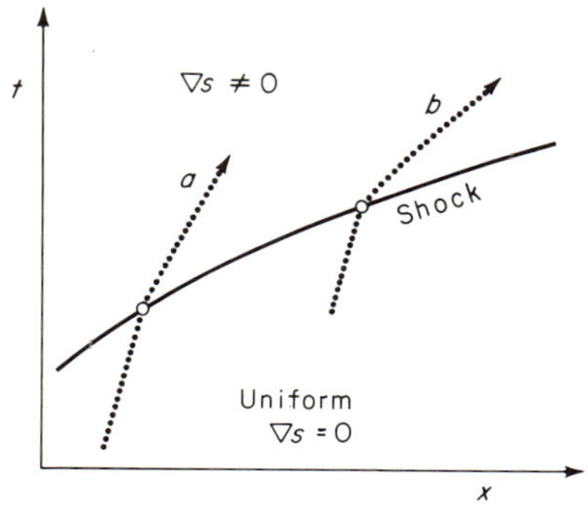

Figure 8.48

nonuniform entropy or acquires nonuniform entropy by crossing a shock of variable strength; e.g., the particles a and b acquire different entropies in Fig. 8.48 because the shock is stronger where particle b crosses. In order for the isentropic model to hold, it is still necessary that viscous dissipation and heat transfer be negligible.[1]

We now seek characteristic forms for this case. The equations of motion (8.1) and (8.2) are rewritten below, together with the isentropic condition:

$$\frac{D\rho}{Dt} + \rho\frac{\partial u}{\partial x} = -\frac{\rho u A'}{A} \tag{8.1}$$

$$\frac{Du}{Dt} + \frac{1}{\rho}\frac{\partial P}{\partial x} = G \tag{8.2}$$

$$\frac{Ds}{Dt} = 0 \tag{8.79}$$

In general *two* thermodynamic quantities are now required to fix the state, e.g., we write $\rho(P,s)$, and two thermodynamic variables will appear in the final characteristic form. We can, however, eliminate derivatives of ρ in favor of derivatives of P as follows: with $\rho = \rho(P,s)$,

$$\frac{D\rho}{Dt} = \left(\frac{\partial\rho}{\partial P}\right)_s \frac{DP}{Dt} + \left(\frac{\partial\rho}{\partial s}\right)_P \frac{Ds}{Dt}$$

which with $Ds/Dt = 0$ is simply

$$\frac{D\rho}{Dt} = \frac{1}{c^2}\frac{DP}{Dt} \tag{8.80}$$

[1] *Except* within the shock. The entropy is of course discontinuous across the shock.

Substituting this in (8.1) and multiplying through by c/ρ and then adding and subtracting (8.1) and (8.2) gives

$$\begin{aligned} \frac{D^+u}{Dt} + \frac{1}{\rho c}\frac{D^+P}{Dt} &= G - \frac{cuA'}{A} \\ \frac{D^-u}{Dt} - \frac{1}{\rho c}\frac{D^-P}{Dt} &= G + \frac{cuA'}{A} \end{aligned} \tag{8.81}$$

where D^+/Dt and D^-/Dt are the derivatives along the characteristics C^+ and C^-, respectively, as before. These equations are of the same form as (8.8); there is, however, a significant difference in that $dP/\rho c$ is no longer an exact differential, i.e., we cannot consider $\rho c = f(P)$, and the Riemann invariants cannot be formed as before. Thus we can only write

$$\begin{aligned} du + \frac{dP}{\rho c} &= G^+\,dt \qquad \text{on } C^+ \\ du - \frac{dP}{\rho c} &= G^-\,dt \qquad \text{on } C^- \end{aligned} \tag{8.82}$$

where $G^\pm = G \mp cuA'/A$, and the required third equation (8.80) can be written

$$dP - c^2\,d\rho = 0 \qquad \text{on } PP \tag{8.83}$$

There are now *three* characteristics (C^+, C^-, and the particle path PP) with three corresponding equations. Except in very special cases, these must be solved numerically. The numerical calculation will be illustrated for the case $A' = G = 0$.† Let the x axis be an initial-data line (Fig. 8.49)

† The nonhomogeneous case is similar to the example in Sec. 8.3.

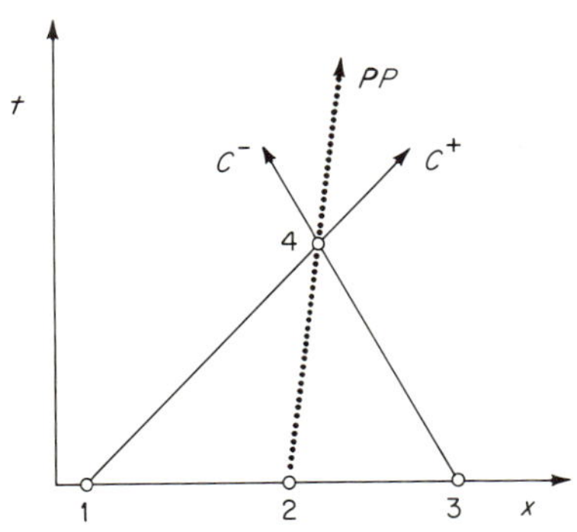

Figure 8.49

and conditions at point 4 to be found. From (8.82) we have, approximately,

$$\rho_1 c_1(u_4 - u_1) + P_4 - P_1 = 0$$

$$\rho_3 c_3(u_4 - u_3) - P_4 + P_3 = 0$$

which can be solved for P_4 and u_4. The location of point 4 has been established as the intersection of characteristics drawn with known initial slopes from the (arbitrary) points 1 and 3. Now the point 2 can be established by laying out the particle path from point 4 with $dx/dt = u_4$. Then (8.83) is written

$$P_4 - P_2 - c_2^2(\rho_4 - \rho_2) = 0$$

which, with P_4 now known, gives ρ_4. Then the velocity and thermodynamic state at point 4 are established.

An alternative method is suitable to a digital computer. This consists in solving (8.1), (8.2), and (8.80) for the three partial time derivatives $\partial\rho/\partial t$, $\partial u/\partial t$, and $\partial P/\partial t$. With the space derivatives known along an initial-data line $t = t_0$, these time derivatives can then be calculated. Then conditions at a later time $t_0 + \Delta t$ can be found by marching forward, for example, $P(t_0 + \Delta t) = P(t_0) + \Delta t\, \partial P/\partial t$. But the calculation will be numerically unstable if initial data for the calculation of a given mesh point do not at least span the domain of dependence of characteristics. This condition can be met by making the time increment Δt sufficiently small relative to the space increment Δx (see, for example, *Richtmyer* [1957, p. 195]).

The numerical method of characteristics can be used for problems involving shocks (see Fig. 8.50). Suppose initial data, including the velocity U of the shock at point 2, are given along the x axis. Then

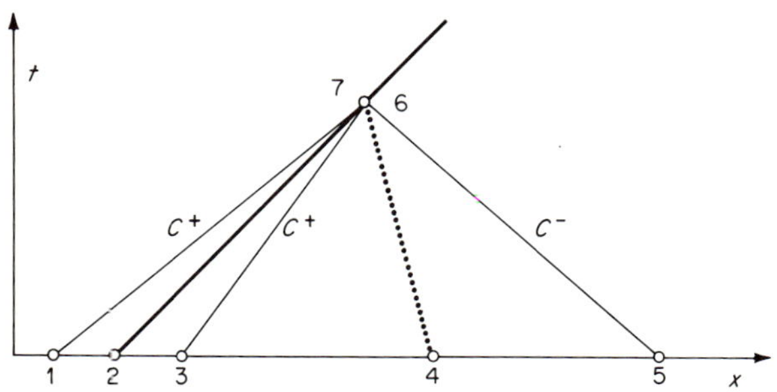

Figure 8.50

conditions at the later congruent points 6 and 7, upstream and downstream of the shock respectively, can be found as follows.

Points 6 and 7 are chosen on the shock path projected with its initial slope at 2. Conditions at 6 are then fully established by the procedure already described under Fig. 8.49. Between points 1 and 7 we have $\rho_1 c_1(u_7 - u_1) + P_7 - P_1 = 0$, which is of the form $P_7(u_7)$. With upstream conditions known, the shock relations give another equation $P_7(u_7)$ [see, for example, Eq. (7.20)] and simultaneous solution gives P_7 and u_7, and all further quantities are given from the shock relations, including the "new" shock velocity U.

8.11 Viscous effects in waves of finite amplitude

For progressive waves in one dimension, the relative magnitudes of the terms in the momentum equation were estimated in Sec. 4.6 to be, for a gas,

$$\underset{(1)}{\frac{\partial u}{\partial t}} + \underset{(S_0)}{u\frac{\partial u}{\partial x}} + \underset{(1)}{\frac{1}{\rho}\frac{\partial P}{\partial x}} = \underset{(\Lambda/\lambda)}{\frac{4}{3}\frac{\mu'}{\rho}\frac{\partial^2 u}{\partial x^2}}$$

Evidently, there is a range of amplitudes and characteristic wavelengths for which the viscous term is at least of the same order of magnitude as the nonlinear convective term, i.e., for small amplitude and small wavelengths. It is just the viscous effect which prevents the eventual shock formation in acoustic waves that is predicted by the classical nonlinear (inviscid) theory.

In general, the effect of heat conduction is negligible in gases to just about the same extent as the effect of viscosity. We therefore lump heat conduction together with viscous action under the heading *viscous effects*.

Consider a simple C^+ wave, i.e., a wave traveling in the $+x$ direction. The inviscid acoustic theory predicts for such a wave

$$\frac{\partial u}{\partial t} + c_0\frac{\partial u}{\partial x} = 0 \tag{8.84}$$

That is, the velocity u has constant amplitude on a (linear) C^+ characteristic. The viscous effects lead to higher-order terms; a derivation will lead to the following equation (in which the smallest terms have been dropped):

$$\frac{\partial u}{\partial t} + c_0\frac{\partial u}{\partial x} = \frac{\delta}{2}\frac{\partial^2 u}{\partial x^2} \tag{8.85}$$

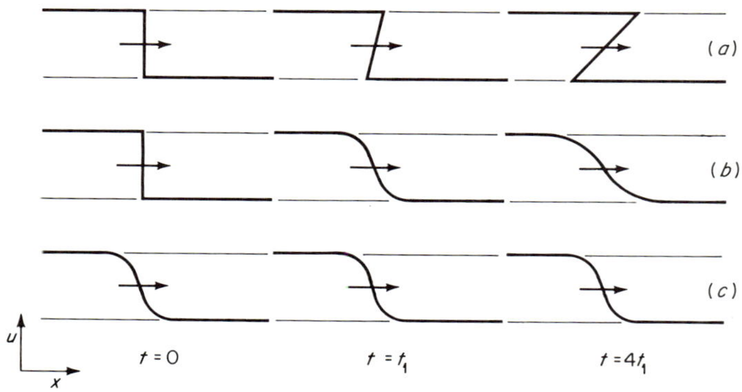

Figure 8.51
Evolution in time of a (shock) step discontinuity: (*a*) classical nonlinear inviscid theory; (*b*) linear viscous theory; and (*c*) stationary form predicted by viscous nonlinear theory. (*After Lighthill* [1956].)

where δ is the diffusivity defined in Eq. (4.165). For an observer traveling with the wave, the appropriate spatial variable is ξ

$$\xi \equiv x - c_0 t$$

In terms of this variable, (8.85) becomes

$$\frac{\partial u}{\partial t} = \frac{\delta}{2}\frac{\partial^2 u}{\partial \xi^2} \tag{8.86}$$

which is just the *heat equation*. We have already discussed the solution of this equation for an initial step discontinuity in Sec. 3.6; this solution gives the behavior shown in Fig. 8.51*b*.

Consider now a simple wave of "finite" amplitude with essentially nonlinear behavior. The inviscid result that u and c (and therefore the sum $u + c$) do not vary along a C^+ characteristic for such a wave can be written

$$\frac{\partial v}{\partial t} + v\frac{\partial v}{\partial x} = 0 \tag{8.87}$$

where $v \equiv u + c$ is the local wave velocity. The approximate form of this equation including viscous effects is

$$\frac{\partial v}{\partial t} + v\frac{\partial v}{\partial x} = \frac{\delta}{2}\frac{\partial^2 v}{\partial x^2} \tag{8.88}$$

This is called the *Burgers equation* (see, for example, *Lighthill* [1956]) and is useful in dealing with nonstationary waves. The corresponding stationary (step) wave was discussed in Sec. 7.10 and is shown in Fig. 8.51*c*.

We will not discuss the viscous effects further. The interested reader is referred to *Lighthill* [1956]. The more general work of *Coleman et al.* [1965] contains an interesting prediction of a critical amplitude above which waves will steepen to form shocks and below which they will not.

Problems

8.1 Derive the continuity equation for viscous compressible flow in an axisymmetric pipe of constant radius R in the form

$$\int_0^R \left(\frac{\partial}{\partial x}\rho u + \frac{\partial \rho}{\partial t}\right) r\,dr = 0$$

where u and v are respectively the axial and radial velocity components and

$$\rho = \rho(r,x,t) \qquad u = u(r,x,t)$$
$$u(R,x,t) = 0 \qquad v(R,x,t) = 0$$

and the radial velocity component v is in general nonzero. May the above integrand be set identically equal to zero?

8.2 Consider a standing normal shock in a steady purely one-dimensional flow. Sketch, on an xt diagram, particle paths and both families of characteristics in the neighborhood of the shock.

8.3 At time $t = 0$ a progressive one-dimensional acoustic wave is given by $S = S^0 \sin(2\pi x/\lambda)$. As an academic but interesting exercise, calculate the distance Δx at which characteristics first intersect to form a shock. Use the inviscid nonlinear theory for a perfect gas and find Δx in terms of S^0, γ, and λ. Calculate this distance for a representative acoustic situation.

Answer $\dfrac{\lambda}{\pi(\gamma + 1)S^0}$

8.4 A lightweight frictionless piston of length L and density ρ is held in a long constant-area tube and released at $t = 0$. The perfect gas within the tube is initially stationary at pressure P_0 with sound speed c_0. The exterior of the tube is at zero pressure. Find the approximate ordinary differential equation for the free-piston velocity $\dot{X} = V(t)$ in terms of the given quantities. If possible, solve the differential equation. Assume that $V < c_0$, and use the binomial expansion if required.

Answer $V \approx \frac{c_0}{\gamma}(1 - e^{-\gamma P_0 t/\rho L c_0})$

8.5 A tube of constant cross-sectional area contains stagnant perfect gas and is closed on one end by a movable piston. At time $t = 0$ the piston is advanced impulsively into the gas (compressive motion) with velocity $\dot{X}$. Find the pressure on the face of the piston in terms of $\dot{X}$ and the properties of the stagnant gas:

(*a*) From shock conditions
(*b*) From the acoustic condition $\delta P = \rho c\, \delta u$
(*c*) From the condition that the characteristic crossing the shock is approximately invariant

Show that the above answers are common in the limit as $\dot{X} \to 0$.

8.6 A horizontal tube contains stationary air at 1 atm and 300 K. The left end of the tube is closed by a movable piston, which at time zero is moved impulsively at a speed $\dot{X} = 400$ ft/s. Find the pressure on the face of the piston:

(*a*) If the piston motion is to the left
(*b*) If the piston motion is to the right

Answer (*a*) 0.600 atm; (*b*) 1.604 atm

8.7 A very long tube of constant area contains uniform stationary gas at P_0, c_0 and is closed by a piston. Starting at time zero, the piston is withdrawn with displacement $X = -\frac{1}{2}(at^2)$, where a is a constant. Find the normalized pressure P/P_0 on the face of the piston as a function of time. Assume a perfect gas.

8.8 Stationary air at 300 K and 1 atm is bounded by a stationary steel plate. A shock surface, parallel to the plate and with pressure ratio $P_2/P_1 = 1.5$, travels toward the plate. Find the flow conditions in *both* media in the xt neighborhood of the shock "collision" with the plate.

8.9 A plane shock front in air with pressure ratio $P_2/P_1 = 1.2$ is reflected from a parallel rigid wall. Find the pressure behind the reflected shock:

(*a*) From linear (acoustic) theory
(*b*) From nonlinear (shock) theory

Answer (*a*) $1.4000P_1$; (*b*) $1.4334P_1$

8.10 A cylinder is closed at the ends by a piston and a rigid wall. Initially stagnant perfect gas is compressed by impulsively moving the piston inward with constant velocity. Using the Rankine-Hugoniot relation and Eq. (7.21), show that

$$\eta_2 = \frac{\mu\eta_1 - 1}{\eta_1 + \mu - 2}$$

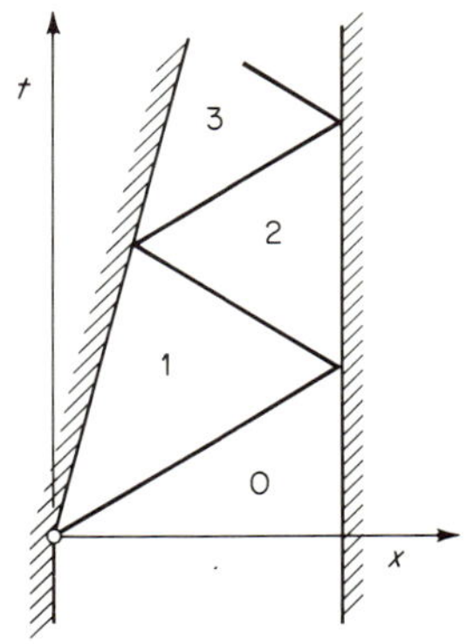

where $\eta_2 = \rho_2/\rho_1$, $\eta_1 = \rho_1/\rho_0$, and $\mu = (\gamma + 1)/(\gamma - 1)$. For such a compression process, is the equation $P\rho^{-\gamma} = $ const expected to be valid in any sense as the number of reflections increases? If not, how should it be modified?

8.11 Air at 200 ft/s, 300 K, and 1 atm flows out of a cylindrical tube into the atmosphere. A shock of pressure ratio $P_2/P_1 = 2$ moves downstream in the tube. Find the flow conditions in the xt neighborhood of the shock collision with the open end; in particular, find the outflow velocity.

8.12 A plane detonation wave with velocity U is initiated at the closed end of a rigid tube at time $t = 0$. The Chapman-Jouguet condition applies to this case, and the sound speed behind the detonation front is $c_1 = U/2$. Find u/U and c/U as functions of the similarity variable x/Ut. Show the flow on a wave diagram. Assume that the detonation products behave as a perfect gas with $\gamma = 2$.

8.13 A cylinder of length L is closed at the ends by a piston and a fixed wall and contains uniform stationary fluid of sound speed c_0. If the piston is given a motion $X = X_0 \cos \omega t$, specify conditions on X_0, ω, L, and c_0 such that the resulting inviscid motion is described by

(*a*) Linear acoustics
(*b*) Spatially uniform properties
(*c*) Nonlinear equations (method of characteristics)

8.14 A simple shock tube consists of a constant-area tube with a diaphragm mounted at the *end*. The outside pressure is P_a. The tube is initially pressurized with perfect gas of sound speed c_1 and pressure $P_1 > P_a$. The diaphragm is ruptured (instantaneously disappears) at $t = 0$.

(*a*) Find the minimum value of P_1/P_a such that the initial outflow will be sonic.
(*b*) Carefully sketch the wave diagram for this case.

Answer $\left(\dfrac{\gamma + 1}{2}\right)^{2\gamma/(\gamma - 1)}$

8.15 Consider the expansion into a vacuum shown in Fig. 8.19*b* and suppose that a fixed rigid wall is located at $x = -L$. A shock will be formed when the leading edge of the expansion ($C^{\pm}$) reflects from the wall. Find the initial velocity of this shock and the temperature behind it.

Answer $+c_0$; $\dfrac{2\gamma}{\gamma - 1} T_0$

8.16 A cylindrical mass m is dropped from rest into the cylindrical chamber (containing gas) shown. Assuming no leakage (perfect fit) and no friction, find the distance h at which the mass is brought to rest. Also suggest a means for finding the corresponding time interval Δt. Justify and use the simplest calculation model possible. Assume isentropic compression and a perfect gas with $\gamma = 2$.

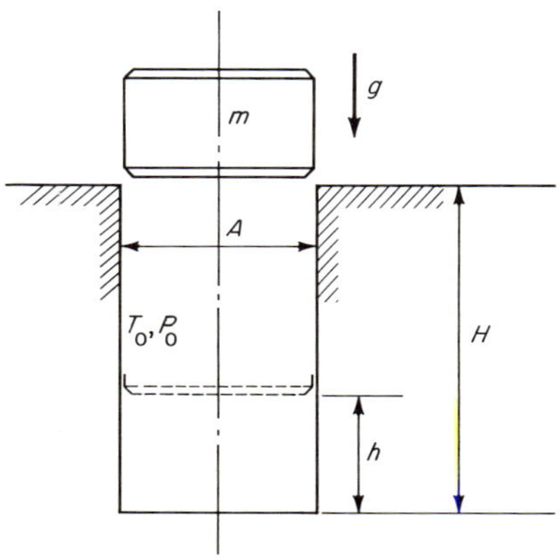

8.17 Briefly discuss the formula $\Delta P = \rho c \, \Delta u$, as it may apply to:

(*a*) Simple acoustic continuous waves
(*b*) Simple nonlinear continuous waves
(*c*) Shocks

8.18 A piston is moved impulsively from rest in the cylinder sketched. The right-hand end of the cylinder is open to the atmosphere. Find the resulting gas motion for as long a time as practicable in the cylinder via the numerical method of characteristics. In particular, show the wave diagram with normalized coordinates x/L and $c_1 t/L$. The piston velocity is $\dot{X} = 0.375 c_1$, and the gas is perfect, with $\gamma = 1.40$.

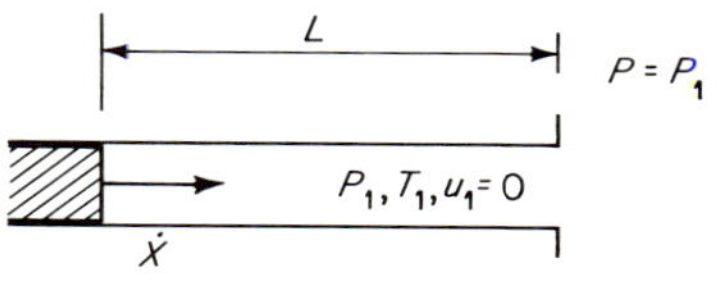

nine
steady supersonic flow in two dimensions

9.1 Introduction

Flows characterized by only two space dimensions may be either plane or axisymmetric. For example, flow over a long airfoil of uniform cross section may lie approximately in parallel planes while flow through a nozzle of circular (and variable) cross section may lie in radial planes passing through the axis of symmetry. We will refer to the first example as *plane flow* and the second as *axisymmetric flow* (strictly, an axisymmetric flow may have a tangential or swirl component of velocity, as indeed some real flows do, but this case is not considered here).

An essential feature of steady supersonic flow is the presence of *Mach waves*, which form restricted regions of influence. Further, stationary shocks (which will in general be oblique to the flow) may occur. It will be found that *two-dimensional supersonic flow is qualitatively analogous to one-dimensional unsteady flow* (Fig. 9.1). In particular, the wave diagrams are similar, with correspondence in the steady/unsteady flows respectively between streamline/particle path, oblique shock/moving shock, and sound wave/Mach wave. While the analogy is striking and helpful, it is only qualitative; e.g., the wave diagrams of Fig. 9.1*a* and *b* cannot in general be geometrically similar. It is only the *linearized* flow discussed in Sec. 5.4 which is formally analogous to a corresponding acoustic motion. The

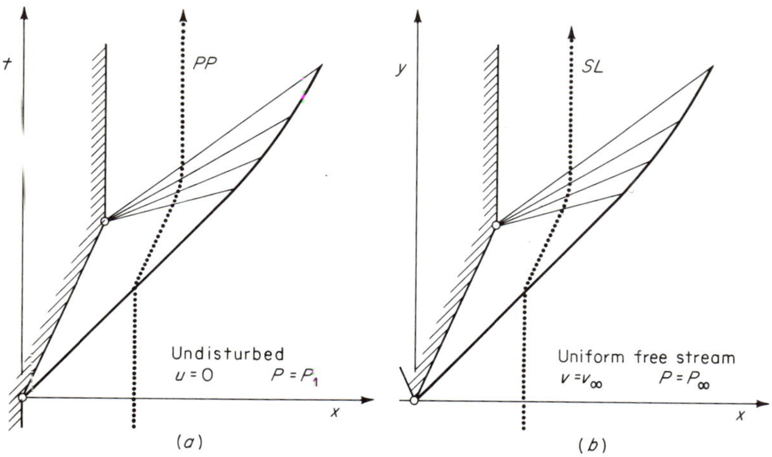

Figure 9.1
Wave diagrams: (*a*) unsteady flow resulting from impulsive starting and stopping of a piston; (*b*) supersonic steady flow about a body with a wedge nose.

linearized theory of supersonic flow given in that section is generalized, in this chapter, to include large-amplitude changes in the flow.

It is thus necessary to develop the theory of two-dimensional supersonic flow from the beginning. The treatment will be reasonably brief, however, since most of the problems have already been considered in another guise. As before, the treatment is confined to inviscid flow.

9.2 The Prandtl-Meyer function

Discontinuous turning of a supersonic stream may be accomplished via a sequence of standing oblique shocks, as shown in Fig. 9.2*a*. If the number of shocks is increased without limit while keeping the total turning angle

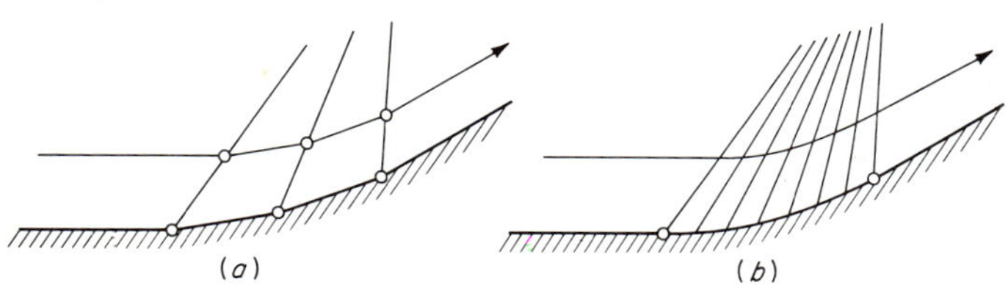

Figure 9.2
Compressive supersonic turning.

fixed, an isentropic turning is achieved in the limit, as mentioned in Sec. 7.6. A finite turning angle then occurs via an infinite number of infinitesimal shocks, as indicated in Fig. 9.2*b*.

To exploit this conceptual device, we begin with the conditions for an oblique shock of infinitesimal strength; in this case we are justified in retaining only the first term in the weak-shock expansions, such as (7.54), viz.,

$$[\theta] = -\frac{\sqrt{M_1^2 - 1}}{M_1^2} \frac{[w]}{c_1} \tag{9.1}$$

where $[\theta]$ is the turning angle across the shock. From the invariance of v,

$$u_2^2 - u_1^2 = (w_2^2 + v^2) - (w_1^2 + v^2) = w_2^2 - w_1^2$$

Omitting second-order terms $[u]^2$ and $[w]^2$, this gives

$$u_1[u] = w_1[w]$$

and with $w_1/u_1 = \sin\beta \approx \sin\mu_1 = 1/M_1$,

$$\frac{[u]}{u_1} = \frac{1}{M_1^2}\frac{[w]}{c_1}$$

Substituting this into (9.1) gives the desired result

$$\frac{[u]}{u_1} = -\frac{[\theta]}{\sqrt{M_1^2 - 1}} \tag{9.2}$$

In the limit of infinitesimal strength, the oblique shock becomes a *Mach wave* and (9.2) becomes exact, with $[\theta] \rightarrow d\theta$ and $[u] \rightarrow du$,

$$d\theta = \pm\sqrt{M^2 - 1}\,\frac{du}{u} \tag{9.3}$$

where the signs $(+)$ and $(-)$ correspond respectively to the lower and upper branches of the shock polar; e.g., for Fig. 7.20 in the derivation above, the upper branch was implicitly considered. With the infinitesimal turning taking place across a Mach wave m^+ or m^-, the signs are sorted out in Fig. 9.3. Because the shock has infinitesimal strength, we now must admit the possibility of rarefaction,[1] with $du > 0$. This means that $d\theta$ can be either positive or negative in every case, and continuous turning via an infinite sequence of Mach waves can be either compressive ($du < 0$) or expansive ($du > 0$).

[1] In this discussion we presume of course that $\Gamma > 0$. Otherwise an ordinary shock is expansive, as discussed in Chap. 7.

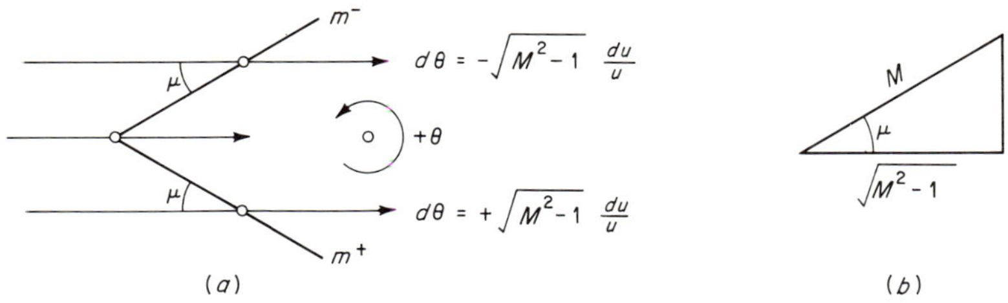

Figure 9.3

The two Mach waves, m^+ and m^-, will later be identified as *characteristics*, just like the sound-wave paths in one-dimensional unsteady flow.

The integral of (9.3) represents continuous isentropic turning through the total angle $\Delta\theta$. We define the quantity $d\omega = \pm d\theta$,

$$d\omega = \sqrt{M^2 - 1}\,\frac{du}{u} = \pm d\theta \tag{9.4}$$

where the plus sign holds across an m^+ wave and the minus sign across an m^- wave. It is to be shown that this is an exact differential of the *Prandtl-Meyer function*[1] ω. Making use of (5.11), viz.,

$$\frac{du}{u} = \frac{dM/M}{1 + (\Gamma - 1)M^2}$$

Eq. (9.4) becomes

$$d\omega = \frac{\sqrt{M^2 - 1}}{1 + (\Gamma - 1)M^2}\,\frac{dM}{M} \tag{9.5}$$

The thermodynamic function Γ can be considered known as a function of M via the energy equation (5.4), which can be written $-2(h - h_0)/c^2 = M^2$. Then (9.5) is integrable to give $\omega(M)$. The Prandtl-Meyer function may be considered another dimensionless measure of the flow speed, alternative to the Mach number. The function is real and has physical significance only for supersonic flow, $M \geq 1$.

For the important and simple case of the *perfect gas* the function Γ is just a constant, $\Gamma = (\gamma + 1)/2$. Equation (9.5) integrates to

$$\omega(M) = \sqrt{\frac{\gamma + 1}{\gamma - 1}}\tan^{-1}\sqrt{\frac{\gamma - 1}{\gamma + 1}(M^2 - 1)} - \tan^{-1}\sqrt{M^2 - 1} \tag{9.6}$$

[1] After *Ludwig Prandtl*, the originator of the boundary-layer concept, who apparently was the first to study simple waves in supersonic flow (1907), and *Theodor Meyer*, a student of Prandtl's, who developed the theory (1908).

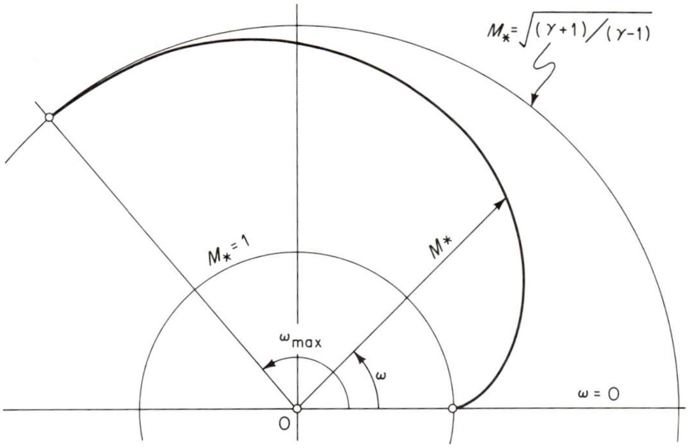

Figure 9.4
Hodograph diagram showing the Prandtl-Meyer function for a perfect gas with $\gamma = 1.40$.

where the constant of integration has been taken as zero, so that $\omega(1) = 0$. This function is tabulated in the isentropic-flow tables; its use will be illustrated below. Because M may go to infinity, it is more convenient for graphical display of (9.6) to rewrite it in terms of the normalized Mach number M_* as given by (5.62). The result is

$$\omega(M_*) = \alpha \tan^{-1} \sqrt{\frac{M_*^2 - 1}{\alpha^2 - M_*^2}} - \tan^{-1} \alpha \sqrt{\frac{M_*^2 - 1}{\alpha^2 - M_*^2}} \tag{9.7}$$

where $\alpha \equiv \sqrt{(\gamma + 1)/(\gamma - 1)}$ is the maximum value of M_*. The maximum value of ω is then

$$\omega_{\max} = \frac{\pi}{2}\left(\sqrt{\frac{\gamma + 1}{\gamma - 1}} - 1\right) \tag{9.8}$$

Equation (9.7) is the formula for an epicycloid in polar coordinates (generated by a wheel rolling on a wheel), as shown in Fig. 9.4, the *hodograph diagram*. This figure represents the variation of flow angle with Mach number in continuous supersonic turning [note that by Eq. (9.4) $\Delta\theta = \pm\Delta\omega$].

EXAMPLE 9.1 SUPERSONIC TURNING

A uniform free stream of air has Mach number $M_1 = 2.5$ and pressure $P_1 = 1$ atm. Find the downstream conditions following a 20° turn. Consider three

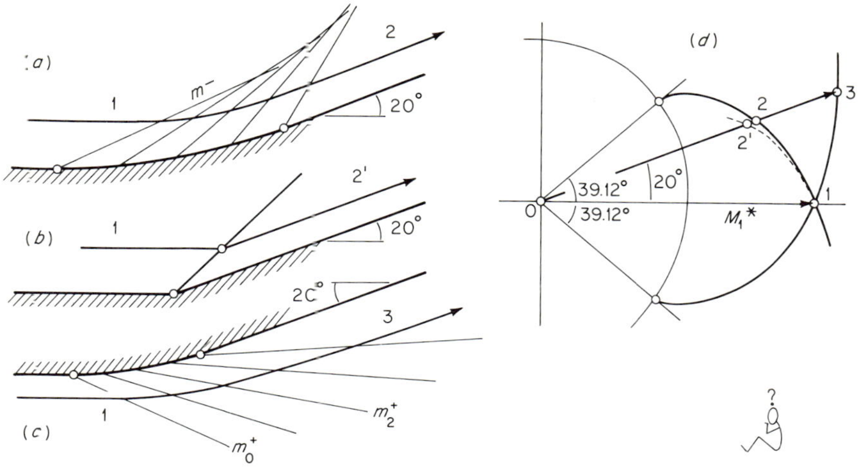

Figure 9.5
(*a*) Continuous compressive turning; (*b*) turning via a shock; (*c*) continuous expansive turning; and (*d*) hodograph diagram.

different cases: (*a*) compressive and continuous, (*b*) compressive and discontinuous, (*c*) expansive and continuous.

Possible physical situations corresponding to the respective cases are shown in Fig. 9.5*a*, *b*, and *c*.

For the compressive case *a*, Eq. (9.4) gives $\Delta\omega = -\Delta\theta = -20°$. Similarly for case *c*, $\Delta\omega = +\Delta\theta = +20°$. Then from the isentropic-flow tables for air with γ assumed 1.40, $\omega_1 = 39.12°$, giving $\omega_2 = 19.12°$ and $\omega_3 = 59.12°$ and corresponding Mach numbers, pressure, etc. Conditions across the oblique shock (weak solution) are found from shock charts (or tables), giving the values shown in Table 9.1.

The hodograph curves are shown in Fig. 9.5*d* with the curve of Fig. 9.4 rotated (and flopped over in the case of process $1 \rightarrow 2$) so that the initial flow direction 1 is aligned with the vector M_1^* in the hodograph diagram and the curves represent the processes $\Delta\theta = \pm\Delta\omega$. The vectors $0 \rightarrow 2$ and $0 \rightarrow 3$

Table 9.1

Field	ω, Degrees	M	M_*	P
1	39.12	2.500	1.826	1.000
2	19.12	1.745	1.507	3.234
2′	.....	1.646	1.452	3.211
3	59.12	3.538	2.070	0.212

represent M_2^* and M_3^*, respectively. The shock polar for the process $1 \rightarrow 2'$ is shown dotted.

Each of the Mach waves is drawn at the local Mach angle μ to the flow. The flow angle θ is known at the wall; this gives ω, which in turn gives M and μ. For example, the wave m_2^+ in Fig. 9.5*c* is drawn from a point on the wall where the local slope is $\theta = 10°$; then $\omega = 39.12° + 10°$, giving $M = 2.967$ and $\mu = 19.7°$.

The preceding has been based on the idea of turning via a sequence of oblique shocks. By letting the number of discontinuities increase without limit, a heuristic scheme for continuous turning was developed. To put this on a strong foundation and to approach more general problems, the method of characteristics is needed: this is developed in the following section.

9.3 Method of characteristics

The problem of putting a given set of equations into characteristic form has a formal solution; i.e., given a set of equations, there are conventional *recipes* for converting to characteristic form. A general recipe for problems with two independent variables will be given in the following section; here we use a more direct and physical approach.

The equations of motion may be taken to be the continuity equation, irrotationality condition, and momentum equation, respectively,

$$\nabla \cdot (\rho \mathbf{u}) = 0 \tag{9.9}$$

$$\nabla \times \mathbf{u} = 0 \tag{9.10}$$

$$\nabla\left(\frac{u^2}{2}\right) + \frac{1}{\rho}\nabla P = 0 \tag{9.11}$$

(Of course, all of the relations already found for steady isentropic flow, e.g., in Sec. 5.2, will still be applicable.) The two independent variables are the space coordinates (which we will label x_1 and x_2, with x_2 corresponding to r in the case of axisymmetric flow). With homentropic flow, there is, in effect, only one thermodynamic variable. There are thus three dependent variables (unknowns) consisting of the two velocity components and the thermodynamic variable.

The basic idea is to seek rates of change along the physical *waves* of the problem. In one-dimensional unsteady flow it was found that certain

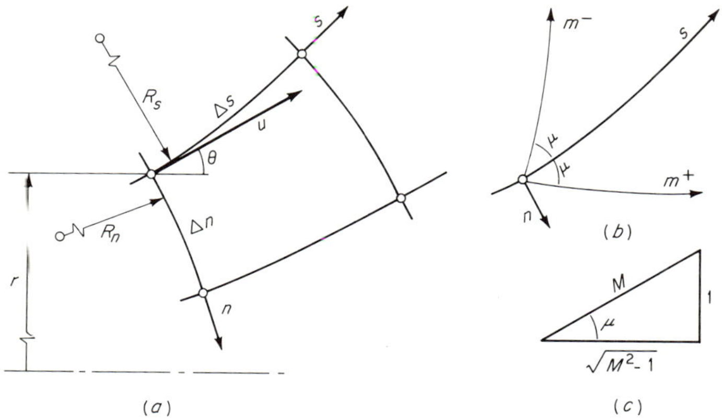

Figure 9.6
(*a*) Streamline coordinates; (*b*) Mach waves intersecting streamline; and (*c*) Mach triangle.

quantities, the Riemann invariants, were constant along sound-wave trajectories. In this problem we seek a similar result for the *Mach waves.*

Following somewhat the derivation of *Liepmann and Roshko* [1957, chap. 7], we will write the equations of motion in a system of streamline coordinates, as shown in Fig. 9.6. The curvilinear coordinates consist of the streamline s and the normal n to the streamlines and lie in the plane of the motion.[1] The velocity $\mathbf{u}$ is everywhere in the local direction of s and at angle θ to a fixed reference direction; we take u and θ to be dependent variables. The axisymmetric case is included by placing the point of interest a distance r from the axis of symmetry (aligned with $\theta = 0$); the plane flow case will follow by letting $r \to \infty$.

From the geometry of the figure it is not difficult to show that

$$\frac{1}{R_n} = \frac{1}{\Delta n}\frac{\partial \Delta n}{\partial s} = -\frac{\partial \theta}{\partial n} \tag{9.12}$$

$$\frac{1}{R_s} = \frac{1}{\Delta s}\frac{\partial \Delta s}{\partial n} = \frac{\partial \theta}{\partial s} \tag{9.13}$$

From Stokes' theorem and Eq. (9.10), the circulation Γ_c about the curvilinear rectangle shown is zero,

$$\int (\nabla \times \mathbf{u}) \cdot \mathbf{n}\, dA = \oint \mathbf{u} \cdot d\mathbf{l} = \Gamma_c = 0$$

[1] This is actually a bastard system in which distances are not stretched; e.g., the equation of an adjacent streamline is *not* given by $n =$ const.

Evaluating the line integral,

$$u\,\Delta s - \left[u\,\Delta s + \Delta n \frac{\partial}{\partial n}(u\,\Delta s)\right] = 0$$

where second-order terms have been omitted. Note that integration along the normals makes no contribution because $\mathbf{u} \cdot d\mathbf{l}$ is identically zero. With (9.13) the above becomes

$$\frac{\partial u}{\partial n} + u\frac{\partial \theta}{\partial s} = 0 \tag{9.14}$$

which is the first equation of motion.

The continuity equation is written in the simple form

$$\rho u r\,\Delta n = \text{const}$$

Differentiated logarithmically with respect to s and with the use of (9.12) this becomes

$$\frac{1}{\rho}\frac{\partial \rho}{\partial s} + \frac{1}{u}\frac{\partial u}{\partial s} - \frac{\partial \theta}{\partial n} = -\frac{\sin\theta}{r} \tag{9.15}$$

where $\sin\theta = \partial r/\partial s$. The momentum equation in the s direction is

$$u\frac{\partial u}{\partial s} + \frac{1}{\rho}\frac{\partial P}{\partial s} = 0$$

With the homentropic transformation $dP = c^2\,d\rho$, combining with (9.15) gives

$$(M^2 - 1)\frac{1}{u}\frac{\partial u}{\partial s} + \frac{\partial \theta}{\partial n} = \frac{\sin\theta}{r} \tag{9.16}$$

This is the second equation of motion.

With the Mach angle defined by $\tan\mu = 1/\sqrt{M^2 - 1}$, Eqs. (9.14) and (9.16) can be rewritten as

$$\tan\mu\,\frac{\sqrt{M^2 - 1}}{u}\frac{\partial u}{\partial n} + \frac{\partial \theta}{\partial s} = 0 \tag{9.17}$$

$$\frac{\sqrt{M^2 - 1}}{u}\frac{\partial u}{\partial s} + \tan\mu\,\frac{\partial \theta}{\partial n} = \frac{\tan\mu\,\sin\theta}{r} \tag{9.18}$$

The Prandtl-Meyer function has been defined by $d\omega = \sqrt{M^2 - 1}\;du/u$. Thus, for example, the first term in (9.18) can be written $\partial\omega/\partial s$, whereby a derivative of the velocity is replaced by a derivative of a unique function of the velocity. Thus, the equations of motion (9.17) and (9.18) become

$$\tan\mu \frac{\partial \omega}{\partial n} + \frac{\partial \theta}{\partial s} = 0 \tag{9.19}$$

$$\frac{\partial \omega}{\partial s} + \tan\mu \frac{\partial \theta}{\partial n} = \frac{\tan\mu \sin\theta}{r} \tag{9.20}$$

These can be added and subtracted to give the *characteristic equations*

$$\begin{aligned} \left(\frac{\partial}{\partial s} + \tan\mu \frac{\partial}{\partial n}\right)(\theta + \omega) &= +\frac{\tan\mu \sin\theta}{r} \\ \left(\frac{\partial}{\partial s} - \tan\mu \frac{\partial}{\partial n}\right)(\theta - \omega) &= -\frac{\tan\mu \sin\theta}{r} \end{aligned} \tag{9.21}$$

These equations can be interpreted as follows. Let dm^+ be an increment of length along the Mach wave m^+ in Fig. 9.6*b*. Then the derivative of any function $F(s,n)$ along m^+ can be written by the chain rule

$$\frac{dF}{dm^+} = \frac{\partial F}{\partial s}\frac{ds}{dm^+} + \frac{\partial F}{\partial n}\frac{dn}{dm^+}$$

and from the geometry of the figure, $ds/dm^+ = \cos\mu$ and $dn/dm^+ = \sin\mu$. Then

$$\frac{dF}{dm^+} = \cos\mu\left(\frac{\partial F}{\partial s} + \tan\mu \frac{\partial F}{\partial n}\right)$$

in which the quantity in parentheses is exactly in the form of the derivative in the first equation of the set (9.21).

A similar result obtains along m^-, and (9.21) can be written in a form analogous to the one-dimensional unsteady characteristic equations (8.8), viz.,

$$\boxed{\begin{aligned} \frac{d}{dm^+}(\theta + \omega) &= +\frac{\sin\mu \sin\theta}{r} \\ \frac{d}{dm^-}(\theta - \omega) &= -\frac{\sin\mu \sin\theta}{r} \end{aligned}} \tag{9.22}$$

These are derivatives along the characteristic lines m^+ and m^- and in this form are given per unit length of the characteristic path (unlike the

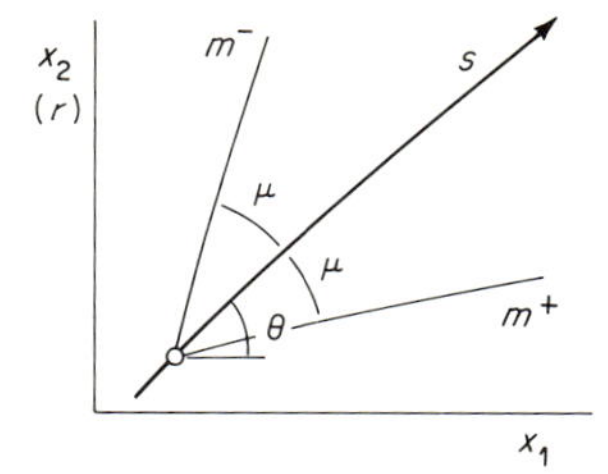

Figure 9.7

unsteady one-dimensional case, where derivatives are of course per unit time). The slope of the characteristics m^+ and m^- respectively is given by $\tan(\theta - \mu)$ and $\tan(\theta + \mu)$, as sketched in Fig. 9.7.[1] These slopes are comparable to $u \pm c$ in the one-dimensional case. To complete the analogy we compare the "invariants" $\omega \pm \theta$ and $u \pm F$; because ω may be considered, via (9.5), uniquely dependent on the thermodynamic state, for example, $\omega(P)$, it is analogous to F; θ and u are analogous because they both represent slopes of particle paths in the respective planes of the independent variables (note, however, that $\theta = 0$ conventionally represents a horizontal path line while $u = 0$ represents a vertical path line). Thus we have a direct analogy

$$\theta \pm \omega \leftrightarrow u \pm F$$

For *plane flow*, the distance $r \to \infty$, and the right-hand sides vanish in the characteristic equations. Then we have the simple result

$$\begin{aligned} \theta + \omega &= \text{const} \qquad \text{on characteristic } m^+ \\ \theta - \omega &= \text{const} \qquad \text{on characteristic } m^- \end{aligned} \tag{9.23}$$

[1] Unfortunately (and unavoidably) this violates the symmetry of the sign convention in that m^+ is associated with $\theta - \mu$, for example, instead of $\theta + \mu$.

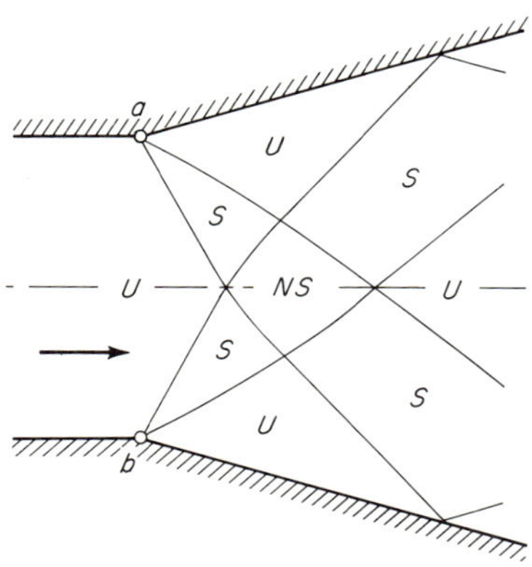

Figure 9.8

which are analogous to the integrals of (8.13). Just as in the one-dimensional case, the existence of *uniform regions* will imply the existence of adjacent *simple wave regions* with straight characteristic lines carrying constant properties. Centered rarefactions, called *Prandtl-Meyer fans,* will arise at a discontinuity in the flow angle, as at a sharp corner of a solid boundary, since the flow angle θ must be multiple-valued at such a point. In Fig. 9.8 a fan is centered at points a and b. The initial supersonic flow is uniform (U) and is divided by characteristics into downstream simple (S), nonsimple (NS), and uniform (U) regions. A flow field of this kind is worked out in detail in Example 9.5 (page 460).

Reflection Conditions at a Boundary

The contact surface separating two supersonic streams is shown in Fig. 9.9 (compare Fig. 8.33). As before, this may be considered a problem in the reflection of known incident characteristics. The matching conditions (7.65) and (7.67) are simply $u_1 \parallel u_2$ and $P_1 = P_2$, which may be conveniently expressed

$$P_1(\omega_1) = P_2(\omega_2) \tag{9.24}$$

$$\theta_1 = \theta_2 \tag{9.25}$$

With the incident characteristics m_2^- and m_1^+ carrying known values $\theta_2 - \omega_2$ and $\theta_1 + \omega_1$, respectively, we have the known quantity

$$(\theta_1 + \omega_1) - (\theta_2 - \omega_2) = \omega_1(P) + \omega_2(P) \tag{9.26}$$

This can in principle be solved for the pressure, which in turn gives ω_1, ω_2, and finally θ.

Note that the incident and reflected characteristics necessarily have equal angles μ of incidence and reflection.

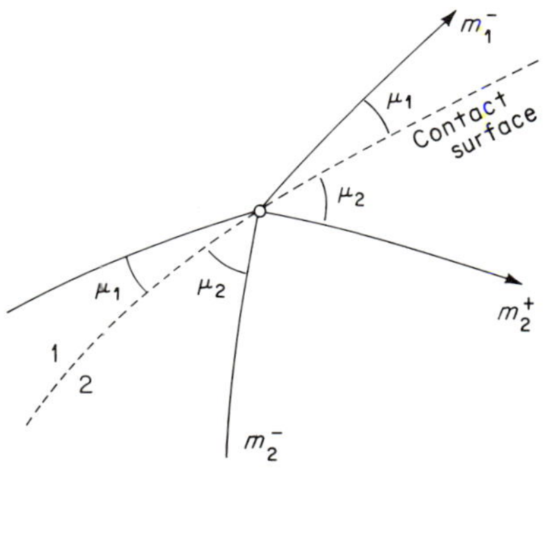

Figure 9.9
Contact surface between two different substances, 1 and 2, flowing at different supersonic speeds.

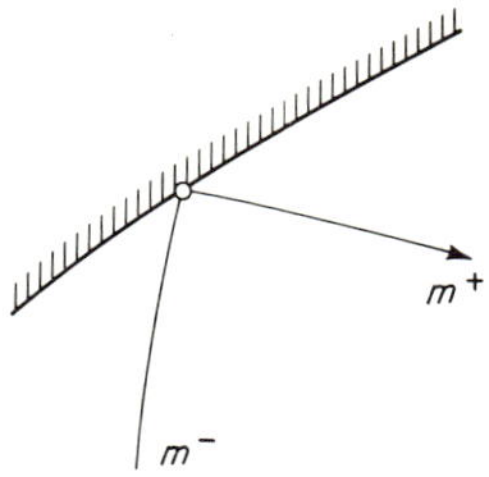

Figure 9.10
Rigid wall.

Two special cases of practical importance are the *rigid wall* and the *free surface*, or constant-pressure boundary. For the rigid wall (see Fig. 9.10) with $\theta - \omega$ known on the incident characteristic m^- and θ_w known, we have

$$\begin{aligned} \omega &= \theta_w - (\theta - \omega) \\ \theta + \omega &= 2\theta_w - (\theta - \omega) \end{aligned} \tag{9.27}$$

and similar relations if the incident characteristic is m^+. A free surface occurs when the supersonic stream is bounded by stagnant fluid at constant pressure (see Fig. 9.11), as when a nozzle discharges into a stagnant atmosphere. With constant pressure P_a on the boundary we have $\omega(P_a) = \omega_a = $ const; with $\theta - \omega$ known,

$$\begin{aligned} \theta &= (\theta - \omega) + \omega_a \\ \theta + \omega &= (\theta - \omega) + 2\omega_a \end{aligned} \tag{9.28}$$

and similar relations if the incident characteristic is m^+. Application of these conditions will be illustrated in following examples.

EXAMPLE 9.2 SUPERSONIC TURNING

A uniform flow at Mach number $M_1 = 1.775$ and $P_1 = 1$ atm follows a wall which is initially straight, followed by a 16° expansive turn formed by a circular arc, followed by a straight section (see Fig. 9.12). Sketch the flow pattern and

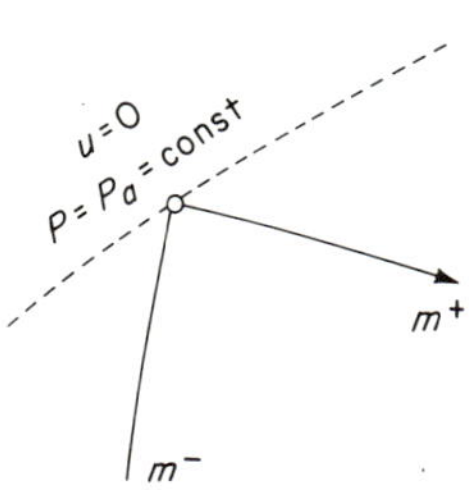

Figure 9.11
Free surface.

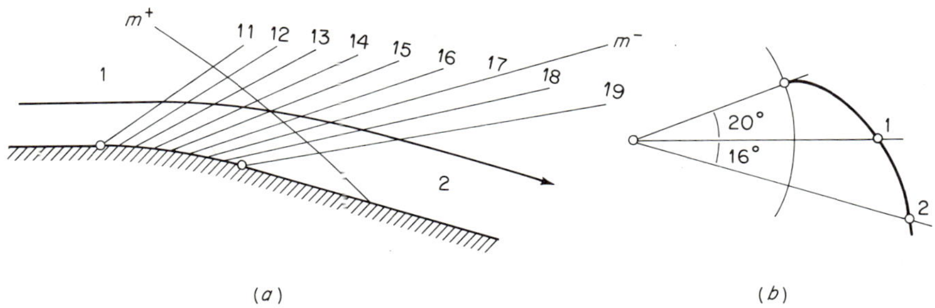

Figure 9.12
Continuous expansive turn: (*a*) physical plane; (*b*) hodograph diagram.

determine the final Mach number M_2 and pressure P_2 (perfect gas, $\gamma = 1.40$, plane flow).

The entire field is covered by characteristics m^+ originating in uniform region 1. Hence, we have simple waves.

$$\theta + \omega = \omega_1 \qquad \text{everywhere}$$

$$\theta - \omega = \text{const} = 2\theta - \omega_1 \qquad \text{on } m^-$$

therefore

$$\theta = \text{const} \qquad \omega = \omega_1 - \theta = \text{const} \qquad \text{on } m^-$$

Thus, m^- characteristics have constant ω and θ and are straight lines. Characteristics have been drawn in the figure for 2° increments in θ; the angle with the

Table 9.2

m^-	θ	ω	M	μ	$\theta + \mu$
11	0	20	1.775	34.3	34.3
12	−2	22	1.844	32.8	30.8
13	−4	24	1.915	31.5	27.5
14	−6	26	1.986	30.2	24.2
15	−8	28	2.059	29.1	21.1
16	−10	30	2.134	27.9	17.9
17	−12	32	2.210	26.9	14.9
18	−14	34	2.289	25.9	11.9
19	−16	36	2.369	25.0	9.0

horizontal is $\theta + \mu$. Thus, for example, on characteristic 17, $\theta = -12°$, and the streamlines crossing this characteristic have constant inclination of $-12°$. The final conditions are

$$M_2 = 2.369$$

$$P_2 = \frac{0.0718}{0.1808} = 0.3970 \text{ atm}$$

The relative distance d of the streamline from the wall is

$$\frac{d_2}{d_1} = \frac{A_2/A_*}{A_1/A_*} = \frac{2.3356}{1.4123} = 1.654$$

Table 9.2 lists the values along each characteristic m^-.

EXAMPLE 9.3 PRANDTL-MEYER FAN

Conditions are the same as in the preceding problem, except that the wall turn is formed by a sharp convex corner.

The physically acceptable solution, with Mach waves running downstream from the corner, is shown in Fig. 9.13*a*. The calculations from the preceding example and the corresponding values (Table 9.2) are applicable. The difference is that the characteristics here intersect at the point of discontinuity in θ, forming a *Prandtl-Meyer fan.*

The solution sketched in Fig. 9.13*b*, with upstream-running Mach waves, is formally constructed in the same way. With $\theta - \omega = -\omega_1 =$ const everywhere, we have, for example, the following values on characteristics:

On m_1^+: $\theta = 0°$ $\omega = 20°$ $\mu = 34.3°$ $M = 1.775$

On m_2^+: $\theta = -16°$ $\omega = 4°$ $\mu = 55.2°$ $M = 1.218$

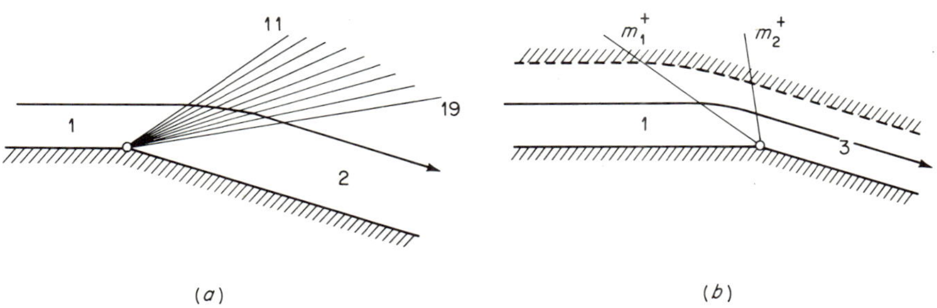

Figure 9.13
(*a*) Prandtl-Meyer fan, or centered rarefaction; (*b*) compressive Prandtl-Meyer fan.

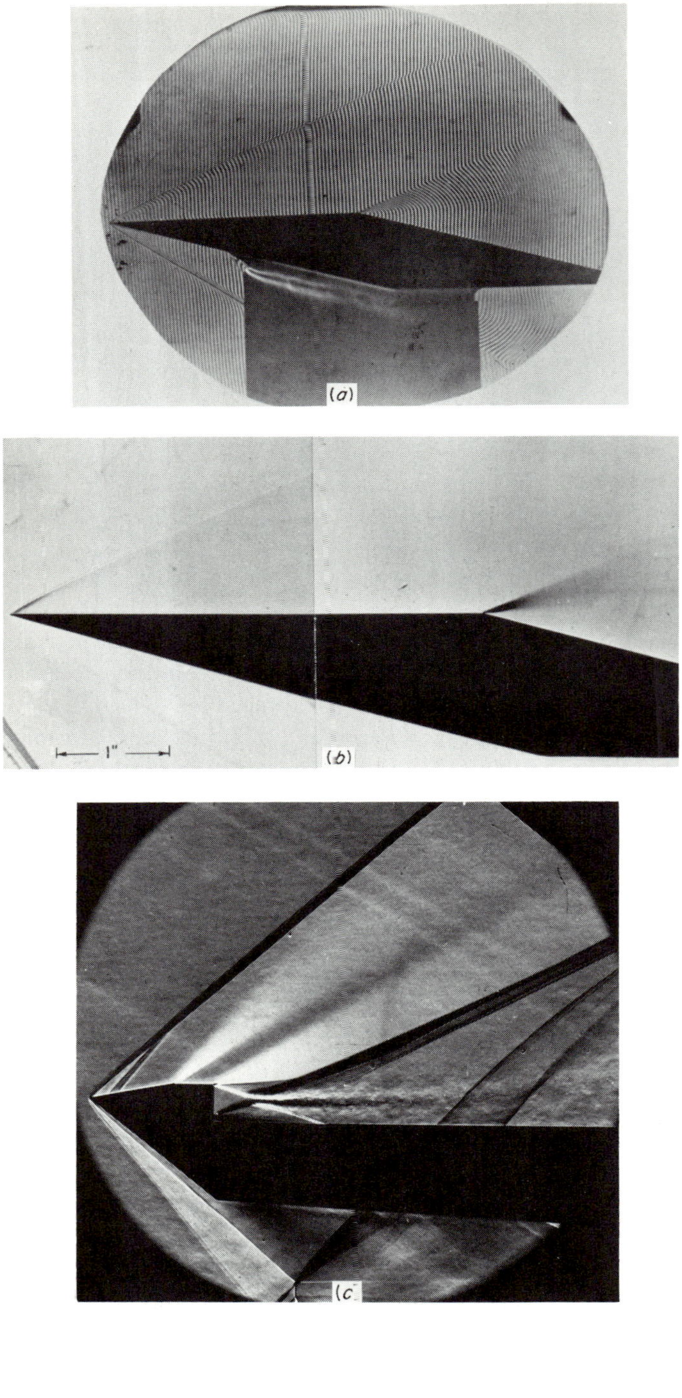
(a)
1"
(b)

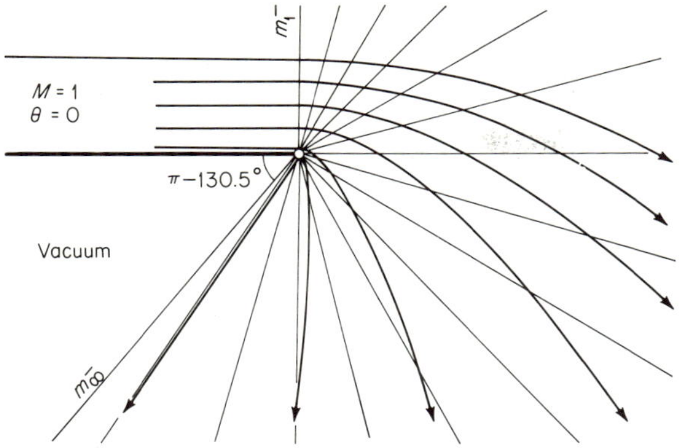

Figure 9.15
Expansion of an initially uniform sonic flow into a vacuum (perfect gas, $\gamma = 1.40$).

and the pressure in region 3 corresponding to $M = 1.218$ is $0.4030/0.1808 = 2.23$ atm. Physically, such a solution could arise only by deflection via an upper wall contoured as a streamline of the centered-wave solution, as sketched. Solutions of this kind are used, for example, in the design of supersonic diffusers.

All the flows considered in this section are *reversible*, provided that shocks are absent. For example, the solution sketched in Fig. 9.13*b* holds for the reversed flow from right to left, giving an expansion instead of a compression (in this case, the reversed flow is the one more in accord with usual experience). Formally, this property of reversibility follows from the invariance of Eqs. (9.9) and (9.11) under time reversal (which merely gives a change in the sign of **u**).

Photographs of Prandtl-Meyer fans are shown in Fig. 9.14.

EXAMPLE 9.4 EXPANSION OF A SONIC FLOW INTO A VACUUM

The ultimate simple wave corresponds to expansion from $M = 1$ to $M = \infty$, that is, to zero pressure. Such a wave is pictured in Fig. 9.15. The initially uniform flow is bounded by a plane wall of zero thickness, such that a turning angle of 180° is geometrically possible. With $\theta + \omega = \text{const} = 0$ everywhere, $\theta = -\omega$ and the minimum value of θ is just $-\omega_{\max}$. Then the complete expansion is represented by the (inverted) hodograph line of Fig. 9.4. The maximum turning angle for a perfect gas from Eq. (9.8) gives the values shown in Table 9.3. For $\gamma < 5/4$, $\omega(\infty) > \pi$, and a complete expansion is geometrically

---

Figure 9.14
Prandtl-Meyer fans (see also Fig. 10.9*a*). (*a*) Flow around sharp corner, interferogram. (*b*) Same as (*a*), schlieren picture. (*Courtesy of I. I. Glass, University of Toronto.*) (*c*) Flow around sharp corner and expansion at tail, visible on upper part of two-dimensional body. (*Courtesy of I. S. Donaldson, University of Manchester.*)

Table 9.3

γ	$\omega(\infty)$, Degrees
$5/4 = 1.25$	180
$4/3 \approx 1.333$	148.12
$7/5 = 1.40$	130.45
$5/3 \approx 1.667$	90

impossible. The value $\omega(\infty)$ is achieved along the limiting characteristic m_∞^-, where $M = \infty$, $P = \rho = 0$, and from (5.61) $u = \sqrt{2/(\gamma - 1)}\, c_0$. Because the Mach angle μ is zero along this characteristic, it is necessarily also a streamline bounding the flow.

A flow of the kind pictured can be physically identified as *flow into a vacuum* (and can be observed in modified form at the edge of a rocket nozzle exhausting into vacuum). This is, of course, the analog of unsteady expansion into a vacuum (compare Fig. 8.19). The rarefaction near the tail of the wave (along the characteristics in the third quadrant near m_∞^-) is extreme; along the vertical characteristic (with $\theta + \mu = -90°$), for example, $M \approx 6.9$ and $\rho/\rho_0 \approx 10^{-3}$. Near the center of the rarefaction wave, the accelerations become very large, with large mean free path. The flow is thus nonequilibrium in this region, just as in the case of unsteady expansion.

EXAMPLE 9.5 SHARP-CORNERED NOZZLE

Design a sharp-cornered nozzle to isentropically accelerate a uniform flow at $M_A = 1.2177$ to a uniform flow at $M_C = 2.0000$. This is to be accomplished with no net change in flow direction, i.e., such that $\theta_C = \theta_A$ (perfect gas, $\gamma = 1.40$, plane flow).

The full wave diagram is given in Fig. 9.16. The nozzle has been split along the plane of symmetry (centerline), which acts as a rigid plane boundary. The upper boundary is constructed as a *free streamline* so that the simple wave between B and C is preserved. The initial expansion fan is terminated on $\omega + \theta = 26.38°$, that is, the ultimate value of ω in region C. The area ratio for the expansion is given by

$$\frac{A_C}{A_A} = \frac{A_C/A_*}{A_A/A_*} = \frac{1.6875}{1.0358} = 1.629$$

It may be noted that any intermediate streamline constructed on this diagram gives a possible nozzle profile.

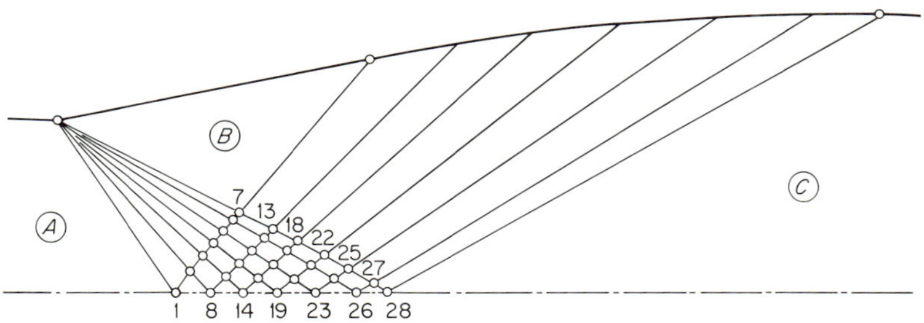

Figure 9.16

Table 9.4

M	Point	$\omega + \theta$	$\omega - \theta$	ω	θ	μ	$\theta + \mu$	$\theta - \mu$
1.2177	1	4	4	4	0	55.2	55.2	−55.2
	2	8	4	6	2	50.6	52.6	−48.6
	3	12	4	8	4	47.1	51.1	−43.1
	4	16	4	10	6	44.2	50.2	−38.2
	5	20	4	12	8	41.7	49.7	−33.7
	6	24	4	14	10	39.5	49.5	−29.5
1.611	7	26.38	4	15.19	11.19	38.4	49.6	−27.2
	8	8	8	8	0	47.1	47.1	−47.1
	9	12	8	10	2	44.2	46.2	−42.2
	10	16	8	12	4	41.7	45.7	−37.7
	11	20	8	14	6	39.5	45.5	−33.5
	12	24	8	16	8	37.6	45.6	−29.6
1.679	13	26.38	8	17.19	9.19	36.6	45.8	−27.4
	14	12	12	12	0	41.7	41.7	−41.7
	15	16	12	14	2	39.5	41.5	−37.5
	16	20	12	16	4	37.6	41.6	−33.6
	17	24	12	18	6	35.9	41.9	−29.9
1.747	18	26.38	12	19.19	7.19	34.9	42.1	−27.7
	19	16	16	16	0	37.6	37.6	−37.6
	20	20	16	18	2	35.9	37.9	−33.9
	21	24	16	20	4	34.3	38.3	−30.3
1.816	22	26.38	16	21.19	5.19	33.5	38.7	−28.3
	23	20	20	20	0	34.3	34.3	−34.3
	24	24	20	22	2	32.8	34.8	−30.8
1.886	25	26.38	20	23.19	3.19	32.1	35.3	−28.9
	26	24	24	24	0	31.5	31.5	−31.5
1.957	27	26.38	24	25.19	1.19	30.8	32.0	−29.6
2.000	28	26.38	26.38	26.38	0	30.0	30.0	−30

EXAMPLE 9.6 UNDEREXPANDED JET

Find the flow field of the jet issuing from an underexpanded (choked) converging nozzle, with the exit pressure twice that of the stationary atmosphere (perfect gas, $\gamma = 1.40$, plane flow).

The flow field in this case (and the axisymmetric one as well) shows an interesting and well-known repetitive pattern somewhat like a string of sausages. The calculated wave diagram is shown in Fig. 9.17. For computational convenience the exit Mach number in uniform field E has been taken as 1.01 (rather than unity). The expansion to uniform field A across a Prandtl-Meyer fan is governed by the condition that the boundary pressure is P_a. Additional expansion to uniform field I is governed by the condition that θ be zero along the plane of symmetry. Thus, we have the following values:

E:	$P = 0.522P_0$	$\omega = 0.04°$	$\theta = 0$
A:	$P = 0.261P_0$	$\omega = 12.79°$	$\theta = 12.75°$
I:		$\omega = 25.54°$	$\theta = 0$
A':		$\omega = 12.79°$	$\theta = -12.75°$

Note that $\omega = 12.79°$ holds everywhere on the free (constant-pressure) boundary. The characteristic net is constructed in the upper half-plane, starting at point 1 with the conditions of field E. Table 9.5 is a skeleton table of values. A weak oblique shock is formed at the point where two m^- characteristics first intersect (above point 78). The calculation of flows including weak shocks will be briefly discussed in Sec. 9.5. The flow is *almost* symmetric about a median transverse plane (even in the absence of shocks, it will not be perfectly symmetric). A new "cell" is formed downstream of the shock and almost

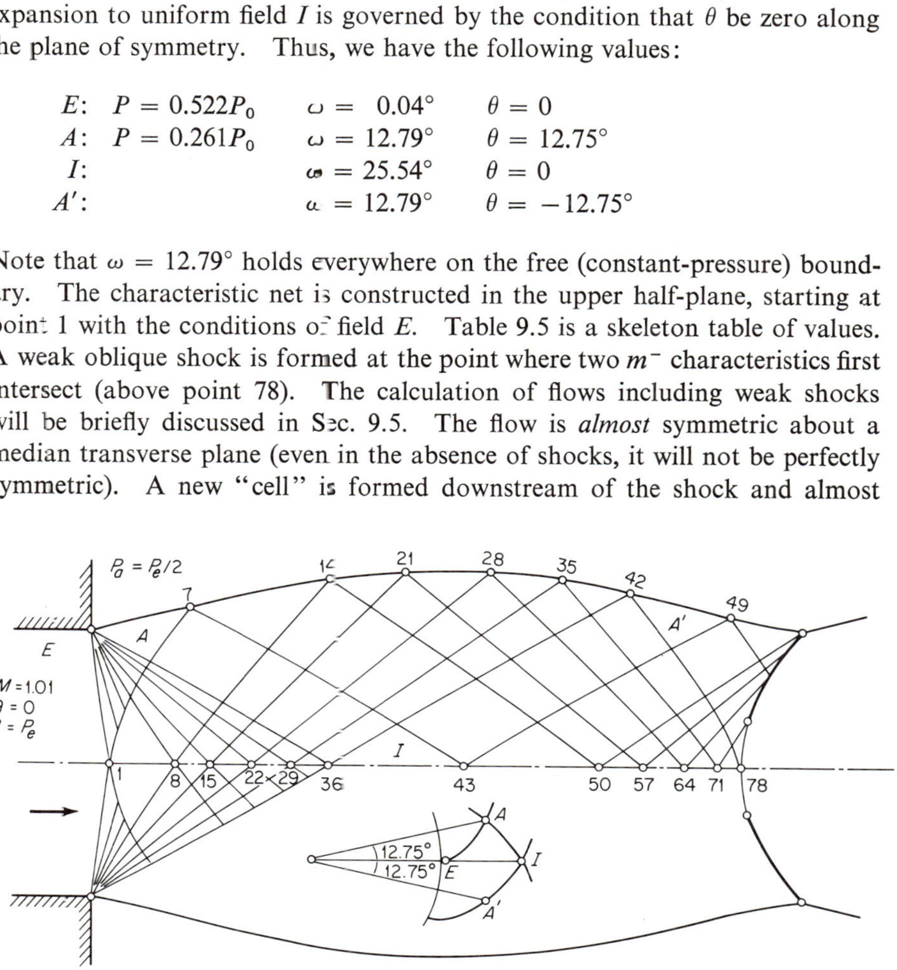

Figure 9.17
Underexpanded jet.

Table 9.5

Point	$\omega + \theta$	$\omega - \theta$	M	θ	μ	$\theta + \mu$	$\theta - \mu$
1	0.04	0.04	0.04	0	81.9	81.0	−81.9
7	25.54	0.04	12.79	12.75	40.8	53.6	−28.0
8	4.79	4.79	4.79	0	53.2	53.2	−53.2
14	20.79	4.79	12.79	8.00	40.8	48.8	−32.8
15	9.79	9.79	9.79	0	44.5	44.5	−44.5
21	15.79	9.79	12.79	3.00	40.8	43.8	−37.8
22	15.79	15.79	15.79	0	37.8	37.8	−37.8
28	9.79	15.79	12.79	−3.00	40.8	37.8	−43.8
29	20.79	20.79	20.79	0	33.7	33.7	−33.7
35	4.79	20.79	12.79	−8.00	40.8	32.8	−48.8
36	25.54	25.54	25.54	0	30.5	30.5	−30.5
42	0.04	25.54	12.79	−12.75	40.8	28.0	−53.6
43	25.54	25.54	25.54	0	30.5	30.5	−30.5
49	0.04	25.54	12.79	−12.75	40.8	28.0	−53.6
50	20.79	20.79	20.79	0	33.7	33.7	−33.7
57	15.79	15.79	15.79	0	37.8	37.8	−37.8
64	9.79	9.79	9.79	0	44.5	44.5	−44.5
71	4.79	4.79	4.79	0	53.2	53.2	−53.2
78	0.04	0.04	0.04	0	81.9	81.9	−81.9

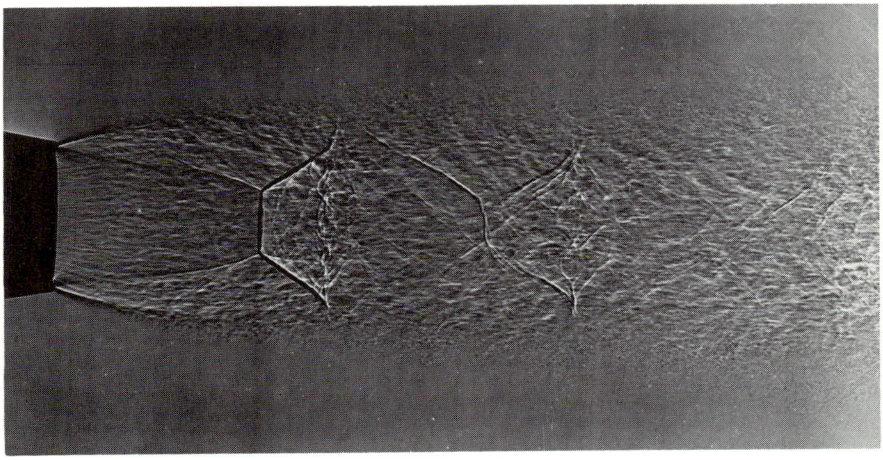

Figure 9.18
Supersonic periodic jet from a choked converging axisymmetric nozzle, $P_e/P_a = 3$.

reproduces the first. Experimental results agree quite well with the inviscid calculation, at least in the first cell or two.[1]

A photograph of an axisymmetric periodic jet of this type is shown in Fig. 9.18.

9.4 General treatment of the method of characteristics for two independent variables

Here we digress from the specific topic of steady supersonic flow to briefly discuss the method of characteristics in general. In particular, a method is developed for putting a set of differential equations in two independent variables into characteristic form.

We have already considered two problems which are suitable to the method of characteristics, viz., unsteady one-dimensional flow (independent variables x, t) and steady supersonic flow (independent variables x_1, x_2). With a restriction to homentropic flows, only two dependent variables (u, P and ω, θ, respectively, say) appear in each of these problems. In general, however, we will consider problems with any number of dependent variables and two independent variables. For example, in unsteady one-dimensional isentropic flow, it is convenient to take u, P, ρ as dependent variables.

Given n first-order equations with n dependent variables and two independent variables, it is desired to find n equations in characteristic form. The procedure given here is due to *Courant and Friedrichs* [1948, chap. 2]. The variables and equations are listed symbolically:

2 independent variables:	x, y	
n dependent variables:	u^k	$k = 1, 2, \ldots, n$
n equations:	$L_i = 0$	$i = 1, 2, \ldots, n$

The given equations L_i are of the general form

$$L_i = a_{ik}\frac{\partial u^k}{\partial x} + b_{ik}\frac{\partial u^k}{\partial y} + c_i = 0 \tag{9.29}$$

where the Einstein convention of summation on repeated indices, introduced in Chap. 1, is used.[2] The coefficients a, b, and c may depend upon

[1] See, for example, E. S. Love and C. E. Grigsby, *NACA Res. Mem.* L54L31, 1955.

[2] For example, in the third equation $L_3 = 0$, the first term becomes

$$a_{31}\frac{\partial u^{(1)}}{\partial x} + a_{32}\frac{\partial u^{(2)}}{\partial x} + \cdots + c_{3n}\frac{\partial u^{(n)}}{\partial x}$$

The superscript k is *not* an exponent.

x, y, and the u_k. These equations are said to be *quasi-linear* because products or powers of the highest-order derivatives (in this case all derivatives are first order) do not appear. It would appear that most differential equations of physical interest are quasi-linear.

We now seek n characteristic directions C. Specifically, we seek a revised form of (9.29) such that *differentiation is in only one direction*, i.e., *along a characteristic curve*. The idea of differentiation along a certain curve is simply illustrated by the material derivative, which is, in one dimension,

$$\frac{D}{Dt} = \frac{\partial}{\partial t} + u\frac{\partial}{\partial x}$$

This represents a derivative with respect to time along a particle path PP in the xt plane, as indicated in Fig. 9.19. For example, the material derivative of some function $f = f(x,t)$ is

$$\frac{Df}{Dt} = \lim_{\Delta t \to 0} \frac{\Delta f}{\Delta t}$$

Suppose that it is desired to find the rate of change of f with respect to the distance l along the curve PP [because x and t have *different* dimensions, a length $\delta l = \sqrt{(\delta x)^2 + (\delta t)^2}$ has grotesque dimensions; to avoid this distraction, consider that x and t have been normalized, i.e., are dimensionless]. Then

$$\frac{Df}{Dt} = \frac{df}{dl}\left(\frac{dl}{dt}\right)_{PP}$$

With $(dt/dl)_{PP} \equiv t_l$, this can be rewritten

$$\frac{df}{dl} = t_l\frac{Df}{Dt} = t_l\left(\frac{\partial}{\partial t} + u\frac{\partial}{\partial x}\right)f$$

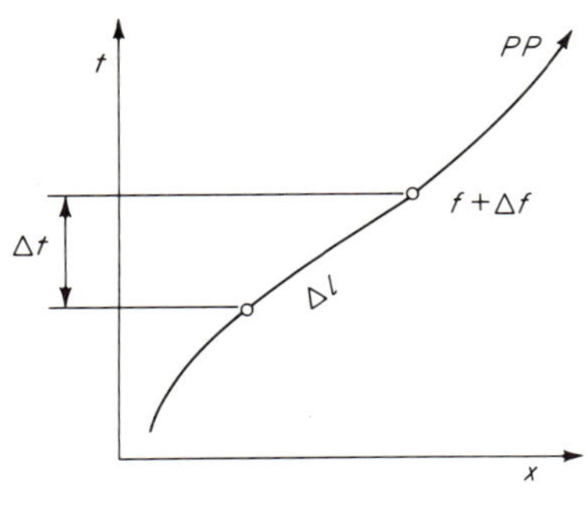

Figure 9.19

We say that either Df/Dt or df/dl represents a derivative along the curve PP. The essential point is that $dx/dt = u$ in both cases; i.e., the ratio of the $\partial/\partial x$ coefficient to the $\partial/\partial t$ coefficient is the same in both cases. Thus, the derivative operator

$$\mathscr{L} \equiv a\frac{\partial}{\partial x} + b\frac{\partial}{\partial y}$$

represents differentiation along a curve C, with

$$\left(\frac{dy}{dx}\right)_C = \frac{b}{a} \tag{9.30}$$

In general, consider a curve C given in parametric form by $x(\sigma)$ and $y(\sigma)$ (see Fig. 9.20). Then with $df = \partial f/\partial x\, dx + \partial f/\partial y\, dy$,

$$\left(\frac{df}{d\sigma}\right)_C = \frac{dx}{d\sigma}\frac{\partial f}{\partial x} + \frac{dy}{d\sigma}\frac{\partial f}{\partial y}$$

or

$$\left(\frac{df}{d\sigma}\right)_C = f_\sigma = x_\sigma\frac{\partial f}{\partial x} + y_\sigma\frac{\partial f}{\partial y}$$

represents differentiation of f along the curve C, which has slope

$$\left(\frac{dy}{dx}\right)_C = \frac{y_\sigma}{x_\sigma} \tag{9.31}$$

A linear combination $\lambda_i L_i$ of Eqs. (9.29) of the form

$$\lambda_i L_i = T_k u_\sigma{}^k + R = 0 \tag{9.32}$$

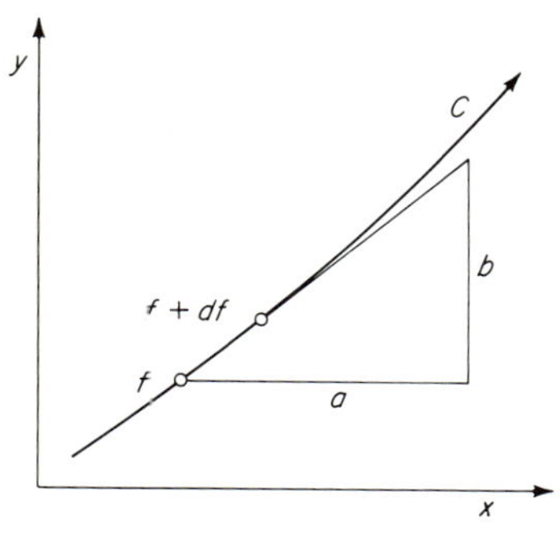

Figure 9.20

would thus be in characteristic form, because there is only one derivative operator $d/d\sigma$ corresponding to a particular characteristic. Written out, $\lambda_i L_i$ is

$$\lambda_i a_{ik} \frac{\partial u^k}{\partial x} + \lambda_i b_{ik} \frac{\partial u^k}{\partial y} + \lambda_i c_i = 0 \tag{9.33}$$

or

$$\left(\lambda_i a_{i1} \frac{\partial u^{(1)}}{\partial x} + \lambda_i b_{i1} \frac{\partial u^{(1)}}{\partial y}\right) + \left(\lambda_i a_{i2} \frac{\partial u^{(2)}}{\partial x} + \lambda_i b_{i2} \frac{\partial u^{(2)}}{\partial y}\right) + \cdots + \lambda_i c_i = 0$$

In order that *each* of the derivatives in parentheses be in the same direction, we require, as in (9.30), that

$$\frac{\lambda_i b_{i1}}{\lambda_i a_{i1}} = \frac{\lambda_i b_{i2}}{\lambda_i a_{i2}} = \cdots = \frac{\lambda_i b_{ik}}{\lambda_i a_{ik}} = \frac{y_\sigma}{x_\sigma} \qquad \text{no sum}$$

Thus, there are n equations

$$\lambda_i a_{ik} y_\sigma = \lambda_i b_{ik} x_\sigma \tag{9.34}$$

which must be satisfied. Let

$$\frac{y_\sigma}{x_\sigma} = \frac{dy}{dx} \equiv \zeta$$

Then (9.34) can be written

$$\lambda_i(a_{ik}\zeta - b_{ik}) = 0 \tag{9.35}$$

which are n homogeneous equations for the unknown λ's. In order to have nontrivial solutions, the coefficient determinant must vanish

$$|a_{ik}\zeta - b_{ik}| = 0 \tag{9.36}$$

This gives rise to an nth-order polynomial for ζ. We then have n roots and designate them ζ_l; if they are all real, as we will suppose, the system of equations (9.29) is said to be *totally hyperbolic* and the method of characteristics will provide a convenient means for solution. The n characteristic directions are given by

$$\zeta_l = \left(\frac{dy}{dx}\right)_l$$

EXAMPLE 9.7 CHARACTERISTIC DIRECTIONS FOR HOMENTROPIC UNSTEADY ONE-DIMENSIONAL FLOW

Let us find the characteristic directions for this problem via the above procedure. For constant area and no body force, the equations of motion (8.2) and (8.3) are

$$\begin{aligned} u\frac{\partial u}{\partial x} + \frac{1}{\rho}\frac{\partial P}{\partial x} + \frac{\partial u}{\partial t} + 0 &= 0 \\ c\frac{\partial u}{\partial x} + \frac{u}{\rho c}\frac{\partial P}{\partial x} + 0 + \frac{1}{\rho c}\frac{\partial P}{\partial t} &= 0 \end{aligned} \tag{9.37}$$

These have been written in the form of (9.29). By virtue of the homentropic condition, ρ and c may be considered to be known functions of P; there are thus only two unknowns, u and P. The determinant (9.36) is

$$\begin{vmatrix} u\zeta - 1 & \dfrac{1}{\rho}\zeta \\ c\zeta & \dfrac{u}{\rho c}\zeta - \dfrac{1}{\rho c} \end{vmatrix} = 0$$

Multiplying the second column by ρc gives

$$\begin{vmatrix} u\zeta - 1 & c\zeta \\ c\zeta & u\zeta - 1 \end{vmatrix} = 0$$

Expanding the determinant, we obtain

$$(u\zeta - 1)^2 - (c\zeta)^2 = 0$$

Solving,

$$\zeta = \frac{dt}{dx} = \frac{1}{u + c}, \frac{1}{u - c}$$

which are of course the characteristic directions found in Sec. 8.2.

EXAMPLE 9.8 CHARACTERISTIC DIRECTIONS FOR ISENTROPIC UNSTEADY ONE-DIMENSIONAL FLOW

With the isentropic condition $Ds/Dt = 0$, the pressure and density are related via $DP/Dt = c^2\, D\rho/Dt$. Then the momentum and continuity equations are exactly as in Example 9.7. The full set of equations is then

$$u\frac{\partial u}{\partial x} + \frac{1}{\rho}\frac{\partial P}{\partial x} + 0 + \frac{\partial u}{\partial t} + 0 + 0 = 0$$

$$c\frac{\partial u}{\partial x} + \frac{u}{\rho c}\frac{\partial P}{\partial x} + 0 + 0 + \frac{1}{\rho c}\frac{\partial P}{\partial t} + 0 = 0$$

$$0 + u\frac{\partial P}{\partial x} - c^2 u\frac{\partial \rho}{\partial x} + 0 + \frac{\partial P}{\partial t} - c^2\frac{\partial \rho}{\partial t} = 0$$

where the last equation is just $DP/Dt = c^2\, D\rho/Dt$. There are now three unknowns, u, P, ρ. The determinant (9.36) is now

$$\begin{vmatrix} u\zeta - 1 & \frac{1}{\rho}\zeta & 0 \\ c\zeta & \frac{u}{\rho c}\zeta - \frac{1}{\rho c} & 0 \\ 0 & u\zeta - 1 & -c^2 u\zeta + c^2 \end{vmatrix} = 0$$

After cleaning up by multiplying the second column by ρc, the third row by $1/\rho c$, and the third column by ρ/c, this becomes

$$\begin{vmatrix} u\zeta - 1 & c\zeta & 0 \\ c\zeta & u\zeta - 1 & 0 \\ 0 & u\zeta - 1 & u\zeta - 1 \end{vmatrix} = 0$$

On expansion, this gives the cubic equation

$$(u\zeta - 1)^3 - c^2\zeta^2(u\zeta - 1) = 0$$

with roots

$$\zeta = \frac{dt}{dx} = \frac{1}{u + c}, \frac{1}{u - c}, \frac{1}{u}$$

which are the characteristic directions, as discussed in Sec. 8.10.

Having determined the characteristic directions, it remains to write the n equations in characteristic form. The linear combination (9.33), rewritten for convenience, is

$$\lambda_i a_{ik}\frac{\partial u^k}{\partial x} + \lambda_i b_{ik}\frac{\partial u^k}{\partial y} + \lambda_i c_i = 0$$

It was found that the λ_i must satisfy

$$\lambda_i a_{ik} y_\sigma = \lambda_i b_{ik} x_\sigma$$

With this substitution, (9.33) can be written

$$\lambda_i b_{ik}\left(x_\sigma \frac{\partial u^k}{\partial x} + y_\sigma \frac{\partial u^k}{\partial y}\right) + \lambda_i c_i y_\sigma = 0 \tag{9.38}$$

or

$$\lambda_i a_{ik}\left(x_\sigma \frac{\partial u^k}{\partial x} + y_\sigma \frac{\partial u^k}{\partial y}\right) + \lambda_i c_i x_\sigma = 0 \tag{9.39}$$

The combination in parentheses is exactly the derivative of u^k along a characteristic, i.e.,

$$u_\sigma{}^k = x_\sigma \frac{\partial u^k}{\partial x} + y_\sigma \frac{\partial u^k}{\partial y}$$

Rewriting (9.38) and (9.39), together with (9.35), we have finally

$$\begin{aligned} \lambda_i(a_{ik}\zeta - b_{ik}) &= 0 \qquad n \text{ equations} \\ \lambda_i(a_{ik}u_\sigma{}^k + c_i x_\sigma) &= 0 \qquad 1 \text{ equation} \\ \lambda_i(b_{ik}u_\sigma{}^k + c_i y_\sigma) &= 0 \qquad 1 \text{ equation} \end{aligned} \tag{9.40}$$

which is in the form of a set of $n + 2$ equations for the multipliers λ_i. It is necessary that any nth-order determinant of coefficients vanish:

$$\begin{vmatrix} \text{Coefficients from} \\ \text{any } n \text{ equations} \\ \text{of (9.40)} \end{vmatrix} = 0$$

If the first n equations are selected, the characteristic directions are obtained. Any other combination leads to the characteristic equations of motion. We proceed to an example, which is the subject of this chapter.

EXAMPLE 9.9 STEADY IRROTATIONAL SUPERSONIC FLOW IN TWO DIMENSIONS

The characteristic equations for this problem can be found via the above procedure. For steady flow, the potential equation (5.19) can be written

$$\mathbf{u} \cdot [(\mathbf{u} \cdot \nabla)\mathbf{u}] - c^2 \nabla \cdot \mathbf{u} = 0 \tag{9.41}$$

Table 9.6

Cylindrical Coordinates	Indicial	Eq. (9.29)
z	x_1	x
r	x_2	y
u_z	u_1	$u^{(1)}$
u_r	u_2	$u^{(2)}$

while the irrotationality condition is

$$\nabla \times \mathbf{u} = 0 \tag{9.42}$$

These equations will be written explicitly in cylindrical coordinates, which will be adaptable to either axisymmetric or plane flow. With coordinates r, θ, z the velocity is

$$\mathbf{u} = \mathbf{e}_z u_z + \mathbf{e}_r u_r$$

the component u_θ in the θ direction being zero. Reference to Appendix E yields the vector formulas, with $u_\theta = \partial/\partial\theta = 0$,

$$\nabla = \mathbf{e}_z \frac{\partial}{\partial z} + \mathbf{e}_r \frac{\partial}{\partial r}$$

$$\nabla \cdot \mathbf{u} = \frac{\partial u_z}{\partial z} + \frac{1}{r}\frac{\partial}{\partial r} r u_r$$

$$\nabla \times \mathbf{u} = r\mathbf{e}_\theta \left(\frac{\partial u_r}{\partial z} - \frac{\partial u_z}{\partial r}\right)$$

These formulas have been written out in the familiar r, θ, z symbols for convenience. It will, however, be more practical to use an indicial notation for position and velocity, let the coordinates be x_1, x_2 and the velocity components be u_1, u_2. We then have the equivalent notations shown in Table 9.6. The coordinate system is shown in Fig. 9.21.

Written in the same form as (9.29), the equations of motion (9.41) and (9.42) are now

$$\begin{aligned} (c^2 - u_1^2)\frac{\partial u_1}{\partial x_1} - u_1 u_2 \frac{\partial u_2}{\partial x_1} - u_1 u_2 \frac{\partial u_1}{\partial x_2} + (c^2 - u_2^2)\frac{\partial u_2}{\partial x_2} + \frac{c^2 u_2}{x_2} &= 0 \\ 0 \quad - \quad \frac{\partial u_2}{\partial x_1} + \quad \frac{\partial u_1}{\partial x_2} + \quad 0 \quad + \; 0 &= 0 \end{aligned} \tag{9.43}$$

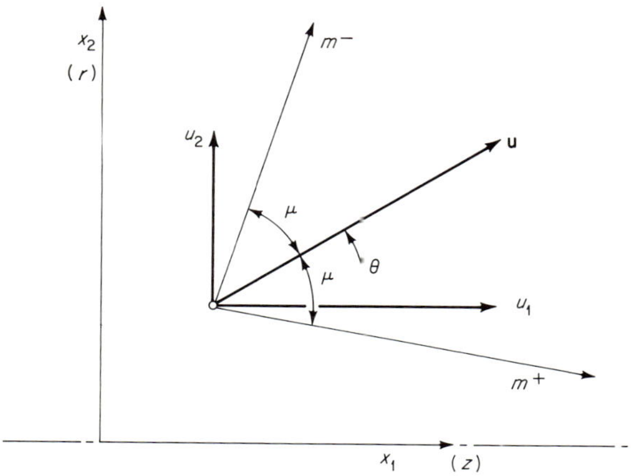

Figure 9.21
Coordinate system for two-dimensional flow.

As in Sec. 9.3, the case of plane flow is recovered by letting $x_2 \to \infty$ in the last term of the first equation, that is, $r \to \infty$.

The sound speed c can be considered a known function of u via the energy equation $h + u^2/2 = \text{const}$, together with the homentropic condition. There are thus only two unknowns, u_1 and u_2.

The coefficient matrix for Eqs. (9.40) is

$$\begin{bmatrix} (c^2 - u_1^2)\zeta + u_1u_2 & -1 \\ -u_1u_2\zeta - (c^2 - u_2^2) & -\zeta \\ (c^2 - u_1^2)u_{1\sigma} - u_1u_2u_{2\sigma} + \dfrac{c^2u_2}{x_2}x_{1\sigma} & -u_{2\sigma} \\ -u_1u_2u_{1\sigma} + (c^2 - u_2^2)u_{2\sigma} + \dfrac{c^2u_2}{x_2}x_{2\sigma} & u_{1\sigma} \end{bmatrix} \tag{9.44}$$

The characteristic directions are found by setting the determinant of the first two rows to zero, yielding the two roots

$$\begin{aligned} \zeta^+ &= \frac{-u_1u_2 + c\sqrt{u_1^2 + u_2^2 - c^2}}{c^2 - u_1^2} = \left(\frac{dx_2}{dx_1}\right)^+ \\ \zeta^- &= \frac{-u_1u_2 - c\sqrt{u_1^2 + u_2^2 - c^2}}{c^2 - u_1^2} = \left(\frac{dx_2}{dx_1}\right)^- \end{aligned} \tag{9.45}$$

The roots or characteristic directions are real only if

$$u_1^2 + u_2^2 = u^2 \geq c^2$$

That is, characteristics exist, and the system of equations is hyperbolic only if the flow is supersonic.

The velocity components are expressed in terms of the flow angle θ by

$$u_1 = u \cos \theta$$
$$u_2 = u \sin \theta$$

and the Mach angle μ is defined by

$$M = \frac{u}{c} = \csc \mu$$

Introducing these relations, (9.45) becomes

$$\zeta = \frac{-\sin \theta \cos \theta \pm \sin \mu \cos \mu}{\sin^2 \mu - \cos^2 \theta}$$

After trigonometric reduction

$$\begin{aligned} \zeta^+ &= \tan(\theta - \mu) \leftrightarrow m^+ \\ \zeta^- &= \tan(\theta + \mu) \leftrightarrow m^- \end{aligned} \tag{9.46}$$

The characteristic directions lie along *Mach waves* m^+ and m^-, as we have already found in Sec. 9.3.

The characteristic curves m^+ and m^- are given by, say,

$$x_1 = x_1(\alpha) \qquad x_2 = x_2(\alpha) \qquad \text{on } m^+$$
$$x_1 = x_1(\beta) \qquad x_2 = x_2(\beta) \qquad \text{on } m^-$$

That is, the parameter σ is replaced by either α or β depending on the characteristic direction.

To find the characteristic forms of the equations of motion, the determinant formed by any two rows from the matrix (9.44) (other than the first two rows, which give the characteristic directions) is set equal to zero. Selecting the first and third rows gives

$$-[(c^2 - u_1^2)\zeta + 2u_1u_2]u_{2\sigma} + (c^2 - u_1^2)u_{1\sigma} + \frac{c^2u_2}{x_2}x_{1\sigma} = 0$$

This becomes two equations by substituting the distinct values of ζ and making the corresponding assignment of σ. Thus,

$$-[(c^2 - u_1^2)\zeta^+ + 2u_1u_2]u_{2\alpha} + (c^2 - u_1^2)u_{1\alpha} + \frac{c^2u_2}{x_2}x_{1\alpha} = 0$$

$$-[(c^2 - u^2)\zeta^- + 2u_1u_2]u_{2\beta} + (c^2 - u_1^2)u_{1\beta} + \frac{c^2u_2}{x_2}x_{1\beta} = 0$$

With the substitution of (9.45) these reduce to[1]

$$
\begin{aligned}
u_{1\alpha} + \zeta^- u_{2\alpha} + \frac{c^2 u_2}{x_2(c^2 - u_1{}^2)} x_{1\alpha} &= 0 \qquad \text{on } m^+ \\
u_{1\beta} + \zeta^+ u_{2\beta} + \frac{c^2 u_2}{x_2(c^2 - u_1{}^2)} x_{1\beta} &= 0 \qquad \text{on } m^-
\end{aligned}
\tag{9.47}
$$

Note that the coefficient of the second term is the slope of the *opposite* characteristic; e.g., for the first equation (along m^+) the coefficient is ζ^-.

Formally, this completes the development. Equations (9.47) are numerically integrable along characteristics m^+ and m^-. It is, however, more convenient to recast these equations into a form with dependent variables ω (the Prandtl-Meyer function) and θ (the flow angle). Using $u_1 = u \cos \theta$ and $u_2 = u \sin \theta$, the derivatives in (9.47) become

$$
\begin{aligned}
u_{1\sigma} &= u_\sigma \cos \theta - \theta_\sigma u \sin \theta \\
u_{2\sigma} &= u_\sigma \sin \theta + \theta_\sigma u \cos \theta
\end{aligned}
\qquad \sigma = \alpha, \beta
$$

The Prandtl-Meyer function ω was defined in Eq. (9.4) as

$$
d\omega = \sqrt{M^2 - 1}\,\frac{du}{u} = \cot \mu \frac{du}{u}
$$

Then the above derivatives become

$$
\begin{aligned}
u_{1\sigma} &= \frac{u}{\cot \mu} \omega_\sigma \cos \theta - \theta_\sigma u \sin \theta \\
u_{2\sigma} &= \frac{u}{\cot \mu} \omega_\sigma \sin \theta + \theta_\sigma u \cos \theta
\end{aligned}
$$

Equations (9.47) become the *two* equations

$$
\omega_\sigma(\cos \theta + \zeta \sin \theta) + \theta_\sigma \cot \mu(\zeta \cos \theta - \sin \theta) + \frac{\sin \mu \cos \mu \sin \theta}{\sin^2 \mu - \cos^2 \theta} \frac{x_{1\sigma}}{x_2} = 0 \tag{9.48}
$$

with corresponding values

$$
\begin{aligned}
\sigma = \alpha &\leftrightarrow \zeta = \zeta^- \\
\sigma = \beta &\leftrightarrow \zeta = \zeta^+
\end{aligned}
$$

[1] For plane flow, corresponding to $x_2 \to \infty$, Eqs. (9.47) can also be written in terms of u_1, u_2 only. For example, the first equation becomes $du_1/du_2 = -\zeta$, which is just the equation for the Prandtl-Meyer curve in the hodograph plane.

These can be considerably simplified. Let m^+ be the distance measured along m^+ and m^- the distance measured along m^-. Then from Fig. 9.21,

$$\frac{dx_1}{dm^+} = \cos(\theta - \mu) \qquad \frac{dx_1}{dm^-} = \cos(\theta + \mu)$$

Hence

$$x_{1\alpha} = m_\alpha^+ \cos(\theta - \mu) \qquad x_{1\beta} = m_\beta^- \cos(\theta + \mu)$$

With this substitution, and after trigonometric reduction,[1] Eqs. (9.48) become

$$\frac{d}{d\alpha}(\theta + \omega) - \frac{m_\alpha^+ \sin\mu \sin\theta}{x_2} = 0$$

$$\frac{d}{d\beta}(\theta - \omega) + \frac{m_\beta^- \sin\mu \sin\theta}{x_2} = 0$$

and with the identities

$$\frac{d}{d\alpha} = \frac{dm^+}{d\alpha}\frac{d}{dm^+} \qquad \frac{d}{d\beta} = \frac{dm^-}{d\beta}\frac{d}{dm^-}$$

we obtain finally

$$\frac{d}{dm^+}(\theta + \omega) = +\frac{\sin\mu \sin\theta}{x_2} \qquad \text{on } m^+$$

$$\frac{d}{dm^-}(\theta - \omega) = -\frac{\sin\mu \sin\theta}{x_2} \qquad \text{on } m^-$$

These are exactly Eqs. (9.22), found in Sec. 9.3.

9.5 Flow with shocks

Where two Mach waves (characteristics) of the same family intersect, an oblique shock necessarily forms, in analogy to shock formation in unsteady one-dimensional flow. Such formation occurs in a compression wave, where characteristics tend to converge, as shown in Fig. 9.22 [compare Example 9.1 (page 447)].

In general, a flow field containing shocks will not be homentropic, not even piecewise, because shocks of variable strength impart different entropies along different streamlines. In the figure, the shock forms downstream of a uniform region. From the point of formation upward, the strength increases, as reflected by the increasing shock angle β, or *shock curvature* (in the case shown the shock initially has infinitesimal strength

[1] A useful identity here is $\cos^2\theta - \sin^2\mu = \cos(\theta + \mu)\cos(\theta - \mu)$.

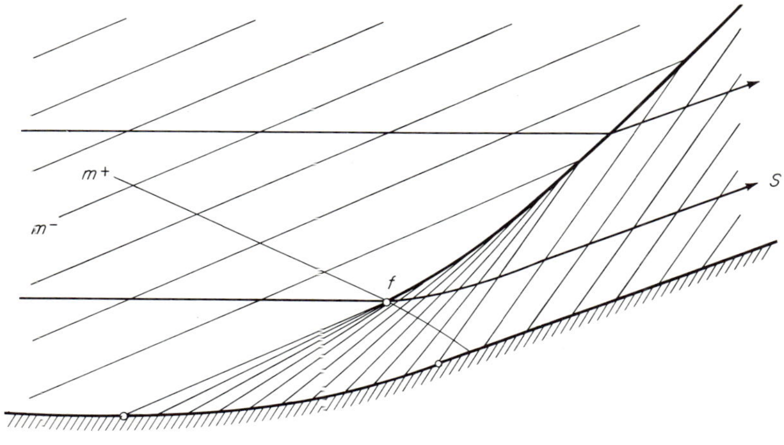

Figure 9.22
Formation of a shock at the intersection of two characteristics, point *f*.

and lies at the Mach angle of the upstream region). Above the streamline passing through point *f* the flow field is thus nonhomentropic, the entropy increasing upward. By the Crocco-Vazsonyi theorem, Eq. (2.44), the flow above streamline *S* is then *rotational*, as already discussed in Example 2.7 (page 74). We summarize the above observations by saying the *flow downstream of a curved shock front is rotational.*[1] This result is due to Hadamard (1903).

In the case illustrated in Fig. 9.22, the shock is formed from a simple wave. Any characteristics m^+ lying above the one shown necessarily cross the shock and in general will suffer a change in the carried invariant $\theta + \omega$; the region above m^+ is thus strictly nonsimple.

From the standpoint of computability of a flow, the occurrence of curved shocks is a misfortune because the relatively simple methods developed in Sec. 9.4 for irrotational flows are no longer applicable. The method of characteristics will appear in revised form, with the streamlines appearing as a third set of characteristics—again in analogy to one-dimensional isentropic flow. (For a treatment of this method, see, for example, *von Mises* [1958, chap. 5].) A further difficulty is the necessity for applying the oblique shock conditions at each point on the shock front.

[1] It is however *conceivable* that a shock front of precisely varied strength could in very special circumstances change a rotational flow to an irrotational one, but this has no apparent physical interest.

If, however, the shocks are straight, there is no particular difficulty. More generally, if they are not straight but may be considered *weak*, then the entropy change and rotation can be neglected. We consider this case now. In accord with Sec. 9.2 the shock is treated as a spatially concentrated simple wave with

$$[\omega] = -|[\theta]| \tag{9.49}$$

An alternative and equivalent statement is that the invariant $\theta \pm \omega$ on the crossing characteristic is unchanged across the shock.

The shock path can be established by the approximate condition that the upstream and downstream characteristics intersect the shock front at equal angles. From Fig. 9.23, these angles are

$$\begin{aligned} \varepsilon_1 &\equiv \beta - \mu_1 \\ \varepsilon_2 &\equiv \mu_2 + [\theta] - \beta \end{aligned} \tag{9.50}$$

Then we wish to show that $\varepsilon_1 \approx \varepsilon_2$. Define the error δ by

$$\varepsilon_1 = \varepsilon_2 + \delta \tag{9.51}$$

By (9.50),

$$\delta = 2(\beta - \mu_1) + \mu_1 - \mu_2 + [\theta] \tag{9.52}$$

The condition that the shock is weak can be expressed $[\theta] \ll 1$. From (7.53) and (7.54) $\beta - \mu_1$ can be obtained as a power series in $[\theta]$, and a

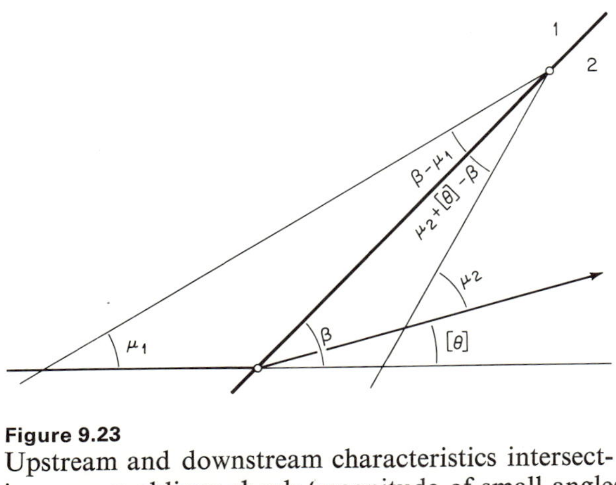

Figure 9.23
Upstream and downstream characteristics intersecting an m^- oblique shock (magnitude of small angles is exaggerated).

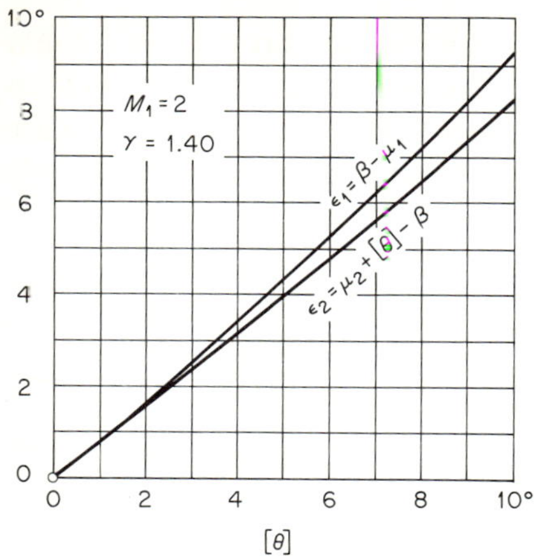

Figure 9.24
The angle ε between a characteristic and an oblique shock front vs. turning angle $[\theta]$, for a perfect gas with $\gamma = 1.40$, $M_1 = 2$.

similar series can be developed for $\mu_1 - \mu_2$. Omitting the derivation, the result is

$$\delta = K[\theta]^2 \tag{9.53}$$

$$K = \frac{M_1^4}{(M_1^2 - 1)^{3/2}} \left[\frac{2M_1^2(\Gamma_1 - 1)^2 - \frac{1}{2}\Gamma_1^2 + 4\Gamma_1 - 2}{2(M_1^2 - 1)} - \frac{1}{4}(\Gamma_1^2 + 6\Gamma_1 + 4) - \frac{1}{2}\frac{c_1^2\Gamma_1}{3v_1T_1}\left(\frac{\partial T}{\partial P}\right)_s + \frac{1}{12}\frac{c_1^6}{v_1^4}\left(\frac{\partial^3 v}{\partial P^3}\right)_s \right]$$

That is, the error involved in setting $\varepsilon_1 = \varepsilon_2$ is proportional to the turning angle squared, i.e., to the square of the shock strength. This result is analogous to that already found in Eq. (8.44) for unsteady flow. Numerical results for a perfect gas are indicated by Fig. 9.24.

EXAMPLE 9.10 THE FAR FIELD FOR FLOW ABOUT A TWO-DIMENSIONAL AIRFOIL

Consider the flow about a slender diamond-shaped airfoil, as shown in Fig. 9.25. Only the upper half of the flow field is shown. For small turning angles, the shocks are everywhere weak, and the field may be considered homentropic.

The flow in region 4 has the same properties as the flow in region 1; in particular, $\theta_4 = \theta_1 = 0$ and $\mu_4 = \mu_1$ (this is a consequence of the condition $P_4 = P_1$, which with homentropic simple waves gives $\omega_4 = \omega_1$ and $\theta_4 = \theta_1$). Thus the Mach waves (m^- characteristics) have the same inclination in regions 1 and 4.

Since the characteristics in the centered rarefaction fan make the same angle with the shock fronts as the (parallel) characteristics coming from the uniform regions, the shock front is necessarily in the form of a *parabola* above points a and b. (That is, a parabola has the "optical" property that rays originating at the focus are reflected at equal angles of incidence and reflection to form parallel rays, and this property is reproduced here by the shock front and characteristics.)

The approximations involved in this argument improve with distance from the airfoil, i.e., as the shocks become weaker. Thus, we say that *the shock fronts far from the airfoil assume the form of a parabola.* The axis of the parabola is of course inclined at angle μ_1 to the free stream.[1]

The angle ε between the shock front and the free-stream characteristics is proportional to the inverse square root of the distance along the axis of the parabola, $\varepsilon \propto 1/\sqrt{l}$. If y is the (vertical) distance from the airfoil, it follows that $\varepsilon \propto 1/\sqrt{y}$. Since the shock strength varies as ε, for example, $[\theta] \propto \varepsilon$, we have finally

$$\text{Shock strength} \propto \frac{1}{\sqrt{y}} \tag{9.54}$$

[1] The presence of a rounded leading edge (to prevent separation) will modify these conclusions somewhat (see *Lighthill* [1954]).

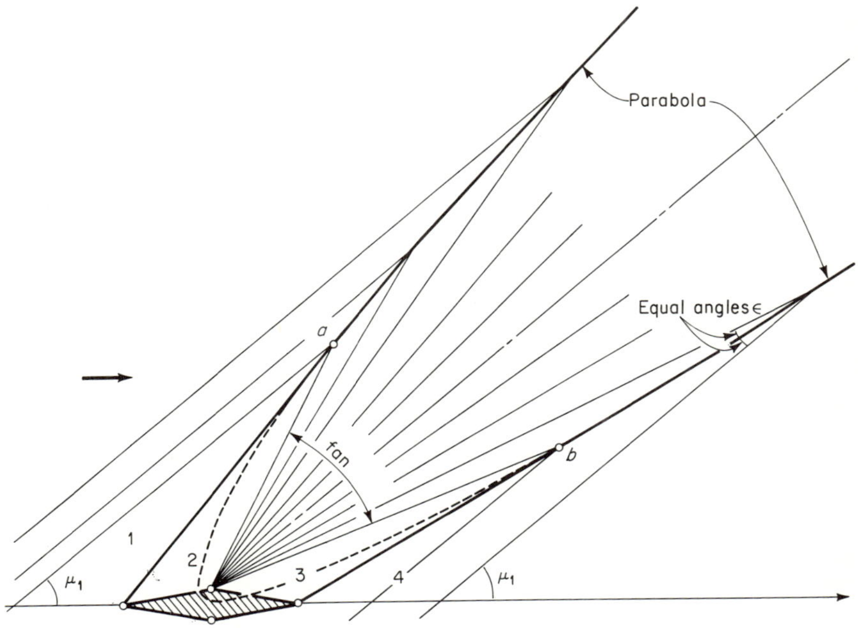

Figure 9.25
Upper wave pattern for streaming flow about a two-dimensional airfoil.

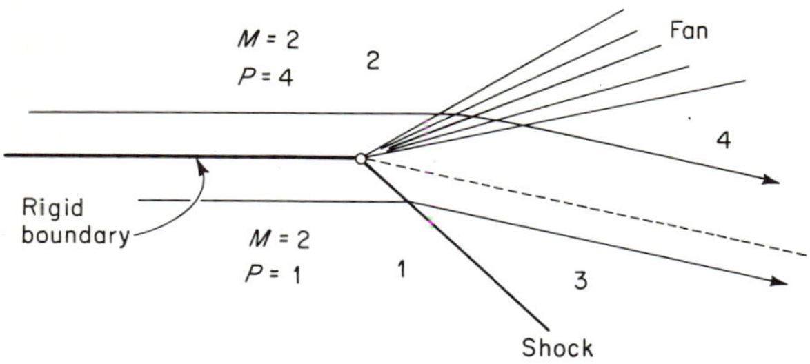

Figure 9.26

in agreement with the result found for plane waves in Example 8.3. For "real" three-dimensional bodies, the corresponding result is

$$\text{Shock strength} \propto \frac{1}{y^{3/4}} \tag{9.55}$$

as discussed by *Lighthill* [1954], and also in agreement with the result found for cylindrical waves in Example 8.3.

EXAMPLE 9.11 SUDDEN EXPANSION

This is the analog of the discontinuity in pressure at the diphragm of a shock tube, as discussed in Sec. 8.9. Two parallel *supersonic* streams at different pressures suddenly confront each other at the terminus of a separating boundary, i.e., a thin plate, as shown in Fig. 9.26.

The pressures and flow directions can be matched across the contact surface as shown in the figure. Both fluids are assumed to be perfect gases with $\gamma = 1.40$, and the solution curves are plotted in Fig. 9.27. The $P(\theta)$ states down-

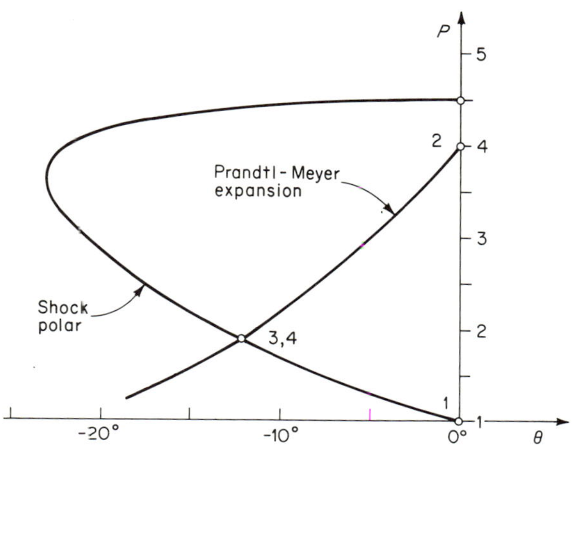

Figure 9.27

stream of the Prandtl-Meyer fan are found from the isentropic flow tables, with $\theta_4 = \omega_2 - \omega_4$. The $P(\theta)$ states downstream of the shock are found directly from the oblique-shock tables. By trial, the final numerical solution is

$$\theta_3 = \theta_4 = -12.16^\circ$$
$$P_3 = P_4 = 1.904$$
$$M_3 = 1.560 \qquad M_4 = 2.476$$

Problems

9.1 A steady supersonic flow expands from Mach number $M_1 = 2$ and pressure P_1 to pressure $P_2 = P_1/2$ across a centered rarefaction, e.g., at the discharge of an underexpanded nozzle. Find the Mach number M_2 and flow direction θ_2. Assume inviscid irrotational flow of a perfect gas, $\gamma = 1.40$.

9.2 At what initial Mach number M_1 does a 90° turn via simple waves correspond to maximum expansion to $M_2 = \infty$? Assume perfect gas, $\gamma = 1.40$.

9.3 At $T_0 = 300$ K and $P_0 = 1$ atm, nitrogen has a mean free path $\Lambda_0 = 65.4$ nm. For steady isentropic flow from these reservoir conditions to Mach number unity, followed by a simple-wave expansion, find the Mach number M and flow angle θ at which the mean free path is equal to 1 cm.

Answer $M \approx 25$; $\theta \approx 119°$

9.4 Consider a two-dimensional duct carrying a perfect gas ($\gamma = 1.40$) with uniform conditions and $M_1 = 2$. Design a 10° turning elbow to achieve a uniform downstream state 2 for each of the following cases:

(*a*) $M_2 > M_1 \qquad s_2 = s_1$
(*b*) $M_2 < M_1 \qquad s_2 = s_1$
(*c*) $M_2 < M_1 \qquad s_2 > s_1$
(*d*) $M_2 = M_1 \qquad s_2 = s_1$

In each case, find numerical values for M, P, and the duct area (compared to that for state 1). Discuss the possibility of separation.

9.5 The linearized momentum and continuity equations for unsteady homentropic one-dimensional flow can be written

$$\rho_0 \frac{\partial u}{\partial t} + c_0{}^2 \frac{\partial \rho}{\partial x} = 0$$

$$\frac{\partial \rho}{\partial t} + \rho_0 \frac{\partial u}{\partial x} = 0$$

Find the characteristic directions and characteristic equations (note that these are the equations corresponding to the *wave equation*).

9.6 A plane Chapman-Jouguet detonation wave advances into solid explosive. If the explosive has a plane boundary with the atmosphere perpendicular to the detonation front, sketch the wave and streamline pattern at this boundary for a reference frame moving with the detonation wave.

9.7 Find the *lift coefficient* $C_L \equiv 2F_L/\rho_\infty u_\infty^2 L$ for a flat plate airfoil with $M_\infty = 2$, $\alpha = 8°$. The fluid is a perfect gas with $\gamma = 1.40$. Compare with the results from linear theory.

9.8 A two-dimensional supersonic inlet diffuser is shown in the sketch. Design the actual duct profile for a perfect gas with $\gamma = 1.40$. Note that the number of possible geometries is infinite. By a suitable choice of wave pattern, however, the calculation is greatly simplified.

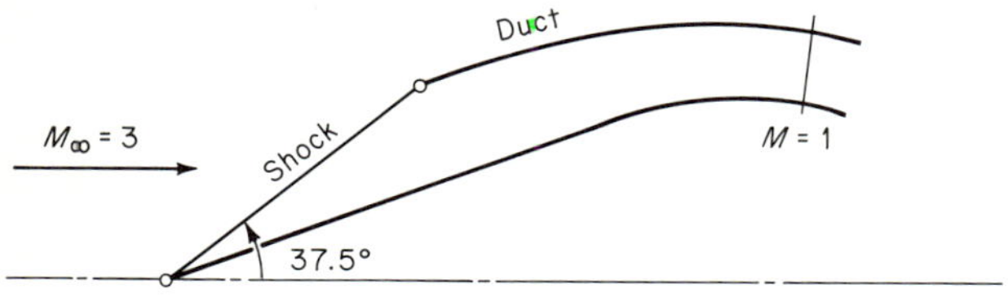

ten

self-similar motions

10.1 Introduction

A motion which preserves its own geometry—in space or time or both—is said to be *self-similar*. In the simplest and most important cases, such motions are described by a single independent variable, the similarity variable.

In Sec. 3.6, the formation of a similarity variable via dimensional arguments was discussed, and was illustrated by the *Rayleigh problem*. Other examples of self-similar motions already encountered are the *centered rarefaction* in unsteady flow and the *Prandtl-Meyer expansion*.

In this chapter we consider several self-similar motions in gas-dynamics. Problems which can be formulated in this way invariably involve simple geometries and simple boundary conditions. The few cases for which such formulation is possible are important for the physical insight they bring and as "exact" references for comparison with the more usual approximate calculations.

10.2 Prandtl-Meyer expansion

As a fairly easy reintroduction to the subject, the Prandtl-Meyer expansion will be worked out from similarity arguments. This is, in fact, how the

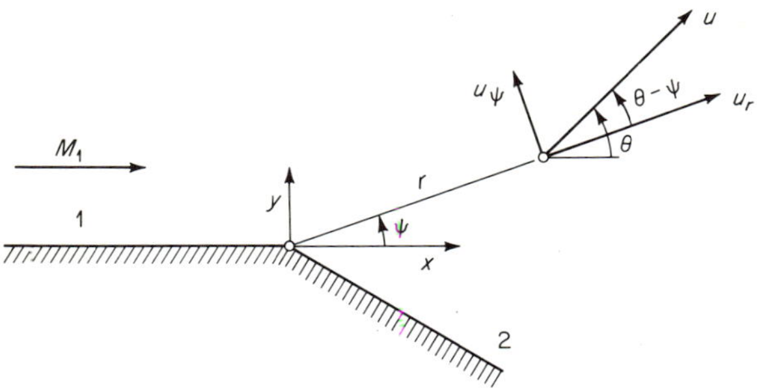

Figure 10.1
Coordinate system for Prandtl-Meyer expansion.

problem was first approached by two of the most famous figures in twentieth-century fluid mechanics, *Prandtl* [1907] and *G. I. Taylor* [1942].

The coordinate system and wall geometry are shown in Fig. 10.1. The upstream boundary conditions 1 are specified by the velocity u and the thermodynamic state, as given by P_1, ρ_1, say. *No characteristic length appears* in either the geometry or the boundary conditions.

The equations of motion for steady irrotational flow can be written

$$\nabla \times \mathbf{u} = 0 \tag{10.1}$$

$$\nabla \cdot (\rho \mathbf{u}) = 0 \tag{10.2}$$

$$h + \frac{u^2}{2} = \text{const} \tag{10.3}$$

The boundary conditions can be written

$$u = u_1 \qquad P = P_1 \qquad \rho = \rho_1 \qquad \theta = 0 \qquad \text{far upstream}$$

$$\theta = \theta_2 \qquad \text{far downstream}$$

Since no characteristic length is available for putting these equations into nondimensional form, we conclude that the spatial variables x, y can appear only in the dimensionless combination y/x. An equivalent assertion is that only the polar angle $\psi = \tan^{-1}(y/x)$, and not the radial distance r, enters as a spatial variable.[1] Then all the dependent variables depend only on ψ: $P = P(\psi)$, $\mathbf{u} = \mathbf{u}(\psi)$, etc. This is a considerable simplification!

[1] The symbol ψ is used here as the polar coordinate, instead of the more usual θ, because we wish to use θ as the local streamline inclination, consistent with the notation of Chap. 9. The above argument is inapplicable to subsonic flow, because there must be *some* downstream geometry of dimension L, the effect of which is propagated upstream in subsonic flow. We have thus implicitly assumed that the flow is *supersonic*, $M > 1$.

The equations of motion now become, in polar coordinates,

$$\frac{du_r}{d\psi} - u_\psi = 0 \tag{10.4}$$

$$\rho u_r + \frac{d}{d\psi}\rho u_\psi = 0 \tag{10.5}$$

$$h + \frac{u^2}{2} = \text{const} \tag{10.3}$$

Incidentally, the irrotationality condition (10.4) is identical to the momentum equation for the r direction.

Because the upstream and downstream boundary conditions are given in terms of the streamline inclination θ, it will be convenient to rewrite these equations in terms of the flow speed u and the angle θ. From the geometry of Fig. 10.1,

$$\begin{aligned} u_r &= u\cos(\theta - \psi) \\ u_\theta &= u\sin(\theta - \psi) \end{aligned} \tag{10.6}$$

Substituting these into (10.4) yields

$$\frac{d\theta}{d\psi} + \frac{1}{u}\frac{du}{d\psi}\cot(\psi - \theta) = 0 \tag{10.7}$$

The continuity equation (10.5) can be written such that the density ρ appears only as the term $\frac{1}{\rho}\frac{d\rho}{d\psi}$. This term is rewritten

$$\frac{d\rho}{d\psi} = \frac{dh}{d\psi}\left(\frac{\partial\rho}{\partial h}\right)_s = -u\frac{du}{d\psi}\left(\frac{\partial\rho}{\partial h}\right)_s$$

where the energy equation (10.3) was used to evaluate $dh/d\psi$. From the Gibbs equation, $dh = T\,ds + v\,dP$, we find $(\partial\rho/\partial h)_s = \rho/c^2$, so that finally

$$\frac{1}{\rho}\frac{d\rho}{d\psi} = -\frac{u}{c^2}\frac{du}{d\psi}$$

Making use of this and (10.6), the continuity equation (10.5) reduces to

$$\cot(\psi - \theta)\frac{d\theta}{d\psi} + (M^2 - 1)\frac{1}{u}\frac{du}{d\psi} = 0 \tag{10.8}$$

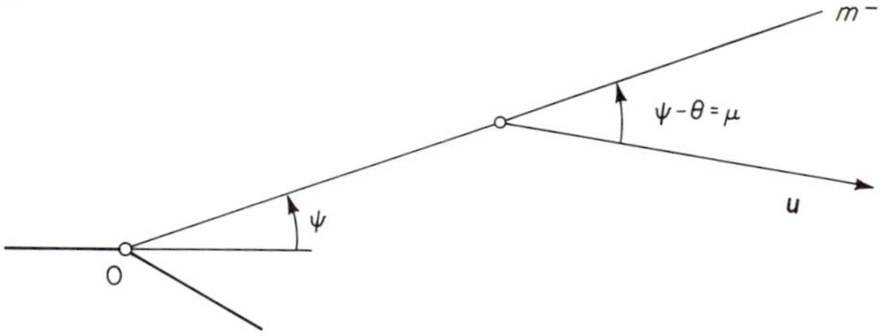

Figure 10.2

The equations of motion are now (10.7) and (10.8). Solving (10.7) for $d\theta/d\psi$ and substituting into (10.8) gives

$$\cot(\psi - \theta) = \pm\sqrt{M^2 - 1} \tag{10.9}$$

Since $\sqrt{M^2 - 1}$ is just the cotangent of the Mach angle μ, (10.9) can be written $\psi - \theta = \pm\mu$. The upstream boundary conditions can be expressed as $\theta = 0$, $\mu = \mu_1$: selecting the positive root of (10.9) gives the admissible upstream condition $\psi_1 = +\mu_1$; selecting the negative root gives the inadmissible condition $\psi_1 = -\mu_1$; that is, ψ cannot be negative. We therefore take the positive root of (10.9), corresponding to

$$\psi - \theta = \mu(M) \tag{10.10}$$

Thus, a radial line from the origin is necessarily a *Mach wave* or characteristic m^- (Fig. 10.2). It follows that the velocity component u_ψ, perpendicular to the radial line, is necessarily sonic.

With (10.9), the equations of motion (10.7) and (10.8) reduce to

$$\frac{d\theta}{d\psi} + \sqrt{M^2 - 1}\,\frac{1}{u}\frac{du}{d\psi} = 0 \tag{10.11}$$

This has a solution

$$\theta + \int \sqrt{M^2 - 1}\,\frac{du}{u} = \text{const}$$

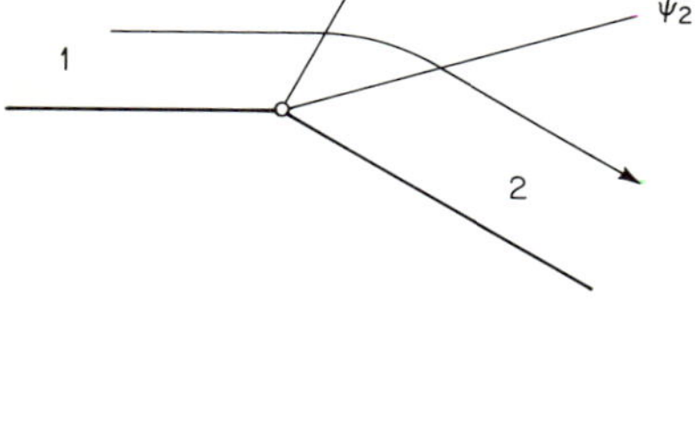

Figure 10.3

The integral is the Prandtl-Meyer function $\omega(M)$ already discussed in Sec. 9.2. With $\theta_1 = 0$, the solution can be written

$$\theta + \omega(M) = \omega(M_1) \tag{10.12}$$

Equations (10.10) and (10.12) represent the solution of the problem. With θ_2 known, the expansion terminates at $\omega(M_2) = \omega(M_1) - \theta_2$, which gives M_2. Then from (10.10) we have the boundaries of the expansion region,

$$\psi_1 = \mu(M_1)$$

$$\psi_2 = \mu(M_2) + \theta_2$$

as shown in Fig. 10.3. Outside the expansion region ($\psi > \psi_1$ or $\psi < \psi_2$) the similarity formulation still holds, because flow properties are still constant on lines ψ = const, but this is trivial because the flow properties are *uniform* within these regions.

0.3 Conical flow

Certain steady supersonic flows depend only on the spherical angle ψ from an axis of symmetry. These flows are called *conical* because all properties are constant on the surface of a cone ψ = const.

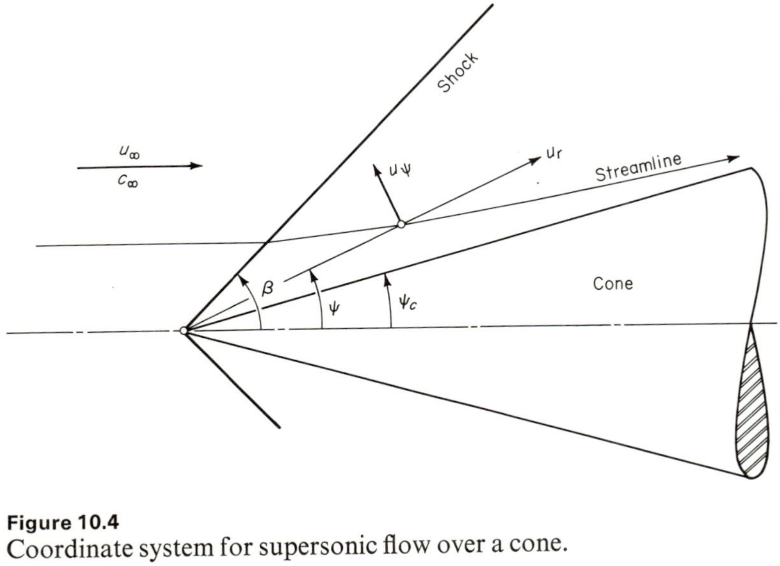

Figure 10.4
Coordinate system for supersonic flow over a cone.

The case of greatest practical importance is that of *supersonic flow over a cone*, which we discuss here (see Fig. 10.4).

Assuming that the flow is everywhere supersonic (and inviscid), no characteristic length L will appear in the problem. Even if the conical body is truncated some distance L downstream (as at the junction of the cone with a cylinder, for example), the effect of this geometry will not propagate upstream. In this respect the problem is like the Prandtl-Meyer flow discussed in Sec. 10.2. We thus conclude that the flow variables depend only on the spherical-coordinate angle ψ and not on the radial distance r.

Written in spherical coordinates, with dependence only on the angle ψ, the equations of irrotationality (equivalently, radial momentum), continuity, and energy are, respectively,

$$\frac{du_r}{d\psi} - u_\psi = 0 \tag{10.13}$$

$$\frac{d}{d\psi}(\rho u_\psi \sin\psi) + 2\rho u_r \sin\psi = 0 \tag{10.14}$$

$$h + \frac{u^2}{2} = \text{const} \tag{10.15}$$

where u_ψ and u_r are the velocity components. Note that these are similar to Eqs. (10.3) to (10.5) for plane self-similar flow.

The density can be eliminated by the use of the energy equation, as before:

$$\frac{1}{\rho}\frac{d\rho}{d\psi} = -\frac{u}{c^2}\frac{du}{d\psi} = -\frac{1}{2c^2}\frac{d}{d\psi}(u_r^2 + u_\psi^2)$$

Substituting this into the continuity equation (10.14) yields

$$u_\psi\left(u_r\frac{du_r}{d\psi} + u_\psi\frac{du_\psi}{d\psi}\right) - c^2\frac{du_\psi}{d\psi} - c^2 u_\psi \cot\psi - 2c^2 u_r = 0$$

Substituting (10.13) into this and rearranging yields finally

$$\frac{d^2u_r}{d\psi^2} + u_r = \frac{c^2(u_r + u_\psi \cot\psi)}{u_\psi^2 - c^2} \tag{10.16}$$

which with the irrotationality condition

$$\frac{du_r}{d\psi} = u_\psi \tag{10.17}$$

and the energy equation, $h(c) + u^2/2 = \text{const}$, forms the basis for a solution. In the case of a *perfect gas*, the latter may be written

$$c^2 + \frac{\gamma - 1}{2} u^2 = \frac{\gamma + 1}{2} c_*^2 = \text{const} \tag{10.18}$$

The boundary conditions at the cone surface and at the shock front may be written, respectively,

$$\psi = \psi_c \qquad u_\psi = 0 \tag{10.19}$$

$$\psi = \beta \qquad u_\psi = -w_2 \qquad u_r = v \tag{10.20}$$

The integration of (10.16) and (10.17) must in general be carried out numerically. One possibility is to start the integration at the cone surface $\psi = \psi_c$, with some arbitrary assumed value for u_r. In this case neither the free-stream Mach number nor the shock angle β is known in advance; these values are determined only by finding the value of ψ at which (10.20) is satisfied for *some* Mach number M_∞, as we discuss below in more detail. This procedure is the one followed by *Kopal* [1947].

A second possibility is to start the calculation from some arbitrary assumed shock angle β, with an assumed value for M_∞. In this case the cone angle ψ_c is not known in advance but is determined in the course of the calculation as the angle at which u_ψ finally vanishes.

For the *perfect gas*, the shock boundary conditions (10.20) can be written in explicit form. From Fig. 7.17, $\tan \beta = w_1/v$; substituting the value of w_1 from (7.46) yields

$$\tan \beta = \frac{c_*^2 - \dfrac{\gamma - 1}{\gamma + 1} v^2}{v w_2}$$

Rewriting this in the notation of the conical flow, per Eq. (10.20), gives

$$\tan \beta = -\frac{c_*^2 - \dfrac{\gamma - 1}{\gamma + 1} u_r^2}{u_r u_\psi} \tag{10.21}$$

A second condition can be obtained by solving (5.62) for M_1^2,

$$M_1^2 = \frac{2}{(\gamma + 1)(c_*^2/u_1^2) - (\gamma - 1)}$$

where $c_*/u_1 = 1/M_1^*$. Inserting the shock relation $\cos\beta = v/u_1$, this can be written

$$M_1{}^2 = \frac{2v^2}{(\gamma+1)c_*{}^2\cos^2\beta - (\gamma-1)v^2}$$

Finally, rewriting this in terms of the conical-flow notation,

$$M_\infty{}^2 = \frac{2u_r{}^2}{(\gamma+1)c_*{}^2\cos^2\beta - (\gamma-1)u_r{}^2} \tag{10.22}$$

Suppose the calculation is started at the surface of the cone $\psi = \psi_c$, as done by Kopal. At this surface $u_\psi = 0$. As ψ increases, the magnitude of u_ψ increases until at some angle $\psi = \beta$ Eq. (10.21) is satisfied, fixing the shock angle. Then M_∞ is determined from (10.22). In practice, it is useful to normalize the equations with respect to the (unknown in advance) constant c_*. Then c_* can be determined after M_∞ is known by means of (5.57), which yields[1]

$$\left(\frac{c_*}{c_\infty}\right)^2 = \frac{2}{\gamma+1} + \frac{\gamma-1}{\gamma+1}M_\infty{}^2 \tag{10.23}$$

The conical-flow field was first calculated by Busemann, who used a graphical technique in the hodograph, that is, $u\theta$, plane. Later, the differential equations were numerically integrated by *Taylor and Maccoll* [1933]. An extensive set of numerical results has been tabulated by *Kopal* [1947]. A general discussion of conical flows, including examples other than supersonic flow over a cone, is given by *Ferri* [1954].

Special Features of the Flow

For any given cone angle ψ_c, a conical solution can be found only above a certain minimum Mach number $(M_\infty)_{\min}$. Below this Mach number, the shock front is detached from the cone and the flow is rotational and not self-similar. Conversely, for any given value of M_∞ there is a maximum cone angle ψ_c for which a conical solution can be found.[2] This is shown in Fig. 10.5, with the corresponding maximum wedge angle ψ_w shown for

[1] The normalizing constant actually used by Kopal was

$$u_{\max} = \sqrt{\frac{\gamma+1}{\gamma-1}}\,c_*$$

[2] At $M_\infty = \infty$, the maximum cone angle permitting an attached shock is 57.52° (for $\gamma =$ 1.40).

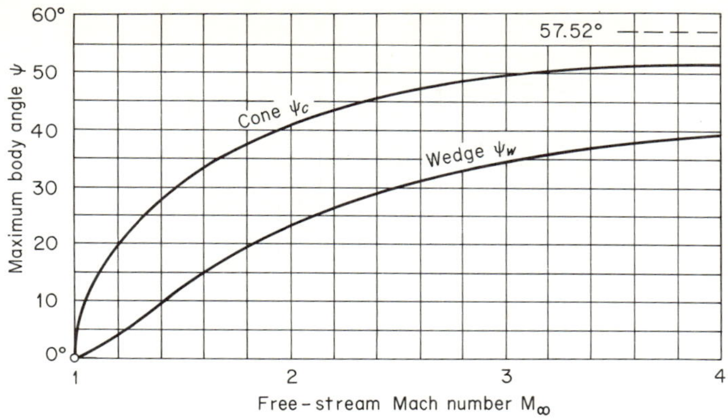

Figure 10.5
Maximum body angle (ψ_c or ψ_w) permitting an attached shock; perfect gas, $\gamma = 1.405$.

comparison (the maximum wedge angle is of course the maximum turning angle across a shock at any given Mach number).

The shock front is necessarily conical, and the conditions behind the shock front are necessarily uniform.

For oblique shocks in general, we know that there are two possible solutions for any given values of M_1 and turning angle [θ]. These are the *weak solution* and the *strong solution*, as discussed in Sec. 7.5. Now suppose, as is normally the case, that the conical shock is weak in the above sense and that the flow behind the shock front is supersonic. The pressure increases toward the cone surface, i.e., with decreasing ψ, however, and the Mach number therefore decreases toward the cone surface.[1] Therefore the Mach number decreases toward the cone surface and is a maximum just behind the shock front.

Thus, a flow which is initially supersonic just behind the shock front may either remain supersonic or go over to a subsonic flow at some surface ψ = const (the *sonic cone* or *sonic line*). In the latter case, the flow field is mixed supersonic and subsonic. Finally, a flow which is initially subsonic behind the shock will necessarily remain subsonic. The three possible cases (pure supersonic, mixed, and pure subsonic) are delimited in Fig. 10.6.

[1] That the pressure increases toward the cone can be seen from the momentum equation for the direction normal to the streamlines, assuming the streamline curvature to be everywhere outward, as indicated in Fig. 10.4; i.e., the normal pressure gradient produces the curvature.

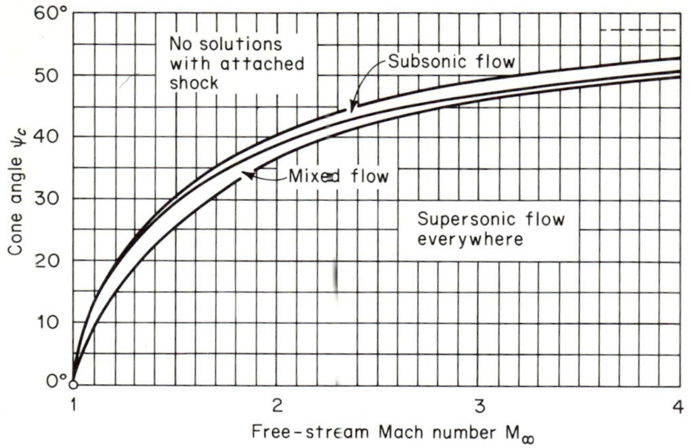

Figure 10.6
The possible flow regimes for conical flow; perfect gas, $\gamma = 1.405$. (*Data from Kopal* [1947].)

A particular flow field which is mixed subsonic and supersonic is shown in Fig. 10.7. The outer part of the field (above the sonic line) is supersonic, and the Mach waves are shown.

There is reason to be suspicious of conical solutions which lead to either pure subsonic or mixed flows, since disturbances can propagate

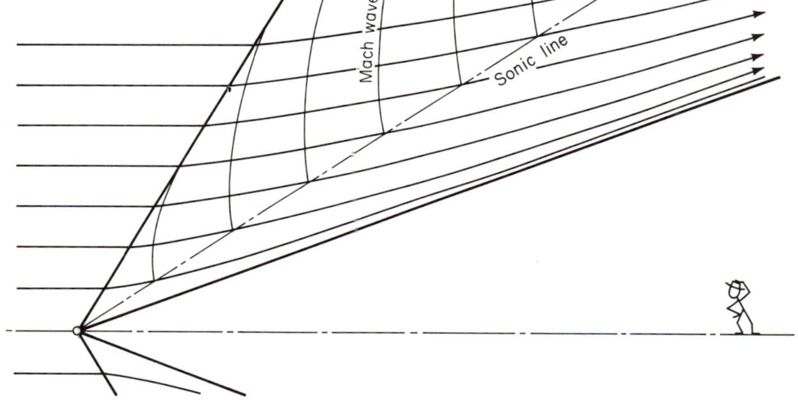

Figure 10.7
Mixed subsonic and supersonic conical flow. $M_\infty = 1.31$, $\gamma = 1.40$. (*After Taylor and Maccoll* [1933].)

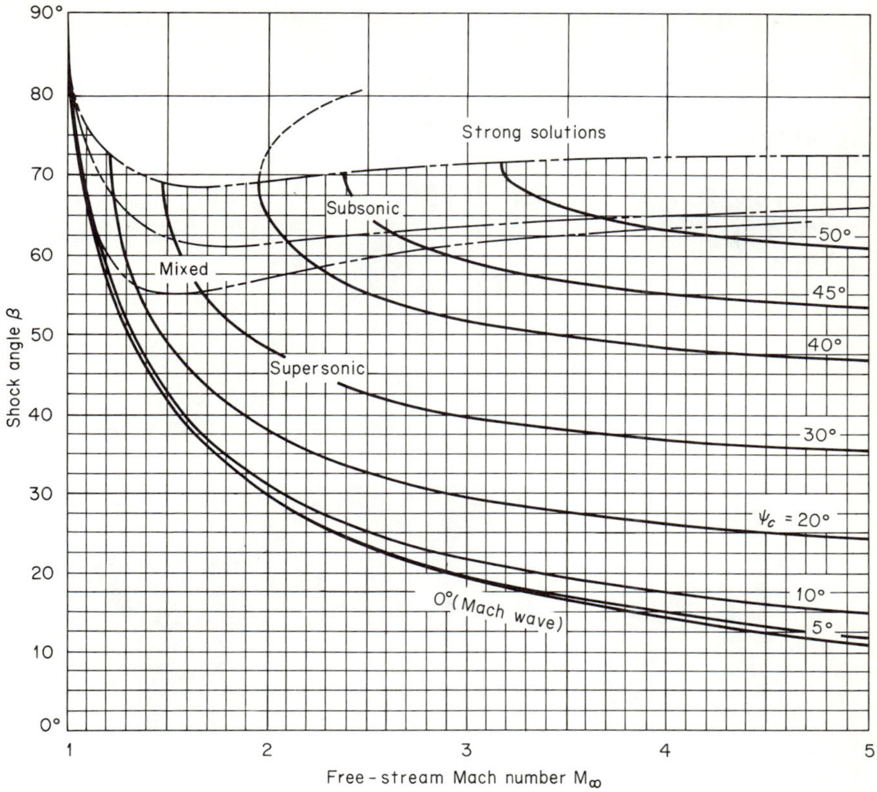

Figure 10.8
Shock-front angle as a function of Mach number; perfect gas, $\gamma = 1.405$. (*Data from Kopal* [1947].)

upstream in these cases and the argument about the absence of a characteristic length loses its force. That is, all cones must terminate somewhere, and the effect of the terminus will propagate upstream. Nevertheless, there is ample experimental evidence that the mixed solution, at least, does occur on finite cones (see *Solomon* [1955]). This case is in fact often cited as an example of *smooth* transition from supersonic to subsonic flow (such shock-free transition was once believed to be impossible).

The shock-front angle β is shown as a function of M_∞ and ψ_c in Fig. 10.8. The strong solutions, corresponding to the larger shock angles, have apparently never been observed. Nor does it seem likely that they will be, since they depend on a strong pressure buildup via downstream flow obstruction (see also the discussion in Sec. 7.5).

Figure 10.9
Conical flows. (*a*) Free flight in air, $M_\infty = 2$, $\psi_c = 10°$. (*b*) $M_\infty = 1.51$, $\psi_c = 20°$. (*c*) Detached shock at $M_\infty = 3$, $\psi_c = 60°$. Flow field is not conical in this case. (*Courtesy of W. Deveikis, NASA.*) (*d*) $M_\infty = 5.3$, $\psi_c = 10°$, shadowgraph picture. (*Courtesy of J. J. Ginoux, von Kármán Institute.*) (*e*) Same as (*d*), schlieren picture.

Experimental results generally confirm the predictions of the conical-flow theory. The conical-flow field has even appeared in quite unexpected places (see the frontispiece in *Birkhoff* [1960]). Photographs of conical flows are shown in Fig. 10.9.

10.4 Intense explosion

The problem of a strong spherical explosion, which we now consider, is one of several spherically symmetric self-similar motions.

A perfect-gas atmosphere is initially uniform and at rest. At time zero a large quantity E of energy is instantaneously released at some point in the gas. It is desired to find the resulting gas motion, which is assumed to have spherical symmetry.

The released energy E appears in the gas as kinetic and internal energy. A strong spherical shock is expected to propagate radially from the origin, the point of energy release. Let the conditions in the undisturbed stationary gas be indicated by subscript zero, for example, $\rho = \rho_0$. Let the conditions immediately behind the shock be indicated by subscript s, for example, $\rho = \rho_s$, and let the shock velocity be U. Assuming that the shock is *strong* [as defined by (7.58), which can be represented by the single condition $(U/c_0)^2 \gg 1$], the shock conditions (7.61) to (7.63) are

$$P_s = \frac{2}{\gamma + 1} \rho_0 U^2 \tag{10.24}$$

$$u_s = \frac{2}{\gamma + 1} U \tag{10.25}$$

$$\rho_s = \frac{\gamma + 1}{\gamma - 1} \rho_0 \tag{10.26}$$

By virtue of these limiting forms of the shock conditions, the only property of the undisturbed atmosphere which enters the calculation is the density ρ_0. No other property, such as the pressure P_0, plays a role (in effect, the pressure P_0 is zero).

The only constant dimensional parameters which appear in the problem are thus ρ_0 and E. The dimensions of these quantities are respectively M/L^3 and ML^2/T^2; it is thus impossible to form a characteristic length (or a characteristic time). In consequence, it will be necessary to form a similarity variable.

The list of independent variables is ρ_0, E, r, t, while the dependent variables are u, P, ρ. We have assumed that the motion is spherically symmetric and the velocity u purely radial. By examining the dimensions of the independent variables we find only one dimensionless combination of the form

$$\eta \equiv \frac{r}{\left(\dfrac{E}{\rho_0}\right)^{1/5} t^{2/5}} \tag{10.27}$$

This is the similarity variable for the problem.

The Shock Radius $R(t)$

Let the instantaneous radius of the shock front be R. This quantity is expected to depend on E, ρ_0, and t,

$$R = R(E,\rho_0,t)$$

The only nondimensional combination that can be formed from these variables is $R(E/\rho_0)^{-1/5}t^{-2/5}$. The law of shock propagation is thus necessarily

$$\frac{R}{\left(\dfrac{E}{\rho_0}\right)^{1/5} t^{2/5}} = \eta_0 = \text{const} \tag{10.28}$$

Then $R \propto t^{2/5}$. The principal objective of the subsequent analysis is to find the numerical value of η_0, which will turn out to be very close to unity.

An alternative similarity variable r/R (used by Taylor) differs from (10.27) only by a constant factor.

By differentiation of (10.28), the shock velocity U is simply

$$U = \frac{dR}{dt} = \frac{2}{5}\frac{R}{t} \tag{10.29}$$

A remarkable and famous verification of these formulas was given by measurements of the first atomic explosion in New Mexico in 1945, as shown in Fig. 10.10. The prediction corresponding to (10.28) was first obtained by G. I. Taylor in a paper written in 1941, publication being delayed by security restrictions until 1950. Taylor was also able to obtain a numerical solution for the gas motion behind the shock (*Taylor* [1950]).

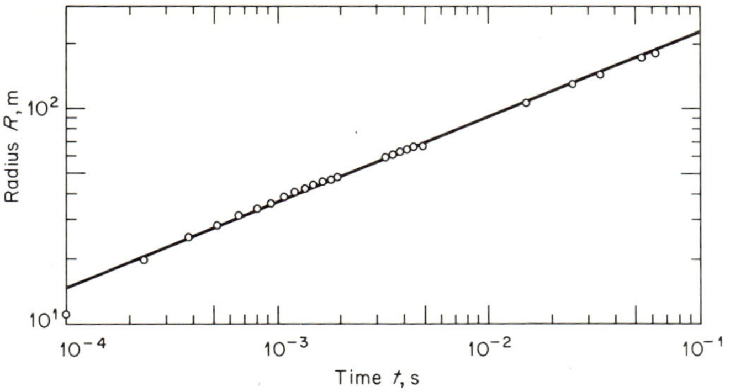

Figure 10.10
Experimental results from the New Mexico explosion compared to the theoretical curve $R \propto t^{2/5}$. (*Data from Taylor* [1950].)

Sedov, in Russia, published in 1946 an analytical solution for the same problem, and these results are included in his well-known book on similarity (*Sedov* [1959]). A recent study of this problem is that of *Laumbach and Probstein* [1969].

Self-similar Formulation of Equations

The dependent variables u, P, and ρ will be put into nondimensional form. There are, however, no constant parameters with dimensions of velocity or pressure; we therefore normalize u with respect to r/t and P with respect to $\rho_0 r^2/t^2$. The nondimensional dependent variables can then be written

$$\hat{u} \equiv \frac{5(\gamma+1)}{4}\frac{u}{r/t}$$

$$\hat{P} \equiv \frac{25(\gamma+1)}{8}\frac{P}{\rho_0 r^2/t^2} \tag{10.30}$$

$$\hat{\rho} \equiv \frac{\gamma-1}{\gamma+1}\frac{\rho}{\rho_0}$$

These are all functions of η. The coefficients involving γ have been introduced, with the help of hindsight, in order to make the boundary conditions (10.24) to (10.26) numerically neat. In formulating the problem in this way, we follow *Landau and Lifshitz* [1959, chap. 10].

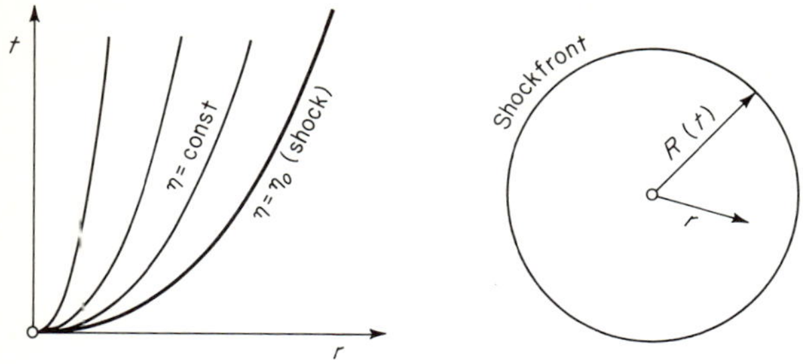

Figure 10.11
The shock path along $\eta = \eta_0$.

In terms of the dimensionless variables, the boundary conditions (10.24) to (10.26) become

$$\hat{u} = 1$$
$$\hat{P} = 1 \qquad \text{at } \eta = \eta_0 \tag{10.31}$$
$$\hat{\rho} = 1$$

These apply at the shock front (see Fig. 10.11).

The gas flow behind the shock front is described by the equations of motion

$$\frac{\partial u}{\partial t} + u\frac{\partial u}{\partial r} + \frac{1}{\rho}\frac{\partial P}{\partial r} = 0 \tag{10.32}$$

$$\frac{\partial \rho}{\partial t} + \rho\frac{\partial u}{\partial r} + u\frac{\partial \rho}{\partial r} + \frac{2\rho u}{r} = 0 \tag{10.33}$$

$$\left(\frac{\partial}{\partial t} + u\frac{\partial}{\partial r}\right)\ln\frac{P}{\rho^{\gamma}} = 0 \tag{10.34}$$

The last equation is the condition that the specific entropy of a particular fluid particle be constant in time. The flow is *not* homentropic, as the strength of the shock varies with time, so that different fluid particles acquire different entropies in crossing the shock.

With the introduction of the dimensionless variables $\hat{u}$, $\hat{P}$, and $\hat{\rho}$ the preceding differential equations become

$$\hat{\rho}(2\hat{u} - \gamma - 1)\frac{d\hat{u}}{d\eta} + (\gamma - 1)\frac{d\hat{P}}{d\eta} = \frac{1}{2\eta}\left[\hat{\rho}\hat{u}(5\gamma + 5 - 4\hat{u}) - 4(\gamma - 1)\hat{P}\right] \tag{10.35}$$

$$\left(\hat{u} - \frac{\gamma + 1}{2}\right) \frac{1}{\hat{\rho}} \frac{d\hat{\rho}}{d\eta} + \frac{d\hat{u}}{d\eta} = -\frac{3\hat{u}}{\eta} \tag{10.36}$$

$$\frac{d}{d\eta} \ln \frac{\hat{P}}{\hat{\rho}^\gamma} = \frac{1}{\eta} \frac{5(\gamma + 1) - 4\hat{u}}{2\hat{u} - (\gamma + 1)} \tag{10.37}$$

The reduction to a set of ordinary differential equations with appropriate boundary conditions (10.31) indicates that the proposed similarity solution is tenable.

Integration

It remains to integrate the equations. One integral can be obtained from the following argument of Landau and Lifshitz. First of all, the energy within the spherical region bounded by the shock is constant and equal to E, the initial internal energy content of the gas (before explosion) being negligible. Furthermore, the energy contained within any spherical surface which moves such that η = const is also constant in time [the arguments leading to Eq. (10.28) apply, where E is the energy confined to the subsphere of radius r].

Such a spherical surface, of instantaneous radius r, has a radial velocity $2r/5t$, from (10.29). From the constancy of the energy, the first term in the integral energy balance (1.81) vanishes. With $\mathbf{T} = -P\mathbf{n}$ and negligible heat flux $\mathbf{q}$ and body force $\mathbf{G}$, this balance gives

$$4\pi r^2 \rho \left(u - \frac{2}{5}\frac{r}{t}\right)(e + \tfrac{1}{2}u^2) = -4\pi r^2 P u$$

With $e = \dfrac{1}{\gamma - 1}\dfrac{P}{\rho}$, this reduces to

$$\frac{\hat{P}}{\hat{\rho}} = \frac{\hat{u}^2(\gamma + 1 - 2\hat{u})}{2\gamma\hat{u} - \gamma - 1} \tag{10.38}$$

which is the desired integral and may be used in place of one of the differential equations[1] (10.35) to (10.37).

By retaining (10.36) to (10.38) as the equations of motion, the integration can be completed. Equation (10.38) substituted into (10.37) gives a relation involving only derivatives of $\hat{\rho}$ and $\hat{u}$. This can be combined

[1] If Eq. (10.38) is differentiated, it is found to be a combination of (10.35) and (10.36).

with (10.36) to give an equation in which only $\hat{u}$ and its derivative appear. Then, with the boundary condition (10.31), integration gives

$$\left(\frac{\eta_0}{\eta}\right)^5 = \hat{u}^2\left[\frac{5(\gamma + 1) - 2(3\gamma - 1)\hat{u}}{7 - \gamma}\right]^{k_1}\left[\frac{2\gamma\hat{u} - \gamma - 1}{\gamma - 1}\right]^{k_2} \tag{10.39}$$

$$\hat{\rho} = \left[\frac{2\gamma\hat{u} - \gamma - 1}{\gamma - 1}\right]^{k_3}\left[\frac{5(\gamma + 1) - 2(3\gamma - 1)\hat{u}}{7 - \gamma}\right]^{k_4}\left[\frac{\gamma + 1 - 2\hat{u}}{\gamma - 1}\right]^{k_5}$$

where

$$k_1 \equiv \frac{13\gamma^2 - 7\gamma + 12}{(3\gamma - 1)(2\gamma + 1)}$$

$$k_2 \equiv \frac{-5(\gamma - 1)}{2\gamma + 1}$$

$$k_3 \equiv \frac{3}{2\gamma + 1} \tag{10.40}$$

$$k_4 \equiv \frac{13\gamma^2 - 7\gamma + 12}{(2 - \gamma)(3\gamma - 1)(2\gamma + 1)}$$

$$k_5 \equiv \frac{1}{\gamma - 2}$$

Equation (10.38) gives the pressure $\hat{P}$.

Only the value of η_0, corresponding to the position of the shock, remains to be determined. From the conservation of the explosion energy E we have

$$E = \int_0^R \left(\tfrac{1}{2}\rho u^2 + \frac{P}{\gamma - 1}\right)4\pi r^2\, dr \tag{10.41}$$

in which $\tfrac{1}{2}\rho u^2$ and $P/(\gamma - 1)$ are, respectively, the kinetic energy and internal energy per unit volume for the ideal gas within the spherical volume bounded by the shock. Written in terms of the nondimensional quantities this integral is

$$1 = \frac{32\pi\eta_0{}^5}{25(\gamma^2 - 1)}\int_0^1 (\hat{\rho}\hat{u}^2 + \hat{P})\xi^4\, d\xi \tag{10.42}$$

where $\xi \equiv \eta/\eta_0$. Evaluation gives η_0 as a function of γ. Representative values are

γ	η_0
1.34	1.000
1.40	1.033
1.67	1.153

A very complete listing of numerical results for this problem, as well as the cylindrical and plane explosion cases, is given by *Sedov* [1959].

Conditions behind the shock, as given by the solutions above, are shown for air ($\gamma = 1.40$) in Fig. 10.12. The density falls off very rapidly behind the shock, in such a way that the average of ρ/ρ_s is $1/6$, that is, atmospheric density. The pressure ratio P/P_s approaches a uniform value near the center. Temperature ($T \propto P/\rho$) increases rapidly toward the center, where it becomes infinite.

The low-density region left behind by the expanding spherical shock is subject to buoyant rise (Archimedean lift) in a gravitational field, producing the notorious mushroom cloud. Taylor makes an approximate calculation for the velocity of rise for the New Mexico explosion, based on the simple equation for the rise velocity

$$U = \tfrac{2}{3}\sqrt{gr_0}$$

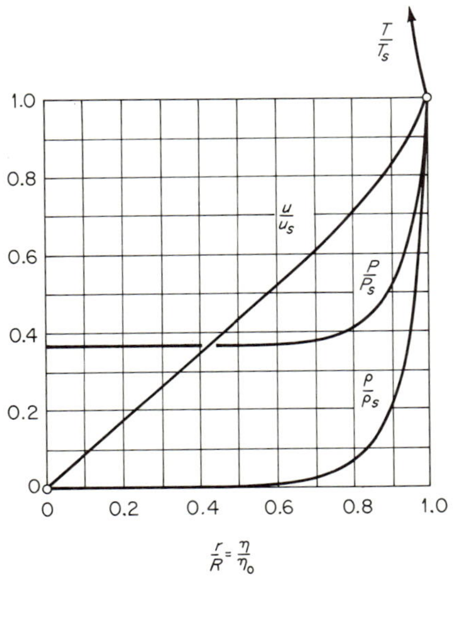

Figure 10.12
Flow conditions within the expanding blast wave ($\gamma = 1.40$).

where r_0 is the radius of the low-density region. The agreement with experimental results is good.

The apparent accuracy of the preceding similarity calculation is somewhat surprising, because in reality the gas can hardly be expected to be perfect. In particular, for air, the ratio of specific heats γ will change from its initial value of 1.40 in the course of the high-temperature processes within the blast.

10.5 Laminar boundary layer behind a shock wave

This is one of the simplest of the compressible boundary-layer flows. The problem will be treated here briefly, yielding only a glimpse of the strange and wonderful transformations which have been found useful in compressible boundary-layer theory.

A boundary layer is normally a thin layer adjacent to a solid surface, within which viscous forces and heat conduction play an essential role. The tangential component of velocity varies from some free-stream value at the outer edge of the layer to zero at the wall, by virtue of the no-slip condition. It is this rapid variation in the tangential velocity which is the dominant feature of boundary layers. Boundary layers in *compressible* fluids are complicated by nonnegligible dissipation (because of the large flow speeds and gradients), variable viscosity (even in perfect gases), and, of course, a variable fluid density (the variation being mainly a consequence of the variation in temperature).

General references to this subject are *Stewartson* [1964] and *Schlichting* [1968, chap. 13]. The main reference for the problem at hand is *Mirels* [1955, 1956], and the material given here more or less follows his work.

Problem Description

Consider a plane shock front advancing into stationary fluid, with a plane wall perpendicular to the shock front, as shown in Fig. 10.13*a*. A practical example of this is a shock propagating down a tube, e.g., the main shock separating regions 1 and 2 in the shock tube (see, for example, Figs. 8.36 and 8.37). In such a case it is assumed that the thickness of the boundary layer is small compared to the tube radius, so that the flow is effectively plane. The requirement that the fluid ahead of the shock be

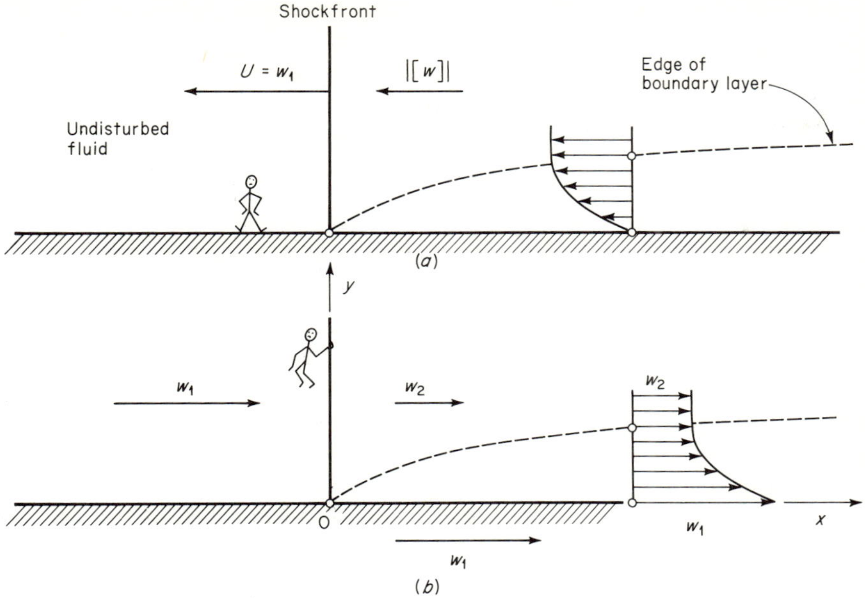

Figure 10.13
Boundary-layer development behind the shock: (*a*) flow as seen by stationary observer, and (*b*) observer moving with the shock.

stationary is essential, because only then will there be no boundary layer at the wall in front of the shock.[1]

The effect of the passage of the shock is to impulsively set the fluid into motion with respect to the wall. From the point of view of an observer traveling with the shock, as shown in Fig. 10.13*b*, this motion can be considered *steady*. It is therefore a great advantage to consider the motion with respect to this moving reference frame.

The Boundary-layer Equations

Because the fluid motion in the boundary layer is dominated by plane laminar shearing, the Navier-Stokes equations assume a greatly simplified approximate form.

[1] This may be related to the observation that plane shock waves are often observed in *unsteady* flow in tubes, while stationary plane shocks are seldom observed in *steady* flow in tubes. In the latter case the shock cannot extend to the wall but interacts with an existing boundary layer to produce relatively complicated shock patterns, such as the *lambda shock*.

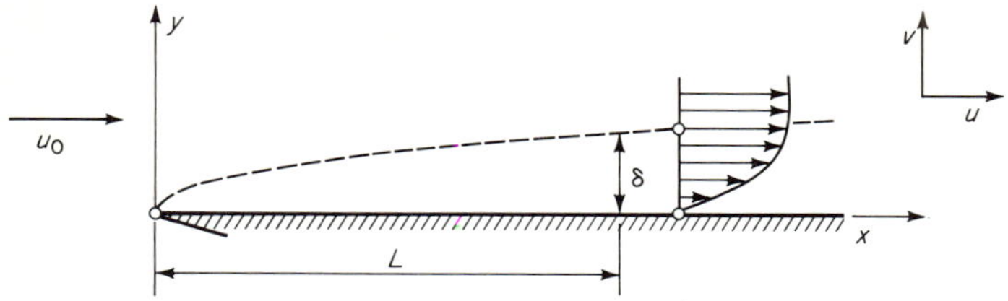

Figure 10.14
Geometry for the boundary layer on a flat plate in uniform streaming flow.

Consider a plane flow with spatial coordinates x and y and corresponding velocity components u and v (this notation conforms to normal practice in boundary-layer theory). For steady flow, the x component of the momentum equation (1.59) is, written out in full,

$$\rho\left(u\frac{\partial u}{\partial x} + v\frac{\partial u}{\partial y}\right) + \frac{\partial P}{\partial x} = \frac{\partial}{\partial x}\left[\left(\tfrac{4}{3}\mu + \mu_v\right)\frac{\partial u}{\partial x} + \left(\mu_v - \tfrac{2}{3}\mu\right)\frac{\partial v}{\partial y}\right] + \frac{\partial}{\partial y}\left[\mu\left(\frac{\partial v}{\partial x} + \frac{\partial u}{\partial y}\right)\right] \qquad (10.43)$$

Several of the terms in this equation are small enough to be negligible, in comparison to the other terms. We proceed to estimate the order of magnitude of individual terms.

The classical boundary layer on a flat plate is shown in Fig. 10.14. We will use this problem as a basis for the magnitude estimates. The boundary-layer equations thus found will be in the somewhat specialized form appropriate to *steady flow in two dimensions with zero pressure gradient*. An external streaming flow with uniform velocity u_0 is brought to rest at the plate surface. The velocity change occurs over a distance of order δ in the y direction and over a distance of order L in the x direction. The velocities and their derivatives are then estimated to have the following magnitudes:

$$u \sim u_0 \qquad \frac{\partial u}{\partial x} \sim \frac{u_0}{L} \qquad \frac{\partial^2 u}{\partial x^2} \sim \frac{u_0}{L^2} \qquad \frac{\partial^2 u}{\partial x\,\partial y} \sim \frac{u_0}{\delta L}$$

$$v \sim u_0\frac{\delta}{L} \qquad \frac{\partial u}{\partial y} \sim \frac{u_0}{\delta} \qquad \frac{\partial^2 u}{\partial y^2} \sim \frac{u_0}{\delta^2} \qquad \frac{\partial^2 v}{\partial x\,\partial y} \sim \frac{u_0}{\delta^2}$$

Note that the estimate for v can be obtained from the (approximate) continuity equation $\partial u/\partial x + \partial v/\partial y = 0$, with $\partial u/\partial x \sim u_0/L$ and $\partial v/\partial y \sim v/\delta$.

We now assume that the boundary layer is thin, $\delta \ll L$. Thus, for example, $\partial^2 u/\partial x^2$ is negligible in comparison with $\partial^2 u/\partial y^2$. The convective derivatives on the left-hand side of (10.43) are each of order $u_0{}^2/L$. The derivatives on the right-hand side (treating viscosity as constant in the estimate) are all of order u_0/L^2 except for the last, which is of order u_0/δ^2 and which will be the only one retained. In this problem the free-stream pressure is constant, and the pressure variation across the boundary layer is negligible, as discussed below. The term $\partial P/\partial x$ can therefore be dropped. Equation (10.43) then becomes, with the corresponding magnitude estimates,

$$\underset{(\rho u_0{}^2/L)}{\rho\left(u\frac{\partial u}{\partial x} + v\frac{\partial u}{\partial y}\right)} = \underset{(\mu u_0/\delta^2)}{\frac{\partial}{\partial y}\left(\mu\frac{\partial u}{\partial y}\right)} \tag{10.44}$$

This is the celebrated boundary-layer equation of Ludwig Prandtl. The two sides of this equation must certainly be of the same order of magnitude! Then equating the magnitude estimates gives, with $x = L$,

$$\delta \sim \sqrt{\frac{\mu x}{\rho u_0}} \tag{10.45}$$

that is, δ increases as the square root of x. This can also be written

$$\frac{\delta}{x} \sim \sqrt{\frac{1}{\mathrm{Re}_x}} \tag{10.46}$$

where $\mathrm{Re}_x \equiv \rho u_0 x/\mu$. For the flow of air at normal pressure and temperature, with $u_0 \sim c$, we find $\mathrm{Re} \sim 10^7\ \mathrm{m}^{-1}$; that is, for $x = 1$ m, $\mathrm{Re}_x = 10^7$. The condition $\delta \ll x$ is thus very well fulfilled, except for a region quite close to the origin. We shall refer to the distance δ as the *boundary-layer thickness* [compare the thickness found for the *Rayleigh problem*, Eq. (3.64)].

Note that in the boundary-layer approximation (10.44) all terms involving the bulk viscosity drop out.

Estimating each of the terms in the *momentum equation for the y direction*, this equation reduces to simply

$$\frac{\partial P}{\partial y} \approx 0 \tag{10.47}$$

That is, the pressure does not vary over the thickness of the boundary layer. Since P = const in the free-stream region, for the problems of interest here, it follows that $P \approx$ const everywhere.

Estimating the magnitude of the various terms in the energy equation (1.74) in similar fashion, this equation reduces to

$$\rho\left(u\frac{\partial h}{\partial x} + v\frac{\partial h}{\partial y}\right) = \mu\left(\frac{\partial u}{\partial y}\right)^2 + \frac{\partial}{\partial y}\left(\kappa\frac{\partial T}{\partial y}\right) \tag{10.48}$$

The reader should observe that (10.44) and (10.48) are somewhat similar in form, at least in the case of a perfect gas with $h \propto T$.

The boundary-layer thickness δ has been rather crudely defined as the distance over which the velocity changes from zero (at the wall) to approximately its free-stream value u_0. We can arbitrarily define δ to be, for example, the distance y at which $u = 0.99u_0$, as was done essentially in the Rayleigh problem. A boundary-layer thickness for *temperature* can be defined in a corresponding way and in general will have a value different from the velocity boundary-layer thickness. We therefore designate these two distances by different symbols, δ_v and δ_t, respectively,

δ_v = velocity boundary-layer thickness

δ_t = temperature boundary-layer thickness

The final boundary-layer equation is just the continuity equation (1.58), for steady flow,

$$\frac{\partial}{\partial x}\rho u + \frac{\partial}{\partial y}\rho v = 0 \tag{10.49}$$

The order-of-magnitude estimates for the classical flat-plate problem are also applicable to the problem at hand, the boundary layer behind a shock. The velocity change u_0 across the boundary layer is simply replaced by the jump $[w]$, which is known from the strength of the shock. The (approximate) boundary-layer equations given above are therefore applicable.

Problem Statement

We consider the motion from the viewpoint of an observer moving with the shock front, as shown in Fig. 10.13*b*. The flow is steady in this reference frame.

The fluid is assumed to be a *perfect gas*. This allows us to put $h = h(T)$, with constant specific heat c_p. The viscosity μ and conductivity

κ must, however, be considered to vary with temperature. Then the boundary-layer equations already found can be rewritten as

$$\frac{\partial}{\partial x}\rho u + \frac{\partial}{\partial y}\rho v = 0 \tag{10.49}$$

$$\rho\left(u\frac{\partial u}{\partial x} + v\frac{\partial u}{\partial y}\right) = \frac{\partial}{\partial y}\left(\mu\frac{\partial u}{\partial y}\right) \tag{10.50}$$

$$\rho c_p\left(u\frac{\partial T}{\partial x} + v\frac{\partial T}{\partial y}\right) = \mu\left(\frac{\partial u}{\partial y}\right)^2 + \frac{\partial}{\partial y}\left(\kappa\frac{\partial T}{\partial y}\right) \tag{10.51}$$

The unknowns in these equations can be considered to be u, v, ρ, and T (for a perfect gas the viscosity and thermal conductivity are functions of the temperature only). The pressure is, however, everywhere constant at the free-stream value; thus $\rho \propto 1/T$, and there is in effect only one thermodynamic variable (in this respect the situation is like inviscid homentropic gasdynamics, in that one thermodynamic variable fixes the thermodynamic state). We thus have three equations and three unknowns.

The external boundary conditions are fixed by the (specified) strength of the shock. In addition we assume an *isothermal wall* fixed at the upstream temperature T_1. Denoting, as usual, conditions ahead of the shock by subscript 1 and behind the shock by subscript 2, the boundary conditions in the shock-fixed reference frame are, for $u(x,y)$, $v(x,y)$, and $T(x,y)$,

$$\begin{aligned} u(x,0) &= w_1 \qquad & u(x,\infty) &= w_2 \qquad & v(x,0) &= 0 \\ T(x,0) &= T_1 \qquad & T(x,\infty) &= T_2 \end{aligned} \tag{10.52}$$

The boundary conditions at $y = \infty$, for example, $u = w_2$, are expected to be very nearly satisfied[1] at $y = \delta$. For a very weak shock, $w_2/w_1 \to 1$ and $T_2/T_1 \to 1$. For a very strong shock, $w_2/w_1 \to (\gamma - 1)/(\gamma + 1)$ and $T_2/T_1 \to \infty$.

The density in the free stream is $\rho_2 = P_2/RT_2$, while the density of the cold gas at the wall is fixed at $\rho_w = P_2/RT_1$. For fairly strong shocks $T_2 \gg T_1$, and the relatively cold wall acts like a sink for mass, somewhat analogous to condensation on the wall. This cooling effect is opposed by the viscous heating, represented by the term $\mu(\partial u/\partial y)^2$ in (10.51).

[1] As a famous student of the subject has remarked, infinity may mean about 1 cm in fluid mechanics.

Stream Function

It is useful and conventional in many two-dimensional flows to introduce a stream function $\psi(x,y)$, defined by

$$\rho u = \rho_w \frac{\partial \psi}{\partial y} \qquad \rho v = -\rho_w \frac{\partial \psi}{\partial x} \tag{10.53}$$

(the selection of the constant density ρ_w on the right side is arbitrary; ρ_2, for example, would serve as well). Substituting the definition (10.53) into (10.49) shows that continuity is satisfied identically, and the continuity equation is thus removed from further consideration. This is analogous to the identical satisfaction of the irrotationality condition by a velocity potential.

The general properties of the stream function are discussed, for example, by *Karamcheti* [1966, chap. 4]. The quintessential property is that *a line* $\psi(x,y) = \text{const}$ *is a streamline.* For example, we may take $\psi = 0$ to be the wall.

Dorodnitsyn-Howarth Transformation

This is the first of our transformations. It has been found that this transformation has the effect of nearly removing the density ρ from the equations and reducing them to essentially incompressible form. We define the new variable $\bar{y}$

$$\bar{y} = \int_0^y \frac{\rho}{\rho_w}\, dy \tag{10.54}$$

where the integration is carried out at constant x. Physically, $\rho_w \bar{y}$ is the fluid mass stored between $y = 0$ and $y = y$, per unit area of the wall. We can describe $\bar{y}$ as a *stretched y coordinate.*

The momentum and energy equations (10.50) and (10.51) can be rewritten in terms of $\psi(\bar{y},x)$ and $T(\bar{y},x)$. By differentiation of (10.54)

$$\left(\frac{\partial \bar{y}}{\partial y}\right)_x = \frac{\rho}{\rho_w}$$

The partial derivative operators become

$$\left.\frac{\partial}{\partial y}\right)_x = \frac{\rho}{\rho_w}\left.\frac{\partial}{\partial \bar{y}}\right)_x$$

$$\left.\frac{\partial}{\partial x}\right)_y = \left.\frac{\partial}{\partial x}\right)_{\bar{y}} + \left(\frac{\partial \bar{y}}{\partial x}\right)_y \left.\frac{\partial}{\partial \bar{y}}\right)_x$$

Henceforth, the subscripts will normally be omitted from partial derivatives, with the understanding that $\partial/\partial x$ will mean differentiation at constant $\bar{y}$. The velocity components become, from (10.53),

$$u = \frac{\partial \psi}{\partial \bar{y}} \qquad v = -\frac{\rho_w}{\rho}\left[\frac{\partial \psi}{\partial x} + u\left(\frac{\partial \bar{y}}{\partial x}\right)_y\right] \tag{10.55}$$

Then (10.50) and (10.51) become, respectively,

$$\frac{\partial \psi}{\partial \bar{y}}\frac{\partial^2 \psi}{\partial x\,\partial \bar{y}} - \frac{\partial \psi}{\partial x}\frac{\partial^2 \psi}{\partial \bar{y}^2} = \frac{\partial}{\partial \bar{y}}\left(\frac{\rho\mu}{{\rho_w}^2}\frac{\partial^2 \psi}{\partial \bar{y}^2}\right) \tag{10.56}$$

$$\frac{\partial \psi}{\partial \bar{y}}\frac{\partial T}{\partial x} - \frac{\partial \psi}{\partial x}\frac{\partial T}{\partial \bar{y}} = \frac{\rho\mu}{{\rho_w}^2 c_p}\left(\frac{\partial^2 \psi}{\partial \bar{y}^2}\right)^2 + \frac{\partial}{\partial \bar{y}}\left(\frac{\rho\kappa}{{\rho_w}^2 c_p}\frac{\partial T}{\partial \bar{y}}\right) \tag{10.57}$$

The boundary conditions (10.52) become, in terms of these variables,

$$\frac{\partial \psi}{\partial \bar{y}} = \begin{cases} w_1 & \text{at } y = 0 \\ w_2 & \text{at } y = \infty \end{cases} \qquad \frac{\partial \psi}{\partial x} = 0 \quad \text{at } y = 0 \tag{10.58}$$

$$T(x,0) = T_1 \qquad\qquad T(x,\infty) = T_2$$

Approximations for the Viscosity and Thermal Conductivity

If it were only possible to set $\rho\mu$ = const and $\rho\kappa$ = const, the variable-density and variable-transport properties would completely drop out of Eqs. (10.56) and (10.57). This approximation is in fact much used in aerodynamics and corresponds to the Chapman viscosity law mentioned in Sec. 2.7. We now discuss this approximation, which is based on putting $\mu \propto T$. This can be written

$$\frac{\mu}{\mu_w} = \frac{T}{T_w} \tag{10.59}$$

where $\mu_w(T_w)$ is evaluated from the known viscosity dependence of the gas in question. The comparison of the empirical "law" (10.59) with a known viscosity law is shown schematically in Fig. 10.15 (compare Fig. 2.24). With P = const, (10.59) can be rewritten as

$$\rho\mu = \rho_w\mu_w \tag{10.60}$$

which is the desired approximation.

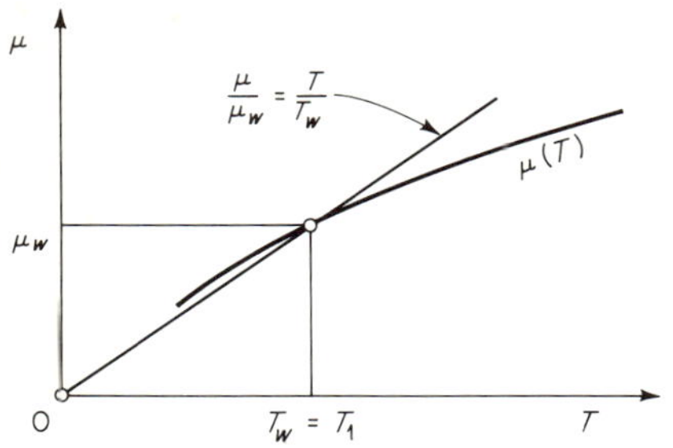

Figure 10.15
The approximation $\mu \propto T$.

Since the Prandtl number, $\mathrm{Pr} \equiv \mu c_p/\kappa$, is independent of temperature to good approximation and $c_p = \text{const}$, it follows that the further approximation $\rho\kappa = \text{const}$ is consistent with (10.60). From the definition of the Prandtl number

$$\frac{\rho\kappa}{c_p} = \frac{\rho\mu}{\mathrm{Pr}}$$

Then from (10.60)

$$\frac{\rho\kappa}{c_p} = \frac{\rho_w\mu_w}{\mathrm{Pr}} = \text{const} \tag{10.61}$$

With the approximate expressions (10.60) and (10.61), the momentum and energy equations (10.56) and (10.57) become finally

$$\frac{\partial\psi}{\partial\bar{y}}\frac{\partial^2\psi}{\partial x\,\partial\bar{y}} - \frac{\partial\psi}{\partial x}\frac{\partial^2\psi}{\partial\bar{y}^2} = \nu_w\frac{\partial^3\psi}{\partial\bar{y}^3} \tag{10.62}$$

$$\frac{\partial\psi}{\partial\bar{y}}\frac{\partial T}{\partial x} - \frac{\partial\psi}{\partial x}\frac{\partial T}{\partial\bar{y}} = \frac{\nu_w}{c_p}\left(\frac{\partial^2\psi}{\partial\bar{y}^2}\right)^2 + \frac{\nu_w}{\mathrm{Pr}}\frac{\partial^2 T}{\partial\bar{y}^2} \tag{10.63}$$

where $\nu_w \equiv \mu_w/\rho_w$ is a constant kinematic viscosity.

Self-similarity

It is remarkable that the equations of motion (10.62) and (10.63) and boundary conditions (10.58) are exactly in the form which would be obtained for an *incompressible* fluid with constant transport properties. In fact, the problem as it is now described is identical to the celebrated *Blasius problem* for the steady incompressible boundary layer on a flat plate (shown in Fig. 10.14) except that in the Blasius problem the first boundary condition in (10.58) reads $\partial\psi/\partial\bar{y} = 0$; that is, the wall is stationary.

It should also be remarked that the momentum equation (10.62) is now decoupled from the energy equation (10.63). That is, the momentum equation contains only one unknown ψ and can be solved, with appropriate boundary conditions, independent of any solution (or lack of solution) for the energy equation.

Similarity variables for the present problem cannot be constructed on the basis of dimensional arguments alone. The Blasius problem, however, has a famous solution for the velocity boundary layer in terms of the similarity variables

$$\eta \equiv \frac{y}{\sqrt{2\nu x/u_0}}$$

$$f(\eta) \equiv \frac{\psi}{\sqrt{2u_0\nu x}}$$

It is therefore natural in the present situation to try for a solution in terms of

$$\eta \equiv \frac{\bar{y}}{\sqrt{2\nu_w x/w_2}} \tag{10.64}$$

$$f(\eta) \equiv \frac{\psi}{\sqrt{2w_2\nu_w x}} \tag{10.65}$$

This choice of similarity variables can be rationalized in several different ways. For example, consider the Rayleigh problem discussed in Chap. 3. In that problem we found a similarity variable of the form $y/\sqrt{\nu t}$, where t is the time from the initiation of motion. In the present problem, from the standpoint of a fluid particle in the free stream, the boundary-layer development occurs from the time that particle passes the origin, i.e., over a time $t = x/u_0$. Thus if we replace t by x/u_0, the similarity variable becomes $y/\sqrt{\nu x/u_0}$ instead of $y/\sqrt{\nu t}$. This happens to be the "correct" variable. (We comment further on the general problem of selecting similarity variables in Sec. 10.6.)

The similarity formulation can be tested by determining whether the differential equations and boundary conditions are expressible purely in terms of the (dimensionless) similarity variables. We will need a non-dimensional temperature, assumed to depend only on η, which can be taken to be

$$\tilde{T}(\eta) \equiv \frac{T}{T_2} \tag{10.66}$$

This quantity is equal to T_1/T_2, known from the shock conditions, at the relatively cold isothermal wall and is equal to unity in the free stream, far from the wall.

With the substitutions (10.64) and (10.65) the equations of motion (10.62) and (10.63) become

$$f''' + ff'' = 0 \tag{10.67}$$

$$\tilde{T}'' + (\mathrm{Pr})f\tilde{T}' = -(\gamma - 1)(\mathrm{Pr})M_{2n}^{\,2}(f'')^2 \tag{10.68}$$

where a prime indicates differentiation with respect to η and $M_{2n} = w_2/c_2$ is known from the shock conditions. The differential equations are thus in a self-similar form. Likewise, the boundary conditions (10.58) reduce, for $f(\eta)$ and $\tilde{T}(\eta)$, to

$$\begin{aligned} f'(0) &= \frac{w_1}{w_2} \qquad f'(\infty) = 1 \qquad f(0) = 0 \\ \tilde{T}(0) &= \frac{T_1}{T_2} \qquad \tilde{T}(\infty) = 1 \end{aligned} \tag{10.69}$$

The similarity formulation is thus valid.

The momentum equation (10.67) is formally identical to that found for the (incompressible) Blasius problem, though the boundary condition at the wall is different. This equation can be solved first, by numerical methods, and the result used in the solution of the (linear) energy equation (10.68).[1] These calculations are given by *Mirels* [1955].

Velocity and Temperature Profiles

The velocity components are, from (10.55), (10.64), and (10.65),

$$\begin{aligned} u &= w_2 f'(\eta) \\ v &= -w_2 \frac{\rho}{\rho_w} \sqrt{\frac{\nu_w}{2w_2 x}} \left[f + 2xf' \left(\frac{\partial \eta}{\partial x} \right)_y \right] \end{aligned} \tag{10.70}$$

In order to actually draw a velocity profile or temperature profile, it is necessary to find how y (as opposed to the stretched coordinate $\bar{y}$) varies with η. From the definitions (10.54) and (10.64)

$$\sqrt{\frac{2\nu_w x}{w_2}}\,\eta = \int_0^y \frac{\rho}{\rho_w}\,dy$$

[1] Analytical solutions are also possible; see, for example, *Evans* [1968].

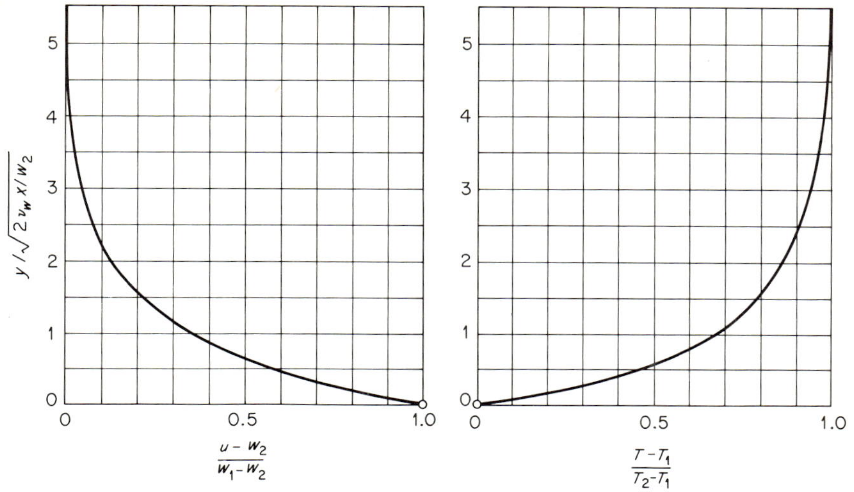

Figure 10.16
Velocity and temperature profiles for $\gamma = 1.40$ and shock density ratio $\rho_2/\rho_1 = w_1/w_2 = 4$, corresponding to a shock Mach number $M_{1n} = 3.162$.

Differentiating,

$$\left(\frac{\partial\eta}{\partial y}\right)_x = \frac{1}{\sqrt{2\nu_w x/w_2}}\frac{\rho}{\rho_w}$$

With $dy = (\partial y/\partial\eta)_x\, d\eta$ at constant x,

$$y = \int_0^\eta \left(\frac{\partial y}{\partial\eta}\right)_x d\eta = \sqrt{\frac{2\nu_w x}{w_2}}\int_0^\eta \frac{\rho_w}{\rho}\, d\eta$$

Finally, with $\rho_w/\rho = T/T_w = T/T_1$,

$$\frac{y}{\sqrt{2\nu_w x/w_2}} = \frac{T_2}{T_1}\int_0^\eta \tilde{T}\, d\eta \tag{10.71}$$

and the integral is known from the solution for $\tilde{T}(\eta)$.

The velocity and temperature profiles corresponding to a particular value of the shock strength are shown in Fig. 10.16. It is remarkable that the temperature profile is very nearly a mirror image of the velocity profile. This similarity of profiles has some basis in theory, related to similarity between the energy and momentum equations, but we will not discuss the point here.

The assumption of an isothermal wall, made near the outset, is not strictly reproducible in experiments since there is necessarily heat transfer

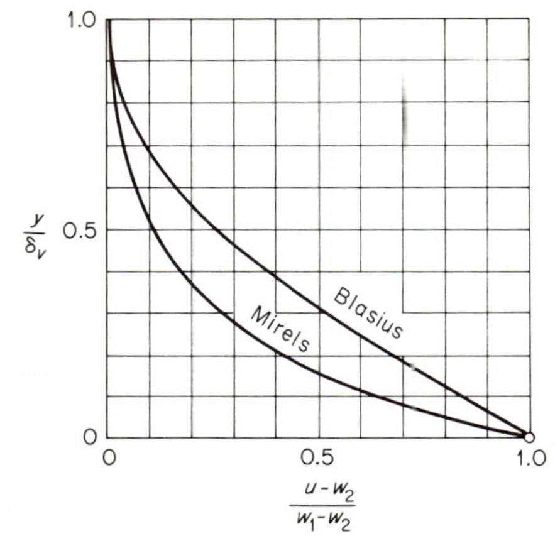

Figure 10.17
Comparison between the Blasius velocity profile and that of Mirels (for the case $\gamma = 1.40$, $w_1/w_2 = 4$).

to the wall. Investigation shows, however, that the increase in wall temperature is very small, normally less than 1 percent (*Mirels* [1956]). This conclusion follows mainly from the smallness of the gas thermal conductivity compared to that of the solid wall material, e.g., iron. Only at enormously large gas density and temperature is there reason to question the isothermal assumption for this particular problem.

The velocity boundary-layer thickness δ_v is typically of order

$$\delta_v \sim 5\sqrt{\frac{2\nu_w x}{w_2}}$$

where δ_v is defined as the distance y at which 99 percent of the velocity change has taken place. A comparison between the compressible boundary layer considered here and the classical incompressible Blasius boundary layer is shown in Fig. 10.17.

At some distance x behind the shock the Reynolds number, for example, $\text{Re} = w_2 x/\nu_w$, becomes sufficiently large for the boundary layer to become *turbulent*. The preceding theory is then of course inapplicable, and one of the semiempirical descriptions for turbulent boundary layers must be resorted to.

10.6 Closure

There are several other self-similar motions of interest in gasdynamics, and we close with a listing of some of them, together with references:

1 Impulsive piston withdrawal (Chap. 8)
2 Impulsive piston advance (Chap. 8)
3 Cylinder or sphere expanding uniformly into stationary gas (*Lighthill* [1948])
4 Expanding gas behind a detonation front (*Taylor* [1950])
5 Spherical implosion (*Zel'dovich and Raizer* [1967, chap. 12])
6 Shock in an exponential atmosphere (*Zel'dovich and Raizer* [1967, chap. 12])
7 Viscous flow of a perfect gas in a converging or diverging channel (a generalization of Hamel's solution for incompressible fluids; *Byrkin* [1969]).

This list is not exhaustive. In particular, self-similar motions involving two or more independent variables are not mentioned. Certain problems which are more or less variations of the above are not mentioned.

In addition to Zel'dovich and Raizer, the books of *Stanyukovich* [1960] and *Sedov* [1959] contain extensive treatments of self-similar motions.

Remarks on the Choice of a Similarity Variable

A similarity solution for any given problem is possible only if allowed by both the equations and the boundary conditions. There are several more or less orderly procedures by which one may seek a suitable similarity variable:

1 By dimensional arguments alone. This technique has been emphasized here; in some cases, however, it will not point to any definite similarity variable, and another strategem must be used.
2 By nondimensionalizing one of the space variables with respect to a variable significant distance. Examples are the Taylor problem (similarity variable r/R) and the Blasius problem (similarity variable y/δ). The dependence of the selected significant distance with time or space must be found from dimensional arguments, for example, $R \propto t^{2/5}$ (Taylor), or magnitude estimates, for example, $\delta \propto x^{1/2}$ (Blasius), or otherwise. This formulation serves to emphasize the self-preserving (similar) nature of the motion.
3 By simply trying a similarity variable written in a general form. For example, if x and y are independent variables, a similarity variable $\eta = yx^m$ can be formulated and tested by substituting into the equations of motion and boundary conditions.
4 By making use of the theory of groups, as discussed by *Birkhoff* [1960] and *Hansen* [1964].

Problems

10.1 Consider the (idealized) fall of a stone from rest, given by the equation $y = \frac{1}{2}gt^2$. In what sense is this a self-similar motion? Find the similarity variable and find the above solution for the differential equation of motion by similarity arguments.

10.2 The blast wave from an intense spherical explosion travels a distance R_0 from the source in time t_0. What is the time to travel a distance $2R_0$ from the source?

10.3 For the intense-explosion problem, find an explicit expression for P_s in terms of E and R. Compare the resulting attenuation law with the attenuation of a spherical acoustic wave.

10.4 Consider the possible similarity solution for impulsive (constant-velocity) piston withdrawal in a constant-area duct. The equations of motion are, for homentropic flow,

$$\frac{\partial u}{\partial t} + u\frac{\partial u}{\partial x} + \frac{2}{\gamma - 1}c\frac{\partial c}{\partial x} = 0$$

$$\frac{2}{\gamma - 1}\frac{\partial c}{\partial t} + \frac{2}{\gamma - 1}u\frac{\partial c}{\partial x} + c\frac{\partial u}{\partial x} = 0$$

Assume a single independent variable $\eta = x/c_0 t$ and dependent variables $c = c_0\psi(\eta)$, $u = c_0\mu(\eta)$. Find the ordinary differential equations of motion in the form

$$(\cdot)\mu' + (\cdot)\psi' = 0 \qquad (\cdot)\psi' + (\cdot)\mu' = 0$$

and show that $[2/(\gamma - 1)]^2\psi'^2 = \mu'^2$.

eleven
analogs in compressible flow

11.1 Introduction

An *analogy* is a correspondence in some functional particulars between two systems; the systems are said to be *analogs* of each other. We could say, for example, that a glider is an analog of the gull.

A formal statement of meaning, more useful for our purposes, is that *two physical systems are analogs of each other if they are both described by the same nondimensional equations.* The analogous systems may or may not resemble each other physically. For example, certain acoustical enclosures can be represented by analogous electrical circuits with no visual resemblance.

Analogs are useful when experiments can be performed with relative convenience and economy on one system, the results being used to predict the performance of a second analogous system. It may seem peculiar that *acoustics* is generally studied in electrical engineering departments at universities; this follows at least partly from the convenience of the electrical-network analogy to acoustics.

A second advantage of some analogies lies in their accessibility to observation, such observation leading to better understanding of the analogous system. The shallow-water analogy to compressible flow is of this type. While the formal analogy itself is somewhat imperfect, direct observation of the shallow-water flow seems to be helpful in gaining a qualitative understanding of a compressible-gas flow.

11.2 Shallow-water flow

The motion of a shallow pool of liquid under the influence of gravity is analogous to the motion of a compressible fluid (a gas, in particular) in two dimensions. Surface waves on the liquid are analogous to density waves in the gas. The liquid motion is sometimes referred to as the *hydraulic analog*.

A convenient experimental basin can be formed by a nearly horizontal sheet of glass with suitable sides and elevated from the floor of the laboratory. Such a device is called a *water table*. Because experiments and splashing around in general are relatively inexpensive activities, the water table can be called the poor man's wind tunnel.

The Shallow-water Equations

The geometry for the problem is shown in Fig. 11.1. The free upper surface of the liquid is located at $x_3 = h(x_1,x_2,t)$. This surface is considered to be at constant atmospheric pressure P_a. The bottom surface is located at the horizontal plane $x_3 = 0$.

We assume that the liquid can be treated as incompressible, with negligible surface tension and constant viscosity. Then the momentum equations (1.66) for the three velocity components are

$$\rho \frac{Du_1}{Dt} + \frac{\partial P}{\partial x_1} = \mu \nabla^2 u_1 \tag{11.1}$$

$$\rho \frac{Du_2}{Dt} + \frac{\partial P}{\partial x_2} = \mu \nabla^2 u_2 \tag{11.2}$$

$$\rho \frac{Du_3}{Dt} + \frac{\partial P}{\partial x_3} = -\rho g + \mu \nabla^2 u_3 \tag{11.3}$$

According to the standard approximations in shallow-water theory, the motion is treated as inviscid ($\mu \to 0$) with negligible vertical velocity ($u_3 \to 0$) and negligible vertical acceleration ($Du_3/Dt \to 0$). The last approximation is justified if the vertical acceleration is small compared to gravity g. We will now accept these approximations but return later to scrutinize them.

The x_3 momentum equation now becomes simply the *hydrostatic* balance,

$$\frac{\partial P}{\partial x_3} = -\rho g$$

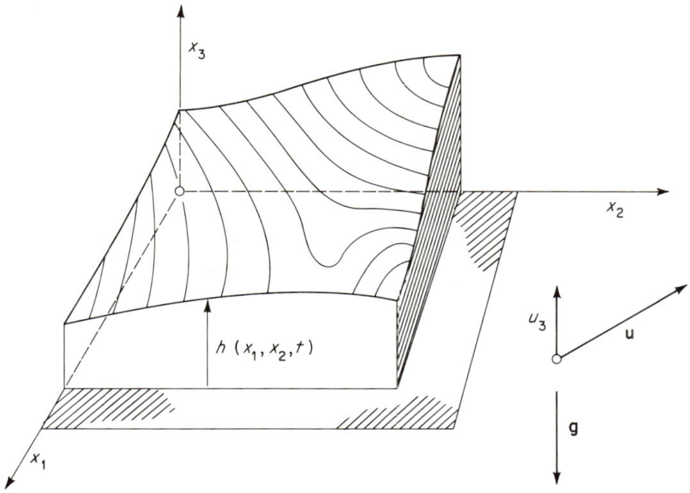

Figure 11.1
Geometry for shallow-water flow.

with the integral

$$P = \rho g(h - x_3) + P_a \tag{11.4}$$

where we have used the boundary condition $P = P_a$ at $x_3 = h(x_1,x_2,t)$. The pressure derivatives in (11.1) and (11.2) are now, from (11.4),

$$\frac{\partial P}{\partial x_1} = \rho g \frac{\partial h}{\partial x_1} \qquad \frac{\partial P}{\partial x_2} = \rho g \frac{\partial h}{\partial x_2}$$

Substituting these into (11.1) and (11.2) and taking the vector sum of the resulting equations yields

$$\frac{D\mathbf{u}}{Dt} + g\,\nabla h = 0 \tag{11.5}$$

where

$$\mathbf{u} \equiv \mathbf{e}_1 u_1 + \mathbf{e}_2 u_2 \tag{11.6}$$

and the operators D/Dt and ∇ now apply only to the *two-dimensional* x_1x_2 space; i.e.,

$$\nabla = \mathbf{e}_1 \frac{\partial}{\partial x_1} + \mathbf{e}_2 \frac{\partial}{\partial x_2}$$

$$\frac{D}{Dt} = \frac{\partial}{\partial t} + u_1 \frac{\partial}{\partial x_1} + u_2 \frac{\partial}{\partial x_2} = \frac{\partial}{\partial t} + \mathbf{u} \cdot \nabla$$

(This convention will hold in all of the following.) The approximate equation (11.5) is the momentum equation for shallow-water flow.

The free surface $x_3 = h$ moves upward, from the point of view of a fluid particle, at just the component velocity u_3,

$$u_3 = \frac{Dh}{Dt} = \frac{\partial h}{\partial t} + \mathbf{u} \cdot \nabla h \qquad \text{at } x_3 = h \tag{11.7}$$

This kinematical condition, which does not involve any approximations, can be obtained formally from a formula which was first given by Lord Kelvin (see *Serrin* [1959, sec. 8]). We shall make use of this in the following.

The continuity equation for the incompressible liquid is just

$$\frac{\partial u_1}{\partial x_1} + \frac{\partial u_2}{\partial x_2} + \frac{\partial u_3}{\partial x_3} = 0$$

In the notation adopted this is written

$$\nabla \cdot \mathbf{u} + \frac{\partial u_3}{\partial x_3} = 0 \tag{11.8}$$

Multiplying by dx_3 and integrating from $x_3 = 0$, where $u_3 = 0$, to $x_3 = h$, where u_3 is given by (11.7), yields

$$\frac{\partial h}{\partial t} + \mathbf{u}_h \cdot \nabla h + h\overline{\nabla \cdot \mathbf{u}} = 0 \tag{11.9}$$

where $\mathbf{u}_h$ means $\mathbf{u}(x_3 = h)$ and $\overline{\nabla \cdot \mathbf{u}}$ is the spatial average,

$$\overline{\nabla \cdot \mathbf{u}} \equiv \frac{1}{h}\int_0^h \nabla \cdot \mathbf{u}\, dy$$

Equation (11.9) is exact. At this point it is necessary to introduce the final shallow-water approximation, that $\mathbf{u}$ can be replaced by an "effective" average value; i.e., we put $\mathbf{u} = \mathbf{u}(x_1,x_2,t)$, leaving out the dependence on x_3. Then (11.9) becomes simply

$$\frac{\partial h}{\partial t} + \nabla \cdot (h\mathbf{u}) = 0 \tag{11.10}$$

Equations (11.5) and (11.10) are the desired shallow-water equations.

Analogy with Homentropic Two-dimensional Flow of a Gas

The momentum equation, appropriate to inviscid flow of a gas, is

$$\frac{D\mathbf{u}}{Dt} + \frac{1}{\rho}\nabla P = 0 \tag{11.11}$$

For homentropic flow $\nabla P = c^2 \nabla \rho$; for a perfect gas $c^2 = \gamma P/\rho$. With $P\rho^{-\gamma} = P_0\rho_0^{-\gamma} = \text{const}$, the momentum equation can then be written

$$\frac{D\mathbf{u}}{Dt} + \frac{\gamma P_0}{\rho_0{}^{\gamma}}\rho^{\gamma-2}\,\nabla\rho = 0$$

Then we can form a table of the continuity and momentum equations for shallow-water and compressible gas flow, Table 11.1. We now inquire under what conditions the corresponding equations become identical (in suitable nondimensional form). It is apparent from the two *continuity* equations that the height h is analogous to density ρ in the compressible flow, $h \leftrightarrow \rho$. The two *momentum* equations, however, are in the same form only if the coefficient of $\nabla\rho$ in the compressible-flow equation is constant; this requirement is satisfied only if the ratio of specific heats γ is equal to 2. Since $P \propto \rho^{\gamma}$ and $h \leftrightarrow \rho$, it follows that $h^2 \leftrightarrow P$. Similarly, $h \leftrightarrow T$.

Thus we have a formal and definite analog, with

$$\begin{aligned} \gamma &= 2 \\ h &\leftrightarrow \rho \\ h^2 &\leftrightarrow P \\ h &\leftrightarrow T \end{aligned} \tag{11.12}$$

Unfortunately, no real gas has $\gamma = 2$; the maximum possible value is of course $\gamma = 5/3$. Thus, even if the idealizing assumptions leading to Eqs.

Table 11.1

Shallow Water	Compressible Flow
$\frac{\partial h}{\partial t} + \nabla\cdot(h\mathbf{u}) = 0$	$\frac{\partial \rho}{\partial t} + \nabla\cdot(\rho\mathbf{u}) = 0$
$\frac{D\mathbf{u}}{Dt} + g\,\nabla h = 0$	$\frac{D\mathbf{u}}{Dt} + \frac{\gamma P_0}{\rho_0{}^{\gamma}}\rho^{\gamma-2}\,\nabla\rho = 0$

(11.5) and (11.10) could be justified, the hydraulic analogy to gas flow would still be approximate.

The reader may notice that the correspondence $h \leftrightarrow \rho$ is consistent with intuition. In a two-dimensional geometry, compression of the gas increases the value of ρ, or mass per unit area. Similarly, the mass per unit area of liquid on the water table is proportional to the height h.

Wave Speed of a Small Disturbance

The propagation speed of an infinitesimal gravity wave is the analog of the speed of sound.

We can obtain an expression for the wave speed by comparing the two momentum equations. Writing $\nabla P = c^2 \nabla \rho$ for the gas, the momentum equations (11.5) and (11.11) are

$$\frac{D\mathbf{u}}{Dt} + gh\frac{\nabla h}{h} = 0 \qquad \frac{D\mathbf{u}}{Dt} + c^2\frac{\nabla \rho}{\rho} = 0$$

Comparing the two equations, we find by induction that the shallow-water wave speed is given by $c^2 = gh$,

$$c = \sqrt{gh} \tag{11.13}$$

A more formal route to the same conclusion is to linearize the shallow-water equations (11.5) and (11.10) to obtain a wave equation. With a small disturbance $h - h_0 \ll h_0$ in a shallow pond, at rest with depth h_0 in the undisturbed state, linearization yields

$$\frac{\partial^2 h}{\partial t^2} - gh_0 \nabla^2 h = 0 \tag{11.14}$$

which is the classical wave equation. Thus, the "undisturbed" sound speed is $c_0 = \sqrt{gh_0}$.

Thus, according to the shallow-water theory, waves propagate at speed $c = \sqrt{gh}$. In ordinary experimental situations, such a speed is convenient for "real time" observation; e.g., with $g = 9.80\ \text{m/s}^2$ and $h = 1$ cm, we find $c \approx 0.3$ m/s. That is, wave motions or fluid velocities of the order of the wave speed are slow enough to permit direct observation.

The result (11.13) can be compared to more comprehensive predictions. The effects of surface tension and finite depth are included in a formula first given by Kelvin (1871):

$$c^2 = \left(\frac{g\lambda}{2\pi} + \frac{2\pi\sigma}{\rho\lambda}\right) \tanh \frac{2\pi h}{\lambda} \tag{11.15}$$

Here σ is the surface tension and λ the wavelength. The waves described by such a formula are said to be *dispersive* because the wave speed, or *phase velocity*, depends upon wavelength. Only in the case of relatively long wavelengths ($\lambda > 2\pi h$, $\rho g\lambda^2 \gg 4\pi^2\sigma$) does this formula go over to $c^2 = gh$. Experimental confirmation of (11.15) is given by *Walbridge and Woodward* [1970]. Surface tension or *capillary* waves can normally be observed on the water table as very short wavelength disturbances.

The Froude Number

In analogy to the Mach number, we define the *Froude number F* to be the ratio of the local flow speed to the local wave speed

$$F = \frac{u}{c} = \frac{u}{\sqrt{gh}} \tag{11.16}$$

All the usual formulas of steady gasdynamics can be applied to shallow-water flow, with $M \to F$, $P \to h^2$, and $\gamma = 2$. For example, Eqs. (5.58) and (5.59) can be written for gasdynamics as

$$\frac{P_0}{P} = \left(1 + \frac{\gamma - 1}{2} M^2\right)^{\gamma/(\gamma-1)} \tag{11.17}$$

The shallow-water version of this equation is

$$\frac{h_0}{h} = 1 + \tfrac{1}{2}F^2 \tag{11.18}$$

where h_0 is the stagnation, or reservoir depth.

Just as in gasdynamics, a critical-flow state occurs when the flow speed is equal to the wave speed, i.e., when $F = 1$. In hydraulics, it is conventional to call flows with $F < 1$ *subcritical* and flows with $F > 1$ *supercritical*. A more or less well-known example of supercritical flow is seen when water flows uphill, such a flow being possible because small disturbances do not propagate upstream.[1] Transition from subcritical to

[1] Like most other shallow-water experiments, this one can be performed at the kitchen sink.

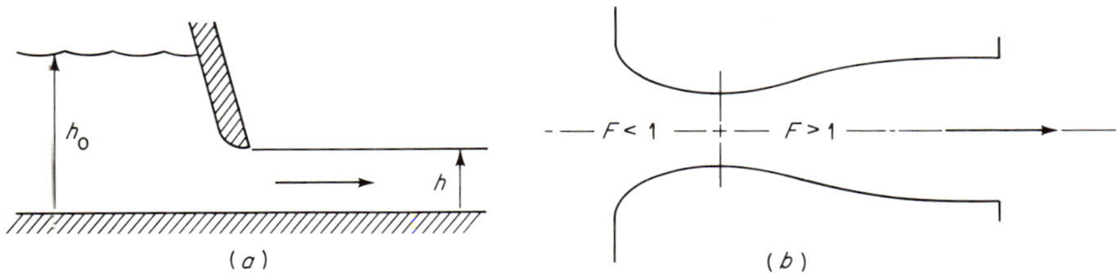

Figure 11.2
Flow from a reservoir: (*a*) under a sluice gate; (*b*) through a variable-width channel, with transition to supercritical flow.

supercritical flow can take place, for example, in a channel shaped like a Laval nozzle (see Fig. 11.2).

Hydraulic Jump

The analog of the shock wave is called the *hydraulic jump*, sketched in Fig. 11.3. This phenomenon can be observed, for example, where tapwater spreads on the horizontal bottom surface of a sink (the "shock front" in this case being circular), or at the base of certain spillways, or as a *tidal bore* moving upstream in certain coastal rivers[1] (see Figs. 11.4 and 11.5).

We now give the shock conditions for a hydraulic jump treated as a normal discontinuity. As before, we adopt the point of view of an observer moving with the discontinuity, with relative normal velocities w_1 and w_2 and corresponding surface elevations h_1 and h_2. The appropriate control volume, which is presumed to be short enough to permit storage terms to be neglected, is shown in Fig. 11.6.

With the relative velocity w assumed uniform at sections 1 and 2, conservation of mass is expressed by

$$w_1 h_1 = w_2 h_2$$

[1] Excellent photographs of tidal bores are given in *Tricker* [1965]. A particularly spectacular form of the hydraulic jump is the front of a powder avalanche in snow; there is reason to expect that the "fluid" in this case is compressible, however.

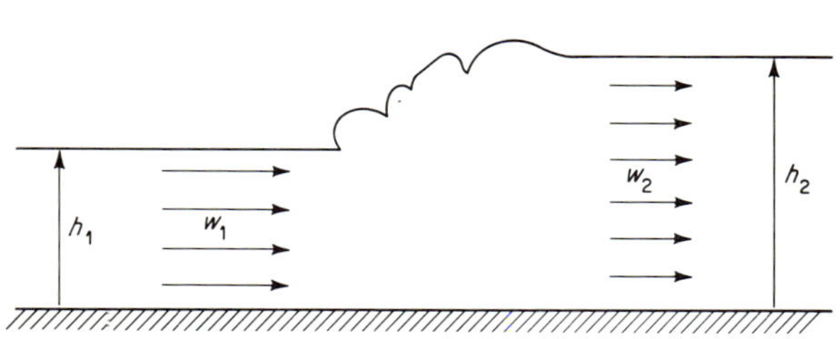

Figure 11.3
Hydraulic jump.

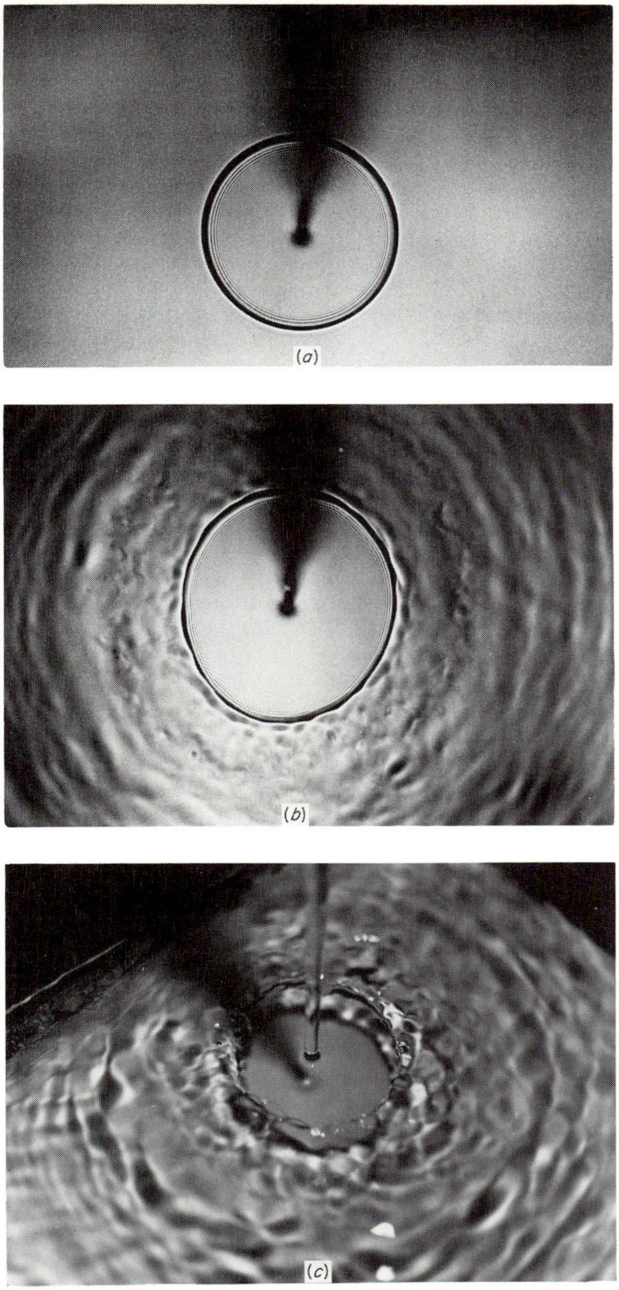

Figure 11.4
Kitchen-sink hydraulic jump: (*a*) shadowgraph of laminar jump; (*b*) shadowgraph of transition jump; and (*c*) view from above, turbulent jump.

Figure 11.5
Petitcodiac tidal bore, New Brunswick.

This can be rewritten in the form

$$\frac{F_{1n}}{F_{2n}} = \left(\frac{h_2}{h_1}\right)^{3/2} \tag{11.19}$$

where $F_{1n} \equiv w_1/c_1$ and $F_{2n} \equiv w_2/c_2$.

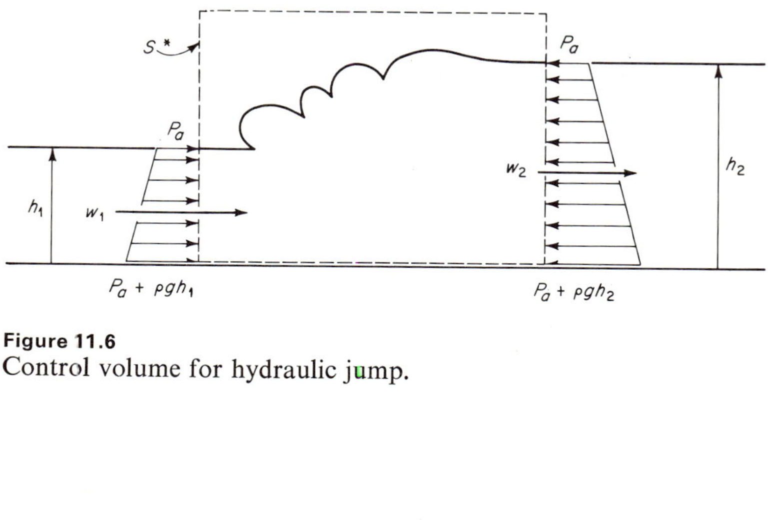

Figure 11.6
Control volume for hydraulic jump.

The momentum balance requires calculation of the net force acting horizontally on the control surface. Consistent with the shallow-water model, the pressure distribution is taken to be hydrostatic. The *net* force acting to the right is then, per unit width,

$$\int_0^{h_1} \rho g x_3 \, dx_3 - \int_0^{h_2} \rho g x_3 \, dx_3 = \frac{\rho g}{2}(h_1^2 - h_2^2)$$

The momentum balance is then

$$\frac{\rho g}{2}(h_1^2 - h_2^2) = \rho w_2^2 h_2 - \rho w_1^2 h_1$$

or

$$w_1^2 h_1 + \tfrac{1}{2} g h_1^2 = w_2^2 h_2 + \tfrac{1}{2} g h_2^2$$

With (11.19) this can be rewritten

$$\frac{F_{1n}^2 + \frac{1}{2}}{F_{1n}^{4/3}} = \frac{F_{2n}^2 + \frac{1}{2}}{F_{2n}^{4/3}} \tag{11.20}$$

Again taking the point of view that the upstream conditions, for example, F_{1n}, are fixed and known, we solve for the downstream conditions. Equations (11.19) and (11.20) then allow the calculation of h_2/h_1 and F_{2n} as functions of F_{1n}; that is, the system is determinate. Thus, no energy condition is required. With the help of (11.19), Eq. (11.20) can be rewritten as a cubic equation in h_2/h_1,

$$\left(\frac{h_2}{h_1} - 1\right)\left[\left(\frac{h_2}{h_1}\right)^2 + \left(\frac{h_2}{h_1}\right) - 2F_{1n}^2\right] = 0$$

The first root is $h_2/h_1 = 1$, corresponding to no discontinuity. The second root is

$$\frac{h_2}{h_1} = -\tfrac{1}{2} + \sqrt{\tfrac{1}{4} + 2F_{1n}^2} \tag{11.21}$$

(the third root is negative and has no apparent physical significance).

While (11.21) has solutions $h_2 < h_1$ for $F_{1n} < 1$, we rule out such solutions on essentially the same grounds that rarefaction shocks were ruled out; i.e., the discontinuity must be dissipative.[1] Thus,

$$h_2 \geq h_1 \qquad F_{1n} \geq 1 \tag{11.22}$$

[1] In thermodynamical terms, the flow across a jump is irreversible. This conclusion will seem inescapable to anyone who has observed a strong hydraulic jump, e.g., at the base of a spillway.

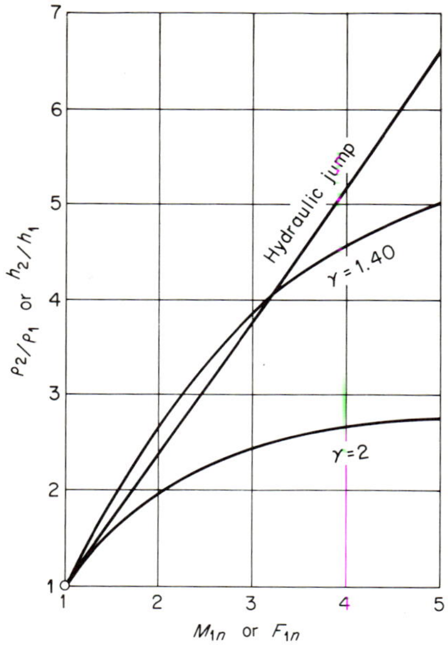

Figure 11.7
Comparison between h_2/h_1 for a hydraulic jump and ρ_2/ρ_1 for a shock.

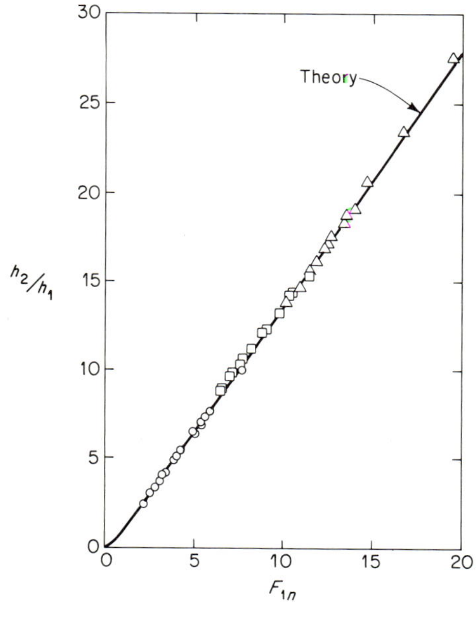

Figure 11.8
Height ratio vs. Froude number. Comparison of experiment and theory [Eq. (11.21)]. Various experimental points correspond to different experiments.

The relative incoming flow must be supercritical. Similarly, the outgoing flow is subcritical, $F_{2n} \leq 1$.

Unfortunately, the result (11.21) is *not* identical to the expression for the density ratio across a normal shock in a perfect gas (even for $\gamma = 2$). From (7.34), the latter expression is

$$\frac{\rho_2}{\rho_1} = \frac{\gamma + 1}{(\gamma - 1) + 2/{M_{1n}}^2} \tag{11.23}$$

Only in the limit $M_{1n} \to 1$, $F_{1n} \to 1$ are the two expressions identical, as they must be, for $\gamma = 2$. For stronger shocks the correspondence is actually better for $\gamma = 1.40$, as shown in Fig. 11.7.

The prediction of Eq. (11.21) is remarkably well confirmed by the experiments of *Peterka* [1963], as shown in Fig. 11.8. These experiments, on a relatively large physical scale, were carried out with a view to the dissipation of energy downstream of large spillways.

The Shallow-water Approximations

We will briefly discuss the nature of the somewhat drastic approximations used in obtaining the shallow-water equations (11.5) and (11.10). Essentially, it has been assumed that vertical accelerations are small, that the viscous force exerted by the horizontal bottom is negligible, and that the vertical variation of $\mathbf{u}$ can be neglected in the continuity equation. We discuss these conditions in turn.

The continuity equation (11.9) is rewritten in the exact form

$$u_3 = h\overline{\nabla \cdot \mathbf{u}}$$

Let u have a variation comparable to its own magnitude over a distance λ. Then we estimate $u_3 \sim h(u/\lambda)$ or

$$\frac{u_3}{u} \sim \frac{h}{\lambda} \tag{11.24}$$

If the fluid motion is wavelike, waves of length λ have a time period λ/c. The vertical acceleration is then of order

$$\frac{Du_3}{Dt} \sim \frac{uh/\lambda}{\lambda/c} = uc\,\frac{h}{\lambda^2}$$

For strong (nonlinear) wave motion, u is of order c, which leads to

$$\frac{Du_3}{Dt} \sim g\,\frac{h^2}{\lambda^2}$$

The vertical acceleration is thus negligible if

$$h^2 \ll \lambda^2$$

This is the basic condition for the validity of the shallow-water model, sometimes written $\lambda \gg h$. As an interesting application, the shallow-water assumptions are fairly well satisfied for *tsunami* ("tidal waves") in the Pacific Ocean. Over large regions of the ocean, the bottom is relatively uniform in depth; between the Hawaiian and Aleutian Islands, for example, the depth is about 18,000 ft. Since the waves in question have very large wavelengths, the shallow-water theory works fairly well in such regions.

The effect of viscous forces on the motion is mainly contained in a modified version of (11.5):

$$\frac{D\mathbf{u}}{Dt} + g\,\nabla h = \nu\frac{\partial^2\mathbf{u}}{\partial {x_3}^2} \tag{11.25}$$

which can be derived from (11.1) to (11.3). The derivative $\partial u/\partial x_3$ is estimated to be of order u/h (see Fig. 11.9). Since there is negligible tangential stress at the free surface, $\partial u/\partial x_3 = 0$ there and the second derivative is of order $\partial^2 u/\partial {x_3}^2 \sim u/h^2$. From (11.25), the viscous term is then negligible only if $g\,\nabla h \gg \nu u/h^2$. Putting this condition in non-dimensional form gives

$$\nabla h \gg \frac{F^2}{\mathrm{Re}_h}$$

where $\mathrm{Re}_h \equiv hu/\nu$. The shallow-water flows commonly of interest for the analog have depths of the order of 1 cm and velocities corresponding to Froude number of order unity; for such cases, assuming the liquid to be

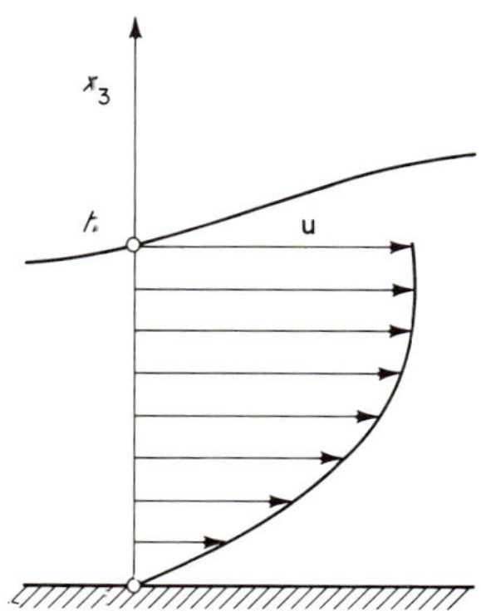

Figure 11.9

water, $\mathrm{Re}_h \sim 3{,}000$ and $F \sim 1$, and the condition for negligible viscous effect is

$$\nabla h \gg \frac{1}{3{,}000}$$

In the modeling of streaming flows, it is normal practice (and a practical necessity!) to incline the water table slightly, so that **g** has a small component in the direction of motion, thus compensating for the effect of friction.

The main approximation involved in the continuity equation appears to be setting $\overline{\nabla \cdot \mathbf{u}} = \nabla \cdot \bar{\mathbf{u}}$. Suppose the motion is just in the x_1 direction; then application of the Leibniz rule for differentiation shows that this is justified only if

$$u \frac{\partial h}{\partial x} \ll h\overline{\nabla \cdot \mathbf{u}}$$

Since $h\overline{\nabla \cdot \mathbf{u}} = u_3$ and the left-hand side is of order u_3 in *steady* flow, the above approximation appears questionable at best.

Recent and detailed treatments of shallow-water flow are given by *Loh* [1969] and *Rajaratnam* [1967].

EXAMPLE 11.1 DAM-BREAK PROBLEM

A plane vertical dam located at $x = 0$ (Fig. 11.10) breaks at some instant. Find the resulting water motion.

This is the analog of gas expansion into a vacuum (impulsive piston withdrawal at infinite speed). The dam is assumed to disappear instantaneously at $t = 0$.

The continuity (11.10) and momentum (11.5) equations for shallow-water flow in one dimension are respectively

$$\frac{\partial h}{\partial t} + u\frac{\partial h}{\partial x} + h\frac{\partial u}{\partial x} = 0 \tag{11.26}$$

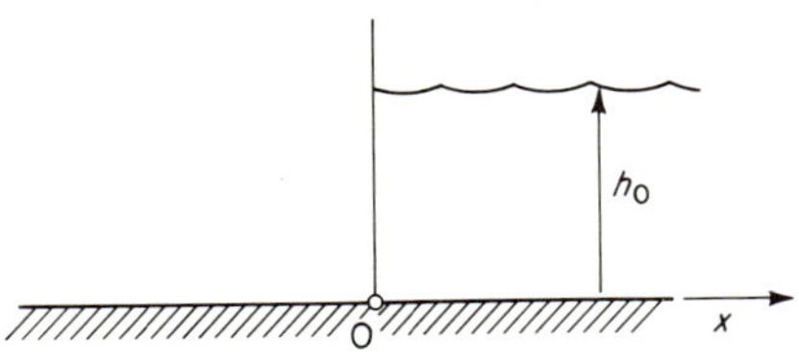

Figure 11.10

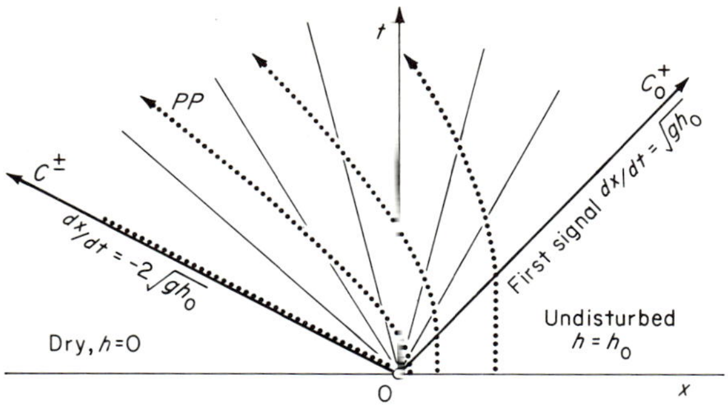

Figure 11.11
Wave diagram for dam-break problem.

and

$$\frac{\partial u}{\partial t} + u\frac{\partial u}{\partial x} + g\frac{\partial h}{\partial x} = 0 \tag{11.27}$$

Equation (11.26) can be made dimensionally homologous to (11.27) by multiplying through by $\sqrt{gh}/h$. Noting that $(\sqrt{gh}/h)\,dh = d(2\sqrt{gh})$, addition and subtraction of the two equations then gives the characteristic form

$$\left[\frac{\partial}{\partial t} + (u \pm \sqrt{gh})\right](u \pm 2\sqrt{gh}) = 0 \tag{11.28}$$

These equations are equivalent to (8.13) written for a perfect gas with $\gamma = 2$, viz.,

$$\left[\frac{\partial}{\partial t} + (u \pm c)\frac{\partial}{\partial x}\right](u \pm 2c) = 0 \tag{11.29}$$

where c is replaced by $\sqrt{gh}$ for the shallow-water flow.

The solution is analogous to that of the gas expanding into a vacuum, shown in Fig. 8.19*b*. In the present problem waves coming *from* the right originate in a uniform region with $h = h_0$, $u = 0$. We therefore have simple C^+ waves with

$$u - 2\sqrt{gh} = -2\sqrt{gh_0} \qquad \text{everywhere} \tag{11.30}$$

The corresponding wave diagram is shown in Fig. 11.11. The slope of a C^+ characteristic is

$$\frac{dx}{dt} = u + c = u + \sqrt{gh} \tag{11.31}$$

and the characteristics are necessarily straight lines intersecting at the origin.

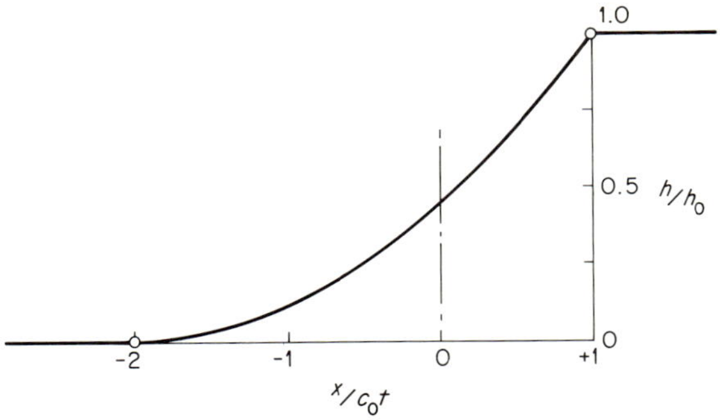

Figure 11.12
Wave profile at time t.

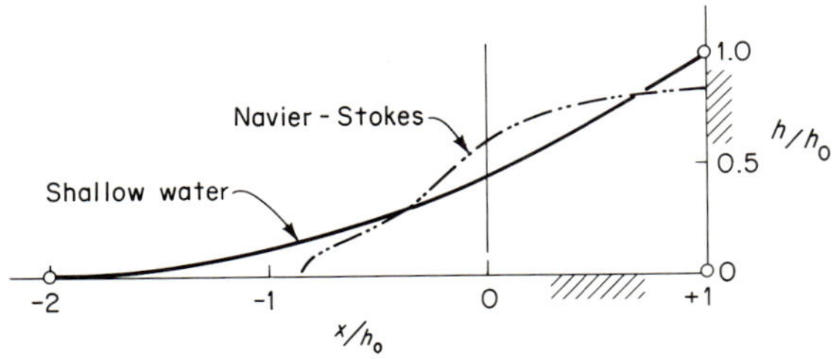

Figure 11.13
Shallow-water profile compared with that computed from the Navier-Stokes equation by *Harlow and Welch* [1965]. The time is $t = h_0/c_0$.

With (11.30) this gives the spatial profile of the wave,

$$\frac{h}{h_0} = \left(\frac{2 + \dfrac{1}{\sqrt{gh_0}}\dfrac{x}{t}}{3}\right)^2 \tag{11.32}$$

as shown in Fig. 11.12. The corresponding velocity distribution is, from (11.30)

$$\frac{u}{\sqrt{gh_0}} = -\frac{2}{3}\left(1 - \frac{1}{\sqrt{gh_0}}\frac{x}{t}\right) \tag{11.33}$$

A comparison between the above profile and one calculated numerically from the full Navier-Stokes equations is shown in Fig. 11.13; the numerical calculation agrees quite well, of course, with experiments. The comparison between the shallow-water profile and that calculated numerically is not very favorable; this is partly in consequence of the (necessarily) early time at which comparison is made, when the shallow-water assumptions are violated (a comparison at later times is not possible, because a fixed back wall at $x = h_0$ is incorporated into the numerical calculation).

11.3 Traffic flow

By a bold assumption, a moving stream of traffic may be treated as a fluid continuum. Specifically, we consider the motion of cars along a long single-lane road (no passing) according to a fluid-mechanical model.

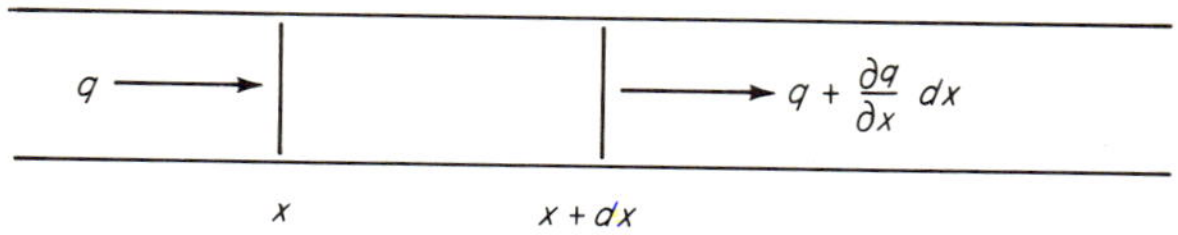

Figure 11.14

The elemental treatment given in the next few pages is that of *Lighthill and Whitham* [1955]. The particular problem is one example of *kinematic waves* or *waves of continuity* (*Wallis* [1969]).

Let the traffic density (cars per mile, say) on a single-lane road be ρ and the traffic flow rate (number of cars passing a fixed point per unit time) be q. If x is the distance coordinate along the road and t is time, the corresponding functions are $\rho(x,t)$ and $q(x,t)$.

The condition that cars are conserved can be expressed as follows. Let x and $x + dx$ be fixed stations; the net outflow rate from the control volume thus formed (Fig. 11.14) is $\partial q/\partial x\, dx$. Equating this to the rate of decrease of the number of cars contained, $-\partial/\partial t\, \rho\, dx$, yields the continuity equation

$$\frac{\partial \rho}{\partial t} + \frac{\partial q}{\partial x} = 0 \tag{11.34}$$

This is, of course, the analog of the one-dimensional continuity equation in gasdynamics. This is perhaps clearer if we write

$$q = \rho u \tag{11.35}$$

where u is the traffic velocity. Then (11.34) is formally identical to the gasdynamic equation (8.1), with A = const.

An additional equation is now required for the unknowns ρ and q. Normally, in fluid mechanics, this is a momentum equation. An alternative possibility is to assume that there is a known function of the form

$$q = q(\rho) \tag{11.36}$$

That is, the local flow rate is fixed by the local density. Some experimental measurements indicate a function of the type shown in Fig. 11.15. The occurrence of no flow under conditions of no traffic and jammed traffic, respectively, can be confirmed by observation![1] It is, however, a strong

[1] Some experiments mentioned by Lighthill and Whitham indicate a value of $q_{\max}$ in the range of 3,000 to 5,000 cars per hour. The value of ρ under jammed conditions can be inferred from the size of the cars.

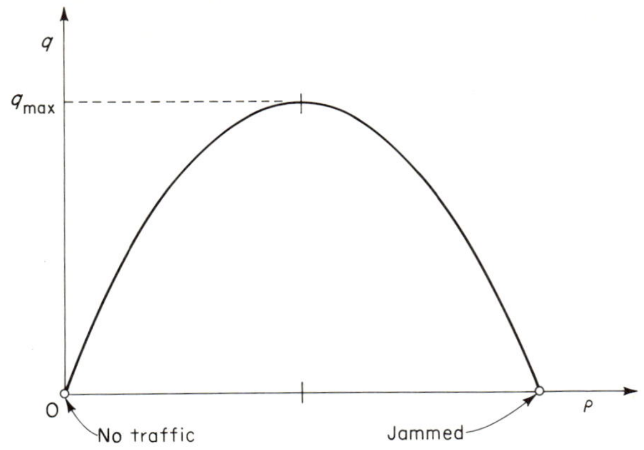

Figure 11.15
The flow-vs.-density curve.

assumption that (11.36) applies even in unsteady traffic conditions. We take the assumption, nonetheless, justifying it partly by the instructive results to follow.

With (11.36), the continuity equation (11.34) can be rewritten

$$\frac{\partial \rho}{\partial t} + c(\rho)\frac{\partial \rho}{\partial x} = 0 \tag{11.37}$$

where

$$c(\rho) \equiv \frac{dq}{d\rho} \tag{11.38}$$

Equation (11.37) represents a complete solution, for any given initial conditions, for the traffic-flow problem, as we now explain. The equation can be rewritten

$$\left(\frac{\partial}{\partial t} + c\frac{\partial}{\partial x}\right)\rho = 0$$

which is analogous, for example, to the first equation (8.13) for one-dimensional gasdynamics or to the statement $Ds/Dt = 0$ for isentropic flow. Taking advantage of our experience with these cases, we find the solution to be simply

$$\rho = \text{const on characteristic lines } \frac{dx}{dt} = c(\rho) \tag{11.39}$$

That is, we have *simple waves* everywhere. The characteristics are necessarily straight lines in the xt plane, because ρ is constant along any given characteristic and c is a unique function of ρ.

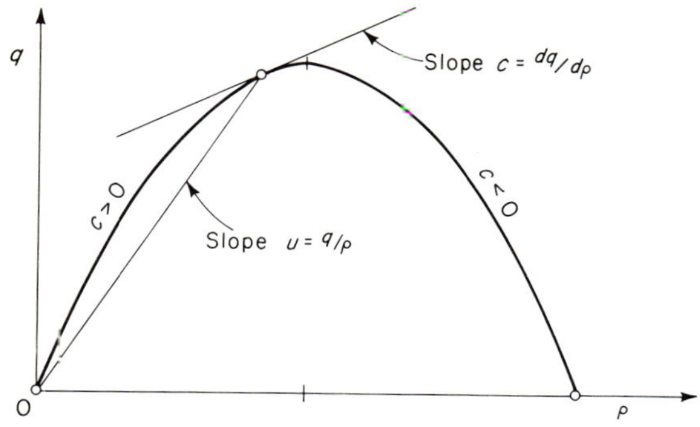

Figure 11.16
Showing wave velocity and flow velocity.

The wave velocity $c(\rho) = dq/d\rho$ can be shown graphically as a tangent on the flow-density curve, Fig. 11.16. It should be noted that c is here the wave velocity with respect to a stationary observer (*not* with respect to the "fluid") and can be positive or negative. The traffic-flow velocity $u = q/\rho$ is represented by the slope of a chord from the origin. If the curve $q(\rho)$ is everywhere convex as shown, $c < u$ everywhere and waves always propagate backward with respect to the traffic.[1]

The *shock conditions* are particularly simple in this case. The conservation of cars (continuity) condition across a shock is just $\rho_1 w_1 = \rho_2 w_2$, where w_1 and w_2 are the velocities relative to the shock front, as before. With the (absolute) shock velocity U, $w_1 = U - u_1$ and $w_2 = U - u_2$ (see Fig. 11.17), $\rho_1 w_1 = \rho_2 w_2$ is rewritten

$$\rho_1(U - u_1) = \rho_2(U - u_2)$$

which gives

$$U = \frac{[\rho u]}{[\rho]} \tag{11.40}$$

[1] There is a nice analogy here with the mechanics of dispersive waves. The velocity u of individual motorcars is analogous to the *phase velocity*, and the wave velocity c is analogous to the *group velocity*. The interested reader can verify that the analogy is in fact precise, with q corresponding to frequency ω and ρ corresponding to wave number k. (A further discussion of phase velocity and group velocity is given in Sec. 11.4.)

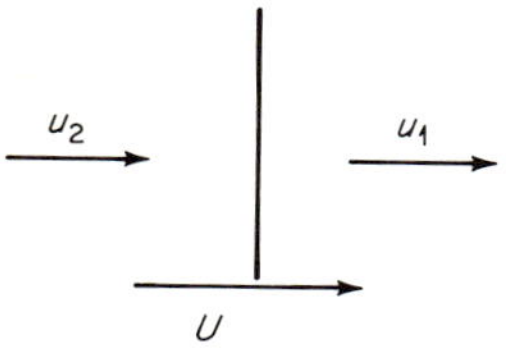

Figure 11.17

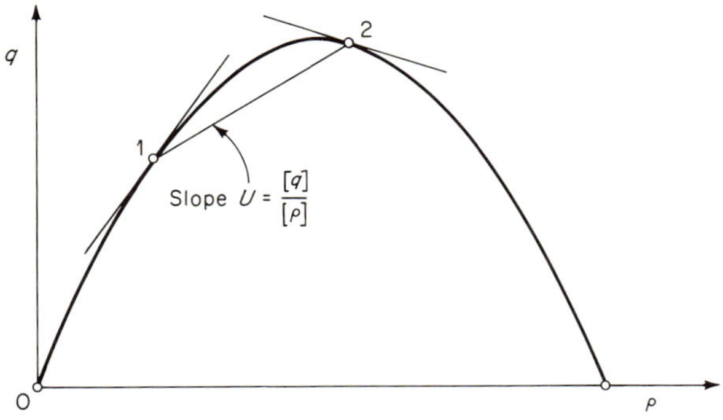

Figure 11.18
Shock path 1 → 2 on flow-density curve.

This equation also applies, of course, to one-dimensional fluid flow. In the traffic notation already adopted, this is written

$$U = \frac{[q]}{[\rho]} \tag{11.41}$$

This is the only shock condition needed. For given upstream condition ρ_1 and given U, the downstream conditions are uniquely fixed. A geometrical interpretation of (11.41) is shown in Fig. 11.18. Note that U is the slope of the chord $1 \to 2$, and that $U = c$ for infinitesimal-strength shocks.

The shocks mentioned above are not intended to be discontinuities of the type shown in Fig. 7.3! That is, individual cars do not actually collide but pass through a quasi-discontinuity in density.

A rarefaction shock is not precluded by (11.41) or by any thermodynamical argument. Rather, we can demonstrate that an initial rarefaction discontinuity will spread out, just as in the case of gasdynamics (as will be shown in the following example).

The impossibility of rarefaction shocks can be put more definitely, as follows. For a shock wave of finite strength to persist, infinitesimal (acoustic) disturbances originating upstream and downstream of the shock *must travel toward the shock front*. In this way, all disturbances tend to be concentrated at the shock front. If, however, the infinitesimal disturbances traveled *away* from the shock front, the shock itself would tend to spread out with them. In gasdynamics this argument is reflected in the stability condition

$$M_{1n} \geq 1 \geq M_{2n}$$

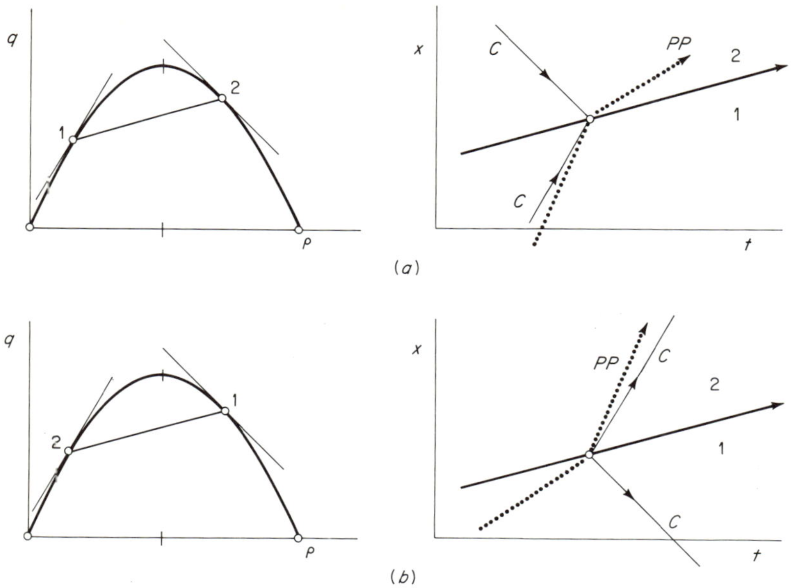

Figure 11.19
(*a*) Stable compression shock with upstream and downstream characteristics pointing into the shock front; (*b*) unstable rarefaction shock with upstream and downstream characteristics pointing away from the shock front.

i.e., that upstream and downstream sound waves are swept toward the shock front. In the present case, the corresponding stability condition is written

$$c_1 \geq U \geq c_2 \tag{11.42}$$

which precludes rarefaction shocks ($\rho_2 < \rho_1$), as illustrated in Fig. 11.19. (In constructing the figure it is convenient to interchange the normal placement of x and t coordinates in the wave diagram, so that lines with known slope dx/dt in the flow-density curve will appear as parallel lines with slope dx/dt in the wave diagram; e.g., the shock front on the wave diagram is drawn parallel to the chord $1 \rightarrow 2$ on the flow-density curve.)

A more formal discussion of the "stability" of a shock front from the above point of view is given by *Landau and Lifshitz* [1959, chap. 9].

EXAMPLE 11.2 WHEN THE LIGHT FINALLY TURNS GREEN

A very long string of traffic is backed up at a traffic light. Find the resulting motion when the light finally turns green.

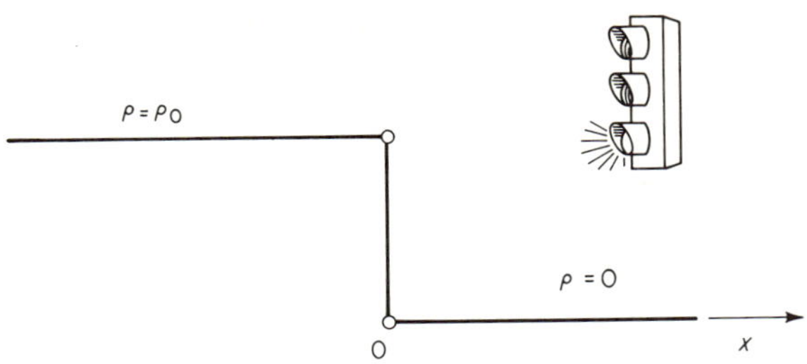

Figure 11.20

The density for $x < 0$ is ρ_0 (jammed density) and for $x > 0$ is zero, with an initial discontinuity at the origin, as shown in Fig. 11.20. By virtue of (11.37), only simple waves with characteristics $dx/dt = c$ will occur. The characteristics will necessarily intersect at the origin, where the density is discontinuous and assumes all possible values $\rho_0 \geq \rho \geq 0$, as shown in Fig. 11.21. The motion is thus still another example of a *centered rarefaction.*

Suppose that the flow-density curve is parabolic; then $q(\rho)$ can be represented by the equation

$$q_{\max} - q = \rho_0 c_0 \left(\frac{\rho}{\rho_0} - \frac{1}{2}\right)^2 \tag{11.43}$$

where $q_{\max} = \rho_0 c_0/4$. The wave speed $c = dq/d\rho$ is then, by differentiation,

$$c = 2c_0\left(\frac{1}{2} - \frac{\rho}{\rho_0}\right) \tag{11.44}$$

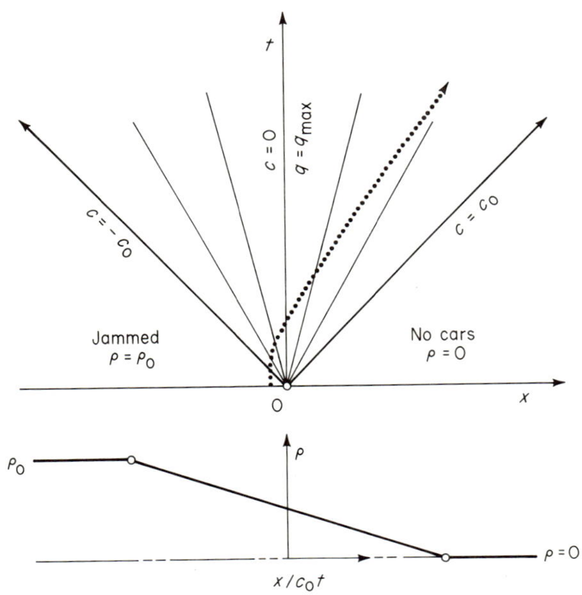

Figure 11.21
Wave diagram and density distribution.

Thus $c_{max} = c_0$ at $\rho = 0$ and $c_{min} = -c_0$ at $\rho = \rho_0$. Setting the slope of the characteristics $c = dx/dt = x/t$ gives the density field

$$\frac{\rho}{\rho_0} = \frac{1}{2}\left(1 - \frac{x}{c_0 t}\right) \tag{11.45}$$

within the fan, as shown in the figure. With $u = q/\rho$, the velocity is then found to be

$$\frac{u}{c_0} = \frac{1}{2}\left(1 + \frac{x}{c_0 t}\right) \tag{11.46}$$

The above results bear at least a qualitative resemblance to the observed flow of traffic away from a stoplight. It is interesting to note that the traffic crossing the intersection ($x = 0$) has $c = 0$ and therefore $q = q_{max}$; that is, the traffic flow across the origin has the maximum possible value.

Additional problems which are essentially gasdynamic in nature are discussed by *Lighthill and Whitham* [1955]. In particular the fate of a *traffic hump*, or region of excess density, is analogous to the motion of a *density hump* in a compressible fluid (as discussed in Sec. 8.5), complete with the eventual formation of a shock wave.

The traffic theory given here should not, of course, be considered too solemnly. It is offered primarily as an example of *kinematic waves* and for whatever entertainment or insight it may lend to gasdynamics.

11.4 The electroacoustical analogy

Acoustical motions in ducts and cavities can be usefully modeled by corresponding electrical networks. To introduce this analogy, and its limitations, we first consider a simple mass-spring analog for wave motion.

A Mechanical Analog for Plane Acoustic Waves

A one-dimensional lattice of springs and discrete masses is shown in Fig. 11.22. We will investigate the conditions under which the motion of this network is analogous to the one-dimensional acoustical motion of a compressible fluid. A model of this kind was used by *Newton* [1686, bk. 2, prop. 47] in his famous calculation of the speed of sound and was suggested as a basis for numerical calculations of compressible fluid motions by *von Neumann* [1944]. An extensive discussion is given by *Brillouin* [1946].

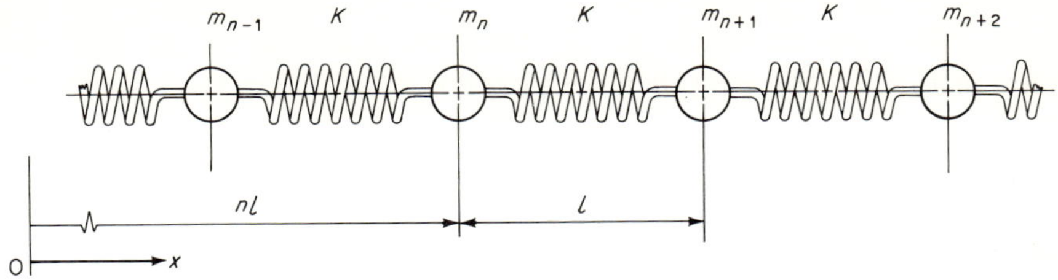

Figure 11.22
One-dimensional mechanical network.

The appropriate values for the spring constant K and mass m can easily be found. Suppose the fluid to be modeled is a *perfect gas*; the elasticity of a prism of fluid of length l and cross-sectional area A is shown schematically in Fig. 11.23. For isentropic state changes, $PV^\gamma = P(Al)^\gamma = \text{const}$. Then logarithmic differentiation gives

$$\frac{dP}{dl} = -\frac{\gamma P}{l}$$

The spring constant K is the negative rate of change of force with distance, $K = -d(PA)/dl$ or

$$K = \frac{\gamma PA}{l} \tag{11.47}$$

The expression for an *arbitrary* simple fluid is obtained by setting $dP = c^2\, d\rho$; with $\rho V = \text{const}$, this gives

$$K = \frac{\rho c^2 A}{l} \tag{11.48}$$

The corresponding fluid mass is just ρV,

$$m = \rho A l \tag{11.49}$$

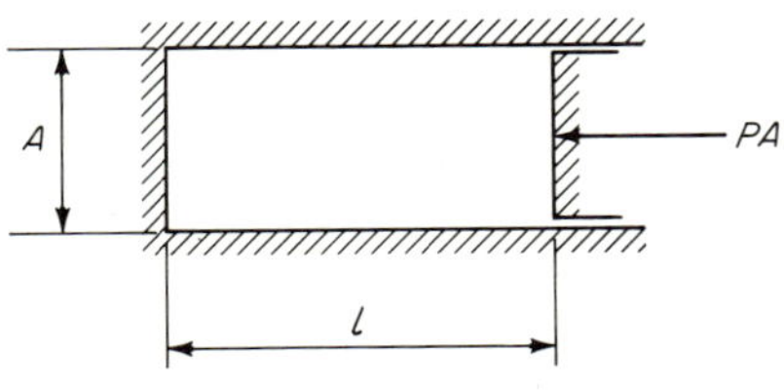

Figure 11.23
The fluid spring.

Let the displacement of the nth discrete mass from its equilibrium position in Fig. 11.22 be ξ_n; then the position x_n of that mass is given by

$$x_n = nl + \xi_n$$

Similarly, $x_{n+1} = (n + 1)l + \xi_{n+1}$, and so on. The equation of motion of the nth mass is, from $ma = F$,

$$m\frac{d^2\xi_n}{dt^2} = K(\xi_{n-1} - \xi_n) - K(\xi_n - \xi_{n+1})$$

where $\xi_{n-1} - \xi_n$ and $\xi_n - \xi_{n+1}$ are respectively the compressions of the left-hand and right-hand springs. Simplifying gives

$$m\frac{d^2\xi_n}{dt^2} = K(\xi_{n-1} + \xi_{n+1} - 2\xi_n) \tag{11.50}$$

A sinusoidal progressive wave in acoustics could be written

$$\xi = B\sin\left[\frac{2\pi}{\lambda}(x - ct)\right] = B\sin(kx - \omega t)$$

where $k \equiv 2\pi/\lambda$ is the wave number and $\omega = 2\pi\nu$ is the angular frequency. With $x = nl$ the (equilibrium) position of the *discrete* masses, the equivalent expression for the present problem would be

$$\xi = B\sin(knl - \omega t)$$

In other words, $\xi(x,t)$ is replaced by $\xi(n,t)$. We will use the complex form

$$\xi_n = Be^{i(knl - \omega t)} \tag{11.51}$$

where as usual only the *real part* of complex solutions will be taken. Such a motion could be generated in the lattice by imposing a sinusoidal motion on the first mass, $n = 0$, with amplitude B.

It is desired to determine whether the equation of motion (11.50) will permit such a progressive-wave solution, i.e., a sinusoidal wave traveling to the right with speed c. Substituting (11.51) into the equation of motion and canceling terms yields

$$-\frac{m\omega^2}{K} = e^{-ikl} + e^{+ikl} - 2$$

Retaining only the real part of the right-hand side gives, after trigonometric reduction,

$$\omega^2 = \frac{4K}{m}\sin^2\frac{kl}{2} \tag{11.52}$$

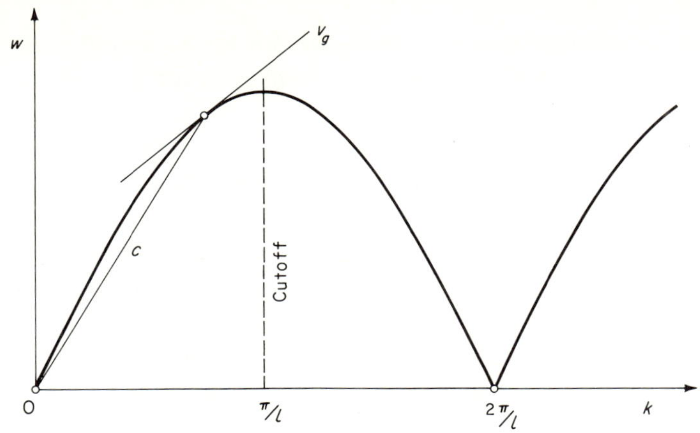

Figure 11.24
Dispersion curve for the one-dimensional lattice.

Thus (11.51) is a solution, and progressive sinusoidal waves will occur, with frequency ω given by this equation. The positive root (of physical interest) is

$$\omega = 2\sqrt{\frac{K}{m}}\left|\sin\frac{kl}{2}\right| \tag{11.53}$$

This function, of the form $\omega = \omega(k)$, is called a *dispersion relation* and is plotted in Fig. 11.24.

The *phase velocity* c, or "speed of sound," for a sinusoidal wave of frequency ν is given by the standard formula

$$c = \nu\lambda = \frac{\omega}{k}$$

i.e., is the slope of the chord in Fig. 11.24. By (11.53) this is

$$c = l\sqrt{\frac{K}{m}}\frac{|\sin(kl/2)|}{kl/2} \tag{11.54}$$

For $kl \ll 1$ (equivalently, $\lambda \gg l$) this becomes simply

$$c = l\sqrt{\frac{K}{m}} \tag{11.55}$$

Putting in the spring constant and mass of the fluid from (11.48) and (11.49) shows that this corresponds exactly to the normal speed of sound in the fluid. Thus, *the analogy is precise provided that the wavelength λ is large compared to the spacing l of the discrete elements*, i.e., for

$$\lambda \gg l \tag{11.56}$$

Although it is not of immediate interest for the acoustic analogy, the behavior of the system at higher wave numbers is interesting. A group of waves will travel not at the phase velocity $c = \omega/k$ but rather at the *group velocity* $v_g = d\omega/dk$, shown in Fig. 11.24.[1] When $kl \ll 1$, consistent with (11.56), c and v_g assume a common stationary value, and there is no distinction between the two wave speeds, as is the case in ordinary acoustics. However, when $k = \pi/l$, the group velocity is *zero* and the solution (11.51) assumes the form

$$\xi_n = Be^{i(n\pi - \omega t)}$$

which has the real part

$$\xi_n = B(-1)^n \cos \omega t \tag{11.57}$$

According to this solution, every pair of discrete masses oscillates as an independent dipole, each particle nodding and bowing as shown in Fig. 11.25. The motion no longer looks like a traveling wave but looks instead like a standing wave. Oscillation at higher frequencies may be shown to decay exponentially with distance; hence, the frequency corresponding to $k = \pi/l$ is called the *cutoff frequency*

$$\nu_c = \frac{1}{\pi}\sqrt{\frac{K}{m}}$$

The lattice itself is called a *low-pass filter*, since it passes only waves of frequency below ν_c.

[1] For a discussion of these distinct wave speeds, see *Brillouin* [1946, chap. 5] or *Fay* [1965, chap. 4]. In music, a group of waves may occur as a *beat*.

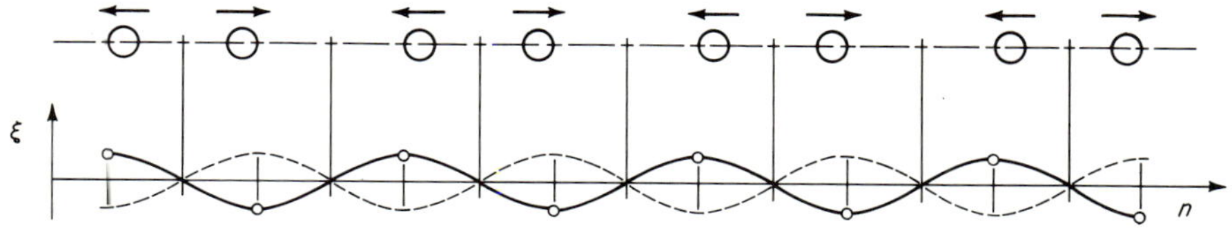

Figure 11.25
Oscillation at the cutoff frequency.

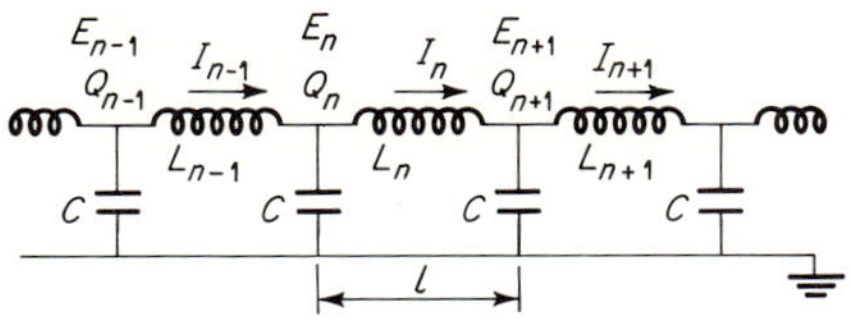

Figure 11.26
Transmission-line circuit.

An Electrical Analog for Plane Acoustic Waves

The electric circuit which is analogous to the one-dimensional mechanical circuit just discussed, i.e., Fig. 11.22, is shown in Fig. 11.26. This is just the circuit for an electric transmission line, where the distributed inductance and capacitance of the line are replaced by lumped parameters, or *discrete* circuit elements. To show that there is indeed an analogy, it is only necessary to show that the systems obey the same differential equations.

We will write the differential equation for the current I_n in a typical inductance L_n. The voltage across the inductance is

$$L\frac{dI_n}{dt} = E_n - E_{n+1} = \frac{Q_n}{C} - \frac{Q_{n+1}}{C}$$

Differentiating

$$L\frac{d^2I_n}{dt^2} = \frac{1}{C}\frac{dQ_n}{dt} - \frac{dQ_{n+1}}{dt}$$

But $dQ_n/dt = I_{n-1} - I_n$, and similarly $dQ_{n+1}/dt = I_n - I_{n+1}$. Then the differential equation becomes

$$L\frac{d^2I_n}{dt^2} = \frac{1}{C}(I_{n-1} + I_{n+1} - 2I_n) \tag{11.58}$$

precisely in the form of (11.50), which is rewritten below for convenience:

$$m\frac{d^2\xi_n}{dt^2} = K(\xi_{n-1} + \xi_{n+1} - 2\xi_n)$$

Thus, there is a definite analogy with $L \leftrightarrow m$, $1/C \leftrightarrow K$, $I \leftrightarrow \xi$ (in acoustics, however, it will turn out to be much more useful to identify I with velocity times the cross-sectional area, that is, with uA).

By analogy with the solution already found for the mechanical system, the transmission line has a low-frequency wave speed

$$c = v_g = l\sqrt{\frac{1}{LC}} \tag{11.59}$$

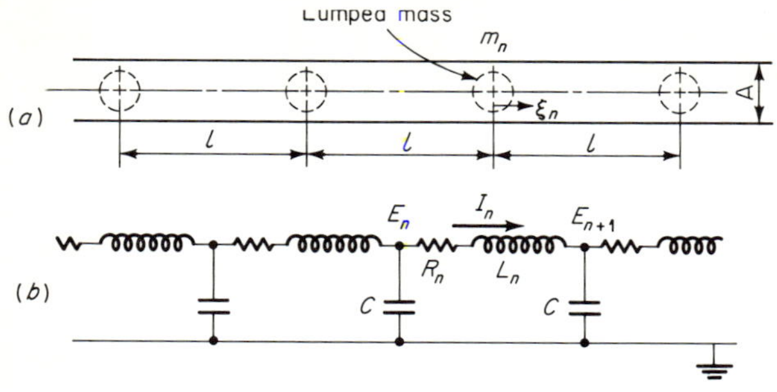

Figure 11.27
(*a*) Lumped-parameter representation of duct; (*b*) equivalent transmission line.

The waves do not travel with the speed of light! The cutoff frequency is

$$\nu_c = \frac{1}{\pi}\sqrt{\frac{1}{LC}} \tag{11.60}$$

For a typical transmission line, this frequency is of course very large compared to 60 Hz.

A more general lumped-constant acoustical model incorporates the effect of *losses*, e.g., sound attenuation in a duct due to the viscous boundary layer. The model for one-dimensional propagation in a duct of cross-sectional area A is shown in Fig. 11.27, together with the analogous transmission-line network. We now discuss this analogy.

The voltage across a typical series combination of resistance R_n and inductance L_n is

$$L\frac{dI_n}{dt} + RI_n = \frac{1}{C}(Q_n - Q_{n+1})$$

Differentiation yields the desired differential equation,

$$L\frac{d^2I_n}{dt^2} = \frac{1}{C}(I_{n-1} + I_{n+1} - 2I_n) - R\frac{dI_n}{dt} \tag{11.61}$$

which differs from (11.58) only by virtue of the added resistance term.

The differential equation for the lumped-parameter acoustical system, i.e., mechanical, is (11.50), with a frictional resistance $r\, d\xi/dt$ opposing the motion added to the force term

$$m\frac{d^2\xi_n}{dt^2} = K(\xi_{n-1} + \xi_{n+1} - 2\xi_n) - r\frac{d\xi_n}{dt} \tag{11.62}$$

where it has been assumed that the acoustic motion is essentially in the axial direction, i.e., that the duct diameter is small compared to the wavelength.

Rather than draw the analogy between current I in (11.61) and displacement ξ in this equation, it is convenient to make use of the *volume velocity* U

$$U \equiv Au = A\frac{d\xi}{dt} \tag{11.63}$$

This quantity will be taken as the analog of current. This identification will be useful because the *volume velocity is conserved at a duct junction in the same sense that current is conserved at an electric junction according to Kirchhoff's law*. This result follows from the condition that the wavelength is very large compared to the duct dimension, so that the *local* flow at a junction has essentially constant density, i.e., may be treated as incompressible, so that ρuA = const is equivalent to U = const. More fundamentally, the result follows from the similarity of conservation of mass to conservation of charge.

Differentiating (11.62), introducing the appropriate value of K from (11.48), and dividing by A yields

$$\frac{m}{A^2}\frac{d^2U_n}{dt^2} = \frac{\rho c^2}{V}(U_{n-1} + U_{n+1} - 2U_n) - \frac{r}{A^2}\frac{dU_n}{dt} \tag{11.64}$$

which is to be compared to the electrical equation (11.61), viz.,

$$L\frac{d^2I_n}{dt^2} = \frac{1}{C}(I_{n-1} + I_{n+1} - 2I_n) - R\frac{dI_n}{dt} \tag{11.61}$$

We can now identify the analogous acoustical and electrical quantities, as discussed in the following section.

Analogous Acoustical and Electrical Quantities

By comparing (11.64) and (11.61) and from our knowledge of the corresponding mechanical model, we make the assignment of analogous quantities shown in Table 11.2. The *acoustic pressure* p will enter as a *driving force* for the acoustic element, just as the potential E acts as a driving force for the electric element.

The electroacoustic analogy in Table 11.2 is often used in acoustics and is called the *classical* or *Kelvin system*. It should be noted that an alternative system of analogy is possible, in which voltage is identified

Table 11.2 Analogous Quantities

Acoustical	Mechanical	Electrical
Inertance $L_a = m/A^2$	Mass m	Inductance L
Acoustic capacitance $C_a = V/\rho c^2$	Compliance $1/K$	Capacitance C
Acoustic resistance $R_a = r/A^2$	Damping constant c	Resistance R
Volume displacement ξA	Displacement ξ	Charge Q
Volume velocity $U = uA$	Velocity u	Current I
Acoustic pressure p	Force F	Potential (voltage) E

with velocity (rather than pressure). This alternative system has the advantage that electric and acoustic circuits are topologically identical; i.e., series and parallel connections are preserved in passing from the electric to the acoustic circuit, and vice versa. We shall not, however, discuss this second system further; a treatment will be found in the book by *Malecki* [1969, chap. 9].

It is conventioned to define an *acoustic impedance* Z_a

$$Z_a = \frac{p}{U} \tag{11.65}$$

which is usefully represented as a complex number, in analogy with the electric impedance Z,

$$Z \equiv \frac{E}{I} \tag{11.66}$$

The acoustic impedance Z_a defined by (11.65) is in general distinct from the *characteristic acoustic impedance* $\rho c = \mathscr{R}$ introduced in Sec. 4.8; in particular, Z_a may be complex, whereas $\mathscr{R}$ is necessarily real.

As a simple example of the impedance of a distributed system, consider progressive plane waves without dissipation in a duct of area A. The acoustic pressure and fluid velocity are related by

$$p = \rho c u$$

where the velocity is considered positive in the direction of propagation. Then from (11.65) we find $Z_a = \rho c/A$, which is entirely real. In *this* case Z_a differs from the characteristic acoustic impedance $\mathscr{R}$ only by a multiplicative constant.

The Helmholtz Resonator

This is one of the simplest possible acoustic devices which can be represented by an electrical analog. Helmholtz resonators, in the form of wine bottles, cider jugs, soda bottles, and the like, have been a source of acoustical amusement for centuries, the resonance being produced by suitable blowing across the mouth of the bottle.

The resonator is sketched in Fig. 11.28. The fluid in the neck of the resonator is considered to move axially, i.e., up and down, like a solid prism, with mass $m = \rho A l$; the *inertance* of the system is thus $L_a = m/A^2 = \rho l/A$. The volume V (which need not be spherical) acts as an elastic reservoir, resisting the downward motion of the neck mass, with acoustic *capacitance* $C_a = V/\rho c^2$. In addition, *resistance* R_a is present in

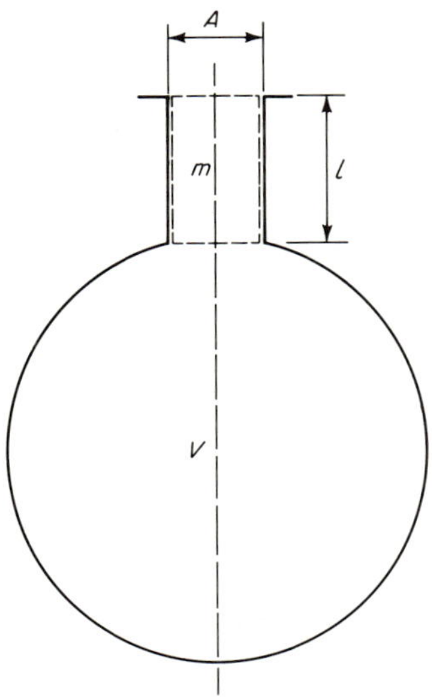

Figure 11.28
Helmholtz resonator.

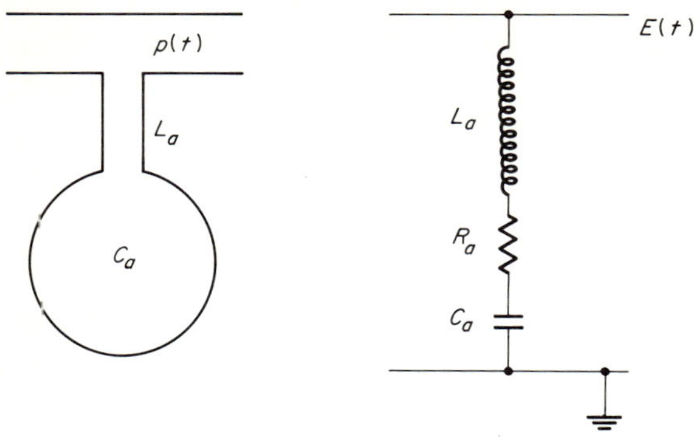

Figure 11.29
Helmholtz resonator on a duct and the equivalent electrical circuit.

various forms, e.g., if the volume is lined with absorbent material and if the mouth of the resonator radiates to space.

The mass m and its associated resistance are thus represented as L_a and R_a in series. Finally the volume velocity U runs into and out of the capacitance (volume) C_a in the same way that current runs into a series capacitor. We thus arrive at the series LRC circuit shown in Fig. 11.29 (more formally, the differential equation for the neck mass motion is found to be identical to that for the current in the LRC circuit; this procedure becomes cumbersome, however, as devices increase in complexity).

The resonant frequency ν_0 of the Helmholtz resonator (in the absence of resistance) can be taken directly from the well-known formula for the series LRC circuit, viz.,

$$\nu_0 = \frac{1}{2\pi\sqrt{L_a C_a}} \tag{11.67}$$

Putting in the physical values $L_a = \rho l/A$ and $C_a = V/\rho c^2$ then gives

$$\nu_0 = \frac{c}{2\pi}\sqrt{\frac{A}{Vl}} \tag{11.68}$$

The presence of resistance tends to smear out ν_0 into a range of frequencies. A resonator mounted on an acoustical line, as in Fig. 11.29, then acts as a simple band-stop filter.[1] To obtain a wider stop band, several resonators can be placed in parallel, i.e., next to each other along the acoustical line.

We will not pursue the electroacoustic analogy any further. Ample treatments can be found in several books on acoustics, e.g., *Beranek* [1954], *Kinsler and Frey* [1950], or *Blitz* [1964].

[1] For a discussion of band-stop filters, see *Brillouin* [1946] or a book on electric-circuit theory.

11.5 An analog for flow with relaxation

As mentioned in Sec. 2.2, the term *relaxation* refers to the departure from thermodynamic equilibrium by some internal property of the fluid. Possible examples of nonequilibrium (relaxing) properties in a gas include the extent of chemical reaction and the level of excitation of vibrational energy in the molecules.

In general, relaxation occurs when changes in the fluid state are so rapid that the relaxing (laggard, unenthusiastic) internal property is unable to follow. A very simple example of this situation is the motion of very fine particles, e.g., dust, suspended in a gas which is subjected to acoustic motion. Depending on the fineness of the particles, their velocity (and temperature) will somewhat lag behind that of the gas. We thus take the particle velocity v to be the relaxing property of the gas mixture and proceed to develop the appropriate theory, which will prove to be formally analogous to that for a gas with thermodynamical relaxation.

Acoustical Motion of a Particle-laden Gas

The particle-laden gas is shown schematically in Fig. 11.30. The particle velocity is $v(x,t)$, and the gas velocity is $u(x,t)$. At the outset we make the following simplifying assumptions:

1 The motion is one-dimensional.

2 Particles are spheres of uniform size, very small compared to the wavelength of sound.

3 Particles collectively occupy negligible volume (but do not have negligible mass or inertia).

4 The fluid-dynamical drag on each particle is given by Stokes' law (appropriate to small Reynolds number and quasi-steady flow, corresponding to sound of sufficiently low frequency).

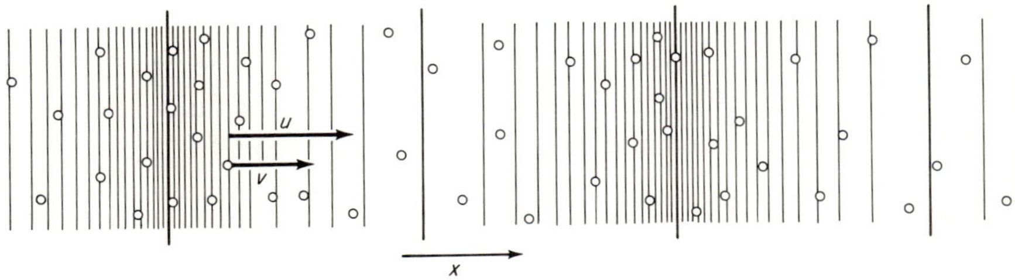

Figure 11.30

5 The heat-transfer rate to the particles is proportional to the temperature difference between the gas and the particle (this is essentially the heat-transfer analog to Stokes' law).

6 The motion is a small perturbation from an undisturbed state. The standard acoustical approximation of neglecting convective derivatives and in general replacing coefficients by their undisturbed values (linearization) is therefore justified.

7 The only viscous effect is the particle drag, and the only heat-conduction effect is the heat transfer between particles and gas.

8 The particle number density is small compared to molecular number density, and Brownian motion can be neglected.

According to Stokes' law, the drag force F_d on the spherical particle is proportional to the relative fluid velocity

$$F_d = 3\pi D\mu(u - v) \tag{11.69}$$

where D is the particle diameter and μ the gas viscosity. The equation of motion for a particle is just $m\,dv/dt = F_d$; with $m = \rho_p \pi D^3/6$ and (11.69), this can be written

$$\frac{dv}{dt} = \frac{u - v}{\tau_m} \tag{11.70}$$

where the *momentum relaxation time* τ_m is given by

$$\tau_m \equiv \frac{\rho_p D^2}{18\mu} \tag{11.71}$$

Equation (11.70) is a typical linear relaxation statement, in which the lagging quantity (particle velocity v in this case) tends to pursue an equilibrium value (u in this case) at a rate proportional to the lag. The relaxation time τ_m is just a measure of the quickness of response. For example, if u is constant and $v(0) = 0$ (as is the case if the particle is released from rest in a uniform streaming flow), Eq. (11.70) has the solution

$$v = u(1 - e^{-t/\tau_m}) \tag{11.72}$$

as shown in Fig. 11.31. Such a particle response could be produced by the passage of a weak shock. In general, however, the fluid velocity is variable in time and space and presents a shifting target for the particle velocity.

For rather small particles, such as particles of cigarette smoke of diameter $D \sim 1\ \mu$m in air, the momentum relaxation time τ_m is of the order of $1\ \mu$s.

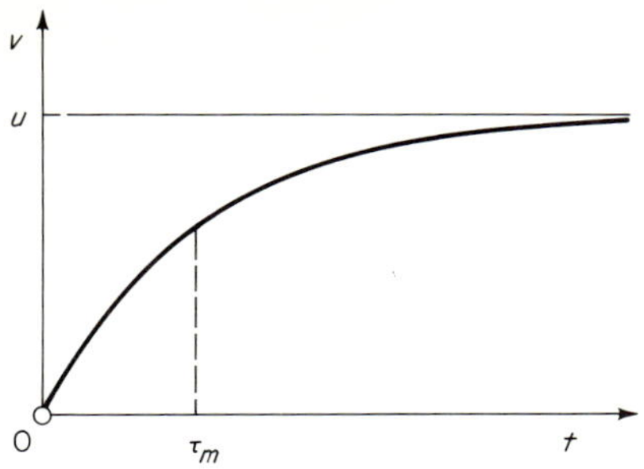

Figure 11.31

Let the gas temperature be denoted by T and the particle temperature by T_p. By a derivation parallel to the above we find a parallel result,

$$\frac{dT_p}{dt} = \frac{T - T_p}{\tau_t} \tag{11.73}$$

where τ_t is a *thermal relaxation time.*[1]

The time derivatives d/dt in (11.70) and (11.73) are, in the notation of fluid mechanics, material derivatives $d/dt = \partial/\partial t + v\,\partial/\partial x$. In the acoustic approximation the convective derivative is neglected, so that we have simply $d/dt = \partial/\partial t$.

Let ρ be the density of the (pure) gas and n the number density of particles. The continuity equations for gas and particles are then, respectively,

$$\frac{\partial \rho}{\partial t} + \rho_0 \frac{\partial u}{\partial x} = 0 \tag{11.74}$$

$$\frac{\partial n}{\partial t} + n_0 \frac{\partial v}{\partial x} = 0 \tag{11.75}$$

The momentum equation for the particles is just (11.70), with $d/dt \to \partial/\partial t$,

$$\frac{\partial v}{\partial t} - \frac{u - v}{\tau_m} = 0 \tag{11.76}$$

[1] At small Reynolds numbers, the Nusselt number $\mathrm{Nu} \equiv hD/\kappa$ for heat transfer is 2. For a gas with a Prandtl number $\mathrm{Pr} = 2/3$, this leads to a very simple relation between τ_m and τ_t,

$$\tau_t = \frac{c_{pp}}{c_{pg}} \tau_m$$

where c_{pp} and c_{pg} are respectively the particle and gas specific heats. That is, a particle with a large specific heat is thermally slow. It may be noted that in assigning a uniform temperature to the particle, we have assumed a large thermal conductivity, $\kappa_{\mathrm{pcl}}/\kappa_{\mathrm{gas}} \gg 1$.

The momentum equation for the gas is just

$$\rho_0 \frac{\partial u}{\partial t} + \frac{\partial P}{\partial x} = -nF_d$$

where $-nF_d$ is the force per unit volume exerted by the particles on the gas; i.e., by Newton's third law, equal and opposite to the drag force on the particles. With F_d/m given by (11.70), this becomes

$$\frac{\partial u}{\partial t} + \frac{1}{\rho_0}\frac{\partial P}{\partial x} + \eta \frac{u - v}{\tau_m} = 0 \tag{11.77}$$

where η is the mass fraction of particles,

$$\eta \equiv \frac{mn_0}{\rho_0} \tag{11.78}$$

This quantity may be, for example, of order unity or less.

The energy equation for the particles may be written simply as (11.73) in the acoustic approximation

$$\frac{\partial T_p}{\partial t} = \frac{T - T_p}{\tau_t} \tag{11.79}$$

Finally, an energy equation for the gas is needed. For pedagogical reasons, it will be helpful to find the exact form of this equation, i.e., without making any acoustic approximations. The rate per unit volume at which heat is transferred from the particles to the gas is $mnc_{pp}(T_p - T)/\tau_t$. The rate per unit volume at which the particles do work on the gas is $-nvF_d$, where the particle velocity v is the velocity of the "surface" separating the two systems. Then the energy equation (1.74) can be written in the form

$$\rho \frac{D}{Dt}\left(h + \frac{u^2}{2}\right) - \frac{\partial P}{\partial t} = mnc_{pp}\frac{T_p - T}{\tau_t} - nvF_d$$

Dividing through by ρ and subtracting from this the product of u with the exact (nonlinearized) form of (11.77) yields

$$\frac{Dh}{Dt} - \frac{1}{\rho}\frac{DP}{Dt} = \eta c_{pp}\frac{T_p - T}{\tau_t} + \eta \frac{(u - v)^2}{\tau_m}$$

By the Gibbs equation the left-hand side is just $T\, Ds/Dt$; thus we finally have the energy equation in the form

$$\frac{Ds}{Dt} = \frac{\eta c_{pp}}{T}\frac{T_p - T}{\tau_t} + \frac{\eta}{T}\frac{(u - v)^2}{\tau_m} \tag{11.80}$$

This is in the form of the general relation (2.16) for the entropy increase. *The effect of particle relaxation is to increase the entropy of the fluid.*[1]

The entropy equation (11.80), elegant though it is, must be abridged in passing to the acoustic approximation

$$\frac{Ds}{Dt} \approx \frac{\partial s}{\partial t}$$

This neglects a term $u\, \partial s/\partial x$. Now $u\, \partial s/\partial x$ and the second term on the right in (11.80) are both of second-order smallness, i.e., of order u^2. Therefore a *consistent* acoustic approximation requires that both terms be neglected, yielding

$$\frac{\partial s}{\partial t} = \frac{\eta c_{pp}}{T}\frac{T_p - T}{\tau_t} \tag{11.81}$$

For convenience, we collect the equations of motion (in acoustic form) below:

$$\frac{\partial \rho}{\partial t} + \rho_0 \frac{\partial u}{\partial x} = 0 \tag{11.74}$$

$$\frac{\partial n}{\partial t} + n_0 \frac{\partial v}{\partial x} = 0 \tag{11.75}$$

$$\frac{\partial v}{\partial t} - \frac{u - v}{\tau_m} = 0 \tag{11.76}$$

$$\frac{\partial u}{\partial t} + \frac{1}{\rho_0}\frac{\partial P}{\partial x} + \eta \frac{u - v}{\tau_m} = 0 \tag{11.77}$$

$$\frac{\partial T_p}{\partial t} - \frac{T - T_p}{\tau_t} = 0 \tag{11.79}$$

$$\frac{\partial s}{\partial t} = \frac{\eta c_{pp}}{T}\frac{T_p - T}{\tau_t} \tag{11.81}$$

The unknowns are n, v, T_p (for the particles) and two thermodynamic variables plus u (for the gas), totaling six. The six equations thus form a complete set.

[1] The dissipation function, proportional to $(u - v)^2$, is always positive. Similarly, the heat-transfer term contains an always positive part; formally,

$$\frac{T_p - T}{T} = \frac{T_p - T}{T_p} + \frac{(T_p - T)^2}{T_p T}$$

The Relation between Pressure and Density

Because the flow is in general nonisentropic, the pressure and density will not be uniquely connected. By the chain rule

$$dP = \left(\frac{\partial P}{\partial \rho}\right)_s d\rho + \left(\frac{\partial P}{\partial s}\right)_\rho ds$$

With $\partial s/\partial t$ given by (11.81) and the thermodynamic identity $(\partial P/\partial s)_v = -(\partial T/\partial v)_s = (\gamma - 1)(\partial T/\partial v)_P$ this becomes

$$\frac{\partial P}{\partial t} = c_0{}^2 \frac{\partial \rho}{\partial t} + \frac{\gamma - 1}{T}\left(\frac{\partial T}{\partial v}\right)_P \eta c_{pp} \frac{T_P - T}{\tau_t} \tag{11.82}$$

where $c_0 = \sqrt{(\partial P/\partial \rho)_s}$ is the normal sound speed in the pure gas.

Wave Equation for the Case of No Thermal Inertia

As a simple and instructive special case, let the heat capacity of the particles be zero. With $c_{pp} \to 0$, it follows that $\tau_t \to 0$ [see the footnote below Eq. (11.73)]. Then from (11.79) it is always true that $T = T_P$, and the flow is isentropic.

Only one relaxation time, viz., τ_m, remains in the problem.

Differentiating the momentum equation (11.77) with respect to time yields

$$\frac{\partial^2 u}{\partial t^2} + \frac{1}{\rho_0}\frac{\partial^2 P}{\partial x\, \partial t} + \frac{\partial}{\partial t}\frac{u - v}{\tau_m} = 0$$

With $dP = c_0{}^2\, d\rho$ and the continuity equation (11.74) this becomes

$$\frac{\partial^2 u}{\partial t^2} - c_0{}^2 \frac{\partial^2 u}{\partial x^2} + \eta \frac{\partial}{\partial t}\frac{u - v}{\tau_m} = 0 \tag{11.83}$$

which is the conventional wave equation with an additional term. Differentiating this again with respect to time and making use of (11.76), (11.77), and (11.74) together with $dP = c_0{}^2\, d\rho$ finally yields an equation in the velocity u only,

$$\tau_m \frac{\partial}{\partial t}\left(\frac{1}{c_f{}^2}\frac{\partial^2 u}{\partial t^2} - \frac{\partial^2 u}{\partial x^2}\right) + \left(\frac{1}{c_e{}^2}\frac{\partial^2 u}{\partial t^2} - \frac{\partial^2 u}{\partial x^2}\right) = 0 \tag{11.84}$$

where

$$c_f = c_0 = \sqrt{\left(\frac{\partial P}{\partial \rho}\right)_s}$$

$$c_e = \frac{c_0}{\sqrt{\eta + 1}} \tag{11.85}$$

are two distinct wave speeds. We call c_f the *frozen sound speed* and c_e the *equilibrium sound speed*, as will be explained below.

Equation (11.84) might be called a general wave equation and is more or less typical of linear wave processes with friction. We proceed to discuss it in some detail.

Suppose the sound is periodic, with period T. If the relaxation time τ_m is large, $\tau_m \gg T$, the first set of terms in (11.84) is dominant and the equation becomes

$$\frac{1}{c_f^2}\frac{\partial^2 u}{\partial t^2} - \frac{\partial^2 u}{\partial x^2} \approx 0$$

where the constant of integration has been set to zero. We thus have ordinary wave motion propagating at the frozen sound speed c_f. The physical interpretation is that the rather large particles (large τ_m) do not follow the motion of the gas at all but are practically stationary. With viscous dissipation neglected, the particles do not influence the sound propagation in any way! The particles are said to be *frozen*, with a fixed state independent of that of the gas.

At the other end of the scale the particles are very small, and the relaxation time is small compared to the time period, $\tau_m \ll T$. Then only the second set of terms survives, and

$$\frac{1}{c_e^2}\frac{\partial^2 u}{\partial t^2} - \frac{\partial^2 u}{\partial x^2} \approx 0$$

This gives ordinary wave motion propagating with the equilibrium sound speed c_e. In this case, the particles follow the gas motion with negligible lag, i.e., are in *equilibrium* with the gas. Suppose the gas is perfect; then the equilibrium sound speed c_e is given, from (11.85), by

$$c_e^2 = \frac{1}{\eta + 1}\left(\frac{\partial P}{\partial \rho}\right)_s = \frac{\gamma P}{(\eta + 1)\rho} = \frac{\gamma P}{mn + \rho} \tag{11.86}$$

The denominator in this expression is just the total density of the mixture; thus, c_e is given by the equation for a perfect gas, provided that the *total* density is used. The sound speed is reduced in comparison to the value for a pure gas by virtue of the added inertia.

The full equation (11.34) can be solved, for pure sinusoidal waves, by standard techniques. Consider the motion induced in the space $x > 0$ by an oscillating piston at $x = 0$; let the piston displacement be given by

$$X = \epsilon \sin \omega t$$

where ϵ is the displacement amplitude and ω is the angular frequency. The corresponding piston velocity is

$$u(0,t) = \epsilon\omega \cos \omega t$$

The piston motion drives simple waves traveling to the right. The solution to this problem is given in detail by *Vincenti and Kruger* [1965, sec. 8.6], who found exactly our equation (11.84) in the framework of vibrational relaxation in a gas. This solution is, in our notation,

$$u(x,t) = \epsilon\omega e^{-\alpha\omega x/c_f} \cos\left[\omega\left(t - \frac{x}{c}\right)\right] \tag{11.87}$$

where the attenuation factor α and the wave speed (phase velocity) c are given by

$$\left.\begin{matrix}\alpha \\ c_f/c\end{matrix}\right\} = \left\{\frac{1}{2(1 + k^2)}\left[\mp(b + k^2) + \sqrt{(1 + k^2)(b^2 + k^2)}\right]\right\}^{1/2} \tag{11.88}$$

where for brevity we have written $k \equiv \omega\tau_m$ and $b \equiv c_f^2/c_e^2 = \eta + 1$.

A plot of the phase velocity c and attenuation factor α is shown in Fig. 11.32. The phase velocity is c_e at very low frequency and c_f at high

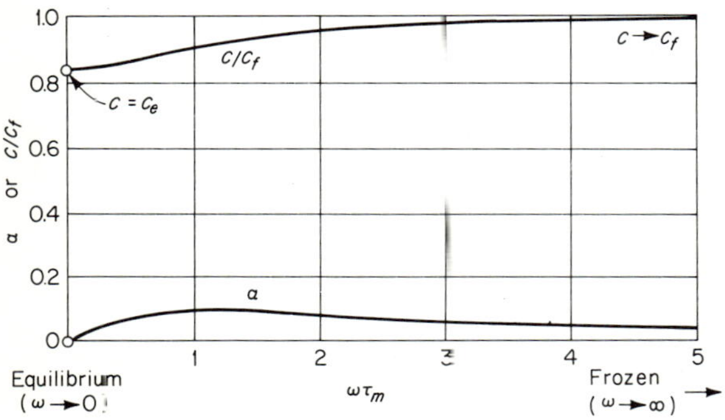

Figure 11.32
Attenuation factor and wave speed vs. nondimensional frequency for the case $\eta = 0.4142$.

frequency, as already suggested. In general, however, the phase velocity depends on frequency and the waves are *dispersive*; i.e., a waveform consisting of a Fourier sum of various frequency components will be progressively distorted because the individual components propagate at different velocities.

The attenuation coefficient is $\alpha\omega/c_f$. At high frequencies $\alpha \to 0$ as $1/\omega$, so that the attenuation coefficient has a finite limiting value. Translating the result of Vincenti and Kruger, this value is

$$\frac{\alpha\omega}{c_f} \to \frac{\eta}{2c_f\tau_m} \tag{11.89}$$

That is, in the frozen limit there is a finite attenuation with distance, as would be expected.

Wave Equation for the General Case

If both thermal and momentum relaxation are allowed, the problem is somewhat more complicated, though not different in principle. We shall not discuss this case in detail.

The following wave equation is obtained:

$$\tau_m{}^2\delta\frac{\partial^2}{\partial t^2}\left(\frac{1}{c_f{}^2}u_{tt} - u_{xx}\right) + \tau_m[1 + \delta(\eta + 1)]\frac{\partial}{\partial t}\left(\frac{1}{c_n{}^2}u_{tt} - u_{xx}\right) + (1 + \eta\delta)\left(\frac{1}{c_e{}^2}u_{tt} - u_{xx}\right) = 0 \tag{11.90}$$

where $\delta \equiv c_{pp}/c_{pg}$ and use has been made of the relation $\tau_t = \tau_m\delta$. The various wave speeds are given by

$$\begin{aligned} c_f{}^2 &= c_0{}^2 = \left(\frac{\partial P}{\partial \rho}\right)_s \\ c_n{}^2 &= \frac{1 + \delta(\eta + 1)}{1 + \delta[(\gamma + 1)\eta + 1]}\,c_0{}^2 \\ c_e{}^2 &= \frac{1 + \eta\delta}{(1 + \gamma\eta\delta)(\eta + 1)}\,c_0{}^2 \end{aligned} \tag{11.91}$$

where γ is the ratio of specific heats in the gas. Note that in the case $\delta = 0$ Eq. (11.90) reduces to the previous wave equation (11.84).

There is an extensive literature on particle-laden gas flows. A detailed treatment, including acoustical motions, is given by *Marble* [1970]. Recent references are given by *Schmitt–von Schubert* [1969].

The standard reference on *molecular* relaxation processes in acoustics is *Herzfeld and Litovitz* [1959].

11.6 Closure

There are many analogies to compressible flow which have not been mentioned.

Probably the most important of these is the *electric-field analogy* with two-dimensional compressible flow, introduced by *Taylor and Scharman* [1928]. This is an elaboration of the well-known analogy for incompressible potential flow and steady-state heat conduction, i.e., for solution of Laplace's equation. It is discussed by *Malavard* [1954].

Several of the remaining compressible-flow analogies have been catalogued by *Murphy et al.* [1963].

references

1686 Newton, I.: *Mathematical Principles of Natural Philosophy*, London (Motte-Cajori translation, University of California Press, Berkeley, 1962).

1845 Stokes, G. G.: "On the Theories of the Internal Friction of Fluids in Motion, and of the Equilibrium and Motion of Elastic Solids," *Trans. Cambridge Phil. Soc.*, **8**: 287–319; reprinted in *Mathematical and Physical Papers*, vol. 1, Johnson Reprint, New York, 1966.

1850 Stokes, G. G.: "On the Effect of the Internal Friction of Fluids on the Motion of Pendulums," *Trans. Cambridge Phil. Soc.*, **9**: 8; reprinted in *Mathematical and Physical Papers*, vol. 3, Johnson Reprint, New York, 1966.

1859 Riemann, G.: "Über die Fortpflanzung ebener Luftwellen von endlicher Schwingungsweite," in *Gesammelte Werke*, Leipzig, 1876; reprinted by Dover New York, 1953.

1889 Tait, P. G.: *Voyage of H.M.S. Challenger*, vol. 2, pt. 4, H.M. Stationery Office, London; also in Tait's *Scientific Papers*, vol. 2, papers 61 and 107, Cambridge University Press, London, 1900.

1894 Rayleigh, Lord: *The Theory of Sound*, Macmillan, London; reprinted by Dover, New York, 1945.

1903 Reynolds, O.: *Papers on Mathematical and Physical Subjects*, vol. 3, *The Sub-mechanics of the Universe*, Cambridge University Press, London.

1907 Prandtl, L.: "Neue Untersuchungen über die stromende Bewegung der Gase und Dampfe," *Physik. Z.*, **8**: 23.

1911 Rayleigh, Lord: "On the Motion of Solid Bodies through Viscous Liquid," *Phil. Mag.*, **21**(6): 697.

1915 Rayleigh, Lord: "The Principle of Similitude," *Nature*, **95**: 66 (March).

1925 Väisälä, V.: "Über die Wirkung der Windschwankungen auf die Pilot Beobachtungen," *Soc. Sci. Fennica, Commentationes Phys. Math.*, (II) **19**: 37.

1928 Mason, W. P.: *Phys. Rev.*, **31**: 283–295.

Taylor, G. I., and C. F. Scharman: "A Mechanical Method for Solving Problems of Flow in Compressible Fluids," *Proc. Roy. Soc. (London)*, **A121**: 194–217; also in Taylor's *Papers*, Cambridge University Press, London, 1963.

1931 Bridgman, P. W.: *Dimensional Analysis*, Yale University Press, New Haven, Conn.

1932 Lamb, H.: *Hydrodynamics*, Cambridge University Press, London; reprinted by Dover, New York, 1945.

1933 Taylor, G. I., and J. W. Maccoll: "The Air Pressure on a Cone Moving at High Speeds," *Proc. Roy. Soc. (London)*, **A139**: 278–297; also in Taylor's *Papers*, Cambridge University Press, London, 1963.

1935 Miller, D. C.: *Anecdotal History of the Science of Sound*, Macmillan, New York.

1937 Hudleston, L. J.: "Intermolecular Forces of Normal Liquids," *Trans. Faraday Soc.*, **33**: 97–103.

1942 Bethe, H. A.: "The Theory of Shock Waves for an Arbitrary Equation of State," *OSRD Rept.* 545.

Kirkwood, J. G., and H. Bethe: "The Pressure Wave Produced by an Underwater Explosion," pt. I, *OSRD Rept.* 588 (Dept. Commerce Bibliography PB 32182).

Kirkwood, J. G., and J. M. Richardson: "The Pressure Wave Produced by an Underwater Explosion," pt. III, *OSRD Rept.* 813 (Dept. Commerce Bibliography PB 32184).

Taylor, G. I., and H. Jones: "Note on the Lateral Expansion behind a Detonation Wave," British Ministry of Supply; collected in Taylor's *Papers*, Cambridge University Press, London, 1963.

1944 Neumann, J. von: "Proposal and Analysis of a New Numerical Method for the Treatment of Hydrodynamical Shock Problems," *Underwater Explosion Research*, vol. 1, Office of Naval Research, Washington, D.C.

1945 Landau, L. D.: "On Shock Waves at Large Distances from the Place of Their Origin," *Prikl. Mat. Mekhan.*, **9**(4): 286–292; also in *Papers*, Pergamon, New York, 1965.

1946 Brillouin, L.: *Wave Propagation in Periodic Structures*, McGraw-Hill, New York; reprinted by Dover, New York, 1953.

Su, G. J.: "Modified Law of Corresponding States for Real Gases," *Ind. Eng. Chem.*, **38**(8): 803–806.

1947 Kopal, Z.: *Tables of Supersonic Flow around Cones*, Department of Electrical Engineering, Massachusetts Institute of Technology, Cambridge, Mass.

Sivian, L. J.: "High Frequency Absorption in Air and Other Gases," *J. Acoust. Soc. Amer.*, **19**(5): 914–916.

1948 Courant, R., and K. O. Friedrichs: *Supersonic Flow and Shock Waves*, Interscience, New York.

Lighthill, M. J.: "The Position of the Shock Wave in Certain Aerodynamic Problems," *Quart. J. Mech. Appl. Math.*, **1**: 309–318.

Stoker, J. J.: "The Formation of Breakers and Bores," *Commun. Pure Appl. Math.*, **1**: 1–87.

1949 Born, M.: *Natural Philosophy of Cause and Chance*, Oxford University Press, London.

Chapman, D. R., and M. W. Rubesin: *J. Aeron. Sci.*, **16**: 547.

Zartman, I. F.: "Ultrasonic Velocities and Absorption in Gases at Low Pressures," *J. Acoust. Soc. Amer.*, **21**(4): 171.

1950 Kinsler, L. E., and A. R. Frey: *Fundamentals of Acoustics*, Wiley, New York.

Lass, H.: *Vector and Tensor Analysis*, McGraw-Hill, New York.

Taylor, G. I.: "The Formation of a Blast Wave by a Very Intense Explosion," *Proc. Roy. Soc.* (*London*), **A201**: 159–174; also in Taylor's *Papers*, Cambridge University Press, London, 1963.

Taylor, G. I.: "The Dynamics of the Combustion Products behind Plane and Spherical Detonation Fronts in Explosives," *Proc. Roy. Soc.* (*London*), **A200**: 235–247; also in Taylor's *Papers*, Cambridge University Press, London, 1963.

1952 Chapman, S., and T. G. Cowling: *The Mathematical Theory of Non-uniform Gases*, Cambridge University Press, London

Lenihan, J. M. A.: "The Velocity of Sound in Air," *Acustica*, **2**: 205–212.

Lighthill, M. J.: "On Sound Generated Aerodynamically," *Proc. Roy. Soc.* (*London*), **A211**: 564.

1953 Parker, J. G., et al.: "Absorption of Sound in Argon, Nitrogen, and Oxygen at Low Pressures," *J. Acoust. Soc. Amer.*, **25**(2): 263.

Shapiro, A. H.: *The Dynamics and Thermodynamics of Compressible Fluid Flow*, 2 vols., Ronald, New York.

Truesdell, C.: "Precise Theory of the Absorption and Dispersion of Forced Plane Infinitesimal Waves According to the Navier-Stokes Equations," *J. Rational Mech. Anal.* **2**(4): 643–741.

1954 Beranek, L. L.: *Acoustics*, McGraw-Hill, New York.

Bergmann, L.: *Der Ultraschall*, Hirzel, Stuttgart.

Bett, K. E., K. Weale, and D. Newitt: "The Critical Evaluation of Compression Data for Liquids and a Revision of the Isotherms of Mercury," *Brit. J. Appl. Phys.*, **5**: 243–251.

Ferri, A.: "Supersonic Flows with Shock Waves," in W. R. Sears (ed.), *General Theory of High Speed Aerodynamics*, vol. 6, Princeton Series in High Speed Aerodynamics and Jet Propulsion, Princeton University Press, Princeton, N.J.

Lighthill, M. J.: "Higher Approximations," in W. R. Sears (ed.), *General Theory of High Speed Aerodynamics*, vol. 6, Princeton Series in High Speed

Aerodynamics and Jet Propulsion, Princeton University Press, Princeton, N.J.

Malavard, L.: "Electrolytic Plotting Tank," in R. W. Ladenburg et al. (eds.), *Physical Measurements in Gas Dynamics and Combustion*, vol. 9, Princeton Series in High Speed Aerodynamics and Jet Propulsion, Princeton University Press, Princeton, N.J.

Shapiro, A. H.: "Free Surface Water Table," in R. W. Ladenburg et al. (eds.), *Physical Measurements in Gas Dynamics and Combustion*, vol. 9, Princeton Series in High Speed Aerodynamics and Jet Propulsion, Princeton University Press, Princeton, N.J.

Smith, A. H., and A. W. Lawson: *J. Chem. Phys.*, **22**: 351.

Truesdell, C.: "Prologue," in L. Euler, *Opera Omnia*, vols. 12 and 13, 2d ser., Swiss Society of Natural Science, Lausanne.

1955 Hilsenrath, J. et al.: "Tables of Thermal Properties of Gases," *Natl. Bur. Std. (U.S.) Circ.* 564.

Lighthill, M. J., and G. B. Whitham: "On Kinematic Waves II: A Theory of Traffic Flow on Long Crowded Roads," *Proc. Roy. Soc. (London)*, **A229**: 317–345.

Mirels, H.: "Laminar Boundary behind Shock Advancing into Stationary Fluid," *NACA Tech. Note* 3401.

Solomon, G. E.: "Transonic Flow past Cone Cylinders," *NACA Rept.* 1242 (supersedes *NACA Tech. Note* 3213).

1956 Lighthill, M. J.: "Viscosity Effects in Waves of Finite Amplitude," in G. K. Batchelor and R. M. Davies (eds.), *Surveys in Mechanics*, Cambridge University Press, London.

Mirels, H.: "Boundary Layer behind Shock or Thin Expansion Wave Moving into Stationary Fluid," *NACA Tech. Note* 3712.

Whitham, G. B.: "On the Propagation of Weak Shock Waves," *J. Fluid Mech.*, **1**: 3.

1957 Liepmann, H. W., and A. Roshko: *Elements of Gasdynamics*, Wiley, New York.

Richtmyer, R. D.: *Difference Methods for Initial Value Problems*, Interscience, New York

1958 Crocco, L.: "One-dimensional Treatment of Steady Gas Dynamics," in H. W. Emmons (ed.), *Fundamentals of Gasdynamics*, vol. 3, Princeton Series in High Speed Aerodynamics and Jet Propulsion, Princeton University Press, Princeton, N.J.

Friedlander, F. G.: *Sound Pulses*, Cambridge University Press, London.

Hayes, W. D.: "The Basic Theory of Gasynamic Discontinuities," in H. W. Emmons (ed.), *Fundamentals of Gasdynamics*, vol. 3, Princeton Series in High Speed Aerodynamics and Jet Propulsion, Princeton University Press, Princeton, N.J.

Mises, R. von: *Mathematical Theory of Compressible Fluid Flow*, Academic, New York.

Polachek, H., and R. Seeger: "Shock Wave Interactions," in H. W. Emmons (ed.), *Fundamentals of Gasdynamics*, vol. 3, Princeton Series in High Speed Aerodynamics and Jet Propulsion, Princeton University Press, Princeton, N.J.

Stever, H. G.: "Condensation Phenomena in High Speed Flows," in H. W. Emmons (ed.), *Fundamentals of Gasdynamics*, vol. 3, Princeton Series in High Speed Aerodynamics and Jet Propulsion, Princeton University Press, Princeton, N.J.

Taylor, G. I., and R. S. Tankin: "Gas Dynamical Aspects of Detonation," in H. W. Emmons (ed.), *Fundamentals of Gasdynamics*, vol. 3, Princeton Series in High Speed Aerodynamics and Jet Propulsion, Princeton University Press, Princeton, N.J.

1959 Greenspan, M.: "Rotational Relaxation in Nitrogen, Oxygen, and Air," *J. Acoust. Soc. Amer.*, **31**(2): 155.

Herzfeld, K. F., and T. A. Litovitz: *Absorption and Dispersion of Ultrasonic Waves*, Academic, New York.

Landau, L. D., and E. M. Lifshitz: *Fluid Mechanics*, trans. from Russian by J. Sykes and W. Reid, Pergamon, New York.

Sedov, L. I.: *Similarity and Dimensional Methods in Mechanics*, M. Holt (ed.), trans. from Russian, Academic, New York.

Serrin, J.: "Mathematical Principles of Classical Fluid Mechanics," in S. Flugge and C. Truesdell (eds.), *Handbuch der Physik*, vol. 8, pt. 1, Springer-Verlag, Berlin.

1960 Birkhoff, G.: *Hydrodynamics*, Princeton University Press, Princeton, N.J.

Eckart, C.: *Hydrodynamics of Oceans and Atmospheres*, Pergamon, New York.

Resnick, R., and D. Halliday: *Physics*, 2 vols., Wiley, New York.

Stanyukovich, K. P.: *Unsteady Motion of Continuous Media*, English trans., M. Holt (ed.), Pergamon, New York.

Truesdell, C., and R. Toupin: "The Classical Field Theories," in S. Flugge (ed.), *Handbuch der Physik*, vol. 3, pt. 1, Springer-Verlag, Berlin.

1961 Morse, P. M., and K. Ingard: "Linear Acoustic Theory," in S. Flugge (ed.), *Handbuch der Physik*, vol. 11, pt. 1, Springer-Verlag, Berlin.

1962 Aris, R.: *Vectors, Tensors, and the Basic Equations of Fluid Mechanics*, Prentice-Hall, Englewood Cliffs, N.J.

Bradley, J. N.: *Shock Waves in Chemistry and Physics*, Wiley, New York.

Hodgman, C. D. (ed.): *Handbook of Chemistry and Physics*, Chemical Rubber, Cleveland.

Lighthill, M. J.: "Sound Generated Aerodynamically," Bakerian Lecture, *Proc. Roy. Soc. (London)*, **A267**: 147–182.

Ryshov, O. S., and G. M. Shefter: "On the Energy of Acoustic Waves Propagating in Moving Media," *Prikl. Math. Mekh.*, **26**: 854; English trans., *Appl. Math. Mech.*, **26**: 1293 (1962).

1963 Gaydon, A. G., and I. Hurle: *The Shock Tube in High Temperature Chemical Physics*, Reinhold, New York.

Murphy, G., J. Shippy, and H. Luo: *Engineering Analogies*, Iowa State University Press, Ames.

Peterka, A. J.: "Hydraulic Design of Stilling Basins and Energy Dissipators," *U.S. Dept. Interior, Bur. Reclamation, Eng. Monograph* 25.

1964 Blitz, J.: *Elements of Acoustics*, Butterworth, London.

Coleman, B. D., and V. Mizel: "Existence of Caloric Equations of State in Thermodynamics," *J. Chem. Phys.*, **40**(4): 1116–1125.

Hansen, A. G.: *Similarity Analyses of Boundary Value Problems in Engineering*, Prentice-Hall, Englewood Cliffs, N.J.

Hirschfelder, J. O., C. Curtiss, and R. Bird: *Molecular Theory of Gases and Liquids*, Wiley, New York.

Mechtly, E. A.: "The International System of Units; Physical Constants and Conversion Factors," *NASA Spec. Publ.* 7012.

Stewartson, K.: *The Theory of Laminar Boundary Layers in Compressible Fluids*, Oxford University Press, London.

1965 Born, M., and E. Wolf: *Principles of Optics*, Pergamon, New York.

Coleman, B., et al.: *Wave Propagation in Dissipative Materials*, Springer, New York.

Davis, W. C., et al.: "Failure of the Chapman-Jouguet Theory for Liquid and Solid Explosives," *Phys. Fluids*, **8**: 2169–2182.

Fay, J. A., *Molecular Thermodynamics*, Addison-Wesley, Reading, Mass.

Harlow, F. H., and J. E. Welch: "Numerical Calculation of Time-dependent Viscous Incompressible Flow of Fluid with Free Surface," *Phys. Fluids*, **8**: 2182–2189.

Kline, S. J.: *Similitude and Approximation Theory*, McGraw-Hill, New York.

Mednikov, E. P.: *Acoustic Coagulation and Precipitation of Aerosols*, Consultants Bureau, New York (trans. from the Russian, U.S.S.R. Academy of Sciences 1963 publication).

Pennelegion, L., R. F. Cash, and D. F. Bedder: "Design and Operating Features of the N.P.L. 6 in. Shock Tunnel," *Aeron. Res. Council R&M* 3449, H.M. Stationery Office, London.

Stull, D. R., et al.: *JANAF Thermochemical Tables* (Joint Army, Navy, Air Force Tables), U.S. Dept. Commerce PB-168-370; see also Addenda to these tables, PB-168-370-1 (1966) and PB-168-370-2 (1967).

Tricker, R.: *Bores, Breakers, Waves and Wakes*, American Elsevier, New York.

Vincenti, W. G., and C. Kruger: *Introduction to Physical Gas Dynamics*, Wiley, New York.

1966 Anon.: *Blasters' Handbook*, E. I. duPont de Nemours and Company, Wilmington.

Attinger, E. O.: "Hydrodynamics of Blood Flow," in V. T. Chow (ed.), *Advances in Hydroscience*, vol. 3, Academic, New York.

Back, L. H., and R. F. Cuffel: "Detection of Oblique Shocks in a Conical Nozzle with a Circular-arc Throat," *AIAA J.*, **4**(12): 2219–2221.

Harlow, F. H., and W. E. Pracht: "Formation and Penetration of High-speed Collapse Jets," *Phys. Fluids*, **9**(10): 1951–1959.

Harris, C. M.: "Absorption of Sound in Air versus Humidity and Temperature," *J. Acoust. Soc. Amer.*, **40**(1): 148–159.

Hayes, W. D., and R. F. Probstein: *Hypersonic Flow Theory*, vol. 1, *Inviscid Flows*, Academic, New York.

Herbert, M., and R. Herd: "Boundary-layer Separation in Propelling Nozzles," *Aeron. Res. Council R&M* 2421, H.M. Stationery Office, London.

Karamcheti, K.: *Principles of Ideal-fluid Aerodynamics*, Wiley, New York.

Pearson, J. M.: *A Theory of Waves*, Allyn and Bacon, Boston.

Reid, R. C., and T. K. Sherwood: *The Properties of Gases and Liquids*, 2d ed., McGraw-Hill, New York.

Zemansky, M. W., and H. C. Van Ness: *Basic Engineering Thermodynamics*, McGraw-Hill, New York.

1967 Batchelor, G. K.: *An Introduction to Fluid Dynamics*, Cambridge University Press, London.

Fox, E. A.: *Mechanics*, Harper & Row, New York.

Meyer, C. A., R. B. McClintock, G. J. Silvestri, and R. C. Spencer, Jr.: *Thermodynamic and Transport Properties of Steam*, American Society of Mechanical Engineers, New York.

Rajaratnam, N.: "Hydraulic Jumps," in V. T. Chow (ed.), *Advances in Hydroscience*, vol. 4, Academic, New York.

Schaaffs, W.: *Molecular Acoustics*, vol. 2, pt. 5 (n.s.) in K.-H. Hellwege and A. M. Hellwege (eds.), *Landolt-Börnstein Encyclopedia of Numerical Data*, Springer-Verlag, Berlin.

Zel′dovich, Y. B., and Y. P. Raizer: *Physics of Shock Waves and High Temperature Hydrodynamic Phenomena*, vol. 2, W. D. Hayes and R. F. Probstein (eds.), Academic, New York.

1968 Bondi, A.: *Molecular Crystals, Liquids, and Glasses*, Wiley, New York.

Evans, H. L.: *Laminar Boundary Layer Theory*, Addison-Wesley, Reading, Mass.

Hayes, W. D.: "Energy Invariant for Geometric Acoustics in a Moving Medium," *Phys. Fluids*, **11**(8): 1654–1656.

Lawrence, R. A., and E. E. Weynand: "Factors Affecting Flow Separation in Contoured Supersonic Nozzles," *AIAA J.*, **6**(6): 1159–1160.

Morse, P. M., and K. Ingard: *Theoretical Acoustics*, McGraw-Hill, New York

Muntz, E. P.: "Molecular Velocity Distribution-function Measurements in a Flowing Gas," *Phys. Fluids*, **11**(1): 64–76.

Schlichting, H.: *Boundary Layer Theory*, 6th ed., trans. from German by J. Kestin, McGraw-Hill, New York.

Truesdell, C.: *Essays in the History of Mechanics*, Springer-Verlag, Berlin.

1969 Byrkin, A. P.: "Exact Solution of the Navier-Stokes Equations for a Compressible Gas," *Prikl. Math. Mekh.*, **33**(1): 582–584.

Coltman, J. W.: "Sound Radiation from the Mouth of an Organ Pipe," *J. Acoust. Soc. Amer.*, **46**(2): 477.

Edwards, D. H.: "A Survey of Recent Work on the Structure of Detonation Waves," 12th International Combustion Symposium, Combustion Institute, Reinhold, New York.

Harlow, F. H.: "Numerical Methods for Fluid Dynamics: An Annotated Bibliography," *Los Alamos Sci. Lab. Rept.* LA-4281.

Keenan, J. H., F. G. Keyes, P. G. Hill, and J. G. Moore: *Steam Tables*, Wiley, New York.

Laumbach, D. D., and R. F. Probstein: "A Point Explosion in a Cold Exponential Atmosphere," *J. Fluid Mech.*, **35**(1): 53–75.

Loh, W. H. T.: "Theory of the Hydraulic Analogy for Steady and Unsteady Gas Dynamics," in W. H. T. Loh (ed.), *Modern Developments in Gas Dynamics*, Plenum, New York.

Malecki, I.: *Physical Foundations of Technical Acoustics*, trans. from Polish by I. Bellert, Pergamon, New York.

Muntz, E. P., and L. N. Harnett: "Molecular Velocity Distribution Measurements in a Normal Shock Wave," *Phys. Fluids*, **12**(10): 2027–2035.

Piercy, J. E.: "Role of the Vibrational Relaxation of Nitrogen in the Absorption of Sound in Air," *J. Acoust. Soc. Amer.*, **46**(3): 602–604.

Rowlinson, J. S.: *Liquids and Liquid Mixtures*, Butterworth, London.

Schafer, Klaus (ed.): *Transportphanomene I*, Landolt-Bornstein (Eigenschaften der Materie in ihren Aggregatzustanden), vol. 2, pt. 5a, Springer-Verlag, Berlin.

Schmitt–von Schubert, B.: "Existence and Uniqueness of Normal Shock Waves in Gas-Particle Mixtures," *J. Fluid Mech.*, **38**(3): 633–655.

Sears, W. R.: "Aerodynamics, Noise, and the Sonic Boom," *AIAA J.*, **7**(4): 577–586.

Wallis, G. B.: *One-dimensional Two-phase Flow*, McGraw-Hill, New York.

1970 Harlow, F. H., and A. A. Amsden: *Fluid Dynamics: An Introductory Text*, (LA–4100) Los Alamos Scientific Laboratory of the University of California, Los Alamos.

Huntley, R.: "A Bel Is Ten Decibels," *Sound and Vibration*, **4**(1): 22.

Klein, M. J.: "Maxwell, His Demon, and the Second Law of Thermodynamics," *Amer. Scientist*, **58**(1): 84–97.

Marble, F. E.: "Dynamics of Dusty Gases," *Ann. Rev. Fluid Mech.*, **2**.

Walbridge, N. E., and L. A. Woodward: "Phase Velocity of Surface Capillary-Gravity Waves," *Phys. Fluids*, **13**(10): 2461–2465.

appendix A
weak-shock conditions

The entropy jump $[s]$ can be found by the method given in Sec. 7.3; retaining terms up to fourth order in $[P]$,

$$[s] = \frac{1}{12T_1}\left(\frac{\partial^2 v}{\partial P^2}\right)_s [P]^3 - \frac{1}{24T_1}\left[\frac{1}{T_1}\left(\frac{\partial T}{\partial P}\right)_s\left(\frac{\partial^2 v}{\partial P^2}\right)_s - \left(\frac{\partial^3 v}{\partial P^3}\right)_s\right][P]^4 \cdots \tag{A.1}$$

where the derivatives are all evaluated at the upstream state 1 (this convention will be used in all that follows). It is helpful to put this equation in the nondimensional form

$$\frac{T_1[s]}{c_1^2} = \frac{\Gamma_1}{6}\Pi^3 - \frac{1}{12}\left[\frac{c_1^2\Gamma_1}{v_1 T_1}\left(\frac{\partial T}{\partial P}\right)_s - \frac{c_1^6}{2v_1^4}\left(\frac{\partial^3 v}{\partial P^3}\right)_s\right]\Pi^4 \cdots \tag{A.2}$$

where Γ is the nondimensional form of the fundamental gasdynamic derivative,

$$\Gamma \equiv \frac{c^4}{2v^3}\left(\frac{\partial^2 v}{\partial P^2}\right)_s \tag{A.3}$$

and is of order unity [for a perfect gas, $\Gamma = (\gamma + 1)/2$]. In highly *exceptional* cases $\Gamma < 0$, with the result that only rarefaction shocks with $[P] < 0$ can satisfy the requirement that $[s] > 0$. The nondimensional pressure jump Π in (A.2) is

$$\Pi \equiv \frac{v_1[P]}{c_1^2} \tag{A.4}$$

By definition, *a weak shock has negligible entropy change.* Then a consequence of (A.2) is that the condition for weakness is that Π be small compared to unity; equivalent conditions can be put in terms of $[w]$, $[v]$, or M_{1n}, as will be seen below. Thus if

$$\left.\begin{array}{c} |\Pi| \\ \dfrac{|[w]|}{c_1} \\ \dfrac{|[v]|}{v_1} \\ |M_{1n} - 1| \end{array}\right\} \ll 1 \tag{A.5}$$

which will necessarily be satisfied mutually, the entropy jump will be negligible and the shock is weak. In practice, it may be reasonable to treat a given shock as isentropic even if the quantities in (A.5) are near unity, simply because the coefficients in (A.2) are generally small. Note, however, that for substances such as liquid water ($c_1{}^2/v_1 \approx 2.2 \times 10^4$ atm) condition (A.5) may be satisfied for shocks with $[P]$ in the neighborhood of 1,000 atm. For perfect gases $c_1{}^2/v_1 = \gamma P_1$, and the condition is far more stringent.

In the following all expressions are given in terms of the nondimensional pressure jump Π, which is thus the independent parameter defining the strength of the shock in question, and are carried out in general only to the third power in this quantity. If expressions are wanted in terms of another parameter, such as $[w]/c_1$, they can be obtained by substitution or reversion.

Retaining only the first term in each of the expansions below gives the equations for an *acoustic discontinuity* (Chap. 4) which propagates with the speed of sound c_1 into the upstream fluid. Such a discontinuity is thus a twice-degenerate shock with $|\Pi| \lll 1$.

In practice, only the first two terms in each of the expansions are usually of interest.

The jump in specific volume v is given by a direct Taylor expansion of $v(P,s)$, retaining terms only up to order $[P]^3$:

$$[v] = \left(\frac{\partial v}{\partial P}\right)_s [P] + \frac{1}{2}\left(\frac{\partial^2 v}{\partial P^2}\right)_s [P]^2 + \frac{1}{6}\left(\frac{\partial^3 v}{\partial P^3}\right)_s [P]^3 + \cdots + \left(\frac{\partial v}{\partial s}\right)_p [s] + \cdots$$

We have also the identities

$$\left(\frac{\partial v}{\partial P}\right)_s = -\frac{v^2}{c^2} \qquad \left(\frac{\partial v}{\partial s}\right)_p = \left(\frac{\partial T}{\partial P}\right)_s$$

Inserting these with Eq. (A.1) into the above gives

$$\frac{-[v]}{v_1} = \Pi - \Gamma_1\Pi^2 - \frac{1}{6}\left[\frac{c_1{}^2\Gamma_1}{v_1T_1}\left(\frac{\partial T}{\partial P}\right)_s + \frac{c_1{}^6}{v_1{}^4}\left(\frac{\partial^3 v}{\partial P^3}\right)_s\right]\Pi^3 \cdots \tag{A.6}$$

The jump in the relative normal velocity $[w]$ is found from continuity and momentum to be

$$-[w] = \sqrt{-[P][v]} \tag{A.7}$$

Substituting Eq. (A.6) into this gives

$$-\frac{[w]}{c_1} = \Pi - \frac{\Gamma_1}{2}\Pi^2 - \frac{1}{4}\left[\frac{c_1{}^2\Gamma_1}{3v_1T_1}\left(\frac{\partial T}{\partial P}\right)_s + \frac{\Gamma_1{}^2}{2} + \frac{c_1{}^6}{3v_1{}^4}\left(\frac{\partial^3 v}{\partial P^3}\right)_s\right]\Pi^3 \cdots \tag{A.8}$$

The jump in sound speed is found by Taylor expansion of $c(P,s)$

$$[c] = \left(\frac{\partial c}{\partial P}\right)_s [P] + \frac{1}{2}\left(\frac{\partial^2 c}{\partial P^2}\right)[P]^2$$

$$+ \frac{1}{6}\left(\frac{\partial^3 c}{\partial P^3}\right)_s [P]^3 + \cdots + \left(\frac{\partial c}{\partial s}\right)_p [s] + \cdots$$

where the derivatives are found to be

$$\left(\frac{\partial c}{\partial P}\right)_s = -\frac{v}{c} + \frac{c^3}{2v^2}\left(\frac{\partial^2 v}{\partial P^2}\right)_s$$

$$\left(\frac{\partial c}{\partial s}\right)_p = \frac{c}{v}\left(\frac{\partial T}{\partial P}\right)_s + \frac{c^3}{2v^2}\left(\frac{\partial^2 T}{\partial P^2}\right)_s$$

and the jump is then

$$\frac{[c]}{c_1} = (\Gamma_1 - 1)\Pi + \frac{1}{2}\left[3\Gamma_1{}^2 + \frac{c_1{}^6}{2v_1{}^4}\left(\frac{\partial^3 v}{\partial P^3}\right)_s\right]\Pi^2$$

$$+ \frac{1}{6}\left[\frac{\Gamma_1 c_1{}^4}{2v_1{}^2T_1}\left(\frac{\partial^2 T}{\partial P^2}\right)_s + \frac{\Gamma_1 c_1{}^2}{v_1T_1}\left(\frac{\partial T}{\partial P}\right)_s + 15\Gamma_1{}^3 - 3\Gamma_1{}^2\right.$$

$$\left. + \frac{c_1{}^6}{2v_1{}^4}(9\Gamma_1 - 1)\left(\frac{\partial^3 v}{\partial P^3}\right)_s + \frac{c_1{}^8}{2v_1{}^5}\left(\frac{\partial^4 v}{\partial P^4}\right)_s\right]\Pi^3 + \cdots \tag{A.9}$$

The series (A.8) can be reverted to give

$$\Pi = -\frac{[w]}{c_1} + \frac{\Gamma_1}{2}\left(\frac{[w]}{c_1}\right)^2 - \frac{1}{4}\left[\frac{c_1{}^2\Gamma_1}{3v_1T_1}\left(\frac{\partial T}{\partial P}\right)_s + \frac{5\Gamma_1{}^2}{2} + \frac{c_1{}^6}{3v_1{}^4}\left(\frac{\partial^3 v}{\partial P^3}\right)_s\right]\left(\frac{[w]}{c_1}\right)^3 \cdots \tag{A.10}$$

By Eq. (7.20) the shock Mach number M_{1n} may be written

$$M_{1n} = \frac{-\Pi}{[w]/c_1} \tag{A.11}$$

and Eq. (A.10) gives directly

$$M_{1n} = 1 - \frac{\Gamma_1}{2}\frac{[w]}{c_1} + \frac{1}{4}\left[\frac{c_1{}^2\Gamma_1}{3v_1T_1}\left(\frac{\partial T}{\partial P}\right)_s + \frac{5\Gamma_1{}^2}{2} + \frac{c_1{}^6}{3v_1{}^4}\left(\frac{\partial^3 v}{\partial P^3}\right)_s\right]\left(\frac{[w]}{c_1}\right)^2 \cdots \tag{A.12}$$

Substitution of (A.8) gives this in terms of Π,

$$M_{1n} = 1 + \frac{\Gamma_1}{2}\Pi + \frac{1}{4}\left[\frac{c_1{}^2\Gamma_1}{3v_1T_1}\left(\frac{\partial T}{\partial P}\right)_s + \frac{3\Gamma_1{}^2}{2} + \frac{c_1{}^6}{3v_1{}^4}\left(\frac{\partial^3 v}{\partial P^3}\right)_s\right]\Pi^2 + \cdots \tag{A.13}$$

Note that $M_{1n} \geq 1$ for both compression shocks ($\Gamma_1 > 0$, $\Pi > 0$) and rarefaction shocks ($\Gamma_1 < 0$, $\Pi < 0$), as is required for shock stability.

The downstream Mach number M_{2n} is

$$M_{2n} \equiv \frac{w_2}{c_2} = \frac{w_1 + [w]}{c_1 + [c]}$$

and substituting (A.8) and (A.9) gives

$$M_{2n} = 1 - \frac{\Gamma_1}{2}\Pi + \frac{1}{4}\left[\frac{c_1{}^2\Gamma_1}{3v_1T_1}\left(\frac{\partial T}{\partial P}\right)_s - \frac{5\Gamma_1{}^2}{2} - \frac{2c_1{}^6}{3v_1{}^4}\left(\frac{\partial^3 v}{\partial P^3}\right)_s\right]\Pi^2 \cdots \tag{A.14}$$

To first power in Π, we have from (A.13) and (A.14) the elegant relation

$$M_{1n} - 1 \approx 1 - M_{2n} \tag{A.15}$$

Note that the downstream flow is always subsonic.

Weak Shock in a Perfect Gas

For a perfect gas the derivatives in the above formulas are

$$\left(\frac{\partial v}{\partial P}\right)_s = -\frac{v}{\gamma P} \qquad \left(\frac{\partial T}{\partial P}\right)_s = \frac{\gamma - 1}{\gamma}\frac{T}{P}$$

$$\left(\frac{\partial^2 v}{\partial P^2}\right)_s = \frac{\gamma + 1}{\gamma^2}\frac{v}{P^2} \qquad \left(\frac{\partial^2 T}{\partial P^2}\right)_s = -\frac{\gamma - 1}{\gamma^2}\frac{T}{P^2}$$

$$\left(\frac{\partial^3 v}{\partial P^3}\right)_s = -\frac{(\gamma + 1)(2\gamma + 1)}{\gamma^3}\frac{v}{P^3}$$

$$\left(\frac{\partial^4 v}{\partial P^4}\right)_s = +\frac{(\gamma + 1)(2\gamma + 1)(3\gamma + 1)}{\gamma^4}\frac{v}{P^4}$$

In addition we have the identities

$$c^2 = \gamma P v = \gamma R T \qquad \Gamma = \frac{\gamma + 1}{2} \qquad \Pi = \frac{[P]}{\gamma P_1}$$

Substitution into Eqs. (A.2), (A.6), (A.8), (A.9), (A.13), and (A.14) then gives the relations below, in which the shock-strength parameter is $[P]/P_1$, as is conventional for a perfect gas,

$$\frac{[s]}{R} = \frac{\gamma + 1}{12\gamma^2}\left(\frac{[P]}{P_1}\right)^3 - \frac{\gamma + 1}{8\gamma^2}\left(\frac{[P]}{P_1}\right)^4 \cdots \tag{A.16}$$

$$\frac{-[v]}{v_1} = \frac{[P]}{\gamma P_1} - \frac{\gamma + 1}{2\gamma^2}\left(\frac{[P]}{P_1}\right)^2 + \frac{(\gamma + 1)^2}{4\gamma^3}\left(\frac{[P]}{P_1}\right)^3 \cdots \tag{A.17}$$

$$\frac{-[w]}{c_1} = \frac{[P]}{\gamma P_1} - \frac{\gamma + 1}{4\gamma^2}\left(\frac{[P]}{P_1}\right)^2 + \frac{3(\gamma + 1)^2}{32\gamma^3}\left(\frac{[P]}{P_1}\right)^3 \cdots \tag{A.18}$$

$$\frac{[c]}{c_1} = \frac{\gamma - 1}{2\gamma}\frac{[P]}{P_1} - \frac{(\gamma - 1)(\gamma + 1)}{8\gamma^2}\left(\frac{[P]}{P_1}\right)^2 + \frac{(\gamma - 1)(\gamma + 1)^2}{16\gamma^3}\left(\frac{[P]}{P_1}\right)^3 \cdots \tag{A.19}$$

$$M_{1n} = 1 + \frac{\gamma + 1}{4\gamma}\frac{[P]}{P_1} - \frac{(\gamma + 1)^2}{32\gamma^2}\left(\frac{[P]}{P_1}\right)^2 \cdots \tag{A.20}$$

$$M_{2n} = 1 - \frac{\gamma + 1}{4\gamma}\frac{[P]}{P_1} + \frac{(\gamma + 1)(7\gamma - 1)}{32\gamma^2}\left(\frac{[P]}{P_1}\right)^2 \cdots \tag{A.21}$$

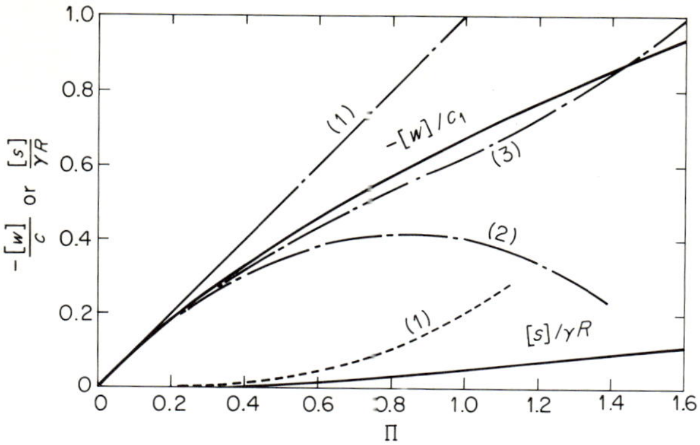

Figure A.1
Velocity jump and entropy jump as functions of pressure jump. The exact results compared to the series expansions including one, two, and three terms, as labeled. (Perfect gas with $\gamma = 1.40$.)

Some idea of the range of applicability of the above series expressions can be obtained from Fig. A.1.

appendix B
basic thermodynamic and vector identities

Thermodynamic Identities

The identities given here are based on the several forms of the Gibbs equation, the definitions of certain quantities such as the specific heats and sound speed, and finally the calculus of derivatives. The derivations, most of which can be found in standard books on thermodynamics, are omitted. For convenience, equations are given in terms of the specific volume v rather than its reciprocal, the density ρ.

The Gibbs Equation

The various forms of the Gibbs equation are

$$de = T\,ds - P\,dv \qquad e \equiv e \tag{B.1a}$$

$$dh = T\,ds + v\,dP \qquad h \equiv e + Pv \tag{B.1b}$$

$$da = -s\,dT - P\,dv \qquad a \equiv e - Ts \tag{B.1c}$$

$$dg = -s\,dT + v\,dP \qquad g \equiv h - Ts \tag{B.1d}$$

In direct consequence we have the following relations, listed in line-for-line correspondence to the above:

$$T = \left(\frac{\partial e}{\partial s}\right)_v \qquad P = -\left(\frac{\partial e}{\partial v}\right)_s \qquad \left(\frac{\partial T}{\partial v}\right)_s = -\left(\frac{\partial P}{\partial s}\right)_v \tag{B.2a}$$

$$T = \left(\frac{\partial h}{\partial s}\right)_p \qquad v = \left(\frac{\partial h}{\partial P}\right)_s \qquad \left(\frac{\partial T}{\partial P}\right)_s = \left(\frac{\partial v}{\partial s}\right)_p \tag{B.2b}$$

$$s = -\left(\frac{\partial a}{\partial T}\right)_v \qquad P = -\left(\frac{\partial a}{\partial v}\right)_T \qquad \left(\frac{\partial s}{\partial v}\right)_T = \left(\frac{\partial P}{\partial T}\right)_v \tag{B.2c}$$

$$s = -\left(\frac{\partial g}{\partial T}\right)_p \qquad v = \left(\frac{\partial g}{\partial P}\right)_T \qquad \left(\frac{\partial s}{\partial P}\right)_T = -\left(\frac{\partial v}{\partial T}\right)_p \tag{B.2d}$$

The identities in the right-hand column are the familiar Maxwell relations.

Definitions

The following quantities are physically important:

Specific heat at constant volume $c_v \equiv \left(\frac{\partial e}{\partial T}\right)_v$

Specific heat at constant pressure $c_p \equiv \left(\frac{\partial h}{\partial T}\right)_p$

Ratio of specific heats $\gamma \equiv \frac{c_p}{c_v}$

Speed of sound $c \equiv \sqrt{\left(\frac{\partial P}{\partial \rho}\right)_s}$

Fundamental gasdynamic derivative $\Gamma \equiv \frac{c^4}{2v^3}\left(\frac{\partial^2 v}{\partial P^2}\right)_s$

No special symbols will be introduced for quantities such as $(\partial v/\partial T)_p$, which have an obvious physical meaning.

Dependence of the Internal Energy and Enthalpy on Pressure

$$\left(\frac{\partial e}{\partial P}\right)_T = -P\left(\frac{\partial v}{\partial P}\right)_T - T\left(\frac{\partial v}{\partial T}\right)_p \tag{B.3}$$

$$\left(\frac{\partial h}{\partial P}\right)_T = v - T\left(\frac{\partial v}{\partial T}\right)_p \tag{B.4}$$

Identities Involving the Specific Heats

$$c_v = T\left(\frac{\partial s}{\partial T}\right)_v \qquad c_p = T\left(\frac{\partial s}{\partial T}\right)_p \tag{B.5}$$

$$c_p - c_v = -T\left(\frac{\partial P}{\partial v}\right)_T \left(\frac{\partial v}{\partial T}\right)_p^2 \geq 0 \tag{B.6}$$

$$\gamma - 1 = \frac{T}{c_p}\left[\frac{c}{v}\left(\frac{\partial v}{\partial T}\right)_p\right]^2 \geq 0 \tag{B.7}$$

$$\left(\frac{\partial s}{\partial v}\right)_p = \frac{c_p}{T}\left(\frac{\partial T}{\partial v}\right)_p = \left(\frac{\partial P}{\partial T}\right)_s \tag{B.8}$$

$$\left(\frac{\partial c_p}{\partial P}\right)_T = -T\left(\frac{\partial^2 v}{\partial T^2}\right)_p \tag{B.9}$$

Isentropic Derivatives

$$\left(\frac{\partial P}{\partial \rho}\right)_s = -v^2\left(\frac{\partial P}{\partial v}\right)_s \tag{B.10}$$

$$\left(\frac{\partial P}{\partial v}\right)_s = \gamma\left(\frac{\partial P}{\partial v}\right)_T \tag{B.11}$$

$$\left(\frac{\partial T}{\partial v}\right)_s = -(\gamma - 1)\left(\frac{\partial T}{\partial v}\right)_p \tag{B.12}$$

$$\left(\frac{\partial T}{\partial v}\right)_s^2 = \frac{(\gamma - 1)c^2 T}{v^2 c_p} \tag{B.13}$$

$$\left(\frac{\partial T}{\partial P}\right)_s = \frac{(\gamma - 1)v^2}{c^2}\left(\frac{\partial T}{\partial v}\right)_p \tag{B.14}$$

Note that isentropic derivatives can be converted to derivatives of the entropy via Maxwell's relations.

Derivatives Related to the Sound Speed and Fundamental Gasdynamic Derivative

$$c^2 \equiv \left(\frac{\partial P}{\partial \rho}\right)_s = -v^2\left(\frac{\partial P}{\partial v}\right)_s > 0 \tag{B.15}$$

$$\Gamma \equiv \frac{c^4}{2v^3}\left(\frac{\partial^2 v}{\partial P^2}\right)_s = \frac{1}{c}\left(\frac{\partial \rho c}{\partial \rho}\right)_s \tag{B.16}$$

$$\left(\frac{\partial^2 P}{\partial v^2}\right)_s = \frac{c^6}{v^6}\left(\frac{\partial^2 v}{\partial P^2}\right)_s = \frac{2c^2}{v^3}\Gamma \tag{B.17}$$

$$\left(\frac{\partial^2 P}{\partial \rho^2}\right)_s = 2vc^2(\Gamma - 1) = \left(\frac{\partial c^2}{\partial \rho}\right)_s \tag{B.18}$$

$$\left(\frac{\partial c}{\partial s}\right)_p = \frac{c}{v}\left(\frac{\partial T}{\partial P}\right)_s + \frac{c^3}{2v^2}\left(\frac{\partial^2 T}{\partial P^2}\right)_s \tag{B.19}$$

$$\left(\frac{\partial c}{\partial P}\right)_s = \frac{v}{c}(\Gamma - 1) \tag{B.20}$$

$$\Gamma - 1 = \frac{c}{v}\left(\frac{\partial c}{\partial P}\right)_T + \frac{cT}{vc_p}\left(\frac{\partial v}{\partial T}\right)_p\left(\frac{\partial c}{\partial T}\right)_p \tag{B.21}$$

$$\left(\frac{\partial^2 c}{\partial P^2}\right)_s = \frac{c^3}{2v^2}\left(\frac{\partial^3 v}{\partial P^3}\right)_s + \frac{3v^2}{c^3}\Gamma^2 \tag{B.22}$$

$$\left(\frac{\partial h}{\partial v}\right)_s = -\frac{c^2}{v} \tag{B.23}$$

$$\left(\frac{\partial^2 h}{\partial v^2}\right)_s = \frac{c^2}{v^2}(2\Gamma - 1) \tag{B.24}$$

Thermodynamic Derivatives for an Ideal Gas

For this important special case we have

$$Pv = RT \tag{B.25}$$

$$c_p - c_v = R \qquad c_p = \frac{\gamma}{\gamma - 1}R \qquad c_v = \frac{1}{\gamma - 1}R \tag{B.26}$$

$$c^2 = \gamma P v = \gamma R T \tag{B.27}$$

$$e, h, c_p, c_v, \gamma = \text{functions of } T \text{ only} \tag{B.28}$$

If γ (and therefore c_p and c_v) is constant, the gas is by definition *perfect*.

A useful rule to keep in mind when evaluating the derivatives for a perfect gas is that the quotient form in the derivative is reproduced in its explicit form, with a multiplicative polynomial in γ; for example,

$$\left(\frac{\partial P}{\partial v}\right)_s = -\gamma\frac{P}{v}$$

If the derivative is with respect to the entropy, however, the quotient form will have the gas constant in place of the entropy, e.g.,

$$\left(\frac{\partial s}{\partial v}\right)_T = \frac{R}{v}$$

For perfect gases, this rule is applicable to derivatives of all orders, e.g.,

$$\left(\frac{\partial^3 v}{\partial P^3}\right)_s = -\frac{(\gamma + 1)(2\gamma + 1)}{\gamma^3}\frac{v}{P^3}$$

but for ideal gases with $\gamma = \gamma(T)$ it is applicable only to first derivatives.

Divergence Theorem (Gauss' Theorem)

Let S be a closed surface, with outward unit normal $\mathbf{n}$, bounding the volume V. Then if $\mathbf{F}(\mathbf{x},t)$ is a *vector* field, the divergence theorem is

$$\int_S \mathbf{n} \cdot \mathbf{F}\, dA = \int_V \nabla \cdot \mathbf{F}\, dV \tag{B.29}$$

where dA is an increment of area on S and dV is an increment of volume in V. This can also be written

$$\int_S n_i F_i\, dA = \int_V F_{i,i}\, dV \tag{B.30}$$

If $F(\mathbf{x},t)$ is a *scalar* field, the divergence theorem is

$$\int_S \mathbf{n} F\, dA = \int_V \nabla F\, dV \tag{B.31}$$

This can be derived directly from (B.29). The corresponding indicial form is

$$\int_S \mathbf{e}_i n_i F\, dA = \int_V \mathbf{e}_i F_{,i}\, dV \tag{B.32}$$

with the component statements

$$\int_S n_i F\, dA = \int_V F_{,i}\, dV \tag{B.33}$$

If $\mathsf{F}(\mathbf{x},t)$ is a *tensor* field, the divergence theorem in dyadic notation is [compare (B.29)]

$$\int_S \mathbf{n} \cdot \mathsf{F}\, dA = \int_V \nabla \cdot \mathsf{F}\, dV \tag{B.34}$$

This can be put in indicial notation from the correspondences $\mathbf{n} \cdot \mathbf{F} \sim n_i F_{ik}$ and $\nabla \cdot \mathbf{F} \sim F_{ik,i}$, yielding [compare (B.30)]

$$\int_S n_i F_{ik}\, dA = \int_V F_{ik,i}\, dV \tag{B.35}$$

Stokes' Theorem

Let S be a surface bounded by the simple closed curve C. Then if $\mathbf{F}(\mathbf{x},t)$ is a continuous vector field, Stokes' theorem is

$$\oint_C \mathbf{F} \cdot \mathbf{t}\, dl = \int_S (\nabla \times \mathbf{F}) \cdot \mathbf{n}\, dA \tag{B.36}$$

where $\mathbf{n}$ is a unit vector normal to S, $\mathbf{t}$ is a unit vector tangential to C, and dl is an increment of length along C.

Determinant Expansion for the Curl

In Cartesian coordinates, if $\mathbf{F}(\mathbf{x},t)$ is a vector field

$$\nabla \times \mathbf{F} = \begin{vmatrix} \mathbf{e}_1 & \mathbf{e}_2 & \mathbf{e}_3 \\ \dfrac{\partial}{\partial x_1} & \dfrac{\partial}{\partial x_2} & \dfrac{\partial}{\partial x_3} \\ F_1 & F_2 & F_3 \end{vmatrix} \tag{B.37}$$

Vector Identities[1]

$$\nabla(uv) = u\,\nabla v + v\,\nabla u \tag{B.38}$$

$$\nabla \cdot (\phi \mathbf{v}) = \phi \nabla \cdot \mathbf{v} + \nabla \phi \cdot \mathbf{v} \tag{B.39}$$

$$\nabla \times (\phi \mathbf{v}) = \phi \nabla \times \mathbf{v} + \nabla \phi \times \mathbf{v} \tag{B.40}$$

$$\nabla \times (\nabla \phi) = 0 \tag{B.41}$$

$$\nabla \cdot (\nabla \times \mathbf{v}) = 0 \tag{B.42}$$

$$\nabla \cdot (\mathbf{u} \times \mathbf{v}) = (\nabla \times \mathbf{u}) \cdot \mathbf{v} - (\nabla \times \mathbf{v}) \cdot \mathbf{u} \tag{B.43}$$

$$\nabla \times (\mathbf{u} \times \mathbf{v}) = (\mathbf{v} \cdot \nabla)\mathbf{u} - \mathbf{v}(\nabla \cdot \mathbf{u}) + \mathbf{u}(\nabla \cdot \mathbf{v}) - (\mathbf{u} \cdot \nabla)\mathbf{v} \tag{B.44}$$

$$\nabla \times (\nabla \times \mathbf{v}) = \nabla(\nabla \cdot \mathbf{v}) - \nabla^2 \mathbf{v} \tag{B.45}$$

[1] From H. Lass, *Vector and Tensor Analysis*. Copyright 1950. McGraw-Hill Book Company. Used by permission.

$$\nabla(\mathbf{u}\cdot\mathbf{v}) = \mathbf{u}\times(\nabla\times\mathbf{v}) + \mathbf{v}\times(\nabla\times\mathbf{u}) + (\mathbf{u}\cdot\nabla)\mathbf{v} + (\mathbf{v}\cdot\nabla)\mathbf{u} \quad (B.46)$$

$$(\mathbf{v}\cdot\nabla)\mathbf{r} = \mathbf{v} \quad (B.47)$$

$$\nabla\cdot\mathbf{r} = 3 \quad (B.48)$$

$$\nabla\times\mathbf{r} = 0 \quad (B.49)$$

$$\nabla\cdot(r^{-3}\mathbf{r}) = 0 \quad (B.50)$$

$$d\mathbf{f} = (d\mathbf{r}\cdot\nabla)\mathbf{f} + \frac{\partial\mathbf{f}}{\partial t}\,dt \quad (B.51)$$

$$d\phi = d\mathbf{r}\cdot\nabla\phi + \frac{\partial\phi}{\partial t}\,dt \quad (B.52)$$

appendix C
constants and conversions

Constants

$$\text{Universal gas constant } \tilde{R} = 8{,}314.3 \text{ J/(kg mol)(K)}$$
$$= 1{,}544.3 \text{ ft}\cdot\text{lb}_f\text{/(lb mol)(°R)}$$

$$\text{Avogadro's number } \tilde{N} = 6.0225 \times 10^{26} \text{ (kg mol)}^{-1}$$

$$\text{Boltzmann's constant } k = \frac{\tilde{R}}{\tilde{N}} = 1.3805 \times 10^{-23} \text{ J/K}$$

$$\text{Planck's constant } h = 6.6256 \times 10^{-34} \text{ J}\cdot\text{s}$$

$$\text{Speed of light in vacuum } c_0 = 2.9979 \times 10^{8} \text{ m/s}$$

$$\text{Standard atmospheric pressure } P_a = 1.01325 \times 10^{5} \text{ N/m}^2$$
$$= 14.696 \text{ lb}_f\text{/in}^2$$

$$\text{Standard acceleration of gravity } g_0 = 9.80665 \text{ m/s}^2$$
$$= 32.1740 \text{ ft/s}^2$$

Defined Conversions (Exact)

$$T(\text{K}) = t(°\text{C}) + 273.15$$
$$T(°\text{R}) = t(°\text{F}) + 459.67$$
$$T(°\text{R}) = 1.8T(\text{K})$$

$$1\ \text{ft} = 0.3048\ \text{m}$$
$$1\ \text{lb}_m = 0.45359237\ \text{kg}$$
$$1\ \text{International Table kcal} = \tfrac{1}{860}\ \text{kWh}$$
$$1\ \text{Hz} = 1\ \text{cycle/second} = 1\ \text{s}^{-1}$$
$$1\ \text{nmi (nautical mile)} = 1.852\ \text{km}$$

Conversion Factors for Units of Mass

$$1\ \text{lb}_m = 0.45359\ \text{kg}$$
$$1\ \text{slug} = 32.174\ \text{lb}_m$$

Conversion Factors for Units of Force

$$1\ \text{lb}_f = 4.4482\ \text{N}$$
$$1\ \text{kg}_f = 9.8066\ \text{N}$$

Conversion Factors for Units of Pressure

$$1\ \text{atm} = 760\ \text{mm Hg} = 1.01325 \times 10^5\ \text{N/m}^2 = 14.696\ \text{lb}_f/\text{in}^2$$
$$1\ \text{bar} = 10^5\ \text{N/m}^2 = 0.98692\ \text{atm}$$
$$1\ \text{tor} = 1\ \text{mm Hg}$$

Conversion Factors for Units of Energy

$$1\ \text{kcal} = 3.9683\ \text{Btu} = 4{,}186.0\ \text{J}$$
$$1\ \text{Btu} = 778.03\ \text{ft}\cdot\text{lb}_f = 1{,}055.04\ \text{J}$$
$$1\ \text{kWh} = 3.6 \times 10^6\ \text{J}$$
$$1\ \text{ft}\cdot\text{lb}_f = 1.3558\ \text{J}$$

Conversion Factors for Units of Viscosity

$$1\ \text{g/(cm)(s)} = 1\ \text{poise} = 10^{-1}\ \text{kg/(m)(s)} = 10^2\ \text{centipoise}$$
$$1\ \text{lb}_m/\text{(ft)(s)} = 1.4882\ \text{kg/(m)(s)}$$
$$1\ \text{stoke} = 1\ \text{cm}^2/\text{s}$$

Miscellaneous Constants

$$\pi = 3.141592653$$
$$e = 2.718281828$$
$$\ln 10 = 2.302585092$$
$$\frac{360}{2\pi} = 1 \text{ rad} = 57.29577951^\circ$$

$\log_{10} 2 = 0.30102$ $\quad \log_{10} 6 = 0.77815$
$\log_{10} 3 = 0.47712$ $\quad \log_{10} 7 = 0.84509$
$\log_{10} 4 = 0.60205$ $\quad \log_{10} 8 = 0.90308$
$\log_{10} 5 = 0.69897$ $\quad \log_{10} 9 = 0.95424$

$\sqrt{2} = 1.4142$ $\quad \sqrt{7} = 2.6458$
$\sqrt{3} = 1.7321$ $\quad \sqrt{8} = 2.8284$
$\sqrt{4} = 2$ $\quad \sqrt{9} = 3$
$\sqrt{5} = 2.2361$ $\quad \sqrt{10} = 3.1623$
$\sqrt{6} = 2.4495$ $\quad \sqrt{11} = 3.3166$

$\sqrt{\pi} = 1.7724$ $\quad 1/\sqrt{\pi} = 0.56419$

appendix D

data tables for compressible flow

Table D.1 Isentropic Flow of a Perfect Gas ($\gamma = 1.40$)†

M	P/P_0	ρ/ρ_0	T/T_0	c/c_0	A/A_*	μ	ω	M_*
0.00	1.0000	1.0000	1.0000	1.0000	∞			0
0.01	0.9999	1.0000	1.0000	1.0000	57.873			0.0100
0.02	0.9997	0.9998	0.9999	1.0000	28.942			0.0219
0.03	0.9994	0.9996	0.9998	0.9999	19.300			0.0329
0.04	0.9989	0.9992	0.9997	0.9998	14.481			0.0438
0.05	0.9983	0.9988	0.9995	0.9998	11.591			0.0548
0.06	0.9975	0.9982	0.9993	0.9996	9.666			0.0657
0.07	0.9966	0.9976	0.9990	0.9995	8.292			0.0766
0.08	0.9955	0.9968	0.9987	0.9994	7.262			0.0876
0.09	0.9944	0.9960	0.9984	0.9992	6.461			0.0985
0.10	0.9930	0.9950	0.9980	0.9990	5.822			0.1094
0.11	0.9916	0.9940	0.9976	0.9988	5.299			0.1204
0.12	0.9900	0.9928	0.9971	0.9986	4.864			0.1313
0.13	0.9883	0.9916	0.9966	0.9983	4.497			0.1422
0.14	0.9864	0.9903	0.9961	0.9980	4.182			0.1531
0.15	0.9844	0.9888	0.9955	0.9978	3.910			0.1640
0.16	0.9823	0.9873	0.9949	0.9974	3.673			0.1748
0.17	0.9800	0.9857	0.9943	0.9971	3.464			0.1857
0.18	0.9776	0.9840	0.9936	0.9968	3.278			0.1965
0.19	0.9751	0.9822	0.9928	0.9964	3.112			0.2074

† The exponential notation for powers of 10 is used in the later parts of this table. This notation is illustrated by, for example, $1.89E - 3 = 1.89 \times 10^{-3} = 0.00189$.

Table D.1 Isentropic Flow of a Perfect Gas ($\gamma = 1.40$) *(Continued)*

M	P/P_0	ρ/ρ_0	T/T_0	c/c_0	A/A_*	μ	ω	M_*
0.20	0.9725	0.9803	0.9921	0.9960	2.964			0.2182
0.21	0.9697	0.9783	0.9913	0.9956	2.829			0.2290
0.22	0.9668	0.9762	0.9904	0.9952	2.708			0.2398
0.23	0.9638	0.9740	0.9895	0.9948	2.597			0.2506
0.24	0.9607	0.9718	0.9886	0.9943	2.496			0.2614
0.25	0.9575	0.9694	0.9877	0.9938	2.403			0.2722
0.26	0.9541	0.9670	0.9867	0.9933	2.317			0.2829
0.27	0.9506	0.9645	0.9856	0.9928	2.238			0.2936
0.28	0.9470	0.9619	0.9846	0.9923	2.166			0.3044
0.29	0.9433	0.9592	0.9835	0.9917	2.098			0.3150
0.30	0.9395	0.9564	0.9823	0.9911	2.035			0.3257
0.31	0.9355	0.9535	0.9811	0.9905	1.977			0.3364
0.32	0.9315	0.9506	0.9799	0.9899	1.922			0.3470
0.33	0.9274	0.9476	0.9787	0.9893	1.871			0.3576
0.34	0.9231	0.9445	0.9774	0.9886	1.823			0.3682
0.35	0.9188	0.9413	0.9761	0.9880	1.778			0.3788
0.36	0.9143	0.9380	0.9747	0.9873	1.736			0.3894
0.37	0.9098	0.9347	0.9733	0.9866	1.696			0.3999
0.38	0.9052	0.9313	0.9719	0.9859	1.659			0.4104
0.39	0.9004	0.9278	0.9705	0.9851	1.623			0.4209
0.40	0.8956	0.9243	0.9690	0.9844	1.590			0.4313
0.41	0.8907	0.9207	0.9675	0.9836	1.559			0.4418
0.42	0.8857	0.9170	0.9659	0.9828	1.529			0.4522
0.43	0.8807	0.9132	0.9643	0.9820	1.501			0.4626
0.44	0.8755	0.9094	0.9627	0.9812	1.474			0.4729
0.45	0.8703	0.9055	0.9611	0.9803	1.449			0.4833
0.46	0.8650	0.9016	0.9594	0.9795	1.425			0.4936
0.47	0.8596	0.8976	0.9577	0.9786	1.402			0.5039
0.48	0.8541	0.8935	0.9559	0.9777	1.380			0.5141
0.49	0.8486	0.8894	0.9542	0.9768	1.359			0.5243
0.50	0.8430	0.8852	0.9524	0.9759	1.340			0.5345
0.51	0.8374	0.8809	0.9506	0.9750	1.321			0.5447
0.52	0.8317	0.8766	0.9487	0.9740	1.303			0.5548
0.53	0.8259	0.8723	0.9468	0.9730	1.286			0.5649
0.54	0.8201	0.8679	0.9449	0.9721	1.270			0.5750
0.55	0.8142	0.8634	0.9430	0.9711	1.255			0.5851
0.56	0.8082	0.8589	0.9410	0.9700	1.240			0.5951
0.57	0.8022	0.8544	0.9390	0.9690	1.226			0.6051
0.58	0.7962	0.8498	0.9370	0.9680	1.213			0.6150
0.59	0.7901	0.8451	0.9349	0.9669	1.200			0.6249

M	P/P_0	ρ/ρ_0	T/T_0	c/c_0	A/A_*	μ	ω	M_*
0.60	0.7840	0.8405	0.9328	0.9658	1.188			0.6348
0.61	0.7784	0.8357	0.9307	0.9647	1.177			0.6447
0.62	0.7716	0.8310	0.9286	0.9636	1.166			0.6545
0.63	0.7654	0.8262	0.9265	0.9625	1.155			0.6643
0.64	0.7591	0.8213	0.9243	0.9614	1.145			0.6740
0.65	0.7528	0.8164	0.9221	0.9603	1.136			0.6837
0.66	0.7465	0.8115	0.9199	0.9591	1.127			0.6934
0.67	0.7401	0.8066	0.9176	0.9579	1.118			0.7031
0.68	0.7338	0.8016	0.9153	0.9567	1.110			0.7127
0.69	0.7274	0.7966	0.9131	0.9555	1.102			0.7223
0.70	0.7209	0.7916	0.9107	0.9543	1.094			0.7318
0.71	0.7145	0.7865	0.9084	0.9531	1.087			0.7413
0.72	0.7080	0.7814	0.9061	0.9519	1.081			0.7508
0.73	0.7016	0.7763	0.9037	0.9506	1.074			0.7602
0.74	0.6951	0.7712	0.9013	0.9494	1.068			0.7696
0.75	0.6886	0.7660	0.8989	0.9481	1.062			0.7789
0.76	0.6821	0.7609	0.8964	0.9468	1.057			0.7883
0.77	0.6756	0.7557	0.8940	0.9455	1.052			0.7975
0.78	0.6691	0.7505	0.8915	0.9442	1.047			0.8068
0.79	0.6625	0.7452	0.8890	0.9429	1.043			0.8160
0.80	0.6560	0.7400	0.8865	0.9416	1.038			0.8251
0.81	0.6495	0.7347	0.8840	0.9402	1.034			0.8343
0.82	0.6430	0.7295	0.8815	0.9389	1.030			0.8433
0.83	0.6365	0.7242	0.8789	0.9375	1.027			0.8524
0.84	0.6300	0.7189	0.8763	0.9361	1.024			0.8614
0.85	0.6235	0.7136	0.8737	0.9347	1.021			0.8704
0.86	0.6170	0.7083	0.8711	0.9333	1.018			0.8793
0.87	0.6106	0.7030	0.8685	0.9319	1.015			0.8882
0.88	0.6041	0.6977	0.8659	0.9305	1.013			0.8970
0.89	0.5977	0.6924	0.8632	0.9291	1.011			0.9058
0.90	0.5913	0.6870	0.8606	0.9277	1.009			0.9146
0.91	0.5849	0.6817	0.8579	0.9262	1.007			0.9233
0.92	0.5785	0.6764	0.8552	0.9248	1.006			0.9320
0.93	0.5721	0.6711	0.8525	0.9233	1.004			0.9407
0.94	0.5658	0.6658	0.8498	0.9219	1.003			0.9493
0.95	0.5595	0.6604	0.8471	0.9204	1.002			0.9578
0.96	0.5532	0.6551	0.8444	0.9189	1.001			0.9663
0.97	0.5469	0.6498	0.8416	0.9174	1.001			0.9748
0.98	0.5407	0.6445	0.8389	0.9159	1.000			0.9833
0.99	0.5345	0.6392	0.8361	0.9144	1.000			0.9917

Table D.1 Isentropic Flow of a Perfect Gas ($\gamma = 1.40$) (*Continued*)

M	P/P_0	ρ/ρ_0	T/T_0	c/c_0	A/A_*	μ	ω	M_*
1.00	0.5283	0.6339	0.8333	0.9129	1.000	90.0	0	1.0000
1.01	0.5221	0.6287	0.8306	0.9113	1.000	81.9	0.04	1.0083
1.02	0.5160	0.6234	0.8278	0.9098	1.000	78.6	0.13	1.0166
1.03	0.5099	0.6181	0.8250	0.9083	1.001	76.1	0.23	1.0248
1.04	0.5039	0.6129	0.8222	0.9067	1.001	74.1	0.35	1.0330
1.05	0.4979	0.6077	0.8193	0.9052	1.002	72.2	0.49	1.0411
1.06	0.4919	0.6024	0.8165	0.9036	1.003	70.6	0.64	1.0492
1.07	0.4860	0.5972	0.8137	0.9020	1.004	69.2	0.80	1.0573
1.08	0.4800	0.5920	0.8108	0.9005	1.005	67.8	0.97	1.0653
1.09	0.4742	0.5869	0.8080	0.8989	1.006	66.6	1.15	1.0733
1.10	0.4684	0.5817	0.8052	0.8973	1.008	65.4	1.34	1.0812
1.11	0.4626	0.5766	0.8023	0.8957	1.010	64.3	1.53	1.0891
1.12	0.4568	0.5714	0.7994	0.8941	1.011	63.2	1.74	1.0970
1.13	0.4511	0.5663	0.7966	0.8925	1.013	62.2	1.94	1.1048
1.14	0.4455	0.5612	0.7937	0.8909	1.015	61.3	2.16	1.1126
1.15	0.4398	0.5562	0.7908	0.8893	1.017	60.4	2.38	1.1203
1.16	0.4343	0.5511	0.7879	0.8877	1.020	59.5	2.61	1.1280
1.17	0.4287	0.5461	0.7851	0.8860	1.022	58.7	2.84	1.1356
1.18	0.4232	0.5411	0.7822	0.8844	1.025	57.9	3.07	1.1432
1.19	0.4178	0.5361	0.7793	0.8828	1.028	57.2	3.31	1.1508
1.20	0.4124	0.5311	0.7764	0.8811	1.030	56.4	3.56	1.1583
1.21	0.4070	0.5262	0.7735	0.8795	1.033	55.7	3.81	1.1658
1.22	0.4017	0.5213	0.7706	0.8778	1.037	55.1	4.06	1.1732
1.23	0.3964	0.5164	0.7677	0.8762	1.040	54.4	4.31	1.1806
1.24	0.3912	0.5115	0.7648	0.8745	1.043	53.8	4.57	1.1879
1.25	0.3861	0.5067	0.7619	0.8729	1.047	53.1	4.83	1.1952
1.26	0.3809	0.5019	0.7590	0.8712	1.050	52.5	5.09	1.2025
1.27	0.3759	0.4971	0.7561	0.8695	1.054	51.9	5.36	1.2097
1.28	0.3708	0.4923	0.7532	0.8679	1.058	51.4	5.63	1.2169
1.29	0.3658	0.4876	0.7503	0.8662	1.062	50.8	5.90	1.2240
1.30	0.3609	0.4829	0.7474	0.8645	1.066	50.3	6.17	1.2311
1.31	0.3560	0.4782	0.7445	0.8628	1.071	49.8	6.45	1.2382
1.32	0.3512	0.4736	0.7416	0.8611	1.075	49.3	6.72	1.2452
1.33	0.3464	0.4690	0.7387	0.8595	1.080	48.8	7.00	1.2522
1.34	0.3417	0.4644	0.7358	0.8578	1.084	48.3	7.28	1.2591
1.35	0.3370	0.4598	0.7329	0.8561	1.089	47.8	7.56	1.2660
1.36	0.3323	0.4553	0.7300	0.8544	1.094	47.3	7.84	1.2729
1.37	0.3277	0.4508	0.7271	0.8527	1.099	46.9	8.13	1.2797
1.38	0.3232	0.4463	0.7242	0.8510	1.104	46.4	8.41	1.2865
1.39	0.3187	0.4418	0.7213	0.8493	1.109	46.0	8.70	1.2932

M	P/P_0	ρ/ρ_0	T/T_0	c/c_0	A/A_*	μ	ω	M_*
1.40	0.3142	0.4374	0.7184	0.8476	1.115	45.6	8.99	1.2999
1.41	0.3098	0.4330	0.7155	0.8459	1.120	45.2	9.28	1.3065
1.42	0.3055	0.4287	0.7126	0.8442	1.126	44.8	9.57	1.3131
1.43	0.3012	0.4244	0.7097	0.8425	1.132	44.4	9.86	1.3197
1.44	0.2969	0.4201	0.7069	0.8407	1.138	44.0	10.15	1.3262
1.45	0.2927	0.4158	0.7040	0.8390	1.144	43.6	10.44	1.3327
1.46	0.2886	0.4116	0.7011	0.8373	1.150	43.2	10.73	1.3392
1.47	0.2845	0.4074	0.6982	0.8356	1.156	42.9	11.02	1.3456
1.48	0.2804	0.4032	0.6954	0.8339	1.163	42.5	11.32	1.3520
1.49	0.2764	0.3991	0.6925	0.8322	1.169	42.2	11.61	1.3583
1.50	0.2724	0.3950	0.6897	0.8305	1.176	41.8	11.91	1.3646
1.51	0.2685	0.3909	0.6868	0.8287	1.183	41.5	12.20	1.3708
1.52	0.2646	0.3869	0.6840	0.8270	1.190	41.1	12.49	1.3770
1.53	0.2608	0.3829	0.6811	0.8253	1.197	40.8	12.79	1.3832
1.54	0.2570	0.3789	0.6783	0.8236	1.204	40.5	13.09	1.3894
1.55	0.2533	0.3750	0.6754	0.8219	1.212	40.2	13.38	1.3955
1.56	0.2496	0.3710	0.6726	0.8201	1.219	39.9	13.68	1.4016
1.57	0.2459	0.3672	0.6698	0.8184	1.227	39.6	13.97	1.4076
1.58	0.2433	0.3633	0.6670	0.8167	1.234	39.3	14.27	1.4135
1.59	0.2388	0.3595	0.6642	0.8150	1.242	39.0	14.56	1.4195
1.60	0.2353	0.3557	0.6614	0.8133	1.250	38.7	14.86	1.4254
1.61	0.2318	0.3520	0.6586	0.8115	1.258	38.4	15.16	1.4313
1.62	0.2284	0.3483	0.6558	0.8098	1.267	38.1	15.45	1.4371
1.63	0.2250	0.3446	0.6530	0.8081	1.275	37.8	15.75	1.4429
1.64	0.2217	0.3409	0.6502	0.8064	1.284	37.6	16.04	1.4487
1.65	0.2184	0.3373	0.6475	0.8046	1.292	37.3	16.34	1.4544
1.66	0.2151	0.3337	0.6447	0.8029	1.301	37.0	16.63	1.4601
1.67	0.2119	0.3302	0.6419	0.8012	1.310	36.8	16.93	1.4657
1.68	0.2088	0.3266	0.6392	0.7995	1.319	36.5	17.22	1.4713
1.69	0.2057	0.3232	0.6364	0.7978	1.328	36.3	17.52	1.4769
1.70	0.2026	0.3197	0.6337	0.7961	1.338	36.0	17.81	1.4825
1.71	0.1996	0.3163	0.6310	0.7943	1.347	35.8	18.10	1.4880
1.72	0.1966	0.3129	0.6283	0.7926	1.357	35.5	18.40	1.4935
1.73	0.1936	0.3095	0.6256	0.7909	1.367	35.3	18.69	1.4989
1.74	0.1907	0.3062	0.6229	0.7892	1.376	35.1	18.98	1.5043
1.75	0.1878	0.3029	0.6202	0.7875	1.386	34.8	19.27	1.5097
1.76	0.1850	0.2996	0.6175	0.7858	1.397	34.6	19.56	1.5150
1.77	0.1822	0.2964	0.6148	0.7841	1.407	34.4	19.86	1.5203
1.78	0.1794	0.2931	0.6121	0.7824	1.418	34.2	20.15	1.5256
1.79	0.1767	0.2900	0.6095	0.7807	1.428	34.0	20.44	1.5308

Table D.1 Isentropic Flow of a Perfect Gas ($\gamma = 1.40$) (*Continued*)

M	P/P_0	ρ/ρ_0	T/T_0	c/c_0	A/A_*	μ	ω	M_*
1.80	0.1740	0.2868	0.6068	0.7790	1.439	33.7	20.73	1.5360
1.81	0.1714	0.2837	0.6041	0.7773	1.450	33.5	21.01	1.5412
1.82	0.1688	0.2806	0.6015	0.7756	1.461	33.3	21.30	1.5463
1.83	0.1662	0.2776	0.5989	0.7739	1.472	33.1	21.59	1.5514
1.84	0.1637	0.2745	0.5963	0.7722	1.484	32.9	21.88	1.5564
1.85	0.1612	0.2715	0.5936	0.7705	1.495	32.7	22.16	1.5614
1.86	0.1587	0.2686	0.5910	0.7688	1.507	32.5	22.45	1.5664
1.87	0.1563	0.2656	0.5884	0.7671	1.519	32.3	22.73	1.5714
1.88	0.1539	0.2627	0.5859	0.7654	1.531	32.1	23.02	1.5763
1.89	0.1516	0.2598	0.5833	0.7637	1.543	31.9	23.30	1.5812
1.90	0.1492	0.2570	0.5807	0.7620	1.555	31.8	23.59	1.5861
1.91	0.1470	0.2542	0.5782	0.7604	1.568	31.6	23.87	1.5909
1.92	0.1447	0.2514	0.5756	0.7587	1.580	31.4	24.15	1.5957
1.93	0.1425	0.2486	0.5731	0.7570	1.593	31.2	24.43	1.6005
1.94	0.1403	0.2459	0.5705	0.7553	1.606	31.0	24.71	1.6052
1.95	0.1381	0.2432	0.5680	0.7537	1.619	30.9	24.99	1.6099
1.96	0.1360	0.2405	0.5655	0.7520	1.633	30.7	25.27	1.6146
1.97	0.1339	0.2378	0.5630	0.7503	1.646	30.5	25.55	1.6193
1.98	0.1318	0.2352	0.5605	0.7487	1.660	30.3	25.83	1.6239
1.99	0.1298	0.2326	0.5580	0.7470	1.674	30.2	26.10	1.6285
2.00	0.1278	0.2300	0.5556	0.7454	1.687	30.0	26.38	1.6330
2.01	0.1258	0.2275	0.5531	0.7437	1.702	29.8	26.65	1.6375
2.02	0.1239	0.2250	0.5506	0.7420	1.716	29.7	26.93	1.6420
2.03	0.1220	0.2225	0.5482	0.7404	1.730	29.5	27.20	1.6465
2.04	0.1201	0.2200	0.5458	0.7388	1.745	29.4	27.48	1.6509
2.05	0.1182	0.2176	0.5433	0.7371	1.760	29.2	27.75	1.6553
2.06	0.1164	0.2152	0.5409	0.7355	1.775	29.0	28.02	1.6597
2.07	0.1146	0.2128	0.5385	0.7338	1.790	28.9	28.29	1.6640
2.08	0.1128	0.2104	0.5361	0.7322	1.806	28.7	28.56	1.6683
2.09	0.1111	0.2081	0.5337	0.7306	1.821	28.6	28.83	1.6726
2.10	0.1094	0.2058	0.5313	0.7289	1.837	28.4	29.10	1.6769
2.11	0.1077	0.2035	0.5290	0.7273	1.853	28.3	29.36	1.6811
2.12	0.1060	0.2013	0.5266	0.7257	1.869	28.1	29.63	1.6853
2.13	0.1043	0.1990	0.5243	0.7241	1.885	28.0	29.90	1.6895
2.14	0.1027	0.1968	0.5219	0.7225	1.902	27.9	30.16	1.6936
2.15	0.1011	0.1946	0.5196	0.7208	1.919	27.7	30.43	1.6977
2.16	0.0996	0.1925	0.5173	0.7192	1.935	27.6	30.69	1.7018
2.17	0.0980	0.1903	0.5150	0.7176	1.953	27.4	30.95	1.7059
2.18	0.0965	0.1882	0.5127	0.7160	1.970	27.3	31.21	1.7099
2.19	0.0950	0.1861	0.5104	0.7144	1.987	27.2	31.47	1.7139

M	P/P_0	ρ/ρ_0	T/T_0	c/c_0	A/A_*	μ	ω	M_*
2.20	0.0935	0.1841	0.5081	0.7128	2.005	27.0	31.73	1.7179
2.21	0.0921	0.1820	0.5059	0.7112	2.023	26.9	31.99	1.7219
2.22	0.0906	0.1800	0.5036	0.7097	2.041	26.8	32.25	1.7258
2.23	0.0892	0.1780	0.5014	0.7081	2.060	26.6	32.51	1.7297
2.24	0.0878	0.1760	0.4991	0.7065	2.078	26.5	32.76	1.7336
2.25	0.0865	0.1740	0.4969	0.7049	2.097	26.4	33.02	1.7374
2.26	0.0851	0.1721	0.4947	0.7033	2.115	26.3	33.27	1.7412
2.27	0.0838	0.1702	0.4925	0.7018	2.134	26.1	33.53	1.7450
2.28	0.0825	0.1683	0.4903	0.7002	2.154	26.0	33.78	1.7488
2.29	0.0812	0.1664	0.4881	0.6986	2.173	25.9	34.03	1.7526
2.30	0.0800	0.1646	0.4859	0.6971	2.193	25.8	34.28	1.7563
2.31	0.0787	0.1628	0.4837	0.6955	2.213	25.7	34.53	1.7600
2.32	0.0775	0.1609	0.4816	0.6940	2.233	25.5	34.78	1.7637
2.33	0.0763	0.1592	0.4794	0.6920	2.254	25.4	35.03	1.7673
2.34	0.0751	0.1574	0.4773	0.6909	2.274	25.3	35.28	1.7709
2.35	0.0740	0.1556	0.4752	0.6893	2.295	25.2	35.53	1.7745
2.36	0.0728	0.1539	0.4731	0.6878	2.316	25.1	35.77	1.7781
2.37	0.0717	0.1522	0.4709	0.6863	2.338	25.0	36.02	1.7817
2.38	0.0706	0.1505	0.4688	0.6847	2.359	24.8	36.26	1.7852
2.39	0.0695	0.1488	0.4668	0.6832	2.381	24.7	36.50	1.7887
2.40	0.0684	0.1472	0.4647	0.6817	2.403	24.6	36.75	1.7922
2.41	0.0673	0.1456	0.4626	0.6802	2.425	24.5	36.99	1.7957
2.42	0.0663	0.1439	0.4606	0.6786	2.448	24.4	37.23	1.7991
2.43	0.0653	0.1424	0.4585	0.6771	2.471	24.3	37.47	1.8025
2.44	0.0643	0.1408	0.4565	0.6756	2.494	24.2	37.71	1.8059
2.45	0.0633	0.1392	0.4544	0.6741	2.517	24.1	37.95	1.8093
2.46	0.0623	0.1377	0.4524	0.6726	2.540	24.0	38.18	1.8126
2.47	0.0613	0.1362	0.4504	0.6711	2.564	23.9	38.42	1.8159
2.48	0.0604	0.1346	0.4484	0.6696	2.588	23.8	38.65	1.8192
2.49	0.0594	0.1331	0.4464	0.6681	2.612	23.7	38.89	1.8225
2.50	0.0585	0.1317	0.4444	0.6667	2.637	23.6	39.12	1.8258
2.51	0.0576	0.1302	0.4425	0.6652	2.661	23.5	39.36	1.8290
2.52	0.0567	0.1288	0.4405	0.6637	2.686	23.4	39.59	1.8322
2.53	0.0559	0.1274	0.4386	0.6622	2.712	23.3	39.82	1.8354
2.54	0.0550	0.1260	0.4366	0.6608	2.737	23.2	40.05	1.8386
2.55	0.0542	0.1246	0.4347	0.6593	2.763	23.1	40.28	1.8417
2.56	0.0533	0.1232	0.4328	0.6578	2.789	23.0	40.51	1.8448
2.57	0.0525	0.1218	0.4309	0.6564	2.815	22.9	40.74	1.8479
2.58	0.0517	0.1205	0.4289	0.6549	2.842	22.8	40.96	1.8510
2.59	0.0509	0.1192	0.4271	0.6535	2.869	22.7	41.19	1.8541

Table D.1 Isentropic Flow of a Perfect Gas ($\gamma = 1.40$) (*Continued*)

M	P/P_0	ρ/ρ_0	T/T_0	c/c_0	A/A_*	μ	ω	M_*
2.60	0.0501	0.1179	0.4252	0.6521	2.996	22.6	41.41	1.8572
2.61	0.0493	0.1166	0.4233	0.6506	2.923	22.5	41.64	1.8602
2.62	0.0486	0.1153	0.4214	0.6492	2.951	22.4	41.86	1.8632
2.63	0.0478	0.1140	0.4196	0.6477	2.979	22.3	42.09	1.8662
2.64	0.0471	0.1128	0.4177	0.6463	3.007	22.3	42.31	1.8692
2.65	0.0464	0.1115	0.4159	0.6449	3.036	22.2	42.53	1.8721
2.66	0.0457	0.1103	0.4141	0.6435	3.065	22.1	42.75	1.8750
2.67	0.0450	0.1091	0.4122	0.6421	3.094	22.0	42.97	1.8779
2.68	0.0443	0.1079	0.4104	0.6406	3.123	21.9	43.19	1.8808
2.69	0.0436	0.1067	0.4086	0.6392	3.153	21.8	43.40	1.8837
2.70	0.0430	0.1056	0.4068	0.6378	3.183	21.7	43.62	1.8865
2.71	0.0423	0.1044	0.4051	0.6364	3.213	21.7	43.84	1.8894
2.72	0.0417	0.1033	0.4033	0.6350	3.244	21.6	44.05	1.8922
2.73	0.0410	0.1022	0.4015	0.6337	3.275	21.5	44.27	1.8950
2.74	0.0404	0.1010	0.3998	0.6323	3.306	21.4	44.48	1.8978
2.75	0.0398	0.0999	0.3980	0.6309	3.338	21.3	44.69	1.9005
2.76	0.0392	0.0989	0.3963	0.6295	3.370	21.2	44.91	1.9032
2.77	0.0386	0.0978	0.3945	0.6281	3.402	21.2	45.12	1.9060
2.78	0.0380	0.0967	0.3928	0.6268	3.434	21.1	45.33	1.9087
2.79	0.0374	0.0957	0.3911	0.6254	3.467	21.0	45.54	1.9114
2.80	0.0368	0.0946	0.3894	0.6240	3.500	20.9	45.75	1.9140
2.81	0.0363	0.0936	0.3877	0.6227	3.534	20.8	45.95	1.9167
2.82	0.0357	0.0926	0.3860	0.6213	3.567	20.8	46.16	1.9193
2.83	0.0352	0.0916	0.3844	0.6200	3.601	20.7	46.37	1.9220
2.84	0.0347	0.0906	0.3827	0.6186	3.636	20.6	46.57	1.9246
2.85	0.0341	0.0896	0.3810	0.6173	3.671	20.5	46.78	1.9271
2.86	0.0336	0.0886	0.3794	0.6159	3.706	20.5	46.98	1.9297
2.87	0.0331	0.0877	0.3777	0.6146	3.741	20.4	47.19	1.9322
2.88	0.0326	0.0867	0.3761	0.6133	3.777	20.3	47.39	1.9348
2.89	0.0321	0.0858	0.3745	0.6119	3.813	20.2	47.59	1.9373
2.90	0.0317	0.0849	0.3729	0.6106	3.850	20.2	47.79	1.9398
2.91	0.0312	0.0840	0.3712	0.6093	3.887	20.1	47.99	1.9423
2.92	0.0307	0.0831	0.3696	0.6080	3.924	20.0	48.19	1.9448
2.93	0.0302	0.0822	0.3681	0.6067	3.961	20.0	48.39	1.9472
2.94	0.0298	0.0813	0.3665	0.6054	3.999	19.9	48.59	1.9497
2.95	0.0293	0.0804	0.3649	0.6041	4.038	19.8	48.78	1.9521
2.96	0.0289	0.0796	0.3633	0.6028	4.076	19.7	49.98	1.9545
2.97	0.0285	0.0787	0.3618	0.6015	4.115	19.7	49.18	1.9569
2.98	0.0281	0.0779	0.3602	0.6002	4.155	19.6	49.37	1.9593
2.99	0.0276	0.0770	0.3587	0.5989	4.194	19.5	49.56	1.9616

M	P/P_0	ρ/ρ_0	T/T_0	c/c_0	A/A_*	μ	ω	M_*
3.00	0.0272	0.0762	0.3571	0.5976	4.235	19.5	49.76	1.9640
3.01	0.0268	0.0754	0.3556	0.5963	4.275	19.4	49.95	1.9662
3.02	0.0264	0.0746	0.3541	0.5951	3.316	19.3	50.14	1.9686
3.03	0.0260	0.0738	0.3526	0.5938	4.357	19.3	50.33	1.9709
3.04	0.0256	0.0730	0.3511	0.5925	4.399	19.2	50.52	1.9731
3.05	0.0253	0.0723	0.3496	0.5913	4.441	19.1	50.71	1.9754
3.06	0.0249	0.0715	0.3481	0.5900	4.483	19.1	50.90	1.9777
3.07	0.0245	0.0707	0.3466	0.5887	4.526	19.0	51.09	1.9799
3.08	0.0242	0.0700	0.3452	0.5875	4.570	18.9	51.28	1.9821
3.09	0.0238	0.0692	0.3437	0.5862	4.613	18.9	51.46	1.9844
3.10	0.0234	0.0685	0.3422	0.5850	4.657	18.8	51.65	1.9866
3.11	0.0231	0.0678	0.3408	0.5838	4.702	18.8	51.83	1.9887
3.12	0.0228	0.0671	0.3393	0.5825	4.747	18.7	52.02	1.9909
3.13	0.0224	0.0664	0.3379	0.5813	4.792	18.6	52.20	1.9931
3.14	0.0221	0.0657	0.3365	0.5801	4.838	18.6	52.39	1.9952
3.15	0.0218	0.0650	0.3351	0.5788	4.884	18.5	52.57	1.9974
3.16	0.0215	0.0643	0.3337	0.5776	4.930	18.4	52.75	1.9995
3.17	0.0211	0.0636	0.3323	0.5764	4.977	18.4	52.93	2.0016
3.18	0.0208	0.0630	0.3309	0.5752	5.025	18.3	53.11	2.0037
3.19	0.0205	0.0623	0.3295	0.5740	5.073	18.3	53.29	2.0057
3.20	0.0202	0.0617	0.3281	0.5728	5.121	18.2	53.47	2.0078
3.21	0.0199	0.0610	0.3267	0.5716	5.170	18.2	53.65	2.0099
3.22	0.0196	0.0604	0.3253	0.5704	5.219	18.1	53.83	2.0119
3.23	0.0194	0.0597	0.3240	0.5692	5.268	18.0	54.00	2.0139
3.24	0.0191	0.0591	0.3226	0.5680	5.319	18.0	54.18	2.0159
3.25	0.0188	0.0585	0.3213	0.5668	5.369	17.9	54.35	2.0179
3.26	0.0185	0.0579	0.3199	0.5656	5.420	17.9	54.53	2.0199
3.27	0.0183	0.0573	0.3186	0.5645	5.472	17.8	54.70	2.0219
3.28	0.0180	0.0567	0.3173	0.5633	5.523	17.8	54.88	2.0239
3.29	0.0177	0.0561	0.3160	0.5621	5.576	17.7	55.05	2.0258
3.30	0.0175	0.0555	0.3147	0.5609	5.629	17.6	55.22	2.0278
3.31	0.0172	0.0550	0.3134	0.5598	5.682	17.6	55.39	2.0297
3.32	0.0170	0.0544	0.3121	0.5586	5.736	17.5	55.56	2.0316
3.33	0.0167	0.0538	0.3108	0.5575	5.790	17.5	55.73	2.0335
3.34	0.0165	0.0533	0.3095	0.5563	5.845	17.4	55.90	2.0354
3.35	0.0163	0.0527	0.3082	0.5552	5.900	17.4	56.07	2.0373
3.36	0.0160	0.0522	0.3069	0.5540	5.956	17.3	56.24	2.0391
3.37	0.0158	0.0517	0.3057	0.5529	6.012	17.3	56.41	2.0410
3.38	0.0156	0.0511	0.3044	0.5517	6.069	17.2	56.58	2.0429
3.39	0.0153	0.0506	0.3032	0.5506	6.126	17.2	56.74	2.0447

Table D.1 Isentropic Flow of a Perfect Gas ($\gamma = 1.40$) (*Continued*)

M	P/P_0	ρ/ρ_0	T/T_0	c/c_0	A/A_*	μ	ω	M_*
3.40	0.0151	0.0501	0.3019	0.5495	6.184	17.1	56.91	2.0465
3.41	0.0149	0.0496	0.3007	0.5484	6.242	17.1	57.07	2.0483
3.42	0.0147	0.0491	0.2995	0.5472	6.301	17.0	57.24	2.0501
3.43	0.0145	0.0486	0.2982	0.5461	6.360	17.0	57.40	2.0519
3.44	0.0143	0.0481	0.2970	0.5450	6.420	16.9	57.56	2.0537
3.45	0.0141	0.0476	0.2958	0.5439	6.480	16.8	57.73	2.0555
3.46	0.0139	0.0471	0.2946	0.5428	6.541	16.8	57.89	2.0572
3.47	0.0137	0.0466	0.2934	0.5417	6.602	16.7	58.05	2.0590
3.48	0.0135	0.0462	0.2922	0.5406	6.664	16.7	58.21	2.0607
3.49	0.0133	0.0457	0.2910	0.5395	6.727	16.7	58.37	2.0624
3.50	0.0131	0.0452	0.2899	0.5384	6.790	16.6	58.53	2.0641
3.51	0.0129	0.0448	0.2887	0.5373	6.853	16.6	58.69	2.0658
3.52	0.0127	0.0443	0.2875	0.5362	6.917	16.5	58.85	2.0675
3.53	0.0126	0.0439	0.2864	0.5351	6.982	16.5	59.00	2.0692
3.54	0.0124	0.0434	0.2852	0.5340	7.047	16.4	59.16	2.0709
3.55	0.0122	0.0430	0.2841	0.5330	7.113	16.4	59.32	2.0726
3.56	0.0120	0.0426	0.2829	0.5319	7.179	16.3	59.47	2.0742
3.57	0.0119	0.0421	0.2818	0.5308	7.246	16.3	59.63	2.0759
3.58	0.0117	0.0417	0.2806	0.5298	7.313	16.2	59.78	2.0775
3.59	0.0115	0.0413	0.2795	0.5287	7.381	16.2	59.94	2.0791
3.60	0.0114	0.0409	0.2784	0.5276	7.450	16.1	60.09	2.0807
3.61	0.0112	0.0405	0.2773	0.5266	7.519	16.1	60.24	2.0823
3.62	0.0111	0.0401	0.2762	0.5255	7.589	16.0	60.40	2.0839
3.63	0.0109	0.0397	0.2751	0.5245	7.659	16.0	60.55	2.0855
3.64	0.0108	0.0393	0.2740	0.5234	7.730	15.9	60.70	2.0871
3.65	0.0106	0.0389	0.2729	0.5224	7.802	15.9	60.85	2.0887
3.66	0.0105	0.0385	0.2718	0.5213	7.874	15.9	61.00	2.0902
3.67	0.0103	0.0381	0.2707	0.5203	7.947	15.8	61.15	2.0918
3.68	0.0102	0.0378	0.2697	0.5193	8.020	15.8	61.30	2.0933
3.69	0.0100	0.0374	0.2686	0.5183	8.094	15.7	61.45	2.0948
3.70	0.00990	0.0370	0.2675	0.5172	8.169	15.7	61.60	2.0963
3.71	0.00977	0.0367	0.2665	0.5162	8.244	15.6	61.74	2.0979
3.72	0.00963	0.0363	0.2654	0.5152	8.320	15.6	61.89	2.0994
3.73	0.00950	0.0359	0.2644	0.5142	8.397	15.6	62.04	2.1009
3.74	0.00937	0.0356	0.2633	0.5132	8.474	15.5	62.18	2.1023
3.75	0.00924	0.0352	0.2623	0.5121	8.552	15.5	62.33	2.1038
3.76	0.00912	0.0349	0.2613	0.5111	8.630	15.4	62.47	2.1053
3.77	0.00899	0.0345	0.2602	0.5101	8.709	15.4	62.62	2.1067
3.78	0.00887	0.0342	0.2592	0.5091	8.789	15.3	62.76	2.1082
3.79	0.00875	0.0339	0.2582	0.5081	8.869	15.3	62.90	2.1096

M	P/P_0	ρ/ρ_0	T/T_0	c/c_0	A/A_*	μ	ω	M_*
3.80	0.00863	0.0335	0.2572	0.5072	8.951	15.3	63.04	2.1111
3.81	0.00851	0.0332	0.2562	0.5062	9.032	15.2	63.19	2.1125
3.82	0.00840	0.0329	0.2552	0.5052	9.115	15.2	63.33	2.1139
3.83	0.00828	0.0326	0.2542	0.5042	9.198	15.1	63.47	2.1153
3.84	0.00817	0.0323	0.2532	0.5032	9.282	15.1	63.61	2.1167
3.85	0.00806	0.0320	0.2522	0.5022	9.366	15.1	63.75	2.1181
3.86	0.00795	0.0316	0.2513	0.5013	9.451	15.0	63.89	2.1195
3.87	0.00784	0.0313	0.2503	0.5003	9.537	15.0	64.03	2.1209
3.88	0.00774	0.0310	0.2493	0.4993	9.624	14.9	64.16	2.1222
3.89	0.00764	0.0307	0.2484	0.4984	9.711	14.9	64.30	2.1236
3.90	0.00753	0.0304	0.2474	0.4974	9.799	14.9	64.44	2.1249
3.91	0.00743	0.0302	0.2464	0.4964	9.888	14.8	64.58	2.1263
3.92	0.00733	0.0299	0.2455	0.4955	9.977	14.8	64.71	2.1276
3.93	0.00723	0.0296	0.2446	0.4945	10.077	14.7	64.85	2.1290
3.94	0.00714	0.0293	0.2436	0.4936	10.156	14.7	64.98	2.1303
3.95	0.00704	0.0290	0.2427	0.4926	10.245	14.7	65.12	2.1316
3.96	0.00695	0.0287	0.2418	0.4917	10.341	14.6	65.25	2.1329
3.97	0.00686	0.0285	0.2408	0.4908	10.434	14.6	65.39	2.1342
3.98	0.00676	0.0282	0.2399	0.4898	10.523	14.6	65.52	2.1355
3.99	0.00667	0.0279	0.2390	0.4889	10.623	14.5	65.65	2.1368
4.00	0.00659	0.0277	0.2380	0.4880	10.72	14.5	65.78	2.1381
4.10	0.00577	0.0252	0.2293	0.4788	11.71	14.1	67.08	2.1505
4.20	0.00506	0.0229	0.2208	0.4699	12.79	13.8	68.33	2.1622
4.30	0.00445	0.0209	0.2129	0.4614	13.95	13.4	69.54	2.1732
4.40	0.00392	0.0191	0.2053	0.4531	15.21	13.1	70.71	2.1837
4.50	0.00346	0.0174	0.1980	0.4450	16.56	12.8	71.83	2.1936
4.60	0.00305	0.0160	0.1911	0.4372	18.02	12.6	72.92	2.2030
4.70	0.00270	0.0146	0.1846	0.4296	19.58	12.3	73.97	2.2119
4.80	0.00239	0.0134	0.1783	0.4223	21.26	12.0	74.99	2.2204
4.90	0.00213	0.0123	0.1724	0.4152	23.07	11.8	75.97	2.2284
5.00	$1.89E-3$	$1.13E-2$	$1.67E-1$	0.4082	$2.500E+1$	11.5	76.92	2.2361
5.10	$1.68E-3$	$1.04E-2$	$1.61E-1$	0.4015	$2.707E+1$	11.3	77.84	2.2433
5.20	$1.50E-3$	$9.62E-3$	$1.56E-1$	0.3950	$2.928E+1$	11.1	78.73	2.2503
5.30	$1.34E-3$	$8.88E-3$	$1.51E-1$	0.3887	$3.165E+1$	10.9	79.60	2.2569
5.40	$1.20E-3$	$8.20E-3$	$1.46E-1$	0.3826	$3.417E+1$	10.7	80.43	2.2631
5.50	$1.07E-3$	$7.58E-3$	$1.42E-1$	0.3766	$3.687E+1$	10.5	81.24	2.2691
5.60	$9.64E-4$	$7.01E-3$	$1.38E-1$	0.3708	$3.974E+1$	10.3	82.03	2.2748
5.70	$8.66E-4$	$6.50E-3$	$1.33E-1$	0.3652	$4.280E+1$	10.1	82.80	2.2803
5.80	$7.79E-4$	$6.02E-3$	$1.29E-1$	0.3597	$4.605E+1$	9.9	83.54	2.2855
5.90	$7.02E-4$	$5.59E-3$	$1.26E-1$	0.3544	$4.951E+1$	9.8	84.26	2.2905

Table D.1 Isentropic Flow of a Perfect Gas ($\gamma=1.40$) (*Continued*)

M	P/P_0	ρ/ρ_0	T/T_0	c/c_0	A/A_*	μ	ω	M_*
6.00	6.33$E-4$	5.19$E-3$	1.22$E-1$	0.3492	5.318$E+1$	9.6	84.96	2.2953
6.10	5.72$E-4$	4.83$E-3$	1.18$E-1$	0.3442	5.708$E+1$	9.4	85.63	2.2998
6.20	5.17$E-4$	4.49$E-3$	1.15$E-1$	0.3393	6.121$E+1$	9.3	86.29	2.3042
6.30	4.68$E-4$	4.19$E-3$	1.12$E-1$	0.3345	6.559$E+1$	9.1	86.94	2.3084
6.40	4.25$E-4$	3.90$E-3$	1.09$E-1$	0.3298	7.023$E+1$	9.0	87.56	2.3124
6.50	3.85$E-4$	3.64$E-3$	1.06$E-1$	0.3253	7.513$E+1$	8.8	88.17	2.3163
6.60	3.50$E-4$	3.40$E-3$	1.03$E-1$	0.3209	8.032$E+1$	8.7	88.76	2.3200
6.70	3.19$E-4$	3.18$E-3$	1.00$E-1$	0.3166	8.580$E+1$	8.6	89.33	2.3235
6.80	2.90$E-4$	2.97$E-3$	9.76$E-2$	0.3124	9.159$E+1$	8.5	89.90	3.3269
6.90	2.65$E-4$	2.78$E-3$	9.50$E-2$	0.3083	9.770$E+1$	8.3	90.44	2.3302
7.00	2.42$E-4$	2.61$E-3$	9.26$E-2$	0.3043	1.041$E+2$	8.2	90.97	2.3333
7.10	2.21$E-4$	2.45$E-3$	9.02$E-2$	0.3004	1.109$E+2$	8.1	91.49	2.3364
7.20	2.02$E-4$	2.30$E-3$	8.80$E-2$	0.2966	1.181$E+2$	8.0	92.00	2.3393
7.30	1.85$E-4$	2.15$E-3$	8.58$E-2$	0.2929	1.256$E+2$	7.9	92.49	2.3421
7.40	1.69$E-4$	2.02$E-3$	8.37$E-2$	0.2893	1.335$E+2$	7.8	92.97	2.3448
7.50	1.55$E-4$	1.90$E-3$	8.16$E-2$	0.2857	1.418$E+2$	7.7	93.44	2.3474
7.60	1.43$E-4$	1.79$E-3$	7.97$E-2$	0.2823	1.506$E+2$	7.6	93.90	2.3499
7.70	1.31$E-4$	1.69$E-3$	7.78$E-2$	0.2789	1.598$E+2$	7.5	94.34	2.3523
7.80	1.21$E-4$	1.59$E-3$	7.59$E-2$	0.2756	1.694$E+2$	7.4	94.78	2.3546
7.90	1.11$E-4$	1.50$E-3$	7.42$E-2$	0.2723	1.795$E+2$	7.3	95.21	2.3569
8.00	1.02$E-4$	1.41$E-3$	7.25$E-2$	0.2692	1.901$E+2$	7.2	95.62	2.3591
8.10	9.45$E-5$	1.33$E-3$	7.08$E-2$	0.2661	2.012$E+2$	7.1	96.03	2.3612
8.20	8.72$E-5$	1.26$E-3$	6.92$E-2$	0.2631	2.128$E+2$	7.0	96.43	2.3632
8.30	8.06$E-5$	1.19$E-3$	6.77$E-2$	0.2601	2.250$E+2$	6.9	96.82	2.3652
8.40	7.45$E-5$	1.13$E-3$	6.62$E-2$	0.2572	2.378$E+2$	6.8	97.20	2.3671
8.50	6.90$E-5$	1.07$E-3$	6.47$E-2$	0.2544	2.511$E+2$	6.8	97.57	2.3689
8.60	6.39$E-5$	1.01$E-3$	6.33$E-2$	0.2516	2.650$E+2$	6.7	97.94	2.3707
8.70	5.92$E-5$	9.56$E-4$	6.20$E-2$	0.2489	2.796$E+2$	6.6	98.29	2.3724
8.80	5.49$E-5$	9.06$E-4$	6.07$E-2$	0.2463	2.948$E+2$	6.5	98.64	2.3740
8.90	5.10$E-5$	8.59$E-4$	5.94$E-2$	0.2437	3.106$E+2$	6.5	98.98	2.3757
9.00	4.74$E-5$	8.15$E-4$	5.81$E-2$	0.2411	3.272$E+2$	6.4	99.32	2.3772
9.10	4.41$E-5$	7.74$E-4$	5.69$E-2$	0.2386	3.445$E+2$	6.3	99.65	2.3787
9.20	4.10$E-5$	7.35$E-4$	5.58$E-2$	0.2362	3.625$E+2$	6.2	99.97	2.3802
9.30	3.82$E-5$	6.98$E-4$	5.47$E-2$	0.2338	3.812$E+2$	6.2	100.28	2.3816
9.40	3.55$E-5$	6.64$E-4$	5.36$E-2$	0.2314	4.008$E+2$	6.1	100.59	2.3830
9.50	3.31$E-5$	6.31$E-4$	5.25$E-2$	0.2291	4.211$E+2$	6.0	100.89	2.3843
9.60	3.09$E-5$	6.01$E-4$	5.15$E-2$	0.2269	4.423$E+2$	6.0	101.19	2.3856
9.70	2.89$E-5$	5.72$E-4$	5.05$E-2$	0.2246	4.644$E+2$	5.9	101.48	2.3869
9.80	2.70$E-5$	5.45$E-4$	4.95$E-2$	0.2225	4.873$E+2$	5.9	101.76	2.3881
9.90	2.52$E-5$	5.19$E-4$	4.85$E-2$	0.2203	5.112$E+2$	5.8	102.04	2.3893

M	P/P_0	ρ/ρ_0	T/T_0	c/c_0	A/A_*	μ	ω	M^*
10.00	$2.36E-5$	$4.95E-4$	$4.76E-2$	0.2182	$5.359E+2$	5.7	102.32	2.3905
11.00	$1.24E-5$	$3.14E-4$	$3.97E-2$	0.1992	$8.419E+2$	5.2	104.80	2.4004
12.00	$6.92E-6$	$2.06E-4$	$3.36E-2$	0.1832	$1.276E+3$	4.8	106.88	2.4080
13.00	$4.02E-6$	$1.40E-4$	$2.87E-2$	0.1695	$1.876E+3$	4.4	108.65	2.4140
14.00	$2.43E-6$	$9.76E-5$	$2.49E-2$	0.1577	$2.685E+3$	4.1	110.18	2.4188
15.00	$1.51E-6$	$6.97E-5$	$2.17E-2$	0.1474	$3.755E+3$	3.8	111.51	2.4227
16.00	$9.73E-7$	$5.08E-5$	$1.92E-2$	0.1384	$5.145E+3$	3.6	112.68	2.4259
17.00	$6.41E-7$	$3.77E-5$	$1.70E-2$	0.1304	$6.921E+3$	3.4	113.71	2.4286
18.00	$4.33E-7$	$2.85E-5$	$1.52E-2$	0.1233	$9.159E+3$	3.2	114.63	2.4308
19.00	$2.98E-7$	$2.18E-5$	$1.37E-2$	0.1169	$1.195E+4$	3.0	115.45	2.4327
20.00	$2.09E-7$	$1.69E-5$	$1.23E-2$	0.1111	$1.538E+4$	2.9	116.20	2.4343
30.00	$1.25E-8$	$2.27E-6$	$5.52E-3$	0.0743	$1.144E+5$	1.9	120.92	2.4427
40.00	$1.69E-9$	$5.42E-7$	$3.12E-3$	0.0558	$4.785E+5$	1.4	123.30	2.4457
50.00	$3.55E-10$	$1.78E-7$	$2.00E-3$	0.0447	$1.455E+6$	1.1	124.73	2.4470
60.00	$9.94E-11$	$7.16E-8$	$1.39E-3$	0.0372	$3.615E+6$	1.0	125.68	2.4478
70.00	$3.38E-11$	$3.32E-8$	$1.02E-3$	0.0319	$7.805E+6$	0.8	126.36	2.4482
80.00	$1.33E-11$	$1.70E-8$	$7.81E-4$	0.0279	$1.521E+7$	0.7	126.87	2.4485
90.00	$5.83E-12$	$9.45E-9$	$6.17E-4$	0.0248	$2.739E+7$	0.6	127.27	2.4487
100.00	$2.79E-12$	$5.58E-9$	$5.00E-4$	0.0224	$4.637E+7$	0.6	127.59	2.4489
∞	0	0	0	0	∞	0	130.5	2.4500

Table D.2 Normal Shock in a Perfect Gas ($\gamma = 1.40$)†

M_{1n}	P_2/P_1	ρ_2/ρ_1	T_2/T_1	c_2/c_1	P_{02}/P_{01}	$\left\|\frac{[w]}{c_1}\right\|$	M_{2n}
1.000	1.000	1.000	1.000	1.000	1.000	0.000	1.0000
1.001	1.002	1.002	1.001	1.000	1.000	0.002	0.9990
1.002	1.005	1.003	1.001	1.001	1.000	0.003	0.9980
1.003	1.007	1.005	1.002	1.001	1.000	0.005	0.9970
1.004	1.009	1.007	1.003	1.001	1.000	0.007	0.9960
1.005	1.012	1.008	1.003	1.002	1.000	0.008	0.9950
1.006	1.014	1.010	1.004	1.002	1.000	0.010	0.9940
1.007	1.016	1.012	1.005	1.002	1.000	0.012	0.9931
1.008	1.019	1.013	1.005	1.003	1.000	0.013	0.9921
1.009	1.021	1.015	1.006	1.003	1.000	0.015	0.9911
1.010	1.023	1.017	1.007	1.003	1.000	0.017	0.9901
1.011	1.026	1.018	1.007	1.004	1.000	0.018	0.9892
1.012	1.028	1.020	1.008	1.004	1.000	0.020	0.9882
1.013	1.031	1.022	1.009	1.004	1.000	0.022	0.9872
1.014	1.033	1.023	1.009	1.005	1.000	0.023	0.9863
1.015	1.035	1.025	1.010	1.005	1.000	0.025	0.9853
1.016	1.038	1.027	1.011	1.005	1.000	0.026	0.9843
1.017	1.040	1.028	1.011	1.006	1.000	0.028	0.9834
1.018	1.042	1.030	1.012	1.006	1.000	0.030	0.9824
1.019	1.045	1.032	1.013	1.006	1.000	0.031	0.9815
1.020	1.047	1.033	1.013	1.007	1.000	0.033	0.9805
1.021	1.050	1.035	1.014	1.007	1.000	0.035	0.9796
1.022	1.052	1.037	1.015	1.007	1.000	0.036	0.9786
1.023	1.054	1.038	1.015	1.008	1.000	0.038	0.9777
1.024	1.057	1.040	1.016	1.008	1.000	0.040	0.9767
1.025	1.059	1.042	1.017	1.008	1.000	0.041	0.9758
1.026	1.061	1.044	1.017	1.009	1.000	0.043	0.9749
1.027	1.064	1.045	1.018	1.009	1.000	0.044	0.9739
1.028	1.066	1.047	1.019	1.009	1.000	0.046	0.9730
1.029	1.069	1.049	1.019	1.010	1.000	0.048	0.9721
1.030	1.071	1.050	1.020	1.010	1.000	0.049	0.9712
1.031	1.073	1.052	1.020	1.010	1.000	0.051	0.9702
1.032	1.076	1.054	1.021	1.011	1.000	0.053	0.9693
1.033	1.078	1.055	1.022	1.011	1.000	0.054	0.9684
1.034	1.081	1.057	1.022	1.011	1.000	0.056	0.9675
1.035	1.083	1.059	1.023	1.011	1.000	0.057	0.9666
1.036	1.086	1.060	1.024	1.012	1.000	0.059	0.9656
1.037	1.088	1.062	1.024	1.012	1.000	0.061	0.9647
1.038	1.090	1.064	1.025	1.012	1.000	0.062	0.9638
1.039	1.093	1.065	1.026	1.013	1.000	0.064	0.9629

† The exponential notation for powers of 10 is used in the later parts of this table. This notation is illustrated by, for example, $1.89E-3 = 1.89 \times 10^{-3} = 0.00189$.

M_{1n}	P_2/P_1	ρ_2/ρ_1	T_2/T_1	c_2/c_1	P_{02}/P_{01}	$\left\lvert\frac{[w]}{c_1}\right\rvert$	M_{2n}
1.00	1.000	1.000	1.000	1.000	1.000	0.000	1.0000
1.01	1.023	1.017	1.007	1.003	1.000	0.017	0.9901
1.02	1.047	1.033	1.013	1.007	1.000	0.033	0.9805
1.03	1.071	1.050	1.020	1.010	1.000	0.049	0.9712
1.04	1.095	1.067	1.026	1.013	1.000	0.065	0.9620
1.05	1.120	1.084	1.033	1.016	1.000	0.081	0.9531
1.06	1.144	1.100	1.039	1.019	1.000	0.087	0.9444
1.07	1.169	1.118	1.046	1.023	1.000	0.113	0.9360
1.08	1.194	1.135	1.052	1.026	0.999	0.128	0.9277
1.09	1.219	1.152	1.059	1.029	0.999	0.144	0.9196
1.10	1.245	1.169	1.065	1.032	0.999	0.159	0.9118
1.11	1.271	1.186	1.071	1.035	0.999	0.174	0.9041
1.12	1.297	1.203	1.078	1.038	0.998	0.189	0.8966
1.13	1.323	1.221	1.084	1.041	0.998	0.204	0.8892
1.14	1.350	1.238	1.090	1.044	0.997	0.219	0.8820
1.15	1.376	1.255	1.097	1.047	0.997	0.234	0.8750
1.16	1.403	1.272	1.103	1.050	0.996	0.248	0.8682
1.17	1.430	1.290	1.109	1.053	0.995	0.263	0.8615
1.18	1.458	1.307	1.115	1.056	0.995	0.277	0.8549
1.19	1.485	1.324	1.122	1.059	0.994	0.291	0.8485
1.20	1.513	1.342	1.128	1.062	0.993	0.306	0.8422
1.21	1.541	1.359	1.134	1.065	0.992	0.320	0.8360
1.22	1.570	1.376	1.141	1.068	0.991	0.334	0.8300
1.23	1.598	1.394	1.147	1.071	0.990	0.347	0.8241
1.24	1.627	1.411	1.153	1.074	0.988	0.361	0.8183
1.25	1.656	1.429	1.159	1.077	0.987	0.375	0.8126
1.26	1.686	1.446	1.166	1.080	0.986	0.389	0.8071
1.27	1.715	1.463	1.172	1.083	0.984	0.402	0.8016
1.28	1.745	1.481	1.178	1.085	0.983	0.416	0.7963
1.29	1.775	1.498	1.185	1.088	0.981	0.429	0.7911
1.30	1.805	1.516	1.191	1.091	0.979	0.442	0.7860
1.31	1.835	1.533	1.197	1.094	0.978	0.456	0.7809
1.32	1.866	1.551	1.204	1.097	0.976	0.469	0.7760
1.33	1.897	1.568	1.210	1.100	0.974	0.482	0.7712
1.34	1.928	1.585	1.216	1.103	0.972	0.495	0.7664
1.35	1.960	1.603	1.223	1.106	0.970	0.508	0.7618
1.36	1.991	1.620	1.229	1.109	0.968	0.521	0.7572
1.37	2.023	1.638	1.235	1.111	0.965	0.533	0.7527
1.38	2.055	1.655	1.242	1.114	0.963	0.546	0.7483
1.39	2.087	1.672	1.248	1.117	0.961	0.559	0.7440

Table D.2 Normal Shock in a Perfect Gas (γ=1.40) (*Continued*)

M_{1n}	P_2/P_1	ρ_2/ρ_1	T_2/T_1	c_2/c_1	P_{02}/P_{01}	$\left\lvert \frac{[w]}{c_1} \right\rvert$	M_{2n}
1.40	2.120	1.690	1.255	1.120	0.958	0.571	0.7397
1.41	2.153	1.707	1.261	1.123	0.956	0.584	0.7355
1.42	2.186	1.724	1.268	1.126	0.953	0.596	0.7314
1.43	2.219	1.742	1.274	1.129	0.950	0.609	0.7274
1.44	2.253	1.759	1.281	1.132	0.948	0.621	0.7235
1.45	2.286	1.776	1.287	1.135	0.945	0.634	0.7196
1.46	2.320	1.793	1.294	1.137	0.942	0.646	0.7157
1.47	2.354	1.811	1.300	1.140	0.939	0.658	0.7120
1.48	2.389	1.828	1.307	1.143	0.936	0.670	0.7083
1.49	2.423	1.845	1.314	1.146	0.933	0.682	0.7047
1.50	2.458	1.862	1.320	1.149	0.930	0.694	0.7011
1.51	2.493	1.879	1.327	1.152	0.927	0.706	0.6976
1.52	2.529	1.896	1.334	1.155	0.923	0.718	0.6941
1.53	2.564	1.913	1.340	1.158	0.920	0.730	0.6907
1.54	2.600	1.930	1.347	1.161	0.917	0.742	0.6874
1.55	2.636	1.947	1.354	1.164	0.913	0.754	0.6841
1.56	2.673	1.964	1.361	1.166	0.910	0.766	0.6809
1.57	2.709	1.981	1.367	1.169	0.907	0.778	0.6777
1.58	2.746	1.998	1.374	1.172	0.903	0.789	0.6746
1.59	2.783	2.015	1.381	1.175	0.899	0.801	0.6715
1.60	2.820	2.032	1.388	1.178	0.895	0.813	0.6684
1.61	2.857	2.049	1.395	1.181	0.891	0.824	0.6655
1.62	2.895	2.065	1.402	1.184	0.888	0.836	0.6625
1.63	2.933	2.082	1.409	1.187	0.884	0.847	0.6596
1.64	2.971	2.099	1.416	1.190	0.880	0.859	0.6568
1.65	3.010	2.115	1.423	1.193	0.876	0.870	0.6540
1.66	3.048	2.132	1.430	1.196	0.872	0.881	0.6512
1.67	3.087	2.148	1.437	1.199	0.868	0.893	0.6485
1.68	3.126	2.165	1.444	1.202	0.864	0.904	0.6458
1.69	3.165	2.181	1.451	1.205	0.860	0.915	0.6431
1.70	3.205	2.198	1.458	1.208	0.856	0.926	0.6405
1.71	3.245	2.214	1.466	1.211	0.852	0.938	0.6380
1.72	3.285	2.230	1.473	1.214	0.847	0.949	0.6355
1.73	3.325	2.247	1.480	1.217	0.843	0.960	0.6330
1.74	3.366	2.263	1.487	1.220	0.839	0.971	0.6305
1.75	3.406	2.279	1.495	1.223	0.835	0.982	0.6281
1.76	3.447	2.295	1.502	1.226	0.830	0.993	0.6257
1.77	3.488	2.311	1.509	1.229	0.826	1.004	0.6234
1.78	3.530	2.327	1.517	1.232	0.822	1.015	0.6210
1.79	3.571	2.343	1.524	1.235	0.817	1.026	0.6188

M_{1n}	P_2/P_1	ρ_2/ρ_1	T_2/T_1	c_2/c_1	P_{02}/P_{01}	$\left\|\frac{[w]}{c_1}\right\|$	M_{2n}
1.80	3.613	2.359	1.532	1.238	0.813	1.037	0.6165
1.81	3.655	2.375	1.539	1.241	0.808	1.048	0.6143
1.82	3.698	2.391	1.547	1.244	0.804	1.059	0.6121
1.83	3.740	2.407	1.554	1.247	0.799	1.070	0.6099
1.84	3.783	2.422	1.562	1.250	0.795	1.080	0.6078
1.85	3.826	2.438	1.569	1.253	0.790	1.091	0.6057
1.86	3.870	2.454	1.577	1.256	0.786	1.102	0.6036
1.87	3.913	2.469	1.585	1.259	0.781	1.113	0.6016
1.88	3.957	2.485	1.592	1.262	0.777	1.123	0.5996
1.89	4.001	2.500	1.600	1.265	0.772	1.134	0.5976
1.90	4.045	2.516	1.608	1.268	0.767	1.145	0.5956
1.91	4.089	2.531	1.616	1.271	0.763	1.155	0.5937
1.92	4.134	2.546	1.624	1.274	0.758	1.166	0.5918
1.93	4.179	2.562	1.631	1.277	0.753	1.177	0.5899
1.94	4.224	2.577	1.639	1.280	0.749	1.187	0.5880
1.95	4.270	2.592	1.647	1.283	0.744	1.198	0.5862
1.96	4.315	2.607	1.655	1.287	0.740	1.208	0.5844
1.97	4.361	2.622	1.663	1.290	0.735	1.219	0.5826
1.98	4.407	2.637	1.671	1.293	0.730	1.229	0.5808
1.99	4.453	2.652	1.679	1.296	0.726	1.240	0.5791
2.00	4.500	2.667	1.688	1.299	0.721	1.250	0.5774
2.01	4.547	2.681	1.696	1.302	0.716	1.260	0.5757
2.02	4.594	2.696	1.704	1.305	0.712	1.271	0.5740
2.03	4.641	2.711	1.712	1.308	0.707	1.281	0.5723
2.04	4.689	2.725	1.720	1.312	0.702	1.292	0.5707
2.05	4.736	2.740	1.729	1.315	0.698	1.302	0.5691
2.06	4.784	2.755	1.737	1.318	0.693	1.312	0.5675
2.07	4.832	2.769	1.745	1.321	0.688	1.322	0.5659
2.08	4.881	2.783	1.754	1.324	0.684	1.333	0.5643
2.09	4.929	2.798	1.762	1.327	0.679	1.343	0.5628
2.10	4.978	2.812	1.770	1.331	0.674	1.353	0.5613
2.11	5.027	2.826	1.779	1.334	0.670	1.363	0.5598
2.12	5.077	2.840	1.787	1.337	0.665	1.374	0.5583
2.13	5.126	2.854	1.796	1.340	0.660	1.384	0.5568
2.14	5.176	2.868	1.805	1.343	0.656	1.394	0.5554
2.15	5.226	2.882	1.813	1.347	0.651	1.404	0.5540
2.16	5.277	2.896	1.822	1.350	0.646	1.414	0.5525
2.17	5.327	2.910	1.831	1.353	0.642	1.424	0.5511
2.18	5.378	2.924	1.839	1.356	0.637	1.434	0.5498
2.19	5.429	2.938	1.848	1.359	0.633	1.444	0.5484

Table D.2 Normal Shock in a Perfect Gas ($\gamma = 1.40$) (*Continued*)

M_{1n}	P_2/P_1	ρ_2/ρ_1	T_2/T_1	c_2/c_1	P_{02}/P_{01}	$\left\lvert\frac{[w]}{c_1}\right\rvert$	M_{2n}
2.20	5.480	2.951	1.857	1.363	0.628	1.455	0.5471
2.21	5.531	2.965	1.866	1.366	0.624	1.465	0.5457
2.22	5.583	2.978	1.875	1.369	0.619	1.475	0.5444
2.23	5.635	2.992	1.883	1.372	0.615	1.485	0.5431
2.24	5.687	3.005	1.892	1.376	0.610	1.495	0.5418
2.25	5.740	3.019	1.901	1.379	0.606	1.505	0.5406
2.26	5.792	3.032	1.910	1.382	0.601	1.515	0.5393
2.27	5.845	3.045	1.919	1.385	0.597	1.525	0.5381
2.28	5.898	3.058	1.929	1.389	0.592	1.535	0.5368
2.29	5.951	3.071	1.938	1.392	0.588	1.544	0.5356
2.30	6.005	3.085	1.947	1.395	0.583	1.554	0.5344
2.31	6.059	3.098	1.956	1.399	0.579	1.564	0.5332
2.32	6.113	3.110	1.965	1.402	0.575	1.574	0.5321
2.33	6.167	3.123	1.974	1.405	0.570	1.584	0.5309
2.34	6.222	3.136	1.984	1.408	0.566	1.594	0.5297
2.35	6.276	3.149	1.993	1.412	0.561	1.604	0.5286
2.36	6.331	3.162	2.002	1.415	0.557	1.614	0.5275
2.37	6.386	3.174	2.012	1.418	0.553	1.623	0.5264
2.38	6.442	3.187	2.021	1.422	0.549	1.633	0.5253
2.39	6.497	3.199	2.031	1.425	0.544	1.643	0.5242
2.40	6.553	3.212	2.040	1.428	0.540	1.653	0.5231
2.41	6.609	3.224	2.050	1.432	0.536	1.663	0.5221
2.42	6.666	3.237	2.059	1.435	0.532	1.672	0.5210
2.43	6.722	3.249	2.069	1.438	0.528	1.682	0.5200
2.44	6.779	3.261	2.079	1.442	0.523	1.692	0.5189
2.45	6.836	3.273	2.088	1.445	0.519	1.702	0.5179
2.46	6.894	3.285	2.098	1.449	0.515	1.711	0.5169
2.47	6.951	3.298	2.108	1.452	0.511	1.721	0.5159
2.48	7.009	3.310	2.118	1.455	0.507	1.731	0.5149
2.49	7.067	3.321	2.128	1.459	0.503	1.740	0.5140
2.50	7.125	3.333	2.138	1.462	0.499	1.750	0.5130
2.51	7.183	3.345	2.147	1.465	0.495	1.760	0.5120
2.52	7.242	3.357	2.157	1.469	0.491	1.769	0.5111
2.53	7.301	3.369	2.167	1.472	0.487	1.779	0.5102
2.54	7.360	3.380	2.177	1.476	0.483	1.789	0.5092
2.55	7.420	3.392	2.187	1.479	0.479	1.798	0.5083
2.56	7.479	3.403	2.198	1.482	0.475	1.808	0.5074
2.57	7.539	3.415	2.208	1.486	0.472	1.817	0.5065
2.58	7.599	3.426	2.218	1.489	0.468	1.827	0.5056
2.59	7.659	3.438	2.228	1.493	0.464	1.837	0.5047

M_{1n}	P_2/P_1	ρ_2/ρ_1	T_2/T_1	c_2/c_1	P_{02}/P_{01}	$\left\lvert\frac{[w]}{c_1}\right\rvert$	M_{2n}
2.60	7.720	3.449	2.238	1.496	0.460	1.846	0.5039
2.61	7.781	3.460	2.249	1.500	0.456	1.856	0.5030
2.62	7.842	3.471	2.259	1.503	0 453	1.865	0.5022
2.63	7.903	3.483	2.269	1.506	0.449	1.875	0.5013
2.64	7.965	3.494	2.280	1.510	0.445	1.884	0.5005
2.65	8.026	3.505	2.290	1.513	0.442	1.894	0.4996
2.66	8.088	3.516	2.301	1.517	0.438	1.903	0.4988
2.67	8.150	3.527	2.311	1.520	0.434	1.913	0.4980
2.68	8.213	3.537	2.322	1.524	0.431	1.922	0.4972
2.69	8.275	3.548	2.332	1.527	0.427	1.932	0.4964
2.70	8.338	3.559	2.343	1.531	0.424	1.941	0.4956
2.71	8.401	3.570	2.354	1.534	0.420	1.951	0.4949
2.72	8.465	3.580	2.364	1.538	0.417	1.960	0.4941
2.73	8.528	3.591	2.375	1.541	0.413	1.970	0.4933
2.74	8.592	3.601	2.386	1.545	0.410	1.979	0.4926
2.75	8.656	3.612	2.397	1.548	0.406	1.989	0.4918
2.76	8.721	3.622	2.407	1.552	0.403	1.998	0.4911
2.77	8.785	3.633	2.418	1.555	0.399	2.007	0.4903
2.78	8.850	3.643	2.429	1.559	0.396	2.017	0.4896
2.79	8.915	3.653	2.440	1.562	0.393	2.026	0.4889
2.80	8.980	3.664	2.451	1.566	0.389	2.036	0.4882
2.81	9.045	3.674	2.462	1.569	0.386	2.045	0.4875
2.82	9.111	3.684	2.473	1.573	0.383	2.054	0.4868
2.83	9.177	3.694	2.484	1.576	0.380	2.064	0.4861
2.84	9.243	3.704	2.496	1.580	0.376	2.073	0.4854
2.85	9.310	3.714	2.507	1.583	0.373	2.083	0.4847
2.86	9.376	3.724	2.518	1.587	0.370	2.092	0.4840
2.87	9.443	3.734	2.529	1.590	0.367	2.101	0.4833
2.88	9.510	3.743	2.540	1.594	0.364	2.111	0.4827
2.89	9.577	3.753	2.552	1.597	0.361	2.120	0.4820
2.90	9.645	3.763	2.563	1.601	0.358	2.129	0.4814
2.91	9.713	3.773	2.575	1.605	0.355	2.139	0.4807
2.92	9.781	3.782	2.586	1.608	0.352	2.148	0.4801
2.93	9.849	3.892	2.598	1.612	0.349	2.157	0.4795
2.94	9.918	3.801	2.609	1.615	0.346	2.167	0.4788
2.95	9.986	3.811	2.621	1.619	0.343	2.176	0.4782
2.96	10.055	3.820	2.632	1.622	0.340	2.185	0.4776
2.97	10.124	3.829	2.644	1.626	0.337	2.194	0.4770
2.98	10.194	3.839	2.656	1.630	0.334	2.204	0.4764
2.99	10.263	3.848	2.667	1.633	0.331	2.213	0.4758

Table D.2 Normal Shock in a Perfect Gas ($\gamma = 1.40$) (*Continued*)

M_{1n}	P_2/P_1	ρ_2/ρ_1	T_2/T_1	c_2/c_1	P_{02}/P_{01}	$\left\lvert \frac{[w]}{c_1} \right\rvert$	M_{2n}
3.00	10.333	3.857	2.679	1.637	0.328	2.222	0.4752
3.01	10.403	3.866	2.691	1.640	0.326	2.231	0.4746
3.02	10.474	3.875	2.703	1.644	0.323	2.241	0.4740
3.03	10.544	3.884	2.714	1.648	0.320	2.250	0.4734
3.04	10.615	3.893	2.726	1.651	0.317	2.259	0.4729
3.05	10.686	3.902	2.738	1.655	0.315	2.268	0.4723
3.06	10.758	3.911	2.750	1.658	0.312	2.278	0.4717
3.07	10.829	3.920	2.762	1.662	0.309	2.287	0.4712
3.08	10.901	3.929	2.774	1.666	0.306	2.296	0.4706
3.09	10.973	3.938	2.786	1.669	0.304	2.305	0.4701
3.10	11.045	3.947	2.799	1.673	0.301	2.315	0.4695
3.11	11.117	3.955	2.811	1.677	0.299	2.324	0.4690
3.12	11.190	3.964	2.823	1.680	0.296	2.333	0.4685
3.13	11.263	3.973	2.835	1.684	0.293	2.342	0.4679
3.14	11.336	3.981	2.848	1.687	0.291	2.351	0.4674
3.15	11.410	3.990	2.860	1.691	0.288	2.360	0.4669
3.16	11.483	3.998	2.872	1.695	0.286	2.370	0.4664
3.17	11.557	4.006	2.885	1.698	0.284	2.379	0.4659
3.18	11.631	4.015	2.897	1.702	0.281	2.388	0.4654
3.19	11.705	4.023	2.909	1.706	0.279	2.397	0.4648
3.20	11.780	4.031	2.922	1.709	0.276	2.406	0.4643
3.21	11.855	4.040	2.935	1.713	0.274	2.415	0.4639
3.22	11.930	4.048	2.947	1.717	0.271	2.425	0.4634
3.23	12.005	4.056	2.960	1.720	0.269	2.434	0.4629
3.24	12.081	4.064	2.972	1.724	0.267	2.443	0.4624
3.25	12.156	4.072	2.985	1.728	0.265	2.452	0.4619
3.26	12.232	4.080	2.998	1.731	0.262	2.461	0.4614
3.27	12.308	4.088	3.011	1.735	0.260	2.470	0.4610
3.28	12.385	4.096	3.023	1.739	0.258	2.479	0.4605
3.29	12.461	4.104	3.036	1.742	0.255	2.488	0.4600
3.30	12.538	4.112	3.049	1.746	0.253	2.497	0.4596
3.31	12.615	4.120	2.062	1.750	0.251	2.507	0.4591
3.32	12.693	4.128	3.075	1.754	0.249	2.516	0.4587
3.33	12.770	4.135	3.088	1.757	0.247	2.525	0.4582
3.34	12.848	4.143	3.101	1.761	0.245	2.534	0.4578
3.35	12.926	4.151	3.114	1.765	0.243	2.543	0.4573
3.36	13.005	4.158	3.127	1.768	0.240	2.552	0.4569
3.37	13.083	4.166	3.141	1.772	0.238	2.561	0.4565
3.38	13.162	4.173	3.154	1.776	0.236	2.570	0.4560
3.39	13.241	4.181	3.167	1.780	0.234	2.579	0.4556

M_{1n}	P_2/P_1	ρ_2/ρ_1	T_2/T_1	c_2/c_1	P_{02}/P_{01}	$\left\lvert\frac{[w]}{c_1}\right\rvert$	M_{2n}
3.40	13.320	4.188	3.180	1.783	0.232	2.588	0.4552
3.41	13.399	4.196	3.194	1.787	0.230	2.597	0.4548
3.42	13.479	4.203	3.207	1.791	0.228	2.606	0.4544
3.43	13.559	4.211	3.220	1.795	0.226	2.615	0.4540
3.44	13.639	4.218	3.234	1.798	0.224	2.624	0.4535
3.45	13.720	4.225	3.247	1.802	0.222	2.633	0.4531
3.46	13.800	4.232	3.261	1.806	0.220	2.642	0.4527
3.47	13.881	4.240	3.274	1.809	0.219	2.652	0.4523
3.48	13.962	4.247	3.288	1.813	0.217	2.661	0.4519
3.49	14.043	4.254	3.301	1.817	0.215	2.670	0.4515
3.50	14.125	4.261	3.315	1.821	0.213	2.679	0.4512
3.51	14.207	4.268	3.329	1.824	0.211	2.688	0.4508
3.52	14.289	4.275	3.342	1.828	0.209	2.797	0.4504
3.53	14.371	4.282	3.356	1.832	0.207	2.706	0.4500
3.54	14.454	4.289	3.370	1.836	0.206	2.715	0.4496
3.55	14.536	4.296	3.384	1.840	0.204	2.724	0.4492
3.56	14.619	4.303	3.398	1.843	0.202	2.733	0.4489
3.57	14.702	4.309	3.412	1.847	0.200	2.742	0.4485
3.58	14.786	4.316	3.426	1.851	0.199	2.751	0.4481
3.59	14.869	4.323	3.440	1.855	0.197	2.760	0.4478
3.60	14.953	4.330	3.454	1.858	0.195	2.769	0.4474
3.61	15.037	4.336	3.468	1.862	0.194	2.777	0.4471
3.62	15.122	4.343	3.482	1.866	0.192	2.786	0.4467
3.63	15.206	4.350	3.496	1.870	0.190	2.795	0.4463
3.64	15.291	4.356	3.510	1.874	0.189	2.804	0.4460
3.65	15.376	4.363	3.525	1.877	0.187	2.813	0.4456
3.66	15.462	4.369	3.539	1.881	0.185	2.822	0.4453
3.67	15.547	4.376	3.553	1.885	0.184	2.831	0.4450
3.68	15.633	4.382	3.567	1.889	0.182	2.840	0.4446
3.69	15.719	4.388	3.582	1.893	0.181	2.849	0.4443
3.70	15.805	4.395	3.596	1.896	0.179	2.858	0.4439
3.71	15.891	4.401	3.611	1.900	0.178	2.867	0.4436
3.72	15.978	4.408	3.625	1.904	0.176	2.876	0.4433
3.73	16.065	4.414	3.640	1.908	0.175	2.885	0.4430
3.74	16.152	4.420	3.654	1.912	0.173	2.894	0.4426
3.75	16.240	4.426	2.669	1.915	0.172	2.903	0.4423
3.76	16.327	4.432	3.684	1.919	0.170	2.912	0.4420
3.77	16.415	4.439	2.698	1.923	0.169	2.921	0.4417
3.78	16.503	4.445	3.713	1.927	0.167	2.930	0.4414
3.79	16.591	4.451	3.728	1.931	0.166	2.938	0.4410

Table D.2 Normal Shock in a Perfect Gas ($\gamma = 1.40$) (*Continued*)

M_{1n}	P_2/P_1	ρ_2/ρ_1	T_2/T_1	c_2/c_1	P_{02}/P_{01}	$\left\lvert \frac{[w]}{c_1} \right\rvert$	M_{2n}
3.80	16.680	4.457	3.743	1.935	0.164	2.947	0.4407
3.81	16.769	4.463	3.757	1.938	0.163	2.956	0.4404
3.82	16.858	4.469	3.772	1.942	0.162	2.965	0.4401
3.83	16.947	4.475	3.787	1.946	0.160	2.974	0.4398
3.84	17.037	4.481	3.802	1.950	0.159	2.983	0.4395
3.85	17.126	4.487	3.817	1.954	0.158	2.992	0.4392
3.86	17.216	4.492	3.832	1.958	0.156	3.001	0.4389
3.87	17.306	4.498	3.837	1.961	0.155	3.010	0.4386
3.88	17.397	4.504	3.862	1.965	0.154	3.019	0.4383
3.89	17.487	4.510	3.878	1.969	0.152	3.027	0.4380
3.90	17.578	4.516	3.893	1.973	0.151	3.036	0.4377
3.91	17.669	4.521	3.908	1.977	0.150	3.045	0.4375
3.92	17.761	4.527	3.923	1.981	0.148	3.054	0.4372
3.93	17.852	4.533	3.939	1.985	0.147	3.063	0.4369
3.94	17.944	4.538	3.954	1.988	0.146	3.072	0.4366
3.95	18.036	4.544	3.969	1.992	0.145	3.081	0.4363
3.96	18.129	4.549	3.985	1.996	0.144	3.090	0.4360
3.97	18.221	4.555	4.000	2.000	0.142	3.098	0.4358
3.98	18.314	4.560	4.016	2.004	0.141	3.107	0.4355
3.99	18.407	4.566	4.031	2.008	0.140	3.116	0.4352
4.000	18.500	4.571	4.047	2.012	0.139	3.125	0.4350
4.100	19.445	4.624	4.205	2.051	0.128	3.213	0.4324
4.200	20.413	4.675	4.367	2.090	0.117	3.302	0.4299
4.300	21.405	4.723	4.532	2.129	0.108	3.390	0.4277
4.400	22.420	4.768	4.702	2.168	$9.95E-2$	3.477	0.4255
4.500	23.458	4.812	4.875	2.208	$9.17E-2$	3.565	0.4236
4.600	24.520	4.853	5.052	2.248	$8.46E-2$	3.652	0.4217
4.700	25.605	4.893	5.233	2.288	$7.81E-2$	3.739	0.4199
4.800	26.713	4.930	5.418	2.328	$7.21E-2$	3.826	0.4183
4.900	27.845	4.966	5.607	2.368	$6.67E-2$	3.913	0.4167
5.000	29.000	5.000	5.800	2.408	$6.17E-2$	4.000	0.4152
5.100	30.178	5.033	5.997	2.449	$5.72E-2$	4.087	0.4138
5.200	31.380	5.064	6.197	2.489	$5.30E-2$	4.173	0.4125
5.300	32.605	5.093	6.401	2.530	$4.91E-2$	4.259	0.4113
5.400	33.853	5.122	6.610	2.571	$4.56E-2$	4.346	0.4101
5.500	35.125	5.149	6.822	2.612	$4.24E-2$	4.432	0.4090
5.600	36.420	5.175	7.038	2.653	$3.94E-2$	4.518	0.4079
5.700	37.738	5.200	7.258	2.694	$3.66E-2$	4.604	0.4069
5.800	39.080	5.224	7.481	2.735	$3.41E-2$	4.690	0.4059
5.900	40.445	5.246	7.709	2.777	$3.18E-2$	4.775	0.4050

M_{1n}	P_2/P_1	ρ_2/ρ_1	T_2/T_1	c_2/c_1	P_{02}/P_{01}	$\left\|\frac{[w]}{c_1}\right\|$	M_{2n}
6.000	41.833	5.268	7.941	2.818	2.97E−2	4.861	0.4042
6.100	43.245	5.289	8.176	2.859	2.77E−2	4.947	0.4033
6.200	44.680	5.309	8.415	2.901	2.58E−2	5.032	0.4025
6.300	46.138	5.329	8.658	2.943	2.42E−2	5.118	0.4018
6.400	47.620	5.347	8.905	2.984	2.26E−2	5.203	0.4011
6.500	49.125	5.365	9.156	3.026	2.11E−2	5.288	0.4004
6.600	50.653	5.382	9.411	3.068	1.98E−2	5.374	0.3997
6.700	52.205	5.399	9.670	3.110	1.86E−2	5.459	0.3991
6.800	53.780	5.415	9.933	3.152	1.74E−2	5.544	0.3985
6.900	55.378	5.430	10.199	3.194	1.63E−2	5.629	0.3979
7.000	57.000	5.444	10.469	3.236	1.54E−2	5.714	0.3974
7.100	58.645	5.459	10.744	3.278	1.44E−2	5.799	0.3968
7.200	60.313	5.472	11.022	3.320	1.36E−2	5.884	0.3963
7.300	62.005	5.485	11.304	3.362	1.28E−2	5.969	0.3958
7.400	63.720	5.498	11.590	3.404	1.20E−2	6.054	0.3954
7.500	65.458	5.510	11.879	3.447	1.13E−2	6.139	0.3949
7.600	67.220	5.522	12.173	3.489	1.07E−2	6.224	0.3945
7.700	69.005	5.533	12.471	3.531	1.01E−2	6.308	0.3941
7.800	70.813	5.544	12.772	3.574	9.51E−3	6.393	0.3937
7.900	72.645	5.555	13.077	3.616	8.98E−3	6.478	0.3933
8.000	74.500	5.565	13.387	3.659	8.49E−3	6.563	0.3929
8.100	76.378	5.575	13.700	3.701	8.03E−3	6.647	0.3925
8.200	78.280	5.585	14.017	3.744	7.59E−3	6.732	0.3922
8.300	80.205	5.594	14.338	3.787	7.19E−3	6.816	0.3918
8.400	82.153	5.603	14.662	3.829	6.81E−3	6.901	0.3915
8.500	84.125	5.612	14.991	3.872	6.45E−3	6.985	0.3912
8.600	86.120	5.620	15.324	3.915	6.11E−3	7.070	0.3909
8.700	88.138	5.628	15.660	3.957	5.80E−3	7.154	0.3906
8.800	90.180	5.636	16.000	4.000	5.50E−3	7.239	0.3903
8.900	92.245	5.644	16.345	4.043	5.23E−3	7.323	0.3901
9.00	94.333	5.651	16.693	4.086	4.96E−3	7.407	0.3898
9.10	96.445	5.658	17.045	4.129	4.72E−3	7.492	0.3895
9.20	98.580	5.665	17.401	4.171	4.49E−3	7.576	0.3893
9.30	100.738	5.672	17.760	4.214	4.27E−3	7.660	0.3891
9.40	102.920	5.679	18.124	4.257	4.06E−3	7.745	0.3888
9.50	105.125	5.685	18.492	4.300	3.87E−3	7.829	0.3886
9.60	107.353	5.691	18.863	4.343	3.68E−3	7.913	0.3884
9.70	109.605	5.697	19.238	4.386	3.51E−3	7.997	0.3882
9.80	111.880	5.703	19.617	4.429	3.35E−3	8.082	0.3880
9.90	114.178	5.709	20.001	4.472	3.19E−3	8.166	0.3878

Table D.2 Normal Shock in a Perfect Gas ($\gamma = 1.40$) (*Continued*)

M_{1n}	P_2/P_1	ρ_2/ρ_1	T_2/T_1	c_2/c_1	P_{02}/P_{01}	$\left\lvert\frac{[w]}{c_1}\right\rvert$	M_{2n}
10.00	116.500	5.714	20.387	4.515	3.04$E-3$	8.250	0.3876
11.00	141.000	5.762	24.471	4.947	1.95$E-3$	9.091	0.3859
12.00	167.833	5.799	28.943	5.380	1.29$E-3$	9.931	0.3847
13.00	197.000	5.828	33.805	5.814	8.77$E-4$	10.769	0.3837
14.00	228.500	5.851	39.055	6.249	6.14$E-4$	11.607	0.3829
15.00	262.333	5.870	44.694	6.685	4.40$E-4$	12.444	0.3823
16.00	298.500	5.885	50.722	7.122	3.21$E-4$	13.281	0.3817
17.00	337.000	5.898	57.138	7.559	2.39$E-4$	14.118	0.3813
18.00	377.833	5.909	63.944	7.996	1.81$E-4$	14.954	0.3810
19.00	421.000	5.918	71.138	8.434	1.39$E-4$	15.789	0.3806
20.00	466.500	5.926	78.722	8.873	1.08$E-4$	16.625	0.3804
30.00	1,049.833	5.967	175.944	13.264	1.45$E-5$	24.972	0.3790
40.00	1,866.500	5.981	312.055	17.665	3.48$E-6$	33.312	0.3786
50.00	2,916.500	5.988	487.055	22.069	1.14$E-6$	41.650	0.3784
60.00	4,199.833	5.992	700.944	26.475	4.61$E-7$	49.986	0.3782
70.00	5,716.499	5.994	953.721	30.882	2.13$E-7$	58.321	0.3782
80.00	7,466.499	5.995	1,245.388	35.290	1.10$E-7$	66.656	0.3781
90.00	9,449.832	5.996	1,575.943	39.698	6.08$E-8$	74.991	0.3781
100.00	11,666.499	5.997	1,945.387	44.107	3.59$E-8$	83.325	0.3781
∞	∞	6	∞	∞	0	∞	0.3780

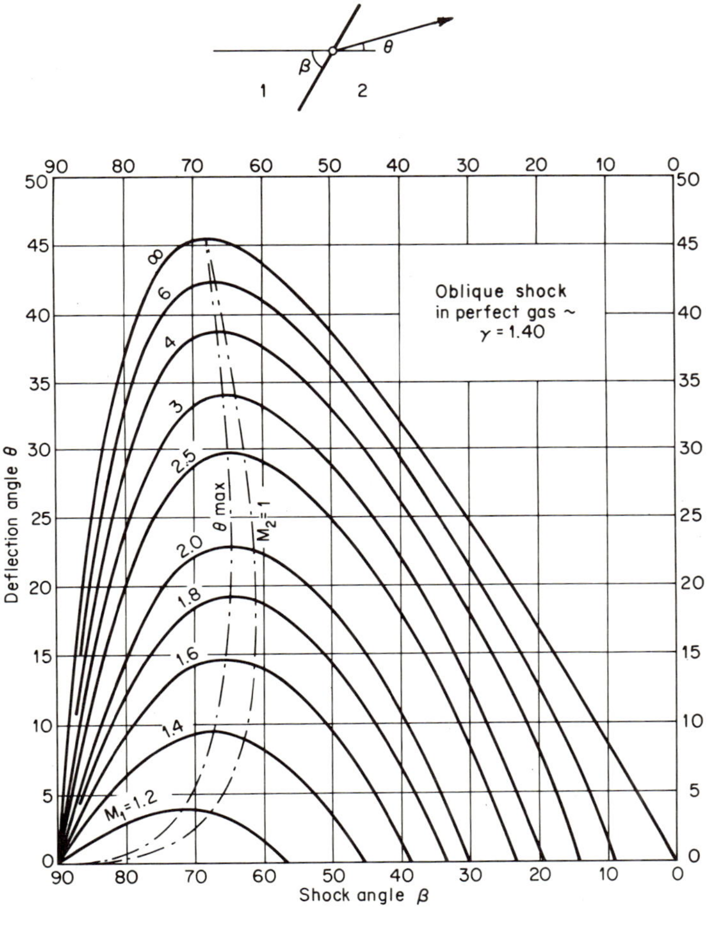

β
θ
1
2
Oblique shock
in perfect gas ~
γ = 1.40
Deflection angle θ
Shock angle β
θ max
$M_2 = 1$
$M_1 = 1.2$
1.4
1.6
1.8
2.0
2.5
3
4
6
∞

Figure D.1

Table D.3 Oblique Shock in a Perfect Gas ($\gamma = 1.40$)

		Weak Solutions			Strong Solutions		
M_1	θ, degrees†	β, degrees	P_2/P_1	M_2	β	P_2/P_1	M_2
1.05	0.0	72.25	1.000	1.050	90.00	1.120	0.953
	(0.56)	79.94	1.080	0.984	79.94	1.080	0.984
1.10	0.0	65.38	1.000	1.100	90.00	1.245	0.912
	1.0	69.81	1.077	1.039	83.58	1.227	0.925
	(1.52)	76.30	1.166	0.971	76.30	1.166	0.971
1.15	0.0	60.41	1.000	1.150	90.00	1.376	0.875
	1.0	63.16	1.062	1.102	85.99	1.369	0.880
	2.0	67.01	1.141	1.043	81.18	1.340	0.901
	(2.67)	73.82	1.256	0.960	73.82	1.256	0.960
1.20	0.0	56.44	1.000	1.200	90.00	1.513	0.842
	1.0	58.55	1.056	1.158	87.04	1.509	0.845
	2.0	61.05	1.120	1.111	83.86	1.494	0.855
	3.0	64.34	1.198	1.056	80.03	1.463	0.876
	(3.94)	71.98	1.353	0.950	71.98	1.353	0.950
1.25	0.0	53.13	1.000	1.250	90.00	1.656	0.813
	1.0	54.88	1.053	1.211	87.66	1.653	0.815
	2.0	56.85	1.111	1.170	85.21	1.644	0.821
	3.0	59.13	1.176	1.124	82.55	1.626	0.832
	4.0	61.99	1.254	1.072	79.39	1.594	0.853
	5.0	66.50	1.366	0.999	74.64	1.528	0.895
	(5.29)	70.54	1.454	0.942	70.54	1.454	0.942
1.30	0.0	50.29	1.000	1.300	90.00	1.805	0.786
	1.0	51.81	1.051	1.263	88.06	1.803	0.787
	2.0	53.48	1.107	1.224	86.06	1.796	0.792
	3.0	55.32	1.167	1.184	83.96	1.783	0.800
	4.0	57.42	1.233	1.140	81.65	1.763	0.812
	5.0	59.96	1.311	1.090	78.97	1.733	0.831
	6.0	63.46	1.411	1.027	75.37	1.679	0.864
	(6.66)	69.40	1.561	0.936	69.40	1.561	0.936
1.35	0.0	47.80	1.000	1.350	90.00	1.960	0.762
	1.0	49.17	1.051	1.314	88.34	1.958	0.763
	2.0	50.64	1.104	1.277	86.65	1.952	0.766

† Figures in parentheses are maximum values.

M_1	θ, degrees†	Weak Solutions β, degrees	Weak Solutions P_2/P_1	Weak Solutions M_2	Strong Solutions β	Strong Solutions P_2/P_1	Strong Solutions M_2
	3.0	52.22	1.162	1.239	84.89	1.943	0.772
	4.0	53.97	1.224	1.199	83.03	1.928	0.781
	5.0	55.93	1.292	1.157	81.00	1.908	0.793
	6.0	58.23	1.370	1.109	78.66	1.877	0.811
	7.0	61.18	1.466	1.052	75.72	1.830	0.839
	8.0	66.92	1.633	0.954	70.03	1.711	0.909
	(8.05)	68.47	1.673	0.931	68.47	1.673	0.931
1.40	0.0	45.59	1.000	1.400	90.00	2.120	0.740
	1.0	46.84	1.050	1.365	88.55	2.119	0.741
	2.0	48.17	1.103	1.330	87.08	2.114	0.743
	3.0	49.59	1.159	1.293	85.57	2.106	0.748
	4.0	51.12	1.219	1.255	83.99	2.095	0.755
	5.0	52.78	1.283	1.216	82.32	2.079	0.764
	6.0	54.63	1.354	1.174	80.49	2.058	0.776
	7.0	56.76	1.433	1.128	78.42	2.028	0.793
	8.0	59.37	1.526	1.074	75.90	1.984	0.818
	9.0	63.19	1.655	1.003	72.19	1.906	0.863
	(9.43)	67.72	1.791	0.927	67.72	1.791	0.927
1.45	0.0	43.60	1.000	1.450	90.00	2.286	0.720
	1.0	44.78	1.050	1.416	88.71	2.285	0.720
	2.0	46.00	1.103	1.381	87.41	2.281	0.723
	3.0	47.30	1.158	1.345	86.08	2.275	0.726
	4.0	48.68	1.217	1.309	84.70	2.265	0.732
	5.0	50.16	1.279	1.272	83.27	2.253	0.739
	6.0	51.76	1.346	1.233	81.74	2.236	0.749
	7.0	53.52	1.419	1.191	80.07	2.213	0.761
	8.0	55.52	1.500	1.146	78.20	2.184	0.778
	9.0	57.89	1.593	1.095	75.98	2.142	0.801
	10.0	61.05	1.711	1.032	73.00	2.076	0.837
	(10.79)	67.10	1.915	0.924	67.10	1.915	0.924
1.50	0.0	41.81	1.000	1.500	90.00	2.458	0.701
	1.0	42.91	1.050	1.466	88.84	2.457	0.702
	2.0	44.07	1.103	1.432	87.67	2.454	0.704
	3.0	45.27	1.158	1.397	86.48	2.448	0.707
	4.0	46.54	1.217	1.362	85.26	2.440	0.711
	5.0	47.89	1.278	1.325	83.99	2.430	0.717

† Figures in parentheses are maximum values.

Table D.3 Oblique Shock in a Perfect Gas ($\gamma = 1.40$) *(Continued)*

M_1	θ, degrees†	Weak Solutions			Strong Solutions		
		β, degrees	P_2/P_1	M_2	β	P_2/P_1	M_2
	6.0	49.33	1.343	1.288	82.66	2.416	0.725
	7.0	50.88	1.413	1.250	81.25	2.398	0.735
	8.0	52.57	1.489	1.208	79.71	2.375	0.748
	9.0	54.47	1.572	1.164	78.00	2.345	0.764
	10.0	56.68	1.666	1.114	76.00	2.305	0.785
	11.0	59.47	1.781	1.056	73.44	2.245	0.817
	12.0	64.36	1.967	0.961	68.79	2.115	0.885
	(12.11)	66.59	2.044	0.921	66.59	2.044	0.921
1.55	0.0	40.18	1.000	1.550	90.00	2.636	0.684
	1.0	41.23	1.051	1.516	88.95	2.635	0.685
	2.0	42.32	1.104	1.482	87.88	2.632	0.686
	3.0	43.45	1.159	1.448	86.80	2.628	0.689
	4.0	44.64	1.217	1.413	85.70	2.621	0.693
	5.0	45.89	1.278	1.378	84.57	2.611	0.698
	6.0	47.22	1.343	1.341	83.39	2.599	0.705
	7.0	48.62	1.411	1.304	82.15	2.584	0.713
	8.0	50.13	1.485	1.265	80.83	2.565	0.723
	9.0	51.78	1.563	1.224	79.40	2.541	0.736
	10.0	53.60	1.649	1.180	77.81	2.511	0.752
	11.0	55.69	1.746	1.132	75.97	2.471	0.772
	12.0	58.24	1.860	1.076	73.69	2.415	0.801
	13.0	61.98	2.018	0.999	70.24	2.316	0.852
	(13.40)	66.17	2.179	0.920	66.17	2.179	0.920
1.60	0.0	38.68	1.000	1.600	90.00	2.820	0.668
	1.0	39.69	1.051	1.566	89.03	2.819	0.669
	2.0	40.73	1.105	1.532	88.06	2.817	0.670
	3.0	41.81	1.160	1.498	87.07	2.812	0.673
	4.0	42.93	1.219	1.464	86.06	2.806	0.676
	5.0	44.11	1.280	1.429	85.03	2.798	0.681
	6.0	45.35	1.345	1.393	83.97	2.787	0.686
	7.0	46.65	1.413	1.357	82.86	2.774	0.693
	8.0	48.03	1.484	1.320	81.69	2.758	0.702
	9.0	49.51	1.561	1.281	80.45	2.738	0.712
	10.0	51.12	1.643	1.240	79.10	2.713	0.725
	11.0	52.89	1.733	1.196	77.61	2.683	0.741

† Figures in parentheses are maximum values.

M_1	θ, degrees†	Weak Solutions β, degrees	P_2/P_1	M_2	Strong Solutions β	P_2/P_1	M_2
	12.0	54.89	1.832	1.148	75.90	2.643	0.761
	13.0	57.28	1.948	1.094	73.82	2.588	0.789
	14.0	60.54	2.097	1.023	70.90	2.500	0.832
	(14.65)	65.83	2.319	0.919	65.83	2.319	0.919
1.65	0.0	37.31	1.000	1.650	90.00	3.010	0.654
	1.0	38.27	1.052	1.616	89.11	3.009	0.654
	2.0	39.27	1.106	1.582	88.20	3.006	0.656
	3.0	40.30	1.162	1.548	87.29	3.003	0.658
	4.0	41.38	1.221	1.514	86.37	2.997	0.661
	5.0	42.50	1.283	1.480	85.42	2.989	0.665
	6.0	43.67	1.348	1.444	84.45	2.980	0.670
	7.0	44.89	1.415	1.409	83.44	2.968	0.676
	8.0	46.18	1.487	1.372	82.39	2.954	0.683
	9.0	47.55	1.563	1.334	81.29	2.937	0.692
	10.0	49.01	1.643	1.295	80.11	2.916	0.703
	11.0	50.58	1.729	1.254	78.83	2.890	0.716
	12.0	52.31	1.822	1.210	77.41	2.859	0.732
	13.0	54.26	1.926	1.163	75.80	2.819	0.752
	14.0	56.54	2.044	1.109	73.87	2.764	0.778
	15.0	59.52	2.192	1.042	71.25	2.681	0.818
	(15.86)	65.55	2.465	0.918	65.55	2.465	0.918
1.70	0.0	36.03	1.000	1.700	90.00	3.205	0.641
	1.0	36.97	1.053	1.666	89.17	3.204	0.641
	2.0	37.93	1.107	1.632	88.33	3.202	0.642
	3.0	38.93	1.164	1.598	87.48	3.199	0.644
	4.0	39.96	1.224	1.564	86.62	3.193	0.647
	5.0	41.03	1.286	1.529	85.75	1.186	0.650
	6.0	42.15	1.351	1.495	84.85	3.178	0.655
	7.0	43.31	1.420	1.459	83.93	3.167	0.660
	8.0	44.53	1.491	1.423	82.97	3.154	0.667
	9.0	45.81	1.567	1.386	81.97	3.139	0.675
	10.0	47.17	1.647	1.348	80.91	3.121	0.684
	11.0	48.61	1.731	1.309	79.78	3.099	0.695
	12.0	50.17	1.822	1.267	78.56	3.072	0.708
	13.0	51.87	1.920	1.223	77.21	3.040	0.724
	14.0	53.77	2.027	1.176	75.67	2.999	0.744

† Figures in parentheses are maximum values.

Table D.3 Oblique Shock in a Perfect Gas ($\gamma = 1.40$) (*Continued*)

M_1	θ, degrees†	Weak Solutions			Strong Solutions		
		β, degrees	P_2/P_1	M_2	β	P_2/P_1	M_2
	15.0	55.99	2.150	1.122	73.84	2.944	0.770
	16.0	58.80	2.300	1.057	71.43	2.863	0.808
	17.0	64.63	2.586	0.932	66.00	2.647	0.905
	(17.01)	65.32	2.617	0.918	65.32	2.617	0.918
1.75	0.0	34.85	1.000	1.750	90.00	3.406	0.628
	1.0	35.75	1.053	1.716	89.22	3.406	0.628
	2.0	36.69	1.109	1.682	88.44	3.404	0.630
	3.0	37.65	1.167	1.648	87.64	3.400	0.631
	4.0	38.65	1.227	1.613	86.84	3.395	0.634
	5.0	39.69	1.290	1.579	86.03	3.389	0.637
	6.0	40.76	1.356	1.544	85.19	3.381	0.641
	7.0	41.87	1.425	1.509	84.34	3.371	0.646
	8.0	43.04	1.497	1.473	83.45	3.360	0.652
	9.0	44.25	1.573	1.437	82.53	3.346	0.659
	10.0	45.53	1.653	1.400	81.57	3.329	0.667
	11.0	46.88	1.737	1.361	80.56	3.310	0.677
	12.0	48.32	1.826	1.321	79.47	3.287	0.688
	13.0	49.87	1.922	1.279	78.29	3.259	0.701
	14.0	51.55	2.025	1.235	76.99	3.225	0.718
	15.0	53.42	2.137	1.187	75.51	3.183	0.738
	16.0	55.59	2.265	1.133	73.76	3.127	0.764
	17.0	58.30	2.420	1.068	71.48	3.046	0.800
	18.0	62.95	2.667	0.965	67.27	2.873	0.877
	(18.12)	65.13	2.775	0.919	65.13	2.775	0.919
1.80	0.0	33.75	1.000	1.800	90.00	3.613	0.617
	1.0	34.63	1.054	1.766	89.27	3.613	0.617
	2.0	35.54	1.110	1.731	88.53	3.611	0.618
	3.0	36.48	1.169	1.697	87.78	3.608	0.619
	4.0	37.44	1.231	1.663	87.03	3.603	0.622
	5.0	38.45	1.295	1.628	86.27	3.597	0.625
	6.0	39.48	1.361	1.593	85.49	3.590	0.628
	7.0	40.56	1.431	1.558	84.69	3.581	0.633
	8.0	41.67	1.504	1.523	83.87	3.570	0.638
	9.0	42.84	1.581	1.486	83.02	3.557	0.644
	10.0	44.06	1.661	1.449	82.13	3.542	0.652

† Figures in parentheses are maximum values.

M_1	θ, degrees†	Weak Solutions β, degrees	Weak P_2/P_1	Weak M_2	Strong Solutions β	Strong P_2/P_1	Strong M_2
	11.0	45.34	1.746	1.412	81.20	3.525	0.660
	12.0	46.69	1.835	1.373	80.22	3.504	0.670
	13.0	48.12	1.929	1.332	79.16	3.480	0.682
	14.0	49.66	2.030	1.290	78.02	3.451	0.696
	15.0	51.34	2.138	1.245	76.76	3.415	0.712
	16.0	53.20	2.257	1.196	75.33	3.371	0.733
	17.0	55.34	2.391	1.142	73.63	3.313	0.759
	18.0	58.00	2.552	1.077	71.43	3.230	0.796
	19.0	62.31	2.797	0.977	67.58	3.064	0.867
	(19.18)	64.99	2.938	0.920	64.99	2.938	0.920
1.85	0.0	32.72	1.000	1.850	90.00	3.826	0.606
	1.0	33.58	1.055	1.815	89.31	3.826	0.606
	2.0	34.47	1.112	1.781	88.61	3.824	0.607
	3.0	35.38	1.172	1.746	87.91	3.821	0.608
	4.0	36.32	1.234	1.711	87.20	3.817	0.611
	5.0	37.30	1.299	1.677	86.48	3.811	0.613
	6.0	38.30	1.367	1.642	85.74	3.804	0.617
	7.0	39.35	1.438	1.607	84.99	3.796	0.621
	8.0	40.43	1.512	1.571	84.23	3.786	0.626
	9.0	41.55	1.590	1.535	83.43	3.774	0.631
	10.0	42.72	1.671	1.498	82.61	3.760	0.638
	11.0	43.94	1.756	1.461	81.75	3.744	0.646
	12.0	45.22	1.845	1.422	80.85	3.725	0.655
	13.0	46.58	1.940	1.383	79.89	3.703	0.665
	14.0	48.02	2.040	1.342	78.86	3.677	0.677
	15.0	49.56	2.146	1.298	77.75	3.646	0.692
	16.0	51.23	2.261	1.252	76.51	3.609	0.709
	17.0	53.09	2.386	1.203	75.11	3.563	0.729
	18.0	55.23	2.528	1.148	73.44	3.502	0.756
	19.0	57.87	2.697	1.082	71.29	3.415	0.793
	20.0	62.10	2.952	0.982	67.55	3.244	0.865
	(20.20)	64.87	3.106	0.920	64.87	3.106	0.920
1.90	0.0	31.76	1.000	1.900	90.00	4.045	0.596
	1.0	32.60	1.056	1.865	89.34	4.044	0.596
	2.0	33.47	1.114	1.830	88.68	4.043	0.597
	3.0	34.36	1.175	1.795	88.01	4.040	0.598

† Figures in parentheses are maximum values.

Table D.3 Oblique Shock in a Perfect Gas ($\gamma = 1.40$) (*Continued*)

		Weak Solutions			Strong Solutions		
M_1	θ, degrees†	β, degrees	P_2/P_1	M_2	β	P_2/P_1	M_2
	4.0	35.28	1.238	1.760	87.34	4.036	0.600
	5.0	36.23	1.304	1.725	86.66	4.031	0.603
	6.0	37.21	1.374	1.690	85.97	4.024	0.606
	7.0	38.22	1.446	1.655	85.26	4.016	0.610
	8.0	39.27	1.521	1.619	84.54	4.007	0.614
	9.0	40.36	1.600	1.583	83.79	3.996	0.620
	10.0	41.49	1.682	1.546	83.02	3.983	0.626
	11.0	42.67	1.768	1.509	82.22	3.968	0.633
	12.0	43.90	1.858	1.471	81.39	3.950	0.641
	13.0	45.19	1.953	1.432	80.50	3.930	0.650
	14.0	46.55	2.053	1.391	79.57	3.907	0.661
	15.0	48.00	2.159	1.349	78.56	3.879	0.674
	16.0	49.55	2.272	1.305	77.47	3.847	0.688
	17.0	51.23	2.393	1.258	76.25	3.807	0.706
	18.0	53.10	2.526	1.208	74.86	3.758	0.727
	19.0	55.24	2.676	1.151	73.21	3.694	0.755
	20.0	57.90	2.856	1.084	71.06	3.601	0.794
	21.0	62.25	3.132	0.979	67.23	3.414	0.869
	(21.17)	64.79	3.280	0.922	64.79	3.280	0.922
1.95	0.0	30.85	1.000	1.950	90.00	4.270	0.586
	1.0	31.68	1.057	1.914	89.37	4.269	0.586
	2.0	32.53	1.116	1.879	88.74	4.267	0.587
	3.0	33.40	1.178	1.844	88.11	4.265	0.589
	4.0	34.31	1.242	1.809	87.47	4.261	0.590
	5.0	35.23	1.310	1.773	86.82	4.256	0.593
	6.0	36.19	1.380	1.738	86.17	4.250	0.596
	7.0	37.18	1.454	1.703	85.50	4.242	0.599
	8.0	38.21	1.530	1.667	84.81	4.233	0.604
	9.0	39.26	1.610	1.630	84.11	4.223	0.609
	10.0	40.36	1.694	1.594	83.38	4.211	0.614
	11.0	41.50	1.781	1.557	82.63	4.197	0.621
	12.0	42.69	1.873	1.519	81.85	4.180	0.628
	13.0	43.93	1.969	1.480	81.03	4.162	0.637
	14.0	45.23	2.069	1.440	80.17	4.140	0.647
	15.0	46.60	2.175	1.398	79.25	4.115	0.658

† Figures in parentheses are maximum values.

M_1	θ, degrees†	Weak Solutions β, degrees	Weak Solutions P_2/P_1	Weak Solutions M_2	Strong Solutions β	Strong Solutions P_2/P_1	Strong Solutions M_2
	16.0	48.06	2.288	1.355	78.26	4.086	0.671
	17.0	49.62	2.408	1.310	77.17	4.051	0.686
	18.0	51.32	2.537	1.262	75.97	4.009	0.705
	19.0	53.21	2.678	1.210	74.59	3.956	0.727
	20.0	55.38	2.838	1.152	72.93	3.887	0.756
	21.0	58.10	3.031	1.082	70.75	3.787	0.796
	22.0	62.86	3.346	0.966	66.53	3.566	0.883
	(22.09)	64.72	3.460	0.923	64.72	3.460	0.923
2.00	0.0	30.00	1.000	2.000	90.00	4.500	0.577
	1.0	30.81	1.058	1.964	89.40	4.500	0.578
	2.0	31.65	1.118	1.928	88.80	4.498	0.578
	3.0	32.51	1.181	1.892	88.20	4.495	0.580
	4.0	33.39	1.247	1.857	87.59	4.492	0.581
	5.0	34.30	1.315	1.821	86.97	4.487	0.584
	6.0	35.24	1.387	1.786	86.34	4.481	0.586
	7.0	36.21	1.462	1.750	85.71	4.474	0.590
	8.0	37.21	1.540	1.714	85.05	4.465	0.594
	9.0	38.25	1.622	1.677	84.39	4.455	0.598
	10.0	39.32	1.707	1.641	83.70	4.444	0.604
	11.0	40.42	1.796	1.603	82.99	4.431	0.610
	12.0	41.58	1.888	1.565	82.26	4.415	0.617
	13.0	42.78	1.986	1.526	81.49	4.398	0.625
	14.0	44.03	2.088	1.487	80.69	4.378	0.634
	15.0	45.35	2.195	1.446	79.83	4.355	0.644
	16.0	46.73	2.308	1.403	78.92	4.328	0.656
	17.0	48.21	2.427	1.359	77.94	4.296	0.669
	18.0	49.79	2.555	1.313	76.86	4.259	0.685
	19.0	51.51	2.692	1.264	75.66	4.214	0.704
	20.0	53.42	2.843	1.210	74.27	4.157	0.728
	21.0	55.65	3.014	1.150	72.59	4.082	0.758
	22.0	58.46	3.223	1.076	70.33	3.971	0.802
	(22.97)	64.67	3.646	0.924	64.67	3.646	0.924
2.10	0.0	28.44	1.000	2.100	90.00	4.978	0.561
	2.0	30.03	1.122	2.026	88.90	4.976	0.562
	4.0	31.72	1.256	1.953	87.78	4.971	0.565
	6.0	33.51	1.402	1.880	86.64	4.961	0.569

† Figures in parentheses are maximum values.

Table D.3 Oblique Shock in a Perfect Gas ($\gamma = 1.40$) *(Continued)*

M_1	θ, degrees†	Weak Solutions			Strong Solutions		
		β, degrees	P_2/P_1	M_2	β	P_2/P_1	M_2
	8.0	35.41	1.561	1.807	85.47	4.946	0.576
	10.0	37.43	1.734	1.733	84.24	4.926	0.585
	12.0	39.59	1.923	1.656	82.94	4.901	0.596
	14.0	41.91	2.129	1.578	81.54	4.867	0.611
	16.0	44.43	2.355	1.495	80.00	4.823	0.630
	18.0	47.21	2.604	1.408	78.26	4.765	0.654
	20.0	50.37	2.885	1.312	76.19	4.685	0.687
	22.0	54.17	3.215	1.202	73.52	4.564	0.735
	24.0	59.77	3.674	1.049	69.11	4.324	0.825
	(24.61)	64.62	4.033	0.927	64.62	4.033	0.927
2.20	0.0	27.04	1.000	2.200	90.00	5.480	0.547
	2.0	28.59	1.127	2.124	88.98	5.478	0.548
	4.0	30.24	1.265	2.049	87.94	5.473	0.550
	6.0	31.98	1.417	1.974	86.89	5.463	0.555
	8.0	33.83	1.583	1.899	85.80	5.450	0.561
	10.0	35.79	1.764	1.823	84.67	5.431	0.569
	12.0	37.87	1.961	1.745	83.49	5.407	0.579
	14.0	40.10	2.176	1.666	82.22	5.376	0.592
	16.0	42.49	2.410	1.583	80.84	5.337	0.609
	18.0	45.09	2.666	1.496	79.31	5.286	0.630
	20.0	47.98	2.949	1.404	77.55	5.218	0.657
	22.0	51.28	3.270	1.301	75.42	5.122	0.694
	24.0	55.36	3.655	1.181	72.56	4.973	0.749
	26.0	62.70	4.292	0.980	66.48	4.581	0.885
	(26.10)	64.62	4.443	0.931	64.62	4.443	0.931
2.30	0.0	25.77	1.000	2.300	90.00	6.005	0.534
	2.0	27.30	1.131	2.221	89.04	6.003	0.535
	4.0	28.91	1.275	2.144	88.07	5.998	0.537
	6.0	30.61	1.434	2.067	87.09	5.989	0.541
	8.0	32.42	1.607	1.990	86.08	5.976	0.547
	10.0	34.33	1.796	1.912	85.03	5.959	0.554
	12.0	36.35	2.002	1.833	83.93	5.936	0.564
	14.0	38.51	2.226	1.751	82.77	5.907	0.576
	16.0	40.82	2.470	1.668	81.51	5.871	0.591
	18.0	43.30	2.736	1.581	80.14	5.824	0.609

† Figures in parentheses are maximum values.

M_1	θ, degrees†	Weak Solutions β, degrees	Weak Solutions P_2/P_1	Weak Solutions M_2	Strong Solutions β	Strong Solutions P_2/P_1	Strong Solutions M_2
	20.0	46.01	3.028	1.489	78.59	5.763	0.633
	22.0	49.03	3.351	1.389	76.77	5.682	0.664
	24.0	52.54	3.722	1.279	74.51	5.565	0.706
	26.0	57.08	4.182	1.143	71.27	5.368	0.774
	(27.45)	64.65	4.874	0.934	64.65	4.874	0.934
2.40	0.0	24.63	1.000	2.400	90.00	6.553	0.523
	2.0	26.12	1.136	2.318	89.10	6.552	0.524
	4.0	27.70	1.286	2.238	88.19	6.547	0.526
	6.0	29.38	1.451	2.159	87.26	6.538	0.530
	8.0	31.15	1.631	2.080	86.31	6.525	0.535
	10.0	33.02	1.829	1.999	85.33	6.509	0.542
	12.0	35.01	2.045	1.918	84.30	6.487	0.551
	14.0	37.11	2.280	1.835	83.22	6.460	0.562
	16.0	39.35	2.535	1.750	82.06	6.425	0.575
	18.0	41.75	2.813	1.661	80.80	6.382	0.592
	20.0	44.34	3.116	1.569	79.40	6.326	0.613
	22.0	47.18	3.448	1.471	77.81	6.253	0.640
	24.0	50.37	3.820	1.364	75.89	6.154	0.675
	26.0	54.19	4.252	1.243	73.40	6.005	0.726
	28.0	59.66	4.838	1.078	69.29	5.713	0.820
	(28.68)	64.71	5.327	0.937	64.71	5.327	0.937
2.50	0.0	23.58	1.000	2.500	90.00	7.125	0.513
	2.0	25.05	1.141	2.416	89.14	7.123	0.514
	4.0	26.61	1.296	2.333	88.28	7.118	0 516
	6.0	28.26	1.468	2.251	87.40	7.110	0.519
	8.0	30.01	1.657	2.169	86.51	7.098	0.524
	10.0	31.85	1.864	2.086	85.58	7.082	0.530
	12.0	33.80	2.090	2.002	84.61	7.061	0.539
	14.0	35.87	2.336	1.917	83.60	7.034	0.549
	16.0	38.06	2.604	1.830	82.52	7.001	0.562
	18.0	40.39	2.895	1.739	81.36	6.960	0.577
	20.0	42.89	3.211	1.646	80.07	6.908	0.596
	22.0	45.60	3.556	1.548	78.63	6.841	0.620
	24.0	48.60	3.936	1.443	76.94	6.753	0.651
	26.0	52.04	4.366	1.327	74.86	6.627	0.693
	28.0	56.34	4.884	1.189	71.95	6.425	0.757
	(29.80)	64.78	5.801	0.940	64.78	5.801	0.940

† Figures in parentheses are maximum values.

Table D.3 Oblique Shock in a Perfect Gas ($\gamma = 1.40$) (*Continued*)

M_1	θ, degrees†	Weak Solutions			Strong Solutions		
		β, degrees	P_2/P_1	M_2	β	P_2/P_1	M_2
2.60	0.0	22.62	1.000	2.600	90.00	7.720	0.504
	2.0	24.07	1.145	2.512	89.19	7.718	0.505
	4.0	25.61	1.307	2.427	88.36	7.714	0.506
	6.0	27.24	1.486	2.342	87.53	7.705	0.510
	8.0	28.97	1.683	2.257	86.67	7.693	0.514
	10.0	30.79	1.900	2.172	85.79	7.678	0.520
	12.0	32.72	2.137	2.085	84.88	7.657	0.528
	14.0	34.75	2.396	1.997	83.92	7.632	0.538
	16.0	36.90	2.677	1.908	82.91	7.600	0.550
	18.0	39.19	2.982	1.815	81.82	7.560	0.564
	20.0	41.62	3.313	1.720	80.63	7.511	0.582
	22.0	44.24	3.672	1.621	79.30	7.448	0.604
	24.0	47.10	4.066	1.516	77.78	7.367	0.631
	26.0	50.31	4.503	1.403	75.96	7.256	0.667
	28.0	54.09	5.007	1.274	73.59	7.091	0.719
	30.0	59.35	5.671	1.106	69.78	6.778	0.811
	(30.81)	64.87	6.297	0.943	64.87	6.297	0.943
2.70	0.0	21.74	1.000	2.700	90.00	8.338	0.496
	2.0	23.17	1.150	2.609	89.22	8.337	0.496
	4.0	24.70	1.318	2.520	88.43	8.332	0.498
	6.0	26.31	1.504	2.432	87.63	8.324	0.501
	8.0	28.02	1.710	2 344	86.82	8 312	0.506
	10.0	29.82	1.937	2.256	85.98	8.297	0.511
	12.0	31.73	2.186	2.167	85.11	8.277	0.519
	14.0	33.74	2.457	2.076	84.20	8.251	0.528
	16.0	35.86	2.752	1.984	83.24	8.220	0.539
	18.0	38.11	3.073	1.889	82.21	8.182	0.553
	20.0	40.50	3.420	1.792	81.10	8.135	0.569
	22.0	43.05	3.796	1.691	79.86	8.075	0.589
	24.0	45.81	5.206	1.585	78.47	7.998	0.615
	26.0	48.85	4.656	1.472	76.83	7.897	0.647
	28.0	52.34	5.163	1.349	74.79	7.753	0.691
	30.0	56.69	5.773	1.202	71.92	7.519	0.759
	(31.74)	64.96	6.814	0.946	64.96	6.814	0.946

† Figures in parentheses are maximum values.

M_1	θ, degrees†	Weak Solutions			Strong Solutions		
		β, degrees	P_2/P_1	M_2	β	P_2/P_1	M_2
2.80	0.0	20.93	1.000	2.800	90.00	8.980	0.488
	2.0	22.35	1.155	2.706	89.25	8.978	0.489
	4.0	23.85	1.329	2.613	88.49	8.974	0.491
	6.0	25.46	1.523	2.522	87.73	8.966	0.494
	8.0	27.15	1.738	2.431	86.95	8.954	0.498
	10.0	28.94	1.975	2.340	86.14	8.939	0.503
	12.0	30.83	2.236	2.248	85.31	8.919	0.510
	14.0	32.82	2.521	2.154	84.44	8.894	0.519
	16.0	34.92	2.831	2.059	83.53	8.864	0.530
	18.0	37.14	3.168	1.961	82.55	8.826	0.543
	20.0	39.49	3.532	1.861	81.50	8.780	0.558
	22.0	41.99	3.927	1.758	80.34	8.722	0.577
	24.0	44.68	4.355	1.651	79.05	8.650	0.600
	26.0	47.61	4.822	1.538	77.55	8.554	0.630
	28.0	50.89	5.340	1.416	75.73	8.424	0.668
	30.0	54.79	5.939	1.278	73.33	8.227	0.724
	32.0	60.43	6.753	1.091	69.21	7.828	0.831
	(32.59)	65.05	7.352	0.949	65.05	7.352	0.949
2.90	0.0	20.17	1.000	2.900	90.00	9.645	0.481
	2.0	21.58	1.160	2.802	89.28	9.643	0.482
	4.0	23.08	1.341	2.706	88.55	9.639	0.484
	6.0	24.67	1.542	2.612	87.81	9.631	0.487
	8.0	26.35	1.766	2.518	87.06	9.619	0.491
	10.0	28.13	2.014	2.423	86.29	9.604	0.496
	12.0	30.01	2.287	2.327	85.49	9.584	0.503
	14.0	31.99	2.586	2.230	84.65	9.560	0.511
	16.0	34.07	2.912	2.132	83.78	9.530	0.521
	18.0	36.27	3.266	2.031	82.85	9.493	0.533
	20.0	38.59	3.650	1.929	81.85	9.448	0.548
	22.0	41.05	4.064	1.823	80.74	9.392	0.566
	24.0	43.67	4.512	1.714	79.54	9.321	0.588
	26.0	46.52	4.998	1.600	78.14	9.231	0.615
	28.0	49.66	5.533	1.479	76.49	9.110	0.650
	30.0	53.28	6.136	1.345	74.39	8.935	0.699
	32.0	57.93	6.879	1.183	71.29	8.635	0.777
	(33.36)	65.15	7.912	0.952	65.15	7.912	0.952

† Figures in parentheses are maximum values.

Table D.3 Oblique Shock in a Perfect Gas ($\gamma = 1.40$) (*Continued*)

M_1	θ, degrees†	Weak Solutions			Strong Solutions		
		β, degrees	P_2/P_1	M_2	β	P_2/P_1	M_2
3.00	0.0	19.47	1.000	3.000	90.00	10.333	0.475
	2.0	20.87	1.166	2.898	89.30	10.322	0.476
	4.0	22.36	1.352	2.799	88.60	10.327	0.477
	6.0	23.94	1.562	2.701	87.88	10.319	0.480
	8.0	25.61	1.795	2.603	87.16	10.307	0.484
	10.0	27.38	2.055	2.505	86.41	10.292	0.489
	12.0	29.25	2.340	2.406	85.64	10.273	0.496
	14.0	31.22	2.654	2.306	84.84	10.248	0.504
	16.0	33.29	2.996	2.204	84.00	10.218	0.514
	18.0	35.47	3.368	2.100	83.11	10.182	0.525
	20.0	37.76	3.771	1.994	82.15	10.137	0.539
	22.0	40.19	4.206	1.886	81.11	10.082	0.556
	24.0	42.78	4.676	1.774	79.96	10.014	0.577
	26.0	45.55	5.184	1.659	78.65	9.927	0.602
	28.0	48.59	5.739	1.537	77.13	9.812	0.635
	30.0	52.02	6.356	1.406	75.24	9.652	0.678
	32.0	56.18	7.081	1.254	72.65	9.399	0.743
	34.0	63.67	8.268	1.003	66.75	8.697	0.908
	(34.07)	65.24	8.492	0.954	65.24	8.492	0.954
3.10	0.0	18.82	1.000	3.100	90.00	11.045	0.470
	2.0	20.21	1.171	2.994	89.32	11.043	0.470
	4.0	21.68	1.364	2.891	88.64	11.039	0.472
	6.0	23.26	1.582	2.789	87.95	11.031	0.474
	8.0	24.93	1.825	2.688	87.24	11.019	0.478
	10.0	26.69	2.096	2.586	86.52	11.004	0.483
	12.0	28.55	2.395	2.484	85.78	10.984	0.490
	14.0	30.51	2.724	2.380	85.00	10.960	0.497
	16.0	32.57	3.083	2.274	84.19	10.930	0.507
	18.0	34.74	3.474	2.167	83.33	10.894	0.518
	20.0	37.02	3.897	2.058	82.42	10.850	0.531
	22.0	39.42	4.354	1.947	81.42	10.795	0.548
	24.0	41.97	4.847	1.833	80.33	10.728	0.567
	26.0	44.69	5.379	1.715	79.09	10.644	0.591
	28.0	47.65	5.956	1.593	77.67	10.533	0.621
	30.0	50.94	6.592	1.462	75.94	10.383	0.661

† Figures in parentheses are maximum values.

M_1	θ, degrees†	Weak Solutions β, degrees	Weak Solutions P_2/P_1	Weak Solutions M_2	Strong Solutions β	Strong Solutions P_2/P_1	Strong Solutions M_2
	32.0	54.80	7.320	1.316	73.66	10.158	0.717
	34.0	60.21	8.277	1.124	69.87	9.717	0.820
	(34.73)	65.34	9.093	0.956	65.34	9.093	0.956
3.20	0.0	18.21	1.000	3.200	90.00	11.780	0.464
	2.0	19.59	1.176	3.090	89.34	11.778	0.465
	4.0	21.06	1.376	2.983	88.68	11.774	0.466
	6.0	22.63	1.602	2.878	88.01	11.766	0.469
	8.0	24.29	1.855	2.773	87.32	11.754	0.473
	10.0	26.05	2.138	2.667	86.62	11.738	0.478
	12.0	27.91	2.451	2.561	85.90	11.719	0.484
	14.0	29.86	2.795	2.453	85.15	11.695	0.491
	16.0	31.92	3.172	2.344	84.37	11.665	0.500
	18.0	34.07	3.583	2.233	83.54	11.629	0.511
	20.0	36.34	4.027	2.121	82.65	11.584	0.524
	22.0	38.72	4.507	2.006	81.70	11.531	0.540
	24.0	41.24	5.024	1.889	80.65	11.464	0.559
	26.0	43.92	5.582	1.770	79.48	11.381	0.581
	28.0	46.81	6.184	1.645	78.13	11.275	0.610
	30.0	50.00	6.843	1.514	76.53	11.131	0.646
	32.0	53.65	7.583	1.371	74.48	10.924	0.697
	34.0	58.35	8.491	1.198	71.41	10.566	0.779
	(35.33)	65.43	9.714	0.959	65.43	9.714	0.959
3.30	0.0	17.64	1.000	3.300	90.00	12.538	0.460
	2.0	19.01	1.181	3.186	89.36	12.537	0.460
	4.0	20.48	1.388	3.075	88.71	12.532	0.462
	6.0	22.04	1.622	2.965	88.06	12.524	0.464
	8.0	23.70	1.886	2.856	87.39	12.512	0.468
	10.0	25.46	2.181	2.747	86.71	12.496	0.473
	12.0	27.31	2.508	2.636	86.01	12.477	0.479
	14.0	29.26	2.869	2.525	85.28	12.452	0.486
	16.0	31.31	3.264	2.412	84.52	12.422	0.495
	18.0	33.46	3.695	2.297	83.72	12.386	0.505
	20.0	35.71	4.162	2.181	82.86	12.342	0.518
	22.0	38.08	4.666	2.064	81.94	12.288	0.533
	24.0	40.57	5.208	1.944	80.93	12.233	0.551
	26.0	43.22	5.792	1.822	79.81	12.141	0.573

† Figures in parentheses are maximum values.

Table D.3 Oblique Shock in a Perfect Gas ($\gamma = 1.40$) (*Continued*)

		Weak Solutions			Strong Solutions		
M_1	θ, degrees†	β, degrees	P_2/P_1	M_2	β	P_2/P_1	M_2
	28.0	46.06	6.421	1.696	78.54	12.036	0.599
	30.0	49.16	7.106	1.564	77.03	11.898	0.634
	32.0	52.67	7.866	1.422	75.15	11.704	0.680
	34.0	56.97	8.762	1.258	72.50	11.390	0.750
	(35.88)	65.52	10.356	0.961	65.52	10.356	0.961
3.40	0.0	17.11	1.000	3.400	90.00	13.320	0.455
	2.0	18.47	1.187	3.281	89.38	13.318	0.456
	4.0	19.93	1.400	3.166	88.74	13.314	0.457
	6.0	21.49	1.643	3.053	88.11	13.305	0.460
	8.0	23.15	1.917	2.940	87.46	13.293	0.463
	10.0	24.90	2.225	2.826	86.79	13.278	0.468
	12.0	26.76	2.566	2.712	86.11	13.258	0.474
	14.0	28.70	2.944	2.596	85.40	13.233	0.481
	16.0	30.75	3.358	2.479	84.66	13.203	0.489
	18.0	32.89	3.810	2.360	83.88	13.167	0.500
	20.0	35.13	4.300	2.241	83.05	13.122	0.512
	22.0	37.49	4.829	2.120	82.16	13.069	0.526
	24.0	39.97	5.398	1.997	81.19	13.003	0.544
	26.0	42.59	6.010	1.872	80.11	12.922	0.565
	28.0	45.39	6.668	1.744	78.89	12.819	0.590
	30.0	48.42	7.380	1.611	77.47	12.685	0.623
	32.0	51.81	8.165	1.469	75.72	12.499	0.665
	34.0	55.84	9.067	1.310	73.36	12.213	0.728
	36.0	61.92	10.331	1.087	68.96	11.582	0.856
	(36.39)	65.60	11.019	0.962	65.60	11.019	0.962
3.50	0.0	16.60	1.000	3.500	90.00	14.125	0.451
	2.0	17.96	1.192	3.377	89.39	14.123	0.452
	4.0	19.42	1.413	3.257	88.77	14.118	0.453
	6.0	20.97	1.664	3.140	88.15	14.110	0.456
	8.0	22.63	1.949	3.022	87.51	14.098	0.459
	10.0	24.38	2.269	2.904	86.86	14.082	0.464
	12.0	26.24	2.626	2.786	86.20	14.062	0.469
	14.0	28.18	3.021	2.666	85.51	14.037	0.476
	16.0	30.23	3.455	2.545	84.78	14.007	0.485
	18.0	32.36	3.928	2.422	84.02	13.970	0.495

† Figures in parentheses are maximum values.

M_1	θ, degrees†	Weak Solutions β, degrees	P_2/P_1	M_2	Strong Solutions β	P_2/P_1	M_2
	20.0	34.60	4.442	2.299	83.22	13.926	0.507
	22.0	36.95	4.997	2.174	82.35	13.872	0.521
	24.0	39.41	5.594	2.048	81.42	13.806	0.537
	26.0	42.01	6.234	1.920	80.38	13.726	0.557
	28.0	44.77	6.923	1.789	79.21	13.624	0.582
	30.0	47.76	7.665	1.655	77.85	13.492	0.613
	32.0	51.05	8.478	1.513	76.21	13.313	0.653
	34.0	54.89	9.397	1.357	74.05	13.046	0.710
	36.0	60.09	10.572	1.159	70.55	12.540	0.811
	(36.87)	65.69	11.703	0.964	65.69	11.703	0.964
3.60	0.0	16.13	1.000	3.600	90.00	14.953	0.447
	2.0	17.48	1.197	3.472	89.40	14.952	0.448
	4.0	18.93	1.425	3.348	88.80	14.947	0.449
	6.0	20.49	1.686	3.226	88.19	14.938	0.452
	8.0	22.14	1.982	3.104	87.57	14.926	0.455
	10.0	23.90	2.315	2.982	86.93	14.910	0.460
	12.0	25.75	2.687	2.859	86.28	14.890	0.465
	14.0	27.70	3.100	2.735	85.60	14.864	0.472
	16.0	29.74	3.554	2.609	84.90	14.834	0.480
	18.0	31.88	4.050	2.483	84.16	14.797	0.490
	20.0	34.11	4.588	2.355	83.37	14.752	0.502
	22.0	36.45	5.170	2.227	82.53	14.698	0.515
	24.0	38.90	5.795	2.097	81.62	14.632	0.532
	26.0	41.48	6.466	1.966	80.62	14.551	0.551
	28.0	44.22	7.186	1.834	79.49	14.450	0.575
	30.0	47.15	7.961	1.697	78.19	14.320	0.604
	32.0	50.38	8.804	1.555	76.64	14.145	0.642
	34.0	54.07	9.746	1.400	74.64	13.892	0.695
	36.0	58.80	10.894	1.215	71.62	13.450	0.781
	(37.31)	65.77	12.407	0.966	65.77	12.407	0.966
3.70	0.0	15.68	1.000	3.700	90.00	15.805	0.444
	2.0	17.03	1.203	3.567	89.41	15.803	0.444
	4.0	18.48	1.438	3.439	88.82	15.798	0.446
	6.0	20.03	1.707	3.312	88.22	15 790	0.448
	8.0	21.69	2.015	3.186	87.61	15.777	0.452
	10.0	23.44	2.361	3.059	86.99	15.761	0.456

† Figures in parentheses are maximum values.

Table D.3 Oblique Shock in a Perfect Gas ($\gamma = 1.40$) (*Continued*)

M_1	θ, degrees†	Weak Solutions			Strong Solutions		
		β, degrees	P_2/P_1	M_2	β	P_2/P_1	M_2
	12.0	25.30	2.750	2.931	86.35	15.740	0.461
	14.0	27.25	3.181	2.803	85.69	15.715	0.468
	16.0	29.29	3.655	2.673	85.00	15.684	0.476
	18.0	31.42	4.174	2.542	84.28	15.646	0.486
	20.0	33.65	4.738	2.410	83.51	15.601	0.497
	22.0	35.99	5.348	2.278	82.69	15.546	0.510
	24.0	38.43	6.003	2.145	81.80	15.480	0.526
	26.0	40.99	6.705	2.011	80.83	15.399	0.545
	28.0	43.71	7.458	1.876	79.74	15.298	0.568
	30.0	46.61	8.266	1.738	78.49	15.169	0.596
	32.0	49.77	9.142	1.594	77.01	14.998	0.632
	34.0	53.35	10.112	1.440	75.14	14.754	0.681
	36.0	57.76	11.260	1.262	72.45	14.352	0.758
	(37.71)	65.85	13.131	0.968	65.85	13.131	0.968
3.80	0.0	15.26	1.000	3.800	90.00	16.680	0.441
	2.0	16.60	1.208	3.662	89.42	16.678	0.441
	4.0	18.05	1.450	3.529	88.84	16.673	0.443
	6.0	19.60	1.729	3.398	88.25	16.664	0.445
	8.0	21.26	2.048	3.267	87.66	16.652	0.448
	10.0	23.02	2.409	3.135	87.05	16.635	0.452
	12.0	24.87	2.813	3.003	86.42	16.614	0.458
	14.0	26.82	3.263	2.870	85.77	16.588	0.464
	16.0	28.87	3.759	2.735	85.09	16.557	0.472
	18.0	31.00	4.302	2.600	84.39	16.519	0.482
	20.0	33.23	4.892	2.464	83.64	16.473	0.493
	22.0	35.56	5.530	2.328	82.84	16.418	0.506
	24.0	37.99	6.216	2.192	81.97	16.351	0.521
	26.0	40.54	6.951	7.055	81.02	16.270	0.540
	28.0	43.24	7.738	1.917	79.97	16.169	0.562
	30.0	46.11	8.581	1.776	78.77	16.040	0.589
	32.0	49.22	9.492	1.631	77.34	15.871	0.624
	34.0	52.70	10.494	1.478	75.57	15.634	0.670
	36.0	56.90	11.654	1.304	73.12	15.259	0.739
	38.0	64.19	13.487	1.029	67.57	14.227	0.913
	(38.09)	65.92	13.876	0.969	65.92	13.876	0.969

† Figures in parentheses are maximum values.

M_1	θ, degrees†	Weak Solutions			Strong Solutions		
		β, degrees	P_2/P_1	M_2	β	P_2/P_1	M_2
3.90	0.0	14.86	1.000	3.900	90.00	17.578	0.438
	2.0	16.20	1.214	3.757	89.43	17.577	0.438
	4.0	17.64	1.463	3.619	88.86	17.571	0.440
	6.0	19.20	1.752	3.483	88.28	17.562	0.442
	8.0	20.85	2.082	3.347	87.70	17.550	0.445
	10.0	22.61	2.457	3.211	87.10	17.533	0.449
	12.0	24.47	2.878	3.074	86.48	17.511	0.455
	14.0	26.42	3.347	2.936	85.84	17.485	0.461
	16.0	28.47	3.865	2.797	85.18	17.453	0.469
	18.0	30.61	4.433	2.657	84.49	17.414	0.478
	20.0	32.83	5.050	2.517	83.75	17.368	0.489
	22.0	35.16	5.717	2.377	82.97	17.312	0.502
	24.0	37.59	6.435	2.237	82.12	17.245	0.517
	26.0	40.13	7.203	2.097	81.20	17.163	0.535
	28.0	42.80	8.026	1.956	80.18	17.061	0.556
	30.0	45.65	8.906	1.813	79.01	16.933	0.583
	32.0	48.72	9.854	1.667	77.64	16.765	0.616
	34.0	52.13	10.890	1.513	75.96	16.533	0.660
	36.0	56.15	12.072	1.343	73.68	16.177	0.724
	38.0	62.09	13.690	1.110	69.50	15.402	0.853
	(38.44)	65.99	14.641	0.970	65.99	14.641	0.970
4.00	0.0	14.48	1.000	4.000	90.00	18.500	0.435
	2.0	15.81	1.219	3.852	89.44	18.498	0.435
	4.0	17.26	1.476	3.709	88.88	18.493	0.437
	6.0	18.81	1.774	3.568	88.31	18.484	0.439
	8.0	20.47	2.117	3.427	87.73	18.471	0.442
	10.0	22.23	2.506	3.287	87.14	18.454	0.446
	12.0	24.10	2.945	3.144	86.54	18.432	0.452
	14.0	26.05	3.434	3.001	85.91	18.405	0.458
	16.0	28.10	3.974	2.857	85.26	18.372	0.466
	18.0	30.24	4.567	2.713	84.58	18.333	0.475
	20.0	32.46	5.212	2.569	83.86	18.286	0.485
	22.0	34.79	5.909	2.425	83.09	18.230	0.498
	24.0	37.21	6.659	2.281	82.26	18.162	0.513
	26.0	39.74	7.463	2.137	81.36	18.079	0.530
	28.0	42.40	8.321	1.994	80.36	17.977	0.551

† Figures in parentheses are maximum values.

Table D.3 Oblique Shock in a Perfect Gas ($\gamma = 1.40$) (*Continued*)

M_1	θ degrees†	Weak Solutions			Strong Solutions		
		β degrees	P_2/P_1	M_2	β	P_2/P_1	M_2
	30.0	45.23	9.240	1.849	79.23	17.848	0.577
	32.0	48.26	10.226	1.701	77.91	17.681	0.609
	34.0	51.61	11.300	1.546	76.30	17.452	0.651
	36.0	55.50	12.510	1.378	74.16	17.110	0.711
	38.0	60.83	14.065	1.164	70.60	16.441	0.820
	(38.77)	66.06	15.426	0.972	66.06	15.426	0.972
6.00	0.0	9.59	1.000	6.000	90.00	41.833	0.404
	2.0	10.91	1.337	5.719	89.53	41.831	0.405
	4.0	12.37	1.761	5.449	89.06	41.822	0.406
	6.0	13.98	2.285	5.182	88.59	41.808	0.408
	8.0	15.73	2.919	4.915	88.11	41.787	0.410
	10.0	17.59	3.668	4.648	87.62	41.760	0.414
	12.0	19.55	4.538	4.382	87.11	41.727	0.418
	14.0	21.61	5.531	4.120	86.60	41.685	0.424
	16.0	23.75	6.647	3.865	86.06	41.635	0.430
	18.0	25.97	7.886	3.618	85.50	41.575	0.438
	20.0	28.26	9.246	3.380	84.92	41.504	0.447
	22.0	30.61	10.723	3.153	84.31	41.419	0.457
	24.0	33.04	12.316	2.935	83.65	41.319	0.469
	26.0	35.54	14.021	2.726	82.95	41.200	0.483
	28.0	38.12	15.836	2.527	82.18	41.056	0.499
	30.0	40.79	17.760	2.335	81.34	40.882	0.518
	32.0	43.59	19.796	2.149	80.40	40.666	0.541
	34.0	46.53	21.950	1.967	79.33	40.393	0.570
	36.0	49.67	24.243	1.786	78.06	40.034	0.605
	38.0	53.14	26.718	1.602	76.48	39.537	0.652
	40.0	57.19	29.501	1.403	74.32	38.765	0.720
	42.0	63.11	33.239	1.141	70.31	37.063	0.859
	(42.44)	66.91	35.376	0.987	66.91	35.376	0.987
8.00	0.0	7.18	1.000	8.000	90.00	74.500	0.393
	2.0	8.50	1.465	7.534	89.56	74.496	0.393
	4.0	10.02	2.093	7.087	89.12	74.482	0.394
	6.0	11.71	2.911	6.642	88.68	74.460	0.396
	8.0	13.56	3.937	6.199	88.23	74.429	0.399
	10.0	15.53	5.185	5.762	87.77	74.387	0.402

† Figures in parentheses are maximum values.

M_1	θ degrees†	Weak Solutions			Strong Solutions		
		β degrees	P_2/P_1	M_2	β	P_2/P_1	M_2
	12.0	17.60	6.660	5.341	87.30	74.334	0.406
	14.0	19.76	8.364	4.939	86.82	74.270	0.411
	16.0	21.98	10.295	4.563	86.33	74.193	0.417
	18.0	24.27	12.450	4.212	85.81	74.101	0.424
	20.0	26.62	14.823	3.887	85.27	73.991	0.433
	22.0	29.02	17.406	3.586	84.70	73.862	0.442
	24.0	31.48	20.194	3.307	84.10	73.710	0.453
	26.0	34.00	23.178	3.048	83.45	73.529	0.466
	28.0	36.58	26.353	2.807	82.76	73.313	0.481
	30.0	39.24	29.713	2.580	82.00	73.052	0.499
	32.0	42.00	33.257	2.366	81.16	72.734	0.520
	34.0	44.87	36.989	2.162	80.21	72.338	0.545
	36.0	47.89	40.928	1.964	79.11	71.832	0.576
	38.0	51.15	45.113	1.769	77.78	71.156	0.615
	40.0	54.77	49.651	1.570	76.10	70.189	0.668
	42.0	59.15	54.868	1.350	73.66	68.587	0.751
	(43.79)	67.28	63.357	0.992	67.28	63.357	0.992
10.00	0.0	5.74	1.000	10.000	90.00	116.500	0.388
	2.0	7.08	1.604	9.298	89.58	116.494	0.388
	4.0	8.65	2.474	8.623	89.15	116.474	0.389
	6.0	10.43	3.657	7.950	88.72	116.442	0.391
	8.0	12.37	5.185	7.289	88.29	116.395	0.393
	10.0	14.43	7.075	6.657	87.84	116.334	0.397
	12.0	16.58	9.332	6.068	87.39	116.258	0.401
	14.0	18.81	11.956	5.529	86.93	116.164	0.406
	16.0	21.09	14.942	5.042	86.44	116.051	0.411
	18.0	23.43	18.280	4.603	85.95	115.916	0.418
	20.0	25.82	21.961	4.207	85.43	115.757	0.426
	22.0	28.25	25.973	3.851	84.88	115.569	0.436
	24.0	30.73	30.302	3.528	84.30	115.348	0.446
	26.0	33.27	34.937	3.235	83.68	115.085	0.459
	28.0	35.86	39.866	2.965	83.02	114.773	0.473
	30.0	38.52	45.078	2.717	82.29	114.398	0.490
	32.0	41.26	50.570	2.485	81.49	113.943	0.510
	34.0	44.11	56.344	2.266	80.59	113.379	0.534
	36.0	47.09	62.417	2.057	79.56	112.665	0.563

† Figures in parentheses are maximum values.

Table D.3 Oblique Shock in a Perfect Gas ($\gamma = 1.40$) (*Continued*)

M_1	θ degrees†	Weak Solutions			Strong Solutions		
		β degrees	P_2/P_1	M_2	β	P_2/P_1	M_2
	38.0	50.27	68.837	1.854	78.33	111.726	0.600
	40.0	53.76	75.723	1.651	76.80	110.417	0.648
	42.0	57.82	83.409	1.435	74.70	108.378	0.719
	44.0	63.74	93.653	1.155	70.75	103.821	0.864
	(44.43)	67.45	99.349	0.995	67.45	99.349	0.995
30.00	0.0	1.91	1.000	30.000	90.00	1,049.831	0.379
	2.0	3.46	3.653	24.149	89.60	1,049.780	0.379
	4.0	5.47	9.386	18.820	89.20	1,049.623	0.381
	6.0	7.69	18.629	14.746	88.79	1,049.359	0.382
	8.0	9.99	31.445	11.865	88.37	1,048.983	0.385
	10.0	12.34	47.807	9.812	87.95	1,048.490	0.388
	12.0	14.72	67.654	8.301	87.53	1,047.869	0.392
	14.0	17.13	90.907	7.151	87.08	1,047.110	0.396
	16.0	19.56	117.474	6.248	86.63	1,046.198	0.402
	18.0	22.01	147.248	5.520	86.16	1,045.114	0.408
	20.0	24.48	180.114	4.920	85.67	1,043.834	0.416
	22.0	26.98	215.950	4.414	85.15	1,042.327	0.425
	24.0	29.51	254.627	3.981	84.61	1,040.555	0.435
	26.0	32.08	296.018	3.604	84.03	1,038.464	0.447
	28.0	34.69	340.000	3.271	83.41	1,035.986	0.461
	30.0	37.36	386.467	2.974	82.73	1,033.027	0.476
	32.0	40.09	435.336	2.703	82.00	1,029.454	0.495
	34.0	42.91	486.582	2.455	81.17	1,025.077	0.517
	36.0	45.84	540.268	2.223	80.23	1,019.598	0.543
	38.0	48.93	596.645	2.003	79.14	1,012.537	0.576
	40.0	52.26	656.354	1.789	77.81	1,003.008	0.618
	42.0	55.98	721.060	1.572	76.09	989.106	0.677
	44.0	60.58	796.423	1.330	73.48	964.921	0.770
	(45.45)	67.75	899.333	0.999	67.75	899.333	0.999

† Figures in parentheses are maximum values.

Table D.4 The Standard Atmosphere†

Height above Sea Level z, m	Pressure P, atm	Density ρ, kg/m³	Temperature T, K	Sound Speed c, m/s	Number Density of Particles n, 1/m³	Height above Sea Level z, ft
0	1.0000	1.2250	288.15	340.29	2.55E+25	0
1,000	0.8870	1.1117	281.65	336.44	2.31E+25	3,281
2,000	0.7846	1.0066	275.15	332.53	2.09E+25	6,562
3,000	0.6920	0.9093	268.66	328.58	1.89E+25	9,842
4,000	0.6085	0.8194	262.17	324.59	1.70E+25	13,120
5,000	0.5334	0.7364	255.68	320.55	1.53E+25	16,400
6,000	0.4660	0.6601	249.19	316.45	1.37E+25	19,680
7,000	0.4057	0.5900	242.70	312.31	1.23E+25	22,970
8,000	0.3519	0.5258	236.22	308.11	1.09E+25	26,250
9,000	0.3040	0.4671	229.73	303.85	9.71E+24	29,530
10,000	0.2615	0.4135	223.25	299.53	8.60E+24	32,810
11,000	0.2240	0.3648	216.77	295.15	7.59E+24	36,090
12,000	0.1915	0.3119	216.65	295.07	6.49E+24	39,370
13,000	0.1636	0.2666	216.65	295.07	5.54E+24	42,650
14,000	0.1399	0.2279	216.65	295.07	4.74E+24	45,930
15,000	0.1195	0.1948	216.65	295.07	4.05E+24	49,210
16,000	0.1022	0.1665	216.65	295.07	3.46E+24	52,490
17,000	0.0873	0.1423	216.65	295.07	2.96E+24	55,770
18,000	0.0747	0.1217	216.65	295.07	2.53E+24	59,050
19,000	0.0638	0.1040	216.65	295.07	2.16E+24	62,340
20,000	0.0546	0.0889	216.65	295.07	1.85E+24	65,610
21,000	0.0467	0.0757	217.58	295.70	1.57E+24	68,900
22,000	0.0399	0.0645	218.57	296.38	1.34E+24	72,180
23,000	0.0342	0.0550	219.57	297.05	1.14E+24	75,460
24,000	0.0293	0.0469	220.56	297.72	9.76E+23	78,740
25,000	0.0252	0.0401	221.55	298.39	8.33E+23	82,020
26,000	0.0216	0.0343	222.54	299.06	7.12E+23	85,300
27,000	0.0186	0.0293	223.54	299.72	6.09E+23	88,580

† Data from *The U.S. Standard Atmosphere, 1962*, U.S. Government Printing Office, Washington, 1962.

1 atm = 1.01325×10^5 n/m² = 14.696 lb_f/in^2
1 kg/m³ = 0.062428 lb_m/ft^3
1 m = 3.2808 ft

Table D.4 The Standard Atmosphere (*Continued*)

Height above Sea Level z, m	Pressure P, atm	Density ρ, kg/m^3	Temperature T, K	Sound Speed c, m/s	Number Density of Particles n, 1/m^3	Height above Sea Level z, ft
28,000	0.0160	0.0251	224.53	300.39	$5.21E+23$	91,860
29,000	0.0137	0.02[illegible]5	225.52	301.05	$4.47E+23$	95,140
30,000	$1.18E-2$	$1.84E-2$	226.51	301.71	$3.83E+23$	98,420
40,000	$2.83E-3$	$4.00E-3$	250.35	317.19	$8.31E+22$	131,200
50,000	$7.87E-4$	$1.03E-3$	270.65	329.80	$2.14E+22$	164,000
60,000	$2.22E-4$	$3.06E-4$	255.77	319.90	$6.36E+21$	196,800
70,000	$5.45E-5$	$8.75E-5$	219.70	295.07	$1.82E+21$	229,700
80,000	$1.02E-5$	$2.00E-5$	180.65	269.44	$4.16E+20$	262,500
90,000	$1.63E-6$	$3.17E-6$	180.65	269.44	$6.59E+19$	295,300
100,000	$2.97E-7$	$4.97E-7$	210.02	290.52	$1.04E+19$	328,100

appendix E
equations in cylindrical and spherical coordinates

Cylindrical coordinates

Coordinates: r, θ, z
Unit vectors: $\mathbf{e}_r, \mathbf{e}_\theta, \mathbf{e}_z$
Velocity components: u_r, u_θ, u_z

Operators

$$\frac{D}{Dt} = \frac{\partial}{\partial t} + u_r \frac{\partial}{\partial r} + \frac{u_\theta}{r}\frac{\partial}{\partial \theta} + u_z \frac{\partial}{\partial z} \tag{E.1}$$

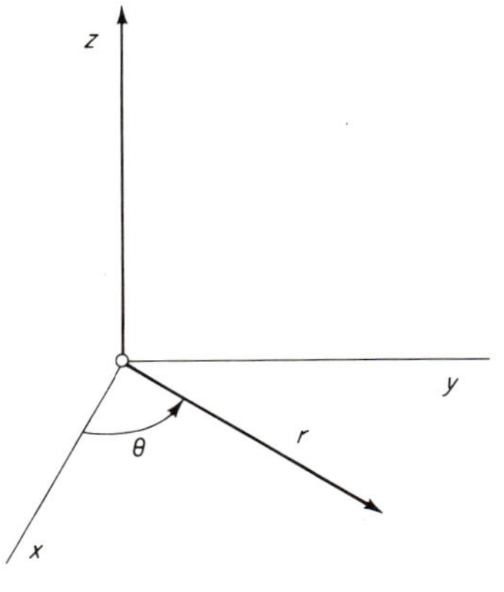

Figure E.1

$$\nabla^2 = \frac{\partial^2}{\partial r^2} + \frac{1}{r}\frac{\partial}{\partial r} + \frac{1}{r^2}\frac{\partial^2}{\partial \theta^2} + \frac{\partial^2}{\partial z^2} \tag{E.2}$$

$$\nabla \cdot \mathbf{u} = \frac{\partial u_r}{\partial r} + \frac{u_r}{r} + \frac{1}{r}\frac{\partial u_\theta}{\partial \theta} + \frac{\partial u_z}{\partial z} \tag{E.3}$$

$$\nabla T = \mathbf{e}_r \frac{\partial T}{\partial r} + \mathbf{e}_\theta \frac{1}{r}\frac{\partial T}{\partial \theta} + \mathbf{e}_z \frac{\partial T}{\partial z} \qquad \text{where } T \text{ is a scalar} \tag{E.4}$$

$$\nabla \times \mathbf{u} = \mathbf{\Omega} = \mathbf{e}_r\left(\frac{1}{r}\frac{\partial u_z}{\partial \theta} - \frac{\partial u_\theta}{\partial z}\right) + \mathbf{e}_\theta\left(\frac{\partial u_r}{\partial z} - \frac{\partial u_z}{\partial r}\right) + \mathbf{e}_z \frac{1}{r}\left(\frac{\partial}{\partial r} r u_\theta - \frac{\partial u_r}{\partial \theta}\right) \tag{E.5}$$

Continuity Equation

$$\frac{\partial \rho}{\partial t} + \frac{\partial}{\partial r}\rho u_r + \frac{1}{r}\frac{\partial}{\partial \theta}\rho u_\theta + \frac{\partial}{\partial z}\rho u_z + \frac{\rho u_r}{r} = 0 \tag{E.6}$$

Viscous Stress Components (Newtonian)

$$\begin{aligned}
\Sigma_{rr} &= 2\mu\left[\frac{\partial u_r}{\partial r}\right] + \mu_v \nabla \cdot \mathbf{u} \\
\Sigma_{\theta\theta} &= 2\mu\left[\frac{1}{r}\frac{\partial u_\theta}{\partial \theta} + \frac{u_r}{r}\right] + \mu_v \nabla \cdot \mathbf{u} \\
\Sigma_{zz} &= 2\mu\left[\frac{\partial u_z}{\partial z}\right] + \mu_v \nabla \cdot \mathbf{u} \\
\Sigma_{r\theta} &= \Sigma_{\theta r} = 2\mu\left[\frac{1}{2}\left(\frac{1}{r}\frac{\partial u_r}{\partial \theta} + r\frac{\partial}{\partial r}\frac{u_\theta}{r}\right)\right] \\
\Sigma_{\theta z} &= \Sigma_{z\theta} = 2\mu\left[\frac{1}{2}\left(\frac{\partial u_\theta}{\partial z} + \frac{1}{r}\frac{\partial u_z}{\partial \theta}\right)\right] \\
\Sigma_{zr} &= \Sigma_{rz} = 2\mu\left[\frac{1}{2}\left(\frac{\partial u_z}{\partial r} + \frac{\partial u_r}{\partial z}\right)\right]
\end{aligned} \tag{E.7}$$

Note that the components of the rate-of-deformation tensor are the quantities in square brackets; for example,

$$D_{zz} = \frac{\partial u_z}{\partial z} \qquad D_{\theta\theta} = \frac{1}{r}\frac{\partial u_\theta}{\partial \theta} + \frac{u_r}{r}$$

That is, the linear stress law (1.37) holds literally if r, θ, z are interpreted as fixed indices.

Momentum Equations

$$\rho\left(\frac{Du_r}{Dt} - \frac{u_\theta^2}{r}\right) = \rho G_r - \frac{\partial P}{\partial r} + \frac{\partial \Sigma_{rr}}{\partial r} + \frac{1}{r}\frac{\partial \Sigma_{r\theta}}{\partial \theta} + \frac{\partial \Sigma_{rz}}{\partial z} + \frac{\Sigma_{rr} - \Sigma_{\theta\theta}}{r} \tag{E.8}$$

$$\rho\left(\frac{Du_\theta}{Dt} + \frac{u_r u_\theta}{r}\right) = \rho G_\theta - \frac{1}{r}\frac{\partial P}{\partial \theta} + \frac{\partial \Sigma_{\theta r}}{\partial r} + \frac{\partial \Sigma_{\theta\theta}}{\partial \theta} + \frac{\partial \Sigma_{\theta z}}{\partial z} + \frac{2\Sigma_{r\theta}}{r} \tag{E.9}$$

$$\rho\frac{Du_z}{Dt} = \rho G_z - \frac{\partial P}{\partial z} + \frac{\partial \Sigma_{zr}}{\partial r} + \frac{1}{r}\frac{\partial \Sigma_{z\theta}}{\partial \theta} + \frac{\partial \Sigma_{zz}}{\partial z} + \frac{\Sigma_{zr}}{r} \tag{E.10}$$

Dissipation Function

The expression (2.8) for the dissipation function Υ holds, with appropriate fixed indices. Thus,

$$\Upsilon = 2\mu(D_{rr}^2 + D_{\theta\theta}^2 + D_{zz}^2 + 2D_{r\theta}^2 + 2D_{\theta z}^2 + 2D_{zr}^2) + (\mu_v - \tfrac{2}{3}\mu)(\nabla \cdot \mathbf{u})^2 \tag{E.11}$$

Energy Equation

$$\rho T\frac{Ds}{Dt} = \Upsilon + \frac{1}{r}\frac{\partial}{\partial r}\left(\kappa r\frac{\partial T}{\partial r}\right) + \frac{1}{r^2}\frac{\partial}{\partial \theta}\left(\kappa\frac{\partial T}{\partial \theta}\right) + \frac{\partial}{\partial z}\left(\kappa\frac{\partial T}{\partial z}\right) \tag{E.12}$$

where the last three terms are $-\nabla \cdot \mathbf{q}$.

Spherical coordinates

Coordinates: r, θ, ϕ
Unit vectors: $\mathbf{e}_r, \mathbf{e}_\theta, \mathbf{e}_\phi$
Velocity components: u_r, u_θ, u_ϕ

Operators

$$\frac{D}{Dt} = \frac{\partial}{\partial t} + u_r\frac{\partial}{\partial r} + \frac{u_\theta}{r\sin\phi}\frac{\partial}{\partial \theta} + \frac{u_\phi}{r}\frac{\partial}{\partial \phi} \tag{E.13}$$

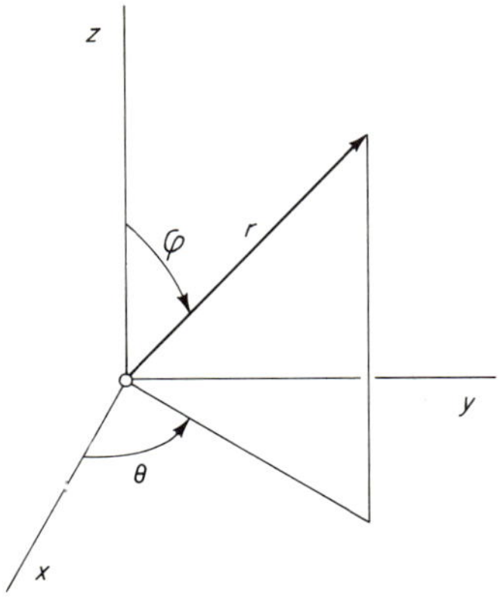

Figure E.2

$$\nabla^2 = \frac{\partial^2}{\partial r^2} + \frac{2}{r}\frac{\partial}{\partial r} + \frac{1}{r^2 \sin^2\phi}\frac{\partial^2}{\partial\theta^2} + \frac{1}{r^2\sin\phi}\frac{\partial}{\partial\phi}\left(\sin\phi\frac{\partial}{\partial\phi}\right) \quad \text{(E.14)}$$

$$\nabla \cdot \mathbf{u} = \frac{\partial u_r}{\partial r} + \frac{2u_r}{r} + \frac{1}{r\sin\phi}\frac{\partial u_\theta}{\partial\theta} + \frac{1}{r\sin\phi}\frac{\partial}{\partial\phi}(u_\phi \sin\phi) \quad \text{(E.15)}$$

$$\nabla T = \mathbf{e}_r\frac{\partial T}{\partial r} + \mathbf{e}_\theta\frac{1}{r\sin\phi}\frac{\partial T}{\partial\theta} + \mathbf{e}_\phi\frac{1}{r}\frac{\partial T}{\partial\phi} \quad \text{where } T \text{ is a scalar} \quad \text{(E.16)}$$

$$\begin{aligned}\nabla \times \mathbf{u} = \mathbf{\Omega} = {} & \mathbf{e}_r\frac{1}{r^2\sin\phi}\left[\frac{\partial}{\partial\phi}(ru_\theta\sin\phi) - \frac{\partial}{\partial\theta}ru_\phi\right] \\ & + \mathbf{e}_\theta\frac{1}{r}\left(\frac{\partial}{\partial r}ru_\theta - \frac{\partial u_r}{\partial\phi}\right) \\ & + \mathbf{e}_\phi\frac{1}{r\sin\phi}\left[\frac{\partial u_r}{\partial\theta} - \frac{\partial}{\partial r}(ru_\theta\sin\phi)\right] \quad \text{(E.17)}\end{aligned}$$

Continuity Equation

$$\frac{\partial\rho}{\partial t} + \frac{1}{r^2}\frac{\partial}{\partial r}r^2\rho u_r + \frac{1}{r\sin\phi}\frac{\partial}{\partial\theta}\rho u_\theta + \frac{1}{r\sin\phi}\frac{\partial}{\partial\phi}(\rho u_\phi \sin_\phi) = 0 \quad \text{(E.18)}$$

Viscous Stress Components (Newtonian)

$$\Sigma_{rr} = 2\mu\left[\frac{\partial u_r}{\partial r}\right] + \mu_v\nabla\cdot\mathbf{u}$$

$$\Sigma_{\theta\theta} = 2\mu\left[\frac{1}{r\sin\phi}\frac{\partial u_\theta}{\partial\theta} + \frac{u_r}{r} + \frac{u_\phi\cot\phi}{r}\right] + \mu_v\nabla\cdot\mathbf{u}$$

$$\Sigma_{\phi\phi} = 2\mu\left[\frac{1}{r}\frac{\partial u_\phi}{\partial \phi} + \frac{u_r}{r}\right] + \mu_v \nabla \cdot \mathbf{u}$$

$$\Sigma_{r\theta} = \Sigma_{\theta r} = 2\mu\left[\frac{1}{2}\left(\frac{1}{r \sin\phi}\frac{\partial u_r}{\partial \theta} + r\frac{\partial}{\partial r}\frac{u_\theta}{r}\right)\right] \tag{E.19}$$

$$\Sigma_{\theta\phi} = \Sigma_{\phi\theta} = 2\mu\left[\frac{1}{2}\left(\frac{\sin\phi}{r}\frac{\partial}{\partial \phi}\frac{u_\theta}{\sin\phi} + \frac{1}{r\sin\phi}\frac{\partial u_\phi}{\partial \theta}\right)\right]$$

$$\Sigma_{\phi r} = \Sigma_{r\phi} = 2\mu\left[\frac{1}{2}\left(\frac{1}{r}\frac{\partial u_r}{\partial \phi} + r\frac{\partial}{\partial r}\frac{u_\phi}{r}\right)\right]$$

Note that the components of the rate-of-deformation tensor are the quantities in square brackets; for example, $D_{rr} = \partial u_r/\partial r$.

Momentum Equations

$$\rho\left(\frac{Du_r}{Dt} - \frac{u_\theta^2 + u_\phi^2}{r}\right) = \rho G_r - \frac{\partial P}{\partial r} + \frac{1}{r\sin\phi}\left[\frac{\sin\phi}{r}\frac{\partial}{\partial r}r^2\Sigma_{rr} + \frac{\partial \Sigma_{r\theta}}{\partial \theta} + \frac{\partial}{\partial \phi}(\Sigma_{r\phi}\sin\phi)\right] - \frac{\Sigma_{\theta\theta} + \Sigma_{\phi\phi}}{r} \tag{E.20}$$

$$\rho\left(\frac{Du_\theta}{Dt} + \frac{u_r u_\theta}{r} + \frac{u_\theta u_\phi \cot\phi}{r}\right) = \rho G_\theta - \frac{1}{r\sin\phi}\frac{\partial P}{\partial \theta} + \frac{1}{r\sin\phi} \times \left[\frac{\sin\phi}{r}\frac{\partial}{\partial r}r^2\Sigma_{\theta r} + \frac{\partial \Sigma_{\theta\theta}}{\partial \theta} + \frac{\partial}{\partial \phi}(\Sigma_{\theta\phi}\sin\phi)\right] + \frac{\Sigma_{\theta r}}{r} + \frac{\Sigma_{\theta\phi}\cot\phi}{r} \tag{E.21}$$

$$\rho\left(\frac{Du_\phi}{Dt} + \frac{u_r u_\phi}{r} - \frac{u_\theta^2 \cot\phi}{r}\right) = \rho G_\phi - \frac{1}{r}\frac{\partial P}{\partial \phi} + \frac{1}{r\sin\phi} \times \left[\frac{\sin\phi}{r}\frac{\partial}{\partial r}r^2\Sigma_{\phi r} + \frac{\partial \Sigma_{\phi\theta}}{\partial \theta} + \frac{\partial}{\partial \phi}(\Sigma_{\phi\phi}\sin\phi)\right] + \frac{\Sigma_{r\phi}}{r} - \frac{\Sigma_{\theta\theta}\cot\phi}{r} \tag{E.22}$$

Dissipation Function

$$\Upsilon = 2\mu(D_{rr}^2 + D_{\theta\theta}^2 + D_{\phi\phi}^2 + 2D_{r\theta}^2 + 2D_{\theta\phi}^2 + 2D_{\phi r}^2) + (\mu_v - \tfrac{2}{3}\mu)(\nabla \cdot \mathbf{u})^2 \tag{E.23}$$

Energy Equation

$$\rho T \frac{Ds}{Dt} = \Upsilon + \frac{1}{r^2}\frac{\partial}{\partial r}\left(r^2 \kappa \frac{\partial T}{\partial r}\right) + \frac{1}{r^2 \sin^2 \phi}\frac{\partial}{\partial \theta}\left(\kappa \frac{\partial T}{\partial \theta}\right) + \frac{1}{r^2 \sin \phi}\frac{\partial}{\partial \phi}\left(\kappa \frac{\partial T}{\partial \phi}\sin \phi\right) \tag{E.24}$$

where the last three terms are $-\nabla \cdot \mathbf{q}$.

appendix F
fluid properties

Table F.1 Atomic Weights

Based on $Carbon^{12} = 12$

Name	Symbol	Atomic Number	Atomic Weight
Argon	Ar	18	39.948
Bromine	Br	35	79.909
Carbon	C	6	12.01115
Cesium	Cs	55	132.905
Chlorine	Cl	17	35.453
Fluorine	F	9	18.9984
Helium	He	2	4.0026
Hydrogen	H	1	1.00797
Krypton	Kr	36	83.80
Mercury	Hg	80	200.59
Neon	Ne	10	20.183
Nitrogen	N	7	14.0067
Oxygen	O	8	15.9994
Radon	Rn	86	222†
Sulfur	S	16	32.064
Xenon	Xe	54	131.30

† Most stable isotope.

Table F.2 Thermal Properties of Gases at 1 Atm and 293.15 K = 20°C = 68°F

Gas	Molecular Weight $\tilde{M}$	Gas Constant R, m²/(s²)(K)	Specific-heat Ratio γ	Density ρ, kg/m³	Sound Speed c, m/s
He	4.0026	2,077.2	1.667	0.1664†	1,007.4†
Ar	39.948	208.13	1.670	1.662	319.0
Xe	131.30	63.32	1.667	5.459†	175.9†
H_2	1.0080	8,248.3	1.406	0.0838	1,303.7
N_2	28.0134	296.80	1.401	1.165	349.1
O_2	31.9988	259.83	1.397	1.332	320.5
Air	28.966	287.03	1.402	1.205	343.3
CO	28.0106	296.83	1.402	1.165	349.1
CO_2	44.0100	188.92	1.297	1.839	266.2
CH_4	16.0431	518.25	1.31	0.594	446.1†
CF_4	92.0367	90.34	1.16	3.673	173.3†

† Values calculated from ideal-gas relations.

Ratio of Specific Heats γ versus Temperature for Gases at Low Pressure

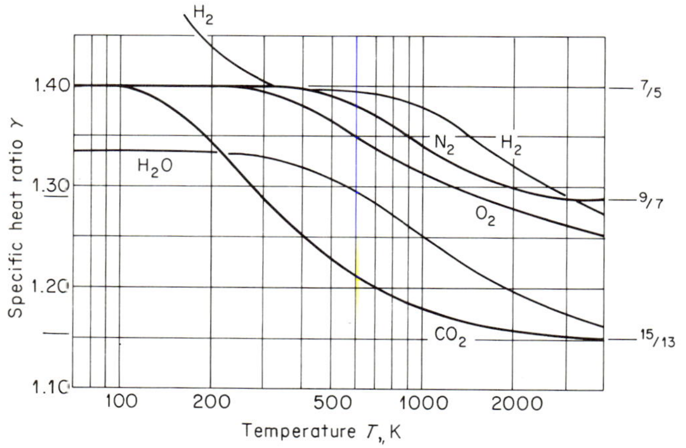

Figure F.1
Ratio of specific heats γ versus T. Note: for monatomic gases, $\gamma = 5/3$ at all values of T.

Viscosity versus Temperature for Gases at Low Pressure

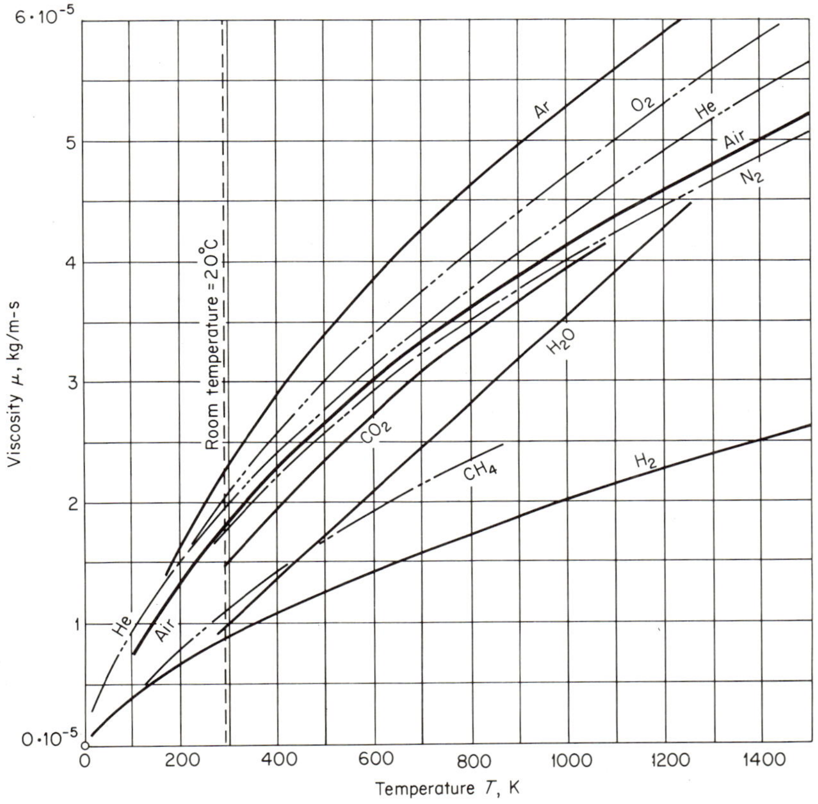

Figure F.2
Viscosity μ versus T.

Viscosity versus Pressure for Gases

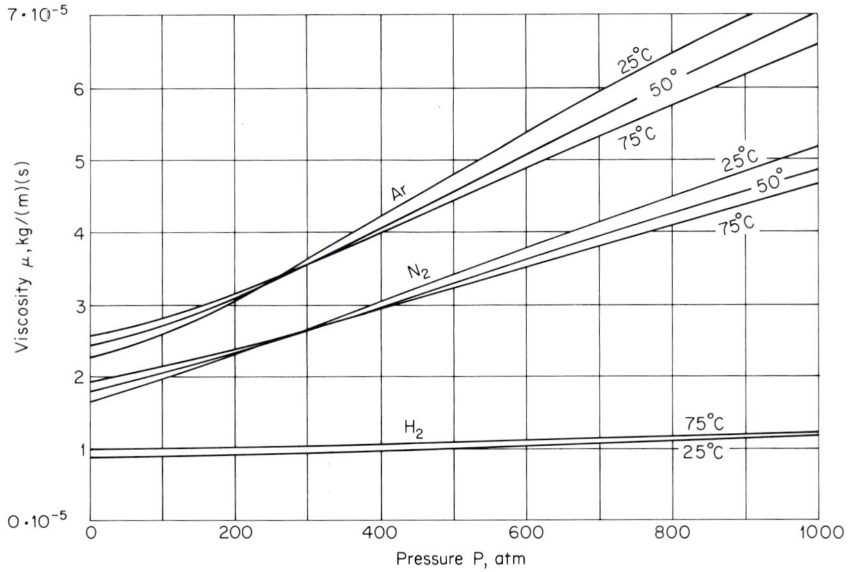

Figure F.3
Viscosity μ versus P.

Prandtl Number versus Temperature for Gases at 1 Atm

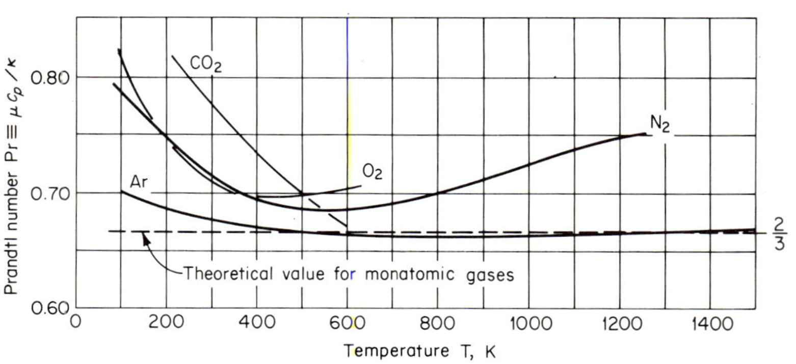

Figure F.4
Prandtl number Pr versus T.

Prandtl Number and Viscosity versus Temperature for Water at 1 Atm

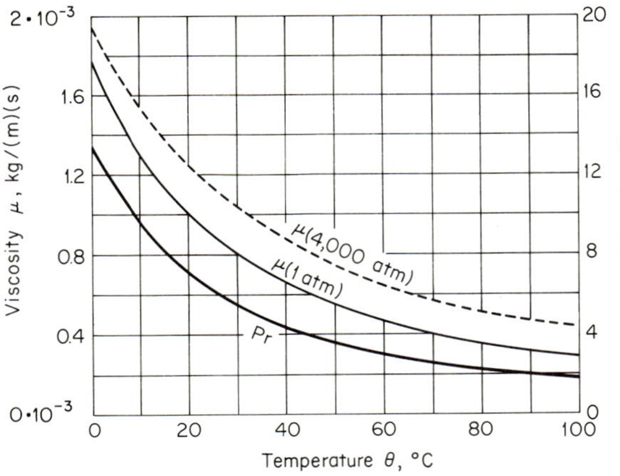

Figure F.5
Prandtl number Pr and viscosity μ versus θ.

Table F.3 Properties of Liquids at 1 Atm and 293.15 K = 20°C = 68°F

Liquid	Density ρ, kg/m^3	Sound Speed c, m/s	Impedance $\rho c \times 10^{-4}$ kg/(m^2)(s)	Thermal Expansion $\frac{1}{v}\left(\frac{\partial v}{\partial T}\right)_p \times 10^4$ K^{-1}	Specific-heat Ratio γ	Specific-heat c_p, J/(kg)(K)	Viscosity $\mu \times 10^3$ kg/(m)(s)	Prandtl Number Pr
Acetone, C_3H_6O	794	1,189	94	14.9	1.416	2,210	0.326	4.5
Benzene, C_6H_6	878	1,326	116	12.4	1.466	1,700	0.652	8.4
Carbon tetrachloride, CCl_4	1,595	938	150	12.4	1.472	840	0.969	7.6
Ethyl alcohol, C_2H_6O	790	1,159	92	11.2	1.208	2,370	1.200	16.7
Glycerin, $C_3H_8O_3$	1,228	1,895	232	5.1	1.117	2,340	1,490	12,500
Kerosene	810	1,315	106			2,000	2.4	32
Mercury, Hg	13,600	1,450	1,970	1.8	1.146	139	1.554	0.025
Water, pure H_2O	998.2	1,484	148	2.1	1.006	4,182	1.002	7.1
Sea, 3.5% H_2O^+	1,025	1,522	156	2.5	1.011	3,900	1.075	7.0

Table F.4 Values of the Fundamental Derivative Γ at 1 Atm and 293.15 K=20°C=68°F†

Substance	Γ
Acetone, C_3H_6O	6.0
Ethyl alcohol, C_2H_6O	6.4
Glycerin, $C_3H_8O_3$	6.1
Methyl alcohol, CH_4O	6.0
n-Propanol, C_3H_8O	6.4
Water, H_2O	4.4
Perfect gas	$(\gamma + 1)/2$
Tait liquid	$(\gamma + 1)/2$

† Values calculated from Eq. (B.21), from data of *Schaaffs* [1967].

Additional Tables and Figures

See also the following in this book:

Table 1.1: Transport Properties page 30
Table 2.2: Physical Constants for Diatomic Molecules page 81
Fig. 2.10: Dissociation of Nitrogen page 82
Table 2.4: Mean Free Path at 1 Atm page 97
Table 2.5: Critical-point Data page 100
Table 2.6: Parameters in the Tait Equation page 102
Fig. 4.3: Sound Speed for Air and for Water page 168
Table 4.1: Sound Speed in Various Substances page 168
Table 4.2: Energy and Momentum Constants page 181
Table 4.3: Characteristic Acoustic Impedance page 186
Table 4.5: Refractive Constants for Gases page 215
Table 4.6: Viscous and Thermal Terms in the Diffusivity δ page 230

Principal References for Fluid Properties

Properties of gases: *Hilsenrath et al.* [1955]

Viscosities: *Schafer* [1969]

Acoustical properties of liquids: *Schaaffs* [1967]

Miscellaneous properties of liquids: *Hodgman* [1962]

Bulk viscosities: *Truesdell* [1953], and others

See also *Stull et al.* [1965]

index

Page references followed by "p." indicate Problems.